INTERNATIONAL UNION OF CRYSTALLOGRAPHY
MONOGRAPHS ON CRYSTALLOGRAPHY

IUCr Monographs on Crystallography

1 *Accurate molecular structures: their determination and importance*
A. Domenicano and I. Hargittai, *editors*

2 *P. P. Ewald and his dynamical theory of X-ray diffraction*
D. W. J. Cruickshank, H. J. Juretschke, and N. Kato, *editors*

3 *Electron diffraction techniques. Volume 1*
J. M. Cowley, *editor*

4 *Electron diffraction techniques, Volume 2*
J. M. Cowley, *editor*

5 *The Rietveld method*
R. A. Young, *editor*

6 *Introduction to crystallographic statistics*
U. Shmueli and G. H. Weiss

7 *Crystallographic instrumentation*
L. A. Aslanov, G. V. Fetisov, and J. A. K. Howard

8 *Direct phasing in crystallography: fundamentals and applications*
C. Giacovazzo

9 *The weak hydrogen bond in structural chemistry and biology*
G. R. Desiraju and T. Steiner

10 *Defect and microstructure analysis by diffraction*
R. L. Snyder, J. Fiala, and H. J. Bunge

11 *Dynamical theory of X-ray diffraction*
A. Authier

12 *The chemical bond in inorganic chemistry*
I. D. Brown

13 *Structure determination from powder diffraction data*
W. I. F. David, K. Shankland, L. B. Mc Cusker, and Ch. Baerlocher, **editors**

14 *Polymorphism in molecular crystals*
J. Bernstein

IUCr Texts on Crystallography

1 *The solid state: from superconductors to superalloys*
A. Guinier and R. Jullien, *translated by* W. J. Duffin

4 *X-Ray charge densities and chemical bonding*
P. Coppens

5 *The basics of crystallography and diffraction, second edition*
C. Hammond

6 *Crystal structure analysis*
W. Clegg, **editor**

7 *Fundamentals of crystallography, second edition*
C. Giacovazzo, **editor**

IUCr Crystallograpic Symposia

1 *Patterson and Pattersons: fifty years of the Patterson function*
J. P. Glusker, B. K. Patterson, and M.Rossi, *editors*

2 *Molecular structure: chemical reactivity and biological activity*
J. J. Stezowski, J. Hung, and M. Shao, *editors*

3 *Crystallographic computing 4: techniques and new technologies*
N. W. Isaacs and M. R. Taylor, *editors*

4 *Organic crystal chemistry*
J. Garbarczyk and D. W. Jones, *editors*

5 *Crystallographic computing 5: from chemistry to biology*
D. Moras, A. D. Podjarny, and J. C. Thierry, *editors*

6 *Crystallographic computing 6: A window on modern crstallography*
H. D. Flack, L. Parkanyi, and K. Simon, *editors*

7 *Correlations, transformations, and interactions in organic crystal chemistry*
D. W. Jones and A. Katrusiak, *editors*

Dynamical Theory of X-Ray Diffraction

ANDRÉ AUTHIER

Université P. et M. Curie, Paris

OXFORD
UNIVERSITY PRESS

This book has been printed digitally and produced in a standard specification in order to ensure its continuing availability

OXFORD
UNIVERSITY PRESS

Great Clarendon Street, Oxford OX2 6DP

Oxford University Press is a department of the University of Oxford.
It furthers the University's objective of excellence in research, scholarship,
and education by publishing worldwide in

Oxford New York

Auckland Cape Town Dar es Salaam Hong Kong Karachi
Kuala Lumpur Madrid Melbourne Mexico City Nairobi
New Delhi Shanghai Taipei Toronto
With offices in
Argentina Austria Brazil Chile Czech Republic France Greece
Guatemala Hungary Italy Japan South Korea Poland Portugal
Singapore Switzerland Thailand Turkey Ukraine Vietnam

Published in the United States
by Oxford University Press Inc., New York

Reprinted 2008

ISBN 978-0-19-852892-0

Preface

For nearly fifty years after P. P. Ewald laid its foundations, the dynamical theory of diffraction remained an elegant construction but of little practical use. The geometrical, or kinematical, theory had the advantage of being much simpler and, although an approximation, it was much better suited for the interpretation of the diffraction patterns used for structure determination purposes or powder analysis. Dynamical theory only worked for a few natural crystals such as calcite, quartz, sodium chloride or diamond that were perfect enough. One actually avoided better crystals to get rid of the extinction problem! The situation changed drastically when highly perfect silicon and germanium crystals were grown for the microelectronic industry. The pioneer observations of single dislocation lines by A. R. Lang, G. Borrmann, U. Bonse and J. B. Newkirk showed that it was possible to characterize crystal perfection and to relate it to the growth history of the sample on the one hand and to its physical properties on the other hand, and it is this double aspect that makes diffraction topography so useful. For the understanding of the diffraction by perfect or nearly perfect crystals, the dynamical theory is compulsory. Until then one had only been interested in the global intensity diffracted by X-rays and not at all in the propagation of the beam inside the crystal. This is however required to explain the formation of defect images on diffraction topographs. The fundamental work of G. Borrmann and N. Kato in the late fifties on anomalous absorption, on X-ray trajectories and on spherical-wave Pendellösung showed the physical reality of the wavefields which had been introduced by P. P. Ewald as a mathematical notion in his thesis. This was the first renewal of dynamical diffraction. These were very exciting times and I was lucky enough to enter the game at that moment and to meet the main actors. By proving experimentally the double refraction of X-rays I added my stone to the construction.

Fascinating developments of dynamical diffraction took place, such as the X-ray interferometer invented by U. Bonse and M. Hart, which enables one to measure accurately the anomalous dispersion corrections and is the ideal tool for nanometre technology, or the standing waves technique for the location of atoms and impurities at surfaces and interfaces suggested by B. W. Batterman.

The extension of the theory to deformed crystals by P. Penning and D. Polder, U. Bonse, S. Takagi and N. Kato came very rapidly while the practical applications of diffraction topography became innumerable. It has proved invaluable for the characterization of high-technology materials such as semiconductors, magnetic, ferroelastic, ferroelectric and piezoelectric materials, crystals for non linear optics, shape-memory alloys, for the understanding of the growth mechanisms not only in synthetic crystals but also in minerals, for the study of the generation and movement of defects during plastic deformation, of the movement of magnetic and ferroelectric domains under an applied field.

The advent of synchrotron radiation and, in particular, of third-generation synchrotron radiation has brought about a second renewal of the dynamical theory and its applications. Exposure times are so short that real-time, *in situ* experiments of crystal growth, plastic deformation and domain movement have become routinely possible. The pulse nature of the radiation enables the stroboscopic study of the strains associated with acoustic vibrations. All the beamlines at a synchrotron source require X-ray optical systems and beam conditioners that are based on the properties of the diffraction by perfect crystals. Thanks to the special properties of third-generation synchrotron radiation such as small source size and very long source-to-sample distances, entirely new experiments based on the properties of coherent beams have become possible. With the recent development of fourth-generation synchrotrons (FELs) time-dependent extensions of the theory have become necessary to interpret the diffraction of femtosecond pulses. For all these reasons the knowledge and understanding of dynamical theory and of its applications have never been as needed as they are now.

One of the beauties of X-ray dynamical theory is that it is entirely based on Maxwell's classic electromagnetic theory; its foundations never had to be reformulated and it is still very actual. It is very similar to the dynamical theories of neutron and electron diffraction because the propagation equation valid for material waves, Schrödinger's equation, is of the same mathematical nature as that deduced for photons from Maxwell's equations. A remarkable aspect of Ewald's work is that it shows the very close relationship between the diffraction of X-rays and the refraction and reflection in usual optics. The fundamental equation of the dynamical theory was already present in his thesis, which was a theory of reflection and refraction, independently of the wavelength. And it is a question relative to this equation which set off Laue's mind in the direction which led to the famous experiment by Friedrich and Knipping.

The book comprises four parts. The first part includes a soft introduction to dynamical theory, some historical considerations, a short reminder of the geometrical theory and a chapter on the interaction of X-rays and matter. The second one gives a detailed treatment of dynamical diffraction of plane and spherical waves by perfect crystals. The third part is devoted to the extensions of the dynamical theory to slightly and highly deformed crystals and the fourth one to a few applications, X-ray optics, standing wave studies of surfaces and interfaces, diffraction topography. Appendix 1 gives a summary of the most frequently used formulae in dynamical theory and Appendix 2 is an unpublished transcript of an evening lecture given by P. P. Ewald at the Summer School on X-ray Topography and Dynamical Theory, which took place in Limoges, France, in August 1975. A list of symbols is given at the end.

My work was done in the Laboratoire de Minéralogie-Cristallographie, Pierre and Marie Curie University in Paris. It owes much to my personal contacts with P. P. Ewald, G. Borrmann, N. Kato and A. R. Lang, to my doctorate students C. Malgrange, F. Balibar, J.-F. Pétroff, M. Sauvage, Y. Epelboin and A. Zarka, to mention the first ones only, and to A. R. Lang, already mentioned, S. Takagi, J. R. Patel and M. Lefeld-Sosnowska who welcomed me in their laboratories. I would also like to thank all the colleagues who were kind enough to read chapters of this book and to send me very useful comments, C. Malgrange, J.-P. Guigay, J. Gronkowski, S.-L. Chang, E. Weckert, V. Holý, J. Härtwig, G. Hildebrandt, D.W.J. Cruickshank, H. Larsen, G. Thorkildsen, V. Dmitrienko.

Peyrat-le-Château A. A.
July 2000

Preface to the second printing

The topics covered by this book are very active and many publications and new developments have appeared since the first printing. It was therefore necessary to take them into account in the present printing and to update the list of references. For the sake of completeness, a number of older references have also been included. In all, more than 280 references have been added. Minor corrections and revisions have also been made throughout the book.

Peyrat-le-Château A. A.
May 2003

To my wife Irena

Dedicated to the memory of P. P. Ewald

Contents

I

Background and basic results

1

Historical developments[1]

1.1 Prologue

P. P. Ewald was born in Berlin in 1888; after a year in Cambridge, UK, he learned mathematics and analysis with D. Hilbert in Göttingen and was fascinated by watching the 'working of a mathematical mind' (Ewald 1968). He then went to Munich to follow A. Pringsheim's lectures which were reputed to present a different approach to mathematics from that offered in Göttingen. Somewhat by chance he went to A. Sommerfeld's Hydrodynamics course. He was immediately attracted to the 'interplay between the mathematical formalism and the physical arguments' which was so vividly described by Sommerfeld and from then on 'his heart was set on Mathematical Physics'. In 1910, he started his thesis. The topic he chose from among those offered to him by Sommerfeld was to investigate whether an anisotropic periodic array of dipoles (oscillating electrons) could produce observable double refraction by the mere fact of this arrangement, even if the resonators themselves were isotropic, and, if this was the case, whether the values obtained were comparable to the measured ones. One of the reasons that Sommerfeld had for proposing such a topic was to obtain supporting evidence for the notion of space lattice and the hypothesis that crystals are triply periodic. The type of arrangement chosen was an orthorhombic one and the crystal Ewald applied his results to was anhydrite ($CaSO_4$). He submitted his thesis in February 1912 and published the main results in 1916 (Ewald 1916*a*,*b*). Although at the time Ewald did not consider X-rays, the theory he developed for solving this problem was valid as well for short wavelengths as for visible light and it turned out to be the basis for his dynamical theory of X-ray diffraction. The first occurrence of the expression *dynamical* to describe the interaction of the dipoles with the radiation is in his thesis.[2]

The problem Ewald had to solve was that of an incident wave falling on a crystalline slab and giving rise to an emergent wave after traversing the crystal. Ewald separated this problem into two parts: the first one of dispersion, the second one of reflection and refraction. He started by studying the field which would be generated by an infinite

[1] The account of the early history is based on the following publications: Bethe and Hildebrandt (1988), Bijvoet *et al.* (1969), Ewald (1938, 1961, 1962, 1965, 1968, 1969, 1979), Hildebrandt (1979*b*,*c*, 1992, 1993, 1995, 1997), Cruickshank *et al.* (1992), Lang (1993, 1995), Laue (1959), Kato (1998), Perutz (1990) and Authier (2003). See also the hitherto unpublished text of a Conference by P. P. Ewald given in the annex to this book: *The early days of dynamical theory*.

[2] '*Dispersion und Doppelbrechung von Elektronengittern*', Inaugural-Dissertation zur Erlangung der Doktorwürde, Königl. Ludwigs-Maximilians-Universität zu München, 16/02/1912, p. 10. The author is grateful to Prof. G. Hildebrandt for pointing it out to him.

three-dimensional array of dipoles, that is not limited by boundaries, when they are excited by a plane wave of frequency ν and with an unknown velocity (Ewald 1916*a*). Each dipole is set in oscillation by that incoming field and emits a spherical wave as shown by Hertz in 1887. This wave, which Ewald calls a 'wavelet', propagates with the velocity of light, $c = \nu/k$ ($k = 1/\lambda$, wave number in vacuum), and contributes to the excitation of the other dipoles. The total wave propagating inside the crystal is the resultant of all these wavelets, it is what Ewald calls the optical field. Due to this interaction with the dipoles, the phase velocity, v, of the resultant field differs from c. There appears therefore an index of refraction, $n = c/v = K/k$ ($K = n/\lambda$, wave number in the medium). The problem is therefore to find all possible values of n; the relation between the wavelength and the frequency is called the *dispersion equation*. Ewald's derivation of this equation which is the basis of his dynamical theory of diffraction was a remarkable feat in mathematics because Fourier transforms were not in use at the time. It is summarized in Section 2.8 with the help of the now well-known properties of Fourier transforms.

In the second part of his study (Theory of refraction and reflection), Ewald (1916*b*) introduces the boundaries of the crystal and an incident wave. He was successful in proving that birefringence could be qualitatively predicted by his model and that it is an intrinsic property of the medium. Although the quantitative agreement was poor (due in part to the fact that the structure of anhydrite it was based on was not correct—he used a P Bravais lattice while anhydrite has an F lattice), this result did confirm that the space lattice hypothesis of crystals was highly probable. It was also very important from an optical viewpoint: it proved that the progressive wave which excites the dipoles in the crystal may be expressed as the sum of two terms, one propagating with velocity c and which cancels out exactly the incident wave and another which satisfies the wave equation for propagation with velocity c/n. This is the so-called Ewald–Oseen extinction theorem (Born and Wolf 1993; Bullough and Hynne 1992) which was also proved by Oseen in 1915 for isotropic media.

1.2 The discovery of X-ray diffraction

The scene for the discovery of X-ray diffraction was set in Munich. Two decisive events which took place in the last decade of the nineteenth century led to it: the derivation of the 230 space groups independently by E. Fedorov and A. Schoenflies around 1890–1891 and, of course, the discovery of X-rays by W. C. Röntgen in Würzburg in November 1895 (Röntgen 1896). This was followed by the discovery of the polarization of X-rays (Barkla 1905) and of the characteristic X-rays (Barkla and Sadler 1909). The big question was that of the nature of X-rays, is it corpuscular or wave? The existence of the photoelectric effect was an argument in favour of the former, a view held by W. H. Bragg, while C. G. Barkla defended the latter. The tentative experiments of diffraction of X-rays by a slit by H. Haga and C. H. Wind (1903) and by B. Walter and R. W. Pohl (1908, 1909) supported the wave theory. The order of magnitude of X-ray wavelengths was determined by two independent ways: at first by Wien in 1905 from the voltage V applied to the tube ($h\nu = eV$) and then by Sommerfeld (1912) on the basis of photometric measurements by Röntgen's chief assistant, P.P. Koch, of Walter and Pohl's plates (Koch 1912). He estimated it to be about 4 Å.

Munich at the beginning of the second decade of the twentieth century was a most propitious place for the discovery which was about to take place. Three institutes played a major role: W. C. Röntgen's Institute of Experimental Physics where W. Friedrich and P. Knipping had prepared their thesis, A. Sommerfeld's Institute of Theoretical Physics where Ewald prepared his thesis as we have seen and which was joined by Laue in 1909, coming from M. Planck's Institute in Berlin, and P. von Groth's Institute of Mineralogy and Crystallography which provided the background insight on the structure of crystals. The spark which set off the chain of events which led to the crucial experiment has been told many times over by Ewald and by Laue: a question by Ewald to Laue in February 1912[3] about the problems he had been working on for his thesis.

Laue did not have a ready answer but asked Ewald what were the distances between the dipoles and what would happen with very short wavelengths. Ewald explained that the distances were not known but probably very small and comparable with the wavelength of X-rays; he added that there were no assumptions in his thesis about the wavelengths of the waves and that his theory of dispersion and refraction should hold for X-rays. He finally showed Laue one of the key results he had obtained (eqn 2.52 in the present book). This immediately gave Laue, who had just written a chapter on Wave optics and gratings in an Encyclopedia edited by A. Sommerfeld, the idea that if the atoms indeed form a lattice they should produce interferences with X-rays similar to the optical interferences produced by gratings, which was verified by the experiment performed by W. Friedrich (then Sommerfeld's assistant) and P. Knipping (then Röntgen's assistant) on 21 April of the same year, despite the scepticism of their respective bosses who feared the diffraction pattern would be smeared out due to the thermal vibrations. In this way, the dynamical theory of diffraction was at the origin of the discovery of the diffraction of X-rays by crystalline matter!

1.3 The geometrical theory of diffraction

The results of Laue, Friedrich and Knipping's work were presented in June and July 1912 at the Bavarian Academy of Sciences. The two papers were published the same year in the *Proceedings of the Academy* and reprinted in 1913 in the *Annalen der Physik*. In the first of these two papers (Friedrich *et al.* 1912), Laue derives the amplitude diffracted by a three-dimensional periodic assembly of atoms by adding the amplitudes of the waves diffracted by each atom, simply taking into account the optical path differences between them, but neglecting the interaction of the propagating waves and matter; it is his 'geometrical theory' of X-ray diffraction. In the second paper (Laue 1912), he gives the geometrical interpretation of the photographic diagram by means of his well-known three conditions which generalize the diffraction condition of a two-dimensional grating and he shows how to index the diagram.

Immediately after hearing an account of these studies by Sommerfeld at a colloquium in Göttingen where he had moved after his thesis, Ewald made the connection with his own work for optical waves and this led him to the introduction of the

[3] See for instance Laue's Nobel lecture (http://www.nobel.se/physics/laureates/1914/laue-lecture.html) or '*Mein Physikalischer Werdegang*' in Laue (1962).

reciprocal lattice in the case of an orthorhombic crystal and of the diffraction sphere, 'Ausbreitungskugel' in his own word (Ewald 1913), now known as the 'Ewald' sphere. The notion of reciprocal lattice was generalized to any type of lattice by Laue (1914) using the concept introduced by Gibbs in 1881. As Laue himself recognized later, it was also very similar to the system of 'polar axes' introduced by Bravais in 1850 but which of course did not include the length relationships.

On the other side of the Channel, in Cambridge, W. L. Bragg took reflections of X-rays from mica plates and, on 11 November 1912, presented them to the Cambridge Philosophical Society, interpreting them as the reflection of the X-rays on families of lattice planes (Bragg and Bragg 1913), deriving Bragg's relation (Bragg 1913). W. H. Bragg applied his experience with ionization chambers to develop the ionization spectrometer and made the first accurate intensity measurements (Bragg 1914) in order to test Debye's theory of the influence of the thermal vibrations (Debye 1913, 1914). Together, W. H. and W. L. Bragg, father and son laid down the bases of crystal structure analysis (Bragg and Bragg 1913*a*,*b*).

While W. L. Bragg introduced the idea of using the amplitude diffracted by one atomic plane as the unit of amplitude, Darwin (1914*a*) calculated the amplitude diffracted by a single atomic plane using Fresnel zones and the intensity diffracted by a family of lattice planes using the same approximation as Laue's and showed immediately (Darwin 1914*b*) the limitations of the geometrical theory which violates the conservation of energy. Compton (1917) followed by Bragg *et al.* (1921) gave expressions of the intensity diffracted by a small crystal rotated in a diffractometer. A summary of the geometrical theory is given in Chapter 3 of the present volume.

Ewald distinguished between *geometrical* and *kinematical* theories (Ewald 1917, 1965, 1969). According to this classification, the geometrical theory corresponds only to the determination of the directions in which the wavelets scattered by the dipoles or the atoms reinforce each other, without cancellation, namely the Ewald construction, while the kinematical theory corresponds to the calculation of the total scattered intensity. But Laue (1959, 1960) preferred to use only the expression *geometrical* theory because the calculation of the intensity is purely geometrical. Indeed, the fine structure of the intensity around a diffraction spot is nothing but the Fourier transform of the shape function of the sample (see Chapter 3).

1.4 Darwin's dynamical theory of diffraction

In the second of his papers, Darwin (1914*b*) took into account the interaction between the waves transmitted and reflected at each atomic plane. The assumption made in the geometrical theory is that the incident wave keeps the same intensity as it propagates through the crystal even though part of it is reflected by each lattice plane it crosses. This of course violates the conservation of energy as is obvious from the intensity formulae (see Chapter 3), but since the fraction which is reflected off is very small the assumption is acceptable for very thin crystals. Darwin's argument is as follows. At each lattice plane, the incident wave generates both a reflected and a transmitted wave. Each one of these in turn generates reflected and transmitted waves each time it crosses

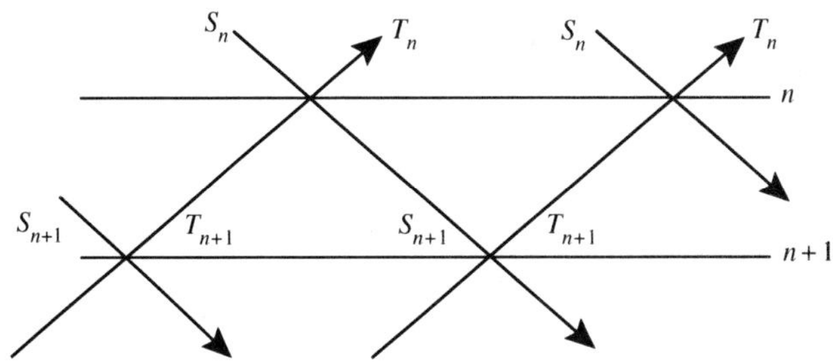

Fig. 1.1. Darwin's theory. S_n: incident amplitude on lattice plane n, T_n: amplitude reflected from lattice plane n.

an atomic plane, and so on (Fig. 1.1). The amplitudes and phases of the transmitted and reflected waves are related and this provides a set of recurrent equations. Let us for example consider the nth and the $(n + 1)$th planes and let T_n and S_n be the amplitudes of the waves respectively incident on the nth plane and reflected from it. Darwin (1914*b*) showed that

$$\begin{aligned} S_n &= \mathrm{i}qT_n + (1 - \mathrm{i}q_0)\exp(-i\phi)S_{n+1}, \\ T_{n+1}\exp(i\phi) &= (1 - \mathrm{i}q_0)T_n - \mathrm{i}\bar{q}\exp(-i\phi)S_{n+1}, \end{aligned} \tag{1.1}$$

where $q = Nd_{hkl}\lambda R|F_{hkl}|/\sin^2\theta_B$ is the ratio of reflected to the incident amplitude on a lattice plane, $\phi = 2\pi d_{hkl}\sin\theta_B/\lambda$ is a phase factor, R is the classical radius of the electron (given in (2.32)), d_{hkl} is the reflecting plane spacing, N the number of scattering units per unit volume, λ the wavelength, θ_B the Bragg angle and F_h the structure factor; $\bar{q}$ and q_0 are given by the same expression as q with F_{hkl} replaced by $F_{\bar{h}\bar{k}\bar{l}}$ and F_{000}, respectively. The imaginary terms are associated with the phase shifts introduced by the diffraction (Darwin 1914*a*). The amplitudes diffracted by a single atomic plane has been discussed recently by Yashiro and Takahashi (2000) and Dub and Litzman (2001).

By solving this set of equations it is possible to obtain the expression of the amplitude reflected at the surface of the crystal. Darwin showed that for a narrow angular range at the centre of the reflection domain this amplitude is pure imaginary. The reflected intensity is therefore equal to 1, this is the total reflection domain. Its width is proportional to the structure factor and is of the order of a few seconds of arc. Furthermore, the centre of the total reflection domain is slightly shifted with respect to the Bragg angle. The integrated intensity, which is measured by the area under the rocking curve, is small despite the existence of the total reflection domain because the rocking curve is so narrow; it is also proportional to the structure factor. In contrast, the integrated intensity predicted by the geometrical theory is much larger because although the width of the rocking curve is very narrow, the peak intensity is very high; it is proportional to the square of the structure factor and to the volume of the crystal (see Section 3.5). The detailed derivation is given in Darwin's original paper

(1914*b*) and in James (1950). A matrix treatment is given in Perkins and Knight (1984). The theory was extended to the case of absorbing crystals by Prins (1930) and Kohler (1933) who showed that, due to absorption, there is no longer a total reflection domain. Darwin's work had been limited to the reflection geometry (Bragg geometry), but it was adapted to the transmission geometry by Borie (1967). This latter treatment is included in Warren (1969). Darwin's calculations are also limited to the case where the scattering plane is perpendicular to the crystal surface. They have been generalized recently for any kind of geometry of the diffraction conditions (Durbin 1995; Yashiro and Takahashi 2000) and to the case where the incidence is grazing or close to $\pi/2$ (Caticha 1994).

Darwin's theory leads to the same expressions of the reflectivity as the Ewald–Laue theory, but is not easily adaptable to the calculation of energy paths and, for that reason, has been seldom used in the later developments of the dynamical theory. It is however very convenient for some calculations as it can be written in matrix form where the elements of a given matrix are related to the partial transmission and partial reflection coefficients for each atomic plane. It is very useful in particular in the case of layered media (Durbin and Follis 1995; Dub and Litzman 1999) or in the case of surface diffraction in order to calculate crystal truncation rod (CTR) diffraction (Nakatani and Takahashi 1994, Takahashi and Nakatani 1995). Chung and Durbin (1995) used this method to calculate the amplitudes of the waves diffracted by a hypothetical layered structure of separate A and B atomic planes arranged in a Fibonacci sequence, $ABAABABA\ldots$, modelling an icosahedral quasicrystal. They could show that such a structure diffracts according to the properties of dynamical diffraction, as observed by Kycia *et al.* (1993).

1.5 Extinction theories

Darwin (1914*a*,*b*) compared his theoretical results with the experimental measurements by Moseley and Darwin (1913) on rock salt and found them in profound disagreement both with respect to the value of the reflected intensity and with respect to the width of the reflection domain which was very much larger than predicted. He was so certain of his theoretical derivation that he concluded (Darwin 1922) that this disagreement was certainly due to the imperfections of the crystal ('warping' or 'cracking'). This led to his model of *extinction* and of diffraction by a *conglomerate of crystalline blocks* (Darwin 1922). This model was further refined by Bragg *et al.* (1926) with the concept of *mosaic* crystal; according to the reminiscences of R.W. James (Ewald 1962, p. 420), the word 'mosaic' was originally suggested by Ewald. They considered that real crystals are made of a mosaic of small blocks more or less misoriented with respect to one another (Fig. 1.2). The geometrical theory applies to the very thin blocks only. When their thickness is larger, their reflectivity becomes closer to that predicted by the dynamical theory and a correction must be applied, this is the *primary extinction* correction which they have shown to be equal to $\tanh(pq)/pq$ where p is the number of lattice planes in the block (see Section 4.11). The incident beam crosses many blocks. If their misalignment is larger than the width of the

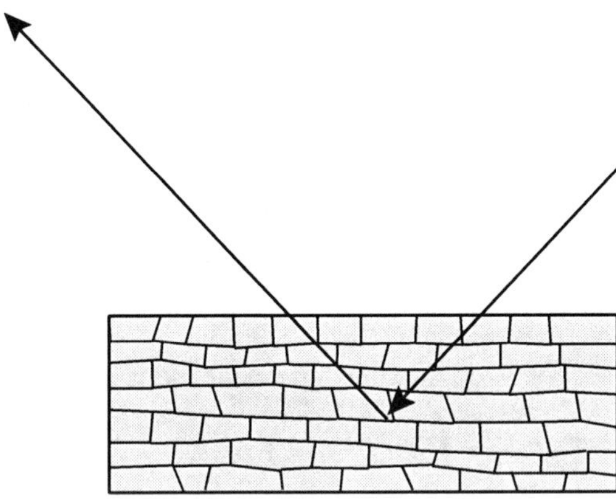

Fig. 1.2. Model of a mosaic crystal.

rocking curve of the individual blocks, they reflect different fractions of the incident beam and the total intensity is the sum of the intensities reflected by the blocks. But if two successive blocks are nearly parallel, part of the incident intensity is reflected off by the first block before it reaches the second one. This is at the origin of the *secondary extinction* correction which is taken into account by an artificial absorption correction. If the crystal is made of a mosaic of widely misoriented very thin crystals, the geometrical theory applies and the crystal is said to be *ideally imperfect*. This model is of course very crude and in general does not correspond physically to the nature of the imperfections in crystals. It may not be too bad if one has a rather homogeneous distribution of subgrains separated by subgrain boundaries, but this is a very particular case. In general, one has a random distribution of defects and long-range strains. One can then try to separate the crystal in *coherent* domains within which phase relationships hold, and reflected amplitudes should be added according to some dynamical theory of diffraction by a deformed crystal; for the contributions due to different domains, intensities, and not amplitudes, should be combined. The size of these coherent domains depends on the diffraction conditions. Very roughly, it will depend on the distance over which the distortion of the lattice planes will vary by an amount larger than the width of the rocking curve; this width depends on the wavelength and the Bragg reflection considered. Different types of coherence lengths can thus be defined according to various criteria. In practice, the notions of primary and secondary extinction do not really have a physical meaning.

The determination of crystal structures and of accurate electron density maps requires, however, good values of the diffracted intensities and, therefore, of the extinction corrections. Many studies were, and still are, devoted to this topic. The mosaic crystal model was improved by Hamilton (1957) and by Zachariasen (1967, 1968) who introduced energy-transfer equations to take into account the coupling between the beams transmitted and diffracted by successive blocks. These equations, called *Hamilton–Darwin Transfer Equations*, are of a form similar to the Takagi–Taupin equations (see Chapters 11 and 14), as discussed by Werner *et al.*

(1986):

$$\begin{aligned} \frac{\partial I_o}{\partial x_o} &= -Q_t I_o + Q_s I_h, \\ \frac{\partial I_h}{\partial x_h} &= -Q_t I_h + Q_s I_o, \end{aligned} \tag{1.2}$$

where $Q_t = Q_s + \mu'$, Q_s is the diffraction cross-section per unit volume, μ' the absorption cross-section per unit volume[4] and x_o and x_h are coordinates along the incident and reflected directions, respectively.

The diffraction cross-section is orientation dependent and is given by

$$Q_s(\theta) = \int \sigma(\theta + \eta) W(\eta)\, \mathrm{d}\eta,$$

where $\sigma(\theta + \eta)$ is the angular distribution of the diffracted intensity due to one mosaic block (*particle-size effect*) and $W(\eta)$ is the angular distribution of the mosaic blocks in the specimen (*mosaic spread*) which can be assumed to be either Gaussian or Lorentzian, according to different authors. Strictly speaking, in order to take into account polarization, eqns (1.2) should apply to the polarization tensor whose diagonal terms are the energies of the σ- and π-polarization states (perpendicular and parallel to the diffraction plane, respectively) and whose complex non-diagonal terms contain their phases (Dmitrienko and Belyakov 1980*a*). Werner (1974) gave the formal solutions of (1.2) in terms of Q_s for a parallepipedic mosaic crystal.

One may distinguish two extreme cases (Zachariasen 1967): type I for which W and the mosaic spread are dominant and type II for which it is σ and the particle-size effect that dominate. That treatment was successful for the secondary extinction in some practical cases but remains unsatisfactory for the reason given above, namely that the mosaic blocks do not correspond in general to reality. It was further refined by Becker and Coppens (1974, 1975) whose formalism is well adapted to many experimental results. Palmer and Jauch (1995) showed for instance that it accounted in a satisfactory way for the wavelength dependence of the extinction-affected observed diffracted intensities in the case of NiF_2 where secondary extinction is dominant. Wilkins (1981) studied numerically the conditions for the presence of primary extinction only to be present in a mosaic block. A comparison of the extinction theories based on the Darwin mosaic model is given in Sabine (1988). More recently, Sabine (1999) proposed a theory in which primary and secondary extinction are treated in a unified way, but which leads to a wavelength dependence which seems less well adapted than the Becker and Coppens treatment (Palmer and Jauch 1995). Tomiyoshi *et al.* (1980) extended the Becker and Coppens treatment to white radiation. Hu (1997*a*) derived solutions of the Hamilton transfer equations in the case of severe extinction and applied them to the case of secondary extinction (Hu 1997*b*).

The statistical dynamical theories constitute another and more modern approach. Their bases were laid down by Kato (1980*a*,*b*), but many authors have contributed

[4] Dmitrienko (1982) has shown that it is possible to introduce the effect of anomalous transmission in the mosaic blocks as a small correction.

since, and a brief review is given in Section 14.5.1. Their principle is to consider the total diffracted intensity to be the sum of two contributions (see Section 14.5.2 for details). The first one is the coherent part. It is calculated using the Takagi–Taupin equations which generalize the fundamental equations of the dynamical theory for any kind of wave and allow for deformations of the crystalline lattice; these equations are amplitude-transfer equations (see Chapter 11). This term corresponds to primary extinction. The second contribution is the incoherent part. It is calculated using intensity-transfer equations of the same type as Hamilton–Darwin's and corresponds to the secondary extinction. These theories are, however, quite sophisticated and are not yet at the stage of being applicable on a routine basis by non-specialists. A comparison of Kato's theory with the Becker and Coppens treatment is given in Becker and Dunstetter (1984).

1.6 Ewald's dynamical theory

Ewald's extinction theorem was published after the discovery of X-ray diffraction (Ewald 1916*b*). Although it was derived for light waves, it proved invaluable for X-rays and the extension of Ewald's dynamical theory to X-rays followed immediately afterwards (Ewald 1917). One of the results obtained by Ewald is the rediscovery of the total reflection domain which is already described in Darwin (1914*b*), and Ewald does refer to Darwin's articles. But his theory has a much wider scope than Darwin's and was far ahead of its time. Of upmost importance is the introduction of the notion of wavefields. This result appears already in his 1913 paper. The optical field which propagates through the crystal and excites the dipoles is a sum of plane waves whose wavevectors can be deduced from one another by translations in reciprocal space:

$$\begin{aligned}\mathbf{E} = {}& \mathbf{E_o} \exp(-2\pi \mathrm{i} \mathbf{K_o} \cdot \mathbf{r}) + \mathbf{E_h} \exp(-2\pi \mathrm{i} \mathbf{K_h} \cdot \mathbf{r}) \\ & + \mathbf{E_g} \exp(-2\pi \mathrm{i} \mathbf{K_g} \cdot \mathbf{r}) + \text{etc.}, \end{aligned} \tag{1.3}$$

where $\mathbf{h} = \mathbf{OH}$ and $\mathbf{g} = \mathbf{OG}$ are reciprocal lattice vectors (Fig. 1.3). The wavefield[5] is characterized by its tiepoint P, which is the common extremity of the wavevectors $\mathbf{K_o} = \mathbf{OP}$, $\mathbf{K_h} = \mathbf{HP}$, $\mathbf{K_g} = \mathbf{GP}$, etc.[6]

The aim of the theory is to find the possible positions of the tiepoint and therefore the wavevectors and the possible values of the index of refraction of the waves propagating in the crystal, as well as their amplitudes E_o, E_h, E_g, etc. The general equations were given for the n-beam case but were applied to the two-beam case only. The shape and the properties of the locus of the tiepoint, the dispersion surface were discussed in this particular case. The equation of the dispersion surface is the dispersion equation. In the general case, the dispersion surface has $2n$ branches and in the two-beam case four branches, two for each direction of polarization (one normal to the $\mathbf{K_o}$, $\mathbf{K_h}$

[5] Expansion (1.3) is nowadays usually called wavefield, sometimes *Ewald wave*. The same concept was introduced in solid state physics by Bloch, 15 years after Ewald (Bloch 1928)—see Section 4.3.

[6] Throughout the book, the wavevectors are oriented *from* the reciprocal lattice points *towards* the tiepoint. This convention follows that introduced by Fues (1939).

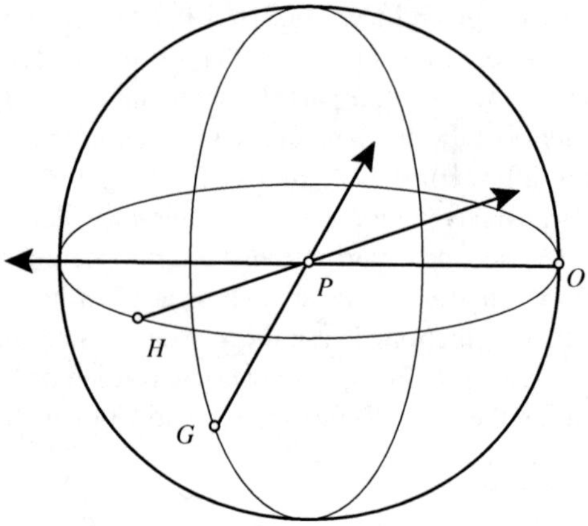

Fig. 1.3. Ewald sphere. O, H and G are reciprocal lattice nodes, and P is the tiepoint of the wavefield. **OP** is the incident wavevector, and **HP** and **GP** are the reflected wavevectors.

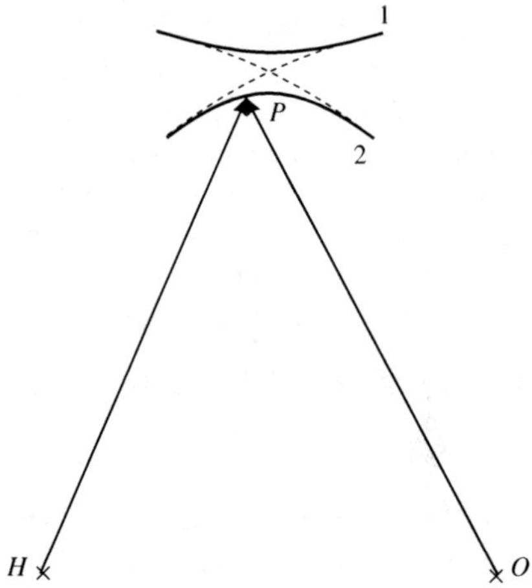

Fig. 1.4. Dispersion surface (solid curve). P: Tiepoint of a wavefield; O and H: reciprocal lattice nodes; the dashed curves are the spheres of centres O and H and radius nk (n, index of refraction).

plane, σ-polarization, the other one in that plane, π-polarization). Figure 1.4 represents schematically that surface for one direction of polarization. After studying the properties of the wavefields inside the crystal, Ewald introduced the boundaries and determined the wavefields generated by the incident beam. Both transmission (Laue) and reflection (Bragg) geometries were considered. The incident beam generates two wavefields for each direction of polarization (*double refraction*). Their tiepoints are situated on different branches of the dispersion surface in the transmission geometry and on the same branch in the reflection geometry.

The theory was limited at that time to the case of a single dipole per unit cell, and the extension to several dipoles per unit cell with the introduction of a factor analogous to the structure factor came later (Ewald 1921, 1925, 1937). Ewald played an important role in the discussions which took place in the 1920s about which expression was correct for the diffracted intensities: proportional to the structure factor or to the square of the structure factor. He insisted that the correct result for perfect crystals was that deduced from the dynamical theory, namely that the intensity should be proportional to the structure factor (Ewald 1925).

The inseparable nature of constituent waves in a wavefield shown by the reciprocity theorem (Ewald 1913) was to be verified much later when anomalous absorption, Pendellösung and double refraction were observed and proved the physical reality of wavefields. Ewald had already predicted the Pendellösung effect: he noted that wavefields associated with different tiepoints would interfere and that this would result in a periodic exchange of energy between the two. He called it Pendellösung in connection with the German verb 'pendeln', to oscillate. He had pointed out that in the transmission geometry the tiepoints of the two wavefields would belong to different branches of the dispersion surface and in the reflection geometry to the same branch. The effect was indeed observed in both cases and will be described in Sections 4.8.7 and 4.9.3, respectively. Ewald recalled much later (Ewald 1965) that he had actually also predicted anomalous absorption which was discovered by Borrmann in 1941 (Borrmann 1941); however, this prediction had not been included in the 1917 paper, but only in one of the questions he had prepared for his *Habilitation* examination and forgotten afterwards!

Ewald's dynamical theory differs from Laue's by the bases which lead up to the fundamental equations. Afterwards, they are the same. The first steps of Ewald's theory are summarized in Section 2.8.

1.7 Early confirmations of the dynamical theory

The first confirmations of the dynamical theory had to do with the rocking curves and the offset of their centre with respect to the Bragg angle predicted by the geometrical theory. The first observations of a deviation from Bragg's law were reported in Siegbahn's laboratory in Lund where X-ray spectroscopy was studied with a high precision, by W. Stenström in his thesis, in 1919, and by Hjalmar (1920). They found that systematic deviations from the Bragg angle occurred for different orders of reflections from sugar and gypsum. Ewald (1920) immediately showed that these variations were to be expected according to the dynamical theory and interpreted quantitatively Hjalmar's measurements. These were followed by more precise measurements by B. Davis and his coworkers (B. Davis and Terrill 1922; B. Davis and Nardroff 1924) who built the first double-crystal diffractometer using calcite crystals. They obtained thus very narrow rocking curve widths with fresh calcite cleavages and showed, on the one hand, the variation of this width with the asymmetry of the reflection (which was to be rediscovered by M. Renninger and is used nowadays daily with the monochromators for synchrotron radiation) and, on the other hand, the increased sensitivity of the deviation from Bragg's angle to the asymmetry of the reflection. This latter effect was also explained by Ewald on the basis of the dynamical theory (Ewald 1924, 1986). Further experiments were performed for his thesis in 1929 by another member of the Swedish school, A. Larsson, in Uppsala, who studied the angle of reflection from mica up to the 11th order and found a remarkable agreement with theory (see Siegbahn 1931).

To conclude this first stage of confirmations of the results of the dynamical theory, let us also mention the measurements made by Renninger (1934), at the time an assistant of Ewalds in Stuttgart, on the cleavage face of a crystallite from a rock salt crystal using the double-crystal technique. He obtained rocking curve widths and integrated intensities which were very close to the values predicted by dynamical theory.

1.8 Laue's dynamical theory

In contrast to Ewald's theory which is microscopic, Laue's reformulation (Laue 1931) is macroscopic. It is based on the assumption that the distribution of the dielectric susceptibility, or polarizability, of the medium for X-rays can be considered as continuous throughout the whole volume of the crystal and proportional to the electron density. Laue had come to this conclusion by observing that it is the whole unit cell which contributes to the diffracted intensities through its Fourier coefficients and not singular points. The validity of this hypothesis was confirmed by Kohler (1935) on the basis of quantum mechanics. In Laue's theory the dielectric susceptibility is considered as a scalar. This is, however, only an approximation for anisotropic crystals, in particular close to an absorption edge (see Section 2.7.1). Moliere (1939*a*) rederived the fundamental equations of dynamical theory using a quantum mechanical treatment of the interactions of X-rays scattered by a crystal; he treated in particular the case where the wavelength is very close to an absorption edge and the tensorial character of the dielectric susceptibility must be taken into account (Moliere 1939*b*). In a third paper he discussed the relations with Laue's theory (Moliere 1939*c*). One of the consequences is the possibility to observe the optical activity of some crystals (Sauvage *et al.* 1983; Siddons *et al.* 1990—see Section 15.3.5.3).

The other important assumption made by Laue is that the surfaces of the crystal can be taken to be mathematical planes. This hypothesis which is natural for light waves is not obvious for X-rays whose wavelengths are of the order of magnitude of the distances between neighbouring atoms, but Laue (1960) concludes that in the absence of any contrary evidence it is the most convenient thing to do, although it is in his opinion the weakest point of the dynamical theory. The ground bases for Laue's dynamical theory are given in Section 2.9.

1.9 *Umweganregung* and *Aufhellung*

If the position of the crystal in the incident beam and the value of the wavelength are such that several reciprocal lattice points besides the origin lie simultaneously on the

Ewald sphere, interaction between the corresponding Bragg reflections occurs. Let us assume for instance that three points lie on the Ewald sphere: O, H and G (Fig. 1.3); $\mathbf{h} = \mathbf{OH}$ and $\mathbf{g} = \mathbf{OG}$ are the corresponding reciprocal lattice vectors. There are therefore three waves in the wavefield and, according to Ewald's reciprocity theorem, **IH** can also be considered as a primary wave which excites the reflections $-\mathbf{h} = \mathbf{HO}$ and $\mathbf{g} - \mathbf{h} = \mathbf{HG}$. If **h** is a strong reflection, there will be a decrease in its intensity due to the interaction between the three waves, which has been called *Aufhellung* by the German school. If **h** is weak, there will be an increase in its intensity, which has been called *Umweganregung* (excitation through a detour, *Umweg*). Such a situation may occur by chance if the incident beam is polychromatic or with a large angular divergence, or if the crystal is bent. White lines due to *Aufhellung* (*Aufhellungslinien*) were first observed by Wagner (1920) and by Berg (1926) on photographs of X-ray reflections from rock salt, NaCl, made with white radiation and by Mayer (1928) with the 400 reflection of diamond (by rotating the crystal around the reciprocal vector **OH**, what is now called a Ψ-scan or Renninger-scan—see Section 9.2.1) and, later, together with dark lines due to *Umweganregung* by Cauchois *et al.* (1937) on bent crystals of quartz. These effects may also be observed in a systematic way, with a Ψ-scan: they will occur every time another node crosses the Ewald sphere. The first one to study *Umweganregung* systematically that way was Renninger (1937*a*,*b*) who measured the forbidden 222 reflection from diamond in an attempt to determine accurately the distribution of electron density. n-beam diffraction is now studied routinely using synchrotron radiation. For pioneer studies, see for instance Chang (1986) and Brown *et al.* (1989) who performed the first Ψ-scan with synchrotron radiation. A review of the more recent work is given in Weckert and Hümmer (1997).

In order to find the positions of these dark and white lines, the geometrical theory of diffraction is enough, but to calculate the reflected intensities in multiple-beam (or 'n-beam') diffraction in general, the dynamical theory is necessary. Two interesting applications may be mentioned. The first one is the enhancement of anomalous absorption (the Borrmann effect) in a three-beam case. It was first observed by Borrmann and Hartwig (1965) and further studied theoretically by Hildebrandt (1967) and experimentally by Uebach and Hildebrandt (1969) and Umeno and Hildebrandt (1975) in the case of simultaneous 111 and $\bar{1}11$ reflections from germanium. This has been called the *super-Borrmann effect* and the explanation is given in Section 9.5. The other application which is now beginning to be of practical importance is the determination of the structure-factor phases. Lipscomb (1949) was the first one to propose to use simultaneous reflections to determine the phases of the reflections, and many studies have been devoted to that topic (see the reviews by Chang 1987, 1992, 1998, 2004; Weckert and Hümmer 1996, 1997). An abridged treatment of the three-beam diffraction is given in Chapter 9. Let us just say here that the analysis of the intensity profiles when the crystal is rotated around the reciprocal lattice vector **OH** and as the reciprocal lattice point G crosses the Ewald sphere enables one to determine what is called a *triplet* phase. It is the sum of the phases of the three simultaneous reflections:

$$\Phi_{3\pm} = \pm[\varphi(\mathbf{g}) + \varphi(\mathbf{h} - \mathbf{g}) - \varphi(\mathbf{h})].$$

Weckert and Hümmer (1997) have shown that the measurement of these triplet phases makes it possible to determine the absolute configuration of enantiomorphs of known structures and to be of great help in the determination of unknown structures, in combination with traditional methods such as direct methods or Maximum Entropy methods. For phases to be determined that way, the simultaneous reflections **h** and **g** must be chosen of comparable intensity because the *Aufhellung* lines and the *Umweganregung* peaks described above do not contain phase information.

n-beam diffraction is also used to characterize the crystal perfection of bulk crystals and epitaxial multilayers (see, for instance, Sasaki *et al.* 1996; Morelhão and Cardoso 1996; Morelhão and Abramof 1999; Morelhão *et al.* 1999).

1.10 The properties of wavefields

The first experimental evidence of the physical existence of Ewald's wavefields is to be found, as noted by Laue (1960, p. 440) and recalled by Hildebrandt (2002), in the light-dark structure of Kossel lines, observed by Borrmann (1936) and explained by Laue (1937) using the properties of the standing waves formed by the wavefields inside the crystal. It was followed by the discovery of anomalous absorption (Borrmann 1941, 1950) and the theoretical determination of the wavefield trajectories (Laue 1952; Kato 1952, 1958; Ewald 1958), which was confirmed by the experimental observation of spherical wave Pendellösung (Kato and Lang 1959) and by that of the double refraction of X-rays, predicted by Borrmann (1955) and observed by Authier (1960). These observations opened up a new area for the dynamical theory.

1.10.1 *Anomalous absorption (the Borrmann effect)*

In 1938, G. Borrmann, who had been studying Kossel lines in Gdansk, was invited by Max v. Laue, then Vice-Director of the Kaiser-Wihelm-Institut for Physics in Berlin-Dahlem, to work with him as his assistant. He was the first one to perform an X-ray diffraction experiment in the transmission geometry (Borrmann 1941). A very divergent incident beam coming from a point focus (wide-angle technique) was sent through thick quartz crystals. The divergence of the beam which was adjusted by means of a slit was of the order of 10–30°, the distance from focus to crystal was a few centimetres and a photographic plate was placed 10–30 cm behind the crystal (Fig. 1.5). The shadow of the slit on the photographic plate is a circle of diameter CD. The rays from the incident beam which satisfy Bragg's condition for a set of lattice planes lie on a cone. The traces on the photographic plate of the cones associated with various reflections are along hyperbolae which are sometimes called pseudo-Kossel lines. Borrmann observed these lines standing out from the background. They correspond to beams which have been partially reflected in the Bragg geometry or in the Laue geometry and it was expected that their intensity should be less than that of the background; this is what is observed for imperfect crystals. However, for a perfect crystal, their intensity is *higher* than that of the background, meaning that the

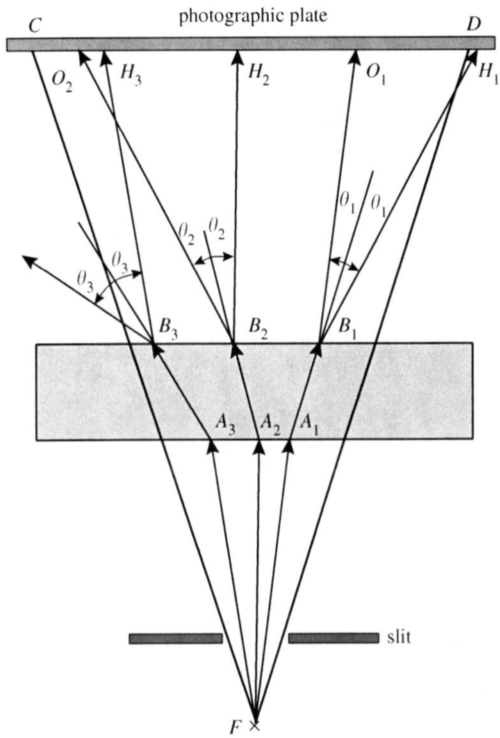

Fig. 1.5. Principle of the setup for wide-angle diagrams (adapted from Borrmann 1964). F: focus of the X-ray tube.

absorption coefficient of the rays which have been Bragg diffracted is smaller than the usual absorption coefficient, this is the *anomalous absorption* or *Borrmann* effect.

Figures 1.6(a)–(c) show three examples from a later publication by Borrmann (1964). They were taken with three crystals of germanium of increasing thicknesses ($t = 0.23$ mm, 0.5 mm and 1.2 mm, respectively) and nickel-filtered Cu$K\alpha$ radiation. Borrmann showed that the diffracted intensity propagated inside the crystal along the lattice planes (for instance along A_1B_1, A_2B_2, A_3B_3 in Fig. 1.5), giving rise on emerging from the crystals to two beams of equal intensity, one along the reflected direction (H_1, H_2, H_3) and one along the incident direction (O_1, O_2). It can be seen that the anomalous absorption effect depends on the nature of the reflection and, for instance, that the reflections with odd indices are absorbed more rapidly than the reflections with even indices. This will be explained in Section 4.8.4.

Laue (1949) interpreted the effect of anomalous absorption by calculating the intensities of the reflected and refracted[7] beams by taking absorption into account and showing that when the Bragg condition is fulfilled there appears indeed an anomalous

[7] Sometimes called *forward diffracted*. See p. 76 (Section 4.5).

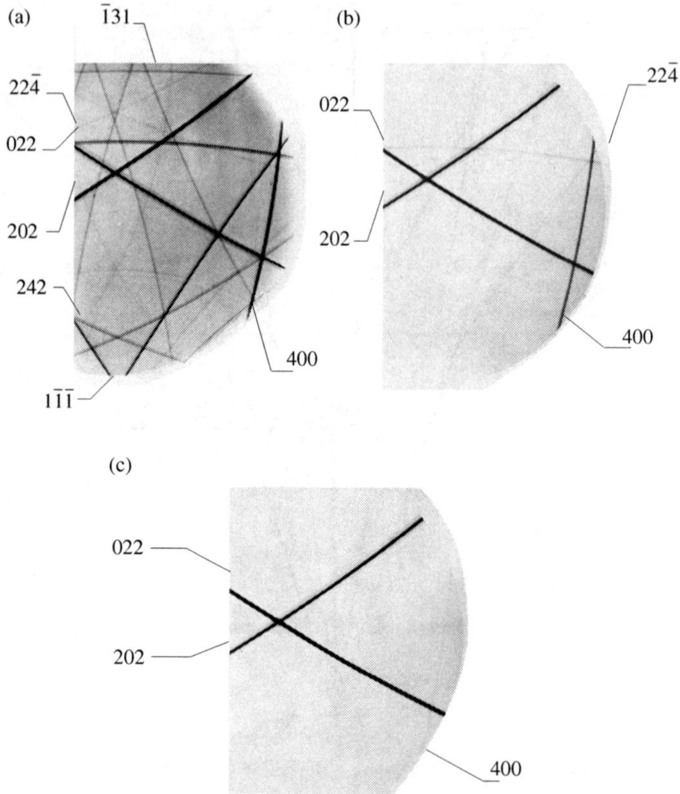

Fig. 1.6. Wide-angle diagrams. (110) germanium crystal, Cu$K\alpha$. Ni filtered, voltage 20 kV, current 1.5 mA. Courtesy G. Borrmann (for the detailed interpretation, see Section 4.8.4). (a) $t = 0.23$ mm; $\mu_o t = 8$; (b) $t = 0.5$ mm; $\mu_o t = 17.5$; (c) $t = 1.2$ mm; $\mu_o t = 42$ (μ_o: normal absorption coefficient).

absorption coefficient which he determined. Borrmann (1950) repeated his early 1941 experiments very carefully with calcite crystals by taking photographs of the reflected and direct beams after different crystal thicknesses. He could estimate the apparent (*anomalous*) absorption coefficient and confirmed Laue's theoretical calculations. Borrmann (1950) gave Laue's simple physical explanation for the phenomenon of anomalous absorption. Let us consider the two-beam case and transmission geometry. The intensity of wavefield (1.3) is

$$|E|^2 = |E_o|^2 + |E_h|^2 + 2|E_o E_h| \cos 2\pi(\mathbf{h} \cdot \mathbf{r} + \psi),$$

where ψ is the phase of E_h/E_o. Laue (1935, 1941) had noted that this expression shows that the two waves interfere and produce a set of standing waves. The term $\cos 2\pi(\mathbf{h} \cdot \mathbf{r} + \psi)$ shows that the nodes lie on planes parallel to the lattice planes and

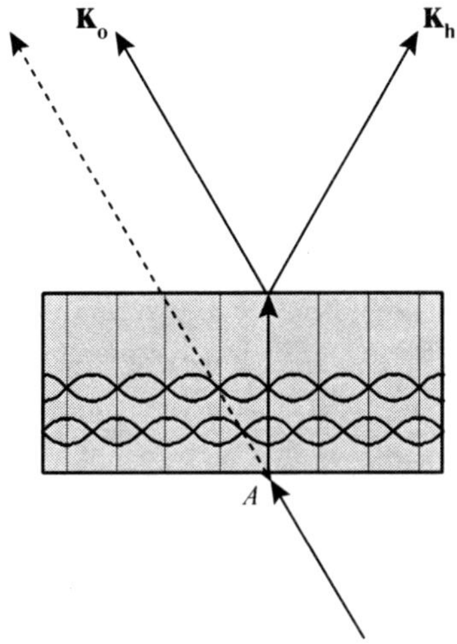

Fig. 1.7. Borrmann effect: the standing wave nodes lie on the atomic planes for branch 1 of the dispersion surface (solid lines). The corresponding wavefields undergo small absorption. For branch 2 (dotted lines), the antinodes lie on the atomic planes and there is a high absorption. In a thick, absorbing, crystal the path of the wavefields is along the lattice planes.

that their periodicity is equal to the distance $1/h = d_{hkl}/n$ where d_{hkl} is the periodicity of the hkl family of lattice planes and n is order of the reflection. The phase ψ is equal to π for wavefields associated with branch 1 of the dispersion surface and to 0 for wavefields associated with branch 2 of the dispersion surface (Fig. 1.4). The nodes of standing waves therefore lie on the atomic planes for a wavefield associated with branch 1 of the dispersion surface while it is the antinodes (maxima of electric field) which lie on the atomic planes for wavefields associated with the other branch of the dispersion surface (Fig. 1.7). Borrmann (1954) pointed out that the former would undergo a very small absorption and penetrate through thick crystals while the latter would be absorbed out very rapidly.

The Borrmann effect was also confirmed independently by Campbell (1951*a*,*b*) and by Schwarz and Rogosa (1954) who measured rocking curves with calcite in transmission with a double-crystal diffractometer. The rocking curves for the refracted beam are asymmetric due to the anomalous absorption and in agreement with those predicted by Laue (1949) (see Chapter 6). The rocking curves for the refracted and the reflected beams through thick absorbing crystals were studied in detail by Brogren and his coworkers (see, for instance, Brogren 1954; Brogren and Adell 1954*a*,*b*). A further confirmation is due to Hirsch (1952) who compared integrated intensities measured on calcite crystals with theoretical values evaluated graphically for the absorbing case (Hirsch and Ramachandran 1950). These integrated intensities were calculated analytically by Kato (1955). The anomalous transmission effect was also found for neutron diffraction by perfect crystals, indirectly by Knowles (1956) who measured the emission of γ-rays from calcite crystals and directly by Sippel *et al.* (1962) who measured directly the anomalous absorption coefficient through an InSb crystal. The latter authors also observed the asymmetry in the rocking curve for the refracted beam (Sippel *et al.* 1964).

It has been noted in Section 1.6 that the dispersion surface has different branches for the two directions of polarization. The theory shows that the anomalous absorption coefficient is smaller for the σ-polarization than for the π-polarization (see Section 4.8.3). It is therefore possible to make an X-ray polarizer by using a sufficiently thick crystal (see Section 15.3.5.1). This was verified by Cole *et al.* (1961).

1.10.2 *Wavefield trajectories*

1. *Poynting vector*

As early as 1952, Kato (1952*a*) showed that, for electrons, the direction of propagation of the wavefields is along the normal to the dispersion surface. For X-rays, it is well known from Maxwell's theory of electromagnetism that the direction of propagation of the energy of an electromagnetic wave is along the Poynting vector, $\mathbf{S} = \mathcal{R}e(\mathbf{E} \wedge \mathbf{H}^*)$, where $\mathcal{R}e(x)$ means real part of x and x^* the complex conjugate of x. Laue (1952) applied this relation to the wavefields inside the crystal: **E** and **H** are then replaced by a sum of terms. The result is that the Poynting vector, averaged over time and over the volume of the unit cell, is proportional to $\left[|D_o|^2\,\mathbf{s_o} + |D_h|^2\,\mathbf{s_h}\right]$ where $\mathbf{D_o}$ and $\mathbf{D_h}$ are the electric displacements of the refracted and reflected waves, respectively, and $\mathbf{s_o}$ and $\mathbf{s_h}$ are unit vectors in the incident and the reflected directions, respectively (see section 5.4). The direction of propagation is common to the two waves in the wavefield. Its orientation is intermediate between those of the incident and reflected directions and depends on the relative weight of the two wave components in the wavefield (Fig. 1.8). In particular, when $D_o = D_h$, which occurs at the center of the reflection domain, the direction of propagation is exactly parallel to that of the lattice planes (Borrmann 1950). Borrmann showed that this is the direction along which the anomalous absorption coefficient is the smallest (Borrmann 1954; Borrmann *et al* 1955). For a thick absorbing crystal (typically if the product μt of the linear absorption coefficient μ by the crystal thickness t is of the order of 10 or more), the beam propagating inside the crystal is restricted to a narrow pencil parallel to the lattice planes (Fig. 1.7): the crystal then acts as a waveguide. The same result was obtained in the reflection geometry (Borrmann *et al.* 1955; Wagner 1956). The spreading out of the propagation directions in a fan was first described by Borrmann *et al.* (1955), hence the expressions Borrmann triangle or Borrmann fan (Fig. 1.10).

It is Kato (1958) who showed that the Poynting vector is normal to the dispersion surface, also in the n-beam case. After he presented his results at the ACA meeting in Pittsburgh in November 1957, Ewald and he had long discussions on the propagation of X-rays in crystals. Ewald pointed out that Kato's calculation involved plane waves and did not provide a physical description of the behaviour of the wavefields. He then proposed a very simple physical proof by substituting wavebundles for plane waves (Ewald 1958). It is well known in optics that the propagation of energy of an

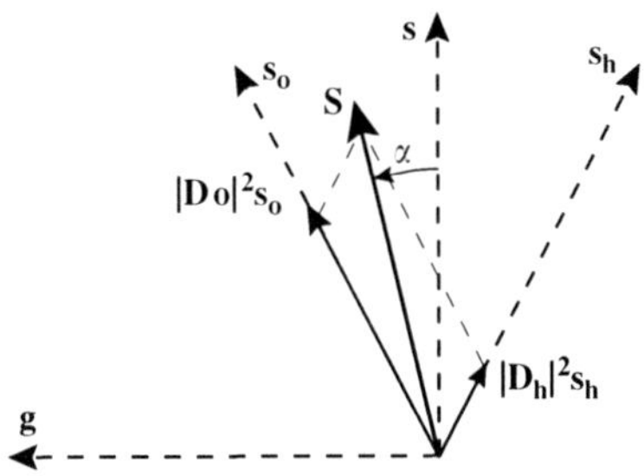

Fig. 1.8. Poynting vector, **S**, of a wavefield. $\mathbf{s_o}$ and $\mathbf{s_h}$ are unit vectors in the incident and reflected directions, respectively; **s** is a unit vector parallel to the lattice planes; **g** is a unit vector parallel to the reciprocal lattice vector.

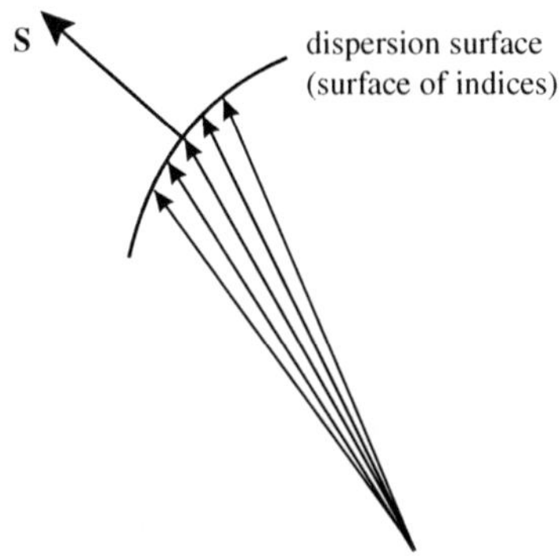

Fig. 1.9. Direction of propagation of a wavepacket, **S**.

electromagnetic wave is along the normal to the surface of indices. This is proved by studying the interferences of the waves in a wavepacket and showing that the direction of propagation as given by the group velocity is along the normal to the locus of the extremity of the wavevectors in the wavepacket (Fig. 1.9)—see Sections 12.4 and 12.5.

Ewald showed that this result can be extended to the propagation of packets of wavefields in a crystal by noting that the dispersion surface plays the same role as the surface of indices; it is the locus of the extremity of the wavevectors. This discussion is justified by the fact that X-ray sources are always divergent; they can be very narrow in the case of a synchrotron radiation source, their structure can be modified by beam conditioners, but they will never be plane waves in the mathematical sense (their Fourier transform is never a Dirac distribution). It was actually the first time that it was realized that the propagation of X-rays should be considered in terms of wavepackets and Ewald's paper provides the physical justification for all the ray tracing studies that were to follow. Wagner (1959*a*,*b*) extended the calculation of the group velocity to the case of electrons and polychromatic waves.

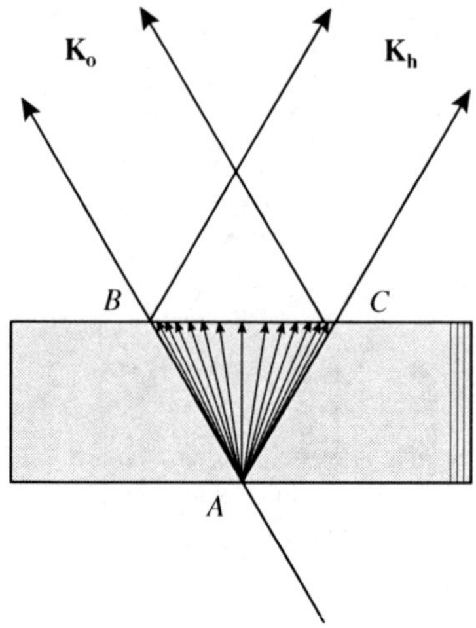

Fig. 1.10. Borrmann triangle. $\mathbf{K_0}$: incident direction; $\mathbf{K_h}$: reflected direction. (After Borrmann 1959*a*.)

2. *The Borrmann triangle (or Borrmann fan)*

The beam coming from a point in the focus of the X-ray tube and incident on the crystal has a divergence which is very much wider than the width of the rocking curve which is of the order of a few seconds of arc. Every point of the dispersion surface is then excited and that the paths of the wavefields span the whole range of directions between the incident and the reflected directions (Fig. 1.10). They are limited inside the crystal within the triangle *ABC* (*AB*, incident direction; *AC*, reflected direction), now called *Borrmann triangle* or *Borrmann fan*. Borrmann (1959*a*) estimated that the wavefields should be more densely distributed along the edges *AB* and *AC*, and Kato (1960*a*) calculated the exact distribution of intensity along the base *BC* of the Borrmann triangle (see Sections 10.4 and 12.7).

3. *Double refraction*

It has been noted in Section 1.6 that an incident plane wave excites four wavefields inside the crystal, two for each direction of polarization. These wavefields propagate inside the crystal along four different directions of propagation which are respectively parallel to the normal to the surface of dispersion at the four corresponding tiepoints. This is what Borrmann (1955, 1959*a*) called the *quadruple refraction* ('*Vierfachbrechung*'). The paths associated with the two directions of polarization are very close to one another and one can only expect to observe a double refraction (Fig. 1.11).

The existence of this property of double refraction for X-rays was first mentioned by Ewald (1938). It is, however, very difficult to prove experimentally: if one could create an incident X-ray plane wave, it would have, by definition, an infinite lateral extension. On the other hand, the beam coming from the X-ray tube has a divergence which is much wider than the width of the rocking curve. This difficulty was overcome by Authier (1960*a*, 1961*a*) in Paris by means of a two-crystal setting (see Fig. 12.14); the

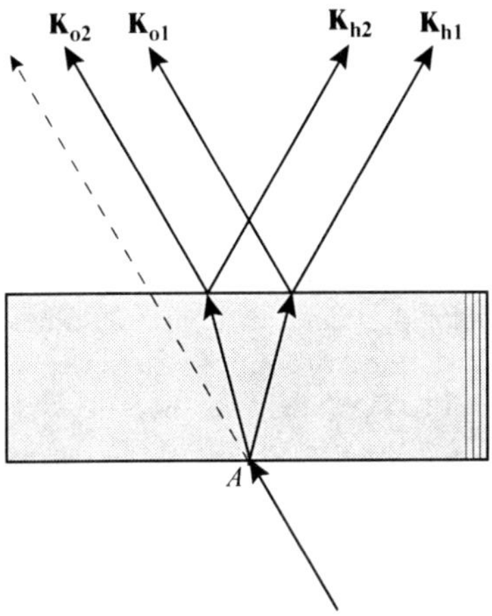

Fig. 1.11. Double refraction. $\mathbf{K}_{o1}$, $\mathbf{K}_{h1}$: wavevectors of wavefield 1; $\mathbf{K}_{o2}$, $\mathbf{K}_{h2}$: wavevectors of wavefield 2.

beam emitted by the X-ray tube falls on a first silicon crystal which is relatively thick and not too absorbing. A fine slit selects a small portion of the reflected beam coming out of the first crystal. This small portion is narrow spatially in direct space and has a very small divergence, much smaller than the width of the rocking curve: it is also very narrow in reciprocal space and can be considered as a 'pseudo-plane' wave with a small lateral extension. This beam is incident on a second silicon crystal where it will generate two wavefields whose paths are sufficiently separated to be observed (Fig. 12.18). The details of the experimental setting and the results are described in Section 12.10. This experiment constituted the most direct proof of the physical reality of the notion of wavefield put forward by Ewald. The same setting was used to record intrinsic rocking curves (Authier 1960*b*; Entin and Assur 1981) and to follow the path of the wavefield excited in the reflection geometry by an incident pseudo-plane wave (Authier 1962). Double refraction was also later directly measured by Lang *et al.* (1986*a*) using a very fine beam from a synchrotron radiation source and a diamond prism.

The refraction effect can also be observed by studying a prism-shaped crystal. As in the case of optics, there is a slight difference in the orientation of the emerging wavevectors after a plane-parallel crystal and a wedge-shaped one. This was first observed in the case of electron diffraction by Cowley and Rees (1947) and Sturkey (1948) in MgO crystallites and discussed by means of dynamical theory by Kato (1949, 1952*a*,*b*). It was first measured in the case of X-rays by Kohra *et al.* (1965) using a silicon crystal (see Section 6.3.2).

1.10.3 *Pendellösung*

1. *Spherical waves*

A. R. Lang, who had been developing topographic techniques for the direct photographic observation of crystal defects (Lang 1957), was joined in Harvard in the fall of 1957 by N. Kato. He had just observed in projection topographs fine thickness fringes along the edges of a silicon wedge-shaped crystal provided by Jim Patel. Kato, who had discussed Pendellösung oscillations in the electron case (Kato 1953), immediately interpreted them as Pendellösung fringes. Section topographs with their hooked-shaped fringes (Fig. 10.6) were more puzzling (Kato and Lang 1959). Kato (1960*b*) guessed that these fringes were not of the type predicted by Ewald, which are due to an incident plane wave, but were due to spherical waves. They result from the interference of wavefields propagating along the same direction of the Borrmann fan but whose tiepoints lie on the extremities of a diameter of the dispersion surface (Fig. 10.7). Kato then extended the dynamical theory to that case (Kato 1961*a*,*b*—see Chapter 10) and showed how the fringes are formed. Spherical-wave Pendellösung fringes were later also observed in the reflection geometry (Uragami 1969). Kato and Lang's fringes were equal-thickness fringes. Using white radiation and a crystal of constant thickness, it is also possible to observe Pendellösung fringes due to the variation of their periods with wavelength (Aristov *et al.* 1977*a*,*b*; Takama *et al.* 1980).

Pendellösung fringes were also observed in neutron diffraction in nearly perfect non-magnetic crystals, in silicon by Sippel *et al.* (1965), Shull (1968), Kikuta *et al.* (1971); Shull and Oberteuffer (1972) and Malgrange *et al.* (1976) and in germanium by Shull and Shaw (1973) and in magnetic crystals ('magnetic Pendellösung'), in yttrium iron garnet (YIG) by Baruchel *et al.* (1982) and in $FeBO_3$ and α-Fe_2O_3 by Kvardakov and Somenkov (1992).

2. *Pseudo-plane waves*

Pendellösung means, as we have said, periodic exchange of energy between two interfering wavefields. If the crystal is wedge shaped thickness fringes are thus expected. These are the so-called *equal-thickness Pendellösung* fringes (Fig. 4.16). But the period of these interferences depends on the orientation of the incident wave on the crystal (see Section 4.8.7). If this orientation is varied, *equal-orientation Pendellösung* fringes are expected (Fig. 4.17). This was first observed by Malgrange and Authier (1965) with the setting used for the production of a pseudo-plane wave and for the observation of the double refraction (Fig. 12.20) and with a wedge-shaped second crystal. In the thinner part of the crystal the paths of the wavefields created by double refraction overlapped and their interferences could be observed photographically. The same setting enabled Lefeld-Sosnowska and Malgrange (1968, 1969) to observe the Pendellösung fringes in the rocking curves from crystals of different thicknesses and with different wavelengths. The angular width of the beam incident on the second crystal was sufficiently small for the agreement of the experimental and theoretical profiles to be quite good.

Pseudo-plane waves can be obtained in various other ways: by means of a triple-crystal diffractometer as proposed by Renninger (1938) and implemented by him (Renninger 1955) or by using a very asymmetric reflection on a first crystal (Renninger 1961; Kohra 1962). The latter technique was used:

(i) In the *transmission geometry*, by Kikuta and Kohra (1968*a*) to observe equal-thickness fringes, by Kikuta and Kohra (1968*b*) and Kohra and Kikuta (1968) to observe fringes in the rocking curves, by Renninger (1968) and Sugii *et al.* (1971) to observe equal-inclination fringes in bent crystals and by Alstrup (1978) to study minute strains in ion-implanted samples.

(ii) In the *reflection geometry*, by Batterman and Hildebrandt (1967, 1968) and by Hashizume *et al.* (1970) to observe fringes in the rocking curve, by Nakayama *et al.* (1970) to observe equal-thickness fringes, and by Renninger (1968) to observe equal-inclination fringes in bent crystals.

Another very interesting method was used by Hart and Milne (1968) who used reflections through two successive slabs in a monolithic block of perfect silicon. The first slab was thin and wedge shaped, the second one thick and plane parallel. Both spherical-wave (after the first crystal) and plane-wave (after the second crystal) equal-thickness Pendellösung fringes were thus obtained and compared. The respective positions of the fringes for the two types of Pendellösung were in good agreement with theory. A slightly different interpretation of that experiment has been given by Indenbom and Slobodetskii (1976).

Nowadays, it is possible to observe plane-wave Pendellösung fringes directly using the beam from a third-generation synchrotron (see Section 15.2). At 150 m from the synchrotron source at ESRF, for instance, the beam received by a point on the sample has a divergence which is much smaller than the width of the rocking curve and it can be approximated by a 'locally plane' wave (Mocella *et al.* 1999, 2000).

1.11 Diffraction by deformed crystals

Dynamical diffraction requires, as we have seen, perfect or nearly perfect crystals. The first tests of the dynamical theory were made with the best crystals available which were natural crystals such as calcite, quartz and diamond. But, as the quality of synthetic crystals increased, the need appeared for characterization techniques which were sensitive to slight imperfections: small density of dislocations or subgrain boundaries, weak long–range strains.

Topography is one of those techniques (see Chapter 17):

(1) *by transmission*, section topography (Lang 1957, 1958), projection topography (Lang 1959*a*,*b*), Borrmann topography (Borrmann *et al.* 1958; Borrmann 1959*b*);

(2) *by reflection*, Berg–Barrett topography (Newkirk 1958*a*,*b*, 1959), double-crystal topography (Bonse and Kappler 1958; Bonse, 1958).

The images of defects on X-ray diffraction topographs could not be interpreted without understanding the path of wavefields in distorted crystals but, at the same time, gave clues for that. Extensions of the dynamical theory were developed for deformed crystals: the ray theories for slightly deformed crystals (Penning and Polder 1961; Kato 1963*d*, 1964*a*,*b*; see Chapter 13) and the generalized theories allowing strong deformations and where the linear fundamental equations are replaced by partial differential equations (Takagi 1962; Taupin 1964; see Chapter 14). The images of dislocations and stacking faults were interpreted (Authier 1967, 1968) and simulated (Balibar and Authier, 1967; Authier *et al.* 1968*a*) by numerical solution of the Takagi–Taupin equations.

Another very important and popular technique of characterization is high-resolution diffractometry, which has become very popular due to the strong interest in semiconductor multilayer devices for microelectronics obtained by vapour deposition, liquid phase epitaxy or ion implantation. This can be done with standard triple-crystal arrangements, for instance, but the sample should be from the same material as that the two other crystals and this limits strongly the applications. This difficulty can be overcome in various ways with multiple-crystal multiple-reflections monochromators usually involving asymmetric cuts, such as the Bartels (1983) three-crystal five-reflection monochromator improved by Fewster (1989) with an eight-reflection setting including an analyser crystal after the sample. The improvement of computer-simulation techniques now makes possible the determination of the strain distribution by interpretation of the rocking curves. Reviews are given in Bowen and Tanner (1998), Holý *et al.* (1999) and Fewster (2000).

1.12 Modern times

The times around the late 1950s and the early 1960s were very exciting not only because the properties of the dispersion surface and of the wavefields were then really understood and verified, 40 odd years after Ewald's founding paper, but also because many applications or extensions of the dynamical theory were developed. In 1960, the number of groups working on dynamical theory was very small and they were located in a handful of countries. It rose sharply and, 10 years later, the number of yearly publications on the topic was nearly 10 times larger. It then remained roughly stable until the years around 1985 when it again rose very fast. By the end of 2002, more than four thousand papers had been published on dynamical theory and its applications.

The increase of studies on dynamical diffraction in the early 1960s is parallel to the development of the transistor and of the LASER. There was a great demand for the characterization of the materials grown for important industrial applications: semiconductors, piezoelectric, pyroelectric, ferroelectric, magnetic, LASER materials. At the same time the existence of these nearly perfect crystals permitted direct applications such as the evaluation of form factors and of anomalous dispersion corrections by measuring Pendellösung fringe distances (Hattori *et al.* 1965*b*), half-widths of Bragg geometry rocking curves (Kikuta *et al.* 1970), interference patterns produced by bicrystals (Hart and Milne 1970) or the fine structure of Laue-geometry rocking curves (Bonse and Teworte 1980); in each of these examples only the first publication is mentioned but many others followed, in particular for the Pendellösung method which is the most popular (see Section 2.7.2). Extremely accurate mapping of tiny variations of the lattice parameter in a crystal became possible (Hart 1968*b*, 1981).

Entire new domains of research appeared: X-ray interferometry invented by Bonse and Hart (1965*a*,*b*) (see Section 15.5), location of atoms at crystal surfaces and interfaces using the fluorescence emitted at the standing waves antinodes, following a suggestion by Batterman (1964*a*, 1969) (see Chapter 16), extension of dynamical diffraction to the cases where the Ewald sphere cannot be approximated by its tangential plane (see Chapter 8).

The real revival of dynamical diffraction which took place in the first half of the 1980s is due to the development of second- and third-generation high brilliance synchrotron sources which have opened entirely new possibilities in X-ray diffraction. They have required X-ray optical systems of many kinds which all operate in the dynamical diffraction regime (see Chapter 15): for the production of beams with a very narrow angular divergence or with a very narrow spectral width around a desired wavelength, for the reduction of harmonics, for the production of polarized X-rays for studies of magnetic diffraction (de Bergevin and Brunel 1972), Compton magnetic scattering (Sakai and Ono 1976) or magnetic dichroism (Schütz *et al.* 1989), etc.

The use of synchrotron radiation as a source started in the 1970s with the first white beam topographs (Tuomi *et al.* 1973, 1974; Hart 1975*a*) and the first specially designed monochromators (Beaumont and Hart 1974; Bonse *et al.* 1976*b*; Hart and Rodrigues 1978—recent developments are described in Chapter 15). With these new sources, the resolution was considerably increased for X-ray topography and diffractometry, the exposure times have been reduced dramatically, down to a few hundredth of a second for white beam experiments at the ESRF (Baruchel *et al.* 1994; Baruchel 1996), thus enabling real-time, *in situ* experiments: crystal growth, plastic deformation and movement of dislocations, movement of magnetic or ferroelectric domains (see Section 17.3.6). In 1976 the first tensile stage (Bowen and Miltat 1976) was built, the first recrystallization experiment (Gastaldi and Jourdan 1976) was made and the first movements of magnetic domains (Tanner *et al.* 1976) were observed. Recent developments are reviewed in Chapter 17. The coming fourth-generation sources (FELs) are opening up the possibility of performing experiments with femtosecond pulses and time-dependent extensions of the dynamical theory are being developed (see Sections 10.10 and 11.1).

Reviews of the classical dynamical theory of X-ray diffraction are given in Zachariasen (1945), Laue (1960), James (1950, 1963), Batterman and Cole (1964), Borrmann (1964), Authier (1970, 2001), Kato (1974), Brümmer and Stephanik (1976), Pinsker (1978) and Chang (2004). An original treatment of the diffraction by deformed crystals is given by Krivoglaz (1996). Many papers on the applications of the dynamical theory can be found in the Proceedings of Schools devoted to that topic (Tanner and Bowen 1980; Weissmann *et al.* 1984; Authier *et al.* 1996) and in the Proceedings of the XTOP meetings: *J. Physics D (Applied Physics)*, 1993, Vol. 26 (ed. J. Baruchel) 1995, Vol. 28 (ed. R. Köhler and H. Klapper), 1999, Vol. 32 (ed. B. K. Tanner), 2001, Vol. 34 (ed. J. Gronkowski), Vol. 36 (ed. J. Baruchel), 2003 and *Il Nuovo Cimento*, 1997, Vol. 19D (ed. C. Bocchi, C. Ferrari, S. Lagomarsino and L. Tapfer).

Reviews of the dynamical theory of neutron diffraction can be found in Rauch and Petrascheck (1978), Klein and Werner (1983), Sears (1979, 1989) and Schlenker and Guigay (1996, 2001).

2

Properties of the electromagnetic field—propagation and scattering

2.1 Maxwell's equations[1]

X-rays are electromagnetic waves of very short wavelength. When they interact with matter they excite the electrons which oscillate and behave like dipoles. The derivation of the wave equation used in both Ewald's and Laue's dynamical theories of X-ray diffraction is based on the properties of the interaction of electromagnetic waves with matter. The electromagnetic field is represented by two vectors, the electric field, **E**, and the magnetic induction, **B**, which are related locally to the electric displacement, **D**, and the magnetic field, **H**, by the *material relations* that describe the reaction of a linear medium to the electromagnetic field:

$$\mathbf{D} = \epsilon\mathbf{E} = \epsilon_0\mathbf{E} + \mathbf{P}, \tag{2.1}$$

$$\mathbf{B} = \mu\mathbf{H} = \mu_0(\mathbf{H} + \mathbf{M}), \tag{2.2}$$

where ϵ and μ are the dielectric constant (or permittivity) and the magnetic permeability of the medium, respectively, ϵ_0 and μ_0 are the electric permittivity and the magnetic permeability of vacuum, **P** is the electric polarization and **M** the magnetization.

In a continuous medium, the space and time derivatives of **E**, **B**, **D**, **H**, of the electric current density, **j** ($\mathbf{j} = \sigma\mathbf{E}$, σ, specific conductivity), and of the local electric charge density ρ are related by Maxwell's equations

$$\operatorname{curl}\mathbf{H} = \frac{\partial\mathbf{D}}{\partial t} + \mathbf{j}, \tag{2.3}$$

$$\operatorname{curl}\mathbf{E} = -\frac{\partial\mathbf{B}}{\partial t}, \tag{2.4}$$

$$\operatorname{div}\mathbf{D} = \rho, \tag{2.5}$$

$$\operatorname{div}\mathbf{B} = 0. \tag{2.6}$$

These equations are complemented by the following boundary conditions at the surface between two neighbouring media, 1 and 2, in the absence of a surface charge

[1] The international system of units SI is used throughout. Useful references are Born and Wolf (1993) and Jackson (1999).

density and of surface current density:

$$\mathbf{E}_{T1} - \mathbf{E}_{T2} = 0, \qquad \mathbf{D}_{N1} - \mathbf{D}_{N2} = 0,$$
$$\mathbf{H}_{T1} - \mathbf{H}_{T2} = 0, \qquad \mathbf{B}_{N1} - \mathbf{B}_{N2} = 0, \tag{2.7}$$

where the indices T and N indicate the tangential and normal components, respectively.

From (2.3) and (2.5) and using the identity $\operatorname{div}\operatorname{curl}\mathbf{X} \equiv 0$ ($\mathbf{X}$, any vector), it follows that

$$\frac{\partial \rho}{\partial t} + \operatorname{div}\mathbf{j} = 0; \tag{2.8}$$

this is the equation of continuity, expressing the conservation of charges.

In the derivation of the expression of the electric field radiated by a dipole it is very useful to make use of the so-called *retarded potentials*. This is done with the help of two sets of quantities, the scalar and vector potentials, and the Hertz vectors.

2.2 The electrodynamic potentials in vacuum

2.2.1 *The vector and scalar potentials*

In vacuum, the dielectric constant and the magnetic permeability are equal to ϵ_0 and μ_0, respectively, and Maxwell's equations (2.3) and (2.5) may be rewritten as

$$\operatorname{curl}\mathbf{B} = \mu_0 \frac{\partial \mathbf{D}}{\partial t} + \mu_0 \mathbf{j}, \tag{2.9}$$

$$\operatorname{div}\mathbf{E} = \frac{\rho}{\epsilon_0}. \tag{2.10}$$

The vector potential $\mathbf{A}$ and the scalar potential ϕ are introduced by setting

$$\mathbf{B} = \operatorname{curl}\mathbf{A}, \tag{2.11}$$

$$\mathbf{E} = -\operatorname{grad}\phi - \frac{\partial \mathbf{A}}{\partial t}. \tag{2.12}$$

These two relations are compatible with Maxwell's equations (2.4) and (2.6). The vector and scalar potentials must be defined in such a way that eqns (2.9) and (2.10) are satisfied. By substituting from (2.11) into (2.9) and using the identity $\operatorname{curl}\operatorname{curl}\mathbf{X} \equiv \operatorname{grad}\operatorname{div}\mathbf{X} - \Delta\mathbf{X}$, we obtain

$$\Delta\mathbf{A} - \epsilon_0\mu_0 \frac{\partial^2 \mathbf{A}}{\partial t^2} - \operatorname{grad}\left(\operatorname{div}\mathbf{A} + \epsilon_0\mu_0 \frac{\partial \phi}{\partial t}\right) = -\mu_0 \mathbf{j}.$$

In a similar way, by substituting from (2.12) into (2.10)

$$\Delta\phi + \frac{\partial(\operatorname{div}\mathbf{A})}{\partial t} = -\frac{\rho}{\epsilon_0},$$

which can also be written in the following way:

$$\Delta\phi - \epsilon_0\mu_0\frac{\partial^2\phi}{\partial t^2} + \frac{\partial(\operatorname{div}\mathbf{A} + \epsilon_0\mu_0\,\partial\phi/\partial t)}{\partial t} = -\frac{\rho}{\epsilon_0}.$$

If one introduces the further relation

$$\operatorname{div}\mathbf{A} + \epsilon_0\mu_0\frac{\partial\phi}{\partial t} = 0, \tag{2.13}$$

the vector and scalar potentials are solutions of the inhomogeneous wave equations:

$$\begin{aligned} \Delta\mathbf{A} - \epsilon_0\mu_0\,\frac{\partial^2\mathbf{A}}{\partial t^2} &= -\mu_0\mathbf{j}, \\ \Delta\phi - \epsilon_0\mu_0\,\frac{\partial^2\phi}{\partial t^2} &= -\frac{\rho}{\epsilon_0}. \end{aligned} \tag{2.14}$$

Equation (2.13) is called the Lorentz condition.

2.2.2 *The retarded potentials*

It can be shown that the following functions are solutions of (2.14) (see, for instance, Born and Wolf 1993):

$$\begin{aligned} \mathbf{A}(\mathbf{r},t) &= \frac{\mu_0}{4\pi}\int\frac{\mathbf{j}(\mathbf{r}',t-|\mathbf{r}-\mathbf{r}'|/c)}{|\mathbf{r}-\mathbf{r}'|}\,\mathrm{d}\tau', \\ \phi(\mathbf{r},t) &= \frac{1}{4\pi\epsilon_0}\int\frac{\rho(\mathbf{r}',t-|\mathbf{r}-\mathbf{r}'|/c)}{|\mathbf{r}-\mathbf{r}'|}\,\mathrm{d}\tau', \end{aligned} \tag{2.15}$$

where $\mathbf{r}$ is the position vector of the observation point, $\mathbf{r}'$ the position vector of volume element $\mathrm{d}\tau'$, and $\mathbf{j}(\mathbf{r}',t-|\mathbf{r}-\mathbf{r}'|/c)$ and $\phi(\mathbf{r}',t-|\mathbf{r}-\mathbf{r}'|/c)$ are the values of $\mathbf{j}$ and ϕ at the time at which each contribution should leave point $\mathbf{r}'$ in order to reach point $\mathbf{r}$ at the required time t. The vector and scalar potentials given by (2.15) are called for that reason the *retarded potentials*. They constitute a particular solution of the wave equations (2.14). The general solution is obtained by adding it to a general solution of the homogeneous equations:

$$\begin{aligned} \Delta\mathbf{A} - \epsilon_0\mu_0\,\frac{\partial^2\mathbf{A}}{\partial t^2} &= 0, \\ \Delta\phi - \epsilon_0\mu_0\,\frac{\partial^2\phi}{\partial t^2} &= 0. \end{aligned}$$

2.3 The electrodynamic potentials in polarized media

We shall assume that the field is propagating in a non-conducting material ($\sigma = 0$) and in a region where there are no charge densities ($\rho = 0$) and no electric current density ($\mathbf{j} = 0$). Substituting the expressions (2.1) and (2.2) of $\mathbf{D}$ and $\mathbf{B}$ into Maxwell's equations (2.3) and (2.4), respectively, we obtain

$$\operatorname{curl} \mathbf{B} = \epsilon_0 \mu_0 \frac{\partial \mathbf{D}}{\partial t} + \mu_0 \mathbf{j}',$$

$$\operatorname{div} \mathbf{E} = \frac{\rho'}{\epsilon_0},$$

where the free current density $\mathbf{j}'$ and the free charge density ρ' are given by

$$\mathbf{j}' = \frac{\partial \mathbf{P}}{\partial t} + \operatorname{curl} \mathbf{M}, \tag{2.16}$$

$$\rho' = -\operatorname{div} \mathbf{P}. \tag{2.17}$$

They obviously satisfy the equation of continuity (2.8).

The vector and scalar potentials are introduced through the same eqns (2.11) and (2.12) and they should satisfy the Lorentz condition (2.13). By analogy with the case of the propagation in vacuum, they are solutions of the wave equations

$$\begin{aligned} \Delta \mathbf{A} - \epsilon_0 \mu_0 \frac{\partial^2 \mathbf{A}}{\partial t^2} &= -\mu_0 \mathbf{j}', \\ \Delta \phi - \epsilon_0 \mu_0 \frac{\partial^2 \phi}{\partial t^2} &= -\frac{\rho'}{\epsilon_0}. \end{aligned} \tag{2.18}$$

According to (2.15), particular solutions of (2.18) are of the form

$$\begin{aligned} \mathbf{A}(\mathbf{r}, t) &= \frac{\mu_0}{4\pi} \int \frac{\partial \mathbf{P}(\mathbf{r}', t - |\mathbf{r} - \mathbf{r}'|/c)/\partial t + \operatorname{curl} \mathbf{M}}{|\mathbf{r} - \mathbf{r}'|} \, \mathrm{d}\tau', \\ \phi(\mathbf{r}, t) &= -\frac{1}{4\pi \epsilon_0} \int \frac{\operatorname{div}' \mathbf{P}(\mathbf{r}', t - |\mathbf{r} - \mathbf{r}'|/c)}{|\mathbf{r} - \mathbf{r}'|} \, \mathrm{d}\tau', \end{aligned} \tag{2.19}$$

where the derivations in div' are over the coordinates of $\mathbf{r}'$.

2.4 Hertz vectors (polarization potentials)

As will be seen in the determination of the electric field radiated by a dipole and in the bases of Ewald's dynamical theory, it is very useful to relate the scalar and vector potentials to another pair of vectors, the Hertz vectors, $\mathbf{\Pi_e}$ and $\mathbf{\Pi_m}$ using Hertz's original notations (Hertz 1889), through the relations

$$\mathbf{A} = \mu_0\left(\frac{\partial \mathbf{\Pi_e}}{\partial t} + \text{curl}\,\mathbf{\Pi_m}\right),$$
$$\phi = -\frac{1}{\epsilon_0}\text{div}\,\mathbf{\Pi_e}. \tag{2.20}$$

The Lorentz condition (2.13) is then automatically satisfied. By substitution of (2.16), (2.17) and (2.20) into (2.18), it is easy to show that the scalar and potential vectors will be solutions of the wave equations (2.18) if the Hertz potentials are solutions of

$$\Delta\Pi_{\mathbf{e}} - \epsilon_0\mu_0\frac{\partial^2 \mathbf{\Pi}_e}{\partial t^2} = -\mathbf{P},$$
$$\Delta\Pi_{\mathbf{m}} - \epsilon_0\mu_0\frac{\partial^2 \mathbf{\Pi_m}}{\partial t^2} = -\mathbf{M}. \tag{2.21}$$

In analogy to the retarded potentials (2.19), particular solutions of (2.21) are of the form

$$\mathbf{\Pi_e}(\mathbf{r}, t) = \frac{1}{4\pi}\int \frac{\mathbf{P}(\mathbf{r}', t - |\mathbf{r} - \mathbf{r}'|/c)}{|\mathbf{r} - \mathbf{r}'|}\,\mathrm{d}\tau',$$
$$\mathbf{\Pi_m}(\mathbf{r}, t) = \frac{1}{4\pi}\int \frac{\mathbf{M}(\mathbf{r}', t - |\mathbf{r} - \mathbf{r}'|/c)}{|\mathbf{r} - \mathbf{r}'|}\,\mathrm{d}\tau'. \tag{2.22}$$

It will be assumed from now on that there is no magnetization and no magnetic interaction ($\mathbf{M} = 0$, $\mu = \mu_0$; $\mathbf{\Pi_m} = 0$). The relations (2.20) reduce to

$$\mathbf{A} = \mu_0\frac{\partial \mathbf{\Pi_e}}{\partial t},$$
$$\phi = -\frac{1}{\epsilon_0}\text{div}\,\mathbf{\Pi_e}. \tag{2.23}$$

The vectors of the electromagnetic field can be expressed directly in terms of the Hertz vector. From (2.2), (2.11) and (2.23), we obtain

$$\mathbf{B} = \mu_0\,\text{curl}\left(\frac{\partial \mathbf{\Pi_e}}{\partial t}\right),$$
$$\mathbf{H} = \text{curl}\left(\frac{\partial \mathbf{\Pi_e}}{\partial t}\right),$$

and from (2.12) and (2.23)

$$\mathbf{E} = \frac{1}{\epsilon_0}\text{grad div}\,\mathbf{\Pi_e} - \mu_0\frac{\partial^2 \mathbf{\Pi_e}}{\partial t^2}.$$

Using the identity

$$\text{grad div}\,\mathbf{X} \equiv \text{curl curl}\,\mathbf{X} + \Delta\mathbf{X}$$

and (2.21), we obtain finally

$$\mathbf{D} = \text{curl curl}\,\mathbf{\Pi_e}. \tag{2.24}$$

2.5 Propagation of an electromagnetic wave in vacuum

Let us consider an electromagnetic wave propagating in vacuum. In a region where there are no charge densities and no current densities, Maxwell's equations reduce to

$$\begin{aligned} \text{curl}\,\mathbf{H} &= \epsilon_0 \frac{\partial \mathbf{E}}{\partial t}, \\ \text{curl}\,\mathbf{E} &= -\mu_0 \frac{\partial \mathbf{H}}{\partial t}, \\ \text{div}\,\mathbf{E} &= 0, \\ \text{div}\,\mathbf{B} &= 0. \end{aligned} \tag{2.25}$$

Using the relation $\epsilon_0\mu_0 = 1/c^2$ where c is the velocity of light, one obtains the equation of propagation of an electromagnetic wave in vacuum by eliminating $\mathbf{H}$ between the first two relations of Maxwell's equations:

$$\Delta\mathbf{E} = \frac{1}{c^2}\frac{\partial^2\mathbf{E}}{\partial t^2}.$$

Its simplest solution is a plane wave,

$$\mathbf{E} = \mathbf{E_o}\exp[2\pi\mathrm{i}(\nu t - \mathbf{k}\cdot\mathbf{r})], \tag{2.26}$$

of which the wave number $k = 1/\lambda$ and the frequency ν are related by $k = \nu/c$.

2.6 Scattering of X-rays by an electron

Let us now consider an X-ray electromagnetic plane wave $\mathbf{E_o}\exp(2\pi\mathrm{i}\nu t)$ incident on an electron. The interaction of an electromagnetic wave with charged particles includes an electric interaction and a magnetic interaction. The magnetic interaction is weak and is neglected in classical treatments of the dynamical theory. It was first observed using conventional sources (de Bergevin and Brunel 1972; Brunel and de Bergevin 1974). The first calculations and the prediction that the effect can be observed are due to Platzmann and Tzoar (1970). The first detailed measurements and comparison with theory are given in de Bergevin and Brunel (1981) where ferro- and ferrimagnetic compounds were studied. The relative intensity changes when the orientation of the direction of magnetization normal to the diffraction plane is reversed are of the order of 10^{-4} to 5×10^{-3} and are therefore quite small, although the experiment made use of the very large anomalous scattering of Cu$K\alpha$. Resonant scattering close to absorption edges of magnetic elements is much more intense (see, for instance, Gibbs *et al.* 1988; Hannon *et al.* 1988). These studies are of course now performed using synchrotron radiation. Reviews of X-ray magnetic scattering studies are given in Lovesey (1987), Blume (1985, 1994), Blume and Gibbs (1988), de Bergevin and Brunel (1991), Lovesey and Collins (1996, 2001) and Laundy *et al.* (1998).

The electric interaction only will be considered here (charge scattering or Thomson scattering). The electric charge of the electron oscillates at the same frequency ν as that of the incoming electromagnetic wave and behaves like an electric dipole. The electromagnetic field radiated by this dipole can be calculated using the Hertz vectors. The rigorous determination of the forces acting on the electron is complicated and must be done using a quantum mechanical treatment. Non-magnetic resonant scattering induces an anisotropy of the dielectric susceptibility (Templeton 1994) which results in the appearance of forbidden reflections (see Section 2.7.1).

However, if we neglect these effects, to a very good approximation justified by the results of the exact calculation, the electron moves under the action of a driving force

$$\mathbf{F} = -e\mathbf{E_o} \exp(2\pi i \nu t),$$

where $-e$ is the charge of the electron and of a linear restoring force; it behaves like an oscillator with a resonant frequency ν_0. Damping can formally be taken into account by a friction term, α, proportional to the velocity of the electron. The equation of motion of the electron can thus be written as

$$m\,\frac{\mathrm{d}^2\mathbf{a}}{\mathrm{d}t^2} = -4\pi^2\nu_0^2 m\mathbf{a} + \alpha\,\frac{\mathrm{d}\mathbf{a}}{\mathrm{d}t} + \mathbf{F},$$

where m is the mass of the electron and $\mathbf{a}$ is the displacement vector of the electron with respect to its mean position.

The solution of the equation of motion is

$$\mathbf{a} = \frac{-e\mathbf{E_o}\exp(2\pi i \nu t)}{4\pi^2 m(\nu_0^2 - \nu^2) - 2\pi i \nu\alpha}.$$

The polarization associated with the dipole can be expressed as a distribution:

$$\mathbf{P}(\mathbf{r}, t) = \mathbf{p}(t)\,\delta(\mathbf{r}), \qquad (2.27)$$

where $\delta(\mathbf{r})$ is a Dirac distribution, the origin is taken at the dipole and

$$\mathbf{p}(t) = -e\mathbf{a} = p(t)\mathbf{n} = \frac{e^2 E_o \exp(2\pi i \nu t)\mathbf{n}}{4\pi^2 m(\nu_0^2 - \nu^2) - 2\pi i \nu\alpha} = \mathbf{P}\exp(2\pi i \nu t), \qquad (2.28)$$

where $\mathbf{n}$ is a unit vector parallel to $\mathbf{E_o}$.

Since we neglect the magnetic interaction, the magnetic Hertz vector, $\Pi_{\mathbf{m}}$, is zero. By substituting from (2.27) into (2.22) and taking into account the properties of Dirac distributions, the associated electric Hertz vector is obtained:

$$\mathbf{\Pi_e} = \frac{1}{4\pi}\frac{\mathbf{p}(t - r/c)}{r}. \tag{2.29}$$

Using (2.24), and if the electromagnetic field radiated by the electron propagates in vacuum, the field vectors are given by

$$\begin{aligned} \epsilon_0 \mathbf{E} = \mathbf{D} &= \text{curl curl } \mathbf{\Pi_e}, \\ \mathbf{H} &= \text{curl } \frac{\partial \mathbf{\Pi_e}}{\partial t}. \end{aligned} \tag{2.30}$$

Using the identity (curl curl $\equiv$ grad div $-$ Δ) and the relation $\Delta \Pi_{\mathbf{e}} - \epsilon_0 \mu_0 \partial^2 \mathbf{\Pi_e}/\partial t^2 = 0$ ($\mathbf{P} = 0$ everywhere except at the origin), (2.30) becomes

$$\epsilon_0 \mathbf{E} = \mathbf{D} = \text{grad div } \mathbf{\Pi_e} - \epsilon_0 \mu_0 \frac{\partial^2 \mathbf{\Pi_e}}{\partial t^2}.$$

Using (2.29) and after performing the derivations, one can express the field vectors in terms of spherical polar coordinates. Neglecting the terms in $1/r^2$ and of higher order, one finds that the radial components of **E** and of **H** are equal to zero. **E** and **H** are therefore perpendicular to **r**; we already know from Maxwell's equations and from (2.25) that in vacuum **H** and **E** are perpendicular. The electromagnetic field is therefore transverse. The value of the electric field is:

$$E = \frac{1}{4\pi\epsilon_0}\frac{\partial^2 p(t - r/c)/\partial t^2}{c^2 r} \sin\psi,$$

where ψ is the angle between the direction of the dipole, **n**, and that of the direction of propagation of the field, **r** (Fig. 2.1).

The $\sin\psi$ term corresponds to the well-known fact that a dipole does not radiate parallel to its direction and is at the origin of the polarization factor in the scattering of X-rays.

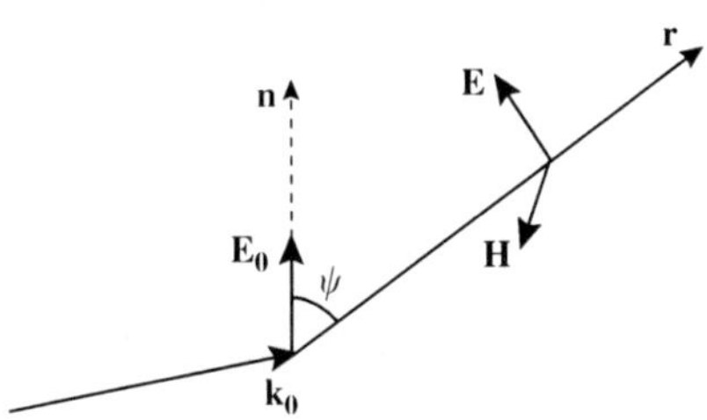

Fig. 2.1. Spherical wave radiated by a dipole.

By substitution from (2.28) we obtain

$$E = -\frac{e^2}{4\pi\epsilon_0 mc^2}\frac{\nu^2}{(\nu_0^2-\nu^2)-\mathrm{i}\alpha\nu/2\pi m}\frac{\exp[2\pi\mathrm{i}\nu(t-r/c)E_o]}{r}\sin\psi.$$

This is a spherical wave of wave number $k = \nu/c$:

$$E = -\frac{e^2}{4\pi\epsilon_0 mc^2}\frac{\nu^2}{(\nu_0^2-\nu^2)-\mathrm{i}\alpha\nu/2\pi m}\frac{\exp[2\pi\mathrm{i}(\nu t-kr)E_o]}{r}\sin\psi. \qquad (2.31)$$

For X-rays and far from an absorption edge, $\nu_0 \ll \nu$, and if we neglect ν_0 and consider the real part only, (2.31) reduces to the well-known J. J. Thomson formula (Thomson 1898). The term

$$R = \frac{e^2}{4\pi\epsilon_o mc^2} \qquad (2.32a)$$

is called the classical radius of the electron. It expresses the strength of the interaction and represents a scattering length. Its value is[2]

$$R = 2.817940 \times 10^{-15}\,\mathrm{m}. \qquad (2.32b)$$

2.7 Polarizability of matter for X-rays

2.7.1 *Elementary dispersion theory*

Let us consider a medium with a continuous distribution of charges. Under the influence of the propagating electromagnetic wave, the local electric charges oscillate in the same way as has been described in Section 2.6 and the medium becomes polarized. If $\rho(\mathbf{r})$ is the electronic density, the dipole moment $\mathrm{d}\mathbf{p}$ of a volume element $\mathrm{d}\tau$ is, by extrapolation of (2.28),

$$\mathrm{d}\mathbf{p}(\mathbf{r}) = \frac{\rho(\mathbf{r})e^2\mathbf{E}(\mathbf{r})}{4\pi^2 m(\nu_0^2-\nu^2)-2\pi\mathrm{i}\nu\alpha}\,\mathrm{d}\tau,$$

where $\mathbf{E}(\mathbf{r})$ is given by (2.26). The polarization of the medium is

$$\mathbf{P}(\mathbf{r}) = \frac{\mathrm{d}\mathbf{p}(\mathbf{r})}{\mathrm{d}\tau} = \frac{\rho(\mathbf{r})e^2\mathbf{E}(\mathbf{r})}{4\pi^2 m(\nu_0^2-\nu^2)-2\pi\mathrm{i}\nu\alpha}. \qquad (2.33)$$

It is related to the electric field through (2.1):

$$\mathbf{D}(\mathbf{r}) = \epsilon\mathbf{E}(\mathbf{r}) = \epsilon_0\mathbf{E}(\mathbf{r}) + \mathbf{P}(\mathbf{r}) = \epsilon_0[1+\chi(\mathbf{r})]\mathbf{E}(\mathbf{r}),$$

[2] CODATA recommended values of the fundamental Physical Constants: 1998 (Mohr and Taylor 2000).

where χ is the dielectric susceptibility, or polarizability.[3] Using (2.33), it follows that

$$\chi(\mathbf{r}) = \frac{\rho(\mathbf{r})e^2}{\epsilon_0[4\pi^2 m(\nu_0^2 - \nu^2) - 2\pi \mathrm{i}\nu\alpha]}.$$

For X-rays and far from an absorption edge, $\nu_0 \ll \nu$, and α is small. The polarizability can then be written as

$$\chi = \chi_n + \chi_a + \mathrm{i}\chi_i = \chi_r + \mathrm{i}\chi_i, \tag{2.34}$$

where χ_a and χ_i are called the anomalous dispersion corrections (classical expressions are given in James 1950) and $\chi_n(\mathbf{r})$ is, using (2.32) and the relation $\lambda = c/\nu$,

$$\chi_n(\mathbf{r}) = -\frac{\rho(\mathbf{r})R\lambda^2}{\pi}, \tag{2.35}$$

Expression (2.35) shows that the polarizability of matter for X-rays is very small, negative, proportional to the electronic density $\rho(\mathbf{r})$ and that it therefore varies with the space coordinates.

In the elementary description, the polarizability is isotropic and represented by a scalar. Actually it is anisotropic, due to several factors such as the anisotropy of the local atomic environment, magnetic ordering, orbital ordering a. s. o., and the anisotropy increases as the wavelength becomes closer to an absorption edge. The susceptibility is therefore a tensor; this induces a birefringence for X-ray scattering, optical activity (see Section 15.3.5.3) and the presence of otherwise forbidden reflections. This is due to the fact that the anisotropy violates the normal extinction rules according to the space group. This effect was discussed by Templeton and Templeton (1980) and studied theoretically by Dmitrienko (1983, 1984); it was first observed experimentally by Templeton and Templeton (1985, 1986) near the bromine K absorption edge. This anisotropy is local and exists even in cubic crystals and icosahedral quasicrystals in sites which do not have a high symmetry. In sites of a high symmetry, the dipole terms do not induce an anisotropy but dipole/quadrupole terms do induce it. The anisotropy of the higher rank susceptibility tensors induces an angular dependence of the absorption coefficient (see Section 2.7.4), forbidden reflections (such as those discovered in hematite by Finkelstein *et al.* (1992) and by Templeton and Templeton (1994) in potassium chromate and in germanium) and interesting polarization effects which are reviewed in Belyakov and Dmitrienko (1989) and in Kirfel and Petcov (1991). For recent discussions, see Dmitrienko *et al.* (1999) and Dmitrienko and Ovchinnikova (2000, 2001) and references therein.

2.7.2 *Fourier expansion of the polarizability*

In a crystalline medium, $\rho(\mathbf{r})$ is a triply periodic function of the space coordinates and so is χ. The polarizabilty and its real and imaginary parts can therefore be expanded

[3] As the physical origin of the susceptibility of the medium for X-rays is its polarization when submitted to the electromagnetic radiation, we shall hereafter refer to χ as the polarizability.

in a Fourier series:

$$\begin{aligned}\chi &= \sum_{\mathbf{h}} \chi_h \exp(2\pi \mathrm{i}\mathbf{h}\cdot\mathbf{r}),\\ \chi_r &= \sum_{\mathbf{h}} \chi_{rh} \exp(2\pi \mathrm{i}\mathbf{h}\cdot\mathbf{r}),\\ \chi_i &= \sum_{\mathbf{h}} \chi_{ih} \exp(2\pi \mathrm{i}\mathbf{h}\cdot\mathbf{r}),\end{aligned} \tag{2.36}$$

where **h** is a reciprocal lattice vector and the summation is extended over all reciprocal lattice vectors.

Neglecting anomalous dispersion, the coefficients of these Fourier expansions are related to the structure factor F_h by taking the Fourier transform of both sides of (2.35):

$$F_h = \int \rho(\mathbf{r}) \exp[-2\pi \mathrm{i}\mathbf{h}\cdot\mathbf{r}]\,\mathrm{d}\tau,$$

and

$$\begin{aligned}\chi_h &= -\frac{R\lambda^2 F_h}{\pi V},\\ \chi_{rh} &= -\frac{R\lambda^2 F_{rh}}{\pi V},\\ \chi_{ih} &= -\frac{R\lambda^2 F_{ih}}{\pi V},\end{aligned} \tag{2.37}$$

where V is the volume of the unit cell. χ_h, χ_{rh}, χ_{ih}, F_h, F_{rh} and F_{ih} are all complex. Using (2.34), and including anomalous dispersion, the structure factor F_h is related to F_{rh} and F_{ih} through

$$F_h = F_{rh} + \mathrm{i}F_{ih} = \sum_j (f_j + f_j' + \mathrm{i}f_j'') \exp[-M_j - 2\pi \mathrm{i}\mathbf{h}\cdot\mathbf{r_j}],$$

f_j is the scattering or form factor of atom j, f_j' and f_j'' are the anomalous dispersion corrections and $\exp(-M_j)$ is the Debye–Waller factor (Debye 1914, 1930; Waller 1923). The summation is over all the atoms in the unit cell. The phase, φ_h, of the structure factor depends of course on the choice of origin of the unit cell. The Fourier coefficients χ_h are dimensionless. Their order of magnitude varies from 10^{-5} to 10^{-7} depending on the wavelength and the structure factor. For instance, it is equal to -9.24×10^{-6} for the 220 reflection of Cu$K\alpha$ on silicon.

The influence of thermal vibrations on the diffraction of X-rays is very well described, for instance, in the books by James (1950), Laue (1960) and Warren (1969). The fact that, in dynamical theory, the structure factor should be multiplied by the Debye–Waller factor was shown theoretically by Laval (1945, 1958), Parthasarathy (1960), Ohtsuki (1964, 1966) and Kambe (1965) and verified experimentally by

Batterman (1962*b* and 1964*b*), Okkerse (1962), Ling and Wagenfeld (1965), Kohra *et al.* (1966) on germanium single crystals and by Hattori *et al.* (1965) and Kyutt and Efimov (1969) on silicon single crystals. This point was taken up again more recently by Chung and Durbin (1999) using numerical simulations of the intensity reflected by an otherwise perfect crystal but whose periodicity is loosened by the thermal vibrations. These simulations are based on an extended Darwin theory (Chung and Durbin 1995; Durbin and Follis 1995). They show a good agreement with the usually accepted behaviour, except at high temperatures and for high-order reflections for which they observed a deviation.

The Debye–Waller factor for silicon was deduced by Aldred and Hart (1973*b*) from high-order structure factors measured from the position of Pendellösung fringes (Aldred and Hart 1973*a*). More recently, highly accurate ($< 0.1\%$) measurements allowed Deutsch *et al.* (1989) to attribute a different Debye–Waller factor to each electronic shell.

The anomalous dispersion corrections are calculated using quantum mechanics and they were first evaluated by Hönl (1933). Several techniques have been used: a non-relativistic multipole approach by Wagenfeld (1975), a relativistic dipole approach by Cromer and Liberman (1970, 1981) completed by a computer program (Cromer 1983), a computer program based on the formalism of the second-order scattering matrix by Kissel *et al.* (1980) (tables are given in Kane *et al.* 1986; Kissel and Pratt 1990), a relativistic multipole approach (Smith 1987; Creagh 1999) and a self-consistent Dirac–Hartree–Fock approach (Chantler 1995). A discussion of the various techniques is given in Chantler (1994) and in Kissel *et al.* (1995). The anomalous dispersion corrections can also be measured using various X-ray techniques:

- *Rocking curves in the Laue geometry* (see, for instance, Teworte and Bonse 1984, Deutsch and Hart 1985*b*, Deutsch *et al.* 1990 and Section 6.4.3)
- *Rocking curves in the Bragg geometry* (see, for instance, Kuribayashi *et al.* 1997 and Subsection 7.1.2)
- *Pendellösung fringes* (see, for instance, Tanemura and Kato 1972, Aldred and Hart 1973*a* and Saka and Kato 1987 and Section 10.5)
- *Powder diffraction* (see, for instance, Suortti *et al.* 1985)
- *X-ray interferometry* (see, for instance, Begum *et al.* 1986 and Section 15.5.2.2)
- *Prism-deviation techniques* (Deutsch and Hart 1984; Fontaine *et al.* 1985; Warburton and Ludwig 1986)
- *Specular reflection* (many papers are devoted to that technique, for a recent example, see Stanglmeier *et al.* 1992)
- *Multilayer interference techniques* (Barbee *et al.* 1984).

The accuracy of experimental structure factor measurements by means of Pendellösung fringes and rocking curves has constantly been improved, reaching values of a few millielectrons per atom in Deutsch and Hart (1985*b*) for silicon and Deutsch *et al.* (1990) for germanium. An analysis of their theoretical and experimental determination is given in Creagh (1985, 1999).

We shall call φ_h, φ_{rh} and φ_{ih} the phase factors of F_h, F_{rh} and F_{ih}, respectively:

$$\begin{aligned} F_h &= |F_h| \exp(i\varphi_h), \\ F_{rh} &= |F_{rh}| \exp(i\varphi_{rh}), \\ F_{ih} &= |F_{ih}| \exp(i\varphi_{ih}), \end{aligned} \tag{2.38}$$

and we shall set

$$\varphi = \varphi_{rh} - \varphi_{ih}. \tag{2.39}$$

It does not depend on the origin of the unit cell, contrary to φ_h. The structure factor corresponding to the $\bar{h}, \bar{k}, \bar{l}$ reflection is

$$F_{\bar{h}} = F_{r\bar{h}} + iF_{i\bar{h}} = F^*_{rh} + iF^*_{ih},$$

where f^* means imaginary conjugate of f. It is important to note that $F_{\bar{h}} \neq F^*_h$.

The following quantities are frequently used in dynamical theory:

$$\begin{aligned} \sqrt{\chi_h \chi_{\bar{h}}} &= \sqrt{|\chi_h \chi_{\bar{h}}|} \exp(i\beta), \\ \sigma^{1/2} &= \frac{\sqrt{\chi_h \chi_{\bar{h}}}}{\chi_{\bar{h}}} = -\sqrt{\left|\frac{\chi_h}{\chi_{\bar{h}}}\right|} \exp(i\beta'), \\ \kappa &= \left|\frac{\chi_{ih}}{\chi_{rh}}\right| = \left|\frac{F_{ih}}{F_{rh}}\right|, \end{aligned} \tag{2.40}$$

where $|\sigma| = |\chi_h/\chi_{\bar{h}}|$ is the departure from Friedel's law (Friedel 1913) in absorbing crystals, that is the ratio of the intensities of the *hkl* and $\bar{h}\bar{k}\bar{l}$ reflections, respectively. Using (2.40), it follows that

$$\begin{aligned} \chi_h \chi_{\bar{h}} &= |\chi_{rh}|^2 (1 - \kappa^2 + 2i\kappa \cos \varphi), \\ \beta + \beta' &= \varphi_h. \end{aligned} \tag{2.41}$$

The following remarks are also often useful:

1. In a centrosymmetric crystal,

$$\varphi = 0 \text{ or } \pi; \qquad \tan \beta = \pm\kappa; \qquad \beta' = \varphi_{rh}; \qquad |\sigma| = 1.$$

2. If the absorption is not too large ($\kappa \ll 1$),[4]

$$\begin{aligned} \beta &= \kappa \cos \varphi / (1 - \kappa^2 \sin^2 \varphi), \\ \sqrt{\chi_h \chi_{\bar{h}}} &= |\chi_{rh}| (1 + i\kappa \cos \varphi), \\ \sigma &= 1 + 2\kappa \sin \varphi. \end{aligned}$$

[4] The case when $\kappa \ll 1$ is not satisfied is discussed by Fukamachi and Kawamura (1993) and Fukamachi *et al.* (2002) (*resonant scattering dynamical theory*).

2.7.3 *Index of refraction*

The index of refraction in the forward direction is given by the relation

$$n = \sqrt{\frac{\epsilon}{\epsilon_0}}$$

and, since $\chi \ll 1$,

$$n \approx 1 + \frac{\chi_o}{2} = 1 - \frac{R\lambda^2 F_o}{2\pi V} = 1 - \frac{Rc^2 F_o}{2\pi \nu^2 V}.$$

It is slightly smaller than 1 and there is specular reflection if a beam of X-rays is incident on a material with a glancing angle smaller than the critical angle $\omega_c = \sqrt{-\chi_{ro}} = \sqrt{F_o R\lambda^2/\pi V}$ (see Section 2.10). For instance, for silicon and Cu$K\alpha$, $\chi_{ro} = -1.5142 \times 10^{-5}$ and $\omega_c = 13.38$ arcmin. From this angle the electronic density F_o/V of a material can be determined.

The phase velocity of an electromagnetic wave is ν/K, where K is the wave number in the medium:

$$K = nk = \frac{\nu}{c} - \frac{F_o Rc}{2\pi V \nu}.$$

It is therefore larger than the velocity of light, c. This is due to a phase shift occurring during the response of the dipoles to the incident electromagnetic wave. The rate at which the energy of the photons is propagated is, however, given by the group velocity

$$\frac{\mathrm{d}\nu}{\mathrm{d}K} = c\left[1 + \frac{F_o Rc^2}{2\pi V \nu^2}\right]^{-1}$$

which is smaller than c, as expected.

2.7.4 *Absorption*

1. *Fourier expansion of the absorption coefficient*: The imaginary part of the polarizability, χ_i, is associated with the attenuation of X-rays propagating through matter. Three main effects contribute to it:

(i) Photoelectric absorption which is taken into account through an absorption coefficient, μ^{PE}.

(ii) Inelastic scattering of the energy in another direction by the Compton effect, which is taken into account through an absorption coefficient, μ^C.

(iii) Inelastic scattering of the energy in another direction by thermal diffuse scattering, which is taken into account through the Debye–Waller factor, and by a very small absorption coefficient, μ^T.

The total linear absorption coefficient is the sum of the three:

$$\mu = \mu^{PE} + \mu^{C} + \mu^{T}.$$

The photoelectric absorption coefficient is the largest component. For instance, Giardina and Merlini (1973) estimated their respective values for silicon and Mo$K\alpha$, at room temperature:

$$\mu = 14.61\,\mathrm{cm}^{-1}; \qquad \mu^{PE} = 14.11\,\mathrm{cm}^{-1};$$
$$\mu^{C} = 0.32\,\mathrm{cm}^{-1}; \qquad \mu^{T} = 0.18\,\mathrm{cm}^{-1}.$$

Experimental values of the linear absorption coefficent of silicon and germanium for Mo$K\alpha$ and Cu$K\alpha$ obtained by different authors are compared in Hildebrandt (1979*a*).

The three phenomena which contribute to the absorption have the same three-dimensional periodicity as the crystalline lattice and μ may be expanded in Fourier series:

$$\mu = \sum_{\mathbf{h}} \mu_h(2\theta) \exp(2\pi \mathrm{i}\mathbf{h} \cdot \mathbf{r})$$

with

$$\mu_h(2\theta) = \mu_h^{PE}(2\theta) + \mu_h^{C}(2\theta) + \mu_h^{T}(2\theta).$$

The Fourier coefficients $\mu_h(2\theta)$ of the absorption coefficient are related to the imaginary parts of the form factors by

$$\begin{aligned} \mu_h(2\theta) &= 2\pi k|\chi_{ih}| = 2R\lambda|F_{ih}|/V \\ &= \frac{2R\lambda}{V}\left|\sum_j f_j'' \exp[-M_j + 2\pi \mathrm{i}\,\mathbf{h} \cdot \mathbf{r_j}]\right| \end{aligned} \tag{2.42a}$$

and the linear absorption coefficient in the forward direction is

$$\mu_o = 2\pi k|\chi_{io}|. \tag{2.42b}$$

2. *Temperature dependence of the absorption coefficient*: The fact that f_j'' should be multiplied by $\exp(-M_j)$ in dynamical diffraction may seem obvious, but it is not. Quite a number of studies have been devoted to the experimental and theoretical proof of that result. This was shown experimentally using anomalous transmission, for instance, by Okkerse (1962), Hildebrandt *et al.* (1973), Batterman (1964*b*) and Ludewig (1969) for germanium, by Ghezzi *et al.* (1971) for copper and zinc, and by Hildebrandt *et al.* (1973) and Giardina and Merlini (1973) for silicon. The first theoretical studies were only concerned with the photoelectric absorption part of the attenuation, and the fact that it is proportional to the Debye–Waller factor was proved by Ohtsuki (1964, 1966) and Kambe (1965). Sano *et al.* (1969) have shown

that it is also true for the absorption coefficient due to Compton scattering, μ_h^C. Afanas'ev and Kagan (1968) discussed the contribution μ_h^T due to inelastic scattering by phonons and showed that its temperature dependence is not through a Debye–Waller factor, but is more complicated. The contribution due to Compton scattering and its temperature dependence were calculated by Giardina and Merlini (1973) and compared to experimental results with a good agreement.

3. *Angular dependence of the absorption coefficient*: By far the main reason for the angular dependence of the absorption coefficient is of course the Debye–Waller factor. But it is easy to understand physically why for each atom j there should also be a small angular dependence of the three contributions to f_j''. They are respectively proportional to the coefficients of the Fourier transform of that part of the electron density which contributes most to the effect. For the photoelectric absorption, it is the core electrons (K- and L-shells mainly): K-shell for germanium and Mo$K\alpha$, L-shell for germanium and Cu$K\alpha$, for instance. Since these electrons are very close to the centre of the atom, the corresponding Fourier transform is very little angle dependent, but it is not a Dirac distribution. Wagenfeld (1966) and Hildebrandt *et al.* (1973, 1975) have shown that the angular dependence of μ_h^{PE} is of the form

$$\mu_h^{PE\perp} = [\tau^D + \tau^Q \cos 2\theta] \exp(-M)$$

for a polarization perpendicular to the diffraction plane, where τ^D and τ^Q are the dipolar and quadrupolar contributions, respectively, and

$$\mu_h^{PE\parallel} = [\tau^D \cos 2\theta + \tau^Q \cos 4\theta] \exp(-M)$$

for polarization parallel to the diffracting plane. For instance, they have obtained, for τ^Q/τ^D, 0.0191 and 0.0463 for silicon and Cu$K\alpha$ and Mo$K\alpha$, respectively, 0.0229 and 0.0254 for germanium and Cu$K\alpha$ and Mo$K\alpha$, respectively.

For Compton scattering and thermal diffuse scattering, on the contrary, the outer electrons play an important role and the angular dependence will be very much like that of the real part of the form factor.

As an example, Fig. 2.2 compares the relative angular dependences of $\exp(-M)$, f_h'' and of the Fourier coefficients of the imaginary part of the polarizability of silicon for Mo$K\alpha$. It confirms that the angular dependence of f_h'' is very small. It is much smaller than that of the Debye–Waller factor. It should be noted that since silicon is face-centred cubic, $|F_{ih}| = 8f_h'' \exp(-M)$ for reflections with even indices and $|F_{ih}| = 4\sqrt{2}f_h'' \exp(-M)$ for reflections with odd indices.

2.8 Ewald's dispersion theory

As has been explained in the prologue of Chapter 1, Ewald considers the interaction of X-rays with a crystal as that of the electromagnetic field with a three-dimensional array of dipoles. Under the influence of the incoming field, each dipole radiates a

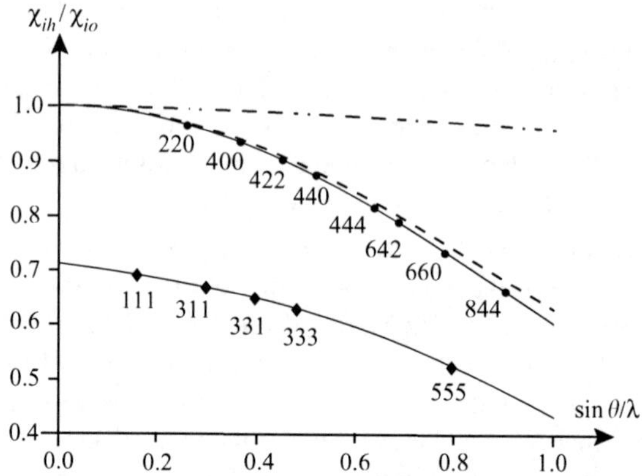

Fig. 2.2. Calculated angular dependence of the imaginary part of the polarizability of silicon for Mo$K\alpha$; σ-polarization; dashed line: Debye–Waller factor (after Aldred and Hart 1973*a,b*); dash-dotted line: f_h''/f_o'' (after Hildebrandt *et al.* 1973). The solid lines correspond to $\exp(-M) f_h''/f_o''$ for the even reflections and to $(\sqrt{2}/2)\exp(-M) f_h''/f_o''$ for the odd reflections.

spherical wave $E \exp[2\pi i \nu t - kr]/r$ which propagates with the velocity of light $c = \nu/k$. The incoming wave which excites each dipole is the superposition of the spherical waves emitted by all the other dipoles and which Ewald calls *wavelets*. This excitation of the dipoles propagates from one dipole to the other as a progressive wave of phase velocity $v = c/n$. It is the *optical field*. The simplest form of propagation is that of a plane wave $E \exp[2\pi i(\nu t - \mathbf{K_o} \cdot \mathbf{r})]$, where $K_o = nk$ and n is the index of refraction. The possible values of n are those for which all the oscillating dipoles constitute a self-consistent system. Ewald (1962, 1965, 1969, 1979) compares it to a system of coupled pendulum-like mechanical resonators. The movement of one will generate the movement of the others and that movement will in turn influence the movement of the first one. If the frequency of the incoming field on this dipole is close to that of its proper frequency, there is a resonance and a strong interaction. The problem is therefore a problem of proper modes and eigenvalues.[5]

The principle of Ewald's original demonstration (Ewald 1916*a*, 1917) can be summarized schematically as follows using the properties of Fourier transforms. Consider an infinite three-dimensional array of dipoles. The position vector of a given dipole is

$$\mathbf{r_{ij}} = \mathbf{r_i} + \mathbf{r_j},$$

where $\mathbf{r_j}$ is the position vector of dipole j in the unit cell and

$$\mathbf{r_i} = u_i \mathbf{a} + v_i \mathbf{b} + w_i \mathbf{c} \tag{2.43}$$

[5] This analogy has been extended by Shevchenko and Pobydaylo (2003) so as to take into account angles of incidence different from Bragg's incidence and deformed crystals.

is the position vector of the origin of a unit cell, **a**, **b** and **c** are the unit vectors of the lattice and u_i, v_i, w_i are three integers. The polarization of this dipole is, according to (2.27) and (2.28),

$$\mathbf{P_{ij}}(\mathbf{r}, t) = \mathbf{p_j}(t)\delta(\mathbf{r} - \mathbf{r_{ij}}) \exp(-2\pi \mathrm{i}\mathbf{K_o} \cdot \mathbf{r}) \exp(2\pi \mathrm{i}\nu t). \tag{2.44}$$

The associated Hertz vector is, according to (2.22),

$$\mathbf{\Pi_{eij}}(\mathbf{r}, t) = \frac{1}{4\pi} \int \frac{\mathbf{P_{ij}}(\mathbf{r}', t - |\mathbf{r} - \mathbf{r}'|/c)}{|\mathbf{r} - \mathbf{r}'|} \,\mathrm{d}\tau'.$$

The resulting Hertz vector is the sum of the Hertz vectors over all the dipoles:

$$\mathbf{\Pi_e}(\mathbf{r}, t) = \frac{\exp(2\pi \mathrm{i}\nu t)}{4\pi} \sum_i \sum_j \int \mathbf{p_j} \frac{\exp(-2\pi \mathrm{i}k|\mathbf{r} - \mathbf{r}'|)}{|\mathbf{r} - \mathbf{r}'|}$$
$$\times \exp(-2\pi \mathrm{i}\mathbf{K_o} \cdot \mathbf{r}')\delta(\mathbf{r}' - \mathbf{r_{ij}}) \,\mathrm{d}\tau',$$

where the sum over j is the sum over all the dipoles in a unit cell and the sum over i is the sum over all the unit cells in the lattice. Let us stress here that k is the wave number of the wavelet emitted by a given dipole and K_o is the wave number of the progressive wave exciting the dipoles. We can interchange the summation and the integration and replace $\mathbf{r_{ij}}$ by its value:

$$\mathbf{\Pi_e}(\mathbf{r}, t) = \frac{\exp(2\pi \mathrm{i}\nu t)}{4\pi} \int \frac{\exp(-2\pi \mathrm{i}k|\mathbf{r} - \mathbf{r}'|)}{|\mathbf{r} - \mathbf{r}'|} \sum_i \left\{ \sum_j \mathbf{p_j} \exp(-2\pi \mathrm{i}\mathbf{K_o} \cdot \mathbf{r_j}) \right\}$$
$$\times \exp(-2\pi \mathrm{i}\mathbf{K_o} \cdot \mathbf{r_i})\delta(\mathbf{r}' - \mathbf{r_i}) \,\mathrm{d}\tau'.$$

This equation can be written, after multiplying the integrand by $\exp(-2\pi \mathrm{i}\mathbf{K_o} \cdot \mathbf{r}')\times$ $\exp(2\pi \mathrm{i}\mathbf{K_o} \cdot \mathbf{r}')$, as

$$\mathbf{\Pi_e}(\mathbf{r}, t) = \frac{\exp(2\pi \mathrm{i}\nu t)}{4\pi} \sum_j \mathbf{p_j} \exp(-2\pi \mathrm{i}\mathbf{K_o} \cdot \mathbf{r_j}) \int F_1 F_2 \exp(-2\pi \mathrm{i}\mathbf{K_o} \cdot \mathbf{r}') \,\mathrm{d}\tau'$$

with

$$F_1 = \frac{\exp(-2\pi \mathrm{i}k|\mathbf{r} - \mathbf{r}'|)}{|\mathbf{r} - \mathbf{r}'|},$$
$$F_2 = \sum_i \delta(\mathbf{r}' - \mathbf{r_i}) \exp[2\pi \mathrm{i}\mathbf{K_o}\cdot(\mathbf{r}' - \mathbf{r_i})].$$

The resultant Hertz vector is seen to be proportional to the Fourier transform of the product $F_1 F_2$. The latter is equal to the convolution of the Fourier transform of these two functions:

$$\mathcal{F}(F_1 F_2) = \mathcal{F}(F_1) * \mathcal{F}(F_2).$$

The Fourier transform of F_1 is calculated by noting that the Fourier transform of a spherical wave $\exp(-2\pi ikr/r)$ is $1/(K^2 - k^2)$ and by changing the integration variable $\mathbf{r}'$ to $\mathbf{r} - \mathbf{r}'$. The Fourier transform of F_2 is calculated by noting that the Fourier transform of an infinite sum of triply periodic Dirac distributions in real space is an infinite sum of triply periodic Dirac distributions in reciprocal space. It follows therefore that

$$\mathcal{F}(F_1) = -\frac{\exp[-2\pi i\mathbf{K_o}\cdot\mathbf{r}]}{K_o^2 - k^2},$$

$$\mathcal{F}(F_2) = \frac{1}{V}\sum_{\mathbf{h}} \delta(\mathbf{K} - \mathbf{K_o} + \mathbf{h}),$$

where the summation is over all reciprocal lattice vectors $\mathbf{h} = h\mathbf{a}^* + k\mathbf{b}^* + l\mathbf{c}^*$ and V is the volume of the unit cell (the reciprocal lattice was introduced by Ewald (1913) for this purpose). The resultant Hertz vector is finally

$$\begin{aligned}\mathbf{\Pi_e}(\mathbf{r}, t) = \frac{-\exp(2\pi i\nu t)}{4\pi V}\sum_{\mathbf{h}} \frac{[\exp -2\pi i\mathbf{K_h}\cdot\mathbf{r}]}{K_h^2 - k^2}\\ \times \sum_j \mathbf{p_j}\exp(-2\pi i\mathbf{K_o}\cdot\mathbf{r_j})\end{aligned} \tag{2.45}$$

with $\mathbf{K_h} = \mathbf{K_o} - \mathbf{h}$.

The vector field associated with this wave is, according to (2.30),

$$\mathbf{E} = \frac{1}{\epsilon_0}\text{curl}\,\text{curl}\,\mathbf{\Pi_e}. \tag{2.46}$$

Equation (2.45) shows that $\mathbf{E}$ is a *superposition* of plane waves having $\mathbf{K_h}$ as wavevectors:

$$\begin{aligned}\mathbf{E} &= \sum_{\mathbf{h}} \mathbf{E_h}\exp(-2\pi i\mathbf{K_h}\cdot\mathbf{r})\exp(2\pi i\nu t)\\ &= \exp(-2\pi i\mathbf{K_o}\cdot\mathbf{r})\exp(2\pi i\nu t)\sum_{\mathbf{h}} \mathbf{E_h}\exp(-2\pi i\mathbf{h}\cdot\mathbf{r}).\end{aligned} \tag{2.47}$$

Ewald called this sum of plane waves, *wavefield*. This is the most important concept in dynamical theory, and as early as 1913 Ewald pointed out that the waves constituting the wavefield are tightly bound together, forming a physical entity (Ewald 1913).

The self-consistency of the problem is ensured by writing that the resulting field $\mathbf{E}$ deduced from $\Pi_\mathbf{e}(\mathbf{r})$ using (2.46) is the same as the field which excites the dipoles. The polarization $\mathbf{p_j}$ in eqn (2.44) is strictly

$$\mathbf{p_j}\exp(2\pi i\nu t) = \alpha_j\mathbf{E}',$$

where α_j is the polarizability of dipole j and $\mathbf{E}'$ is the field created by all the dipoles except dipole j, but Ewald noted that the difference between $\mathbf{E}'$ and $\mathbf{E}$ can be neglected. Using (2.47), it follows that

$$\mathbf{p_j} = \alpha_j \sum_{\mathbf{h}'} \mathbf{E_{h'}} \exp(-2\pi \mathrm{i} \mathbf{K_h} \cdot \mathbf{r_{ij}}), \tag{2.48}$$

since the value of $\mathbf{p_j}$ must be considered at the position $\mathbf{r_{ij}}$ of the dipole. Let us set

$$\chi_{h'} = -\frac{\pi}{V\epsilon_0} \sum_j \alpha_j \exp(2\pi \mathrm{i} \mathbf{h}' \cdot \mathbf{r_j}).$$

Substituting from (2.47) into (2.45) and noting that $\mathbf{h}' \cdot \mathbf{r_i}$ is an integer since $\mathbf{r_i}$ is a direct lattice vector (2.43):

$$\Pi_{\mathbf{e}}(\mathbf{r}, t) = \frac{\epsilon_0}{4\pi^2} \sum_{\mathbf{h}} \left[\frac{\exp(-2\pi \mathrm{i} \mathbf{K_h} \cdot \mathbf{r})}{K_h^2 - k^2} \sum_{\mathbf{h}'} \chi_{h'} \mathbf{E_{h'}} \right]. \tag{2.49}$$

Expression (2.46) is calculated by noting that

$$\operatorname{curl}\operatorname{curl}[\mathbf{E_{h'}} \exp(-2\pi \mathrm{i} \mathbf{K_h} \cdot \mathbf{r})] = -4\pi^2 \mathbf{K_h} \wedge (\mathbf{K_h} \wedge \mathbf{E_{h'}}) \exp(-2\pi \mathrm{i} \mathbf{K_h} \cdot \mathbf{r}).$$

Using the identity $\mathbf{a} \wedge (\mathbf{b} \wedge \mathbf{c}) = (\mathbf{a} \cdot \mathbf{c})\mathbf{b} - (\mathbf{a} \cdot \mathbf{b})\mathbf{c}$, and letting $\mathbf{s_h}$ be a unit vector in the direction of $\mathbf{K_h}$, we obtain

$$\begin{aligned} \mathbf{K_h} \wedge (\mathbf{K_h} \wedge \mathbf{E_{h'}}) &= -K_h^2[\mathbf{E_{h'}} - (\mathbf{s_h} \cdot \mathbf{E_{h'}})\mathbf{s_h}] \\ &= -K_h^2 \mathbf{E_{h'[h]}}, \end{aligned} \tag{2.50}$$

where the notation $\mathbf{E_{h'[h]}}$ means the projection of $\mathbf{E_{h'}}$ on the plane normal to $\mathbf{K_h}$ (Fig. 2.3) and, finally,

$$\operatorname{curl}\operatorname{curl}[\mathbf{E_{h'}} \exp(-2\pi \mathrm{i} \mathbf{K_h} \cdot \mathbf{r})] = 4\pi^2 K_h^2 \mathbf{E_{h'[h]}} \exp[-2\pi \mathrm{i} \mathbf{K_h} \cdot \mathbf{r}], \tag{2.51}$$

using (2.46), (2.49) and (2.51), we obtain the relation expressing the self-consistency of the problem:

$$\mathbf{E} = \sum_{\mathbf{h}} \left[K_h^2 \frac{\exp(-2\pi \mathrm{i} \mathbf{K_h} \cdot \mathbf{r})}{K_h^2 - k^2} \sum_{\mathbf{h}'} \chi_{h'} \mathbf{E_{h'[h]}} \right]$$

and, using the development (2.47) of $\mathbf{E}$,

$$\mathbf{E_h} = \left[\frac{K_h^2}{K_h^2 - k^2} \sum_{\mathbf{h}'} \chi_h' \mathbf{E_{h'[h]}} \right]. \tag{2.52}$$

The set of equations (2.52) is the equivalent of the set of fundamental equations (4.9) in Laue's dynamical theory. The form (2.52) is not exactly correct because as

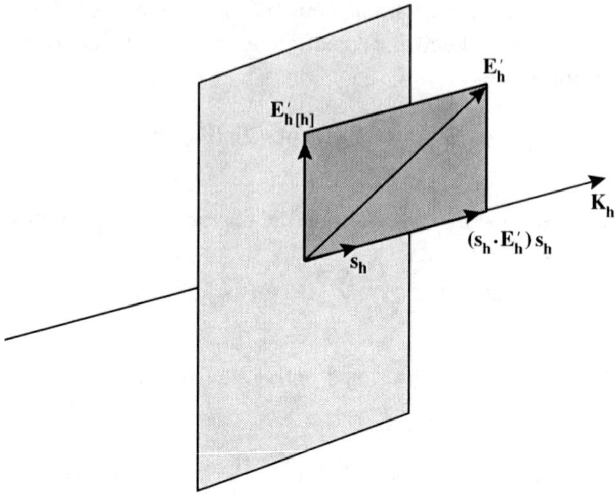

Fig. 2.3. Calculation of the double vector product $\mathbf{K_h} \wedge \mathbf{K_h} \wedge \mathbf{E_{h'}}$: $\mathbf{K_h} \wedge \mathbf{K_h} \wedge \mathbf{E_{h'}} = -K_h^2[\mathbf{E_{h'}} - (\mathbf{s_h} \cdot \mathbf{E_{h'}})]\mathbf{s_h} = -K_h^2\mathbf{E_{h'[h]}}$, where $\mathbf{s_h}$ is a unit vector parallel to $\mathbf{K_h}$.

Ewald himself pointed out (Ewald 1937), the relation $\mathbf{D} = \epsilon_0\mathbf{E}$ which was used in the derivation does not take into account the average polarizability of the medium; the correct equation is given in Section 5.2. The important point, however, is the presence of the terms $1/(K_h^2 - k^2)$ which have been called *resonance factors* by Ewald. They show that the only components of the wavefield which have important values are those for which $(K_h^2 - k^2)$ is the smallest: '*this condition selects from the infinite number of proper harmonics of the space lattice those which, for a given frequency of the X-rays, have the largest intensity*' (from the translation of the Ewald 1913 paper in Cruickshank *et al.* 1992). Ewald (1913) showed that this condition can also be represented geometrically. Relation $\mathbf{K_h} = \mathbf{K_o} - \mathbf{h}$ can be interpreted by saying the wavevectors drawn from the reciprocal lattice points converge on a same point which Ewald later called *tiepoint* (see, for instance, Figs 1.4, 4.3 and 4.6). Those of them for which the resonance factors are very large have the same length and the corresponding reciprocal lattice vectors can be obtained by drawing a sphere centred at the tiepoint and having that length as radius: this is the 'Ewald construction' and the 'Ewald sphere' (see also the comments by Cruickshank 1992). Ewald pointed out also that whichever one of these waves is taken as the 'primary' wave, the same wavefield is generated (the *reciprocity theorem*). Equations (2.52) represent a set of eigenvalue equations. For them to have a non-trivial solution, the associated determinant must be equal to zero. This provides the dispersion equation and the equation of the dispersion surface.

The detailed correspondence between Ewald's and Laue's dynamical theories was worked out by Wagenfeld (1968).

2.9 Propagation equation of an electromagnetic wave in materials in Laue's dynamical theory

2.9.1 *Laue's basic assumption*

In contrast to Ewald's dynamical theory which discusses the interaction of an electromagnetic wave with a distribution of discrete dipoles, Laue's basic assumption (1931, 1960) is to consider that the electric negative and positive charges are distributed in a continuous way throughout the volume of the crystal. Since the crystal must be neutral, they cancel out and the local electric charge and density of current are equal to zero. This corresponds to the assumption made in Section 2.3 and Maxwell's equations become

$$\begin{aligned}
\operatorname{curl}\mathbf{H} &= \frac{\partial \mathbf{D}}{\partial t},\\
\operatorname{curl}\mathbf{E} &= -\frac{\partial \mathbf{B}}{\partial t},\\
\operatorname{div}\mathbf{D} &= 0,\\
\operatorname{div}\mathbf{B} &= 0.
\end{aligned}\tag{2.53}$$

A similar expression to (2.31) would be obtained for the interaction of the electronic wave with a proton. Since the mass of the proton is 1836 times larger than that of the electron, its scattering length is 1836 times smaller and the interaction of electromagnetic waves with the positive charges is neglected as a first approximation. Resonant nuclear scattering of X-rays has, however, been observed for γ- and X-rays and is discussed in Hannon *et al.* (1988), Hannon (1994), Kohn (1994), Kohn and Shvyd'ko (1995), Kohn and Chumakov (2002) and others but will not be taken into account in the present treatment of dynamical theory.

2.9.2 *Propagation equation*

Consider an electromagnetic wave $\mathbf{E}(\mathbf{r})\exp(2\pi i\nu t)$ propagating in a material. The medium is polarized under the influence of the electric field and $\mathbf{E} = \mathbf{D}/\epsilon$, where $\epsilon = \epsilon_0(1+\chi)$ varies with the space coordinates. By eliminating **D**, **H** and **B** in Maxwell's equations (2.53), one obtains a wave equation for **E**:

$$\operatorname{curl}\operatorname{curl}\mathbf{E} - 4\pi^2k^2(1+\chi)\mathbf{E} = 0. \tag{2.54}$$

However, while $\operatorname{div}\mathbf{D}$ is equal to zero, this is not true for $\operatorname{div}\mathbf{E}$, and the electric displacement is in general a more suitable quantity than the electric field because it simplifies the description of the polarization states of the field inside the crystal. It is used in Laue's formulation of the dynamical theory. Since the index of refraction is so close to 1, the boundary conditions (2.7) can be condensed and one can assume that there is continuity of both the normal and the tangential components of **D**. In special cases, however, such as when the angle of incidence of the beam on the crystal is very small and close to the critical angle for specular reflection, this approximation is no longer valid and the boundary conditions for the tangential component electric field must be taken into account separately (see Section 2.10 and Chapter 8).

Since χ is very much smaller than 1, the electric field can be written, to a first approximation, valid for hard X-rays, as

$$\mathbf{E} = \frac{\mathbf{D}}{\epsilon_0(1+\chi)} = \frac{\mathbf{D}}{\epsilon_0}(1-\chi).$$

Using this expression and eliminating $\mathbf{H}$ between the first two relations of Maxwell's equations, one obtains the equation of propagation of an electromagnetic wave in the crystalline medium:

$$\Delta\mathbf{D} + \text{curl curl}\,\chi\mathbf{D} + 4\pi^2k^2\mathbf{D} = 0. \tag{2.55}$$

The solutions of this equation of propagation will be discussed in Chapter 5.

2.10 Specular reflection—Fresnel relations

As mentioned in Section 2.7.3, since the index of refraction for X-rays is smaller than 1, there is total reflection when the angle of incidence on the crystal surface is smaller than the critical angle. The expressions of the reflectivity and of the transmittivity are obtained by applying the boundary conditions.

1. *Boundary conditions for the wavevectors*: Figure 2.4 represents the wavevector $\mathbf{K_o^{(a)}}$ of an incident plane wave, the wavevector $\mathbf{K_o}$ of the refracted wave inside the crystal and the wavevector $\mathbf{K_s}$ of the specularly reflected wave. The wavelengths $K_o^{(a)}$ and K_s are equal to k, wave number in vacuum, and K_o is equal to nk where n is the index of refraction of the medium. The extremities of $\mathbf{K_o^{(a)}}$ and $\mathbf{K_s}$ lie therefore on a sphere of radius k and the extremity of $\mathbf{K_o}$ on a sphere of radius nk. Let **Ix, Iy** and **Iz** be three coordinate axes: **Ix** parallel to the crystal surface and the plane of incidence, **Iy** perpendicular to the plane of incidence and **Iz** perpendicular to the crystal surface. The phase velocity of the three waves must be the same along the crystal surface (the Snell–Descartes refraction law) and the projections $K_{ox}^{(a)}$, K_{ox} and K_{sx} of their wavevectors on the crystal surface must be equal:

$$K_{ox}^{(a)} = K_{ox} = K_{sx}. \tag{2.56}$$

The extremities of the three wavevectors lie on the same normal to the crystal surface. If this normal intersects the sphere of radius nk, there are both a refracted and a specularly reflected wave, if it is tangent to the sphere, the refracted wave is parallel to the crystal surface and the angle of incidence of $\mathbf{K_o^{(a)}}$ is the critical angle, ω_c. If the normal does not intersect the sphere of radius nk, there is no refracted wave and there is total reflection. Actually, if the crystal is absorbing, which is the usual case, n is complex and total reflection does not occur.

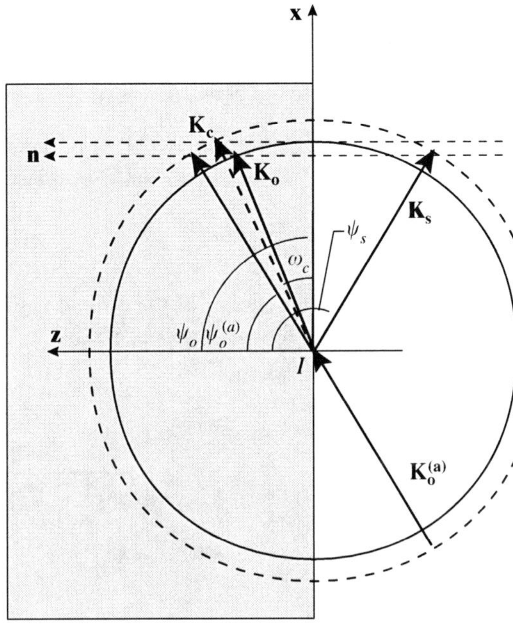

Fig. 2.4. Refraction and specular reflection of a wave at the surface separating two media with different indices of refraction. Broken curve: surface of indices in vacuum; solid curve: surface of indices of the medium; $\mathbf{K_o^{(a)}}$: incident wavevector; $\mathbf{K_o}$: refracted wavevector in the medium; $\mathbf{K_s}$: specularly reflected wavevector; $\mathbf{K_c}$: incident wavevector corresponding to the critical angle, ω_c.

Let us call $\psi_o^{(a)}$, ψ_o and ψ_s the angles between the normal to the crystal surface and the three wavevectors, respectively (Fig. 2.4):

$$\psi_o^{(a)} = (\mathbf{Iz}, \mathbf{K_o^{(a)}}); \qquad \psi_o = (\mathbf{Iz}, \mathbf{K_o}); \qquad \psi_s = (\mathbf{Iz}, \mathbf{K_s}).$$

The x and z components of the three wavevectors are, respectively,

$$\begin{aligned} K_{ox}^{(a)} &= k \sin \psi_o^{(a)}; & K_{oz}^{(a)} &= k \cos \psi_o^{(a)}; \\ K_{ox} &= nk \sin \psi_o; & K_{oz} &= nk \cos \psi_o; \\ K_{sx} &= k \sin \psi_s; & K_{sz} &= k \cos \psi_s. \end{aligned}$$

Using (2.56) and these relations, we obtain the law of refraction:

$$\mathcal{R}e(n) \sin \psi_o = \sin \psi_o^{(a)},$$

where $\mathcal{R}e(n)$ means real part of n. The critical angle corresponds to $\psi_o = \pi/2$. For X-rays, n is very close to 1 and the incidence corresponding to the critical angle is a grazing incidence. Let us set $\Delta\psi = \pi/2 - \psi_o^{(a)}$. The value ω_c of $\Delta\psi$ corresponding to the critical angle is given by

$$\cos\omega_c = \mathcal{R}e(n) = \mathcal{R}e(\sqrt{1+\chi_o}).$$

If we replace the left-hand term by the expansion $1 - \omega_c^2/2$ of the cosine and $\mathcal{R}e(n)$ by its value $1 + \chi_{ro}/2$ (Section 2.7.2), we obtain the value of the critical angle:

$$\omega_c = \sqrt{-\chi_{ro}} = \sqrt{\frac{RF_o}{\pi V}}\,\lambda.$$

Figure 2.5 represents the variations with energy of the critical angle for silicon and Table 15.3 gives the critical angles of a few materials at 10 keV.

Noting that $\psi_s = -\pi - \psi_o^{(a)}$, it follows that

$$\begin{aligned} K_{ox}^{(a)} &= \sqrt{k^2 - [K_{oz}^{(a)}]^2}; & K_{oz}^{(a)} &\approx k\Delta\psi; \\ K_{ox} &= K_{ox}^{(a)}; & K_{oz} &= \sqrt{n^2k^2 - K_{ox}^2}; \\ K_{sx} &= K_{ox}^{(a)}; & K_{sz} &= -K_{oz}^{(a)}; \end{aligned} \qquad (2.57)$$

and

$$\begin{aligned} \cos\psi_o^{(a)} &= \frac{K_{oz}^{(a)}}{k}; & \sin\psi_o^{(a)} &= \frac{K_{ox}^{(a)}}{k}; \\ \cos\psi_o &= \frac{K_{oz}}{nk}; & \sin\psi_o &= \frac{K_{ox}^{(a)}}{nk}; \\ \cos\psi_s &= -\frac{K_{oz}^{(a)}}{k}; & \sin\psi_s &= \frac{K_{ox}^{(a)}}{k}. \end{aligned}$$

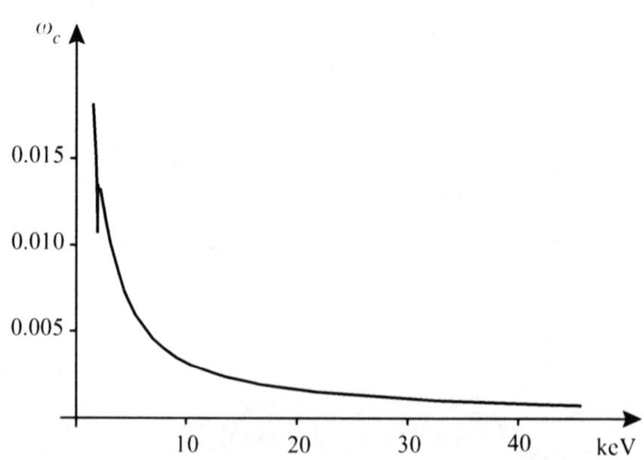

Fig. 2.5. Variation with energy of the critical angle, ω_c, of silicon, in milliradians. The values of the form factor f_o of silicon are taken from Chantler (1995).

2. *Boundary conditions for the field vectors*: According to (2.7) the tangential components of the electric field and of the magnetic field on the crystal surface are the same on each side of the boundary. The electric field of the incident wave, $\mathbf{E}_\mathbf{o}^{(\mathbf{a})}$, is normal to the wavevector $\mathbf{K}_\mathbf{o}^{(\mathbf{a})}$. Its components parallel and perpendicular to the plane of incidence are $\mathbf{E}_\mathbf{o}^{(\mathbf{a})\|}$ (π-component) and $\mathbf{E}_\mathbf{o}^{(\mathbf{a})\perp}$ (σ-component), respectively (Fig. 2.6), and the components of $\mathbf{E}_\mathbf{o}^{(\mathbf{a})}$ are

$$
\begin{aligned}
&x\text{-component:} \quad && E_o^{(a)\|} \cos\psi_o^{(a)} = E_o^{(a)\|} \frac{K_{oz}^{(a)}}{k},\\
&y\text{-component:} \quad && E_o^{(a)\perp},\\
&z\text{-component:} \quad && E_o^{(a)\|} \sin\psi_o^{(a)} = -E_o^{(a)\|} \frac{K_{ox}^{(a)}}{k}.
\end{aligned}
$$

Similar relations hold for the refracted wave (with k replaced by nk) and the reflected wave, and the continuity of the tangential component of the electric field is expressed by

$$
\begin{aligned}
E_o^{\|} \frac{K_{oz}}{n} &= E_o^{(a)\|} K_{oz}^{(a)} - E_s^{\|} K_{oz}^{(a)},\\
E_o^{\perp} &= E_o^{(a)\perp} + E_s^{\perp}.
\end{aligned} \tag{2.58}
$$

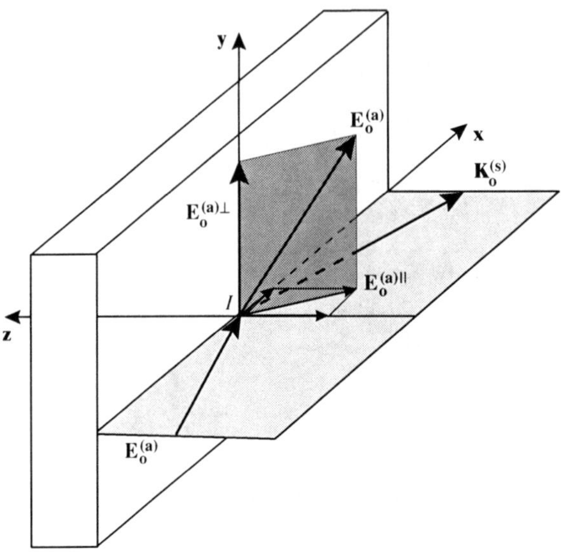

Fig. 2.6. Specular reflection of X-rays—continuity of the tangential component of the electric field.

The magnetic field is obtained using (2.2), (2.4) and (2.26):

$$\mathbf{H}_{\mathbf{o}}^{(\mathbf{a})} = \frac{1}{\mu_0 \nu} \mathbf{K}_{\mathbf{o}}^{(\mathbf{a})} \wedge \mathbf{E}_{\mathbf{o}}^{(\mathbf{a})}. \tag{2.59}$$

Its tangential components are

$$x\text{-component:} \quad \frac{1}{\mu_0 \nu} K_{oz}^{(a)} E_o^{(a)\perp},$$

$$y\text{-component:} \quad \frac{1}{\mu_0 \nu} k E_o^{(a)\parallel}.$$

Similar relations hold for the refracted wave (with k replaced by nk) and the reflected wave, and the continuity of the tangential component of the magnetic field is expressed by

$$\begin{aligned} E_o^{\perp} \frac{K_{oz}}{n} &= E_o^{(a)\perp} K_{oz}^{(a)} - E_s^{\perp} K_{oz}^{(a)}, \\ n E_o^{\parallel} &= E_o^{(a)\parallel} + E_s^{\parallel}. \end{aligned} \tag{2.60}$$

From eqns (2.58) and (2.60), it is easy to obtain the expressions of the π- and σ-components of the refracted and specularly reflected electric field:

$$\begin{aligned} E_s^{\perp} &= \frac{K_{oz}^{(a)} - K_{oz}}{K_{oz}^{(a)} + K_{oz}} E_o^{(a)\perp}; \qquad & E_s^{\parallel} &= \frac{n^2 K_{oz}^{(a)} - K_{oz}}{n^2 K_{oz}^{(a)} + K_{oz}} E_o^{(a)\parallel}; \\ E_o^{\perp} &= \frac{2 K_{oz}^{(a)}}{K_{oz}^{(a)} + K_{oz}} E_o^{(a)\perp}; \qquad & E_o^{\parallel} &= \frac{2n K_{oz}^{(a)}}{n^2 K_{oz}^{(a)} + K_{oz}} E_o^{(a)\parallel}. \end{aligned} \tag{2.61}$$

These relations are called *Fresnel relations*.

3. *Reflectivity and transmittivity*: The intensity of an electromagnetic field is equal to the magnitude of the Poynting vector:

$$\mathbf{S} = \mathcal{R}e(\mathbf{E} \wedge \mathbf{H}^*).$$

Substituting from (2.59), it follows, for the incident beam, that

$$\mathbf{S}_{\mathbf{o}}^{(\mathbf{a})} = \frac{\mathbf{K}_{\mathbf{o}}^{(\mathbf{a})}}{\mu_0 \nu} |E_o^{(a)}|^2$$

and similar relations for the refracted and reflected waves. The energies per unit area of the crystal surface of the three waves are, respectively,

$$\text{incident beam:} \quad |S_o^{(a)}| \cos \psi_o^{(a)} = \frac{|E_o^{(a)}|^2}{\mu_0 \nu} K_{oz}^{(a)},$$

$$\text{reflected beam:} \quad |S_s| \cos \psi_s = \frac{|E_s|^2}{\mu_0 \nu} K_{oz}^{(a)},$$

$$\text{refracted beam:} \quad |S_o| \cos \psi_o = \frac{|E_o|^2}{\mu_0 \nu} \mathcal{R}e(K_{oz}).$$

The reflectivity is

$$R = \frac{|S_s| \cos \psi_s}{|S_o^{(a)}| \cos \psi_o^{(a)}} = \frac{|E_s|^2}{|E_o^{(a)}|^2}$$

and the transmittivity is

$$T = \frac{|S_o| \cos \psi_o}{|S_o^{(a)}| \cos \psi_o^{(a)}} = \frac{|E_s|^2 \mathcal{R}e(K_{oz})}{|E_o^{(a)}|^2 K_{oz}^{(a)}}.$$

Substituting from (2.57) and (2.61), we finally obtain for the π- and the σ-component, respectively,

$$\begin{aligned} R^{\perp} &= \left| \frac{\Delta\psi - \sqrt{\chi_o + \Delta\psi^2}}{\Delta\psi + \sqrt{\chi_o + \Delta\psi^2}} \right|^2; \\ T^{\perp} &= \frac{4\Delta\psi \mathcal{R}(\sqrt{\chi_o + \Delta\psi^2})}{|\Delta\psi + \sqrt{\chi_o + \Delta\psi^2}|^2}; \\ R^{\parallel} &= \left| \frac{(1+\chi_o)\Delta\psi - \sqrt{\chi_o + \Delta\psi^2}}{(1+\chi_o)\Delta\psi + \sqrt{\chi_o + \Delta\psi^2}} \right|^2; \\ T^{\parallel} &= \frac{4(1+\chi_o)\Delta\psi \mathcal{R}(\sqrt{\chi_o + \Delta\psi^2})}{|(1+\chi_o)\Delta\psi + \sqrt{\chi_o + \Delta\psi^2}|^2}. \end{aligned} \tag{2.62}$$

Figure 2.7 represents the variations of the reflectivity and transmittivity of the σ-component with $\Delta\psi$ for silicon and Cu$K\alpha$ radiation, and Fig. 2.8 represents the variations of the reflectivity of silicon with energy at a glancing angle of $\Delta\psi = 3$ mrad (σ-component).

A recent review of the applications of reflectivity studies of interfaces and multilayers is given in Als-Nielsen and McMorrow (2000).

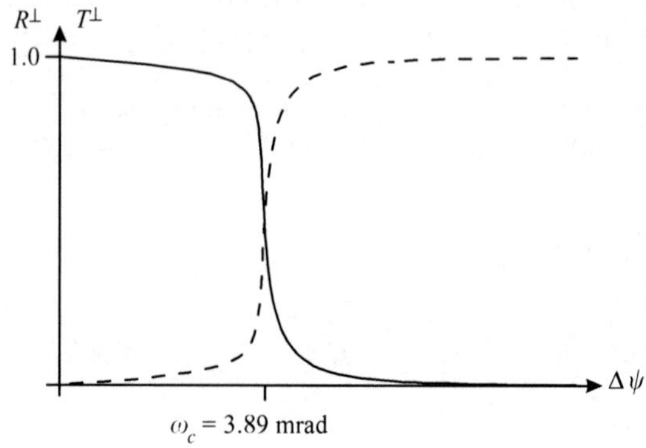

Fig. 2.7. Specular reflection of X-rays: reflectivity, $R^{\perp}$, (solid curve) and transmittivity, $T^{\perp}$, of silicon for Cu$K\alpha$ close to the critical angle, ω_c.

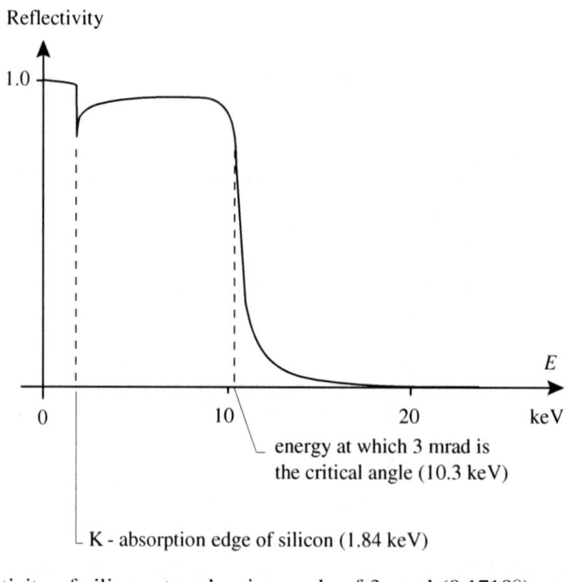

Fig. 2.8. Reflectivity of silicon at a glancing angle of 3 mrad (0.1719°) versus energy. The values of the form factor f_o of silicon are taken from Chantler (1995).

3

Geometrical theory of X-ray diffraction

3.1 Classical scattering by an electron—polarization

Let us consider a polarized electromagnetic plane wave $\mathbf{E_o} \exp(-2\pi \mathrm{i} \mathbf{K_o} \cdot \mathbf{r})$ incident on a free electron. As we have seen in Section 2.6, the coherently scattered wave is a spherical wave whose amplitude is given by J. J. Thomson's formula. If the point of observation, M, is very far from the electron (Fig. 3.1), the spherical wave can be replaced locally by a plane wave tangential to it, $\mathbf{E_h} \exp(-2\pi \mathrm{i} \mathbf{K_h} \cdot \mathbf{r})$ where $\mathbf{K_h}$ is parallel to **OM** and its length is $1/\lambda$. The scattered amplitude can be written as

$$E = E_o \mathcal{A}_e \sin \psi, \tag{3.1}$$

where ψ is the angle between the electric field vector $\mathbf{E_o}$ of the incident wave and the direction of observation of the scattered wave (Fig. 3.1) and

$$\mathcal{A}_e = \frac{R \exp(2\pi \mathrm{i} \nu t)}{r_o}. \tag{3.2}$$

$r_o = OM$ and R is the classical radius of the electron (2.32).

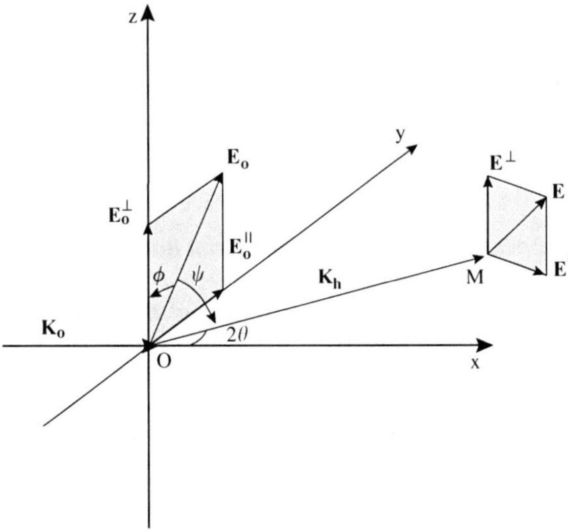

Fig. 3.1. Classical scattering of an electromagnetic wave by a free electron.

Let ϕ be the angle between the incident electric field and the normal to the scattering plane, $\mathbf{K_o}$, $\mathbf{K_h}$ (Fig. 3.1). $\mathbf{E_o}$ can be decomposed into its two components, perpendicular and parallel to that plane, $E_o^{\perp} = E_o \cos\phi$, and $E_o^{\parallel} = E_o \sin\phi$, respectively. The scattered field, $\mathbf{E}$, can be decomposed in a similar way into a component normal to the scattering plane, $\mathbf{E}^{\perp}$, called σ-polarization, and a component parallel to that plane, $\mathbf{E}^{\parallel}$, called π-polarization. The first component is obtained by the scattering of $\mathbf{E_o^{\perp}}$; the angle ψ is equal to $\pi/2$ and

$$E^{\perp} = E_o \cos\phi\, \mathcal{A}_e.$$

In a similar way, the second component is obtained by the scattering of $\mathbf{E_o^{\parallel}}$; the angle ψ is equal to $\pi/2 - 2\theta$ where 2θ is the scattering angle ($\mathbf{K_o}$, $\mathbf{K_h}$) and

$$E^{\parallel} = E_o \sin\phi\, \mathcal{A}_e \cos 2\theta.$$

The ratio of the scattered to the incident amplitude is therefore $E^{\perp}/E_o^{\perp} = \mathcal{A}_e$ for σ-polarization and $E^{\parallel}/E_o^{\parallel} = \mathcal{A}_e \cos 2\theta$ for π-polarization. Introducing the polarization coefficient C equal to 1 or to $\cos 2\theta$ depending on the state of polarization, the ratio of the scattered to the incident amplitude is, in all cases, $C\mathcal{A}_e$. The total scattered field is $\mathbf{E} = \mathbf{E}^{\perp} + \mathbf{E}^{\parallel}$ and its intensity is

$$E^2 = E^{\perp 2} + E^{\parallel 2} = \mathcal{A}_e^2 E_o^2[\cos^2\phi + \sin^2\phi \cos^2 2\theta]. \tag{3.3}$$

The reader will check that the right-hand term is equal to $\mathcal{A}_e^2 E_o^2 \sin^2\psi$, as it should be.

If the incident wave is unpolarized, the intensity of the total scattered field is

$$E^2 = \langle E^{\perp}\rangle^2 + \langle E^{\parallel}\rangle^2 = \mathcal{A}_e^2 E_o^2\left[\frac{1+\cos^2 2\theta}{2}\right], \tag{3.4}$$

since $\langle\cos^2\phi\rangle = \langle\sin^2\phi\rangle = \frac{1}{2}$.

We shall see in Section 5.3 that the equations of the dynamical theory must be solved separately for the two components of polarization.

3.2 Amplitude diffracted by a periodic electron distribution

Let us consider a periodic medium with a continuous distribution of electronic density, $\rho(\mathbf{r})$. The number of electrons in a volume element $\mathrm{d}\tau$ is $\rho(\mathbf{r})\,\mathrm{d}\tau$ and the ratio of the amplitude scattered by volume element $\mathrm{d}\tau$ to the incident amplitude is $\mathcal{A}_e \sin\psi \rho(\mathbf{r})\,\mathrm{d}\tau$. If we take the origin of phases at a plane-wave front passing through a given point O (Fig. 3.2), the phase of the wave scattered by a scattering centre located at P defined by a position vector $\mathbf{r} = \mathbf{OP}$ is $2\pi\mathrm{i}(\mathbf{K_h} - \mathbf{K_o})\cdot\mathbf{r}$ and the amplitude scattered by the volume element around P is

$$\mathcal{A}_e \sin\psi\; \rho(\mathbf{r}) \exp[2\pi\mathrm{i}\Delta\mathbf{K}\cdot\mathbf{r}]\,\mathrm{d}\tau,$$

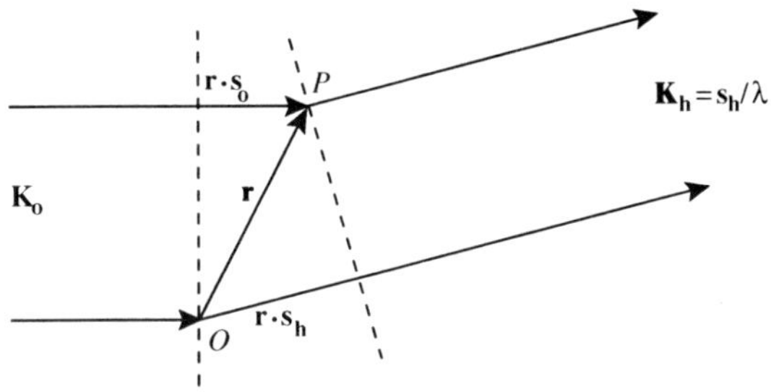

Fig. 3.2. Phase difference between the waves emitted by two scattering centres.

where $\mathbf{\Delta K} = (\mathbf{K_h} - \mathbf{K_o})$ is the diffraction vector. The total amplitude diffracted by the medium is

$$\mathcal{A}(\mathbf{\Delta K}) = \mathcal{A}_e \sin\psi \int \rho(\mathbf{r}) \exp[2\pi \mathrm{i} \mathbf{\Delta K} \cdot \mathbf{r}]\, \mathrm{d}\tau,$$

where the integral is over the whole volume of the diffracting medium and $\mathcal{A}(\mathbf{\Delta K})$ is the distribution of scattered amplitude in reciprocal space.

Let $y(\mathbf{r})$ be the shape function equal to 1 inside the medium and 0 outside. The diffracted amplitude can also be written as

$$A(\mathbf{\Delta K}) = \mathcal{A}_e \sin\psi \int_{-\infty}^{\infty} y(\mathbf{r})\rho(\mathbf{r}) \exp[2\pi \mathrm{i} \mathbf{\Delta K} \cdot \mathbf{r}]\, \mathrm{d}\tau. \tag{3.5}$$

Expression (3.5) shows that the distribution of scattered amplitude in reciprocal space is proportional to the Fourier transform of the product of the shape function by the electronic density. We can therefore write

$$\mathcal{A}(\mathbf{\Delta K}) = \mathcal{A}_e \sin\psi \mathcal{F}[y(\mathbf{r})\rho(\mathbf{r})].$$

The Fourier transform $\mathcal{F}[y(\mathbf{r})\rho(\mathbf{r})]$ is equal to the convolution of the Fourier transforms, $Y(\mathbf{\Delta K})$ of the shape function $y(\mathbf{r})$, and $\mathcal{A}_\infty(\mathbf{\Delta K})$ of the electronic density$\rho(\mathbf{r})$:

$$\begin{aligned}\mathcal{A}(\mathbf{\Delta K}) &= \mathcal{A}_e \sin\psi\ \mathcal{F}[y(\mathbf{r})] * \mathcal{F}[\rho(\mathbf{r})] \\ &= \mathcal{A}_e \sin\psi\, Y(\mathbf{\Delta K}) * \mathcal{A}_\infty(\mathbf{\Delta K})\end{aligned} \tag{3.6}$$

with

$$Y(\mathbf{\Delta K}) = \int_{-\infty}^{\infty} y(\mathbf{r}) \exp[2\pi \mathrm{i} \mathbf{\Delta K} \cdot \mathbf{r}]\, \mathrm{d}\tau,$$

$$\mathcal{A}_\infty(\mathbf{\Delta K}) = \int_{-\infty}^{\infty} \rho(\mathbf{r}) \exp[2\pi \mathrm{i} \mathbf{\Delta K} \cdot \mathbf{r}]\, \mathrm{d}\tau.$$

The electronic density is a periodic function in direct space. It can be expressed as the convolution of the electronic density in the unit cell, $\rho_o(\mathbf{r})$, by a periodic distribution of Dirac distributions:

$$\rho(\mathbf{r}) = \rho_o(\mathbf{r}) * \sum_{i=-\infty}^{\infty} \delta(\mathbf{r} - \mathbf{r_i}),$$

where $\mathbf{r_i} = u_i\mathbf{a} + v_i\mathbf{b} + w_i\mathbf{c}$.

The Fourier transform of an infinite sum of Dirac distributions in direct space is an infinite sum of Dirac distributions in reciprocal space. In a non-orthonormal space we have

$$\mathcal{F}\left[\sum_{i=-\infty}^{\infty} \delta(\mathbf{r} - \mathbf{r_i})\right] = V^{-1} \sum_{h=-\infty}^{\infty} \delta(\mathbf{\Delta K} - \mathbf{h}),$$

where V is the volume of the unit cell and the sum is over all the reciprocal lattice vectors $\mathbf{h}$. It follows that the Fourier transform of the electronic density is

$$\mathcal{A}_\infty(\mathbf{\Delta K}) = \frac{F(\mathbf{\Delta K})}{V} \sum_{h=-\infty}^{\infty} \delta(\mathbf{\Delta K} - \mathbf{h}),$$

where $F(\mathbf{\Delta K})$ is the Fourier transform of $\rho_o(\mathbf{r})$. The distribution of diffracted amplitude is 0 everywhere except when the diffraction vector $\mathbf{\Delta K}$ is equal to a reciprocal lattice vector, that is when the diffraction condition is satisfied. The values of F need only be taken at the reciprocal lattice points and

$$\mathcal{A}_\infty(\mathbf{\Delta K}) = V^{-1} \sum_{h=-\infty}^{\infty} F_{hkl}\, \delta(\mathbf{\Delta K} - \mathbf{h}), \tag{3.7}$$

where

$$\begin{aligned} F_{hkl} &= \int_{-\infty}^{\infty} \rho_o(\mathbf{r}) \exp[-2\pi i\mathbf{h} \cdot \mathbf{r}]\, \mathrm{d}\tau = \int_{\text{unit cell}} \rho(\mathbf{r}) \exp[-2\pi i\mathbf{h} \cdot \mathbf{r}]\, \mathrm{d}\tau \\ &= \sum_j f_j \exp[-2\pi i\mathbf{h} \cdot \mathbf{r}] \end{aligned}$$

is the structure factor. The electronic density, ρ_o, is considered to be the sum of the electronic densities of all the atoms, j, in the unit cell; f_j is the form factor of atom j and is to be multiplied by the Debye–Waller factor $\exp(-M)$ to take into account thermal vibrations (James 1950; Warren 1969).

Substituting from (3.7) into (3.6), it follows finally that

$$\mathcal{A}(\mathbf{\Delta K}) = \frac{A_e \sin\psi}{V} \sum_{\mathbf{h}} F_{hkl}\, Y(\mathbf{\Delta K} - \mathbf{h}), \tag{3.8}$$

which shows that the fine structure of each diffraction spot is given by the variations of the Fourier transform of the shape function.

3.3 Intensity diffracted by a small crystal

Consider a small crystal bathed in the incident beam (Fig. 3.3). For simplicity, we shall assume that it has the shape of a parallelepipedic slab with the big face parallel to the reflecting planes. The Fourier transform of the shape function, $Y(\mathbf{\Delta K})$, of this crystal is

$$V \frac{\sin(\pi X_h N_1)}{\pi X_h} \frac{\sin(\pi X_k N_2)}{\pi X_k} \frac{\sin(\pi X_l N_3)}{\pi X_l},$$

where N_1, N_2 and N_3 are the numbers of unit cells in the three directions perpendicular to the crystal faces, X_h, X_k and X_l are the numerical coordinates of the reciprocal space vector $\mathbf{\Delta K}-\mathbf{h}$ and $\mathbf{h} = h\mathbf{a}^*+k\mathbf{b}^*+l\mathbf{c}^*$ is the reciprocal lattice vector associated with the Bragg reflection considered.

Substituting into (3.8) and using (3.1), (3.2) and (3.4), we obtain for the diffracted intensity, in the case of unpolarized incident radiation:

$$I = I_o \frac{R^2}{r_o^2} \frac{1+\cos^2 2\theta_B}{2} |F_{hkl}|^2 \frac{\sin^2 \pi X_h N_1}{\pi^2 X_h^2} \frac{\sin^2 \pi X_k N_2}{\pi^2 X_k^2} \frac{\sin^2 \pi X_l N_3}{\pi^2 X_l^2}. \tag{3.9}$$

Experimentally, one can record the diffracted intensity in a detector placed in the reflected beam at a distance r_o from the crystal with a wide open entrance slit while the crystal is rotated through the reflection domain around an axis perpendicular to the

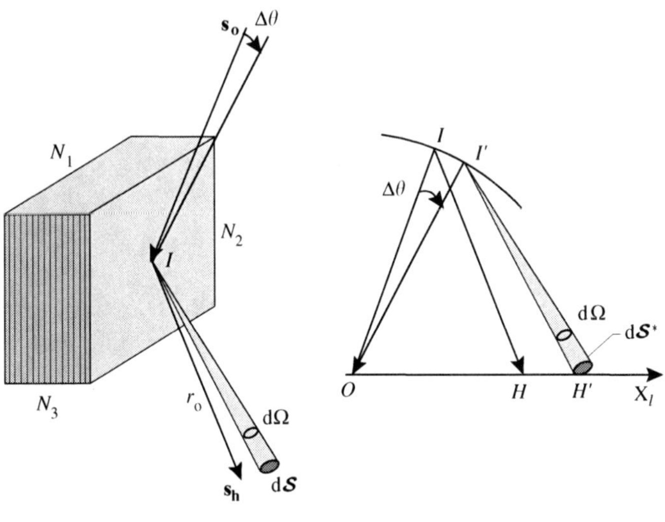

Fig. 3.3. Diffraction by a crystal. Left-hand side: direct space; right-hand side: reciprocal space; I: centre of the Ewald sphere; $d\Omega$: element solid angle diffracted by a point scatterer; $d\mathcal{S}$: surface element of the receptor; $d\mathcal{S}^*$: surface element in reciprocal space.

diffracting plane. Let $\mathbf{s_o}$ and $\mathbf{s_h}$ be unit vectors in the incident and reflected directions satisfying Bragg's condition exactly, $\Delta\theta$ be the angle between the actual incident wavevector and $\mathbf{s_o}$ (Fig. 3.3) and $\mathrm{d}\mathcal{S}$ be the area of the detector; the reflected waves received by the detector are contained in a small cone of solid angle $\mathrm{d}\Omega = \mathrm{d}\mathcal{S}/r_o^2$. The mean reflected direction in reciprocal space is $\mathbf{I'H'}$ ($I'H' = k = 1/\lambda$) and $\mathbf{HH'}$ is the vector $\mathbf{\Delta K}$ of coordinates X_h, X_k and X_l.

We shall now transform eqn (3.9) in order to obtain the expression of the intensity received by the detector as a function of $\Delta\theta$. Let ω be the angular velocity of the crystal. During time $\mathrm{d}t$, the crystal is rotated by $\mathrm{d}(\Delta\theta) = \omega\,\mathrm{d}t$ and the energy received by the detector is

$$\mathrm{d}\mathcal{E} = I\,\mathrm{d}\mathcal{S}\,\mathrm{d}t$$

with $\mathrm{d}\mathcal{S} = r_o^2\,\mathrm{d}\Omega$. This energy can be rewritten, using (3.9), as

$$\begin{aligned}\mathrm{d}\mathcal{E} &= I\,\frac{r_o^2}{\omega}\,\mathrm{d}\Omega\,\mathrm{d}(\Delta\theta)\\ &= \frac{I_o}{\omega}\,R^2\,\frac{1+\cos^2 2\theta_B}{2}\,|F_{hkl}|^2\\ &\quad\times\frac{\sin^2\pi X_h N_1}{\pi^2 X_h^2}\,\frac{\sin^2\pi X_k N_2}{\pi^2 X_k^2}\,\frac{\sin^2\pi X_l N_3}{\pi^2 X_l^2}\,\mathrm{d}\Omega\,\mathrm{d}(\Delta\theta).\end{aligned} \tag{3.10}$$

The elements of integration $\mathrm{d}\Omega\,\mathrm{d}(\Delta\theta)$ can be evaluated in reciprocal space in terms of $\mathrm{d}X_h$, $\mathrm{d}X_k$ and $\mathrm{d}X_l$. We shall take the $\mathbf{OX}_l$ axis of reciprocal space parallel to the reciprocal lattice vector $\mathbf{OH}$ (Fig. 3.3) and the two other axes parallel to the lattice planes.

1. *Evaluation of the element of solid angle* $\mathrm{d}\Omega$: The incident wavevector is $\mathbf{I'O}$ in reciprocal space and the reflected directions are contained in the small cone of solid angle $\mathrm{d}\Omega$ around $\mathbf{I'H'}$. Let $\mathrm{d}\mathcal{S}^*$ be a surface element in reciprocal space normal to that cone in H':

$$\mathrm{d}\mathcal{S}^* = k^2\,\mathrm{d}\Omega.$$

The projection of the surface element $\mathrm{d}\mathcal{S}^*$ on the $\mathbf{OX_h}$, $\mathbf{OX_k}$ plane (parallel to the lattice planes) is $\mathrm{d}\mathcal{S}^*\sin\theta_B$; this projection can be written as a surface element of that plane. Since X_h and X_k are numerical coordinates, it is equal to $\mathrm{d}X_h \times \mathrm{d}X_k \times$ the surface of the basis of the unit cell of the reciprocal lattice normal to $\mathbf{OH}$, namely $V^*/(1/d_{hkl}) = V^* d_{hkl}$ (d_{hkl}, reflecting plane spacing; V^*, volume of the unit cell of the reciprocal lattice):

$$\mathrm{d}\mathcal{S}^*\sin\theta_B = k^2\,\mathrm{d}\Omega\sin\theta_B = \mathrm{d}X_h \times \mathrm{d}X_k \times V^* d_{hkl}. \tag{3.11}$$

2. *Evaluation of* $\mathrm{d}(\Delta\theta)$: It is obtained by calculating the variation $\mathbf{HH'}$ of the reciprocal lattice vector $\mathbf{OH}$ associated with a rotation $(OI, OI') = \mathrm{d}(\Delta\theta)$ of the

direction of incidence. Bragg's law can be written as

$$OH = \frac{1}{d_{hkl}} = \frac{2}{\lambda} \sin \theta_B = 2k \sin \theta_B.$$

It follows, by derivation, that

$$\mathrm{d}(OH) = 2k \cos \theta_B \, \mathrm{d}(\Delta\theta).$$

and, using the normal coordinate of H' along $\mathbf{OX}_l$,

$$\mathrm{d}X_l = \frac{\mathrm{d}(OH)}{OH} = 2kd_{hkl} \cos \theta_B \, \mathrm{d}_{hkl}(\Delta\theta). \tag{3.12}$$

From (3.11) and (3.12), it follows that

$$\mathrm{d}\Omega \, \mathrm{d}(\Delta\theta) = \frac{\lambda^3}{V \sin 2\theta_B} \mathrm{d}X_h \, \mathrm{d}X_k \, \mathrm{d}X_l$$

and, by substitution into (3.10),

$$\mathrm{d}\mathcal{E} = \frac{I_o}{\omega} \frac{\lambda^3 R^2 (1 + \cos^2 2\theta_B)}{2V \sin 2\theta_B} |F_{hkl}|^2 \times \frac{\sin^2 \pi X_h N_1}{\pi^2 X_h^2} \frac{\sin^2 \pi X_k N_2}{\pi^2 X_k^2} \frac{\sin^2 \pi X_l N_3}{\pi^2 X_l^2} \mathrm{d}X_h \, \mathrm{d}X_k \, \mathrm{d}X_l.$$

By integration of this expression over $\mathrm{d}X_h$ and $\mathrm{d}X_k$, one obtains the total energy received by the detector for a given position of the crystal in the incident beam (and therefore for a given value of $\Delta\theta$):

$$\mathrm{d}(\Delta\mathcal{E}) = \frac{I_o}{\omega} \frac{\lambda^3 R^2 (1 + \cos^2 2\theta_B)}{2V \sin 2\theta_B} |F_{hkl}|^2 N_1 N_2 \frac{\sin^2 \pi X_l N_3}{\pi^2 X_l^2} \mathrm{d}X_l. \tag{3.13}$$

3.4 Reflectivity

Let us assume that the crystal is entirely irradiated by the incident beam, and let S be its surface and t its thickness (Fig. 3.4).

The volume of the unit cell can be written as

$$V = \frac{St}{N_1 N_2 N_3} = \frac{Sd_{hkl}}{N_1 N_2}$$

($t = N_3 d_{hkl}$). Substituting in (3.13) and using (3.12), we obtain

$$\mathrm{d}(\Delta\mathcal{E}) = \frac{I_o}{\omega} \frac{\lambda^2 R^2 (N_1 N_2)^2 (1 + \cos^2 2\theta_B)}{2S \sin \theta_B} |F_{hkl}|^2 \times \frac{\sin^2 (2\pi N_3 k d_{hkl} \cos \theta_B \, \Delta\theta)}{(2\pi k d_{hkl} \cos \theta_B \Delta\theta)^2} \mathrm{d}(\Delta\theta).$$

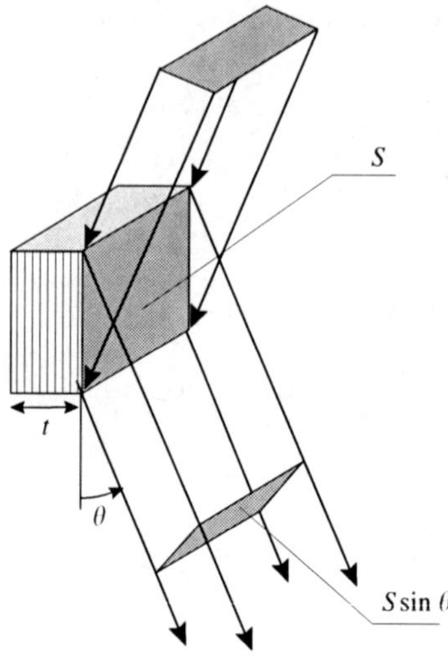

Fig. 3.4. Cross-section of the beam diffracted by a small crystal.

The *reflectivity* is, by definition, the ratio of the energy received per unit area of the detector per unit time divided by the incident intensity. It is a function of $\Delta\theta$:

$$I_h = \frac{\mathrm{d}(\Delta\mathcal{E})\omega}{I_o\,\mathrm{d}(\Delta\theta)\,S\sin\theta_B}$$
$$= \frac{R^2\lambda^2(1+\cos^2 2\theta_B)}{2\sin^2\theta_B} n^2 |F_{hkl}|^2 \left[\frac{\sin(2\pi N_3 k d_{hkl}\cos\theta_B\Delta\theta)}{2\pi k d_{hkl}\cos\theta_B\Delta\theta}\right]^2, \qquad (3.14)$$

where $n = N_1 N_2/S$ is the number of diffracting units per unit surface of the crystal. This is the classical expression of the reflecting power of a small crystal obtained with the geometrical theory (see, for instance, James 1950). Figure 3.5 represents the variations of I_h with $\Delta\theta$ in arbitrary units (*rocking curve*).

The maximum reflectivity, at the centre of the rocking curve, is

$$\frac{R^2\lambda^2 t^2(1+\cos^2 2\theta_B)}{2V^2\sin^2\theta_B}|F_{hkl}|^2$$

($nN_3 = tV^{-1}$). It is proportional to the square of the structure factor and to the square of the crystal thickness. The *Full Width at Half Maximum* (FWHM) is $\approx 0.442946\lambda/t\cos\theta_B$.

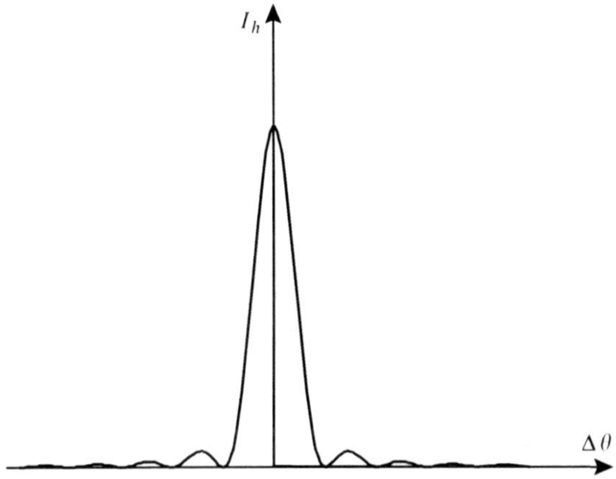

Fig. 3.5. Reflectivity according to the geometrical theory. Arbitrary units.

3.5 Integrated intensity

1. *The crystal is completely irradiated by the beam and is not absorbing*

The *integrated intensity* is the ratio of the total energy received by the detector, while the crystal is rotated, divided by the intensity of the incident beam (see Kalman (1979) for various experimental methods of measuring the integrated intensity). It is obtained by integrating expression (3.13) over dX_l (the same result is obtained using slightly different methods in James 1950; Warren 1969):

$$I_e = \frac{\omega}{I_o}\int d(\Delta\mathcal{E}) = \frac{\lambda^3 R^2}{\sin 2\theta_B}\frac{\Delta V}{V^2}\frac{(1+\cos^2 2\theta_B)}{2}|F_{hkl}|^2, \tag{3.15}$$

where ΔV is the volume of the crystal and V is the volume of the unit cell. I_e has the dimension of an area and so can be regarded as a cross-section. The factor $1/\sin 2\theta_B$ in equation (3.15) and the following equations, (3.16*a*,*b*) and (3.17), is called the Lorentz factor. It takes on a more complicated expression when the axis of rotation of the crystal is not normal to the $\mathbf{K_o}$, $\mathbf{K_h}$ plane (see, for instance, Zachariasen 1945, Rossmanith 1992).

2. *The crystal is larger than the beam and is absorbing*

(i) *Reflection geometry*: We now consider a thick absorbing crystal larger than the cross-section of the incident beam (Fig. 3.6) and a small volume element dV of thickness dz and surface S, situated at depth z from the crystal surface.

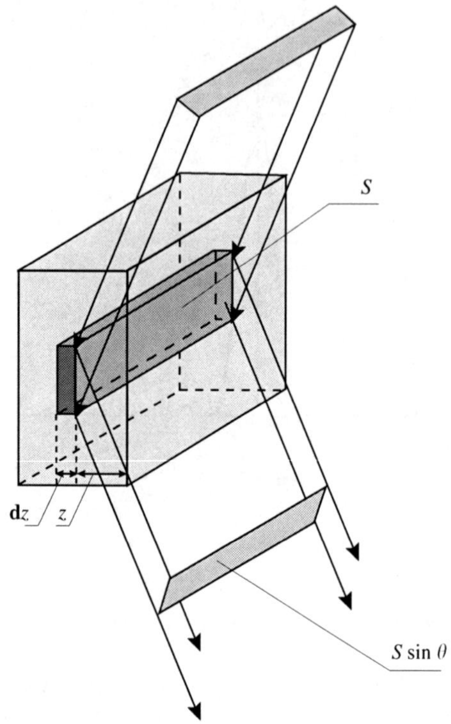

Fig. 3.6. Diffraction of a small crystalline slab at depth z from the surface.

The total path of the X-rays inside the crystal is $2z/\sin\theta_B$. The intensity diffracted by this volume element is, taking into account absorption,

$$\begin{aligned} dI_e &= \frac{\lambda^3 R^2(1+\cos^2 2\theta_B)}{2\sin 2\theta_B}\frac{dV}{V^2}|F_{hkl}|^2 \exp\left(\frac{-2\mu z}{\sin\theta_B}\right) \\ &= \frac{\lambda^3 R^2(1+\cos^2 2\theta_B)}{2\sin 2\theta_B}\frac{S}{V^2}|F_{hkl}|^2 \exp\left(\frac{-2\mu z}{\sin\theta_B}\right) dz. \end{aligned}$$

The total intensity diffracted by the crystal is

$$I_e = \frac{\lambda^3 R^2(1+\cos^2 2\theta_B)}{2\sin 2\theta_B}\frac{S}{V^2}|F_{hkl}|^2 \int_0^t \exp\left(\frac{-2\mu z}{\sin\theta_B}\right) dz,$$

where t is the crystal thickness. The integrated reflectivity, I_{hi}, is here I_e divided by the cross-section $S \sin\theta_B$ of the beam:

$$I_{hi} = \frac{I_e}{S \sin\theta_B} = \frac{\lambda^3 R^2}{\sin 2\theta_B} \frac{1}{V^2} \frac{1}{2\mu} \frac{(1+\cos^2 2\theta_B)}{2} |F_{hkl}|^2 \times \left[1 - \exp\left(\frac{-2\mu t}{\sin\theta_B}\right)\right]. \tag{3.16a}$$

It is a dimensionless number. If the crystal is non-absorbing, (3.16a) is replaced by

$$I_{hi} = \frac{\lambda^3 R^2}{\sin 2\theta_B} \frac{1}{V^2} \frac{t}{\sin\theta_B} \frac{(1+\cos^2 2\theta_B)}{2} |F_{hkl}|^2. \tag{3.16b}$$

(ii) *Transmission geometry*: In a similar way, the integrated reflectivity through a thick, absorbing crystal is

$$I_{hi} = \frac{\lambda^3 R^2}{\sin 2\theta_B} \frac{1}{V^2} \frac{t}{\cos\theta_B} \exp\left(-\frac{\mu t}{\cos\theta_B}\right) \frac{1+\cos^2 2\theta_B}{2} |F_{hkl}|^2. \tag{3.17}$$

The results of the geometrical theory are compared with those obtained with dynamical theory in Section 4.11.

3.6 Mosaic crystals

In the approximation of the 'ideally imperfect' crystal, the crystal is a mosaic of small crystallites (see Fig. 1.2), each one small enough for the approximation of the geometrical theory to be valid, and the misorientations between successive blocks large enough for them to diffract different portions of the incident beam. In such a situation, the rocking curve width depends on the distribution of defects and is in general very much wider than that of a single crystallite; it is a measure of the 'mosaicity' of the crystal. The integrated intensity does not depend on these misorientations and is given by the expressions of Section 3.5. Darwin's extinction theory (see Sections 1.5 and 4.11) is based on this mosaic model.

4

Elementary dynamical theory

4.1 Limitations of the geometrical theory

Darwin (1914*b*) was the first to point out that the geometrical theory cannot be correct because it violates the conservation of energy. Equation (3.15) shows, for instance, that the integrated intensity diffracted by a small crystal entirely bathed in the incident radiation is proportional to its volume. The reason is of course that geometrical theory assumes that the amplitude of the wave received by every diffracting centre in the crystal is the same, thus neglecting the interaction of the wave with matter and the fact that part of its energy has already been reflected away by the layers it has crossed before. The amplitudes diffracted by each scattering centre are added by taking simply into account the interferences due to the differences in optical path. This approximation is equivalent to the Born approximation in quantum mechanics. It works quite well for the diffracted intensities from thin crystals because the interaction of X-rays with matter is very weak. This is also the case for neutrons, but not quite so for high-energy electrons and certainly not so for low-energy electrons. But there is another reason why the geometrical theory is not satisfactory, even for small crystals or for *ideally imperfect* crystals. It does not provide any phase information about the processes inside the crystal. The size of the diffraction spot and of the reflection domain is inversely proportional to the size of the crystal. For an infinite and perfectly periodic crystal, the reflection domain would be reduced to a Dirac distribution and would be represented by a single point in reciprocal space (thermal agitation and local disorder will produce a diffuse spot). This point is the centre of the Ewald sphere and has been called *Laue point* by Ewald. It is situated on the intersection of the two spheres centred at the reciprocal lattice nodes O and H and with radii $k = 1/\lambda$ (Fig. 4.1). If an incident wave such as **OM** does not satisfy Bragg's condition rigorously, there would be no reflected wave, according to the geometrical theory.

It will be shown in Section 4.11 that the results of the dynamical theory tend asymptotically towards those of the geometrical theory when the crystal thickness is much smaller than a certain length, Λ_o (called Pendellösung distance in the transmission geometry and extinction distance in the reflection geometry—Section 4.7.4), which is proportional to the structure factor, F_h, and inversely proportional to the wavelength. The condition of validity of the geometrical theory therefore depends on the diffraction conditions. But, even so, the geometrical theory fails to give a good description of the diffraction by multilayered structures (Underwood and Barbee 1981), or of the surface diffraction at grazing incidence (Vineyard 1982; Colella 1991; Caticha 1993).

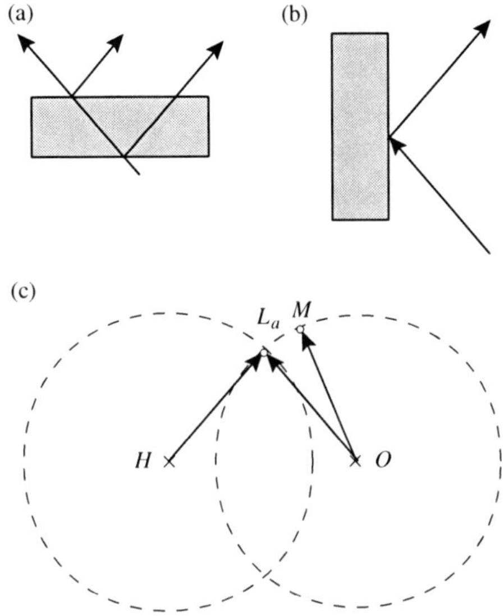

Fig. 4.1. Diffraction according to the geometrical theory. (a) Direct space, Laue geometry. (b) Direct space, Bragg geometry. (c) Reciprocal space; $\mathbf{OH} = \mathbf{h}$: reciprocal lattice vector; L_a: Laue point; $\mathbf{OL_a}$, incident wave ($\overline{OL_a} = k$) and $\mathbf{HL_a}$, reflected wave ($\overline{HL_a} = k$) satisfy Bragg's condition; **OM**: incident wave which does not satisfy Bragg's condition. Wavevectors are oriented from the reciprocal lattice points (see footnote 6 on page 11).

4.2 Introduction of the dispersion surface

Actually, the waves propagating in the medium interact with matter and there is an index of refraction, n. The wave numbers of the waves satisfying the diffraction condition are therefore nk and the point representing the reflection domain in reciprocal space is on the intersection of two spheres with the same centres, but with $nk = (1 + \chi_o/2)k$ as radius (Fig. 4.2) (the expression of the index of refraction is given in Section 2.7.3). This point has been called *Lorentz point* by Ewald. The distance between the Lorentz and the Laue point is $\overline{L_oL_a} = k(1 - n)/\cos\theta_B = k|\chi_o|/2\cos\theta_B = R\lambda F_o/(\pi V 2\cos\theta_B)$. The wave propagating in the incident direction is called the *refracted* wave. Since the index of refraction is smaller than 1 for X-rays (for neutrons it is smaller or larger than 1, depending on the sign of the scattering length), these two spheres lie inside the spheres with radius k. We shall limit ourselves in the present discussion to the case where there are only two reciprocal lattice points simultaneously on the Ewald sphere.

When the crystal is rocked through the reflection domain, it is not possible physically that the reflected beam occurs for one angular position only of the crystal, as it does in the geometrical theory for a semi-infinite crystal. For every position of the crystal within the reflection domain, there is a refracted wave, of wavevector $\mathbf{K_o} = \mathbf{OP}$, and a reflected wave $\mathbf{K_h} = \mathbf{HP}$. For reasons of symmetry, P cannot lie

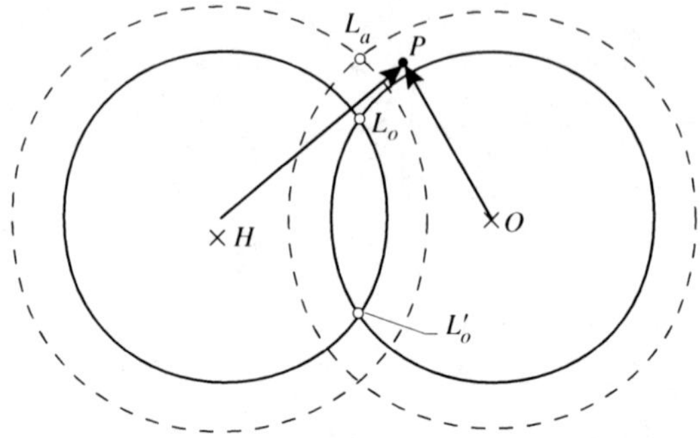

Fig. 4.2. Influence of the index refraction. L_o: Lorentz point; $\overline{OL_o} = \overline{HL_o} = nk$, where n is the index of refraction; **OP**: refracted wave; **HP**: reflected wave; L_o': Lorentz point associated with the reciprocal lattice vector $\mathbf{HO} = -\mathbf{h}$.

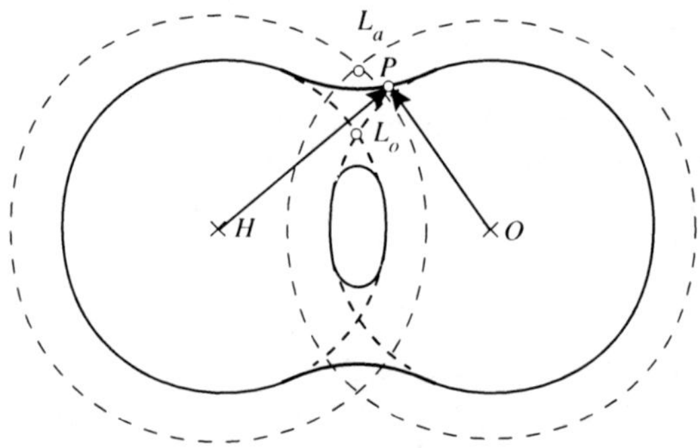

Fig. 4.3. Interaction of the refracted (**OP**) and reflected (**HP**) waves. Solid curve: dispersion surface; P: tiepoint.

on either of the two spheres (there is no reason for it to lie on one of the two spheres rather than on the other one). It must therefore lie on a connecting surface between the two spheres (Fig. 4.3).

Far from the intersection of the two spheres, there is only one wave propagating in the crystal, of wave number nk and wavevector $\mathbf{K_o} = \mathbf{OP}$ or $\mathbf{K_h} = \mathbf{HP}$, and P lies on one of the two spheres. When it lies on the connecting surface, there are two waves and both their wave numbers differ slightly from nk. This surface, which has two sheets, is called the *dispersion surface*. The angular width of the reflection domain is represented in reciprocal space by the size of this connecting surface between the two spheres. It characterizes the strength of the interaction between the waves and matter. It will be shown in Section 4.5 that the diameter of the dispersion surface is proportional to the structure factor (eqn (4.16)). Therefore, the larger the structure factor, the stronger the interaction and the larger the reflection domain. For an infinitely weak reflection, the dispersion surface would be concentrated at the Lorentz point; this would correspond to an 'empty' crystal (no scatterer in the unit cell). The distance of the Lorentz point to the dispersion surface, $|C|\sqrt{\chi_h \chi_{\bar{h}}}/2\cos\theta_B$, is shorter than the distance of the Lorentz point to the Laue point. These two distances could only be equal if the Fourier transform of the polarizability was constant in reciprocal space, that is if the scattering material was constituted by point scatterers located at the nodes of the direct lattice.

It has been noticed in Section 2.7.2 that the Fourier coefficients of the polarizability are of the order of 10^{-5} to 10^{-6}. The size of the dispersion surface is therefore very small with respect to the radius of the Ewald sphere, and the wave numbers K_o and K_h are very close to one another and to k. The common extremity, P, of the two wavevectors has been called *tiepoint* by Ewald to stress that the two waves are closely linked together to form a wavefield; it is called *Anregungspunkt* by Laue. All the properties of X-ray diffraction, X-ray propagation and X-ray interferences in crystals proceed from the properties of the wavefields and can be described in relation to the dispersion surface.[1]

4.3 Analogy with the band theory of solids

The band theory of solids offers many similarities with the dynamical theory of diffraction in that it treats the properties of a wave in a periodic medium. However, in the band theory one studies the possible energies of electrons as a function of their wave number while in diffraction theory the energy is constant and one looks for the possible positions of the wavevectors in reciprocal space. The propagation of a free electron is the analogue of the propagation of a single wave. The kinetic energy of the electron is $\mathcal{E} = \frac{1}{2} mv^2$ and, according to the de Broglie equivalence of its wave and corpuscular natures, $mv = hK$ (h, Planck's constant; K, wave number). The relation between the energy and the wave number of the free electron is, therefore,

$$\mathcal{E} = \frac{h^2 K^2}{2m}. \tag{4.1}$$

To describe the properties of electrons interacting with a medium with a certain effective potential energy $\varphi(\mathbf{r})$, one writes that the wave function of the electron satisfies Schrödinger's equation:

$$-\frac{\hbar^2}{2m}\Delta\Psi + \varphi(\mathbf{r})\Psi = \mathcal{E}\Psi. \tag{4.2}$$

For a free electron, the potential energy is zero:

$$-\frac{\hbar^2}{2m}\Delta\Psi = \mathcal{E}\Psi.$$

The simplest solution is a plane wave of the form

$$\Psi = \Psi_o \exp(-2\pi \mathrm{i}\mathbf{K}\cdot\mathbf{r}).$$

[1] A nice survey of the properties of the dispersion surface is given in Whelan (1986).

Substituting into Schrödinger's equation with a zero potential, we obtain the expression given above for the energy of a free electron.

Bloch (1928) showed that to describe the properties of the electron in a periodic medium, the potential $\varphi(\mathbf{r})$ should be taken as a periodic function, with the periodicity of the direct lattice. In the same way as the polarizability of crystals for X-rays (2.36):

$$\varphi(\mathbf{r}) = \sum_{\mathbf{h}} \varphi_h \exp(2\pi i \mathbf{h} \cdot \mathbf{r}).$$

It was shown by Floquet (1883) that the solutions of linear partial differential equations of the same type as (4.2) with a periodic interaction term are the product of a certain function by a function having the same periodicity as that term. The wave function of the electron is, therefore,

$$\Psi(\mathbf{r}) = \exp(-2\pi i \mathbf{K_o} \cdot \mathbf{r}) \sum_{\mathbf{h}} \Psi_h \exp(2\pi i \mathbf{h} \cdot \mathbf{r}). \tag{4.3}$$

This wave function is known in solid state physics as a Bloch wave. It is exactly the same as the wavefield introduced by Ewald (1917) (eqn (2.47)).

Equation (4.1) shows that the variations of the energy of a free electron with wave number are parabolic. Figure 4.4 shows these variations for an electron in a periodic one-dimensional medium with a vanishingly small potential. The parabola is repeated with the periodicity of the reciprocal lattice. The intersection of two successive parabolae lies on the limit between two Brillouin zones. It corresponds to a Bragg reflection of the wave associated with the electron; the wave function of the electron is then the result of the superposition of the two waves. The corresponding values of the eigenenergy are given by the solutions of (4.2). Their variations are represented by connecting curves between the two parabolae and Fig. 4.5 shows that there are energies which the electron cannot have; they correspond to the energy gap between two energy bands. This situation is very similar to that observed with the dispersion surface in diffraction theory. Stern *et al.* (1969) have shown that the

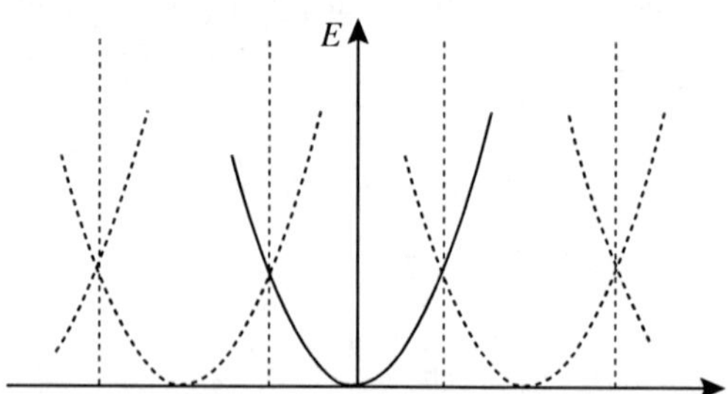

Fig. 4.4. Energy of an electron in a one-dimensional periodic medium, of period d in direct space and with a vanishingly small potential. The dashed vertical lines represent the Brillouin zone boundaries.

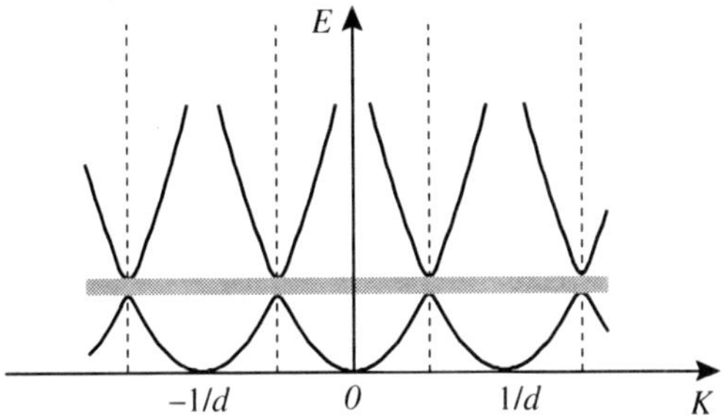

Fig. 4.5. Energy bands in a one-dimensional periodic medium; the shaded area represents the energy gap.

dispersion surface in electron diffraction and the band diagram are cuts of the same dispersion hypersurface in four-dimensional space (energy and the three coordinates of reciprocal space) which defines totally the allowed electronic states in the crystal. The cuts are by two three-dimensional hyperplanes; the three dimensions are the three dimensions of reciprocal space in the former case (the dispersion surface is a constant energy surface), energy and two dimensions of reciprocal space in the latter case. The relationship between total reflection in Bragg diffraction and band gaps in solids was first discussed by Morse (1930) in the case of electron diffraction.

4.4 Propagation equation

The propagation equation of an electromagnetic wave in a medium with a continuous polarizability $\chi(\mathbf{r})$ has been given in Section 2.9 (eqn (2.55)):

$$\Delta\mathbf{D} + \text{curl curl}\,\chi(\mathbf{r})\mathbf{D} + 4\pi^2k^2\mathbf{D} = 0.$$

For electrons and neutrons, the propagation equation is Schrödinger's equation (4.2) which can be rewritten as

$$\Delta\Psi + 4\pi^2k^2[1 + \chi(\mathbf{r})]\Psi = 0, \tag{4.4}$$

where $k = 1/\lambda$ is the wave number in vacuum and $\chi(\mathbf{r})$ is $\varphi(\mathbf{r})/W$ ($\varphi(\mathbf{r})$, potential in the crystal; W, the accelerating voltage) in the case of electron diffraction and $-2mV(\mathbf{r})/h^2k^2$ ($V(\mathbf{r})$, Fermi pseudo-potential) in the case of neutron diffraction. Reviews of the dynamical theory of electron diffraction are given for instance in Kambe and Moliere (1970), Cowley (1992) and in Moodie *et al.* (1997) and for neutron diffraction in Rauch and Petrascheck (1978), Klein and Werner (1983), Sears (1979, 1989) and in Schlenker and Guigay (1996, 2001).

Equations (2.55) and (4.4) are linear, homogeneous, partial differential equations. Any linear combination

$$\mathbf{D} = \sum_j A_j \mathbf{D_j} \tag{4.5a}$$

or

$$\Psi = \sum_j A_j \Psi_j \tag{4.5b}$$

of particular solutions is also a solution. The choice of the solutions and the values of the coefficients A_j are determined by the boundary conditions of the problem.

The interaction term, χ (polarizability for X-rays, potential energy for electrons, Fermi pseudo-potential for neutrons), is a triply periodic function of space coordinates and a particular solution is, in each case, a wavefield expressed by (2.47) or (4.3). The only difference, the curl curl term, comes from the fact that electromagnetic waves are vector waves while the wave function Ψ associated with an electron or a neutron beam is scalar. The derivations are very similar but are more complicated for X-rays because polarization must be taken into account. In this chapter, where the most important results of dynamical theory are presented in a simplified way, the wave equation (4.4) for scalar waves will be used. Its results hold separately for each direction of polarization of an incident X-ray wave in the two-beam case. Polarization is taken into account by multiplying the Fourier coefficients χ_h and $\chi_{\bar{h}}$ of the polarizability by C (equal to 1 for σ-polarization and $\cos 2\theta_B$ for π-polarization).

Expression (4.3) can be rewritten as

$$\Psi(\mathbf{r}) = \sum_{\mathbf{h}} \Psi_h \exp(-2\pi \mathrm{i} \mathbf{K_h} \cdot \mathbf{r}) \tag{4.6}$$

with

$$\mathbf{K_h} = \mathbf{K_o} - \mathbf{h}. \tag{4.7}$$

Equation (4.6) shows that the wavefield is a sum of plane waves whose wavevectors can be deduced from one another by translations in reciprocal space. In the two-beam case to which we shall limit ourselves in this chapter, there are two waves in the wavefield, Ψ_o and Ψ_h. We shall orient the wavevectors *from* the reciprocal lattice points O and H *towards* the tiepoint P: $\mathbf{K_o} = \mathbf{OP}$, $\mathbf{K_h} = \mathbf{HP}$, $\mathbf{h} = \mathbf{OH}$ (Fig. 4.3).

4.5 Fundamental equations of dynamical theory

The solutions of the propagation equation are obtained by substituting the expressions (2.36) of $\chi(\mathbf{r})$ and (4.6) of $\Psi(\mathbf{r})$ into (4.4). It is easy to show that

$$\Delta\Psi(\mathbf{r}) = -4\pi^2 \sum_{\mathbf{h}} K_h^2 \Psi_h \exp[-2\pi \mathrm{i}(\mathbf{K_h} \cdot \mathbf{r})]. \tag{4.8}$$

The product $\chi(\mathbf{r})\Psi(\mathbf{r})$ can be written as

$$\chi(\mathbf{r})\Psi(\mathbf{r}) = \Big[\sum_{\mathbf{g}}\sum_{\mathbf{h}'} \chi_g \Psi_{h'} \exp[2\pi \mathrm{i}(\mathbf{g}+\mathbf{h}')\cdot\mathbf{r}]\Big] \exp[-2\pi \mathrm{i}(\mathbf{K_o}\cdot\mathbf{r})].$$

The vectors $\mathbf{g}$ and $\mathbf{h}'$ are reciprocal lattice vectors and their sum $\mathbf{g}+\mathbf{h}'=\mathbf{h}$ is also a reciprocal lattice vector. It follows, using (4.7) and substituting the summation over $\mathbf{h}$ to the summation over $\mathbf{g}$, that

$$\chi(\mathbf{r})\Psi(\mathbf{r}) = \sum_{\mathbf{h}} \exp[-2\pi \mathrm{i}(\mathbf{K_h}\cdot\mathbf{r})] \sum_{\mathbf{h}'} \chi_{h-h'}\Psi_{h'}. \tag{4.9}$$

Substituting from (4.8) and (4.9) into the equation of propagation (4.4), we finally obtain

$$\sum_{\mathbf{h}} \Big[-4\pi^2 K_h^2 \Psi_h + 4\pi^2 k^2 \Psi_h + 4\pi^2 k^2 \sum_{\mathbf{h}'} \chi_{h-h'}\Psi_{h'}\Big] \times \exp[-2\pi i(\mathbf{K_h}\cdot\mathbf{r})] = 0.$$

This equation with an infinite number of terms can be replaced by an infinite set of equations. Let us multiply both sides by $\exp[2\pi \mathrm{i}(\mathbf{K_h}\cdot\mathbf{r})]$ and integrate over the space coordinates from $-\infty$ to $+\infty$. All the terms in the sum but one vanish, and one obtains the set of equations

$$\Psi_h = \frac{k^2}{K_h^2 - k^2} \sum_{\mathbf{h}'} \chi_{h-h'}\Psi_{h'}. \tag{4.10}$$

These equations have been called *fundamental equations* (*Grundgleichungen*) of dynamical theory by Laue. The corresponding equations for vector waves are (2.52) in Ewald's form of the dynamical theory and (5.4) in Laue's.

One recognizes in (4.10) the resonance factors $1/(K_h^2 - k^2)$ introduced by Ewald (eqn (2.52)). Only those waves in the wavefield for which $K_h \approx k$ have non-negligible amplitudes, namely those for which the Bragg condition is satisfied.

1. *Solution for one wave*: If one resonance term only is very large, (4.10) reduces to

$$\Psi_h = \frac{k^2}{K_h^2 - k^2} \chi_o \Psi_h.$$

For this equation to have a non-trivial solution, we must have

$$K_h^2 = k^2(1+\chi_o)$$

and, since χ_o is very small,

$$K_h = k\left(1 + \frac{\chi_o}{2}\right).$$

The tiepoint P lies on a sphere centred at any reciprocal lattice point (O, H, etc.) and of radius $k(1 + \chi_o/2) = nk$ (Fig. 4.3).

2. *Solution for two beams*: If two resonance terms are very large and two nodes lie simultaneously on the Ewald sphere, (4.10) becomes

$$\Psi_o = \frac{k^2}{K_o^2 - k^2}[\chi_o \Psi_o + \chi_{\bar{h}} \Psi_h],$$

$$\Psi_h = \frac{k^2}{K_h^2 - k^2}[\chi_h \Psi_o + \chi_o \Psi_h],$$

where $\chi_{\bar{h}}$ is the Fourier coefficient of the polarizability corresponding to the $\bar{h}$, $\bar{k}$, $\bar{l}$ reflection. In an infinite crystal, the choice of the wave which is labelled 'o' is arbitrary. In a semi-infinite crystal the logical choice is of that one which has the same tangential component as the incident wave. It is the *refracted wave* and the other one is the *reflected wave*. There is no discontinuity in the properties of the 'o' wave when the crystal is rocked through the reflection domain and the tiepoint, P, moves on the dispersion surface from the region where only one wave is excited in the crystal (the sphere of radius nk centred in O) to the connecting surface where two waves are excited (see Fig. 4.3). It is normally called refracted wave far from the reflection domain and there is no reason not to call it the same way throughout. This is why 'refracted wave' will be preferred to the expression 'forward diffracted wave' which is sometimes used in the literature.

For non-trivial solutions of these linear, homogeneous, equations to exist, the determinant should be equal to zero. The secular equation thus obtained gives a relation between the lengths K_o and K_h of the two wavevectors of the field and hence gives the loci of the tiepoints P and the equation of the dispersion surface. Let us rearrange these equations in the following way:

$$\left[\frac{K_o^2 - k^2}{2k} - \frac{k\chi_o}{2}\right]\Psi_o - \frac{k\chi_{\bar{h}}}{2}\Psi_h = 0,$$
$$-\frac{k\chi_h}{2}\Psi_o + \left[\frac{K_h^2 - k^2}{2k} - \frac{k\chi_o}{2}\right]\Psi_h = 0, \tag{4.11}$$

and let us set

$$X_o = \frac{K_o^2 - k^2}{2k} - \frac{k\chi_o}{2},$$
$$X_h = \frac{K_h^2 - k^2}{2k} - \frac{k\chi_o}{2}. \tag{4.12}$$

An approximation is made here in the classical treatment, which is satisfactory when the tiepoint P is not far from the Lorentz point L_o, namely that the spheres passing through that point may then be approximated by their tangential planes. This approximation is not valid, for instance, when the Bragg angle is close to $\pi/2$ (see Section 5.9) or for very asymmetric reflections (see Chapter 8). Since $K_h \approx k$ and $K_o \approx k$, $K_h^2 - k^2/2k \approx (K_h - k)$, we can write

$$\begin{aligned} X_o &\approx K_o - k\left(1 + \frac{\chi_o}{2}\right), \\ X_h &\approx K_h - k\left(1 + \frac{\chi_o}{2}\right). \end{aligned} \tag{4.13}$$

One may notice that X_o and X_h are the differences between OP and HP, respectively, on the one hand, and the radius of the spheres passing through the Lorentz point on the other hand (Fig. 4.3). They represent the distances of the tiepoint from these spheres and expressions (4.12) are called *resonance errors*. If the crystal is absorbing, the form factors are complex and so are χ_h, X_o and X_h. The dispersion surface is then also a complex surface. The relative distances $(k - K_o)/k$ and $(k - K_h)/k$ from the spheres passing through the Laue point have been called *excitation errors* ('*Anregungsfehlern*') by Laue (1960). Substituting from (4.13) into (4.11), it follows finally that

$$\begin{aligned} X_o\Psi_o - \frac{k\chi_{\bar{h}}}{2}\Psi_h &= 0, \\ -\frac{k\chi_h}{2}\Psi_o + X_h\Psi_h &= 0. \end{aligned} \tag{4.14}$$

The determinant of these equations is $X_oX_h - k^2\chi_h\chi_{\bar{h}}/4$ and the secular equation, which is the equation of the dispersion surface, is

$$X_oX_h = k^2\chi_h\chi_{\bar{h}}/4. \tag{4.15}$$

It shows that the product of the distances of P from the two spheres passing through the Lorentz point is constant and that the dispersion surface is a surface of revolution around axis **OH**. Close to the Lorentz point, these two spheres can be replaced by their tangential planes and their intersections by the diffraction plane can be replaced by two straight lines, unless the reflection is highly asymmetric (see Chapter 8). Equation (4.15) is then the equation of a hyperbola whose asymptotes are these two lines, T_o and T_h (Fig. 4.6).

The angle between the asymptotes is $2\theta_B$. The branch of the hyperbola which lies on the same side of the asymptotes as the Laue point is called branch 1; the other one, branch 2, is on the same side as O and H. With the sign conventions used in this book, X_o and X_h are positive when the tiepoint lies on branch 1 (K_{o1} and $K_{h1} > nk$) and negative when it lies on branch 2 (K_{o2} and $K_{h2} < nk$). If A_{o1} and A_{o2} are the

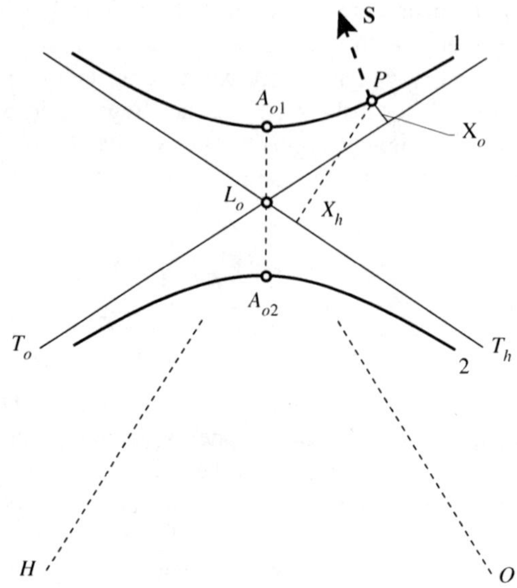

Fig. 4.6. Dispersion surface. P: tiepoint of a wavefield; **S**: Poynting vector of that wavefield; X_o and X_h: coordinates of the tiepoint; A_{o1}, A_{o2}: vertices of the dispersion surface.

vertices of the hyperbola on these two branches, respectively, it follows from (4.15) that the diameter of the dispersion surface is

$$G_B = \overline{A_{o2}A_{o1}} = \frac{k|C|\sqrt{\chi_h \chi_{\bar{h}}}}{\cos\theta_B} = \frac{R\lambda|C|\sqrt{F_h F_{\bar{h}}}}{\pi V \cos\theta_B}, \tag{4.16}$$

where the polarization factor $|C|$ has been included to stress that when polarization is taken into account the dispersion surface has four branches, two for each direction of polarization (see Chapter 5). It is then constituted by two hyperbolae whose diameters are proportional to $|C|$; the interaction of electromagnetic waves with matter is therefore weaker for π-polarization.

The gap $G_B = \overline{A_{o2}A_{o1}}$ between the two branches of the dispersion surface is equivalent to the gap between two successive bands and it is sometimes called for that reason *Bragg gap*; eqn (4.16) shows that it is larger, the larger the structure factor, that is the stronger the interaction of the waves with matter, and the longer the wavelength.

It is important to note that the Fourier coefficient $\chi_{\bar{h}}$ corresponding to the $\bar{h}$, $\bar{k}$, $\bar{l}$ reflection plays the same role as χ_h. This is not surprising and it is due to the interaction of the waves reflected on both sides of each reflecting plane as has been seen in Darwin's dynamical theory. This is the reason for Friedel's law which is only valid in the absence of absorption for non-centrosymmetric crystals. It also fails when simultaneous reflections occur (Miyake and Uyeda 1955).

4.6 Amplitude ratio of the refracted and reflected waves

Equations (4.14) can be written in the following way in order to give the expression of the ratio ξ of the amplitudes of the two waves in the wavefield:

$$\xi = \frac{\Psi_h}{\Psi_o} = \frac{2X_o}{k\chi_{\bar{h}}} = \frac{k\chi_h}{2X_h},$$
$$\xi^2 = \frac{\chi_h}{\chi_{\bar{h}}}\frac{X_o}{X_h}. \tag{4.17}$$

These relations show that the ratio ξ is determined by the position of the tiepoint and some conclusions can immediately be drawn from the simple observation of the dispersion surface.

1. *Phase of the amplitudes*: The ratio ξ is negative for branch 1 and positive for branch 2 (χ_h is negative) and the phase ψ of ξ is, in the non-absorbing case,

$$\psi = \pi + \varphi_h \quad \text{for branch 1},$$
$$\psi = \varphi_h \quad \text{for branch 2},$$

where φ_h is the phase of the structure factor (eqn (2.38)).

This difference of π between the two branches has many important consequences such as the difference in the positions of the nodes of standing waves and, consequently, the difference in anomalous absorption.

2. *Nodes of standing waves*: The development (4.3) of the wave function Ψ can be written as

$$\Psi(\mathbf{r}) = \Psi_o \exp(-2\pi \mathrm{i}\mathbf{K_o}\cdot\mathbf{r})[1 + \xi\exp(2\pi\mathrm{i}\mathbf{h}\cdot\mathbf{r})] \tag{4.18}$$

and the intensity of the wavefield is

$$|\Psi(\mathbf{r})|^2 = |\Psi_o|^2[1 + |\xi|^2 + 2|\xi|\cos\{2\pi(\mathbf{h}\cdot\mathbf{r}) + \psi\}]. \tag{4.19}$$

The nodes of the standing wave system lie on the planes of equation:

$$\mathbf{h}\cdot\mathbf{r} = \mathcal{N} - \frac{\varphi_h}{2\pi} \quad \text{(branch 1);}$$
$$\mathbf{h}\cdot\mathbf{r} = \mathcal{N} + \frac{1}{2} - \frac{\varphi_h}{2\pi} \quad \text{(branch 2),}$$

where $\mathcal{N}$ is an integer. They lie on the planes of maximum electronic density for branch 1 and halfway in between for branch 2 (Figs 1.7 and 5.13).

3. *Direction of propagation*: The direction of propagation of the wavefield is given by the group velocity (Fig. 1.9) or by the Poynting vector. These vectors are normal

to the dispersion surface at the tiepoint characteristic of the wavefield. This is shown in Section 12.5 for the group velocity and in Section 5.4 for the Poynting vector. The expression of the Poynting vector, **S**, is

$$\mathbf{S} \propto \mathbf{s_o}|\Psi_o|^2 + \mathbf{s_h}|\Psi_h|^2 = \Psi_o^2[\mathbf{s_o} + |\xi|^2\mathbf{s_h}] \tag{4.20}$$

and its angle, α, with the direction of the reflecting planes (see Fig. 1.8) is given by

$$\tan\alpha = \frac{1-|\xi|^2}{1+|\xi|^2}\tan\theta_B. \tag{4.21}$$

Figure 4.6 shows the Poynting vector associated with tiepoint P. Its orientation varies from that of the refracted wave, **OP**, to that of the reflected wave, **HP**. On one side of the reflection domain (right-hand side of Fig. 4.6), its orientation is that of **OP** for branch 1, with $\xi = 0$, while it is that of **HP** for branch 2, with $\xi \to \infty$. On the other side of the reflection domain, it is the other way around. In the middle of the reflection domain, when the tiepoint is in A_{o1} or A_{o2}, the propagation direction is parallel to the lattice planes.

4.7 Solutions of plane-wave dynamical theory

4.7.1 *Boundary conditions*

The previous sections have shown which are the possible solutions of the propagation equation for an infinite medium: they are represented geometrically by the dispersion surface. Each one is characterized by its tiepoint whose position determines the direction of propagation of the wavefield and the ratio of the amplitudes of the waves it is constituted of. The choice of the 'o' wave in the expansion of the wavefield is arbitrary. As is shown by Ewald's reciprocity theorem, the presence of any of the waves in the wavefield ensures the existence of the others. In order to find out which are the solutions which exist in a practical case where a beam is incident on a crystal, one has to introduce the boundary conditions, at first for a semi-infinite crystal to determine the wavefields which are created inside the crystal and afterwards by introducing the exit surface to determine the waves which have been refracted or reflected out of the crystal. There are two types of boundary conditions: the conditions for the wavevectors and the conditions for the amplitudes. The former will be considered in the present section and the latter in Sections 4.8 and 4.9.

The condition for the wavevectors comes from the requirement that the phase velocities of the waves should be the same along the crystal–vacuum interface, in agreement with the *Snell–Descartes* law. It demands that the tangential components of the wavevectors should be continuous across the boundary. This condition is satisfied when the difference between the wavevectors on each side of the interface is parallel to the normal to that interface. This is shown geometrically in Fig. 4.7, and

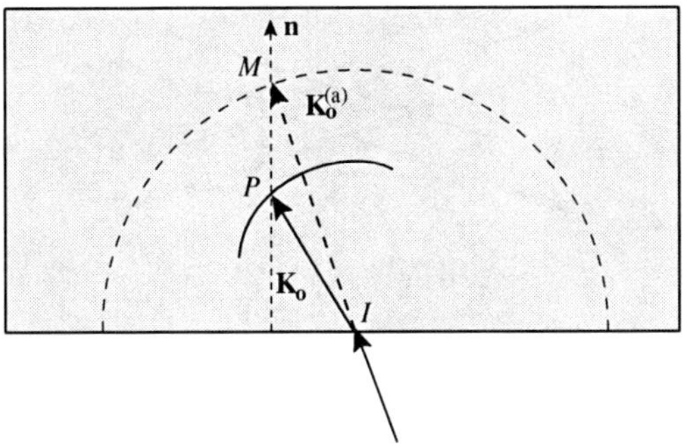

Fig. 4.7. Refraction of a wave at the surface separating two media with different indices of refraction. Broken curve: surface of indices in vacuum; solid curve: surface of indices of the medium; $\mathbf{IM} = \mathbf{K}_o^{(a)}$: incident wavevector; $\mathbf{IP} = \mathbf{K}_o$: refracted wavevector in the medium; M and P lie on a normal, $\mathbf{n}$, to the entrance surface.

formally in (4.22):

$$\mathbf{K_o} - \mathbf{K}_o^{(a)} = \mathbf{IP} - \mathbf{IM} = \overline{MP}\,\mathbf{n}, \tag{4.22}$$

where $\mathbf{n}$ is a unit vector normal to the crystal surface and $\mathbf{K}_o^{(a)}$ is the incident wavevector; $\overline{MP}$ is positive if its orientation is the same as that of $\mathbf{n}$; it is negative here since the index of refraction for X-rays is smaller than 1. A relation similar to (4.22) holds for the reflected waves.

4.7.2 *Departure from Bragg's angle of the incident wave*

The wave number of waves propagating in vacuum in the incident or reflected directions is $k = 1/\lambda$ and the common extremity, M, of their wavevectors $\mathbf{OM} = \mathbf{K}_o^{(a)}$ and $\mathbf{HM} = \mathbf{K}_h^{(a)}$ lies on the spheres of radius k and centred at O and H, respectively, passing through the Laue point, L_a. The intersections of these spheres with the plane of incidence are two circles which can be approximated by their tangents T_o' and T_h' (Fig. 4.8).

Bragg's condition is exactly satisfied according to the geometrical theory of diffraction when M lies at L_a. The departure $\Delta\theta$ from Bragg's incidence of an incident wave is measured by the angle between the corresponding wavevectors $\mathbf{OM}$ and $\mathbf{OL_a}$. As $\Delta\theta$ is very small as compared to the Bragg angle in the general case of X-rays or neutrons, one may write

$$\begin{aligned} \mathbf{K}_o^{(a)} &= \mathbf{OM} = \mathbf{OL_a} + \mathbf{L_aM}, \\ \Delta\theta &= \frac{\overline{L_a M}}{k}. \end{aligned} \tag{4.23}$$

The tangent T_o' is oriented in such a way that $\Delta\theta$ is negative when the angle of incidence is smaller than Bragg's angle.

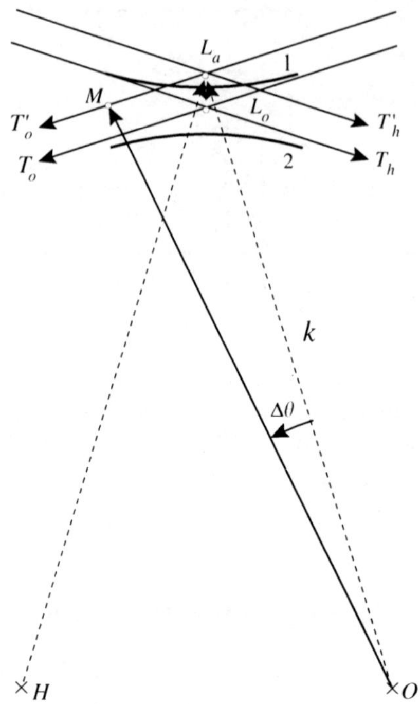

Fig. 4.8. Departure from Bragg's angle of an incident wave: $\Delta\theta = \overline{L_a M}/k$; L_a: Laue point; T'_o, T'_h: tangents to the spheres of centres O and H and of radius k.

4.7.3 *Transmission and reflection geometries*

The boundary condition for the continuity of the tangential component of the wavevectors is applied by drawing from M a line parallel to the normal, **n**, to the crystal surface. The tiepoints of the wavefields excited in the crystal by the incident wave are at the intersections of this line with the dispersion surface. Two different situations may occur:

1. *Transmission geometry, or Laue geometry (Fig. 4.9)*: The reflected wave is directed towards the *inside* of the crystal. The normal to the crystal surface drawn from the excitation point M intersects *both* branches of the dispersion surface and lies in the angle between the asymptotes T_o and T_h which contains the dispersion surface.

Let ψ_o, ψ_h and ψ_n be the angles between the normal to the crystal surface, **n**, oriented towards the inside of the crystal and the incident direction, the reflected direction and the trace **s** of the lattice planes, respectively (Figs 4.9 and 4.11):

$$\psi_o = (\mathbf{n}, \mathbf{s_o}),$$

$$\psi_h = (\mathbf{n}, \mathbf{s_h}),$$

$$\psi_n = (\mathbf{n}, \mathbf{s}).$$

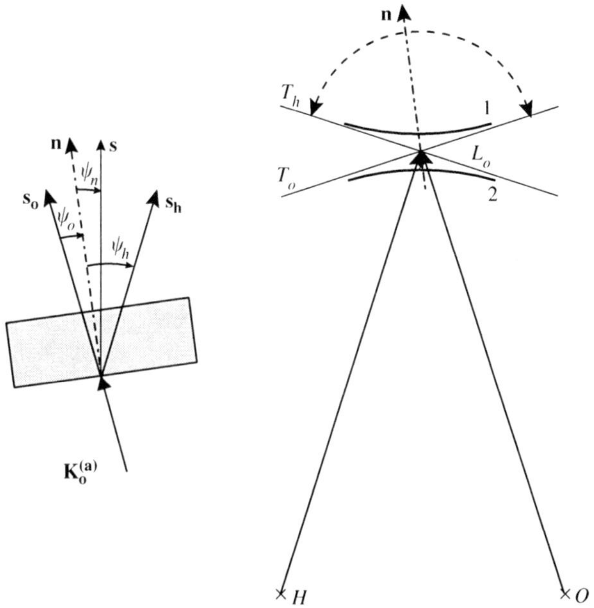

Fig. 4.9. Transmission, or *Laue*, geometry: $-\pi/2 < \psi_h < \pi/2 - 2\theta_B$; $-\pi/2 + 2\theta_B < \psi_o < \pi/2$. The normal **n** to the crystal surface intersects both branches of the dispersion surface at the same time. Inset: direct space.

The directions **s**, $\mathbf{s_o}$ and $\mathbf{s_h}$ are oriented in such a way that

$$(\mathbf{s}, \mathbf{s_o}) = +\theta_B; \qquad (\mathbf{s}, \mathbf{s_h}) = -\theta_B$$

for an h, k, l reflection (for an $\bar{h}, \bar{k}, \bar{l}$ reflection, one should take the opposite sign for θ_B). With these conventions, we have

$$\psi_o = \psi_n + \theta_B,$$
$$\psi_h = \psi_n - \theta_B.$$

Figure 4.9 shows that, in the transmission geometry,

$$-\frac{\pi}{2} + 2\theta_B < \psi_o < \frac{\pi}{2},$$
$$-\frac{\pi}{2} < \psi_h < \frac{\pi}{2} - 2\theta_B,$$
$$-\left(\frac{\pi}{2} - \theta_B\right) < \psi_n < \frac{\pi}{2} - \theta_B.$$

2. *Reflection geometry, or Bragg geometry (Fig. 4.10)*: The reflected wave is directed towards the *outside* of the crystal. The normal to the crystal surface drawn from the excitation point lies in the angle between the asymptotes which does not

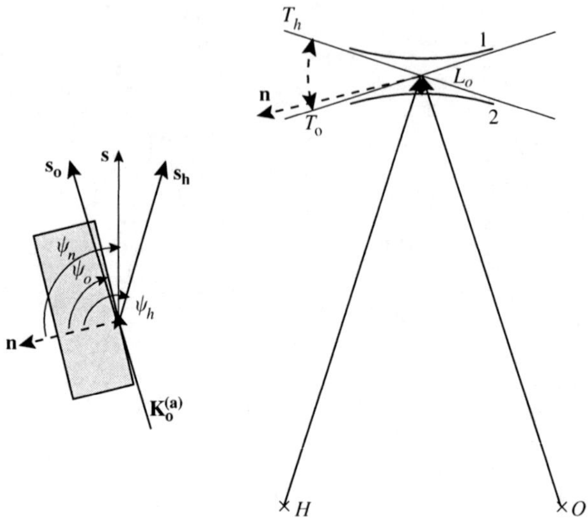

Fig. 4.10. Reflection, or *Bragg*, geometry: $-\pi/2 - 2\theta_B < \psi_h < -\pi/2; -\pi/2 < \psi_o < -\pi/2 + 2\theta_B$. The normal **n** to the crystal surface intersects one branch of the dispersion surface at a time. When the intersection is imaginary, there is total reflection. Inset: direct space.

contain the dispersion surface. It intersects one branch of the dispersion surface at a time and there are three possible situations: the normal to the crystal surface drawn from M intersects either branch 1 or branch 2 or the intersection points are imaginary.

The angular relationships are here, from Fig. 4.10,

$$-\frac{\pi}{2} < \psi_o < -\frac{\pi}{2} + 2\theta_B,$$

$$-\frac{\pi}{2} - 2\theta_B < \psi_h < -\frac{\pi}{2},$$

$$-\frac{\pi}{2} - \theta_B < \psi_n < -\frac{\pi}{2} + \theta_B.$$

3. *Asymmetry ratio*: Let $\gamma_o = \cos\psi_o$ and $\gamma_h = \cos\psi_h$ be the cosines of the angles between the normal to the crystal surface and the incident and reflected directions, respectively. Their ratio, γ, called *asymmetry ratio* is

$$\gamma = \frac{\gamma_h}{\gamma_o} = \frac{\cos(\psi_n - \theta_B)}{\cos(\psi_n + \theta_B)}. \tag{4.24}$$

It will be noted that γ_o, γ_h and γ are all positive in the Laue geometry and that γ_h and the asymmetry ratio γ are negative in the Bragg geometry. Figure 4.23 represents the variations of $\sqrt{|\gamma|}$ with ψ_n in the transmission and in the reflection geometries, respectively.

4.7.4 *Deviation parameter*

The solutions of dynamical theory are best described by introducing a reduced parameter called the *deviation parameter*, η, sometimes called y or w in the literature:

$$\eta = \frac{\Delta\theta \sin 2\theta_B + \chi_o (1-\gamma)/2}{|C| \sqrt{|\gamma|}\sqrt{\chi_h \chi_{\bar{h}}}} = \frac{(\Delta\theta - \Delta\theta_{os})}{\delta_{os}} \tag{4.25}$$

where

$$\begin{aligned} \Delta\theta_{os} &= \frac{\overline{L_a I}}{k} = -\frac{\chi_o(1-\gamma)}{2\sin 2\theta_B}, \\ \delta_{os} &= \frac{C\sqrt{|\gamma|\chi_h\chi_{\bar{h}}}}{\sin 2\theta_B} = \frac{\lambda}{\Lambda_o}\frac{|\gamma_h|}{\sin 2\theta_B}, \\ \Lambda_o &= \frac{\lambda\sqrt{\gamma_o|\gamma_h|}}{C\sqrt{\chi_h\chi_{\bar{h}}}} = \frac{\pi V\sqrt{\gamma_o|\gamma_h|}}{R\lambda C\sqrt{F_h F_{\bar{h}}}}, \end{aligned} \tag{4.26}$$

where the factor $C = 1$ (σ-polarization) or $C = \cos 2\theta_B$ (π-polarization) must be included in the case of X-rays to take polarization into account (see Chapter 5). Point I is at the intersection of the normal to the crystal surface drawn from the Lorentz point, L_o, with T_o' (Fig. 4.11), Λ_o is the *Pendellösung* distance in the transmission geometry and the extinction distance in the reflection geometry. In an absorbing crystal, η, $\Delta\theta_{os}$, δ_{os} and Λ_o are complex.

1. *Darwin width*: The real part of $2\delta_{os}$, $2\delta_{sr}$, is the full width at half-maximum (FWHM) of the rocking curve in the transmission geometry and the width of the total reflection domain in the reflection geometry.
2. *Deviation from Bragg's incidence of the middle of the reflection domain*: The real part of $\Delta\theta_{os}$ is the angle between the wavevector exactly satisfying Bragg's law, $\mathbf{OL_a}$, and the wavevector $\mathbf{OI}$ corresponding to the middle of the reflection domain in dynamical theory. In the *Bragg case*, the asymmetry ratio, γ, is negative and $\Delta\theta_{os}$ is never equal to zero. This deviation from Bragg's angle is due to the refraction effect which is neglected in the geometrical theory. In the *Laue geometry*, $\Delta\theta_{os}$ is equal to zero for symmetric reflections ($\gamma = 1$).

4.7.5 *Determination of the tiepoints*

The coordinates of the tiepoints excited by the incident wave are obtained by looking for the intersection of the dispersion surface, (4.15), with the normal $\mathbf{n}$ to the crystal surface drawn from the excitation point (Fig. 4.11). One finds that

$$X_{oj} = k\frac{S(\gamma_h)}{2\sqrt{|\gamma|}}[\sqrt{\chi_h\chi_{\bar{h}}}]\left[\eta \pm \sqrt{\eta^2 + S(\gamma_h)}\right], \tag{4.27}$$

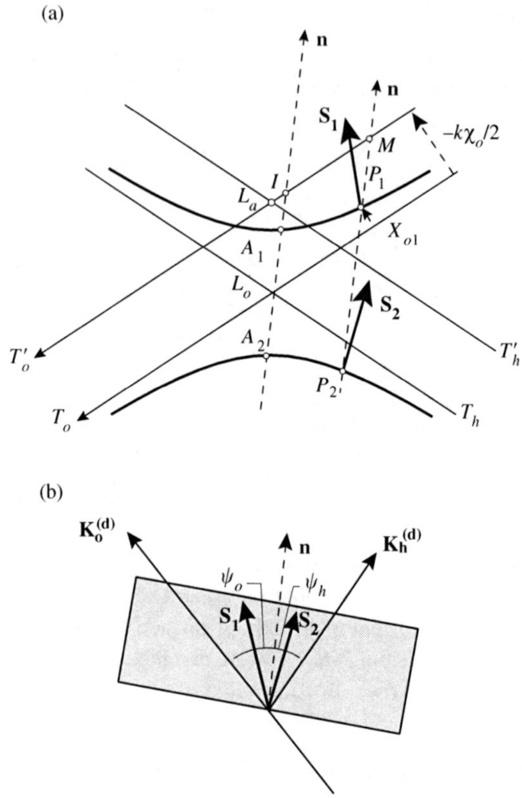

Fig. 4.11. Wavefields created inside the crystal by the incident plane wave. M: excitation point; $\mathbf{n}$: normal to the crystal surface; P_1 and P_2: tiepoints; $\mathbf{S_1}$ and $\mathbf{S_2}$: Poynting vectors; I: middle of the reflection domain. (a) Reciprocal space. (b) Direct space.

where the *plus* sign corresponds to a tiepoint on branch 1 ($j = 1$) and the *minus* sign to a tiepoint on branch 2 ($j = 2$) and $S(\gamma_h)$ means sign of γ_h (+1 in transmission geometry, −1 in reflection geometry). In the reflection geometry, X_o is complex for $|\eta| < 1$ and the intersection of the normal to the entrance surface with the dispersion surface is imaginary.

The ratio ξ_j of the amplitudes of the waves of the corresponding wavefields is related to their coordinates by (4.17) and is

$$\begin{aligned}\xi_j &= \frac{\Psi_{hj}}{\Psi_{oj}} \\ &= \frac{S(\gamma_h)}{\sqrt{|\gamma|}} \frac{\sqrt{\chi_h \chi_{\bar{h}}}}{\chi_{\bar{h}}} \Big[\eta \pm \sqrt{\eta^2 + S(\gamma_h)} \Big]. \end{aligned} \tag{4.28}$$

4.7.6 *Effective absorption coefficient*

In absorbing crystals, absorption is taken into account by introducing an imaginary part to the form factor and to the polarizability (Section 2.7). The wavevectors therefore also have an imaginary part:

$$\mathbf{K_h} = \mathbf{K_h^r} + \mathrm{i}\mathbf{K_h^i}.$$

There is no absorption in vacuum and the incident wavevector is real. Relation (4.7) between the wavevectors of a given wavefield shows that the imaginary part is the same for all of them:

$$\mathbf{K_o^i} = \mathbf{K_h^i}. \tag{4.29}$$

This relation shows that the behaviour towards absorption of all the waves in a given wavefield is the same. Expression (4.19) of the intensity of a wavefield becomes, in the presence of absorption,

$$|\Psi(\mathbf{r})|^2 = \exp(4\pi\mathbf{K_o^i}\cdot\mathbf{r})|\Psi_o|^2\{1 + |\xi|^2 + 2|\xi|\cos 2\pi[(\mathbf{h}\cdot\mathbf{r}) + \psi]\},$$

where ψ is the phase factor of ξ.

Equation (4.22) shows that if the medium is absorbing, the imaginary part of wavevector $\mathbf{K_o}$ is normal to the interface :

$$\mathbf{K_o^i} = \mathcal{I}m(\overline{MP})\mathbf{n}, \tag{4.30}$$

where $\mathcal{I}m(\,)$ means imaginary part of.

The attenuation factor can therefore be written as

$$\exp(4\pi\mathbf{K_o^i}\cdot\mathbf{r}) = \exp\left(-\frac{\mu z}{\gamma_o}\right),$$

where z is the depth of the observation point along the normal $\mathbf{n}$ to the crystal surface, and the *effective* absorption coefficient μ_e is

$$\mu_e = -4\pi\gamma_o K_o^i,$$

where γ_o is defined in Section 4.7.3.

Figure 4.11 shows that

$$\gamma_o\overline{MP} = \tfrac{1}{2}k\chi_o + X_o.$$

It follows that

$$\mu_e = -4\pi\gamma_o K_o^i = \mu_o - 4\pi X_o^i. \tag{4.31}$$

When there is more than one wave in the wavefield, the effective absorption coefficient, μ_e, can differ significantly from the normal value, $\mu_o = -2\pi k\chi_{io}$, defined in Section 2.7.4 (see Sections 4.8.4 and 5.8).

4.8 The diffracted waves in the transmission geometry

4.8.1 *Double refraction*

Let us consider an infinite plane wave incident on a crystal surface of infinite lateral extension in the transmission geometry. The normal to the crystal surface intersects both branches of the dispersion surface, at P_1 and P_2, and two wavefields are excited in the crystal (Fig. 4.11) which have different indices of refraction. This is the *double refraction* effect described in Section 1.10.2 and whose experimental observation is reported in Sections 12.9 and 12.10. As shown in the figure, the Poynting vectors of the two wavefields are both directed towards the inside of the crystal.

4.8.2 *Boundary conditions for the amplitudes at the entrance surface*

The amplitudes of the refracted and reflected waves are Ψ_{o1}, Ψ_{h1}, respectively, for wavefield 1 and Ψ_{o2}, Ψ_{h2}, for wavefield 2. The boundary conditions for the amplitudes at the surface are

$$\Psi_o^{(a)} = \Psi_{o1} + \Psi_{o2},$$
$$0 = \Psi_{h1} + \Psi_{h2},$$

where $\Psi_o^{(a)}$ is the amplitude of the incident wave.

The amplitudes Ψ_{hj} and Ψ_{oj} at the entrance surface can be expressed by means of their ratio ξ_j (4.17):

$$\Psi_{h1} = \frac{\xi_1\xi_2}{\xi_2 - \xi_1}\Psi_o^{(a)}; \qquad \Psi_{h2} = \frac{-\xi_1\xi_2}{\xi_2 - \xi_1}\Psi_o^{(a)};$$
$$\Psi_{o1} = \frac{\xi_2}{\xi_2 - \xi_1}\Psi_o^{(a)}; \qquad \Psi_{o2} = \frac{-\xi_1}{\xi_2 - \xi_1}\Psi_o^{(a)}.$$

By substituting from (4.28), it follows that

$$\begin{aligned} \Psi_{hj} &= \pm\frac{\sqrt{\chi_h\chi_{\bar{h}}}}{2\sqrt{\gamma}\chi_{\bar{h}}}\frac{\Psi_o^{(a)}}{\sqrt{1+\eta^2}}, \\ \Psi_{oj} &= \mp\frac{\eta \mp \sqrt{1+\eta^2}}{2\sqrt{1+\eta^2}}\Psi_o^{(a)}. \end{aligned} \tag{4.32}$$

The amplitudes of the waves are multiplied by the corresponding phase factors, $\exp(-2\pi i\mathbf{K_{hj}}\cdot\mathbf{r})$ and $\exp(-2\pi i\mathbf{K_{oj}}\cdot\mathbf{r})$, respectively, where the origin of $\mathbf{r}$ is taken on the entrance surface.

4.8.3 *Intensities of the reflected and refracted waves*

Using (4.28), (4.31) and (4.32), it can be shown that the *intensities* of the four waves are

$$
\begin{aligned}
|\Psi_{oj}|^2 &= |\Psi_o^{(a)}|^2 \exp\left(\frac{-\mu_{ej} z}{\gamma_o}\right) \frac{\left[\sqrt{1+\eta_r^2} \mp \eta_r\right]^2}{4(1+\eta_r^2)}, \\
|\Psi_{hj}|^2 &= |\Psi_o^{(a)}|^2 \exp\left(\frac{-\mu_{ej} z}{\gamma_o}\right) \left|\frac{F_h}{F_{\bar{h}}}\right| \frac{\gamma^{-1}}{4(1+\eta_r^2)}
\end{aligned}
\tag{4.33}
$$

with

$$
\mu_{ej} = \mu_o \left[\frac{1}{2}(1+\gamma^{-1}) \mp \frac{\frac{1}{2}\eta_r(1-\gamma^{-1}) + \sqrt{\gamma^{-1}}|C\chi_{ih}/\chi_{io}|\cos\varphi}{\sqrt{1+\eta_r^2}}\right],
\tag{4.34}
$$

where $\varphi = \varphi_{rh} - \varphi_{ih}$ is the phase difference between χ_{rh} and χ_{ih} (2.39), η_r is the real part of the deviation parameter η, the upper sign corresponds to branch 1 and the lower sign to branch 2 of the dispersion surface; expression (4.34) is derived in Section 5.8.2. For X-rays, χ_h must be multiplied by the polarization factor ($C = 1$ or $\cos 2\theta_B$).

Figure 4.12 represents the variations of the Poynting vector of the two wavefields across the reflection domain. It can be noted that its orientation varies slowly when the tiepoint is far from the centre of the reflection domain and rapidly when it is

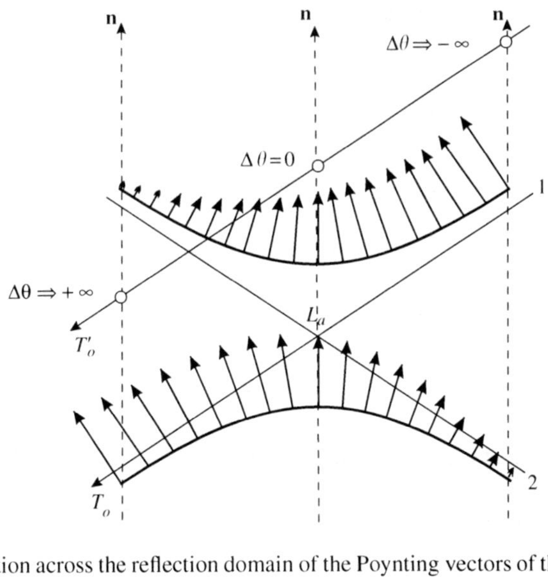

Fig. 4.12. Variation across the reflection domain of the Poynting vectors of the two wavefields excited by an incident plane wave. Symmetric transmission geometry, no absorption.

close to the centre. This is due to the curvature of the dispersion surface and explains the distribution of wavefields in the Borrmann triangle when the incident wave is a spherical wave (see Fig. 1.8, Section 1.10.2 and Chapter 12). The length of the Poynting vector is given by (4.20) where $|\Psi_{oj}|^2$ and $|\Psi_{hj}|^2$ are taken from (4.33). Its variations illustrate the variation of the relative weights of the two wavefields with the deviation parameter η_r (or the departure from Bragg's angle). It can be seen that far from the reflection domain on the small angle side ($\Delta\theta \Rightarrow -\infty$) wavefield 1 is the strongest while on the large angle side ($\Delta\theta \Rightarrow +\infty$) it is wavefield 2. The propagation direction of the wavefields is along the refracted direction when they are the strongest and this situation 'jumps' from one branch of the dispersion surface to the other one across the reflection domain. This is an important property of dynamical theory which also holds in the Bragg case, and occurs as well when a wavefield crosses a highly distorted region in a deformed crystal (the so-called *interbranch scattering*, Section 14.4).

4.8.4 *Anomalous absorption*

Expression (4.34) of μ_{ej} is simpler in the case of a symmetric reflection ($\gamma = 1$):

$$\mu_{ej} = \mu_o \left[1 \mp \frac{|C\chi_{ih}/\chi_{io}| \cos\varphi}{\sqrt{1+\eta_r^2}} \right].$$

The effective absorption coefficient is much smaller than normal for branch 1 wavefields and much larger for branch 2 wavefields. Figure 4.13 shows the variations of the effective absorption coefficient μ_{ej} with η_r for wavefields belonging to branches 1 and 2, respectively, in the case of gallium arsenide, a symmetric 400 reflection and Cu$K\alpha$. The minimum value occurs for $\eta_r = 0$ and is equal to $\mu_o(1 - |C\chi_{ih}/\chi_{io}| \cos\varphi)$; it depends on the nature of the reflection through the structure factor and on the temperature through the Debye–Waller factor to which χ_{ih} is proportional (2.37).

Examples. The effect of anomalous transmission is illustrated in Tables 4.1 and 4.2 which give the values of $|C\chi_{ih}/\chi_{io}|$ and of the corresponding minimum attenuation coefficient $\exp(-\mu_{ej}t/\cos\theta_B)$ (branch 1, $\eta_r = 0$) for germanium crystals of various thicknesses and for two different wavelengths. Germanium is face-centred cubic and $|\chi_{ih}/\chi_{io}| \approx \sqrt{2}/2 \exp(-M)$ for reflections with odd indices and $|\chi_{ih}/\chi_{io}| \approx \exp(-M)$ for reflections with even indices.

1. *Influence of temperature*: Table 4.1 compares the attenuation coefficient for a 1 mm thick germanium crystal, Mo$K\alpha$, 111, 220 and 400 reflections at room temperature and 220 reflection at 5 K.

The Debye–Waller factor becomes closer to 1 as temperature decreases and the anomalous absorption effect becomes more and more important as temperature decreases as is shown by the values given for 220 at 293 and 5 K.

2. *Interpretation of Fig. 1.6*: Figures 1.6(a)–(c) represent wide-angle diagrams by Borrmann (1964) of germanium crystals of thicknesses 0.23 mm, 0.5 mm and 1.2 mm,

Table 4.1 Influence of temperature on anomalous absorption—germanium, 1 mm thick, Mo$K\alpha$, $C = 1$ (reproduced with permission from Ludewig 1969)

hkl	$\lvert\chi_{ih}/\chi_{io}\rvert$	μ_{ej} (cm^{-1}) min	$\exp(-\mu_{ej}t/\cos\theta_B)$ ($t = 1$ mm)
111 (293 K)	0.697	97.0	0.58×10^{-4}
220 (293 K)	0.964	11.5	0.310
220 (5 K)	0.990	3.2	0.72
400 (293 K)	0.930	22.4	0.0989
000	1	$\mu_o = 320$	1.3×10^{-14}

Table 4.2 Variation of the attenuation factor, $\exp(-\mu_{ej}t/\cos\theta_B)$, with the nature of the reflection and crystal thickness—germanium; Cu$K\alpha$; $\mu_o = 341.7\,\text{cm}^{-1}$

hkl	$\lvert C\chi_{ih}/\chi_{io}\rvert$	$\exp(-\mu_{ej}t/\cos\theta_B)$		
		$t = 0.23$ mm	$t = 0.50$ mm	$t = 1.20$ mm
(a) σ-polarization ($C = 1$)				
111	0.698	0.087	0.49×10^{-2}	0.29×10^{-5}
311	0.673	0.056	0.19×10^{-2}	0.25×10^{-6}
333	0.627	0.0156	0.12×10^{-3}	0.38×10^{-11}
220	0.965	0.741	0.522	0.210
400	0.931	0.524	0.245	0.034
440	0.867	0.193	0.028	0.19×10^{-2}
000	1.0	0.39×10^{-2}	0.38×10^{-7}	0.16×10^{-17}
(b) π-polarization ($C = \cos 2\theta_B$)				
111	0.620	0.046	0.13×10^{-2}	0.11×10^{-6}
311	0.398	0.005	0.97×10^{-5}	0.94×10^{-12}
333	0.003	0.15×10^{-4}	0.33×10^{-10}	0.7×10^{-25}
220	0.678	0.064	0.26×10^{-2}	0.6×10^{-6}
400	0.377	0.28×10^{-2}	0.3×10^{-5}	0.6×10^{-13}
440	0.165	0.3×10^{-4}	0.18×10^{-9}	0.4×10^{-23}

respectively. Table 4.2(a) shows that the attenuation factor, $\exp(-\mu_{ej}t/\cos\theta_B)$, is very much larger for Bragg reflections than for no reflection ('000' reflection) and that no intensity propagates outside the Bragg reflections for the crystals of thicknesses 0.5 mm and 1.2 mm. It is also much larger for reflections with even indices than for reflections with odd indices, which explains why the latter are not visible on the diagrams with 0.5 mm and 1.2 mm thicknesses. It is larger for 220 and 400 than for 440 and the latter reflection is no longer visible in Fig. 1.6(c) taken with the 1.2 mm thick crystal.

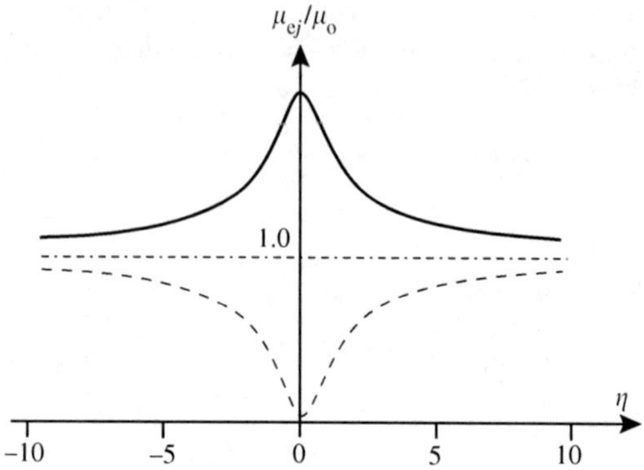

Fig. 4.13. Variation of the effective absorption coefficient with the deviation parameter. GaAs, 400 reflection, Cu$K\alpha$. Solid curve: branch 2; dashed curve: branch 1.

3. *Influence of polarization*: Tables 4.2(a) and (b) compare the attenuation coefficients for σ-polarization ($C = 1$) and π-polarization ($C = \cos 2\theta_B$). It can be seen that the latter is absorbed out much more rapidly than the former. In the case of germanium, 333 and Cu$K\alpha$, $\theta_B = 45.15°$ and π-polarization ($C \approx 0$) is completely absorbed out.

4.8.5 *Boundary conditions at the exit surface*

The incident wave has generated two wavefields in the crystal. They follow different paths inside the crystal. When they reach the exit surface the coupling ceases to exist and they break up into their constituent waves. The wavevectors and amplitudes of the waves outside and inside the crystal are related by the boundary conditions:

1. *Wavevectors*: The wavevectors of the waves outside the crystal are obtained by applying again the condition of the continuity of their tangential components along the crystal surface. The extremities, M_j and N_j, of these wavevectors

$$\mathbf{OM_j} = \mathbf{K_{oj}^{(d)}}; \qquad \mathbf{HN_j} = \mathbf{K_{hj}^{(d)}}$$

lie at the intersections of the spheres of radius k and centred at O and H, respectively (replaced by their asymptotes T'_o and T'_h), with the normal to the crystal exit surface, drawn from P_j ($j = 1$ and 2). If the crystal is plane parallel, N_1 and N_2 are confused (Fig. 4.14) but this is not the case for a wedge-shaped crystal (Fig. 6.4).

The relation between the wavevectors of the emerging wavevectors is

$$\mathbf{HN_j} = \mathbf{HO} + \mathbf{OM_j} + \mathbf{M_jN_j}.$$

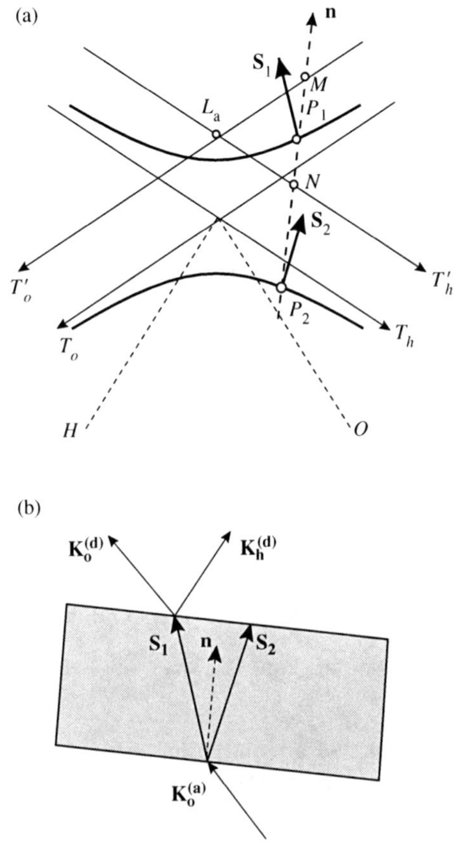

Fig. 4.14. Boundary condition for the wavevectors at the exit surface, plane-parallel crystal. **n**: normal to the entrance and exit surfaces; N and M: tiepoints of the refracted, ($\mathbf{K_o^{(d)}}$), and reflected, ($\mathbf{K_h^{(d)}}$), wavevectors generated by wavefields 1 and 2, respectively. (a) Reciprocal space. (b) Direct space.

It follows, if $\vec{\tau}$ is a unit vector parallel to the crystal surface, that

$$\mathbf{HN_j} \cdot \vec{\tau} = \mathbf{HO} \cdot \vec{\tau} + \mathbf{OM_j} \cdot \vec{\tau}.$$

The difference between the projections of the emerging wavevectors on the exit surface is the projection of the reciprocal lattice vector on the surface.

2. *Amplitudes*: The incident wave is a plane wave and its lateral extension is infinite. Two wavefields arrive at any point of the exit surface which have been generated at different points on the entrance surface. Their constituent waves interfere and generate emerging waves in the refracted and reflected directions (Fig. 4.15). Their respective

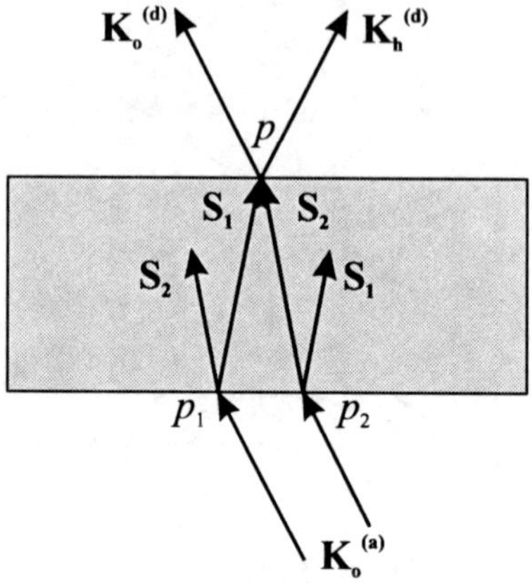

Fig. 4.15. Plane-wave *Pendellösung*: interference of wavefields belonging to branches 1 and 2 of the dispersion surface, respectively, when they reach the exit surface.

amplitudes on the exit surface are given by the boundary conditions:

$$\Psi_o^{(d)} \exp(-2\pi i \mathbf{K_o^{(a)}} \cdot \mathbf{r}) = \Psi_{o1} \exp(-2\pi i \mathbf{K_{o1}} \cdot \mathbf{r}) + \Psi_{o2} \exp(-2\pi i \mathbf{K_{o2}} \cdot \mathbf{r}),$$
$$\Psi_h^{(d)} \exp(-2\pi i \mathbf{K_h^{(a)}} \cdot \mathbf{r}) = \Psi_{h1} \exp(-2\pi i \mathbf{K_{h1}} \cdot \mathbf{r}) + \Psi_{h2} \exp(-2\pi i \mathbf{K_{h2}} \cdot \mathbf{r}), \tag{4.35a}$$

where **r** is the position vector of a point on the *exit surface*, the origin of phases being taken on the *entrance surface*. The phase factors before and after the crystal are discussed in Section 6.4.1.

In a plane-parallel crystal, these expressions reduce to

$$\begin{aligned}\Psi_o^{(d)} &= \Psi_{o1} \exp(-2\pi i \overline{MP_1}\, t) + \Psi_{o2} \exp(-2\pi i \overline{MP_2}\, t),\\ \Psi_h^{(d)} &= \Psi_{h1} \exp(-2\pi i \overline{MP_1}\, t) + \Psi_{h2} \exp(-2\pi i \overline{MP_2}\, t),\end{aligned} \tag{4.35b}$$

where t is the crystal thickness. The case where the exit surface is not parallel to the entrance surface is discussed in Section 6.3.

4.8.6 *Reflectivity*

The intensities of the reflected and refracted beams are obtained by substitution from (4.32) into (4.35b). To simplify, we shall express them here in the non-absorbing case:

$$\begin{aligned}|\Psi_h^{(d)}|^2 &= \frac{1 - \cos[2\pi \overline{P_2 P_1}\, t]}{2\gamma(1+\eta^2)} |\Psi_o^{(a)}|^2,\\ |\Psi_o^{(d)}|^2 &= \frac{1 + 2\eta^2 + \cos[2\pi \overline{P_2 P_1}\, t]}{2(1+\eta^2)} |\Psi_o^{(a)}|^2\end{aligned} \tag{4.36}$$

with

$$\overline{P_2P_1} = \overline{MP_1} - \overline{MP_2} = \frac{\sqrt{1+\eta^2}}{\Lambda_L},$$

where $\Lambda_L = 1/\overline{A_2A_1} = \Lambda_o$ is called the *Pendellösung distance*. Λ_o is given by (4.26) and $\overline{A_2A_1}$ is the diameter of the dispersion surface parallel to the normal to the entrance surface (Fig. 4.11). If we set $\Lambda^{-1} = \sqrt{1+\eta^2}\Lambda_L^{-1} = \overline{P_2P_1}$, expressions (4.36) can be written, finally, as

$$\begin{aligned} |\Psi_h^{(d)}|^2 &= \frac{\sin^2[\pi t \Lambda^{-1}]}{\gamma(1+\eta^2)} |\Psi_o^{(a)}|^2, \\ |\Psi_o^{(d)}|^2 &= \frac{\eta^2 + \cos^2[\pi t \Lambda^{-1}]}{1+\eta^2} |\Psi_o^{(a)}|^2. \end{aligned} \tag{4.37}$$

What is actually measured is the *reflectivity* which, as defined in Section 3.4, is the ratio of the energy of the reflected or refracted beam received per unit area of the detector per unit time by the intensity of the incident beam. The power of a beam is obtained by multiplying its intensity by its cross-section. If the incident beam is limited by a slit of height h and of width $l_o^{(a)}$, its cross-section is $h \times l_o^{(a)}$ (Fig. 4.16); the cross-sections of the refracted and reflected beams are $h \times l_o^{(d)}$ and $h \times l_h^{(d)}$,

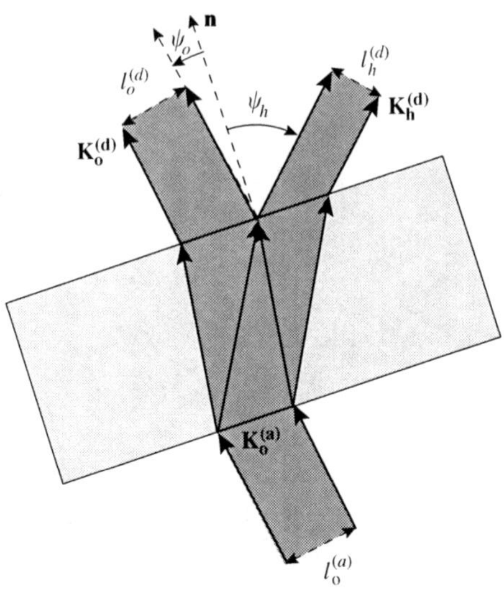

Fig. 4.16. Cross-sections of the incident, $l_o^{(a)}$, refracted, $l_o^{(d)}$, and reflected, $l_h^{(d)}$, waves, respectively.

respectively. For a plane-parallel crystal,

$$l_o^{(d)} = l_o^{(a)},$$
$$l_h^{(d)} = \frac{\gamma_h}{\gamma_o} l_o^{(a)} = \gamma l_o^{(a)}.$$

The reflectivities are therefore, respectively,

$$\begin{aligned} \text{refracted beam:} \quad I_o^{(d)} &= \frac{h l_o^{(d)} |\Psi_o^{(d)}|^2}{h l_o^{(a)} |\Psi_o^{(a)}|^2} = \frac{|\Psi_o^{(d)}|^2}{|\Psi_o^{(a)}|^2}, \\ \text{reflected beam:} \quad I_h^{(d)} &= \frac{h l_h^{(d)} |\Psi_h^{(d)}|^2}{h l_o^{(a)} |\Psi_o^{(a)}|^2} = \gamma \frac{|\Psi_h^{(d)}|^2}{|\Psi_o^{(a)}|^2}. \end{aligned} \tag{4.38}$$

The complete expressions for absorbing crystals are given in Section 6.4.

4.8.7 *Pendellösung*

Expressions (4.37) show that the reflected and refracted intensities are oscillating functions of crystal thickness (*equal-thickness Pendellösung*—Fig. 4.17) and of crystal orientation (*equal-orientation Pendellösung*—Fig. 4.18). Their oscillations are complementary: when one is maximum, the other one is minimum and *vice versa*; there is a periodic exchange of energy between the two beams, as in two coupled pendulums, hence the name given to that effect by Ewald (1917) who also gave a mechanical analogy (Ewald 1965), extended by Shevchenko and Pobydaylo (2003) – see page 44, footnote 5.

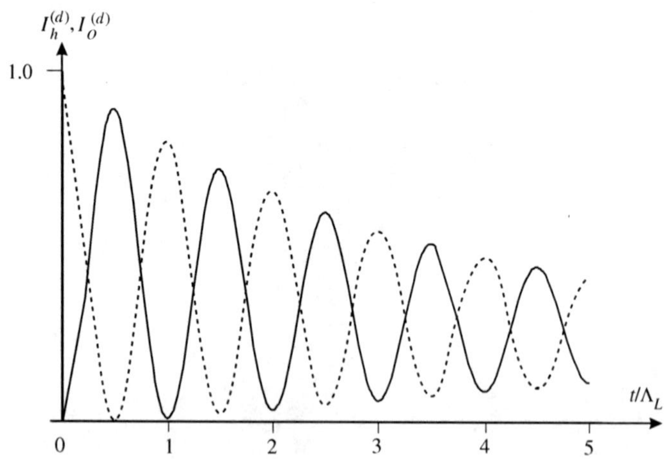

Fig. 4.17. Equal-thickness *Pendellösung* in the transmission case: variations of the reflected intensity, $I_h^{(d)}$ (solid curve), and of the refracted intensity, $I_o^{(d)}$ (dashed curve), with t/Λ_L (t, crystal thickness) for $\eta = 0$; silicon, 220 reflection, Cu$K\alpha$, $\Lambda_L = 15.26\,\mu$m.

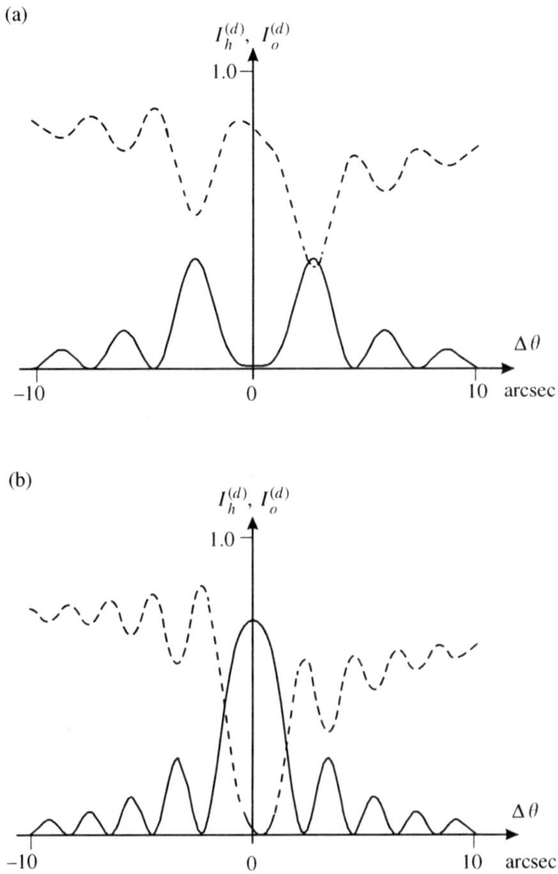

Fig. 4.18. Orientation *Pendellösung* fringes in the transmission case: thin crystals (arbitrary units); silicon, Cu$K\alpha$, 220 reflection. Solid curve: reflected beam. Dashed curve: refracted beam. (a) $t/\Lambda_L = 1.0$ ($\mu_o t = 0.22$). (b) $t/\Lambda_L = 1.5$ ($\mu_o t = 0.33$).

The period of the oscillations of equal-thickness Pendellösung is Λ. Its maximum value is Λ_L. In the symmetric case it is the inverse of the Bragg gap and its expression (4.26) reduces to

$$\Lambda_L = \frac{\lambda}{2\sin\theta_B \delta_{os}} = \frac{d_{hkl}}{\delta_{os}},$$

where d_{hkl} is the reflecting plane spacing. It is quite remarkable that the half-width δ_{os} can thus be interpreted as the ratio of the reflecting plane spacing to the Pendellösung distance.

It is easy to check from (4.37) that in the non-absorbing case $I_o^{(d)} + I_h^{(d)} = 1$, namely that there is conservation of energy. The *FWHM of the rocking curve*, averaged over the Pendellösung oscillations, corresponds, in the non-absorbing case, to $\Delta\eta = 2$, that is to $\Delta\theta = 2\delta_{os}$, where δ_{os} is given by (4.26).

Equal-orientation Pendellösung were first observed by Lefeld-Sosnowska and Malgrange (1968, 1969) as oscillations in rocking curves and were further studied by Bonse *et al.* (1977) and by Bonse and Teworte (1980). They are discussed in Section 6.4.3.

4.8.8 *Integrated intensity*

The integrated intensity is the ratio of total energy recorded in the counter (when the crystal is rocked) to the intensity of the incident beam. It is proportional to the area under the line profile:

$$I_{hi}^{(d)} = \int_{-\infty}^{+\infty} I_h^{(d)}\, \mathrm{d}(\Delta\theta). \tag{4.39}$$

The integration has been performed by Laue (1960) in the zero-absorption case. Using (4.25) and (4.38), the result is

$$I_{hi}^{(d)} = \frac{\pi |C\chi_h|\sqrt{\gamma}}{2\sin 2\theta_B} \int_0^{2\pi t \Lambda_L^{-1}} J_0(z)\, \mathrm{d}z, \tag{4.40}$$

where $J_0(z)$ is the zeroth-order Bessel function.

Figure 4.19 represents the variations of $\int_0^{2\pi t \Lambda_L^{-1}} J_0(z)\, \mathrm{d}z$ with t/Λ_L. It oscillates around 1 when the crystal thickness increases and is equal to 1 when the crystal thickness tends towards infinity. The oscillations are due to Pendellösung (the interferences between branch 1 and branch 2 wavefields). The mean value of the integrated intensity when the crystal thickness is larger than one Pendellösung distance is therefore proportional to $|\chi_h|$ and to the structure factor, $|F_h|$, in contrast to the value of

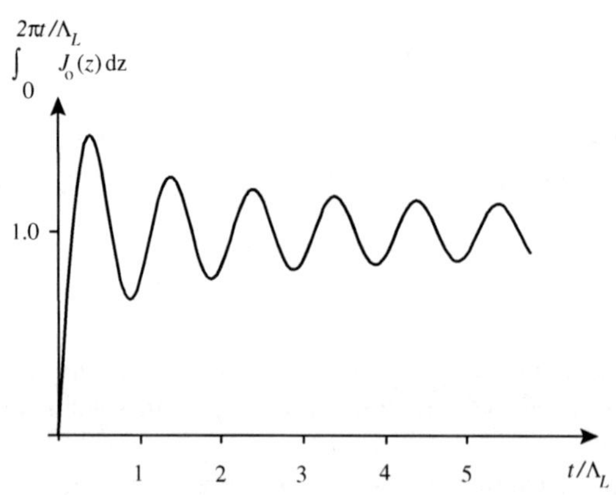

Fig. 4.19. Variations with t/Λ_L of $\int_0^{2\pi t/\Lambda_L} J_0(z)\, \mathrm{d}z$.

the integrated intensity given by the geometrical theory, which is proportional to the square of the structure factor, $|F_h|^2$. When $t/\Lambda_L \Rightarrow 0$,

$$I_{hi}^{(d)} \Rightarrow \frac{\pi^2 C^2 |\chi_h|^2 kt}{2\gamma_o \sin 2\theta_B},$$

which is equivalent to expression (3.17) obtained with the geometrical theory. The geometrical theory is therefore a good approximation when the crystal thickness is much smaller than t/Λ_L.

It will be shown in Section 10.7 that the intensity recorded when traversing the crystal in an incident spherical wave, as is the case on a traverse topograph, is equal to the same expression. Equal-thickness fringes are therefore expected on a traverse topograph of a wedge-shaped crystal. This is how Kato and Lang (1959) observed Pendellösung fringes for the first time.

The integration of (4.39) was performed for absorbing crystals by Kato (1955). The result is given in Section 6.4.

4.9 The diffracted waves in the reflection geometry

4.9.1 *Tiepoints*

The geometrical construction in Fig. 4.20 shows that, in the Bragg case, the normal to the crystal surface drawn from the extremity of the incident wavevector intersects the dispersion surface either at two points of the same branch, P_j' and P_j'', or at imaginary points. Figure 4.20 shows that the Poynting vector is oriented towards the inside of the crystal for one of them (P_1' and P_2', respectively) and towards the *outside* for the other one (P_1'' and P_2'', respectively). The wavefield propagating towards the entrance surface of the crystal, $\mathbf{S}_1''$ or $\mathbf{S}_2''$, can only exist if it has been created by reflection of $\mathbf{S}_1'$ or $\mathbf{S}_2'$ at the back surface of the crystal. The Poynting vectors $\mathbf{S}_j'$ and $\mathbf{S}_j''$ are parallel if P_j' and P_j'' lie on the extremities of a diameter of the dispersion surface ($j = 1$ or 2).

4.9.2 *Thick crystals—total reflection*

In a very thick crystal, the wavefields propagating towards the inside of the crystal are absorbed out before they reach the exit surface of the crystal because there is always a small amount of absorption. One concludes that in the thick crystal case and in the reflection geometry, only one wavefield is excited inside the crystal. It corresponds to branch 1 on the low-angle side of the rocking curve (P_1') and to branch 2 on the high-angle side (P_2'). There is the same 'jump' from one branch of the dispersion surface to the other one across the reflection domain as in the transmission geometry.

The amplitude, $\Psi_h^{(a)}$, of the wave reflected at the crystal surface is obtained by applying the boundary conditions which are particularly simple in this case:

$$\Psi_o = \Psi_o^{(a)}; \qquad \Psi_h^{(a)} = \Psi_h.$$

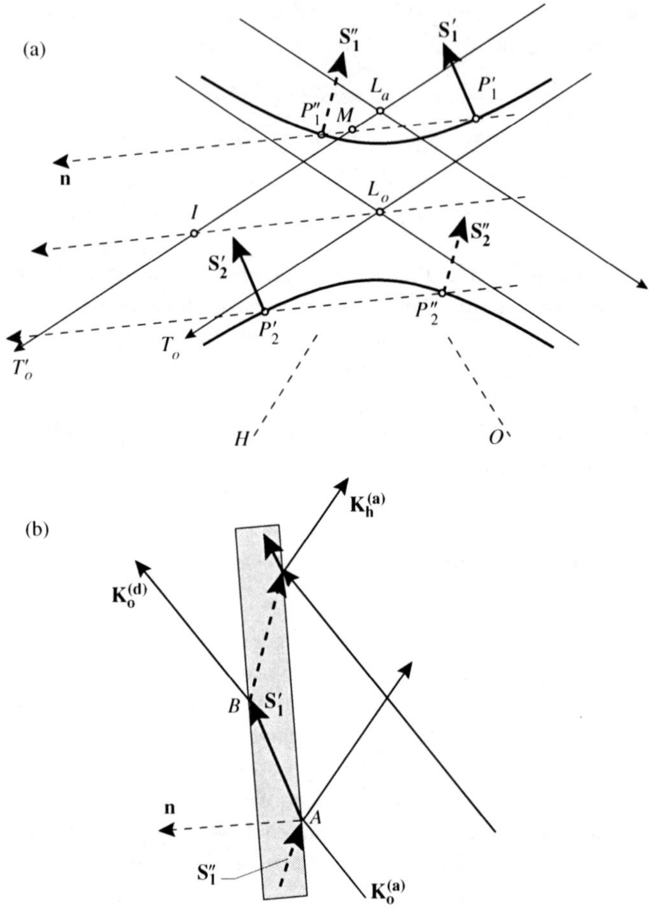

Fig. 4.20. Reflection geometry: wavefields created inside the crystal by the incident plane wave. **n**: normal to the crystal surface; $\mathbf{S}'_1$ and $\mathbf{S}''_1$: Poynting vectors; I: middle of the reflection domain. (a) Reciprocal space. (b) Direct space.

The reflectivity is given by an expression similar to (4.38):

$$I_h^{(a)} = \gamma \frac{|\Psi_h^{(a)}|^2}{|\Psi_o^{(a)}|^2} = |\gamma||\xi_j|^2, \tag{4.41}$$

where the expression of ξ_j is given by (4.28), with $S(\gamma_h) < 0$, and the $\pm$ sign is chosen so that the solution converges:

$$I_h^{(a)} = \left|\eta - S(\eta_r)\sqrt{\eta^2 - 1}\right|^2, \tag{4.42}$$

where η_r is the real part of η.

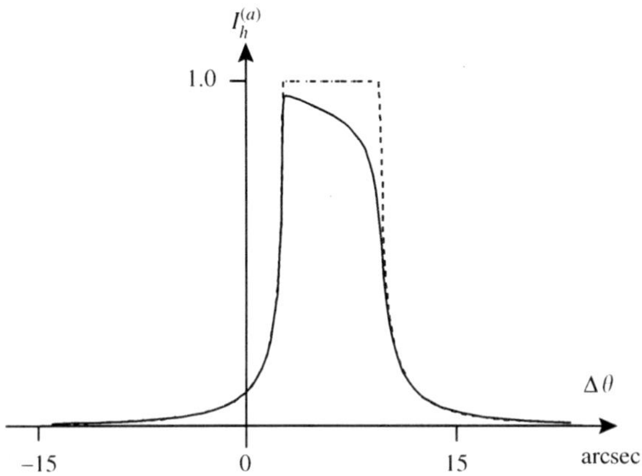

Fig. 4.21. Theoretical rocking curve in the reflection geometry for a thick crystal. Solid curve: GaAs, 400 reflection, Cu$K\alpha$; dashed curve: non-absorbing crystal (Darwin width: $2\delta_{os} = 7.54$ arcsec).

1. *Zero-absorption case, top-hat curve*: The dispersion surface and η are real. When the normal to the entrance surface does not intersect any of the branches of the dispersion surface, that is when $-1 < \eta < +1$,

$$|\xi|^2 = |\gamma|^{-1}, \qquad I_h = 1,$$

and there is *total reflection*. The rocking curve has the well-known top-hat shape (Fig. 4.21).

Far from the total reflection domain, the curve can be approximated by the function $I_h(\text{appr}) = 1/(4\eta^2)$. The width of the total reflection domain is $\Delta\eta = 2$ and its angular width is therefore equal, using (4.25), to $2\delta_{os}$, where δ_{os} is given by (4.26). The properties of this width and the influence of the asymmetry of the reflection are discussed in Section 4.10.

2. *Integrated intensity*: It is defined by (4.39) and was first calculated by Darwin (1914*b*). In the zero-absorption case, it is proportional to the width of the total reflection domain and therefore also to the structure factor, a result stressed by Ewald (1925):

$$I_{hi}^{(a)} = \frac{8\delta_{os}}{3} = \frac{8R\lambda^2|C|\sqrt{|\gamma|}}{\pi V \sin 2\theta_B}\sqrt{F_h F_{\bar{h}}}. \tag{4.43}$$

3. *Penetration depth in the absence of absorption*: Within the total reflection domain, there are two wavefields propagating inside the crystal, with imaginary wavevectors, one towards the inside of the crystal and the other one in the opposite direction, so that they cancel out and, globally, no energy penetrates the crystal.

The effective absorption coefficient of the waves penetrating in the crystal is given by $\mu_e = -4\pi K_o^i \gamma_o$ (4.31). It will be shown in Section 5.5.5 that

$$\mu_e = -4\pi K_o^i \gamma_o = \frac{2\pi \gamma_o \sqrt{1-\eta^2}}{\Lambda_B}, \tag{4.44}$$

where $\Lambda_B = \lambda\sqrt{\gamma_o|\gamma_h|}/\sqrt{\chi_h \chi_{\bar{h}}}$.

The penetration depth z_e is the distance from the crystal surface for which the attenuation factor is equal to $1/e$:

$$\exp\left(-\frac{\mu_e z_e}{\gamma_o}\right) = -1 \quad \Longrightarrow \quad z_e = \frac{\gamma_o}{\mu_e}. \tag{4.45a}$$

It follows, using (4.44), that

$$z_e = \frac{\Lambda_B}{2\pi\sqrt{1-\eta^2}}. \tag{4.45b}$$

It is minimum at the middle of the reflection domain and is then $\Lambda_B/2\pi$. This attenuation effect is called *extinction*, and Λ_B is called the *extinction distance*. It is a specific property due to the existence of wavefields. The resulting propagation direction of energy is parallel to the crystal surface, but with a cross-section equal to zero: it is an *evanescent wave* (see, for instance, Cowan *et al.* 1986). The variations of the penetration depth with angle of incidence are represented in Fig. 5.16 (Section 5.8.4).

4. *Absorbing crystals*: There is no longer a total reflection domain and energy penetrates inside the crystal at all incidence angles although with a very high absorption coefficient within the domain $|\eta_r| \leq 1$. Figure 4.21 gives an example of a rocking curve for a thick absorbing crystal. It was first observed by Renninger (1955). The shape is asymmetric and is due to the anomalous absorption effect: absorption is lower than normal on the low-angle side, which is associated with wavefields belonging to branch 1 of the dispersion surface, and larger than normal on the high-angle side, which is associated with branch 2 wavefields. An analytical expression of the integrated intensity has been given by Afanas'ev and Perstnev (1969).

4.9.3 *Thin crystals*

1. *Boundary conditions*: If the crystal is thin, the wavefield created at the reflecting surface at A and penetrating inside can reach the back surface at B (Fig. 4.20). The *incident* direction there points towards the *outside* of the crystal, while the *reflected* direction points towards the inside. The wavefield propagating along AB therefore generates at B:

(i) *A partially transmitted wave* outside the crystal, $\mathbf{D}_\mathbf{o}^{(\mathbf{d})} \exp(-2\pi \mathrm{i} \mathbf{K}_\mathbf{o}^{(\mathbf{d})} \cdot \mathbf{r})$.

(ii) *A partially reflected wavefield* inside the crystal. The corresponding tiepoints are obtained by applying the usual condition of the continuity of the tangential components of wavevectors. If the crystal is a plane-parallel slab, these points are M and P_j'', respectively, and $\mathbf{K_o^{(d)}} = \mathbf{K_o^{(a)}}$.

The boundary conditions are then written as:

(iii) *Entrance surface*:

$$\Psi'_{oj} + \Psi''_{oj} = \Psi_o^{(a)}; \qquad \Psi'_{hj} + \Psi''_{hj} = \Psi_h^{(a)}$$

($j = 1$ or 2 according to the branch of the dispersion surface to which belongs the tie point).

(iv) *Back surface*:

$$\Psi'_{oj} \exp(-2\pi \mathrm{i} \mathbf{K'_{oj}} \cdot \mathbf{r}) + \Psi''_{oj} \exp(-2\pi \mathrm{i} \mathbf{K''_{oj}} \cdot \mathbf{r}) = \Psi_o^{(d)} \exp(-2\pi \mathrm{i} \mathbf{K_o^{(d)}} \cdot \mathbf{r}),$$

$$\Psi'_{hj} \exp(-2\pi \mathrm{i} \mathbf{K'_{hj}} \cdot \mathbf{r}) + \Psi''_{hj} \exp(-2\pi \mathrm{i} \mathbf{K''_{hj}} \cdot \mathbf{r}) = 0.$$

2. *Rocking curve in the non-absorbing crystal case*: The expressions of the intensity reflected at the entrance surface, $I_h^{(a)}$, and of the intensity refracted through the crystal, $I_o^{(d)}$, are (Laue 1960)

$$I_h^{(a)} = \frac{1 - \cos^2(\pi t/\Lambda)}{\eta^2 - \cos^2(\pi t/\Lambda)}, \tag{4.46a}$$

$$I_o^{(d)} = \frac{\eta^2 - 1}{\eta^2 - \cos^2(\pi t/\Lambda)} \tag{4.46b}$$

with $\Lambda = \Lambda_B/\sqrt{\eta^2 - 1}$.

(i) $|\eta| < 1$: $\sqrt{\eta^2 - 1} = \mathrm{i}\sqrt{1 - \eta^2}$, the cosine terms are replaced by hyperbolic cosines. There is no longer a total reflection domain but there is still the extinction effect as is shown by the hyperbolic cosine term. The maximum height of the rocking curve decreases as the thickness of the crystal decreases.

(ii) $|\eta| > 1$: The oscillations are due to the interference between the wavefields associated with P_1' and P_1'', respectively (or P_2' and P_2'') and were also called *Pendellösung* by Ewald (1917). The *equal-orientation* fringes were observed for the first time by Batterman and Hildebrandt (1967, 1968). The angular positions of the minima of the reflected beam are given by

$$\eta = \mp\sqrt{\frac{K^2\Lambda_B^2}{t^2 + 1}},$$

where K is an integer.

3. *Conservation of energy*: It is obvious from (4.46a) and (4.46b) that $I_h^{(a)} + I_o^{(d)} = 1$ and that there is conservation of energy.

4. *Integrated intensity*: The integrated intensity is

$$I_{hi}^{(a)} = \pi\delta_{os} \tanh\left[\frac{\pi t}{\Lambda_B}\right], \tag{4.47a}$$

where t is the crystal thickness. When this thickness becomes very large, the integrated intensity tends towards

$$I_{hi}^{(a)} = \pi \delta_{os}. \tag{4.47b}$$

This expression differs from (4.43) by the factor π which appears here instead of $\frac{8}{3}$. The reason is as follows. When the crystal thickness is much larger than the extinction distance, but the absorption and $\mu_o t$ are very small so that both wavefields should be taken into account, as in neutron diffraction, expression (4.46a) should be averaged over thickness (see Section 7.3.2). The resulting values of the reflectivity outside the total reflection domain, given by (7.13), are larger than those given by (4.42) for an infinitely thick crystal and the corresponding integrated intensity is also larger.

5. *Absorbing crystals*: The expressions of the reflected and refracted intensities in the case of absorbing crystals are given in Section 7.3.2.

4.10 Influence of the asymmetry on the position and width of the rocking curve and of the angular distribution of the reflected beam

The widths of the rocking curve and the angular distribution of the reflected beam are different when the reflection is asymmetric and this is very important for the properties of monochromators (see Section 15.2).

The FWHM of the rocking curve in the transmission geometry, neglecting Pendellösung (and the width of the total reflection domain in the reflection geometry), and the deviation from Bragg's angle of the middle of the reflection domain are given by (4.26). If we replace χ_h by its value (2.37) and remember that it must be multiplied by the factor C to take into account polarization (Section 4.5), we find

$$\Delta\theta_{os} = -\frac{R\lambda^2 F_o}{2\pi V \sin 2\theta_B}(1-\gamma), \tag{4.48a}$$

$$\begin{aligned} 2\delta_{os} &= \frac{2R\lambda^2}{\pi V \sin 2\theta_B}\sqrt{|\gamma|}\,|C|\sqrt{F_h F_{\bar{h}}} \\ &= \frac{2R}{\pi V}\frac{\lambda d_{hkl}}{\sqrt{1-\lambda^2/4d_{hkl}^2}}\sqrt{|\gamma|}\,|C|\sqrt{F_h F_{\bar{h}}}, \end{aligned} \tag{4.48b}$$

where $d_{hkl} = \lambda/\sin\theta_B$ and the asymmetry ratio, $|\gamma|$, is given by (4.24). The width of the reflection domain can easily be visualized in reciprocal space (Fig. 4.22); there is total reflection when the normal to the crystal surface, **n**, drawn from the extremity M of the incident wavevector intersects the dispersion surface at imaginary points; the corresponding points lie on the segment $I_{o1}I_{o2}$ along the tangent T_o' to the sphere of centre O and radius k where I_{o1} and I_{o2} are at the intersections of T_o' with the

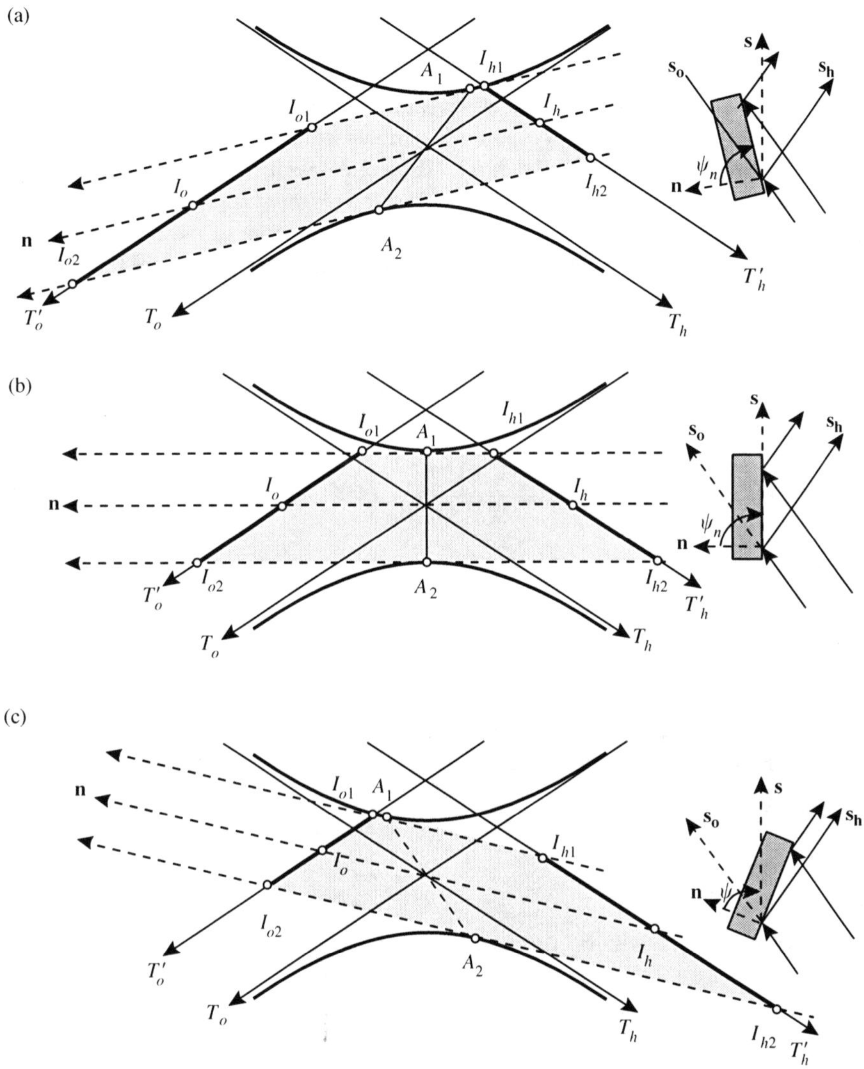

Fig. 4.22. Reflection geometry: variation of the width, $\overline{I_{o1}I_{o2}}$, of the total reflection domain and of the width, $\overline{I_{h1}I_{h2}}$, of the angular distribution of the reflected beam with the asymmetry of the reflection; **n**: normal to the crystal surface; **s**: unit vector parallel to the reflecting planes; $\mathbf{s_o}$: unit vector in the incident direction; insets: direct space. (a) $-\pi/2 - \theta_B < \psi_n < -\pi/2$. (b) $\psi_n = -\pi/2$. (c) $-\pi/2 < \psi_n < -\pi/2 + \theta_B$.

tangents to the two branches of the dispersion surface which are parallel to **n**. The position of the middle of the total reflection domain, I_o, is given by

$$\overline{L_a I_o} = k\Delta\theta_{os}.$$

Figure 4.22 illustrates three cases: glancing incidence (Fig. 4.22(a)), symmetric reflection (Fig. 4.22(b)) and glancing emergence (Fig. 4.22(c)), respectively.

In a similar way, the angular distribution of the *reflected* beam can be represented graphically. The extremities of the wavevectors of the reflected beam lie at the intersection of the normal **n** to the crystal surface drawn from the tiepoint with the tangent, T'_h, to the sphere of centre H and radius k. The angular width of the reflected beam corresponding to total reflection is represented by the segment $I_{h1}I_{h2}$ on T'_h (Fig. 4.22). It can be seen in the figure that it is narrow when the width of the rocking curve is wide, and *vice versa*. It is easy to show that the angular width $2\delta_{hs}$ of the reflected beam and the deviation $\Delta\theta_{hs}$ of its centre from Bragg's angle are deduced from the corresponding expressions (4.26) for the incident beam by replacing $\sqrt{|\gamma|}$ by $1/\sqrt{|\gamma|}$:

$$2\delta_{hs} = \frac{2R\lambda^2}{\pi V \sin 2\theta_B} \frac{1}{\sqrt{|\gamma|}} |C| \sqrt{F_h F_{\bar{h}}}, \tag{4.49a}$$

$$\Delta\theta_{hs} = -\frac{R\lambda^2 F_o}{2\pi V \sin 2\theta_B}\left(1 - \frac{1}{\gamma}\right). \tag{4.49b}$$

The position of the middle of the angular distribution of the reflected beam, I_h, is given by

$$\overline{L_a I_h} = k\Delta\theta_{hs}.$$

Expressions (4.48) and (4.49) show that the factors which influence the width of the rocking curve and of the angular distribution of the reflected beam are

- wavelength
- structure factor
- polarization
- asymmetry.

The last parameter will be discussed here and the first three in Section 7.1.2. Figure 4.23 shows the variations of $\sqrt{|\gamma|}$ and of $1/\sqrt{|\gamma|}$ with the angle between the normal to the crystal surface and the lattice planes, ψ_n, in the reflection and transmission geometries (ψ_n is defined in Section 4.7.3).

The variation of the width of the rocking curve with the asymmetry of the reflection was first observed by Renninger (1961) with a double-crystal arrangement while the influence of asymmetry on the shape of the rocking curve was observed by Bubáková (1962) by means of a triple-crystal arrangement, both with highly perfect germanium crystals.

The asymmetry ratio $\sqrt{|\gamma|}$ takes all values from 0 to ∞ within the angular range of variation of ψ_n. The formula is of course not valid in a very asymmetric geometry (or when the Bragg angle tends towards $\pi/2$) because of the initial approximation to replace the Ewald sphere by its tangential plane. These extreme cases are discussed in Chapter 8 for the very asymmetric geometries and in Section 5.9 for Bragg angles close to $\pi/2$.

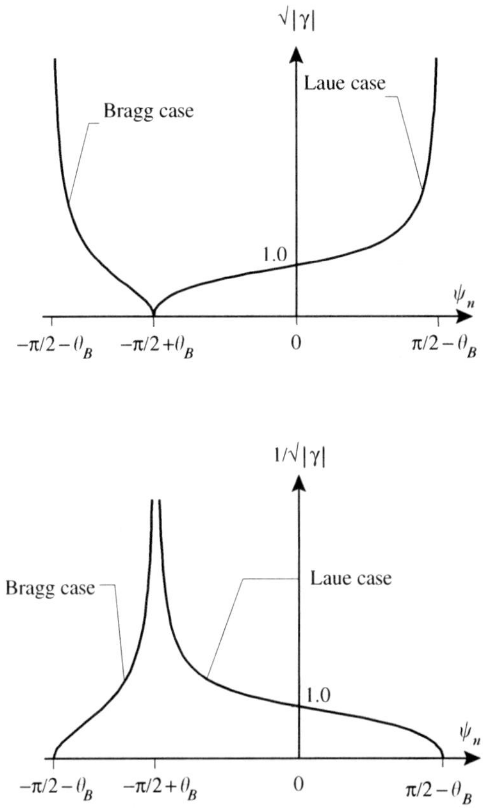

Fig. 4.23. Variations of $\sqrt{|\gamma|}$ and $1/\sqrt{|\gamma|}$ (γ: asymmetry ratio) with the angle ψ_n between the normal to the crystal surface and the lattice planes; silicon, 220 reflection, Cu$K\alpha$ ($\theta_B = 23.67°$).

The effect of the asymmetry makes it possible to vary considerably the widths of the rocking curve and of the reflected beam by using a suitable cut for the crystal surface. This property is used systematically in the design of monochromators (see Section 15.3). Figure 4.24 shows the variations of the shape and position of the rocking curves with the asymmetry of the reflection. It can be seen that when the angle between the incident beam and the crystal surfacee is very small (Fig. 4.22(a)), the width of the rocking curve is very large, while it is the contrary when that angle is very large (Fig. 4.22(c)).

4.11 Comparison with geometrical theory

1. *Reflectivity*: When t/Λ_o tends towards zero ($\Lambda_o \equiv \Lambda_L$ or Λ_B depending on the geometry), the reflectivities (4.37) and (4.46) of the reflected beams for non-absorbing

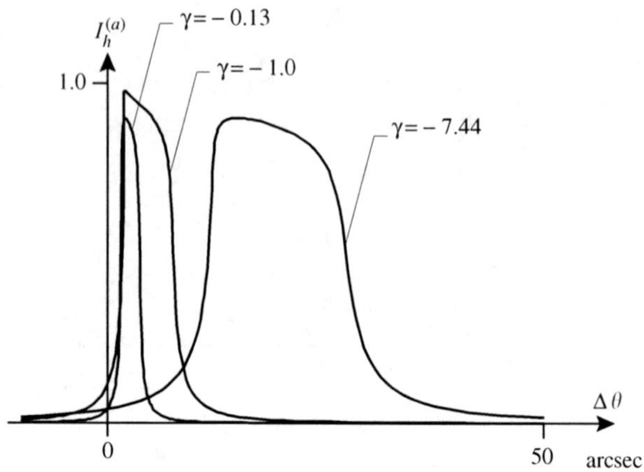

Fig. 4.24. Influence of asymmetry on the theoretical rocking curve of a thick crystal in the reflection geometry; silicon, 220 reflection, $\lambda = 1.65$ Å (7.5 keV): For $\gamma = -1$ (symmetric reflection), $\Delta\theta_{os} = 4.64$ arcsec, $2\delta_{os} = 5.58$ arcsec; for $\gamma = -0.13$ (grazing emergence), $\Delta\theta_{os} = 2.63$ arcsec, $2\delta_{os} = 2.04$ arcsec; for $\gamma = -7.44$ (grazing incidence), $\Delta\theta_{os} = 19.6$ arcsec, $2\delta_{os} = 15.22$ arcsec.

crystals in both the transmission and the reflection geometries tend towards expression (3.14) of the reflectivity obtained with the geometrical theory. The last term of (3.14) can be rewritten, using the notations of dynamical theory, for a symmetric reflection as

$$\left[\frac{\sin 2\pi N_3 kd \cos\theta_B \Delta\theta}{2\pi kd \cos\theta_B \Delta\theta}\right]^2 = \left[\frac{\sin(\pi\eta t/\Lambda_o)}{\pi\eta d/\Lambda_o}\right]^2.$$

Figure 4.25 compares the normalized reflectivities obtained by the geometrical theory and by the dynamical theory in the transmission and reflection geometries, respectively, for $t/\Lambda_o = 0.1$. It can be seen that the reflectivities differ slightly for $|\eta| < 5$. For the approximation of geometrical theory to be valid for the diffracted intensities, the crystal thickness must be smaller than about one tenth of the Pendellösung, or of the extinction distance. It depends therefore on the conditions of the diffraction: the weaker the reflection, the larger is that limiting thickness. From (4.26), it can be seen that

$$\Lambda_o = \frac{\pi V\sqrt{\gamma_o|\gamma_h|}}{R\lambda\sqrt{F_h F_{\bar{h}}}}.$$

The apparent 'perfection' of the crystal can thus be decreased by increasing the Pendellösung distance, for instance by using a higher-order reflection, a shorter wavelength or neutrons instead of X-rays. The high sensitivity to strain of high-energy (short-wavelength) photons added to their strong penetration in heavy materials make them a very good tool for the characterization of bulk materials as shown, for instance,

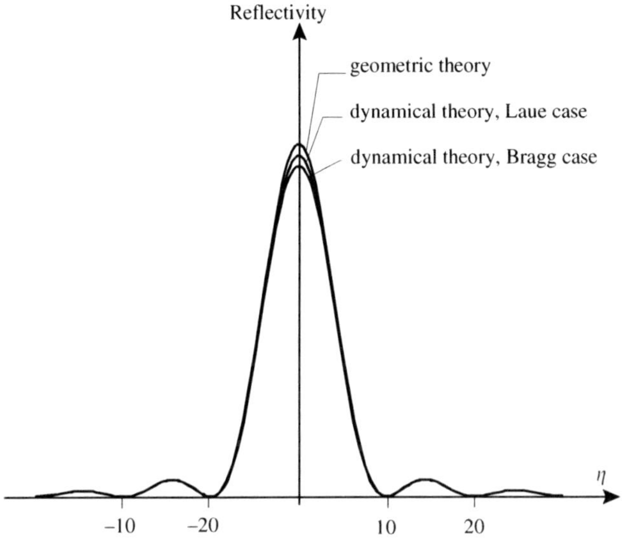

Fig. 4.25. Comparison between the rocking curves obtained with the geometrical theory, the dynamical theory in the transmission geometry and the reflection geometries; $t/\Lambda_o = 0.1$; normalized coordinates.

by Abdul Gani *et al.* (1981), Deutsch (1987), Schneider *et al.* (1988) with high-energy γ-rays from a radioactive source and by Holý *et al.* (1988) with the 160 keV spectrum from a sealed X-ray tube and as is nowadays routine with synchrotron radiation.

2. *Integrated intensity*: Using expression (3.16b) for the integrated intensity given by the geometrical theory, expression (4.47) given by the dynamical theory for thin non-absorbing crystals in the reflection geometry can be written as

$$I_{hi} = I_{hi}^{\text{geom}} \frac{\tanh[\pi t/\Lambda_B]}{[\pi t/\Lambda_B]}. \tag{4.50}$$

The coefficient $\tanh[\pi t/\Lambda_B]/[\pi t/\Lambda_B]$ is the correction given by Darwin (1922) for primary extinction. It tends towards 1 when t/Λ_B tends towards zero. For instance, it is equal to 0.968, 0.992 and 0.999 for $t/\Lambda_B = 0.1$, 0.05 and 0.01, respectively. As in the case of the rocking curve, this shows that for the geometrical theory to be a good approximation, the crystal thickness must be smaller than about one tenth of the extinction distance. For a given crystal thickness, the longer the extinction distance, the better the crystal satisfies the mosaic crystal model.

By deforming a crystal in a controlled way, it is possible to span the whole range between the perfect and the ideally imperfect crystal cases. This was shown, for instance, by Kies *et al.* (1973) who generated vibrations in a cadmium sulphide crystal by means of the acoustoelectric effect and by increasing the amplitude of the vibrations could vary the integrated intensity from the perfect crystal value to the ideally imperfect crystal one given by the geometrical theory; it was also shown by Lukáš *et al.* (1991) who applied a thermal gradient to indium antimonide and silicon crystals. Increasing the degree of imperfection of a crystal is sometimes used in order to heighten the reflectivity of monochromators for high energy synchrotron radiation. Schneider *et al.* (1988, 1989) have for this purpose applied a heat treatment above 1200°C to silicon crystals.

Examples. The fact that the 'apparent' perfection of a crystal depends on the conditions of diffraction is illustrated, for instance, by the following experience by Freund (1973, 1990). He studied the variation with wavelength of the integrated intensity of the 222 reflection of several copper crystals with different degrees of perfection: samples $B1$ to $B4$ were Bridgman crystals with dislocation densities 2×10^4, 4×10^4, 7×10^4 and 2×10^7 dislocations/cm^2, respectively, and sample $C1$ was a dislocation-free Czochralski crystal (Fig. 4.26). The wavelength ranged from 0.03 to 1.66 Å (the short wavelengths were γ-rays). Sample $C1$ behaves like a perfect crystal for wavelengths greater than 0.1 Å but seems quite imperfect for wavelengths smaller than about 0.07 Å. Samples $B1$, $B2$ and $B3$ behave nearly like a perfect crystal for wavelengths above the K absorption edge of copper whereas for a wavelength of about 0.01 Å they behave like an ideal mosaic crystal, with every intermediary stage in between. Sample $B4$ behaves essentially like a mosaic crystal. These results can be interpreted by noting that the extinction distance increases with decreasing wavelength (4.26): the 'coherent' domains in the samples appear therefore smaller and smaller as the wavelength becomes shorter and all the samples behave more and more like imperfect crystals.

4.12 Dynamical diffraction by quasicrystals

Quasicrystals are materials characterized by the absence of translational symmetry but by the presence of rotational symmetry. Since they present no translational symmetry, they can be invariant with respect to a five-fold or a ten-fold rotation axis. Indeed, they were discovered because Shechtman *et al.* (1984) observed sharp diffraction peaks arranged in a pattern presenting icosahedral symmetry with a sample of rapidly quenched Al_6Mn alloy. If such a crystal is set for a Bragg diffraction corresponding to a sharp peak, expression (4.19) holds and a set of standing waves is expected to take place. One therefore expects anomalous transmission to be present as predicted by Berenson and Birman (1986). Chung and Durbin (1995) confirmed this prediction by calculating by means of Darwin's theory the diffraction properties of a hypothetical crystal constituted by the stacking of A and B atomic planes arranged in a Fibonacci sequence *AB AA BA BA* .... The diffraction of X-rays by Fibonacci superlattices was studied by Mikulik *et al.* (1995) using a semi-kinematical approximation of the dynamical theory. At the end of the eighties, a new class of quasicrystals constituted by ternary alloys, Al–Cu–Fe, Al–Cu–Ru, Al–Pd–Mn, etc., were discovered which could be prepared without any sign of disorder (see, for instance, Guryan

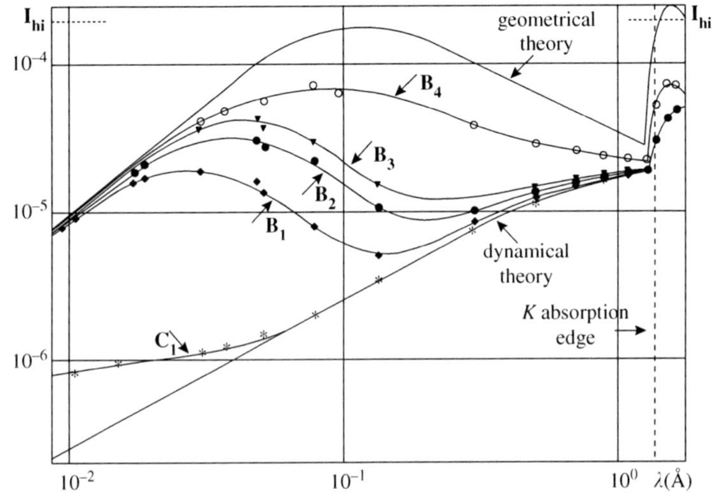

Fig. 4.26. Variation with wavelength of the integrated intensity of the 222 reflection from several copper crystals with different degrees of perfection (Courtesy A. Freund).

et al. 1989; Bancel 1989). With such crystals, Kycia *et al.* (1993, 1995[2]) were able to observe anomalous transmission as well as evidence of Pendellösung fringes. In a quasicrystal, there is no periodic set of lattice planes perpendicular to the diffraction vector, as in a usual crystal; the anomalous transmission observed is therefore not as strong but is nevertheless quite visible; it was further measured in Al-Mn-Pd samples by Härtwig *et al.* (2001). The standing wave pattern was observed by Jach *et al.* (1998, 1999) and Cappello *et al.* (2000). The diffraction diagram of a quasicrystal contains many spots, in particular very many weak spots and n-beam diffraction is very frequent. It has been studied systematically by Lee *et al.* (1993, 1996), Colella *et al.* (1994), Eisenhower *et al.* (1998) and Zhang *et al.* (1998). The Debye-Waller factor of Al-Pd-Mn was determined by Colella *et al.* (2001) from measurements made for several reflections between room temperature and below 20 K. The crystal perfection of quasicrystals was studied with synchrotron white beam topography by Wang *et al.* (1994), Baruchel *et al.* (1994, 2000), Gastaldi *et al.* (1995, 1999, 2003) and Mancini *et al.* (1998*b*) and with synchrotron monochromatic beam topography by Reinier *et al.* (1998) and Klein *et al.* (2001).

[2] *http://www.chess.cornell.edu/Newsletter_1995/quasicrystal.html.*

II

Advanced dynamical theory

5

Properties of wavefields

5.1 Relations between the field vectors

The field vectors satisfy Maxwell's equations (2.53) and the propagation equations (2.54) for the electric field and (2.55) for the electric displacement. The interaction term they contain, the polarizability χ, is triply periodic and its Fourier expansion is given by (2.36). As in the case of scalar waves (Sections 4.3 and 4.4), their general solutions are linear combinations of particular solutions which have the same periodicity as χ. The expansion of the electric field is given in (2.47) and those of the electric displacement and of the magnetic field are, respectively,[1]

$$\mathbf{D} = \sum_{\mathbf{h}} \mathbf{D_h} \exp(-2\pi \mathrm{i} \mathbf{K_h} \cdot \mathbf{r}) = \exp(-2\pi \mathrm{i} \mathbf{K_o} \cdot \mathbf{r}) \sum_{\mathbf{h}} \mathbf{D_h} \exp(2\pi \mathrm{i} \mathbf{h} \cdot \mathbf{r}),$$
$$\mathbf{H} = \sum_{\mathbf{h}} \mathbf{H_h} \exp(-2\pi \mathrm{i} \mathbf{K_h} \cdot \mathbf{r}) \tag{5.1}$$

with $\mathbf{K_h} = \mathbf{K_o} - \mathbf{h}$. The wavevectors are oriented *from* the reciprocal lattice points *towards* the tiepoint (see footnote 6 page 11).

By substitution of the developments (2.47) and (5.1) into Maxwell's equations and into the material relation (2.1), we obtain equations where a sum of an infinite number of terms is equal to zero. It has been shown in Section 4.5 that in equations of that type each term is equal to zero. The following relations can thus be proven, using (2.50) and (4.9):

$$\text{from div}\,\mathbf{B} = 0 \quad \Longrightarrow \quad \mathbf{K_h} \cdot \mathbf{B_h} = 0, \tag{5.2a}$$

$$\text{from div}\,\mathbf{D} = 0 \quad \Longrightarrow \quad \mathbf{K_h} \cdot \mathbf{D_h} = 0, \tag{5.2b}$$

$$\text{from curl}\,\mathbf{H} = \frac{\partial \mathbf{D}}{\partial t} \quad \Longrightarrow \quad \mathbf{D_h} = -\frac{1}{\nu} \mathbf{K_h} \wedge \mathbf{H_h}, \tag{5.2c}$$

$$\text{from curl}\,\mathbf{E} = -\frac{\partial \mathbf{B}}{\partial t} \quad \Longrightarrow \quad \mu_0 \mathbf{H_h} = \mathbf{B_h} = \frac{1}{\nu} \mathbf{K_h} \wedge \mathbf{E_h} \tag{5.2d}$$

$$\Longrightarrow \quad \mathbf{H_h} = \frac{\nu}{K_h^2} \mathbf{K_h} \wedge \mathbf{D_h}, \tag{5.2e}$$

[1] For simplification, the exponential factor including the time will be left out in the equations of the present chapter.

$$\text{from } \mathbf{D} = \epsilon_0[1+\chi]\mathbf{E} \quad \Longrightarrow \quad \mathbf{D_h} = \epsilon_0\Big[\mathbf{E_h} + \sum_{\mathbf{h}'} \chi_{h-h'}\mathbf{E_{h'}}\Big] \tag{5.2f}$$

$$\Longrightarrow \quad \mathbf{E_h} \approx \epsilon_0^{-1}\Big[\mathbf{D_h} - \sum_{\mathbf{h}'} \chi_{h-h'}\mathbf{D_{h'}}\Big], \tag{5.2g}$$

where $\mathbf{B_h}$ is a component of the expansion of the magnetic induction.

These relations show that the components of electric displacement, $\mathbf{D_h}$, are transverse waves: $\mathbf{D_h} \perp \mathbf{K_h} \perp \mathbf{H_h}$ (Fig. 5.1) while this is not the case for the components of the electric field. This is due to the fact that div $\mathbf{D} = 0$, but div $\mathbf{E} \neq 0$. The 'h' component of the electric field is, however, very nearly parallel to the corresponding component of the electric displacement because the Fourier components of the polarizability are so small. From (5.2c) and (5.2d), and using (2.50), it is easy to deduce that

$$\mathbf{E_{h[h]}} = \frac{k^2}{\epsilon_0 K_h^2}\mathbf{D_h}, \tag{5.3}$$

where $\mathbf{E_{h[h]}}$ is the component of $\mathbf{E_h}$ on the plane normal to $\mathbf{K_h}$ (Fig. 5.1).

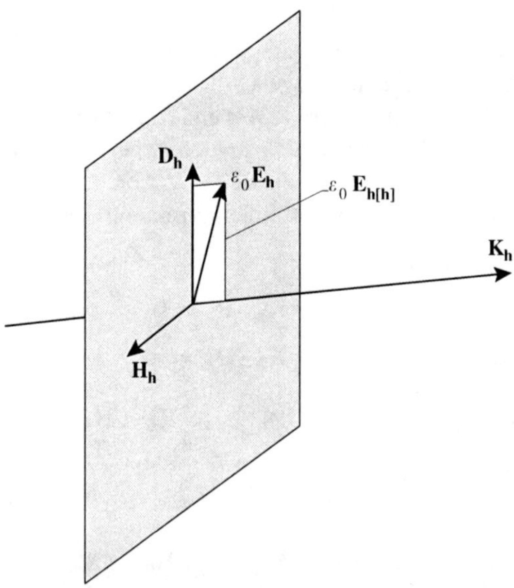

Fig. 5.1. Respective orientations of the electric field, $\mathbf{E_h}$, the electric displacement, $\mathbf{D_h}$, the magnetic field, $\mathbf{H_h}$, and the wavevector $\mathbf{K_h}$. $\mathbf{D_h}$ is a transverse wave, $\mathbf{E_h}$ is not.

5.2 Fundamental equations of the dynamical theory

The fundamental equations of dynamical theory are obtained by substituting the expansions (2.36) and (2.47) or (5.1) into the propagation equations (2.54) or (2.55), respectively. Using the relations (2.51) and (4.9), one obtains, *without* approximation for the electric field

$$\mathbf{D_h} = \frac{K_h^2}{K_h^2 - k^2} \sum_{\mathbf{h'}} \chi_{h-h'} \mathbf{D_{h'[h]}}, \tag{5.4}$$

$$\frac{K_h^2 \mathbf{E_{h[h]}} - k^2 \mathbf{E_h}}{k^2} = \sum_{\mathbf{h'}} \chi_{h-h'} \mathbf{E_{h'}}. \tag{5.5}$$

Either of the two sets of equations can be used to derive the equations of dynamical theory: Laue (1931, 1960) used the electric displacement and Miyake and Ohtsuki (1974) used the electric field. These equations are infinite sets of linear homogeneous equations and there is no rigorous solution. However, as was first noticed by Ewald in 1913, only those terms for which the resonance factors $1/(K_h^2 - k^2)$ take important values contribute significantly to the solution. As we have seen in Section 4.5, these terms are associated with the reciprocal lattice points which are very close to the Ewald sphere. If there are p such points, it is therefore enough to consider sets of p equations and it is of course sufficient to solve one set. They are *vector* equations and each one is equivalent to a set of $3p$ *scalar* equations. This is a severe complication with respect to the case of electron or neutron diffraction which imply scalar waves. The fact that the components of the electric displacement are transverse waves introduces a simplification because it provides further information about their orientation. The set of $3p$ scalar equations for the components of the electric displacement are thus reduced to a set of $2p$ equations with $2p$ unknowns. The same thing can be done, with an approximation, for the components of the electric field. In the two-beam case the solution can be decomposed into two solutions, one for σ-polarization and the other for π-polarization. When p is larger than 2, the solution is more complicated, and is discussed in Chapter 9.

The unknowns of the sets of $2p$ equations are the amplitudes of the waves, and for non-trivial solutions to exist, the determinant associated with either set must be zero. The secular equation thus obtained relates the lengths of the wavevectors of the wavefield. It is the equation of the dispersion surface, of degree $4p$, the *dispersion equation*.

If the crystal is absorbing, which is the usual case in practice, the Fourier components of the polarizability have an imaginary part, as is shown in Section 2.7. The amplitudes of the field vectors and the wavevectors also have an imaginary part and the sets of equations (5.4) and (5.5) must be solved as sets of complex equations. They may be decomposed into two subsets: one for the real part and the other for the imaginary part. To a first-order approximation, when the imaginary parts are small, the subsets for the real part are identical to eqns (5.4) or (5.5), where the form factor is replaced by its real part in χ_h and $\mathbf{K_h}$ and $\mathbf{D_h}$ (or $\mathbf{E_h}$) are replaced by their real parts $\mathbf{K_{hr}}$ and $\mathbf{D_{hr}}$ (or $\mathbf{E_{hr}}$).

5.3 The dispersion equation in the two-beam case

The *two-beam* case corresponds to a situation where two reciprocal lattice nodes only lie simultaneously on the Ewald sphere ($p = 2$). A wavefield can then be written as

$$\begin{aligned} \mathbf{D} &= \mathbf{D_o} \exp(-2\pi \mathrm{i} \mathbf{K_o} \cdot \mathbf{r}) + \mathbf{D_h} \exp(-2\pi \mathrm{i} \mathbf{K_h} \cdot \mathbf{r}), \\ \mathbf{E} &= \mathbf{E_o} \exp(-2\pi \mathrm{i} \mathbf{K_o} \cdot r) + \mathbf{E_h} \exp([-2\pi \mathrm{i} \mathbf{K_h} \cdot \mathbf{r}), \end{aligned} \tag{5.6}$$

and the two sets of fundamental equations become

$$\begin{aligned} \mathbf{D_o}(K_o^2 - k^2) &= K_o^2(\chi_o \mathbf{D_{o[o]}} + \chi_{\bar{h}} \mathbf{D_{h[o]}}), \\ \mathbf{D_h}(K_h^2 - k^2) &= K_h^2(\chi_h \mathbf{D_{o[h]}} + \chi_o \mathbf{D_{h[h]}}), \\ K_o^2 \mathbf{E_{o[o]}} - k^2 \mathbf{E_o} &= k^2(\chi_o \mathbf{E_o} + \chi_{\bar{h}} \mathbf{E_h}), \\ K_h^2 \mathbf{E_{h[h]}} - k^2 \mathbf{E_h} &= k^2(\chi_h \mathbf{E_o} + \chi_o \mathbf{E_h}). \end{aligned}$$

One may note that since the electric displacement is a transverse wave,

$$\mathbf{D_{o[o]}} = \mathbf{D_o}; \qquad \mathbf{D_{h[h]}} = \mathbf{D_h}.$$

The fundamental equations are solved separately for the two directions of polarization, normal and parallel to the scattering plane.

1. *σ-polarization*: $\mathbf{D_o}$, $\mathbf{D_h}$, $\mathbf{E_o}$ and $\mathbf{E_h}$ are perpendicular to the scattering plane and

$$\mathbf{D_{o[h]}} = \mathbf{D_o}; \qquad \mathbf{D_{h[o]}} = \mathbf{D_h}; \qquad \mathbf{E_{o[o]}} = \mathbf{E_o}; \qquad \mathbf{E_{h[h]}} = \mathbf{E_h}.$$

2. *π-polarization*: Figure 5.2 shows that we have rigorously for the electric displacement and *approximately* for the electric field (since div $\mathbf{E}$ is not very different from 0):

$$\mathbf{D_{o[h]}} = D_o \cos 2\theta_B \frac{\mathbf{D_h}}{D_h}; \qquad \mathbf{D_{h[o]}} = D_h \cos 2\theta_B \frac{\mathbf{D_o}}{D_o},$$

$$\mathbf{E_{o[o]}} = \mathbf{E_o}; \qquad \mathbf{E_{h[h]}} = \mathbf{E_h}.$$

If we introduce the polarization coefficient, $C = 1$ for σ-polarization and $C = \cos 2\theta_B$ for π-polarization, the two sets of fundamental equations can be written, after projection on $\mathbf{D_o}$, $\mathbf{D_h}$, $\mathbf{E_o}$ and $\mathbf{E_h}$, respectively, for equations (5.7a), (5.7b),

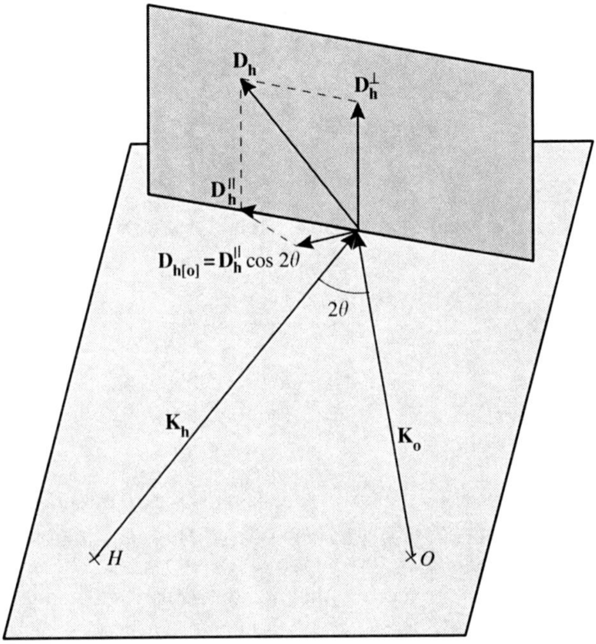

Fig. 5.2. Polarization coefficient: $\mathbf{D}_\mathbf{h}^\perp$, σ-polarization ($C = 1$); $\mathbf{D}_\mathbf{h}^\parallel$, π-polarization ($C = \cos 2\theta_B$); $\mathbf{D}_{\mathbf{h}[\mathbf{o}]}^\parallel = D_h^\parallel \cos 2\theta_B$, projection of $\mathbf{D}_\mathbf{h}^\parallel$ on the plane normal to $\mathbf{K}_\mathbf{o}$.

(5.8a) and (5.8b):

$$\left[\frac{K_o^2 - k^2}{K_o^2} - \chi_o\right] D_o - C\chi_{\bar{h}} D_h = 0, \tag{5.7a}$$

$$-C\chi_h D_o + \left[\frac{K_h^2 - k^2}{K_h^2} - \chi_o\right] D_h = 0, \tag{5.7b}$$

$$[K_o^2 - k^2(1 + \chi_o)]E_o - k^2 C\chi_{\bar{h}} E_h = 0, \tag{5.8a}$$

$$-k^2 C\chi_h E_o + [K_h^2 - k^2(1 + \chi_o)]E_h = 0. \tag{5.8b}$$

The denominator, K_o^2 or K_h^2, in eqns (5.7) can be replaced, to a small approximation, by k^2 and that set then takes the same form as the set for the components of electric field:

$$\begin{aligned} &[K_o^2 - k^2(1 + \chi_o)]D_o - k^2 C\chi_{\bar{h}} D_h = 0, \\ &-k^2 C\chi_h D_o + [K_h^2 - k^2(1 + \chi_o)]D_h = 0. \end{aligned} \tag{5.9}$$

The set (5.9) has also the same form as the set (4.14) obtained in the case of scalar waves. The equation of dispersion is obtained by setting the determinant equal to zero.

It is the same whether it is deduced using the electric field, the electric displacement or a scalar wave, and is identical to eqn (4.15):

$$X_o X_h = \frac{k^2}{4} C^2 \chi_h \chi_{\bar{h}},$$

where X_o and X_h have been defined by (4.12) and can be written as

$$\begin{aligned} X_o &= \frac{K_o^2 - k^2(1+\chi_o)}{2k}, \\ X_h &= \frac{K_h^2 - k^2(1+\chi_o)}{2k}. \end{aligned} \tag{5.10a}$$

The dispersion surface consists of two sheets, one for each direction of polarization; each sheet is a fourth-degree surface, connecting the spheres of radius nk centred at O and H and corresponding to the propagation of a single wave. The intersection of the dispersion surface with the scattering plane is represented schematically in Figs 5.3 and 5.4. When the angle of incidence of the beam on the crystal surface is quite different from the critical angle for specular reflection or far from $\pi/2$, the dispersion surface can be approximated by a second-degree surface: close to their intersection, the two spheres can be replaced by their tangential planes T_o and T_h (Fig. 5.4), X_o and X_h by the distances (4.13) of the tiepoint from T_o and T_h, and the intersection of the dispersion surface with the scattering plane consists of two sets of two hyperbolae of diameter $k|C|\sqrt{\chi_h \chi_{\bar{h}}}/\cos\theta_B$ ($C = 1$ or $\cos 2\theta_B$) (4.16), one around each of the two Lorentz points, L_o and L'_o. The former corresponds to the h, k, l reflection and the latter to the $-h, -k, -l$ reflection.

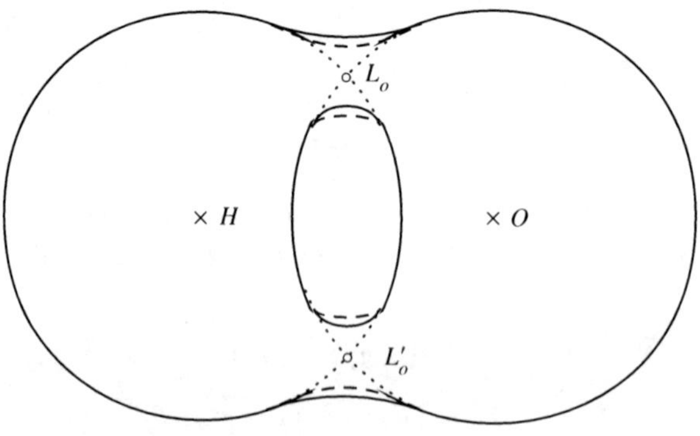

Fig. 5.3. Dispersion surface. Solid line: σ-polarization ($C = 1$); dashed line: π-polarization ($C = \cos 2\theta_B$). L_o: Lorentz point for the hkl reflection; L'_o: Lorentz point for the $\bar{h}\bar{k}\bar{l}$ reflection.

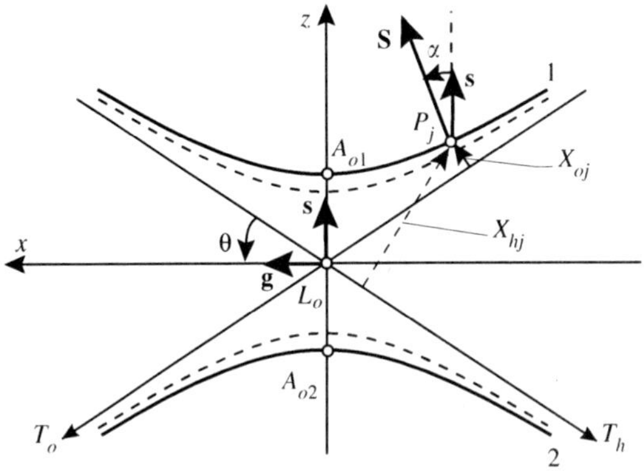

Fig. 5.4. Dispersion surface for the σ-polarization (solid curve) and for the π-polarization (dashed curve), close-up; **S**: Poynting vector; P_j: tiepoint of a wavefield; X_{oj} and X_{hj}: coordinates of the tiepoint; A_{o1}, A_{o2}: vertices of the dispersion surface; **s**, **g** unit vectors parallel and perpendicular to the lattice planes, respectively.

The relation between the amplitude ratio ξ_j of the reflected to the refracted wave and the coordinates X_{oj} and X_{hj} of the tiepoint is given by (4.17)

$$\xi_j = \frac{D_{hj}}{D_{oj}} = \frac{2X_{oj}}{k\chi_{\bar{h}}}. \tag{5.10b}$$

A generalized solution of system (5.8) for any Bragg angle has been proposed by De Caro *et al.* (1997) but it is only valid for $\sigma-$polarization (in which case system (5.8) is without approximation) and a general computation method for solving system (5.7) has been given by Huang and Dudley (2003); the exact formulation for π-polarization is given in Cho *et al.* (2004). Highly asymmetric reflections Highly asymmetric reflections for which the shape of the fourth-degree surface is needed are discussed in Chapter 8 and the situation occurring when the Bragg angle is close to $\pi/2$ in Section 5.9. The dispersion surface when more than two beams are excited (more than two reciprocal lattice points on the Ewald sphere) will be discussed in Chapter 9. A survey of the various approximations used in dynamical theory is given in Holý and Fewster (2003). In absorbing crystals, the dispersion surface is complex - see Authier (2008).

5.4 Poynting vector of the wavefields

The direction of propagation of an electromagnetic wave is given by its Poynting vector. The time-averaged Poynting vector is

$$\mathbf{S} = \mathcal{R}e(\mathbf{E} \wedge \mathbf{H}^*), \tag{5.11a}$$

where $\mathcal{R}e(x)$ means real part of x and x^* is the complex conjugate of x. Using expansion (2.47) of the electric field, (5.1) of the magnetic field, (4.9) and (4.29), it follows that

$$\mathbf{S} = \exp[4\pi \mathbf{K}^{\mathbf{i}}_{\mathbf{o}} \cdot \mathbf{r}]\, \mathcal{R}e\left(\sum_{\mathbf{h}} \sum_{\mathbf{h}'} \mathbf{E}_{\mathbf{h}} \wedge \mathbf{H}^*_{\mathbf{h}'} \exp[2\pi \mathrm{i}(\mathbf{h} - \mathbf{h}') \cdot \mathbf{r}] \right).$$

The sum on the right-hand side contains constant factors such as $\mathbf{E}_{\mathbf{h}} \wedge \mathbf{H}^*_{\mathbf{h}'}$ and rapidly oscillating factors $\exp[2\pi \mathrm{i}(\mathbf{h} - \mathbf{h}') \cdot \mathbf{r}]$. The average Poynting vector is obtained by integration over the unit cell. The attenuation factor $\exp(4\pi \mathbf{K}^{\mathbf{i}}_{\mathbf{o}} \cdot \mathbf{r})$ is slowly varying and may be taken as constant in the integration and

$$\langle \mathbf{S} \rangle = \exp(4\pi \mathbf{K}^{\mathbf{i}}_{\mathbf{o}} \cdot \mathbf{r}) \sum_{\mathbf{h}} \mathcal{R}e(\mathbf{E}_{\mathbf{h}} \wedge \mathbf{H}^*_{\mathbf{h}}).$$

If we replace $\mathbf{H}_{\mathbf{h}}$ by its expression (5.2e) and $\mathbf{E}_{\mathbf{h}}$ by its approximate value $\mathbf{D}_{\mathbf{h}}/\epsilon_0$, we obtain finally

$$\langle \mathbf{S} \rangle \approx \frac{c}{\epsilon_0} \exp(4\pi \mathbf{K}^{\mathbf{i}}_{\mathbf{o}} \cdot \mathbf{r})[\mathbf{s}_{\mathbf{o}} |D_o|^2 + \mathbf{s}_{\mathbf{h}} |D_h|^2], \tag{5.11b}$$

where $\mathbf{s}_{\mathbf{o}}$ and $\mathbf{s}_{\mathbf{h}}$ are unit vectors in the incident and reflected directions, respectively.

The geometrical construction of the Poynting vector is represented in Fig. 1.8. Let $\mathbf{s}$ and $\mathbf{g}$ be unit vectors parallel and perpendicular to the lattice planes, respectively (Fig. 5.4). The projections of eqn (5.11b) on these vectors are

$$\begin{aligned} \langle \mathbf{S} \rangle \cdot s &= \langle S \rangle \cos\alpha = \frac{c}{\epsilon_0} \exp(4\pi \mathbf{K}^{\mathbf{i}}_{\mathbf{o}} \cdot \mathbf{r})[|D_o|^2 + |D_h|^2] \cos\theta_B, \\ \langle \mathbf{S} \rangle \cdot g &= \langle S \rangle \sin\alpha = \frac{c}{\epsilon_0} \exp(4\pi \mathbf{K}^{\mathbf{i}}_{\mathbf{o}} \cdot \mathbf{r})[|D_o|^2 - |D_h|^2] \sin\theta_B. \end{aligned} \tag{5.11c}$$

From these equations it follows easily that

$$p = \frac{\tan\alpha}{\tan\theta_B} = \frac{|D_o|^2 - |D_h|^2}{|D_o|^2 + |D_h|^2} = \frac{1 - |\xi|^2}{1 + |\xi|^2}, \tag{5.12a}$$

where $\alpha = (\mathbf{s}, \mathbf{S})$ is the angle between the Poynting vector and the lattice planes (Fig. 5.4) and ξ is the amplitude ratio D_h/D_o. The variations of p with ξ_r are represented in Fig. 5.5. When ξ is zero or $\pm\infty$, the tiepoint is very close to the asymptotes to the dispersion surface, and $\alpha = \pm\theta_B$. When the tiepoint is at one of the apices of the dispersion surface, $\xi_r = \pm 1$ and $\alpha = 0$.

Conversely, it is possible to deduce the ratio ξ from the slope of the wavefield:

$$|\xi|^2 = \frac{1 - p}{1 + p} = \frac{\sin(\theta_B - \alpha)}{\sin(\theta_B + \alpha)}. \tag{5.12b}$$

In order to show that the Poynting vector is normal to the dispersion surface at the tiepoint, we shall represent the hyperbola with respect to a set of axes $\mathbf{L_o x}$ and $\mathbf{L_o z}$ respectively parallel to the reciprocal lattice vector and to the lattice planes (Fig. 5.4); its equation is

$$-x^2 \sin^2\theta_B + z^2 \cos^2\theta_B = \frac{k^2}{4} C^2 \chi_h \chi_{\bar{h}}$$

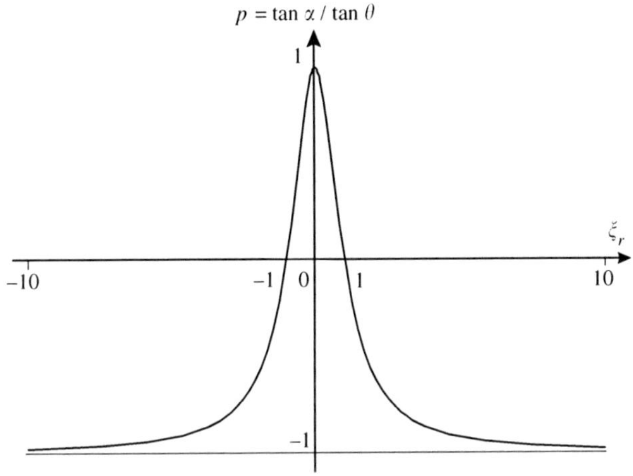

Fig. 5.5. Variations of the slope, p, of a wavefield with the real part of the ratio, ξ_r, of the amplitudes of the two waves which constitute it.

with

$$x = \frac{X_o - X_h}{2\sin\theta_B},$$

$$z = \frac{X_o + X_h}{2\cos\theta_B}.$$

The slope of the normal to the hyperbola at P with respect to the lattice planes is

$$\tan\alpha = -\frac{\mathrm{d}z}{\mathrm{d}x} = -\frac{x\sin^2\theta_B}{z\cos^2\theta_B} = \frac{X_h - X_o}{X_h + X_o}\tan\theta_B. \tag{5.13}$$

By substitution from the expressions (4.17) of X_o and X_h in terms of ξ, this expression becomes identical with (5.12a), thus proving that the Poynting vector is normal to the dispersion surface. The variations of its orientation across the reflection domain are illustrated in Fig. 4.12. The Poynting vector is parallel to the group velocity as will be shown in Section 12.5.

5.5 Determination of the tiepoints—geometrical interpretation of the deviation parameter

5.5.1 *Boundary condition for the wavevectors*

The tiepoints are obtained by applying the condition of the continuity of the tangential component of the wavevectors. This is done by looking for the intersections with the

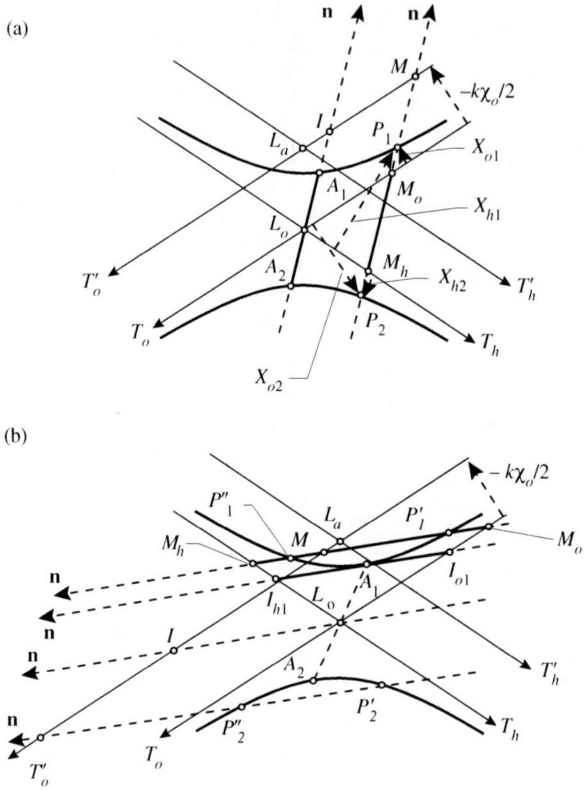

Fig. 5.6. Determination of the tiepoints of the wavefields created inside the crystal by the incident plane wave; M: excitation point; **n**: normal to the crystal surface. (a) Transmission geometry; $\overline{A_2A_1}$: inverse of the Pendellösung distance; geometrical interpretation of the deviation parameter: $\eta = \overline{M_oM_h}/\overline{A_2A_1}$. (b) Reflection geometry; $\overline{I_{o1}I_{h1}}$: inverse of the extinction distance. Geometrical interpretation of the deviation parameter: $\eta = \overline{M_oM_h}/\overline{I_{01}I_{h1}}$.

dispersion surface of the normal to the crystal surface drawn from the extremity of the wavevector in vacuum, **OM**. They lie on different branches of the dispersion surface in the transmission geometry (P_1, and P_2, Fig. 5.6(a)) and on the same branch in the reflection geometry (P'_1, P''_1, or P'_2, P''_2, Fig. 5.6(b)).

Let M_o and M_h be the intersections of that normal with the tangents T_o and T_h to the spheres centred in O and H and with radius nk, respectively. The coordinates X_{oj} and X_{hj} are the distances of the tiepoint to T_o and T_h; they are illustrated in Fig. 5.6(a) only, for clarity, but the derivation is the same for both geometries. They are related to $\overline{M_oP_j}$ and $\overline{M_hP_j}$ through

$$\frac{X_{oj}}{\gamma_o} = \overline{M_oP_j}; \qquad \frac{X_{hj}}{\gamma_h} = \overline{M_hP_j},$$

where γ_o and γ_h are the cosines of the angles ψ_o and ψ_h between the normal to the crystal surface and incident and reflected directions, respectively (see Section 4.7.3). From these relations, it follows that

$$\frac{X_{oj}}{\gamma_0} - \frac{X_h}{\gamma_h} = \overline{M_o M_h}.$$

It can be seen in the figure that

$$\overline{M_o M_h} = \overline{L_o M_o} \frac{\sin 2\theta_B}{\gamma_h}.$$

It has been shown in Section 4.7.2 that the departure from Bragg's incidence, $\Delta\theta$, is equal to $\overline{L_a M}/k$, where L_a is the Laue point. It has also been shown that the incident wavevector corresponding to the middle of the reflection domain is **OI**, where I is at the intersection of the normal to the crystal surface drawn from the Lorentz point L_o with the tangent T'_o to the sphere centred in O and with radius k and we have set $\Delta\theta_{os} = \overline{L_a I}/k$ (eq. 4.24). Noting that

$$\overline{L_o M_o} = \overline{IM} = \overline{L_a M} - \overline{L_a I}$$

and that, therefore,

$$\overline{M_o M_h} = k(\Delta\theta - \Delta\theta_{os}) \frac{\sin 2\theta_B}{\gamma_h}, \tag{5.14}$$

we obtain the equation of the normal to the crystal surface, drawn from M:

$$\frac{X_o}{\gamma_o} - \frac{X_h}{\gamma_h} = k(\Delta\theta - \Delta\theta_{os}) \frac{\sin 2\theta_B}{\gamma_h}. \tag{5.15}$$

Angle $\Delta\theta - \Delta\theta_{os}$ is the angle between the incident wavevector and the wavevector corresponding to the middle of the reflection domain.

5.5.2 *Deviation from Bragg's angle of the middle of the reflection domain*

The fact that the wavevector corresponding to the middle of the reflection domains is different for the geometrical and dynamical theories of diffraction is due to the fact that the effect of refraction is neglected in the geometrical theory of diffraction and was observed very early on (see Section 1.7). The angle between the corresponding directions, $\Delta\theta_{os}$, is called the *deviation from Bragg's angle of the middle of the reflection domain* and it is easy to relate it to the index of refraction. Let X_{oI} and X_{hI} be the distances from I to T_o and T_h (Fig. 5.7). The distance between T_o and T'_o (and between T_h and T'_h) is the difference between the radii nk and k and is $-k\chi_o/2$ (χ_o is negative).

It follows from the figure that

$$X_{oI} = \frac{-k\chi_o}{2} = \gamma_o \overline{L_o I},$$

$$X_{hI} = \frac{-k\chi_o}{2} - \overline{L_a I} \sin 2\theta_B = \gamma_h \overline{L_o I}.$$

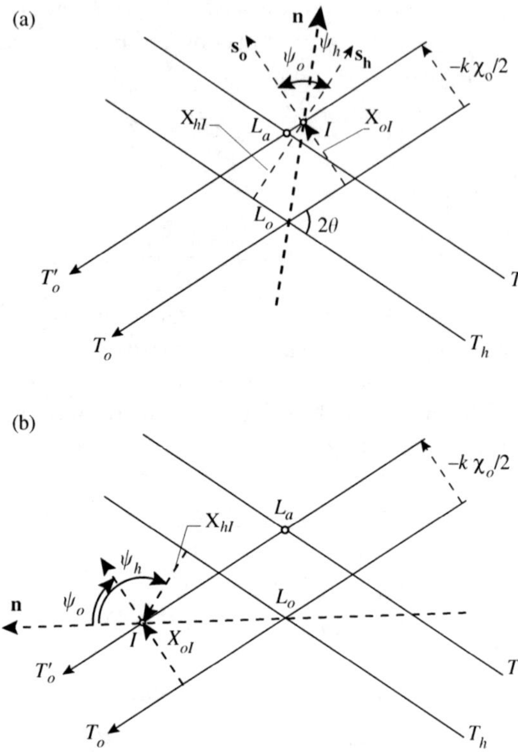

Fig. 5.7. Deviation from Bragg's angle of the middle of the reflection domain: I is the excitation point of the wave corresponding to the middle of the reflection domain; it differs from the position given by the geometrical theory, L_a, centre of the Ewald sphere, by $k\Delta\theta_{os} = \overline{L_a I}$. (a) Laue, transmission geometry. (b) Bragg, reflection geometry.

By elimination of $\overline{L_o I}$ between these two relations we obtain

$$\Delta\theta_{os} = \frac{\overline{L_a I}}{k} = -\frac{\chi_o(1-\gamma)}{2\sin 2\theta_B}, \tag{5.16}$$

where $\gamma = \gamma_h/\gamma_o$ is the asymmetry ratio. In the Laue geometry (Fig. 5.7(a)), γ is positive and the deviation from Bragg's angle of the middle of the reflection domain is zero for a symmetric reflection ($\gamma = 1$). In the Bragg geometry (Fig. 5.7(b)), γ is negative and there is always a deviation from Bragg's angle, even for a symmetric reflection.

5.5.3 *Coordinates of the tiepoint*

The coordinates of the tiepoint are obtained by combining the equation of the dispersion surface, (4.15), and the equation of the normal to the crystal surface, (5.15).

Equation (4.15) can be rewritten as

$$\frac{X_{oj}X_{hj}}{\gamma_o\gamma_h} = \frac{S(\gamma_h)}{4}\frac{k^2C^2\chi_h\chi_{\bar{h}}}{\gamma_o|\gamma_h|} = \frac{S(\gamma_h)}{4\Lambda_o^2}, \tag{5.17a}$$

where $S(\gamma_h)$ means sign of γ_h and

$$\Lambda_o^{-1} = \frac{k|C|\sqrt{\chi_h\chi_{\bar{h}}}}{\sqrt{\gamma_o|\gamma_h|}}, \tag{5.17b}$$

which, as was mentioned in Section 4.7.4, is called *Pendellösung distance* in the transmission geometry and *extinction distance* in the reflection geometry.

Let us set

$$\eta = (\Delta\theta - \Delta\theta_{os})\frac{\Lambda_o \sin 2\theta_B}{\lambda|\gamma_h|} = \Lambda_o S(\gamma_h)\overline{M_o M_h}. \tag{5.18a}$$

This is the *deviation parameter* introduced in Section 4.7.4:

$$\eta = \frac{\Delta\theta - \Delta\theta_{os}}{\delta_{os}}, \tag{5.18b}$$

where

$$\delta_{os} = \frac{\lambda|\gamma_h|}{\Lambda_o \sin 2\theta_B} \tag{5.18c}$$

is half the Darwin width.

Equation (5.15) can be rewritten, using (5.18), as

$$\frac{X_{oj}}{\gamma_o} - \frac{X_{hj}}{\gamma_h} = S(\gamma_h)\frac{\eta}{\Lambda_o}. \tag{5.19}$$

From (5.17a) and (5.19) it is easy to obtain the expressions of the coordinates of the tiepoint in terms of the deviation parameter:

$$\begin{aligned} X_{oj} &= \frac{\gamma_o S(\gamma_h)}{2\Lambda_o}\left[\eta \pm \sqrt{\eta^2 + S(\gamma_h)}\right], \\ X_{hj} &= \frac{|\gamma_h|}{2\Lambda_o}\left[-\eta \pm \sqrt{\eta^2 + S(\gamma_h)}\right]. \end{aligned} \tag{5.20}$$

1. *Transmission geometry*: γ_h is positive and (5.20) takes the form

$$\begin{aligned} X_{oj} &= \frac{\gamma_o}{2\Lambda_o}\left[\eta \pm \sqrt{\eta^2 + 1}\right], \\ X_{hj} &= \frac{\gamma_h}{2\Lambda_o}\left[-\eta \pm \sqrt{\eta^2 + 1}\right], \end{aligned} \tag{5.21a}$$

where the plus sign corresponds to branch 1 ($j = 1$) and the minus sign to branch 2 ($j = 2$). The slope (5.13) of the Poynting vector is, in the symmetrical case,

$$\tan\alpha = \frac{\mp\eta}{\sqrt{\eta^2+1}}\tan\theta_B. \tag{5.21b}$$

2. *Reflection geometry*: γ_h is negative and (5.20) takes the form

$$\begin{aligned} X_{oj} &= -\frac{\gamma_o}{2\Lambda_o}\left[\eta \pm \sqrt{\eta^2-1}\right], \\ X_{hj} &= -\frac{\gamma_h}{2\Lambda_o}\left[-\eta \pm \sqrt{\eta^2-1}\right]. \end{aligned} \tag{5.22a}$$

Here, the plus and minus signs are associated with solutions belonging to the *same* branch of the dispersion surface: $j =' 1$ or $j =' 2$ for the plus sign and $j ='' 1$ or $j ='' 2$ for the minus sign. The directions of propagation of the corresponding wavefields are given by the slopes (5.13) of their respective Poynting vectors. One finds, in the symmetrical case, that

$$\tan\alpha = \frac{\mp\sqrt{\eta^2-1}}{\eta}\tan\theta_B. \tag{5.22b}$$

For branch 1, η is negative: $j =' 1$ (*plus* sign) corresponds to a tiepoint for which the slope of the Poynting vector is positive, which means propagation towards the inside of the crystal, and $j ='' 1$ (*minus* sign) corresponds to a wavefield propagating from the bottom of the crystal towards the outside (Fig. 4.19). For branch 2, η is positive, $j =' 2$ corresponds to a wavefield propagating towards the outside of the crystal and $j ='' 2$ to a wavefield propagating towards the inside of the crystal. The same result holds for an asymmetric reflection.

When $\eta = \pm 1$, $\tan\alpha = 0$ and the propagation direction is parallel to the lattice planes. In a non-absorbing crystal, the intersections are complex for $-1 < \eta < +1$.

5.5.4 *Deviation parameter, Pendellösung distance and FWHM in the transmission geometry*

1. *Geometrical interpretations*: It has been shown in Section 4.8.6 that the period of the interferences between the two wavefields generated by an incident plane wave is the inverse of

$$\Lambda^{-1} = \overline{P_2P_1} = \overline{MP_1} - \overline{MP_2},$$

where, using (5.20),

$$\begin{aligned} \overline{MP_j} &= \frac{X_{oj} + k\chi_o/2}{\gamma_o} \\ &= \frac{k\chi_o}{2\gamma_o} + \frac{S(\gamma_h)}{2\Lambda_o}\eta \pm \frac{S(\gamma_h)}{2\Lambda_o}\sqrt{\eta^2 + S(\gamma_h)}. \end{aligned} \tag{5.23}$$

In the transmission geometry, $S(\gamma_h) = 1$ and

$$\overline{P_2P_1} = \frac{\sqrt{\eta^2+1}}{\Lambda_o}. \tag{5.24}$$

When $\eta = 0$, the normal to the crystal surface passes through L_o, and P_1 and P_2 are the vertices A_1 and A_2 of a diameter of the dispersion surface (Fig. 5.6(a)). The geometrical interpretation of the *Pendellösung* distance Λ_o (5.17b) in the Laue geometry is therefore $(\overline{A_2A_1})^{-1}$ which will be called Λ_L hereafter:

$$\Lambda_L = (\overline{A_2A_1})^{-1} = \Lambda_o \quad \text{(Laue geometry)}. \tag{5.25}$$

It follows that

$$\Lambda = \frac{\Lambda_L}{\sqrt{\eta^2+1}}; \qquad \sqrt{\eta^2+1} = \frac{\overline{P_2P_1}}{\overline{A_2A_1}} = \frac{\Lambda_L}{\Lambda}. \tag{5.26}$$

Using (5.18), it is possible to give a geometrical interpretation of the deviation parameter in the transmission geometry:

$$\eta = \frac{\overline{M_oM_h}}{\overline{A_2A_1}}. \tag{5.27}$$

From the definition (5.18) of η, it can be seen that $\eta = \pm 1$ corresponds to values of $\Delta\theta - \Delta\theta_o$ equal to $\pm\delta_{os}$, half the FWHM. It is easy to find the corresponding excitation points; let us draw the tangents to the dispersion surface at the intersections A_1 and A_2 of the dispersion surface with the normal to the entrance surface of the crystal drawn from L_o. They intersect T_o and T_h at I_{o1}, I_{h1}, I_{o2} and I_{h2}, respectively (Fig. 5.8(a)).

The lines $I_{h2}I_{o1}$ and $I_{o2}I_{h1}$ are also parallel to the normal to the entrance surface and intersect T_o' at I_1 and I_2. We have

$$-\overline{I_{o1}I_{h2}} = \overline{I_{o2}I_{h1}} = \overline{A_2A_1}$$

and, according to (5.27), I_1 and I_2 are the excitation points corresponding to $\eta = -1$ and $\eta = +1$, respectively.

Relation (4.26) between the half-width, δ_{os}, and the Pendellösung distance, Λ_L, can be obtained geometrically from Fig. 5.8(a). We have

$$2k\delta_{os} = \overline{I_1I_2} = \overline{I_{o1}I_{o2}} = 2\overline{L_oI_{o1}},$$
$$\Lambda_L^{-1} = \overline{A_2A_1} = \overline{I_{h2}I_{o1}}.$$

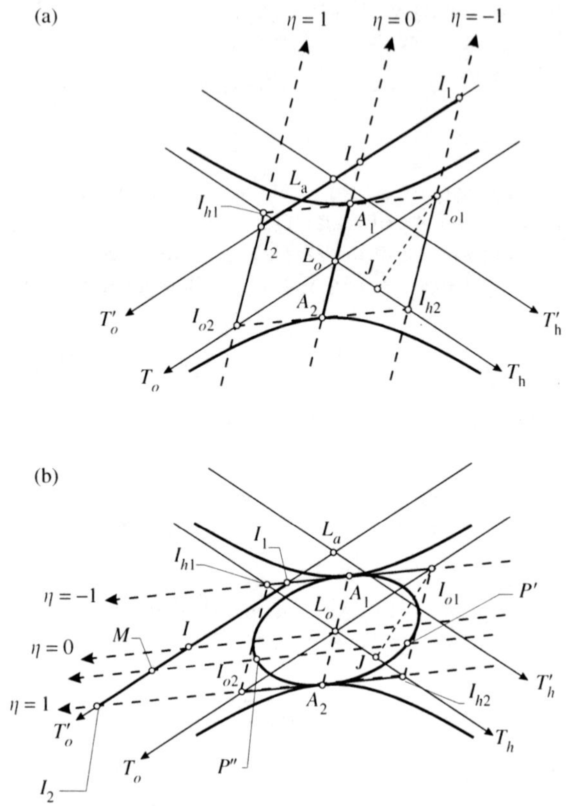

Fig. 5.8. Geometrical interpretation of the width, $\overline{I_1 I_2}$, of the reflection domain: I_1, I_2 positions of the excitation point corresponding to $\eta = \pm 1$. (a) Transmission geometry. (b) Reflection geometry.

Let J be the projection of I_{o1} on T_h, $\overline{JI_{o1}}$ is equal both to $\overline{L_o I_{o1}} \sin 2\theta_B$ and to $\overline{I_{h2} I_{o1}} \gamma_h$ and the relation between δ_{os} and Λ_L follows:

$$\delta_{os} = \Lambda_L^{-1} \frac{\gamma_h}{k \sin 2\theta_B}.$$

2. *Variations with the asymmetry of the reflection*: From Figs 5.6(a) and 5.8(a), it is clear that the Pendellösung distance (5.17b) and the half-width of the rocking curve (4.26), $\delta_{os} = \sqrt{|\gamma| \chi_h \chi_{\bar{h}}} / \sin 2\theta_B$, vary strongly with the angle ψ_n between the lattice planes and the normal to the crystal surface. The variations of $\sqrt{|\gamma|}$ are represented in Fig. 4.22. The variations of Λ_L are represented in Fig. 5.9(a) in the case of silicon 220, Mo$K\alpha$ ($\theta_B = 10.64°$), Cu$K\alpha$ ($\theta_B = 23.65°$) and 333, Cu$K\alpha$ ($\theta_B = 47.45°$). It can be seen that it is maximum for symmetric reflections and that

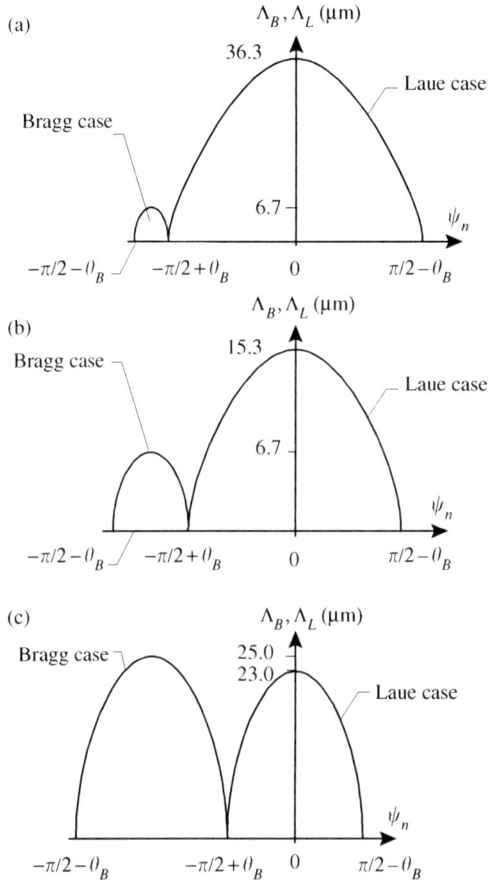

Fig. 5.9. Variations of the Pendellösung distance, Λ_L (Laue case), and of the extinction distance, Λ_B (Bragg case), with the asymmetry of the reflection; ψ_n, angle between the lattice planes and the normal to the entrance surface. (a) Silicon, 220 reflection, Mo$K\alpha$, $\theta_B = 10.64°$. (b) Silicon, 220 reflection, Cu$K\alpha$, $\theta_B = 23.65°$. (c) Silicon, 333 reflection, Cu$K\alpha$, $\theta_B = 47.45°$.

it goes down to zero for very asymmetric reflections. The value of the maximum increases with decreasing structure factors.

3. *Variations with the wavelength*: The variations of the Darwin width are discussed in Section 7.1.2 in the reflection geometry and are represented in Fig. 7.4. The expression (5.17b) of the Pendellösung distance can be written, using (2.37), as

$$\Lambda_L = \frac{\pi V \sqrt{\gamma_o |\gamma_h|}}{R\lambda |C| \sqrt{F_h F_{\bar{h}}}}$$

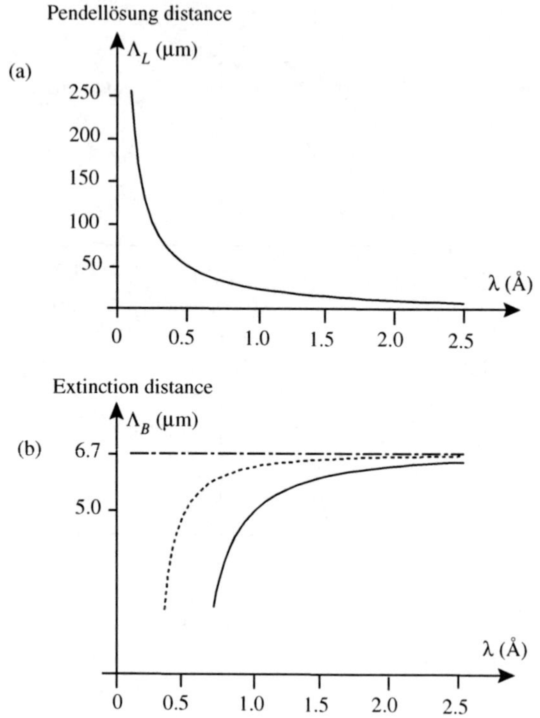

Fig. 5.10. Variations of the Pendellösung distance and of the extinction distance with wavelength. (a) Pendellösung distance; symmetric reflection, silicon, 220 reflection. (b) Extinction distance; silicon, 220 reflection; dash-dotted line: symmetric reflection ($\psi_n = -90°$); dashed line, $\psi_n = -85°$, solid line: $\psi_n = -80°$.

and, using the relations $\gamma_o = \cos(\psi_n + \theta_B)$ and $\gamma_h = \cos(\psi_n - \theta_B)$ (Section 4.7.3),

$$\Lambda_L = \frac{\pi V\sqrt{\cos^2\psi_n - \sin^2\theta_B}}{R\lambda|C|\sqrt{F_h F_{\bar{h}}}} = \frac{\pi V\sqrt{\cos^2\psi_n - \lambda^2/4d_{hkl}^2}}{R\lambda|C|\sqrt{F_h F_{\bar{h}}}},$$

where d_{hkl} is the reflecting-plane spacing.

The Pendellösung distance decreases with increasing wavelength and its variations are represented in Fig. 5.10(a) for a symmetric reflection ($\psi_n = 0$).

5.5.5 *Deviation parameter, extinction distance, penetration depth and Darwin width in the reflection geometry*

1. *Geometrical interpretation of the deviation parameter*: Expression (5.22a) of the coordinates of the tiepoints in the reflection geometry shows that, when $\eta = \pm 1$, P_1'

and P_1'', and P_2' and P_2'', respectively, merge (Fig. 5.8(b)). The normal to the entrance surface is then tangent to the dispersion surface at one of the extremities of a diameter, A_1A_2. Points M_o and M_h defined in Fig. 5.6(b) occupy then the positions I_{o1} and I_{h1} or I_{o2} and I_{o2}, respectively. The length in direct space $(\overline{I_{o1}I_{h1}})^{-1}$ is called the *extinction distance*, Λ_B. Applying the definition (5.18) of the deviation parameter shows that when $\eta = -1$, it is possible to give a geometrical interpretation of the value of Λ_o in the Bragg geometry:

$$\Lambda_o \text{ (Bragg geometry)} = (\overline{I_{o1}I_{h1}})^{-1} = \Lambda_B. \tag{5.28}$$

The interpretation of the deviation parameter is, therefore, in the reflection geometry,

$$\eta = \frac{\overline{M_oM_h}}{\overline{I_{o1}I_{h1}}} \tag{5.29}$$

(to be compared with (5.27) for the transmission geometry).

2. *Penetration depth*: It was shown in Section 4.9.2 that the penetration depth of the beam in the crystal within the total reflection domain is limited by extinction and is directly related to the imaginary part of the wavevectors. These can be written as

$$\mathbf{K_{oj}} = \mathbf{OP_j} = \mathbf{OM} + \mathbf{MP_j},$$
$$\mathbf{K_{hj}} = \mathbf{HP_j} = \mathbf{HM} + \mathbf{MP_j}.$$

When $|\eta| < 1$, they are complex. Using (5.23), it follows that

$$\mathbf{K_{oj}} = \mathbf{OM} + \left(\frac{k\chi_o}{2\gamma_o} - \frac{\eta}{2\Lambda_B} \mp \mathrm{i}\frac{\sqrt{1-\eta^2}}{2\Lambda_B}\right)\mathbf{n}$$

and a similar expression for $\mathbf{K_{hj}}$. The sign must be chosen so that the wave propagates towards the inside of the crystal.

If the crystal is *non-absorbing*, the imaginary part of the wavevectors is simply

$$K_{oj}^i = K_{hj}^i = -\frac{\sqrt{1-\eta^2}}{2\Lambda_B}$$

and the penetration depth is, using (4.44) and (4.45),

$$z_e = \frac{-1}{4\pi K_o^i} = \frac{\Lambda_B}{2\pi\sqrt{1-\eta^2}}. \tag{5.30a}$$

The minimum penetration distance is $z_e = \Lambda_B/2\pi$ and this is why Λ_B is called the extinction distance. It is possible to give a geometrical interpretation of this length by introducing a new surface, complementary of the dispersion surface (Authier 1986). From (5.23) and Fig. 5.6(b), it can be seen that, outside the domain of total reflection,

$$\overline{P_j'P_j''} = \frac{\sqrt{\eta^2-1}}{\Lambda_B}$$

(j = 1 or 2). It is a chord of the dispersion surface. Let us now consider the surface whose intersection with the diffraction plane is an ellipse tangent to the dispersion surface at A_1 and A_2 on the one hand, and $\overline{I_{h2}I_{o1}}$ and $\overline{I_{o2}I_{h1}}$ on the other hand (Fig. 5.8(b)). If the excitation point, M, lies within the total reflection domain, $\overline{I_1 I_2}$, the normal to the crystal surface drawn from M intersects that surface at P' and P''. The quantity $\sqrt{1-\eta^2}/\Lambda_B$ is then the chord $\overline{P'P''}$ of that ellipse and $z_e = \overline{P'P''}/2\pi$.

In the presence of absorption, the imaginary part of the wavevectors is, for the wave penetrating towards the inside of the crystal,

$$K^i_{oj} = K^i_{hj} = -\frac{\mu_{ej}}{4\pi\gamma_o} = \frac{k\chi_{io}}{2\gamma_o} - \mathcal{I}m\left[\frac{\eta + \mathrm{i}\sqrt{1-\eta^2}}{2\Lambda_B}\right]$$

and the penetration depth is, using (4.45a),

$$z_e = \frac{1}{\mu_o/\gamma_o + 2\pi\mathcal{I}m[(\eta + \sqrt{\eta^2 - 1})/\Lambda_B]}. \tag{5.30b}$$

The variations of the penetration depth with the angle of incidence are represented in Fig. 5.16. Its detailed expression is discussed in Section 5.8.4.

3. *Darwin width*: The same derivation applies as in the transmission geometry, using Fig. 5.8(b) this time and $|\gamma_h|$ instead of γ_h.

4. *Variation with the asymmetry of the reflection*: As in the transmission geometry, the Darwin width, the extinction distance (and therefore the penetration depth) are strongly dependent on the angle between the lattice planes and the normal to the crystal surface. The variations of the asymmetry ratio in the expression of the Darwin width are represented in Fig. 4.22 and its influence on the rocking curves is discussed in Section 4.10. The variations of the extinction distance with the asymmetry are represented in Fig. 5.9(b) for three values of the Bragg angle. As for the Pendellösung distance, the extinction distance and the penetration depth go down to zero for very asymmetric reflections. This property is put to advantage in Grazing Incidence Diffraction (Chapter 8) and in reflection topography (Chapter 17).

5. *Variations with wavelength*. The variations of the Darwin width with wavelength are discussed in Section 7.1.2. The expression of the extinction distance is deduced from (5.17b) in the same way as it was done for the Pendellösung distance:

$$\Lambda_B = \frac{\pi V\sqrt{\lambda^2/4d^2_{hkl} - \cos^2\psi_n}}{R\lambda|C|\sqrt{F_h F_{\bar{h}}}}.$$

For a symmetric reflection ($\psi_n = -\pi/2$) it reduces to

$$\Lambda_B = \frac{\pi V}{2Rd_{hkl}|C|\sqrt{F_h F_{\bar{h}}}}$$

and does not depend on the wavelength. Figure 5.10(b) shows the variations of Λ_B with wavelength for two asymmetric positions of the crystal.

5.5.6 *Index of refraction for dynamical diffraction*

The index of refraction of a wave is, by definition,

$$n = \frac{K_o}{k}.$$

Very far from the Bragg diffraction condition, when there is only one wave propagating in the crystal, it is (Section 2.7.3)

$$n = 1 + \frac{\chi_o}{2}.$$

Under the conditions of Bragg diffraction, one has

$$K_{oj} = \overline{OP_j} = k + \gamma_o \overline{MP_j}$$

and

$$n = 1 + \frac{\gamma_o \overline{MP_j}}{k},$$

where $\overline{MP_j}$ is given by (5.23). Far from the centre of the diffraction domain ($|\eta| \gg 1$, $\Delta\theta - \Delta\theta_{os} \gg \delta_{os}$), one may expand the square root in (5.23) and it follows that

$$\overline{MP_j} \approx \frac{k\chi_o}{2\gamma_o} + \frac{S(\gamma_h)}{2\Lambda_o}\eta \pm \frac{S(\gamma_h)|\eta|}{2\Lambda_o} \pm \frac{|\eta|}{2\Lambda_o \eta^2}.$$

When $\eta \Rightarrow -\infty$, the intensity of type 2 wavefields tends towards zero and type 1 wavefields merge with the incident beam. When $\eta \Rightarrow +\infty$, it is the opposite: it is the type 2 wavefields which merge into the incident beam. For these wavefields it follows that

$$\overline{MP_j} \approx \frac{k\chi_o}{2\gamma_o} - \frac{1}{2\Lambda_o \eta}$$

and, using (5.17b) and (5.18), the diffraction correction to the index of refraction is

$$\Delta n = n - \left(1 + \frac{\chi_o}{2}\right) \approx -\frac{|C|^2 \chi_h \chi_{\bar{h}}}{2(\Delta\theta - \Delta\theta_{os}) \sin 2\theta_B}.$$

It induces a phase shift in the refracted wave of wavevector $\mathbf{K_o}$ (also called forward-diffracted beam) which has been measured directly using an interferometer and synchrotron radiation by Hirano and Momose (1996).

5.6 The deviation parameter in absorbing crystals

When the crystal is absorbing, the polarizability has an imaginary part, and χ_o and χ_h are complex (Section 2.7). The quantities δ_{os}, $\Delta\theta_{os}$, Λ_B and Λ_L are also complex. So is the deviation parameter, η, which can be written as

$$\eta = \frac{\Delta\theta \sin 2\theta_B + \chi_o(1-\gamma)/2}{\sqrt{|\gamma|}|C|\sqrt{\chi_h\chi_{\bar{h}}}} = \eta_r + \mathrm{i}\eta_i, \tag{5.31}$$

where

$$\begin{aligned} \eta_r &= \frac{\Delta\theta - (\Delta\theta_{os})_r}{\delta_{sr}} \\ &= \frac{\Delta\theta \sin 2\theta_B \cos\beta + (1-\gamma)(\chi_{ro}\cos\beta + \chi_{io}\sin\beta)/2}{\sqrt{|\gamma|}|C|\sqrt{|\chi_h\chi_{\bar{h}}|}}, \\ \eta_i &= A\eta_r + B \end{aligned} \tag{5.32}$$

with

$$\begin{aligned} A &= -\tan\beta, \\ B &= \frac{(1-\gamma)\chi_{io}}{2\cos\beta\sqrt{|\gamma|}|C|\sqrt{|\chi_h\chi_{\bar{h}}|}}. \end{aligned} \tag{5.33}$$

δ_{sr} and $(\Delta\theta_{os})_r$ are the real parts of δ_{os} and $\Delta\theta_{os}$, respectively, β is the phase angle of $\sqrt{\chi_h\chi_{\bar{h}}}$ (2.41). Parameter B is equal to zero for a symmetric reflection in the transmission geometry. It is to be noted that expressions (5.32) and (5.33) include no approximation relative to the value of the absorption and are valid even if the absorption is very large. If the real part of the scattering factor is very small, β becomes close to $\pi/2$, and $|A|$ and $|B|$ take very large values.

5.7 Amplitude ratio of the refracted and reflected waves

From (5.9), (5.10), (5.17b), (5.20) and (2.40), one deduces the amplitude ratio of the reflected and refracted waves:

$$\begin{aligned} \xi_j &= \frac{D_{hj}}{D_{oj}} = \frac{2X_{oj}}{kC\chi_{\bar{h}}} = \frac{S(C)S(\gamma_h)}{\sqrt{|\gamma|}}\frac{\sqrt{\chi_h\chi_{\bar{h}}}}{\chi_{\bar{h}}}\left[\eta \pm \sqrt{\eta^2 + S(\gamma_h)}\right] \\ &= -\frac{S(C)S(\gamma_h)}{\sqrt{|\gamma|}}\sqrt{\left|\frac{\chi_h}{\chi_{\bar{h}}}\right|}\exp(\mathrm{i}\beta')\left[\eta \pm \sqrt{\eta^2 + S(\gamma_h)}\right]. \end{aligned} \tag{5.34}$$

We shall set the ψ_j phase of ξ_j:

$$\xi_j = |\xi_j|\exp(\mathrm{i}\psi_j).$$

The amplitude ratio ξ_j is characteristic of the tiepoint and is independent of the orientation of the normal to the crystal surface. It is therefore very often useful to reverse (5.34) and to express the deviation parameter in terms of the amplitude ratio and of the asymmetry ratio γ. Noting that

$$\eta \pm \sqrt{\eta^2 + S(\gamma_h)} = -\frac{S(\gamma_h)}{\eta \mp \sqrt{\eta^2 + S(\gamma_h)}},$$

it is easy to show that

$$\eta = \frac{S(C)}{2\sqrt{|\gamma|}} \frac{\chi_{\bar{h}}}{\sqrt{\chi_h \chi_{\bar{h}}}} \left[\gamma \xi_j - \frac{\chi_h}{\chi_{\bar{h}}} \frac{1}{\xi_j} \right]. \tag{5.35}$$

5.7.1 *Phase of the amplitude ratio in the transmission geometry*

The asymmetry ratio γ is positive and

$$\xi_j = \frac{S(C)}{\sqrt{\gamma}} \frac{\sqrt{\chi_h \chi_{\bar{h}}}}{\chi_{\bar{h}}} \left[\eta \pm \sqrt{\eta^2 + 1} \right]. \tag{5.36}$$

The imaginary part of η is small and $\sqrt{1+\eta^2}$ can be expanded as

$$\sqrt{1+\eta^2} = \sqrt{1+\eta_r^2} + \mathrm{i} \frac{\eta_i \eta_r}{\sqrt{1+\eta_r^2}}. \tag{5.37}$$

It follows that

$$\xi_j = -\frac{S(C)}{\sqrt{\gamma}} \sqrt{\left| \frac{\chi_h}{\chi_{\bar{h}}} \right|} \exp(\mathrm{i}\beta') \left(\eta_r \pm \sqrt{1+\eta_r^2} \right) \left[1 \pm \mathrm{i} \frac{\eta_i}{\sqrt{1+\eta_r^2}} \right].$$

This expression shows that the phase of ξ_j varies across the reflection domain, but this is a small variation only, which does not occur for non-absorbing crystals. When $\eta_r \Rightarrow \pm\infty$, $\eta_i \Rightarrow A\eta_r = -\tan\beta\eta_r$ and the last term tends towards $\exp(\pm \mathrm{i}\beta)$ (for $\beta \ll 1$). The variations of the phase are summarized in Table 5.1, using the relation $\beta' + \beta = \varphi_h$ (2.41).

Table 5.1 Variations of the phase ψ_j of the amplitude ratio—transmission geometry

η_r	$-\infty$	0	$+\infty$
Branch 1	$\pi + \varphi_h$	$\pi + \varphi_h - \beta$	$\pi + \varphi_h - 2\beta$
Branch 2	$\varphi_h - 2\beta$	$\varphi_h - \beta$	φ_h

Example. For GaAs, 400 and Cu$K\alpha$, we have

$$\chi_{rh} = -1.674 \times 10^{-5}; \qquad \chi_{ih} = -7.778 \times 10^{-7}$$

and $2\beta = 0.0928\,\text{rad} = 5.317°$.

5.7.2 *Phase of the amplitude ratio in the reflection geometry*

The asymmetry ratio γ is negative and

$$\xi_j = -\frac{S(C)}{\sqrt{|\gamma|}} \frac{\sqrt{\chi_h \chi_{\bar{h}}}}{\chi_{\bar{h}}} \left[\eta \pm \sqrt{\eta^2 - 1}\right]. \tag{5.38}$$

The situation is very different from the transmission case because, on the one hand, when η_r is small, the approximation used above is not valid and, on the other hand, the tiepoint of the wavefield propagating towards the inside of the crystal lies on a different branch of the dispersion surface for $\eta_r < -1$ and $\eta_r > -1$. Let us set

$$Z \exp(\mathrm{i}\psi') = \eta \pm \sqrt{\eta^2 - 1}, \tag{5.39}$$

where the sign must be chosen in such a way that Z converges (see Chapter 7). It follows that

$$\xi_j = \frac{S(C)}{\sqrt{|\gamma|}} \sqrt{\left|\frac{\chi_h}{\chi_{\bar{h}}}\right|} Z \exp \mathrm{i}(\pi + \beta' + \psi').$$

1. *No absorption*: The deviation parameter is real; when $\eta \leq -1$, the tiepoint of the wavefield propagating towards the inside of the crystal is P_1', on branch 1 (Fig. 5.6(b)), corresponding to the plus sign in (5.34), and $\psi' = \pi$, $\psi = \pi + \varphi_h$ (φ_h is the phase of the structure factor); when $|\eta| < 1$, the tiepoint is imaginary and can be represented by P' on the new surface shown in Fig. 5.8(b), ψ' varies progressively from π to 0, being equal to $\pi/2$ for $\eta = 0$, as shown by the dashed curve in Fig. 5.11; when $\eta \geq +1$, the tiepoint of the wavefield propagating towards the inside of the crystal is P_2'', on branch 2 (Fig. 5.6(b)), corresponding to the minus sign in (5.34) and $\psi' = 0$, $\psi = \varphi_h$.
2. *Absorbing crystals*: There is no longer a total reflection domain and ψ' varies continuously when η_r varies from $-\infty$ to $+\infty$. When $\eta_r \Rightarrow -\infty$, the same approximation holds as in the transmission case, and $\psi' \Rightarrow \pi + \beta$, $\psi \Rightarrow \pi + \varphi_h + \beta$; when $\eta_r \Rightarrow +\infty$, $\psi' \Rightarrow \beta$, $\psi \Rightarrow \varphi_h + \beta$. The variations of the ψ' are represented in Fig. 5.11.

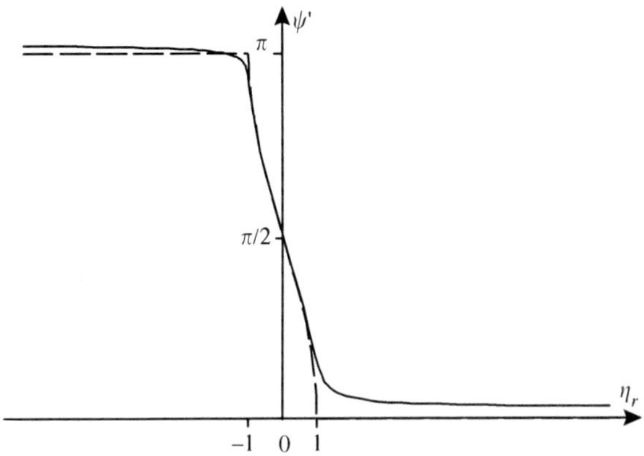

Fig. 5.11. Variations of the phase of $\eta \pm \sqrt{\eta^2 - 1}$. Solid curve: with absorption; dashed curve: without absorption; GaAs, 111 reflection, $\lambda = 1.195$ Å.

5.8 Anomalous absorption

5.8.1 *Effective absorption coefficient in the transmission geometry*

The effective absorption coefficient is given by (4.31):

$$\mu_{ej} = -4\pi \gamma_o K^i_{oj} = \mu_o - 4\pi X^i_{oj}$$

($j = 1$ or 2), where X^i_{oj} is the imaginary part of X_{oj} given by (5.21):

$$X_{oj} = k \frac{|C|}{2\sqrt{\gamma}} \sqrt{\chi_h \chi_{\bar{h}}} \big[\eta \pm \sqrt{1 + \eta^2}\big].$$

Using approximate expression (5.37), it follows that

$$X_{oj} = k \frac{|C|}{2\sqrt{\gamma}} \sqrt{|\chi_h \chi_{\bar{h}}|} \exp(\mathrm{i}\beta)\big(\eta_r \pm \sqrt{1 + \eta_r^2}\big)\left[1 \pm \mathrm{i}\frac{\eta_i}{\sqrt{1 + \eta_r^2}}\right]$$

and, using (5.32),

$$X^i_{oj} = \frac{k|\chi_{io}|}{2} \times \left[\frac{1}{2}(1 - \gamma^{-1}) \pm \frac{(1/2)\eta_r(1 - \gamma^{-1}) + \sqrt{\gamma^{-1}}|C|\sqrt{|\chi_h \chi_{\bar{h}}|}\sin\beta/|\chi_{io}|}{\sqrt{1 + \eta_r^2}}\right].$$

When the absorption is not too large, far from an absorption edge, $\sqrt{|\chi_h \chi_{\bar{h}}|}\sin\beta$ can be approximated by $|\chi_{ih}|\cos\varphi$, where $\varphi = \varphi_{rh} - \varphi_{ih}$ is the phase difference between

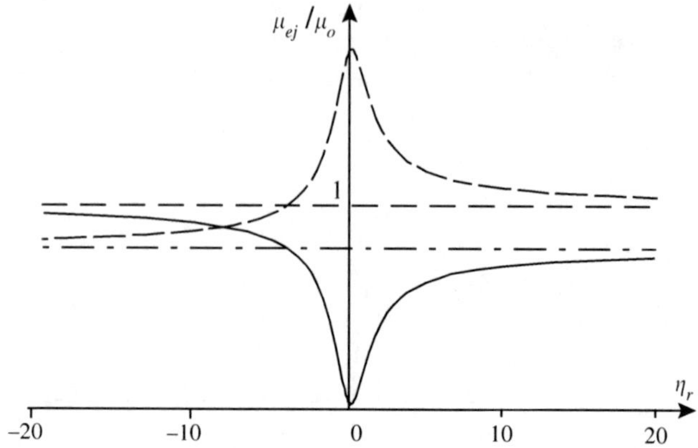

Fig. 5.12. Variations of the effective absorption coefficient, μ_{ej}, along the incident direction; transmission geometry; silicon, 220 reflection, Cu$K\alpha$; asymmetric reflection ($\theta_B = 23.65°$, angle between the normal to the crystal surface and the lattice planes: 15°, $\gamma = 1.27$); solid curve: branch 1; dashed curve: branch 2.

χ_{rh} and χ_{ih}. Remembering that $\mu_o = 2\pi k|\chi_o|$ (2.42b) and using (4.31), we finally obtain, for the effective absorption coefficient,

$$\begin{aligned}\mu_{ej} &= \mu_o - 4\pi X^i_{oj} \\ &= \mu_o\left[\frac{1}{2}(1+\gamma^{-1}) \mp \frac{(1/2)\eta_r(1-\gamma^{-1}) + \sqrt{\gamma^{-1}}|C||\chi_{ih}/\chi_{io}|\cos\varphi}{\sqrt{1+\eta_r^2}}\right].\end{aligned} \tag{5.40}$$

The variations of the effective absorption coefficient are represented in Fig. 4.13 for a symmetric reflection and in Fig. 5.12 for an asymmetric reflection in silicon (220, Cu$K\alpha$ radiation) and an angle of 15° between the normal to the crystal surface and the lattice planes ($\gamma = 1.27$).

The value of the minimum absorption coefficient depends slightly on the asymmetry, it decreases with increasing values of γ. The effective absorption coefficient has been calculated to give the absorption along the incident direction: $\mu_{ej} = -4\pi\gamma_o K^i_{oj}$ (4.34). Far from the reflection domain ($\eta_r \Rightarrow \pm\infty$), the effective absorption coefficient tends towards the normal absorption coefficient:

1. *Branch 1*: For $\eta_r \Rightarrow -\infty$, the propagation direction of the wavefields is nearly parallel to the incident direction and $\mu_{ej} \Rightarrow \mu_0$; for $\eta_r \Rightarrow +\infty$, $\mu_{ej} \Rightarrow \mu_0/\gamma$, the propagation direction of the wavefields is nearly parallel to the reflected direction and the absorption coefficient to be considered is $-4\pi\gamma_h K^i_{oj} = \mu_{ej}\gamma \Rightarrow \mu_0$.
2. *Branch 2*: The situation is similar, for $\eta_r \Rightarrow -\infty$ the propagation is along the reflected direction and the associated absorption coefficient is

$-4\pi\gamma_h K^i_{oj} = \mu_{ej}\gamma \Rightarrow \mu_0$, for $\eta_r \Rightarrow +\infty$ the propagation direction of the wavefields is nearly parallel to the incident direction and $\mu_{ej} \Rightarrow \mu_0$.

5.8.2 *Absorption coefficient in the propagation direction*

Wagner introduced the absorption coefficient along the direction of propagation of the wavefields, μ_S (Borrmann and Wagner 1955; Wagner 1956); The attenuation coefficient of the beam through the crystal may also be written as

$$\exp(-\mu_{ej}t/\gamma_o) = \exp(-\mu_S \overline{Ap}),$$

where $\overline{Ap}$ is the actual path of the wavefield thorough the crystal, along the direction of the Poynting vector, **S** (Fig. 5.13),

$$\mu_S = \mu_{ej}\frac{\cos a}{\gamma_o}$$

is the absorption coefficient along the Poynting vector and $a = (\mathbf{n}, \mathbf{S})$ is the angle between the normal to the entrance surface and the Poynting vector (Fig. 5.13).

Using (5.40), the expression of μ_S can be written as

$$\mu_S = \mu_o \cos a \left[\frac{1}{2\gamma_o} \frac{\sqrt{1+\eta_r^2} \mp \eta_r}{\sqrt{1+\eta_r^2}} + \frac{1}{2\gamma_h} \frac{\sqrt{1+\eta_r^2} \pm \eta_r}{\sqrt{1+\eta_r^2}} \mp \frac{|C|\cos\varphi}{\sqrt{\gamma_o\gamma_h}\sqrt{1+\eta_r^2}} \left| \frac{\chi_{ih}}{\chi_{io}} \right| \right]. \tag{5.41a}$$

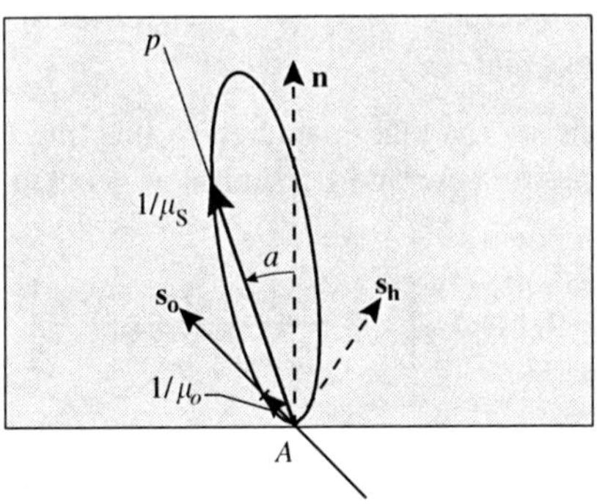

Fig. 5.13. Variation of the absorption length, $1/\mu_S$, in the direction of propagation for tiepoints lying on branch 1 of the dispersion surface; a: angle between the lattice planes and the direction of propagation; $\mathbf{s_o}$: unit vector in the incident direction; $\mathbf{s_h}$: unit vector in the reflected direction; dotted line: trace of the lattice planes; $1/\mu_o$: normal absorption length (when there is no Borrmann effect).

This expression can be transformed by introducing the amplitude ratio $D_h/D_o = \xi$ whose magnitude is given by (4.28), neglecting η_i:

$$\sqrt{\gamma}\left|\frac{D_h}{D_o}\right| = |\eta_r \pm \sqrt{1+\eta_r^2}|,$$

$$\frac{1}{\sqrt{\gamma}}\left|\frac{D_o}{D_h}\right| = |-\eta_r \pm \sqrt{1+\eta_r^2}|$$

with the approximation $\sqrt{|\chi_h/\chi_{\bar{h}}|} \approx 1$ used by Laue (1960). It follows that

$$\mu_{\mathbf{S}} = \mu_o \cos a\left[\frac{|D_o|^2+|D_h|^2}{\gamma_o|D_o|^2+\gamma_h|D_h|^2} \mp |C|\cos\varphi\left|\frac{\chi_{ih}}{\chi_{io}}\right|\frac{|D_oD_h|}{\gamma_o|D_o|^2+\gamma_h|D_h|^2}\right].$$

Noticing that the scalar product of the Poynting vector and the normal, **n**, to the entrance surface is

$$\langle\mathbf{S}\rangle\cdot n = \langle\mathbf{S}\rangle\cos a \propto \langle\mathbf{S}\rangle[\gamma_o|D_o|^2+\gamma_h|D_h|^2],$$

and using (5.11b), one obtains finally that

$$\mu_{\mathbf{S}} = \mu_o\frac{\cos\alpha}{\cos\theta_B}\left[1\mp|C|\left|\frac{\chi_{ih}}{\chi_{io}}\right|\cos\varphi\sqrt{1-\frac{\tan^2\alpha}{\tan^2\theta_B}}\right], \qquad (5.41\text{b})$$

where α is the angle between the Poynting vector and the lattice planes. The variations of $\mu_{\mathbf{S}}$ with the direction of propagation of the wavefields are shown in Fig. 5.13 for branch 1 of the dispersion surface (top sign). It is to be noted that expression (5.41b) does not depend on the asymmetry of the reflection and could have been obtained directly from (5.41a) by setting $\gamma_o = \gamma_h = \cos\theta_B$ and using (5.21b).

5.8.3 *Discussion of anomalous absorption—properties of the standing wavefield*

It is easier to see which are the main parameters affecting the effective absorption coefficient in the symmetric case and transmission geometry ($\gamma = 1$); expression (5.40) becomes

$$\mu_{ej} = \mu_o\left[1\mp\frac{|C||\chi_{ih}/\chi_{io}|\cos\varphi}{\sqrt{1+\eta_r^2}}\right]. \qquad (5.42)$$

These parameters are:

- the branch on which the tiepoint lies
- the value of the deviation parameter, η_r
- the polarization
- the structure factor, through the nature of the reflection, the value of $\cos\varphi$, the temperature.

As has been seen in Section 1.10.1, Borrmann interpreted the phenomenon of anomalous absorption by the set of standing waves created by the interferences of the reflected and refracted waves of a given wavefield: the probability that a photon should be photoelectrically absorbed by an atom is proportional to the number of photons incident on that atom, that is on the intensity of the electric field. The effect of each of the parameters just mentioned can be understood by means of the properties of the standing wavefield. The intensity of a wavefield is

$$|D|^2 = |\mathbf{D_o} + \mathbf{D_h} \exp 2\pi i\mathbf{h} \cdot \mathbf{r}|^2.$$

The scalar product $\mathbf{D_o} \cdot \mathbf{D_h}$ is equal to $D_o D_h$ when $\mathbf{D_o}$ and $\mathbf{D_h}$ are perpendicular to the diffraction plane (σ-polarization) and to $D_o D_h \cos 2\theta_B$ when they are parallel to the diffraction plane (π-polarization). It follows that the intensity of the wavefield is

$$\begin{aligned} |D_j|^2 &= |D_{oj}|^2[1 + |\xi_j|^2 + 2C|\xi_j|\cos(2\pi\mathbf{h} \cdot \mathbf{r} + \psi_j)] \\ &= [|D_{oj}|^2 + |D_{hj}|^2]\left[1 + \frac{2C|\xi_j|}{1 + |\xi_j|^2}\cos(2\pi\mathbf{h} \cdot \mathbf{r} + \psi_j)\right], \end{aligned} \tag{5.43}$$

where the variations of the phase of ξ_j, ψ_j, across the reflection domain are given, in the transmission geometry, by Table 5.1. These variations are very small and can be neglected here. It follows that

$$|D_j|^2 = [|D_{oj}|^2 + |D_{hj}|^2]\left[1 \mp \frac{2C|\xi_j|}{1 + |\xi_j|^2}\cos(2\pi\mathbf{h} \cdot \mathbf{r} + \varphi_h)\right], \tag{5.44}$$

where the minus sign corresponds to tiepoints lying on branch 1 and the plus sign to tiepoints lying on branch 2.

This expression shows that the maxima and minima lie on families of planes $\mathbf{h} \cdot \mathbf{r} =$ constant in direct space whose spacing d_{hkl} is the lattice spacing d divided by the order of the reflection (it is called the reflecting-plane spacing). The product $\mathbf{h} \cdot \mathbf{r}$ can therefore be written as

$$\mathbf{h} \cdot \mathbf{r} = N + \frac{\Delta d}{d_{hkl}}, \tag{5.45}$$

where Δd is the distance of the extremity of the position vector from the origin of the unit cell along the normal to the reflecting planes (Fig. 5.14) and N is an integer.

1. *Branch on which the tiepoint lies*: The minima (nodes), for branch 1, and the maxima (antinodes), for branch 2, lie on the planes

$$\mathbf{h} \cdot \mathbf{r} = N + \frac{\Delta d}{d_{hkl}} = N - \frac{\varphi_h}{2\pi}.$$

We shall choose the origin in such a way that $\varphi_h = 0$, namely on the planes along which the electronic density is maximum. The photoelectric absorption is minimum for branch 1 and maximum for branch 2 (Fig. 5.15).

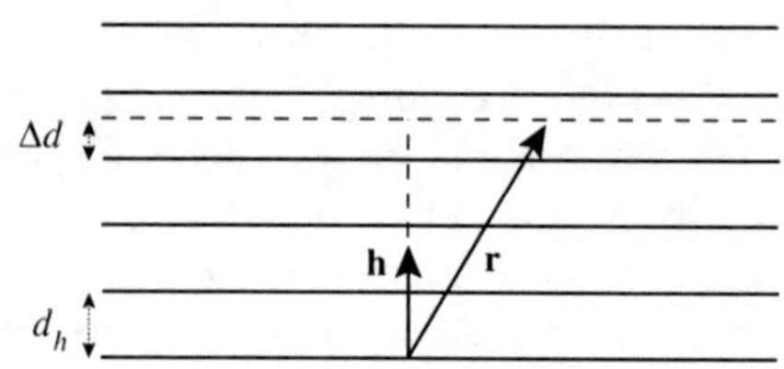

Fig. 5.14. Position of a point with respect to the direct lattice. **r**: position vector in direct space; **h**: reciprocal lattice vector; $d_{hkl} = d/n$: reflecting-plane spacing (d, lattice spacing; n, order of the reflection); $\mathbf{h} \cdot \mathbf{r} = (Nd_{hkl} + \Delta d)/d_{hkl} = \Delta d/d_{hkl}$.

2. *Value of the deviation parameter*: The intensity of the wavefield is minimum when the cosine is equal to -1 and is

$$|D_j|^2 = [|D_{oj}|^2 + |D_{hj}|^2]\left[1 - \frac{2C|\xi_j|}{1+|\xi_j|^2}\right].$$

This value depends on ξ_j and therefore also on η_r. It is minimum when $|\xi_j| = 1$, that is at the centre of the reflection domain; the intensity of the electric field along the planes of maximum electronic density and the anomalous absorption are therefore minimum for $\eta_r = 0$, as shown by (5.42) (Fig. 5.15(a)).

3. *Polarization*: The minimum value of $|D_j|^2$, for $\eta_r = 0$, is $1 - C$. It is equal to 0 for σ-polarization ($C = 1$) and to $1 - \cos 2\theta_B$ for π-polarization: the Borrmann effect will be less marked for a polarization parallel to the diffraction plane, as shown also by (5.42) (Fig. 5.15(b)).

4. *Influence of the nature of the reflection*: Three types of effects affect the value of the effective absorption coefficient:

(i) *Size of the volume where the absorption takes place*: The minimum effective absorption coefficient is, when $\eta_r = 0$ and $C = 1$,

$$\mu_{ej} = \mu_o[1 - |\chi_{ih}/\chi_{io}|\cos\varphi].$$

It is never zero because $|\chi_{ih}/\chi_{io}|$ is never equal to 1. This is due to the fact that the region of space where the photoelectric absorption takes place is not reduced to a point and that its Fourier transform therefore decreases with $\sin\theta_B/\lambda$ (Fig. 2.2). The first reason is that this region corresponds to the inner shells of the electron distribution around the atom (the K- and L-shells) and the second one is thermal agitation. The intensity of the electric field is slightly larger where the absorption takes place than at the maximum of the electronic density and μ_{ej} cannot be 0. As temperature and thermal agitation decrease, the average volume where absorption takes place also decreases and so does the effective anomalous absorption coefficient (see Table 4.1).

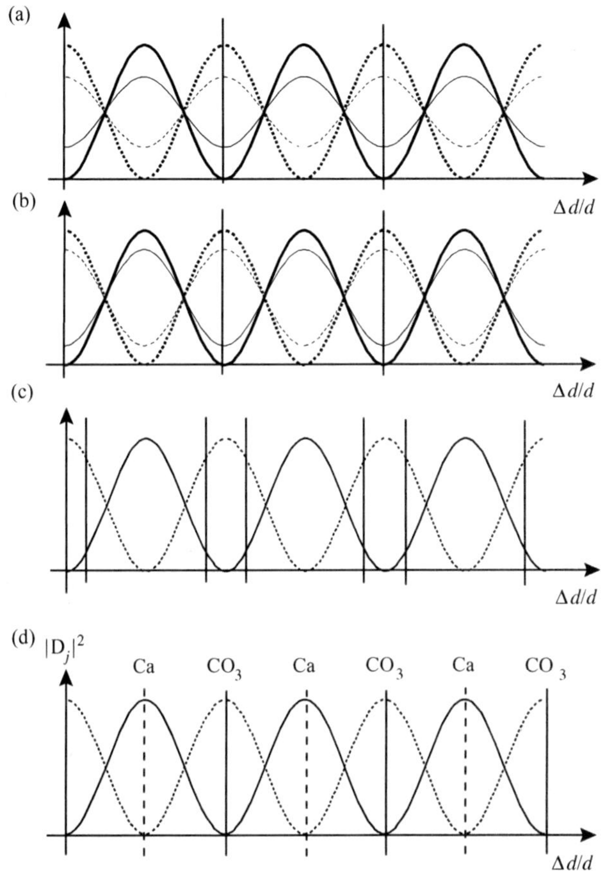

Fig. 5.15. Variations of the intensity of the wavefield with the position in the direct lattice. Solid curves: branch 1; dashed curves: branch 2; vertical lines: atomic planes. (a) Silicon, 220 reflection; Cu$K\alpha$; thick curves: $\eta_r = 0$, $|\xi| = 1$, $C = 1$; thin curves: $\eta_r = 2.27$, $|\xi| = 3.73$, $C = 1$. (b) Silicon, 220 reflection; Cu$K\alpha$; thick curves: $\eta_r = 0$, $|\xi| = 1$, $C = 1$ (σ-polarization); thin curves: $\eta_r = 0$, $|\xi| = 1$, $C = \cos 2\theta_B = 0.678$ (π-polarization). (c) Silicon, 111 reflection. (d) Calcite, $CaCO_3$, 222 reflection.

(ii) *Nature of the reflection*: χ_{ih} is given by (2.37):

$$\chi_{ih} = -\frac{R\lambda^2}{\pi V} F_{ih}$$
$$= -\frac{R\lambda^2}{\pi V} \sum_j f_j'' \exp[-M_j + 2\pi \mathrm{i}\, \mathbf{h} \cdot \mathbf{r_j}].$$

The ratio $|\chi_{ih}/\chi_{io}|$ is only close to 1 if the waves diffracted by all the atoms in the unit cell are in phase. For instance, silicon is face-centred cubic with one atom at the nodes of the lattice and one at half the tetrahedral sites; the positions of these atoms are, taking the origin at a node of the cubic cell, (0, 0, 0) and $(\frac{1}{4}, \frac{1}{4}, \frac{1}{4})$, respectively (Fig. 16.4). The structure of silicon is similar to that of gallium arsenide (Fig. 7.9), where the gallium and arsenic atoms are replaced by silicon atoms; silicon is therefore centrosymmetric but gallium arsenide is not. The values of the structure factor of silicon for the 111 and 220 reflections are, respectively,

$$F_{111} = 4f_{\text{Si}}[1 + \exp 3\pi \text{i}/2]; \quad |F_{111}| = 4\sqrt{2} f_{\text{Si}},$$
$$F_{220} = 8 f_{\text{Si}}.$$

The ratio $|\chi_{ih}/\chi_{io}|$ is close to $\sqrt{2}/2$ for 111 and the Borrmann effect is much smaller than for 220, as shown in Tables 4.1 and 4.2. This result can be understood easily by looking at the position of the standing wave pattern with respect to that of the diffracting planes. The structure of silicon can be described as a stacking of pairs of (111) planes, the distance between the two planes in a pair being one fourth of the periodicity along [111] (Fig. 16.4); the planes along which $\varphi_h = 0$ lie halfway between these planes. This is where $|D_j|^2$ is minimum (Fig. 5.15(c)) and not on any of the 111 planes; the photoelectric absorption is therefore larger than for 220.

(iii) *Phase difference between the Fourier coefficients of the real and imaginary parts of the polarizability* (φ): In a non-centrosymmetric crystal, $\varphi \neq 0$ or π and $|\cos\varphi| < 1$. Again, the intensity of electric field cannot be zero simultaneously on all the atomic planes and the Borrmann effect is smaller than for a centrosymmetric crystal. In this latter case an interesting situation occurs if $\varphi = \pi$ and $\cos\varphi = -1$. The result is that the roles of branches 1 and 2 are interchanged: absorption is larger than normal for branch 1, and smaller than normal for branch 2. The phase shift of π is due to the fact that the maxima of the total electronic density do not coincide with the maxima of the electronic density of the inner shells where the absorption takes place. An example was given by Cowley (1964) with the 222 reflection on calcite, $CaCO_3$. The structure of calcite can be described by a stacking of successive layers of Ca^{++} and CO_3^{--} ions parallel to (111). There are two molecular groups per unit cell and the structure factor is

$$F_h = \left[f_{\text{Ca}} + f_{\text{C}} \exp -2\pi \text{i}\left(\frac{h+k+l}{4}\right)\right]\left[1 + \exp -2\pi \text{i}\left(\frac{h+k+l}{2}\right)\right]$$
$$+ 2f_{\text{O}}\left[\cos 2\pi\left(\frac{k}{2} + \frac{3l}{4}\right) + \cos 2\pi\left(\frac{h}{2} + \frac{3k}{4}\right) + \cos 2\pi\left(\frac{3h}{4} + \frac{l}{2}\right)\right].$$

For the 222 reflection,

$$F_{222} = 2(f_{\text{Ca}} - f_{\text{C}} - 3f_{\text{O}}).$$

Table 5.2 Values of F_{r222} and F_{i222} of calcite

	Mo$K\alpha$	Cu$K\alpha$
$+F_{r222}$	-13.94	-13.70
$+F_{i222}$	0.72	2.68

The numerical values of the real and imaginary parts are given in Table 5.2 for Mo$K\alpha$ and Cu$K\alpha$; since they are of opposite signs, $\cos\varphi = -1$. The reason for this phase shift is that the maxima of the total electronic density lie on the Ca^{++} planes and the maxima of the electronic density of the inner shells lie on the CO_3^{--} planes. The minima of $|D_j|^2$ lie on these latter planes for branch 2 (Fig. 5.15(d)) and it is therefore for branch 2 that the absorption is anomalously small.

On the contrary, the structure factor of calcite for the 211 reflection on the cleavage planes is

$$F_{222} = 2(f_{\mathrm{Ca}} + f_{\mathrm{C}} + f_{\mathrm{O}})$$

and $\cos\varphi = 1$.

Other examples for which $\cos\varphi = -1$ and the roles of branches 1 and 2 are interchanged are exhibited by the 200 reflection on InAs (Blech and Meieran 1969) and on GaAs for a wavelength just above the absorption edge of Ga (Fukamachi *et al.* 1996; Negishi *et al.* 1998). The effect is observed through the reversal of the asymmetry of the rocking curves for the refracted beam in the Laue geometry and for the reflected beam in the Bragg geometry (see Section 6.4.3).

The properties of the standing wave pattern are further discussed in Section 7.2 and in Chapter 16.

5.8.4 *Anomalous absorption in the reflection geometry—penetration depth*

As seen in Section 5.5.5, the effective absorption coefficient is

$$\mu_{ej} = -4\pi\gamma_o K_{oj}^i = \mu_o - 4\pi X_{oj}^i,$$

where X_{oj} is given by (5.22a). We shall only consider here those wavefields which propagate towards the inside of the crystal. Using (5.30b) and (5.39), it follows that

$$\begin{aligned}\mu_{ej} &= \mu_o + 2\pi\gamma_o \mathcal{I}m\left[\frac{\eta + \sqrt{\eta^2 - 1}}{\Lambda_B}\right] \\ &= \mu_o + 2\pi k|C|\sqrt{|\chi_h\chi_{\bar{h}}|}Z\frac{\sin(\psi' + \beta)}{\sqrt{|\gamma|}},\end{aligned} \qquad (5.46)$$

where β is the phase angle of $\sqrt{\chi_h\chi_{\bar{h}}}$ (eqn (2.40)).

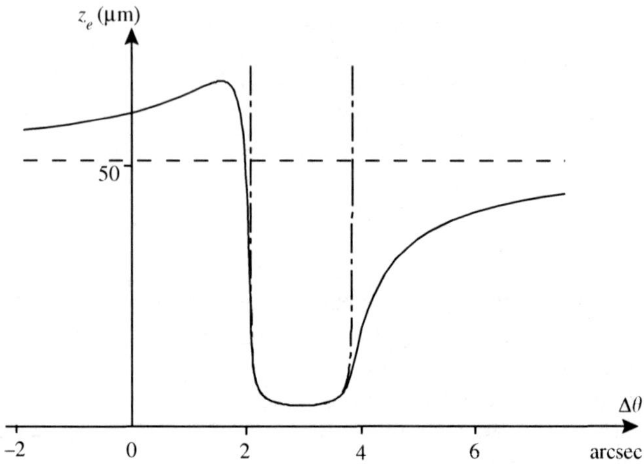

Fig. 5.16. Variations with incidence angle of the penetration depth, z_e, in the reflection geometry. Solid curve: with absorption; dash-dotted curve: without absorption; horizontal line: normal absorption depth; silicon, 333 reflection, Cu$K\alpha$.

The variations of the penetration depth $z_e = \gamma_o/\mu_{ej}$ with the angle of incidence are represented in Fig. 5.16 for a symmetric reflection. The effective absorption coefficient is smaller than the normal absorption coefficient when $\psi' > \pi - \beta$. This occurs for tiepoints lying on branch 1 of the dispersion surface ($\eta_r < -1$, Fig. 5.11) which, as in the transmission geometry and for the same reasons, is associated with anomalously low absorption. When $\psi' < \pi - \beta$, absorption is larger than normal. Within the Bragg gap, absorption is essentially due to extinction; outside the Bragg gap, and for positive values of η_r, the tiepoints lie on branch 2 of the dispersion surface which is associated with anomalously large absorption. The anomalous absorption effect can be understood, as in the transmission geometry, by reference to the position of the nodes of standing waves (see Section 7.2).

5.9 Dispersion surface when the Bragg angle is close to $\pi/2$

5.9.1 *Deviation from Bragg's angle and Darwin width*

When the Bragg angle is close to $\pi/2$, the approximation that the spheres of radii k or nk and centres O and H can be replaced by their tangential planes is no longer valid and neither is the expression of the deviation from Bragg's angle nor that of the Darwin width. Several authors have studied this situation both theoretically and experimentally (for instance, Kohra and Matsushita 1972; Brümmer *et al.* 1979; Caticha and Caticha-Ellis 1982, 1990*a*,*b*; Caticha 1994; Kushnir and Suvorov 1986; Hashizume and Nakahata 1988; Stepanov and Kondrashkina 1991; De Caro and Tapfer 1997; De Caro *et al.* 1997; Kikuta *et al.* 1998; Shvyd'ko *et al.* 1998; Nikulin

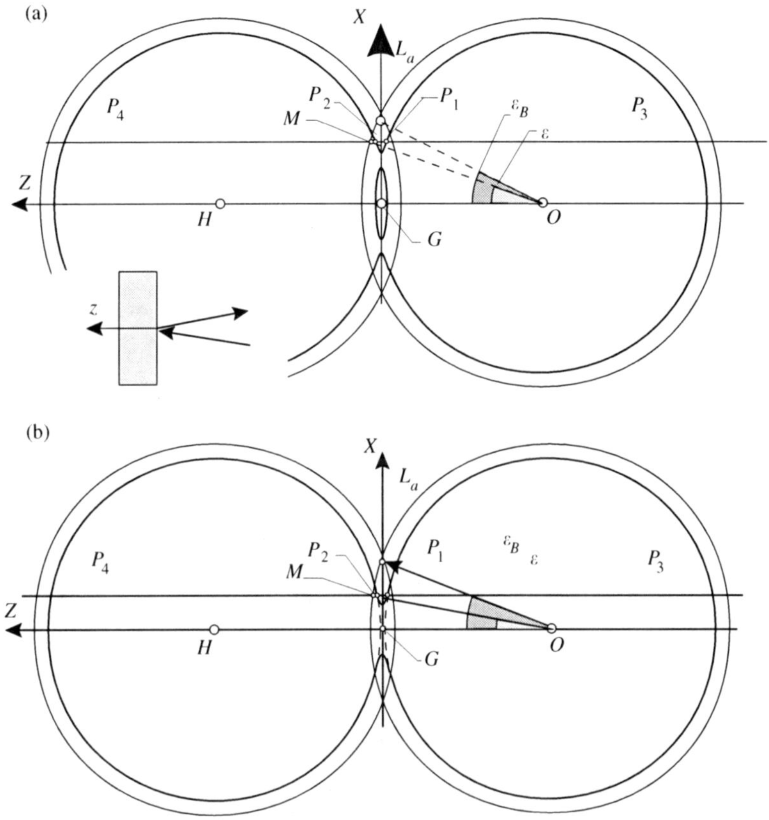

Fig. 5.17. Dispersion surface when the Bragg angle is close to $\pi/2$; L_a: Laue point. Inset: direct space. (a) $\varepsilon > \sqrt{-\chi_o + |C|\sqrt{\chi_h \chi_{\bar{h}}}}$. (b) $\sqrt{-\chi_o - |C|\sqrt{\chi_h \chi_{\bar{h}}}} < \varepsilon < \sqrt{-\chi_o + |C|\sqrt{\chi_h \chi_{\bar{h}}}}$.

et al. 2000, 2001. n-beam diffraction is frequent at such Bragg angles (see Kohn *et al.* 1999; Sutter *et al.* 2001; and Nikulin *et al.* 2003).

Let $\varepsilon_B = \pi/2 - \theta_B$ (Fig. 5.17(a)) where θ_B satisfies Bragg's law, $2d_{hkl} \sin\theta_B = \lambda$. One has, approximately,

$$2d_{hkl}\left(1 - \frac{\varepsilon_B^2}{2}\right) = \lambda, \tag{5.47a}$$

$$\frac{h}{2}\left(1 + \frac{\varepsilon_B^2}{2}\right) = k. \tag{5.47b}$$

For simplicity, we shall limit ourselves to a symmetric reflection geometry. As usual, **OM** is the incident wavevector, and $\mathbf{K_{oj}} = \mathbf{OP_j}$ and $\mathbf{K_{hj}} = \mathbf{HP_j}$ are the wavevectors of the waves associated with tiepoint P_j. The angle of incidence, θ, can be written as

$$\theta = \frac{\pi}{2} - \varepsilon,$$

where ε is the angle between **OM** and **OH** (Fig. 5.17). Let G be the midpoint of **OH**, **GZ** and **GX** be two coordinate axes perpendicular and parallel to the lattice planes, respectively (Fig. 5.17) and X and Z be the coordinates of P_j. One has

$$K_{oj}^2 = \left(\frac{h}{2} + Z\right)^2 + X^2,$$
$$K_{hj}^2 = \left(\frac{h}{2} - Z\right)^2 + X^2,$$

where $h/2 = OG = OH/2$.

It can be seen in Fig. 5.17 that the abscissa, X, of M and P_j satisfies

$$\frac{2X}{h} = \varepsilon. \tag{5.48}$$

The equation of the dispersion surface is the secular equation of the set (5.9):

$$[K_{oj}^2 - k^2(1 + \chi_o)][K_{hj}^2 - k^2(1 + \chi_o)] = k^4 C^2 \chi_h \chi_{\bar{h}}.$$

Substituting the expression of K_{oj}^2 and K_{hj}^2, it follows that

$$\left[\left(\frac{h}{2} + Z\right)^2 + X^2 - k^2(1 + \chi_o)\right]\left[\left(\frac{h}{2} - Z\right)^2 + X^2 - k^2(1 + \chi_o)\right] = k^4 C^2 \chi_h \chi_{\bar{h}}. \tag{5.49}$$

The apices of the dispersion surface lie on **GZ** and their ordinates are determined by putting $Z = 0$. It follows that

$$X^2 = k^2(1 + \chi_o \pm |C|\sqrt{\chi_h \chi_{\bar{h}}}) - \frac{h^2}{4},$$

which is also written, using (5.47b) and (5.48), as

$$X^2 = \frac{h^2}{4}(\varepsilon_B^2 + \chi_o \pm |C|\sqrt{\chi_h \chi_{\bar{h}}}), \tag{5.50a}$$

$$\varepsilon^2 = (\varepsilon_B^2 + \chi_o \pm |C|\sqrt{\chi_h \chi_{\bar{h}}}). \tag{5.50b}$$

Three cases must be distinguished:

1. $\varepsilon_B > \sqrt{-\chi_o + |C|\sqrt{\chi_h \chi_{\bar{h}}}}$: The dispersion surface has the usual shape represented schematically in Fig. 5.17(a) and eqn (5.49) has four roots. The angular

width of the two total reflection domains is given by

$$2\delta_{oc} = \frac{\Delta X}{h/2} = \left[\sqrt{\varepsilon_B^2 + \chi_o + |C|\sqrt{\chi_h \chi_{\bar{h}}}} - \sqrt{\varepsilon_B^2 + \chi_o - |C|\sqrt{\chi_h \chi_{\bar{h}}}}\right] \quad (5.51)$$

and the deviation from Bragg's angle of the middle of the total reflection domains is given by

$$\Delta\theta_{oc} = \varepsilon_B - \sqrt{\varepsilon_B^2 + \chi_o}. \quad (5.52)$$

2. $\sqrt{-\chi_o - |C|\sqrt{\chi_h \chi_{\bar{h}}}} < \varepsilon_B < \sqrt{-\chi_o + |C|\sqrt{\chi_h \chi_{\bar{h}}}}$: The shape of the dispersion surface is that represented in Fig. 5.17(b), and eqn (5.49) has two roots only. There is only one total reflection domain, centred around $\varepsilon = 0$ and the width of the single total reflection domain is

$$2\delta_{oc} = \frac{\Delta X}{h/2} = 2\sqrt{\varepsilon_B^2 + \chi_o + |C|\sqrt{\chi_h \chi_{\bar{h}}}}. \quad (5.53)$$

3. $\varepsilon_B < \sqrt{-\chi_o - |C|\sqrt{\chi_h \chi_{\bar{h}}}}$: There is no reflection.

Figure 5.18 compares the variations of the Darwin width and of the deviation from Bragg's angle calculated with the extended theory ($2\delta_{oc}$ and $\delta\theta_{oc}$, respectively) and with the conventional dynamical theory ($2\delta_{os}$ and $\delta\theta_{os}$, respectively) for germanium, the 440 reflection and $\lambda = 2\,\text{Å}$ ($E = 6.2\,\text{keV}$). It can be seen that the two theories give approximately the same result for $\varepsilon_B > 1°$.

The values of ε_B at the limits between cases 1 and 2 and 2 and 3 are, respectively, $\varepsilon_B = \sqrt{-\chi_o + |C|\sqrt{\chi_h \chi_{\bar{h}}}}$ and $\varepsilon_B = \sqrt{-\chi_o - |C|\sqrt{\chi_h \chi_{\bar{h}}}}$. The Darwin width is maximum at the limit between cases 1 and 2 and is then $2\delta = \sqrt{2|C|\sqrt{\chi_h \chi_{\bar{h}}}}$. When $\varepsilon_B = \sqrt{-\chi_o}$, the two spheres of radii nk and centres O and H are tangent ($OH = 2nk$).

5.9.2 *Dispersion surface*

After rearranging the terms, eqn (5.49) of the dispersion surface can be written as

$$\left[Z^2 + X^2 - \frac{h^2}{4}(\varepsilon_B^2 + \chi_o)\right]^2 - h^2 Z^2 = k^4 C^2 \chi_h \chi_{\bar{h}}. \quad (5.54)$$

The equation obtained by setting the right-hand term equal to zero:

$$\left[Z^2 + X^2 - \frac{h^2}{4}(\varepsilon_B^2 + \chi_o)\right]^2 - h^2 Z^2 = 0$$

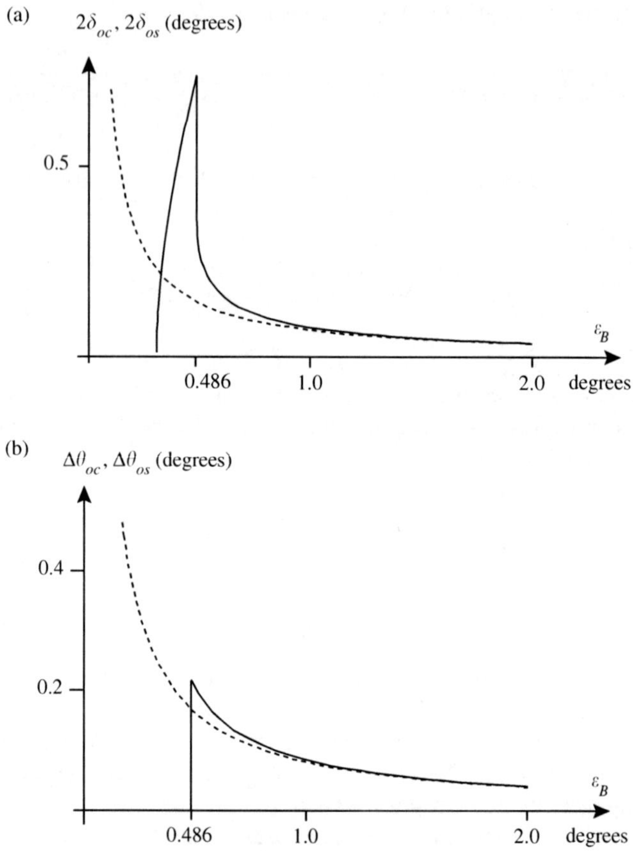

Fig. 5.18. Deviations from conventional dynamical theory when the Bragg angle, θ_B, is close to $\pi/2$. Solid line: extended theory, dashed line: conventional theory; $\varepsilon_B = \pi/2 - \theta_B$; germanium, 440 reflection, $\lambda = 2$ Å. (a) Variations of the Darwin width, $2\delta_{oc}$ (extended theory), $2\delta_{os}$ (conventional theory). (b) Deviation from Bragg's angle, $\Delta\theta_{oc}$ (extended theory), $\Delta\theta_{os}$ (conventional theory).

is the equation of the two circles centred at O and H and with radius nk. Its four roots are

$$Z_1 = -\tfrac{1}{2}\left[h - \sqrt{h^2(1 + \varepsilon_B^2 + \chi_o) - 4X^2}\right],$$

$$Z_2 = +\tfrac{1}{2}\left[h - \sqrt{h^2(1 + \varepsilon_B^2 + \chi_o) - 4X^2}\right],$$

$$Z_3 = -\tfrac{1}{2}\left[h + \sqrt{h^2(1 + \varepsilon_B^2 + \chi_o) - 4X^2}\right],$$

$$Z_4 = +\tfrac{1}{2}\left[h + \sqrt{h^2(1 + \varepsilon_B^2 + \chi_o) - 4X^2}\right],$$

and eqn (5.54) of the dispersion surface can also be written as

$$(Z - Z_1)(Z - Z_2)(Z - Z_3)(Z - Z_4) = k^4 C^2 \chi_h \chi_{\bar{h}}.$$

Its four roots correspond to the intersections with the dispersion surface of the normal **MZ** to the entrance surface drawn, as usual, from the extremity M of the incident wavevector (Fig. 5.17). These intersections are the four tiepoints, P_1, P_2, P_3 and P_4, represented in the figure. Two of these four points, P_3 and P_4, actually lie on the two circles and their ordinates are Z_3 and Z_4. Knowing two of the roots of (5.54) enables to reduce its degree to two. We are looking for the ordinates of P_1 and P_2; they are very small as compared to Z_3 and Z_4 and the factor $(Z - Z_3)(Z - Z_4)$ can be replaced by $(Z_3 Z_4)$. After doing this and neglecting the higher-order terms, one finds, using (5.47b) and (5.48), that

$$Z_j = \pm\frac{h}{4}\sqrt{\left[\varepsilon_B^2 + \chi_o - 4\frac{X^2}{h^2}\right]^2 - C^2 \chi_h \chi_{\bar{h}}}, \tag{5.55a}$$

$$Z_j = \pm\frac{h}{4}\sqrt{[\varepsilon_B^2 - \varepsilon^2 + \chi_o]^2 - C^2 \chi_h \chi_{\bar{h}}} \tag{5.55b}$$

with $j = 1$ or 2. It is easy to check that when $Z_j = 0$, X takes the values given by (5.50).

As usual, there is total reflection if **MZ** intersects the dispersion surface at imaginary points; outside the domain of total reflection one wavefield only is generated if the crystal is thick, that one propagating towards the inside of the crystal. If the crystal is thin, the two wavefields are excited, giving rise to Pendellösung oscillations (see, for instance, Caticha and Caticha-Ellis 1990*a*; Kikuta *et al.* 1998).

The amplitude ratio of the reflected and refracted waves is, using (4.17),

$$\xi_j = \frac{4[Z_j^2 + hZ_j + (h^2/4)(\varepsilon^2 - \varepsilon_B^2 + \chi_o)]}{h^2 C \chi_{\bar{h}}}.$$

Using (5.50), it is easy to check that $|\xi_j| = 1$ when $Z_j = 0$.

5.9.3 *Penetration depth*

The minimum penetration depth is obtained in the middle of the total reflection domain:

1. $\varepsilon_B > \sqrt{-\chi_o + |C|\sqrt{\chi_h \chi_{\bar{h}}}}$: The dispersion surface has the usual shape described by Fig. 5.17(a). The middle of the total reflection domain corresponds to

$$X_j = \frac{h}{2}\sqrt{\chi_o + \varepsilon_B^2}.$$

The value of Z is then, using (5.55),

$$Z = -\mathrm{i}\frac{h}{4}|C|\sqrt{\chi_h \chi_{\bar{h}}}.$$

The minimum penetration depth is, remembering that $h = 2k$,

$$z_e = -\frac{1}{4\pi K_o^i} = -\frac{1}{4\pi\, \mathcal{I}m(Z)} = \frac{\lambda}{2\pi |C| \sqrt{\chi_h \chi_{\bar{h}}}}.$$

Since $\theta \approx \pi/2$, one has $\sqrt{\gamma_o|\gamma_h|} \approx 1$ and $z_e = \Lambda_B/2\pi$, in agreement with the usual result (5.30), far from $\theta \approx \pi/2$.

2. $\sqrt{-\chi_o - |C|\sqrt{\chi_h\chi_{\bar{h}}}} < \varepsilon_B < \sqrt{-\chi_o + |C|\sqrt{\chi_h\chi_{\bar{h}}}}$: The dispersion surface has the shape described by Fig. 5.17(b). The middle of the total reflection domain corresponds to $X = 0$. The corresponding value of Z is

$$Z = -\mathrm{i}\frac{h}{4}\sqrt{C^2 \chi_h \chi_{\bar{h}} - (\varepsilon_B^2 + \chi_o)^2}$$

and the minimum penetration depth is $z_e = -1/4\pi\, \mathcal{I}m(Z)$.

As ε_B varies from $\sqrt{-\chi_o - |C|\sqrt{\chi_h\chi_{\bar{h}}}}$ to $\sqrt{-\chi_o}$ and from $\sqrt{-\chi_o}$ to $\sqrt{-\chi_o + |C|\sqrt{\chi_h\chi_{\bar{h}}}}$, $-\mathcal{I}m(Z)$ varies from 0 to $(h/4)|C|\sqrt{\chi_h\chi_{\bar{h}}}$ to 0 and the penetration depth from ∞ to $\Lambda_B/2\pi$ to ∞.

5.9.4 *Applications*

Diffraction at Bragg angles close to $\pi/2$ has aroused much interest because of the large Darwin width. It is for that reason less sensitive to a small density of imperfections and to a small degree of mosaicity. This has been applied to the study of standing waves (Woodruff *et al.* 1987, 1988; Hashizume *et al.* 1992 and Grehk *et al.* 2001—see Sections 16.2.2 and 16.6). Because of the small bandwidth $\Delta\lambda/\lambda = \tan\varepsilon_B \Delta\theta$, diffraction at Bragg angles close to $\pi/2$ has also been applied to high-resolution X-ray measurements (Graeff and Materlik 1982; Dorner *et al.* 1986), to the use of crystals as resonant cavities for X-rays (Caticha *et al.* 1989; Colella and Luccio 1984), to the design of back-reflection monochromators for synchrotron radiation (Giles and Cusatis 1991; Baron *et al.* 2001) and Fabry–Perot interferometers (Steyerl and Steinhauser 1979; Caticha and Caticha-Ellis 1990*b*).

6

Intensities of plane waves in the transmission geometry

6.1 Boundary conditions for the amplitudes at the entrance surface

A plane wave incident on a semi-infinite crystal in the transmission geometry excites two wavefields. The coordinates X_{oj} and X_{hj} of their tiepoints P_1 and P_2 (Fig. 6.1) are given by (5.21a) in terms of the deviation parameter η. We shall limit ourselves in this chapter to the case where the normal to the crystal surface is parallel to

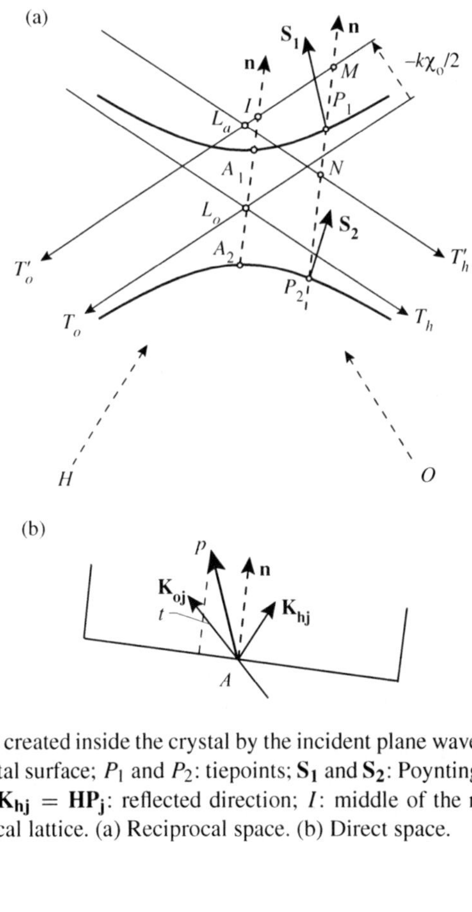

Fig. 6.1. Wavefields created inside the crystal by the incident plane wave. M: excitation point; **n**: normal to the crystal surface; P_1 and P_2: tiepoints; $\mathbf{S_1}$ and $\mathbf{S_2}$: Poynting vectors; $\mathbf{K_{oj}} = \mathbf{OP_j}$: refracted direction; $\mathbf{K_{hj}} = \mathbf{HP_j}$: reflected direction; I: middle of the reflection domain; O: origin of the reciprocal lattice. (a) Reciprocal space. (b) Direct space.

the diffraction plane. The limits of the orientation of the incident wavevector in transmission geometry are given in Fig. 4.9.

The amplitudes of the waves in the two wavefields are D_{o1}, D_{h1}, D_{o2} and D_{h2} and their ratios are given by (5.36). Their values at the entrance surface of the crystal are related to the amplitudes of the waves in vacuum by Maxwell's boundary equations (2.7), in the absence of a surface charge density and of surface current density:

$$\mathbf{E}_{T1} - \mathbf{E}_{T2} = 0, \qquad \mathbf{D}_{N1} - \mathbf{D}_{N2} = 0,$$
$$\mathbf{H}_{T1} - \mathbf{H}_{T2} = 0, \qquad \mathbf{B}_{N1} - \mathbf{B}_{N2} = 0,$$

where the indices T and N indicate the tangential and normal components, respectively.

The polarizability is in principle represented by a tensor (see Section 2.7) but, unless very close to an absorption edge, it can be approximated by a scalar. Under these circumstances **E** and **D** are proportional and the boundary conditions (2.7) show that all their components are continuous across the boundary between the inside and the outside of the crystal. Since the magnetic induction **B** and the magnetic field **H** are also proportional, the four conditions (2.7) reduce to two independent conditions, one for **D** or **E** and one for **H** or **B**. Unless when the incidence or the emergence angle is grazing (see Chapter 8), the conditions on **D** or **E** and **B** or **H** are redundant. The condition on **D** will be used in the present chapter and in Chapter 7.

Let $\mathbf{D}_\mathbf{o}^{(\mathbf{a})}$ be the amplitude of the incident wave, $\mathbf{r}^{(\mathbf{a})}$ a vector joining a point on a given wave plane before the crystal to a point A on the crystal surface. In the general case, one has to consider also outside the crystal a wave of wave vector $\mathbf{K}_\mathbf{h}^{(\mathbf{a})} = \mathbf{K}_\mathbf{o}^{(\mathbf{a})} - \mathbf{h}$ and amplitude $\mathbf{D}_\mathbf{h}^{(\mathbf{a})}$ (Laue 1960, p. 359). The expressions of these waves along the crystal surface are, respectively:

$$\mathcal{D}_\mathbf{o}^{(\mathbf{a})} = \mathbf{D}_\mathbf{o}^{(\mathbf{a})} \exp\left(-2\pi \mathrm{i}\left[\mathbf{K}_\mathbf{o}^{(\mathbf{a})} \cdot (\mathbf{r}^{(\mathbf{a})} + \mathbf{r}_\mathbf{s})\right]\right),$$
$$\mathcal{D}_\mathbf{h}^{(\mathbf{a})} = \mathbf{D}_\mathbf{h}^{(\mathbf{a})} \exp\left(-2\pi \mathrm{i}\left[\mathbf{K}_\mathbf{h}^{(\mathbf{a})} \cdot (\mathbf{r}^{(\mathbf{a})} + \mathbf{r}_\mathbf{s})\right]\right),$$

where $\mathbf{r}_\mathbf{s}$ is a position vector along the crystal surface. If the direction cosine, γ_h, of the reflected wave is > 0, which is the case in the transmission geometry (Fig. 6.1), $\mathbf{D}_\mathbf{h}^{(\mathbf{a})} = 0$ and this wave is virtual, but its phase factor carries the information of the optical path of the incident wave, $\mathbf{D}_\mathbf{o}^{(\mathbf{a})}$. Similarly, the refracted and reflected waves are, in the crystal surface:

$$\mathcal{D}_\mathbf{oj} = \mathbf{D}_\mathbf{oj} \exp\left(-2\pi \mathrm{i}\left[\mathbf{K}_\mathbf{o}^{(\mathbf{a})} \cdot \mathbf{r}^{(\mathbf{a})}\right]\right) \exp\left(-2\pi \mathrm{i}\left[\mathbf{K}_\mathbf{oj} \cdot \mathbf{r}_\mathbf{s}\right]\right),$$
$$\mathcal{D}_\mathbf{hj} = \mathbf{D}_\mathbf{hj} \exp\left(-2\pi \mathrm{i}\left[\mathbf{K}_\mathbf{h}^{(\mathbf{a})} \cdot \mathbf{r}^{(\mathbf{a})}\right]\right) \exp\left(-2\pi \mathrm{i}\left[\mathbf{K}_\mathbf{hj} \cdot \mathbf{r}_\mathbf{s}\right]\right),$$

where the first exponential term in each expression corresponds to the optical path of the wave till it reaches the crystal surface. The boundary condition for the electric induction field is:

$$\mathcal{D}_{\mathbf{o}}^{(\mathbf{a})} + \mathcal{D}_{\mathbf{h}}^{(\mathbf{a})} = \sum_{j=1,2} \{\mathcal{D}_{\mathbf{oj}} + \mathcal{D}_{\mathbf{hj}}\} \tag{6.1a}$$

Due to the continuity of the tangential components of the wavevectors, one has:

$$\mathbf{K}_{\mathbf{o}}^{(\mathbf{a})} \cdot \mathbf{r}_{\mathbf{S}} = \mathbf{K}_{\mathbf{o}1} \cdot \mathbf{r}_{\mathbf{S}} = \mathbf{K}_{\mathbf{o}2} \cdot \mathbf{r}_{\mathbf{S}},$$

$$\mathbf{K}_{\mathbf{h}}^{(\mathbf{a})} \cdot \mathbf{r}_{\mathbf{S}} = \mathbf{K}_{\mathbf{h}1} \cdot \mathbf{r}_{\mathbf{S}} = \mathbf{K}_{\mathbf{h}2} \cdot \mathbf{r}_{\mathbf{S}}.$$

Let us multiply both sides of the boundary condition (6.1a) by $\exp\left(+2\pi\mathrm{i}\left[\mathbf{K}_{\mathbf{o}}^{(\mathbf{a})} \cdot \mathbf{r}_{\mathbf{s}}\right]\right)$ and let us integrate over the surface of the crystal, which has been assumed to be infinite. Remembering that $\mathbf{r}^{(\mathbf{a})}$ is a constant, it follows that:

$$\mathbf{D}_{\mathbf{o}}^{(\mathbf{a})} = \mathbf{D}_{\mathbf{o}1} + \mathbf{D}_{\mathbf{o}2} \tag{6.1b}$$

By multiplying by $\exp\left(+2\pi\mathrm{i}\left[\mathbf{K}_{\mathbf{h}}^{(\mathbf{a})} \cdot \mathbf{r}_{\mathbf{s}}\right]\right)$ and integrating, it follows in the same way that:

$$\mathbf{D}_{\mathbf{h}}^{(\mathbf{a})} = \mathbf{D}_{\mathbf{h}1} + \mathbf{D}_{\mathbf{h}2} \tag{6.1c}$$

By projecting both sides of (6.1a) and (6.1b) on the normal to the diffracting plane and on the diffraction plane, it is easy to show that:

$$\mathbf{D}_{\mathbf{o}}^{(\mathbf{a})\perp} = \mathbf{D}_{\mathbf{o}1}^{\perp} + \mathbf{D}_{\mathbf{o}2}^{\perp}; \quad \mathbf{D}_{\mathbf{o}}^{(\mathbf{a})\parallel} = \mathbf{D}_{\mathbf{o}1}^{\parallel} + \mathbf{D}_{\mathbf{o}2}^{\parallel},$$

$$\mathbf{D}_{\mathbf{h}}^{(\mathbf{a})\perp} = \mathbf{D}_{\mathbf{h}1}^{\perp} + \mathbf{D}_{\mathbf{h}2}^{\perp}; \quad \mathbf{D}_{\mathbf{h}}^{(\mathbf{a})\parallel} = \mathbf{D}_{\mathbf{h}1}^{\parallel} + \mathbf{D}_{\mathbf{h}2}^{\parallel}.$$

All the derivations of dynamical theory are the same for σ- and π-polarization, the only difference being that the Fourier coefficients of the polarizability, χ_h, are multiplied by $C = 1$ in the former case and by $C = \cos 2\theta_B$ in the latter. It will not be necessary, therefore, to use the superscripts $\perp$ or $\parallel$.

6.2 Amplitudes of the refracted and reflected waves

Let us now consider the inside of the crystal. The amplitudes of the refracted and reflected waves are obtained by solving the system formed by (6.1b,c) and

$$\xi_1 = \frac{D_{h1}}{D_{o1}},$$

$$\xi_2 = \frac{D_{h2}}{D_{o2}},$$

where ξ_j is given by (5.36). In the transmission geometry, $\mathbf{D}_\mathbf{h}^{(\mathbf{a})} = 0$, and one finds:

$$D_{h1} = D_o^{(a)} \frac{\xi_1 \xi_2}{\xi_2 - \xi_1}; \qquad D_{h2} = D_o^{(a)} \frac{-\xi_1 \xi_2}{\xi_2 - \xi_1};$$
$$D_{o1} = D_o^{(a)} \frac{\xi_2}{\xi_2 - \xi_1}; \qquad D_{o2} = D_o^{(a)} \frac{-\xi_1}{\xi_2 - \xi_1}. \tag{6.2a}$$

Substituting from (5.36), it follows that

$$D_{hj} = \pm \frac{\sqrt{\chi_h \chi_{\bar{h}}}}{2\sqrt{\gamma}\chi_{\bar{h}}} \frac{1}{\sqrt{1+\eta^2}} D_o^{(a)},$$
$$D_{oj} = \frac{\sqrt{1+\eta^2} \mp \eta}{2\sqrt{1+\eta^2}} D_o^{(a)}, \tag{6.2b}$$

where the top sign corresponds to $j = 1$ (branch 1) and the bottom sign to $j = 2$.

The position vector of point A on the crystal surface is $\mathbf{r}^{(\mathbf{a})}$ and let $\mathbf{r}$ be the position vector of a point p at depth z inside the crystal (Fig. 6.1b). The four waves arriving at p are

$$\mathcal{D}_{\mathbf{hj}} = D_{hj} \exp(-2\pi \mathrm{i}[\mathbf{K}_{\mathbf{hj}} \cdot (\mathbf{r} - \mathbf{r}^{(\mathbf{a})}) + \mathbf{K}_\mathbf{h}^{(\mathbf{a})} \cdot \mathbf{r}^{(\mathbf{a})}]),$$
$$\mathcal{D}_{\mathbf{oj}} = D_{oj} \exp(-2\pi \mathrm{i}[\mathbf{K}_{\mathbf{oj}} \cdot (\mathbf{r} - \mathbf{r}^{(\mathbf{a})}) + \mathbf{K}_\mathbf{o}^{(\mathbf{a})} \cdot \mathbf{r}^{(\mathbf{a})}]), \tag{6.3a}$$

where (Fig. 6.1)

$$\mathbf{K}_{\mathbf{hj}} = \mathbf{HP}_\mathbf{j} = \mathbf{HO} + \mathbf{K}_\mathbf{o}^{(\mathbf{a})} + \mathbf{MP}_\mathbf{j},$$
$$\mathbf{K}_{\mathbf{oj}} = \mathbf{OP}_\mathbf{j} = \mathbf{K}_\mathbf{o}^{(\mathbf{a})} + \mathbf{MP}_\mathbf{j}.$$

$\mathbf{MP}_\mathbf{j}$ is parallel to the normal to the crystal surface and the four waves can also be written as

$$\mathcal{D}_{\mathbf{hj}} = D_{hj} \exp(-2\pi \mathrm{i}[\overline{MP}_j\, z + \mathbf{K}_\mathbf{h}^{(\mathbf{a})} \cdot \mathbf{r}]),$$
$$\mathcal{D}_{\mathbf{oj}} = D_{oj} \exp(-2\pi \mathrm{i}[\overline{MP}_j\, z + \mathbf{K}_\mathbf{o}^{(\mathbf{a})} \cdot \mathbf{r}]), \tag{6.3b}$$

where $\overline{MP}_j$ is given by (5.23).

The expressions of the amplitudes of the reflected and the refracted waves are given by (4.35b). Figures 6.2(a)–(c) represent the variations with the deviation parameter of the corresponding reflectivities in the non-absorbing case and for $\mu_o t = 0.5$ and 2.5, respectively (μ_o, normal absorption coefficient; t, crystal thickness). Far from the reflection domain, $|D_{h1}|^2$ and $|D_{h2}|^2$ tend towards zero, as is normal, while

$$|D_{o1}|^2 \gg |D_{o2}|^2 \quad \text{for } \eta_r \Rightarrow -\infty \quad (|D_{o1}|^2 \Rightarrow 1 \text{ in the non-absorbing case}),$$
$$|D_{o1}|^2 \ll |D_{o2}|^2 \quad \text{for } \eta_r \Rightarrow +\infty \quad (|D_{o2}|^2 \Rightarrow 1 \text{ in the non-absorbing case}).$$

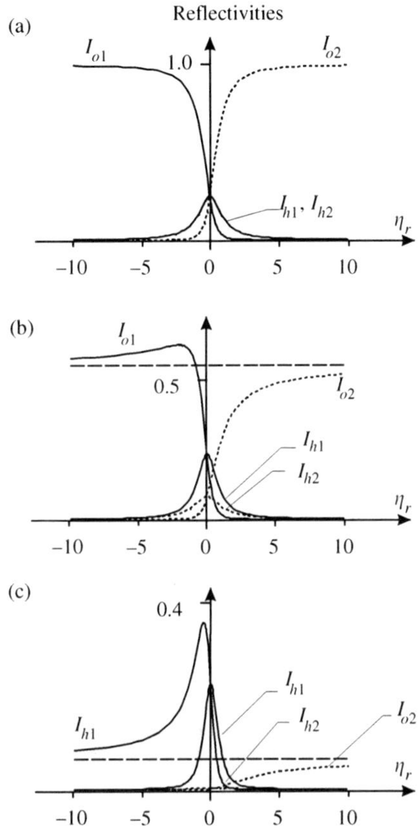

Fig. 6.2. Variations with the deviation parameter of the reflectivities of the 'o' and 'h' components of wavefields 1 and 2. Solid curves: branch 1; dashed curves: branch 2. (a) No absorption, I_{h1} and I_{h2} coincide. (b) Silicon, 220 reflection, Cu$K\alpha$, $\mu_o t = 0.5$. (c) Silicon, 220 reflection, Cu$K\alpha$, $\mu_o t = 2.5$.

This result shows, as was noticed in Section 4.8.3, that it is the wavefields whose tiepoints lie on branch 1 which are the more intense on the low-angle side of the rocking curve, while it is the wavefields associated with branch 2 on the high-angle side, despite the fact that absorption is stronger; this is due to the fact that their propagation direction becomes closer to that of the incident beam as $\eta_r \Rightarrow +\infty$ (Fig. 4.12). In the middle of the reflection domain ($\eta_r = 0$), the amplitudes $|D_{h1}|$ and $|D_{h2}|$, $|D_{o1}|$ and $|D_{o2}|$ are equal, but this is not true for the intensities of the waves because of the anomalous absorption effect. It will be shown in Section 12.10 that the variations of the intensities of the four waves can be measured directly in a single experiment.

Conservation of the energy: When the crystal is not absorbing, the total energy of the four waves inside the crystal must be equal to the energy of the incident wave.

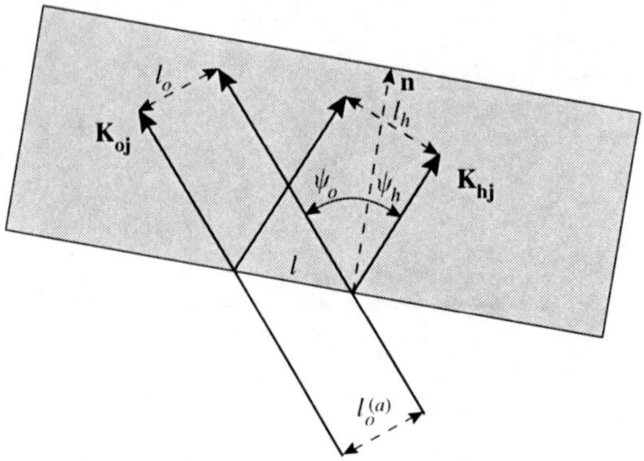

Fig. 6.3. Cross-sections of the incident, refracted and reflected waves, $l_o^{(a)}, l_o, l_h$, respectively.

Let W_o be the energy of the incident wave crossing per second a rectangular area of height h and of width $l_o^{(a)}$ (Fig. 6.3). The intensity of the incident beam is given by the Poynting vector (5.11a) and its energy is

$$W_o = \frac{c}{\epsilon_0} l_o^{(a)} h |D_o^{(a)}|^2. \tag{6.4}$$

The total energy of the four waves created at the entrance surface is

$$W = \frac{c}{\epsilon_0}\left[l_o h \sum_{j=1,2} |D_{oj}|^2 + l_h h \sum_{j=1,2} |D_{hj}|^2 \right], \tag{6.5}$$

where l_o and l_h are the widths of the refracted and reflected beams, respectively.

Let l be the width of the trace of the incident beam on the crystal surface. It can be seen in Fig. 6.3, neglecting the refraction effect, that

$$l_o^{(a)} = l_o = \gamma_o l, \qquad l_h = \gamma_h l,$$

where $\gamma_o = \cos\psi_o$ and $\gamma_h = \cos\psi_h$ have been defined in Section 4.7.3. Using these relations, expressions (6.4) and (6.5) become

$$\begin{aligned} W_o &= \frac{c}{\epsilon_0} l h \gamma_o |D_o^{(a)}|^2, \\ W &= \frac{c}{\epsilon_0} l h \left[\gamma_o \sum_{j=1,2} |D_{oj}|^2 + \gamma_h \sum_{j=1,2} |D_{hj}|^2 \right]. \end{aligned} \tag{6.6}$$

Substituting from (4.33) into the second of equations (6.6), and replacing $|\psi_{mj}|^2$ by $|D_{mj}|^2$ $(m = o, h)$, we obtain, at the surface of the crystal $(z = 0)$,

$$W = \frac{c}{\epsilon_0} l h \gamma_o \left[\frac{\left(\eta + \sqrt{1+\eta^2}\right)^2 + \left(\eta - \sqrt{1+\eta^2}\right)^2 + 2}{4(1+\eta^2)} \right] |D_o^{(a)}|^2,$$

which is readily seen to be equal to W_o. This demonstrates the conservation of energy.

6.3 Boundary conditions for the wavevectors at the exit surface

6.3.1 *Condition for the existence of two outgoing waves*

When a wavefield propagating within the crystal reaches the exit surface, it breaks up into its two constituent waves. If the directions of their wavevectors (both of them) lie towards the outside of the crystal, there are two emerging waves; this is the so-called *Laue–Laue* geometry (Fig. 4.14 for a plane-parallel crystal, Fig. 6.4 for a wedge-shaped crystal). If, on the contrary, one of the wavevectors points towards the inside of the crystal, there is one emerging wave only and the other one is partially reflected; this is the so-called *Laue–Bragg* geometry (Fig. 6.5).

Let $\mathbf{n}'$ be a unit vector parallel to the normal to the exit surface, pointing towards the outside of the crystal (Figs 6.4 and 6.5). By analogy with the definitions of

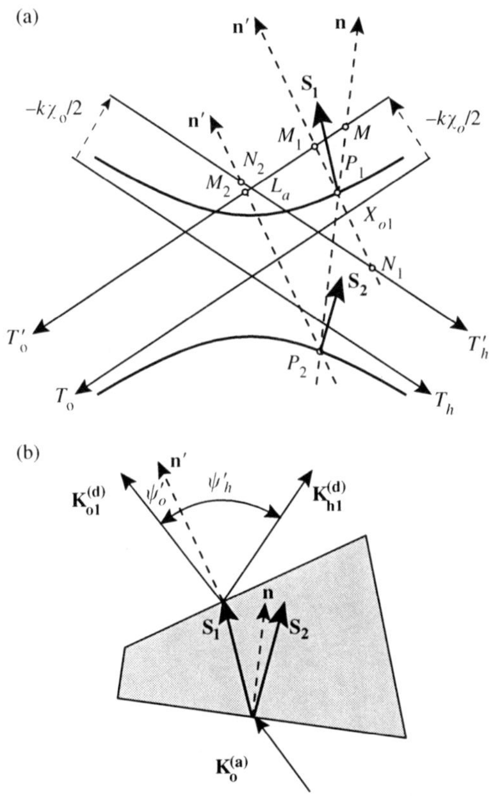

Fig. 6.4. Boundary condition for the wavevectors at the exit surface; wedge-shaped crystal, Laue–Laue geometry. $\mathbf{n}$: normal to the entrance surface; $\mathbf{n}'$: normal to the exit surface; N_1 and M_1, N_2 and M_2: tiepoints of the reflected ($\mathbf{K}_\mathbf{h}^{(\mathbf{d})}$) and refracted ($\mathbf{K}_\mathbf{o}^{(\mathbf{d})}$) wavevectors generated by wavefields 1 and 2, respectively. (a) Reciprocal space. (b) Direct space.

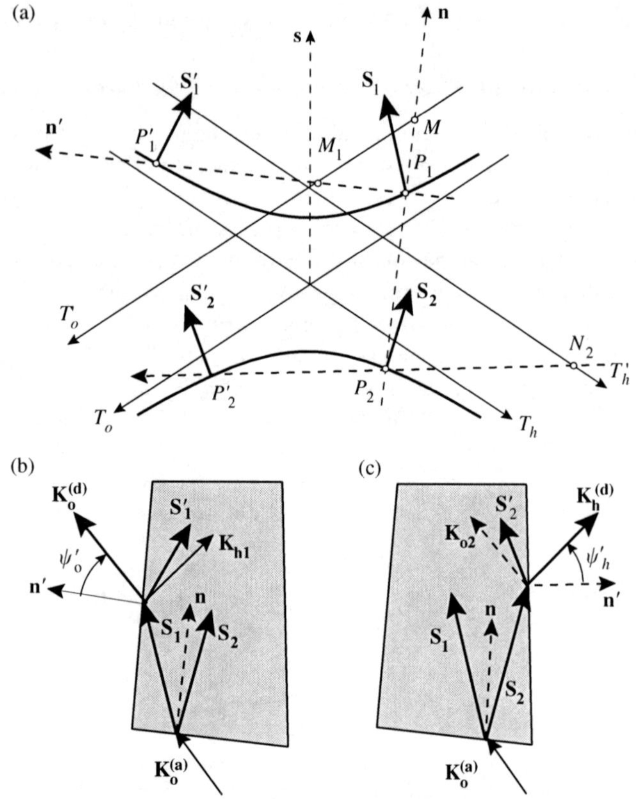

Fig. 6.5. Boundary condition for the wavevectors at the exit surface; wedge-shaped crystal, Laue–Bragg geometry. **n**: normal to the entrance surface; **n**′: normal to the exit surface. (a) Reciprocal space; M excitation point of the incident wave; **s**: trace of the lattice planes. (b) Direct space, the reflected wave points towards the inside of the crystal, $\mathbf{S}_1'$ is partially reflected (tiepoint: P_1'); M_1 is the extremity of the refracted wave, $\mathbf{K}_\mathbf{o}^{(\mathbf{d})}$, outside the crystal. (c) Direct space, the refracted wave points towards the inside of the crystal, $\mathbf{S}_2'$ is partially reflected (tiepoint: P_2'); N_2 is the extremity of the reflected wave, $\mathbf{K}_\mathbf{h}^{(\mathbf{d})}$, outside the crystal.

Section 4.7.3, we shall set

$$\begin{aligned} \psi_o' &= (\mathbf{n}', \mathbf{s_o}); & \gamma_o' &= \cos\psi_o'; \\ \psi_h' &= (\mathbf{n}', \mathbf{s_h}); & \gamma_h' &= \cos\psi_h'; \\ \psi' &= (\mathbf{n}', \mathbf{s}); & \gamma' &= \frac{\gamma_h'}{\gamma_o'}. \end{aligned} \tag{6.7}$$

The conditions for the existence of two outgoing waves are

$$\begin{aligned} -\frac{\pi}{2} + 2\theta_B < \psi'_o < \frac{\pi}{2}; \qquad & \sin 2\theta_B > \gamma'_o > 0; \\ -\frac{\pi}{2} < \psi'_h < \frac{\pi}{2} - 2\theta_B; \qquad & 0 < \sin 2\theta_B < \gamma'_h. \end{aligned} \tag{6.8}$$

6.3.2 *Wavevectors of the outgoing waves (Laue–Laue geometry)*

The incident wave has generated four wavefields inside the crystal, two for the σ-polarization and two for the π-polarization. If condition (6.8) is satisfied, there will be eight outgoing waves, four with wavevectors parallel to the incident direction and four with the wavevector in the reflected direction. The exact orientation of these wavevectors is obtained by applying again the condition of the continuity of their tangential components along the crystal surface. The derivation is analogous for the two directions of polarization and there is no need to specify for which one it is done. The extremities, M_j and N_j, of the wavevectors of the outgoing waves, $\mathbf{OM_j} = \mathbf{K_{oj}^{(d)}}$ and $\mathbf{HN_j} = \mathbf{K_{hj}^{(d)}}$, lie at the intersections of the spheres of radius k and centred at O and H, respectively (replaced by their asymptotes T'_o and T'_h), with the normal to the crystal exit surface, drawn from P_j ($j = 1$ and 2) (see Fig. 4.14 for a plane-parallel crystal and Fig. 6.4 for a wedge-shaped crystal). The equation of this normal is

$$\frac{X_h - X_{hj}}{X_o - X_{oj}} = \gamma', \tag{6.9}$$

where X_o, X_{oj}, X_h and X_{hj} are coordinates normal to T_o and T_h taken as axes; X_{oj} and X_{hj} are given by (5.21a). The equations of T'_o and T'_h relative to these same axes are

$$T'_o:\;\; X_o = -\tfrac{1}{2}k\chi_o; \qquad T'_h:\;\; X_h = -\tfrac{1}{2}k\chi_o;$$

from these two sets of equations the coordinates of M_j and N_j are easily deduced:

$$M_j \quad \begin{cases} X_o = -\tfrac{1}{2}k\chi_o \\ X_h = X_{hj} - \gamma'\left(\tfrac{1}{2}k\chi_o + X_{oj}\right) \end{cases}$$

$$N_j \quad \begin{cases} X_o = X_{oj} - \gamma'^{-1}\left(\tfrac{1}{2}k\chi_o + X_{hj}\right) \\ X_h = -\tfrac{1}{2}k\chi_o \end{cases}$$

Let $\Delta\theta_{osj}$ and $\Delta\theta_{hsj}$ be the angles between the angles of the outgoing wavevectors $\mathbf{OM_j}$ and $\mathbf{HN_j}$, respectively, on the one hand, and the wavevectors which satisfy Bragg's condition according to the geometric theory, $\mathbf{OL_a}$ and $\mathbf{HL_a}$, on the other hand:

$$\Delta\theta_{osj} = (\mathbf{OL_a}, \mathbf{OM_j}); \qquad \Delta\theta_{hsj} = (\mathbf{HL_a}, \mathbf{HN_j}).$$

These may be expressed in terms of the coordinates of the tiepoints P_j:

$$\Delta\theta_{osj} = \frac{\overline{L_a M_j}}{k} = \frac{1}{k\sin 2\theta_B}\Big[\gamma' X_{oj} - X_{hj} + \frac{1}{2}k\chi_o(\gamma' - 1)\Big],$$

$$\Delta\theta_{hsj} = \frac{\overline{L_a N_j}}{k} = \frac{1}{k\sin 2\theta_B}\Big[\gamma'^{-1} X_{hj} - X_{oj} + \frac{1}{2}k\chi_o(\gamma'^{-1} - 1)\Big].$$

It may be noticed that $\Delta\theta_{oj}$ and $\Delta\theta_{hj}$ are proportional and of opposite signs ($\Delta\theta_{osj} = -\gamma'\Delta\theta_{hj}$). They are simply opposite if the exit surface is perpendicular to the lattice planes ($\gamma' = 1$, $\Delta\theta_{osj} = -\Delta\theta_{hsj}$). In practice, it is useful to express these angles in terms of the deviation parameter. It follows, using (5.21a), that

$$\Delta\theta_{osj} = \frac{1}{2\sin 2\theta_B}\Bigg\{|C|\sqrt{\frac{\chi_h\chi_{\bar{h}}}{\gamma}}\big[\eta(\gamma' + \gamma) \pm \sqrt{1+\eta^2}\,(\gamma' - \gamma)\big] + \chi_o(\gamma' - 1)\Bigg\}. \qquad (6.10)$$

The emerging wavevectors $\mathbf{K}_{o1}^{(d)} = \mathbf{OM}_1$ and $\mathbf{K}_{o2}^{(d)} = \mathbf{OM}_2$ have different directions, as well as $\mathbf{K}_{h1}^{(d)} = \mathbf{HN}_1$ and $\mathbf{K}_{h2}^{(d)} = \mathbf{HN}_2$ (Fig. 6.4). The angle between them is

$$\Delta\theta_{os1} - \Delta\theta_{os2} = \frac{|C|\sqrt{\chi_h\chi_{\bar{h}}}}{\sqrt{\gamma}}(\gamma' - \gamma)\sqrt{1+\eta^2}. \qquad (6.11)$$

It is of course equal to zero for a plane-parallel slab ($\gamma' = \gamma$).

Example: Kohra *et al.* (1965) have checked expression (6.10). For this, they used a two-crystal setting in which the first crystal, C_1, has a bevelled lateral face and the second crystal, C_2, set in the reflection geometry, serves as an analyser (Fig. 6.6).

Two measurements were made successively, the first one with the reflected beam coming out of the exit surface parallel to the entrance surface (Fig. 6.6(a)) and the second one with the beam coming out through the lateral bevelled face (Fig. 6.6(b)). The thickness of the first crystal was such that in both cases only the wavefields belonging to branch 1 and corresponding to values of η close to 0 were transmitted. The angle between the positions of the analyser crystal at maximum reflection in the two cases is, according to (6.10),

$$\Delta\theta_{os1} = -\frac{(|\chi_o| - |\chi_h|)\sin\psi'_n}{2\cos\theta_B\cos(\psi'_n + \theta_B)},$$

where ψ'_n is the angle between the bevelled face and the normal to the entrance surface. This experiment provides a method for a direct measurement of the refraction of X-rays. Kohra *et al.* used silicon, Mo$K\alpha$ and 220 for which $\theta_B = 10.65°$; ψ'_n

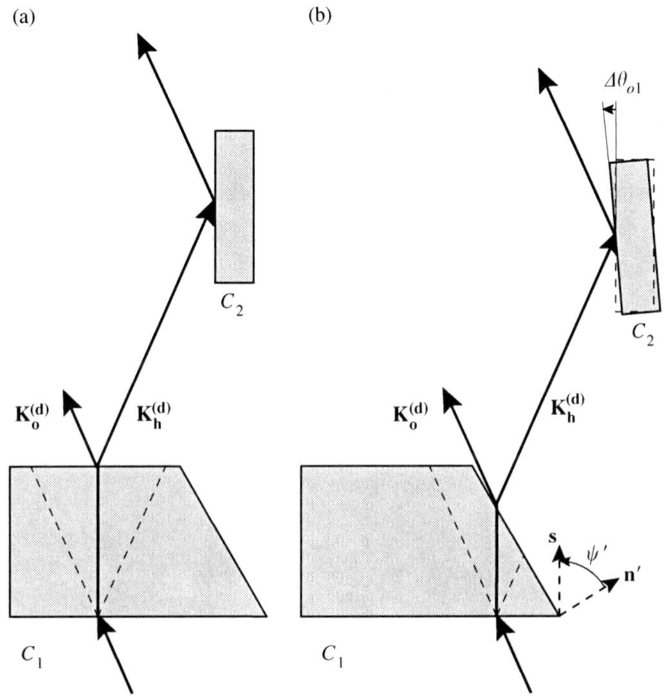

Fig. 6.6. Measurement of the refraction effect in a wedge-shaped crystal; silicon, 220 reflection, Mo$K\alpha$; C_1: sample crystal, C_2: analyser. (a) The exit surface is parallel to the entrance surface, $\mu_o t = 18.9$. (b) The exit surface is inclined, $\mu_o t = 9.5$. Adapted from Kohra *et al.* (1965).

was equal to 16.22°; the measured value of $\Delta\theta_{os1}$ was 0.94 ± 0.07 arcsec, in good agreement with an expected theoretical value of 1.08 arcsec.

6.3.3 *Laue–Bragg geometry*

The Laue–Bragg geometry corresponds to a situation where the orientation of the exit surface is such that one of the wavevectors of a wavefield reaching it points towards the inside of the crystal. This is at the origin of a partial reflection on that exit surface (Fig. 6.5). It may be either the wavevector of the reflected wave, $\mathbf{K_{hj}}$ (Fig. 6.5(b)), or that of the refracted wave, $\mathbf{K_{oj}}$ (Fig. 6.5(c)). In either case, the normal to the exit surface drawn from the tiepoint of the wavefield intersects the *same* branch of the dispersion surface (at P_1' in the example of Fig. 6.5(b) and P_2' in the example of Fig. 6.5(c)). The associated Poynting vector ($\mathbf{S}_1'$ or $\mathbf{S}_2'$, respectively, in these two examples) points towards the inside of the crystal. It is easy to find the orientation of the reflected wavefield. Let X_{hj}' and X_{oj}' be the coordinates of the tiepoints P_j' of the reflected wavefield ($j = 1$ or 2). They satisfy on the one hand eqn (6.9) of the normal to the exit surface drawn from P'_j,

$$\frac{X'_{hj} - X_{hj}}{X'_{oj} - X_{oj}} = \gamma',$$

and the equation of the dispersion surface on the other hand:

$$X'_{hj} X'_{oj} = X_{hj} X_{oj}.$$

The result is

$$X'_{hj} = -\gamma' X_{oj}; \qquad X'_{oj} = -\gamma'^{-1} X_{hj}. \tag{6.12}$$

The slope of the Poynting vector of a wavefield with respect to the lattice planes is given by (5.13) in terms of the coordinates of its tiepoint:

$$\tan \alpha'_j = \frac{X'_{hj} - X'_{oj}}{X'_{hj} + X'_{oj}} \tan \theta_B.$$

Substituting from (6.12), we obtain the slope of the reflected wavefield in terms of the coordinates of the tiepoint of the incident wavefield:

$$\tan \alpha'_j = -\frac{X_{hj} - \gamma'^2 X_{oj}}{X_{hj} + \gamma'^2 X_{oj}}. \tag{6.13}$$

They are of opposite signs, as expected. If $\gamma' = 1$ (the exit surface is parallel to the lattice planes), the incident and reflected wavefields are of course simply symmetric to one another.

Such partial reflections on a lateral face of a crystal have been studied in particular by Borrmann and Lehmann (1963), Lehmann and Borrmann (1967), Lang *et al.* (1986*b*, 1990) and Snigirev and Suvorov (1993), who investigated interferences between partially reflected wavefields and wavefields that were not partially reflected (see Section 12.12.4).

6.4 Rocking curves of the reflected and refracted beams

6.4.1 *Boundary conditions for the amplitudes at the exit surface*

Using an argument similar to that used in Section **6.1**, it can be shown that there is continuity of the amplitudes of the reflected and refracted waves at the exit surface:

$$D^{(d)}_{hj} = D_{hj}; \quad D^{(d)}_{oj} = D_{oj}.$$

Taking into account the phase factors, the expressions of the four waves after the crystal surface are, in the general case:

$$\begin{aligned}
\mathcal{D}_{hj}^{(d)} &= D_{hj} \exp\left[-2\pi \mathrm{i}\mathbf{K}_{\mathbf{h}}^{(\mathbf{a})} \cdot \mathbf{r}^{(\mathbf{a})}\right] \exp\left[-2\pi \mathrm{i}\mathbf{K}_{\mathbf{hj}} \cdot \left(\mathbf{r}^{(\mathbf{d})} - \mathbf{r}^{(\mathbf{a})}\right)\right] \\
&\quad \times \exp\left[-2\pi \mathrm{i}\mathbf{K}_{\mathbf{hj}}^{(\mathbf{d})} \cdot \left(\mathbf{r} - \mathbf{r}^{(\mathbf{d})}\right)\right], \\
\mathcal{D}_{oj}^{(d)} &= D_{oj} \exp\left[-2\pi \mathrm{i}\mathbf{K}_{\mathbf{o}}^{(\mathbf{a})} \cdot \mathbf{r}^{(\mathbf{a})}\right] \exp\left[-2\pi \mathrm{i}\mathbf{K}_{\mathbf{oj}} \cdot \left(\mathbf{r}^{(\mathbf{d})} - \mathbf{r}^{(\mathbf{a})}\right)\right] \\
&\quad \times \exp\left[-2\pi \mathrm{i}\mathbf{K}_{\mathbf{oj}}^{(\mathbf{d})} \cdot \left(\mathbf{r} - \mathbf{r}^{(\mathbf{d})}\right)\right],
\end{aligned} \tag{6.14a}$$

where $\mathbf{r}^{(\mathbf{a})}$, $\mathbf{r}^{(\mathbf{d})}$ and $\mathbf{r}$ are the position vectors of points on the entrance surface, the exit surface and outside the crystal, respectively. The three exponentials in the expression of each wave correspond to the optical paths before, inside and after the crystal, respectively. The consideration of these phase factors is particularly important when studying the propagation of ultrashort pulses or time-dependent diffraction, as is possible with fourth-generation synchrotrons (Malgrange and Graeff 2003).

Given that $\mathbf{OP_j} = \mathbf{OM} + \mathbf{MP_j}$ and $\mathbf{HP_j} = \mathbf{HM} + \mathbf{MP_j}$ (Fig. 6.1a), the expressions (6.14a) of the reflected and refracted waves reduce to, for a *plane-parallel* crystal:

$$\begin{aligned}
\mathcal{D}_{hj} &= \sum \mathcal{D}_{hj}^{(d)} = \sum \left\{D_{hj} \exp\left[-2\pi \mathrm{i}\overline{MP_j}\, t\right]\right\} \Phi_h(\mathbf{r}), \\
\mathcal{D}_{oj} &= \sum \mathcal{D}_{oj}^{(d)} = \sum \left\{D_{oj} \exp\left[-2\pi \mathrm{i}\overline{MP_j}\, t\right]\right\} \Phi_o(\mathbf{r}).
\end{aligned} \tag{6.14b}$$

where t is the crystal thickness and

$$\begin{aligned}
\Phi_h(\mathbf{r}) &= \exp\left[-2\pi \mathrm{i}\mathbf{K}_{\mathbf{h}}^{(\mathbf{a})} \cdot \mathbf{r}^{(\mathbf{d})}\right] \exp\left[-2\pi \mathrm{i}\mathbf{K}_{\mathbf{hj}}^{(\mathbf{d})} \cdot \left(\mathbf{r} - \mathbf{r}^{(\mathbf{d})}\right)\right], \\
\Phi_o(\mathbf{r}) &= \exp\left[-2\pi \mathrm{i}\mathbf{K}_{\mathbf{o}}^{(\mathbf{a})} \cdot \mathbf{r}\right],
\end{aligned}$$

are phase factors.

Expression (5.23) of $\overline{MP_j}$ in terms of the deviation parameter η can be rearranged in terms of the departure from Bragg's angle, $\Delta\theta$. The relation between η and $\Delta\theta$ is given by (5.18) which can also be written, using (4.26), as

$$\frac{\eta}{\Lambda_L} = k(\Delta\theta - \Delta\theta_{os})\frac{\sin 2\theta_B}{\gamma_h} = k\Delta\theta\frac{\sin 2\theta_B}{\gamma_h} + \frac{k\gamma_o}{2}\left(\frac{1}{\gamma_h} - \frac{1}{\gamma_o}\right).$$

It follows that

$$\overline{MP_j} = \frac{k\chi_o}{4}(\gamma_o^{-1} + \gamma_h^{-1}) + \frac{k\Delta\theta \sin 2\theta_B}{2\gamma_h} \pm \frac{\sqrt{\eta^2 + 1}}{2\Lambda_L}. \tag{6.15}$$

Substituting from (6.15) into (6.14) we obtain, for the amplitudes of the emerging waves,

$$\mathcal{D}_h^{(d)} = D_{h1}\Phi_h(\mathbf{r})\exp\Big(-2\pi\mathrm{i}\Big[\frac{k\chi_o t}{4}(\gamma_o^{-1}+\gamma_h^{-1})+\frac{k\Delta\theta t\sin 2\theta_B}{2\gamma_h}\Big]\Big)$$
$$\times\left\{\exp\left(-2\pi\mathrm{i}\frac{t\sqrt{\eta^2+1}}{2\Lambda_L}\right)-\exp\left(+2\pi\mathrm{i}\frac{t\sqrt{\eta^2+1}}{2\Lambda_L}\right)\right\},$$
$$\mathcal{D}_o^{(d)} = \Phi_o(\mathbf{r})\exp\Big(-2\pi\mathrm{i}\Big[\frac{k\chi_o t}{4}(\gamma_o^{-1}+\gamma_h^{-1})+\frac{k\Delta\theta\sin 2\theta_B}{2\gamma_h}\Big]\Big)$$
$$\times\left\{D_{o1}\exp\left(-2\pi\mathrm{i}\frac{t\sqrt{\eta^2+1}}{2\Lambda_L}\right)+D_{o2}\exp\left(+2\pi\mathrm{i}\frac{t\sqrt{\eta^2+1}}{2\Lambda_L}\right)\right\}. \quad (6.16a)$$

6.4.2 *Reflectivity*

Substituting from (6.2b), and using (4.38), it follows, for the reflectivities, that

$$I_h^{(d)} = \left|\frac{\chi_h}{\chi_{\bar{h}}}\right|\exp\left[-\frac{\mu_o}{2}\left(\frac{1}{\gamma_o}+\frac{1}{\gamma_h}\right)t\right]\frac{|\mathcal{E}_1-\mathcal{E}_2|^2}{4|1+\eta^2|},$$
$$I_o^{(d)} = \exp\left[-\frac{\mu_o}{2}\left(\frac{1}{\gamma_o}+\frac{1}{\gamma_h}\right)t\right]\frac{|\eta[\mathcal{E}_1-\mathcal{E}_2]+\sqrt{1+\eta^2}\,[\mathcal{E}_1+\mathcal{E}_2]|^2}{4|1+\eta^2|} \quad (6.16b)$$

with $\mathcal{E}_1 = \exp(\pi i t/\Lambda)$; $\mathcal{E}_2 = \exp(-\pi i t/\Lambda)$; $\Lambda = \Lambda_L/\sqrt{1+\eta^2}$.

In absorbing crystals, the deviation parameter and the Pendellösung distance are complex and the reflectivities (6.16b) are valid whatever the value of the imaginary part of the scattering factor. They can also be expressed explicitly in terms of η_r and the following expressions are valid when χ_{ih}/χ_{rh} is not large.

One finds, using (5.17b), (5.23), (5.31), (5.37), (6.2b) and (6.16a):

$$I_h^{(d)} = \left|\frac{\chi_h}{\chi_{\bar{h}}}\right|\frac{\exp[-(\frac{\mu_o}{2})(\gamma_o^{-1}+\gamma_h^{-1})t]}{2(1+\eta_r^2)}\left\{\cosh\left[\mu_o t\left[\frac{1}{2}(\gamma_o^{-1}-\gamma_h^{-1})\eta_r\right.\right.\right.$$
$$\left.\left.\left.+|C|\left|\frac{\chi_{ih}}{\chi_{io}}\right|\frac{\cos\varphi}{\sqrt{\gamma_o\gamma_h}}\right]\frac{1}{\sqrt{1+\eta_r^2}}\right]-\cos\left[2\pi t\left(\frac{1}{\Lambda_o}\right)_r\sqrt{1+\eta_r^2}\right]\right\},$$
$$I_o^{(d)} = \left|\frac{\chi_h}{\chi_{\bar{h}}}\right|\frac{\exp[-(\mu_o/2)(\gamma_o^{-1}+\gamma_h^{-1})t]}{2(1+\eta_r^2)}$$
$$\times\left\{\cosh\left[2v+\mu_o t\left[\frac{1}{2}(\gamma_o^{-1}-\gamma_h^{-1})\eta_r+|C|\left|\frac{\chi_{ih}}{\chi_{io}}\right|\frac{\cos\varphi}{\sqrt{\gamma_o\gamma_h}}\right]\frac{1}{\sqrt{1+\eta_r^2}}\right]\right.$$
$$\left.-\cos\left[2\pi t\left(\frac{1}{\Lambda_o}\right)_r\sqrt{1+\eta_r^2}-\frac{2\eta_i}{\sqrt{1+\eta_r^2}}\right]\right\}, \quad (6.16c)$$

where $v = \operatorname{arcsinh}\eta_r$ and $(1/\Lambda_o)_r$ is the real part of $1/\Lambda_o$.

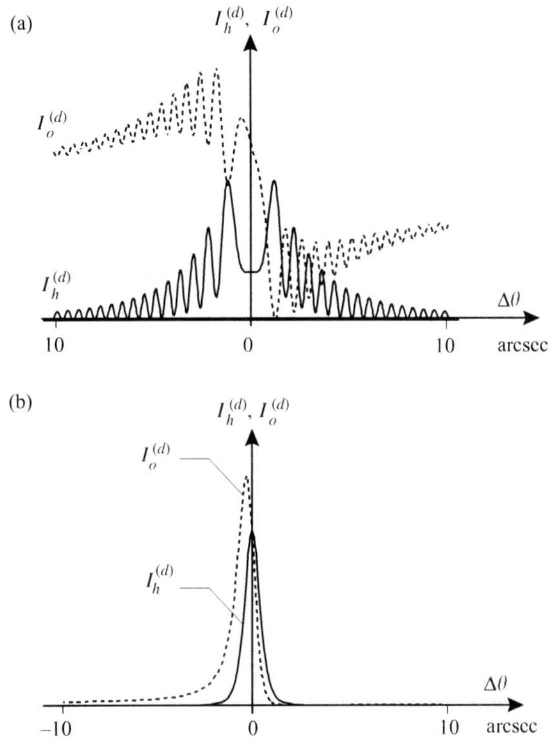

Fig. 6.7. Rocking curves in the transmission case. Silicon, 220, Cu$K\alpha$ ($\Lambda_L = 15.3\,\mu$m). Solid curve: reflected beam. Dashed curve: refracted beam. (a) $t/\Lambda_L = 5$ ($\mu_o t = 1.09$). (b) $t/\Lambda_L = 25$ ($\mu_o t = 5.46$).

6.4.3 *Properties of the rocking curves*

Figures 4.18(a) and (b) represent the variations of $I_h^{(d)}$ and $I_o^{(d)}$ for thin crystals ($t/\Lambda_L = 1.0$ and 1.5) and Figs 6.7(a) and (b) for thicker crystals where absorption is important ($t/\Lambda_L = 5.0$, $\mu_o t = 1.09$ and $t/\Lambda_L = 25.0$, $\mu_o t = 5.45$, respectively). The oscillatory structure of the rocking curves in the transmission geometry was first observed by Lefeld-Sosnowska and Malgrange (1968, 1969) and later by Bonse *et al.* (1977). It was used by Bonse and Teworte (1980), Teworte and Bonse (1984), to measure directly the structure factor for X-rays. The same technique was extended to the case of high-energy γ-rays by Dewey *et al.* (1994). Deutsch and Hart (1985*a* and *b*) and Deutsch *et al.* (1990) improved the accuracy significantly by using a multiparameter pattern fitting of theoretical and experimental rocking curves, without any length measurement.

As thickness increases, the period of the Pendellösung fringes in η_r decreases and, because of the increasing values of $\mu_o t$, their contrast decreases till it completely disappears, as in Fig. 6.7(b). At the same time the rocking curve for the refracted beam becomes strongly asymmetric: the reflectivity is larger for negative values of the deviation parameter which correspond to tiepoints on branch 1 of the dispersion surface and to anomalously low absorption, while for positive values of the deviation parameter, the tiepoints lie on branch 2 and absorption is higher than normal. When $\mu_o t > 5$ or 6, and for a symmetric reflection, the maximum of I_o occurs to a good approximation for a value of the deviation parameter given by

$$\eta_r = -2 \Big/ \left(\frac{\mu_o t}{\cos\theta_B} |C| \left| \frac{\chi_{ih}}{\chi_{io}} \right| \cos\varphi + 2 \right).$$

The observation of the asymmetry of the refracted beam rocking curve by Campbell (1951*a*,*b*), Rogosa and Schwarz (1952), Brogren (1954), Brogren and Adell (1954*a*,*b*), among others, served as early confirmations of the Borrmann effect. The sign of the asymmetry depends on the value of φ, and its reversal for wavelengths just below and just above the absorption edge of Ga in the case of the 200 reflection on gallium arsenide is due to the variation of φ from 0 to π (Fukamachi *et al.* 1996; Negishi *et al.* 1998).

The rocking curve is always symmetric for the reflected beam. For a symmetric reflection ($\gamma = 1$) and for values of t/Λ_L for which Pendellösung oscillations are negligible, $I_o^{(d)} = I_h^{(d)}$ for $\eta_r = 0$.

Width of the rocking curve: in an absorbing crystal, the Full Width at Half Maximum (FWHM) corresponds, when neglecting Pendellösung, to $\Delta\eta_r = 2$ and is given by (4.26):

$$\delta_{os} = \sqrt{\gamma}\, \frac{\sqrt{|\chi_h \chi_{\bar{h}}|}}{\sin 2\theta_B}.$$

The various factors influencing the value of δ_{os} are discussed in Sections 4.10 and 7.1. The FWHM decreases when the thickness increases, because of the effect of absorption. When $\mu_o t$ is large, wavefields whose tiepoints belong to branch 2 can be neglected and, in the case of a symmetric reflection ($\gamma = 1$), the reflectivity (6.16b) reduces to, using (5.42),

$$I_h^{(d)} = \left| \frac{\chi_h}{\chi_{\bar{h}}} \right| \exp\left[-\frac{\mu_o t}{\cos\theta_B} \right] \frac{\exp[\zeta_s (1+\eta^2)^{-1/2}]}{1+\eta^2},$$

where $\zeta_s = 2\pi k |C \chi_{ih} \cos\varphi| t / \cos\theta_B$. The corresponding profile is very narrow and it can be shown that it can be replaced, with a very good approximation, for $\mu_o t \geq 6$, by

$$I_h^{(d)} = \left| \frac{\chi_h}{\chi_{\bar{h}}} \right| \exp\left[-\frac{\mu_o t}{\cos\theta_B} \right] \exp(\zeta_s) \exp\left(-\frac{\zeta_s \eta^2}{2\delta_{os}^2} \right).$$

Its FWHM is, therefore,

$$2\delta_{os}\sqrt{\frac{2\log 2}{\zeta_s}},$$

where log 2 is the natural logarithm of 2. For instance, for silicon, Cu$K\alpha$, 220 and $\mu_o t = 6$, the FWHM is $0.467 \times 2\delta_{os}$.

6.5 Integrated intensity

The integrated intensity has been calculated by Kato (1955) in the absorbing case:

$$I_{hi}^{(d)} = A\Big[\int_0^{(2\pi t/\Lambda_L)} J_0(z)\,\mathrm{d}z - 1 + I_0(\zeta)\Big], \tag{6.17}$$

where

$$A = \frac{\pi|C\chi_h|\sqrt{\gamma}}{2\sin 2\theta_B}\left|\frac{\chi_h}{\chi_{\bar{h}}}\right|\exp\Big[-\frac{\mu_o t}{2}\Big(\frac{1}{\gamma_o}+\frac{1}{\gamma_h}\Big)\Big],$$

$$\zeta = \frac{\mu_o t}{\sqrt{\gamma_o\gamma_h}}\sqrt{|C|^2\left|\frac{\chi_{ih}}{\chi_{io}}\right|^2\cos^2\varphi + \frac{\gamma_h-\gamma_o}{4\gamma_o\gamma_h}},$$

and $I_0(\zeta)$ is the modified Bessel function of zeroth order.

Figure 6.8 compares the variations of the integrated intensity with t/Λ_L given by (6.17) with those given by the geometrical theory (3.17). They are identical for $t/\Lambda_L < 0.1$. As the crystal thickness becomes larger, the value given by the geometrical theory, I_{hi}^{geom}, increases rapidly above that given by the dynamical theory, I_{hi}, but it is damped because of absorption and it reaches a maximum. It then decreases more rapidly than I_{hi}, because the latter is less absorbed on account of the Borrmann effect. For a certain value of $\mu_o t$, I_{hi}^{geom} becomes smaller than the value given by the dynamical theory as was noticed, and checked experimentally, by Parthasarathy *et al.* (1963).

Expression (6.17) can be approximated:

- when ζ is small, by

$$I_{hi}^{(d)} = A\Big[\int_0^{2\pi t/\Lambda_L} J_0(z)\,dz + \frac{\zeta^2}{4} + \frac{\zeta^4}{64} + \cdots\Big]$$

- when ζ is large, by

$$I_{hi}^{(d)} = A\frac{\exp(\zeta)}{\sqrt{2\pi\zeta}}\Big[1 + \frac{1}{8\zeta} + \frac{9}{128\zeta^2} + \cdots\Big].$$

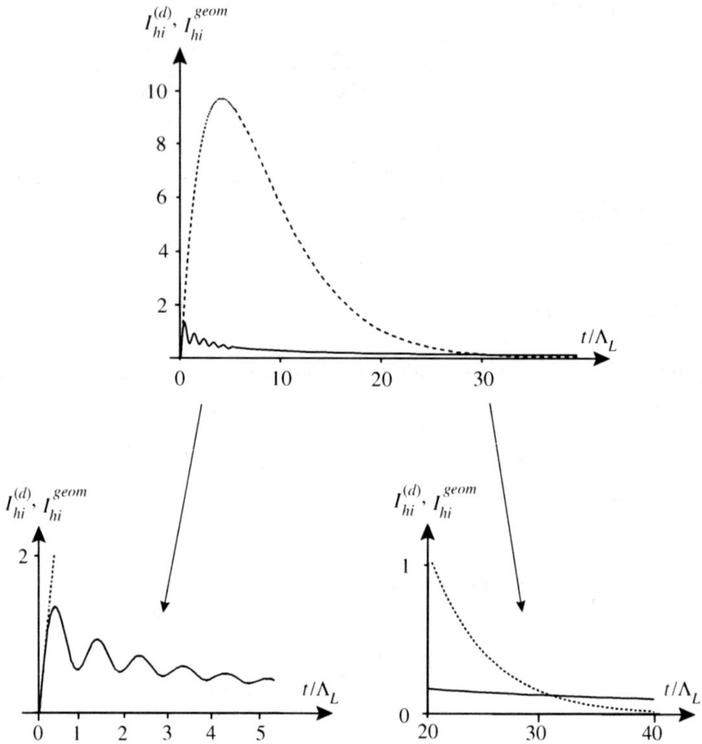

Fig. 6.8. Variations of the integrated intensity with crystal thickness in the transmission geometry. Solid curve: $I_{hi}^{(d)}$ (dynamical theory); dashed curve: I_{hi}^{geom} (geometrical theory); Silicon, 220 reflection, Cu$K\alpha$, $\Lambda_L = 15.25\,\mu\text{m}$). The two bottom drawings are enlargements of two portions of the top drawing.

Stephenson (1997) compared the integrated reflectivities of diamond, silicon and germanium for various asymmetries of the 111 and 220 reflections in order to optimize the choice of Laue-geometry monochromators and showed that perfect thin ($\geq 50\,\mu$m) diamond crystals diffract higher integrated reflectivities than those of equivalent silicon or germanium crystals.

7

Intensities of plane waves in the reflection geometry

7.1 Thick absorbing crystals

7.1.1 *Reflectivity*

As has been shown in Section 4.9.2, only one wavefield needs to be considered in the reflection geometry when the crystal thickness is very large (Fig. 4.20). The expressions of the amplitude D_{hj} and of the reflectivity I_h are given by (4.28) or (5.38) and (4.41), respectively:

$$D_{hj} = \frac{1}{\sqrt{|\gamma|}}\sqrt{\left|\frac{\chi_h}{\chi_{\bar{h}}}\right|}\exp(\mathrm{i}\beta')\big[\eta \pm \sqrt{\eta^2 - 1}\big]D_o^{(a)},$$

$$I_h^{(a)} = \gamma\frac{|D_h^{(a)}|^2}{|D_o^{(a)}|^2} = |\gamma||\xi_j|^2 = \left|\frac{\chi_h}{\chi_{\bar{h}}}\right|\left|\eta \pm \sqrt{\eta^2 - 1}\right|^2. \tag{7.1}$$

The deviation parameter is complex and its expression is given in Section 5.6. As has been said in Section 5.7.2, $\sqrt{\eta^2 - 1}$ has a very large imaginary part when $|\eta_r| \leq 1$ and it is not possible to use the same approximations as in the Laue case. The sign in (7.1) must be chosen in such a way that $|\xi|$ is smaller than 1. If $I_h^{(a)}$ is calculated with a computer program using complex numbers such as Mathematica©, the following expression gives the correct solution:

$$I_h^{(a)} = \left|\frac{\chi_h}{\chi_{\bar{h}}}\right|\left|\eta - S(\eta_r)\sqrt{\eta^2 - 1}\right|^2.$$

It is also possible to calculate I_h analytically using the parameter Z and the phase ψ' introduced in Section 5.7.2. It follows that

$$\xi_j = |\xi_j|\exp(\mathrm{i}\psi) = \frac{1}{\sqrt{|\gamma|}}\sqrt{\left|\frac{\chi_h}{\chi_{\bar{h}}}\right|}Z\exp\mathrm{i}(\beta' + \psi'),$$

$$I_h^{(a)} = \left|\frac{\chi_h}{\chi_{\bar{h}}}\right|Z^2. \tag{7.2}$$

Let us set

$$\rho^2\exp(2\mathrm{i}\omega) = \eta^2 - 1, \tag{7.3}$$

where there are two solutions for ω, ω' and $\omega'' = \omega' + \pi$, the correct value being that for which Z is smaller than 1. Using (7.3), Z^2 can be written as

$$Z^2 = |\eta|^2 + \rho^2 + \eta^\star \rho \exp(\mathrm{i}\omega_j) + \eta\rho \exp(-\mathrm{i}\omega_j),$$

where j designates one of the two solutions for ω_j. Let us set

$$L = |\eta|^2 + \rho^2,$$

$$M = \eta^\star \rho \exp(\mathrm{i}\omega_j) + \eta\rho \exp(-\mathrm{i}\omega_j).$$

It can easily be shown that

$$M = 2\rho(\eta_r \cos\omega_j + \eta_i \sin\omega_j),$$

$$M^2 = L^2 - 1$$

and, therefore, that

$$Z = \sqrt{L \pm \sqrt{L^2 - 1}},$$

where the $\pm$ sign corresponds to the two solutions for ω_j. The correct solution is the one for which Z is less than 1:

$$Z = \sqrt{L - \sqrt{L^2 - 1}}. \tag{7.4}$$

The correct solution for ω_j can also be found by noting that M must be negative. Table 7.1 shows the variations with the real part of the deviation parameter of $\eta_i = A\eta_r + B$, $\tan 2\omega$, ω', ω'' and the values of M, M' and M'', corresponding to ω' and ω'', respectively. It shows that the correct solution for ω is ω'. The corresponding values of the phases ψ' and ψ are also given. The variations of ψ' with η_r are represented in Fig. 5.9.

The expression of the reflected intensity in the absorbing Bragg geometry was first given by Prins (1930). The way of representing it given here was introduced by Hirsch and Ramachandran (1950) and expression (7.4) was first given by Fingerland (1971). Analytical expressions of the integrated intensity have been given by Afanas'ev and Persinev for both strong and weak absorption.

Table 7.1 Variations of the phase ω_j of $\pm\sqrt{\eta^2 - 1}$ and of ψ' and ψ with η_r

η_r	$-\infty$	$-B/A$	-1	0	$+1$		$+\infty$
$\eta_i = A\eta_r + B$	>0	0	<0	B		<0	
$\tan 2\omega$	$2A/1 - A^2$	0	$\pi/2$	π	$3\pi/2$		$2A$
ω'	$-\beta$	0	$\pi/4$	$\pi/2$	$3\pi/4$		$\pi + A$
ω''	$\pi - \beta$	π	$5\pi/4$	$3\pi/2$	$7\pi/4$		A
M'	<0	$-B/A$	<0	B		>0	
M''	>0	B/A	>0	$-B$		<0	
ψ'	$\pi + \beta$	π		$\pi/2$			β
$\psi = \psi' + \beta'$	$\pi + \beta + \beta' = \varphi_h$	$\pi + \beta'$		$\pi/2 + \beta'$			φ_h

7.1.2 *Shape of the rocking curves*

There is no longer a total reflection domain and energy penetrates inside the crystal at all incidence angles although with a very high absorption coefficient within the domain $|\eta_r| \leq 1$, as has been shown in Section 5.8.3. Figure 4.21 shows the typical shape of the rocking curves for a thick absorbing crystal in the reflection geometry. It was first observed by Renninger (1955) on calcite using a triple-crystal arrangement and further studied by Bubáková (1962) and Bubáková *et al.* (1962*a*) on germanium and by Bubáková *et al.* (1962*b*) on silicon, also by means of a triple-crystal diffractometer, and by Okkerse (1963) on germanium, using consecutive reflections in the transmission and reflection geometries on the same perfect crystal. The main characteristics of the curve are its shape, its width and the height of the maximum.

1. *Shape*: The shape of the curve is asymmetric and was studied by Bucksch *et al.* (1967), Fingerland (1971) and Wilkins (1978*a*). The asymmetry is due to the anomalous absorption effect: on the small-angle side, the reflected intensity is associated with wavefields belonging to branch 1 of the dispersion surface which undergo an absorption smaller than normal, while on the high-angle side it is associated with wavefields belonging to branch 2 with an absorption higher than normal. The amount of asymmetry depends on the asymmetry ratio γ and on the values of the Fourier coefficients of the imaginary part of the polarizability, χ_{io} and χ_{ih}, through the coefficients A and B in the expression of the imaginary part of the deviation parameter (defined in (5.33)). In the case of a centrosymmetric crystal and of a symmetric reflection, they can be approximated by

$$A = -\frac{\chi_{ih}}{\chi_{rh}}; \qquad B = -\frac{\chi_{io}}{\chi_{rh}}. \tag{7.5}$$

The ratio A/B can therefore be approximated by the ratio χ_{ih}/χ_{io} of the h and o components of the Fourier coefficients of the imaginary part of the polarizability. The variation of this ratio with $\sin\theta_B/\lambda$ was discussed in Section 2.7.4.

Figure 7.1 shows the variations of the shape of the rocking curve with the values of B and A/B. B is proportional to the normal absorption coefficient, μ_o, and A/B to the angular dependence of the Fourier coefficients of the imaginary part of the polarizability. For $A/B = 1$ (a purely theoretical hypothesis), the imaginary part of the deviation parameter is $\eta_i = A(\eta_r + 1)$ and, at the low-angle side of the total reflection domain ($\eta_r = -1$), η_i and Z are 0 and 1, respectively, as in the non-absorbing case. The asymmetry is highest for $A/B = 1$ and increases with absorption, as can be seen in the figure.

Examples. Figure 7.2 shows the rocking curves of the 400 reflection of a gallium arsenide crystal for two wavelengths, 1.0447 Å (just below the absorption edge of arsenic) for which absorption is very large ($91.3\,\text{mm}^{-1}$) and 1.2312 Å for which the absorption is much smaller ($18.4\,\text{mm}^{-1}$); the values of A/B are practically equal ($\approx$0.9174) in the two cases. The rocking curve is much more asymmetric in the former case than in the latter and this is due to anomalous absorption.

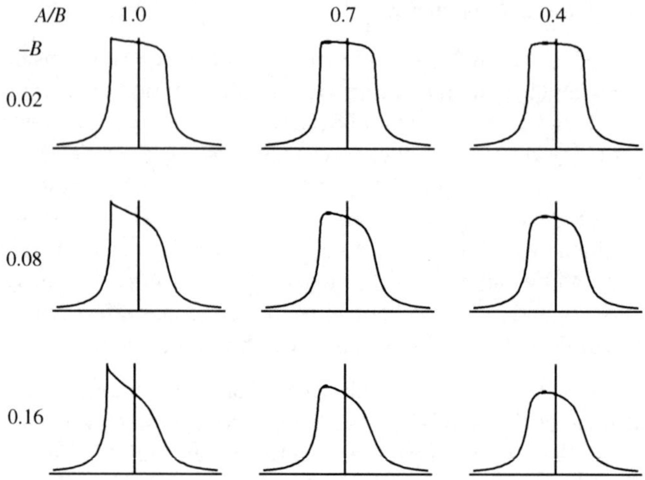

Fig. 7.1. Variation of the shape of the rocking curve in the reflection geometry for thick crystals with parameters A and B, defined by (5.33), and approximated by (7.5).

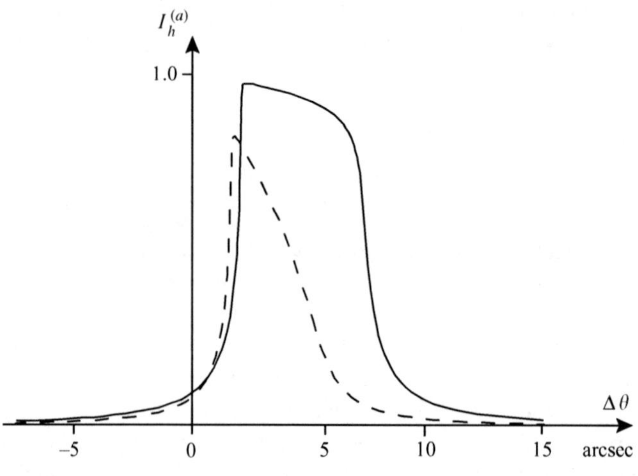

Fig. 7.2. Calculated rocking curves in the reflection geometry for thick crystals and two different absorption coefficients. GaAs, 400 reflection. Solid curve: $\lambda = 1.2312$ Å, $\mu_o = 18.4\,\text{mm}^{-1}$, $A/B = 0.917$; dashed curve: $\lambda = 1.0447$ Å, $\mu_o = 91.3\,\text{mm}^{-1}$, $A/B = 0.917$. For the same value of A/B, the asymmetry of the curve is much more pronounced when the absorption is larger.

In a second example, Fig. 7.3 compares the rocking curves of two reflections of a face-centred cubic material, one with even indices and the other one with odd indices, for the same wavelength (220 and 311, silicon and Cu$K\alpha$ radiation for Fig. 7.3(a); 400 and 111, GaAs and $\lambda = 0.8211$ Å for Fig. 7.3(b)). The ratio A/B is quite larger

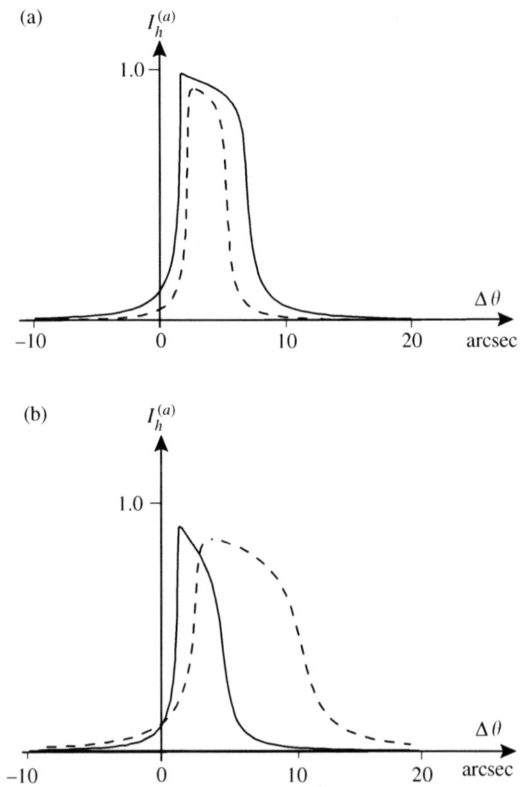

Fig. 7.3. Calculated rocking curves in the reflection geometry for thick crystals, influence of the value of the ratio A/B. (a) Silicon, Cu$K\alpha$, $\mu_o = 14.32\,\text{mm}^{-1}$; solid curve: 220 reflection, $A/B = 0.97$; dashed curve: 111 reflection, $A/B = 0.68$. (b) GaAs, $\lambda = 0.8211$ Å, $\mu_o = 47.9\,\text{mm}^{-1}$; solid curve: 400 reflection, $A/B = 0.917$; dashed curve: 111 reflection, $A/B = 0.70$. Note that in both cases, (a) and (b), for the same absorption, the asymmetry of the curve is much more pronounced when A/B is larger.

in the former case than in the latter and the asymmetry is clearly more pronounced. Figure 7.3(a) shows an example where the absorption is small and Fig. 7.3(b) one where the absorption is large.

2. *Width*: The width $2\delta_{os}$ is given by (4.48b):

$$2\delta_{os} = \frac{2R}{\pi V}\frac{\lambda d_{hkl}}{\sqrt{1-\lambda^2/4d_{hkl}^2}}\sqrt{|\gamma|}|C|\sqrt{F_h F_{\bar{h}}}.$$

(i) It increases with increasing wavelength, as is shown in Fig. 7.4, which represents its variations with λ for silicon and three different reflections, 111, 220 and 400, and as is illustrated in Fig. 7.5 comparing the rocking curves of the 400

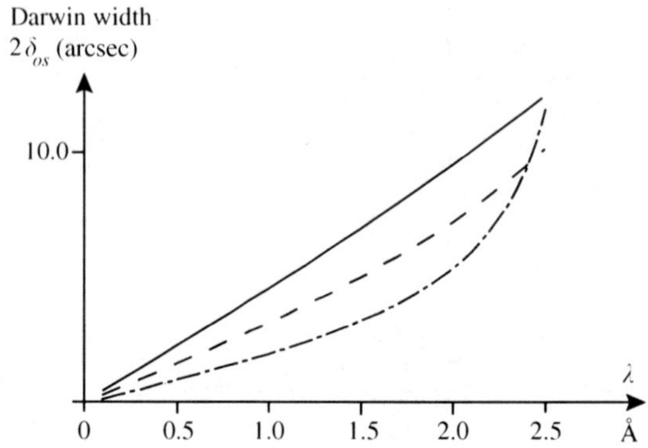

Fig. 7.4. Variations of the Darwin width, $2\delta_{os}$, with wavelength. Silicon; solid curve: 111 reflection; dashed curve: 220 reflection, dash-dotted curve: 400 reflection.

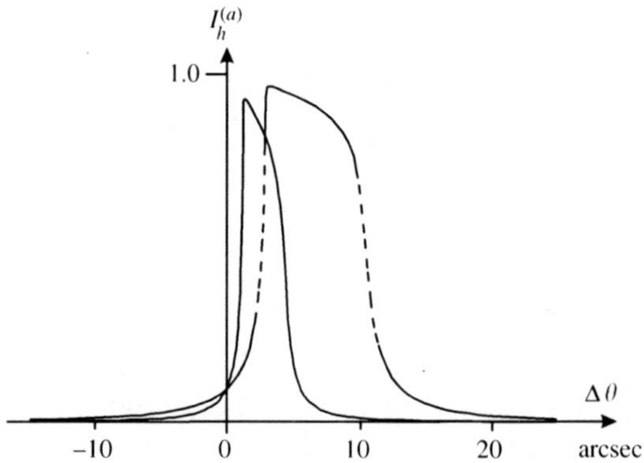

Fig. 7.5. Calculated rocking curves in the reflection geometry for thick crystals and two different wavelengths. GaAs, 400 reflection. Solid curve: Mo$K\alpha$, $\mu_o = 32.3\,\text{mm}^{-1}$, $A/B = 0.917$; dashed curve: Cu$K\alpha$, $\mu_o = 34.6\,\text{mm}^{-1}$, $A/B = 0.917$.

reflection from GaAs for Mo$K\alpha$ and Cu$K\alpha$. The very narrow Darwin widths at very short wavelengths provide a very high sensitivity to small strains which is used for the characterization of materials (see Section 4.11). This is also useful to separate split Bragg peaks when two phases are present near a phase transition as was shown, for instance, in quartz by Bastie and Dolino (1985)—see also the references therein.

(ii) It is larger, the stronger, the reflection as is illustrated by Figs 7.3(a) and (b).

(iii) It is larger for σ-polarization than for π-polarization, as illustrated in Fig. 7.6.

(iv) It varies with the asymmetry of the reflection as has been shown in Section 4.10 (see Fig. 4.23).

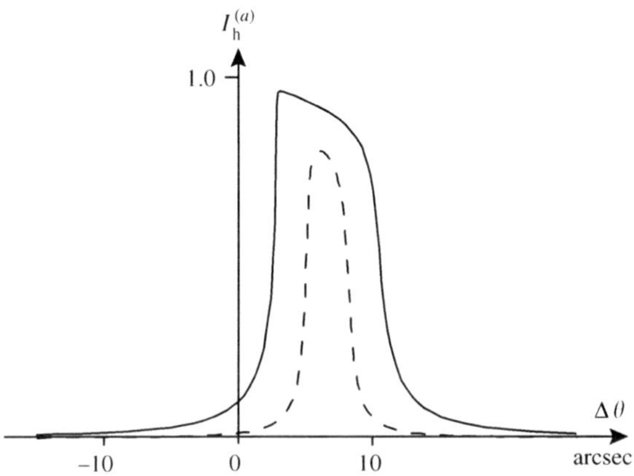

Fig. 7.6. Calculated rocking curves in the reflection geometry for thick crystals, influence of polarization. GaAs, 400 reflection, Cu$K\alpha$. Solid curve: σ-polarization, $A/B = 0.917$; dashed curve: π-polarization, $A/B = 0.37$.

3. *Height of the maximum*: It is possible to show that the maximum of the reflectivity occurs for

$$\eta_r = -\frac{A}{B}\frac{1+B^2}{1+A^2}$$

and is

$$I_h^{(a)} = \frac{(\sqrt{1+B^2} - \sqrt{B^2 - A^2})^2}{1+A^2}.$$

4. *Departure from Friedel's law in non-centrosymmetric crystals*: The reflectivity is proportional to $|\chi_h/\chi_{\bar{h}}|$ which is different from 1 in non-centrosymmetric crystals. This effect is illustrated in Fig. 7.7 which compares the rocking curves of the 111 (surface occupied by gallium atoms) and $\bar{1}\bar{1}\bar{1}$ (surface occupied by arsenic atoms) reflections of gallium arsenide for a wavelength slightly below the absorption edge of gallium for which the absorption is very high ($|\chi_h/\chi_{\bar{h}}| = 1.184$ and 0.844, respectively).

5. *Very large values of the imaginary part of the scattering factor*: The expressions given in this section for the reflected intensity are valid even for very large values of $|\chi_i|$. Figure 7.8 shows the variation of the shape of the rocking curve for increasing values of $|\chi_{ih}/\chi_{rh}|$ and constant values of $|\chi_h|$ and of $|\chi_{ih}/\chi_{io}|$. It can be seen that there is a rocking curve for $\chi_{rh} = 0$ and that it is symmetric. This was first noticed by Kato (1992).

The rocking curves for high values of χ_{hi} were further studied by Fukamachi and Kawamura (1993), Fukamachi *et al.* (1995, 2002) and Kanamaru *et al.* (1995). The

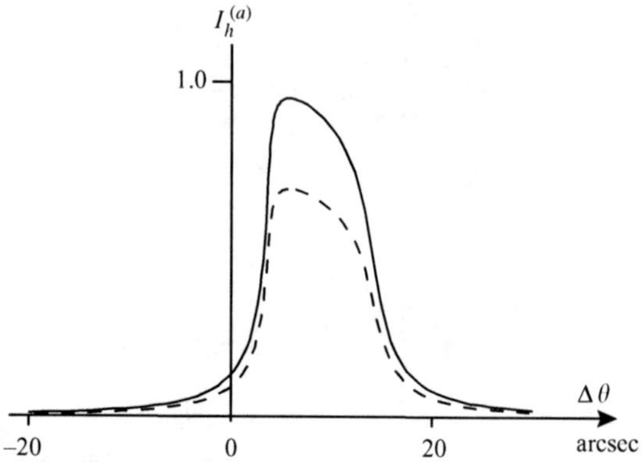

Fig. 7.7. Calculated rocking curves in the reflection geometry, Friedel pair. GaAs, $\lambda = 1.1953$ Å. Solid curve: 111 reflection, surface occupied by gallium atoms, $|\chi_h/\chi_{\bar{h}}| = 1.184$; dashed curve: $\bar{1}\bar{1}\bar{1}$ reflection, surface occupied by arsenic atoms, $|\chi_h/\chi_{\bar{h}}| = 0.844$.

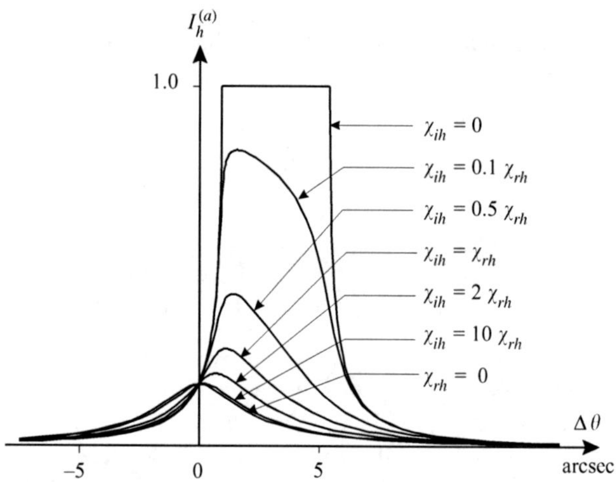

Fig. 7.8. Variation of the shape of the rocking curve in the reflection geometry with increasing values of the imaginary part of the scattering factor. The curves have been calculated for $|\chi_o/\chi_h| = 0.71$.

formation of X-ray standing waves when the real part of the scattering factor vanishes was discussed by Negishi *et al.* (1999).

Fukamachi *et al.* (1993) measured the integrated intensity of the 844 reflection very close to the absorption edge of germanium using synchrotron radiation and obtained satisfactory agreement with the theoretical calculations even for vanishing values of χ_{rh}. Kuribayashi *et al.* (1997) recorded the 111 rocking curves of germanium and gallium arsenide very close to the absorption edges of germanium, gallium and arsenic, respectively, with synchrotron radiation and a six-reflections monochromator. Using fitting procedures with theoretical curves they could measure the values of the anomalous dispersion corrections.

The fact that the rocking curve becomes symmetrical when $\chi_{rh} = 0$ can easily be understood. It follows in that case that

$$\chi_h = \mathrm{i}\chi_{ih}; \qquad \chi_{\bar{h}} = \mathrm{i}\chi_{ih}^{\star};$$
$$\chi_h\chi_{\bar{h}} = -|\chi_{ih}|^2; \qquad \sqrt{\chi_h\chi_{\bar{h}}} = \mathrm{i}|\chi_{ih}|.$$

Using these relations, the deviation parameter becomes

$$\eta = \frac{\chi_{io}}{|\chi_{ih}|}\frac{(1-\gamma)/2}{\sqrt{|\gamma|}C} - \mathrm{i}\frac{\Delta\theta \sin 2\theta_B}{\sqrt{|\gamma|}C|\chi_{ih}|}.$$

This shows that η_r is constant and that η_i changes sign with $\Delta\theta$. The quantity L in (7.4) remains therefore constant when $\Delta\theta$ changes sign and the rocking curve is symmetric. Its shape will also remain unchanged for a Friedel pair of reflections from a non-centrosymmetric crystal.

7.2 Standing waves

The intensity of the wavefield penetrating in the crystal is given by (5.43):

$$|D|^2 = |D_o|^2[1 + |\xi|^2 + 2C|\xi|\cos(2\pi\mathbf{h}\cdot\mathbf{r} + \psi)],$$

where ψ is the phase of ξ and the subscript j has been dropped.

The position of the nodes and antinodes of standing waves are given by (5.45); they are, respectively,

$$\text{nodes:} \quad \frac{\Delta d}{d_{hkl}} = N + 0.5 - \frac{\psi}{2\pi},$$
$$\text{antinodes:} \quad \frac{\Delta d}{d_{hkl}} = N - \frac{\psi}{2\pi}.$$

The variations of the phase with η_r are represented in Table 7.1 and in Fig. 5.9.

As an example, the variations of $\Delta d/d_{hkl}$ will be represented in the case of gallium arsenide, the 111 reflection and $\lambda = 1.195$ Å. Gallium arsenide has the zinc blende structure; gallium lies at the nodes of a face-centred cubic lattice and arsenic occupies half the tetrahedral sites (position $\frac{1}{4}, \frac{1}{4}, \frac{1}{4}$). The structure can be described as a compact packing $AA'BB'CC'$ of $\{111\}$ planes occupied alternately by arsenic and gallium atoms, the distance between the A and A' planes being one fourth of that between the A' and B planes (Fig. 7.9).

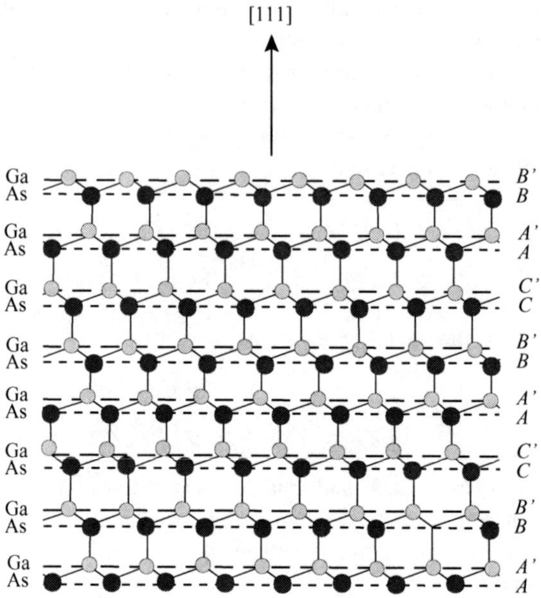

Fig. 7.9. Structure of gallium arsenide. It is constituted by a stacking of (111) double layers occupied alternately by arsenic and gallium atoms; projection on a $(1\bar{1}0)$ plane.

The structure factor of gallium arsenide is

$$F_h = 4\Big[f_{\mathrm{Ga}} \exp\Big(-2\pi \mathrm{i}\frac{h+k+l}{8}\Big) + f_{\mathrm{As}} \exp\Big(+2\pi \mathrm{i}\frac{h+k+l}{8}\Big)\Big]$$

with the origin located at mid-distance between a gallium and an arsenic atom along a [111] direction, that is at mid-distance between an A' and a B plane. For a 111 reflection, the structure factor is

$$F_{111} = -2\sqrt{2}[f_{\mathrm{As}} + f_{\mathrm{Ga}} + \mathrm{i}(f_{\mathrm{As}} - f_{\mathrm{Ga}})],$$

where the dispersion corrections are not explicitly included, for simplicity. The surface of the crystal is then an arsenic plane, with the normal oriented from the A towards the A' planes. The phase of the structure factor, for the origin which has been chosen, is

$$\varphi_h = \pi + \arctan\left[\frac{f_{\mathrm{As}} - f_{\mathrm{Ga}}}{f_{\mathrm{As}} + f_{\mathrm{Ga}}}\right]$$

and, since $f_{\mathrm{Ga}} = 26.67$ and $f_{\mathrm{As}} = 28.17$ for 111, φ_h is not 0 on the planes situated at mid-distance between the arsenic A and gallium A' layers (or B and B'), but slightly offset ($\Delta d/d_{hkl} = \pm 0.5 - 0.0044$).

The $\bar{1}\bar{1}\bar{1}$ reflection corresponds to a surface occupied by gallium atoms.

1. $\eta_r = -\infty$: Far from the reflection domain, on the low-angle side, $\psi \Rightarrow \pi + \varphi_h$. The tiepoints lie on branch 1 of the dispersion surface and the position of the nodes is

$$\frac{\Delta d}{d_{hkl}} = N - \varphi_h.$$

They are slightly offset towards the surface of the crystal with respect to the positions situated halfway between the arsenic A layer and the gallium A' layers (or B and B') and lie on the maxima of electronic density; the positions of the nodes as a function of η_r are represented by a solid line in Fig. 7.10.

It is interesting to note that the asymptotic position of the nodes far from the total reflection domain is directly related to the phase of the structure factor (Bedzyk and Materlik 1985; Authier 1986).

The antinodes lie at the positions $\Delta d/d_{hkl} = N + 0.5 - \varphi_h$ corresponding to the minima of electronic density and are slightly offset with respect to the mid-distance

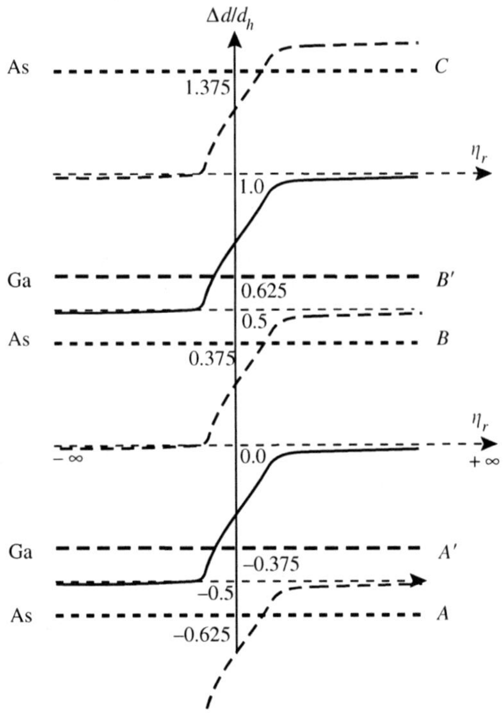

Fig. 7.10. Variations of the relative position of the nodes and antinodes of standing waves with the deviation parameter in the reflection geometry. Symmetrical 111 reflection on gallium arsenide, $\lambda = 1.195$ Å. Solid curves: nodes; dashed curves: antinodes. The thick dotted and dash-dotted straight lines show the positions of the arsenic and gallium (111) planes. The ordinate axis is oriented towards the inside of the crystal from the surface.

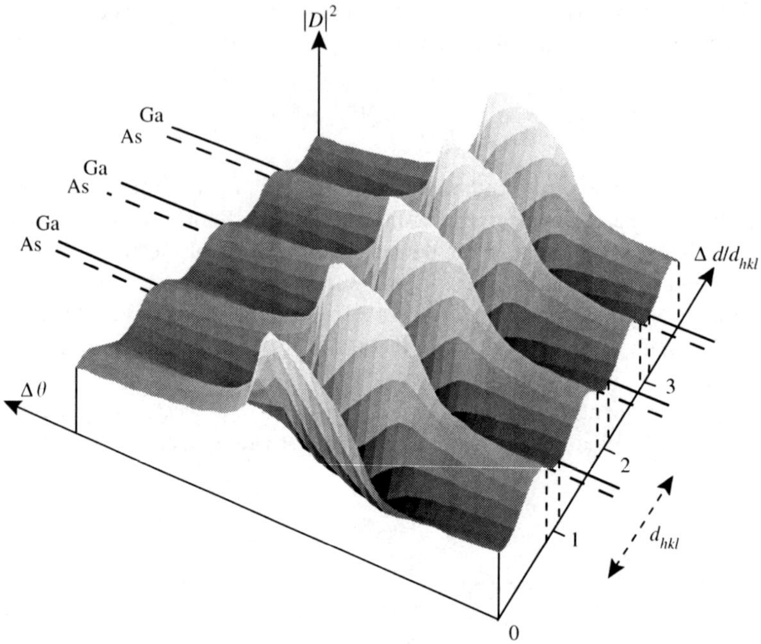

Fig. 7.11. Variations of the intensity of the standing wavefield $|D|^2$ with position $\Delta d/d_{hkl}$ within the unit cell and with angle of incidence $\Delta\theta$. GaAs, 111 reflection, $\lambda = 1.195\,\text{Å}$; the thick dashed and thick solid lines show the positions of the arsenic and gallium planes, respectively. On the low-angle side of the rocking curve, the nodes lie on the maxima of electronic density and on the high-angle side on the minima of electronic density.

between the gallium A' and the arsenic B layers; the positions of the antinodes as a function of η_r are represented by a dashed line in Fig. 7.10.

2. $\infty < \eta_r < -1$: The tiepoint remains on branch 1, ψ is close to $\pi + \varphi_h$.

3. $-1 < \eta_r < 1$: ψ varies from $\pi + \varphi_h$ to φ_h and the nodes move progressively to a position situated halfway between a gallium A' layer and an arsenic A layer; they have been shifted by $d_{hkl}/2$ (Fig. 7.10). Similarly, the antinodes move by the same amount.

4. $1 < \eta_r < +\infty$: The tiepoints lie on branch 2. The nodes and antinodes are now on the maxima and minima of electronic density, respectively.

Figure 7.11 represents the variations of the intensity of the wavefields with the position $\Delta d/d_{hkl}$ and with the angle of incidence for the same example of gallium arsenide and a 111 reflection. The shift of the nodes and antinodes by $d_{hkl}/2$ across the reflection domain is clearly to be seen.

7.3 Thin crystals

7.3.1 *Boundary conditions for the amplitudes*

In the reflection geometry, when the crystal is thin, the wavefields created at the entrance surface propagate till the exit surface where they are backreflected towards the entrance surface. The corresponding tiepoints are P'_j and P''_j and belong to the same branch of the dispersion surface ($j = 1, 2$; Fig. 4.20). The boundary conditions for the amplitudes at the entrance and exit surfaces have been given in Section 4.9.3:

1. *Entrance surface*:

$$D_{o'} + D_{o''} = D_o^{(a)}, \qquad D_{h'} + D_{h''} = D_h^{(a)}. \tag{7.6}$$

2. *Back surface*:

$$\begin{aligned} &D_{o'} \exp(-2\pi \mathrm{i} \mathbf{K_{o'}} \cdot \mathbf{r}) + D_{o''} \exp(-2\pi \mathrm{i} \mathbf{K_{o''}} \cdot \mathbf{r}) \\ &\quad = D_o^{(d)} \exp(-2\pi \mathrm{i} \mathbf{K_o^{(d)}} \cdot \mathbf{r}), \\ &D_{h'} \exp(-2\pi \mathrm{i} \mathbf{K_{h'}} \cdot \mathbf{r}) + D_{h''} \exp(-2\pi \mathrm{i} \mathbf{K_{h''}} \cdot \mathbf{r}) = 0, \end{aligned} \tag{7.7}$$

where $\mathcal{D}_{\mathbf{o}}^{(\mathbf{d})} = \mathbf{D_o^{(d)}} \exp(-2\pi \mathrm{i} \mathbf{K_o^{(d)}} \cdot \mathbf{r})$ is the emerging wave at the back surface.

Due to the continuity of the tangential components of wavevectors, it follows (Fig. 4.19) that

$$\mathbf{K_{o'}} = \mathbf{K_o^{(a)}} + \mathbf{MP'_j}; \qquad \mathbf{K_{o''}} = \mathbf{K_o^{(a)}} + \mathbf{MP''_j},$$

$$\mathbf{K_{h'}} = \mathbf{K_h^{(a)}} + \mathbf{MP'_j}; \qquad \mathbf{K_{h''}} = \mathbf{K_h^{(a)}} + \mathbf{MP''_j}.$$

Using these relations, (7.7) can be written as

$$\begin{aligned} &D_{o'} \exp\left(-2\pi \mathrm{i} \overline{MP'_j}\, t\right) + D_{o''} \exp\left(-2\pi \mathrm{i} \overline{MP''_j}\, t\right) = D_o^{(d)}, \\ &D_{h'} \exp\left(-2\pi \mathrm{i} \overline{MP'_j}\, t\right) + D_{h''} \exp\left(-2\pi \mathrm{i} \overline{MP''_j}\, t\right) = 0 \end{aligned} \tag{7.8}$$

with, according to (5.23) and (7.3),

$$\begin{aligned} \overline{MP'_j} &= \frac{k\chi_o}{2\gamma_o} - \frac{\eta}{2\Lambda_B} - \frac{\sqrt{\eta^2 - 1}}{2\Lambda_B} \\ &= \frac{k\chi_o}{2\gamma_o} - \frac{\eta}{2\Lambda_B} - \frac{\rho}{2|\Lambda_B|} \exp \mathrm{i}(\beta + \omega'), \\ \overline{MP''_j} &= \frac{k\chi_o}{2\gamma_o} - \frac{\eta}{2\Lambda_B} + \frac{\sqrt{\eta^2 - 1}}{2\Lambda_B} \\ &= \frac{k\chi_o}{2\gamma_o} - \frac{\eta}{2\Lambda_B} - \frac{\rho}{2|\Lambda_B|} \exp \mathrm{i}(\beta + \omega''), \end{aligned} \tag{7.9}$$

where ω' and ω'' are the two solutions of (7.3), with $\omega'' = \omega' + \pi$.

7.3.2 *Reflectivity*

The system constituted by (7.6) and (7.8) is solved by substituting from $\xi_j = D_{hj}/D_{oj}$:

$$
\begin{aligned}
D_{o'j} &= \frac{-\xi''_j \exp\left(-2\pi \mathrm{i}\,\overline{MP''_j}\,t\right)}{\xi'_j \exp\left(-2\pi \mathrm{i}\,\overline{MP'_j}\,t\right) - \xi''_j \exp\left(-2\pi \mathrm{i}\,\overline{MP''_j}\,t\right)} D_o^{(a)},\\
D_{o''j} &= \frac{-\xi'_j \exp\left(-2\pi \mathrm{i}\,\overline{MP'_j}\,t\right)}{\xi'_j \exp\left(-2\pi \mathrm{i}\,\overline{MP'_j}\,t\right) - \xi''_j \exp\left(-2\pi \mathrm{i}\,\overline{MP''_j}\,t\right)} D_o^{(a)},
\end{aligned}
\tag{7.10}
$$

where the prime and double prime correspond to the two solutions of (7.3). The amplitude D_{hj} is obtained by multiplying D_{oj} by ξ_j. Using (7.6) and (7.8), we obtain, for the reflected wave at the entrance surface and for the refracted wave at the back surface,

$$
\begin{aligned}
D_h^{(a)} &= \frac{\xi'_j \xi''_j \left[\exp\left(-2\pi \mathrm{i}\,\overline{MP'_j}\,t\right) - \exp\left(-2\pi \mathrm{i}\,\overline{MP''_j}\right)t\right]}{\xi'_j \exp\left(-2\pi \mathrm{i}\,\overline{MP'_j}\,t\right) - \xi''_j \exp\left(-2\pi \mathrm{i}\,\overline{MP''_j}\,t\right)} D_o^{(a)},\\
D_o^{(d)} &= \frac{(\xi'_j - \xi''_j) \exp\left(-2\pi \mathrm{i}\left[\overline{MP'_j} + \overline{MP''_j}\right]t\right)}{\xi'_j \exp\left(-2\pi \mathrm{i}\,\overline{MP'_j}\,t\right) - \xi''_j \exp\left(-2\pi \mathrm{i}\,\overline{MP''_j}\,t\right)} D_o^{(a)}.
\end{aligned}
$$

Substituting from (5.38) and (7.9) it follows finally, for the reflectivities, that

$$
\begin{aligned}
I_h^{(a)} &= |\gamma| \left|\frac{D_h^{(a)}}{D_o^{(a)}}\right|^2 = \left|\frac{\chi_h}{\chi_{\bar h}}\right| \frac{|\mathcal{E}_1 - \mathcal{E}_2|^2}{|\eta(\mathcal{E}_1 - \mathcal{E}_2) + \sqrt{\eta^2 - 1}(\mathcal{E}_1 + \mathcal{E}_2)|^2},\\
I_o^{(d)} &= \left|\frac{D_o^{(d)}}{D_o^{(a)}}\right|^2 = \frac{4|\eta^2 - 1| \exp[-(\mu_o/2)(1/\gamma_o + 1/\gamma_h)t]}{|\eta(\mathcal{E}_1 - \mathcal{E}_2) + \sqrt{\eta^2 - 1}(\mathcal{E}_1 + \mathcal{E}_2)|^2}
\end{aligned}
\tag{7.11}
$$

with $\mathcal{E}_1 = \exp(\pi i t/\Lambda)$; $\mathcal{E}_2 = \exp(-\pi i t/\Lambda)$; $\Lambda = \Lambda_B/\sqrt{\eta^2 - 1}$.

Alternatively, the following expressions do not involve complex quantities:

$$
\begin{aligned}
I_h^{(a)} &= \left|\frac{\chi_h}{\chi_{\bar h}}\right| \frac{\cosh 2b - \cos 2a}{L \cosh 2b + \sqrt{L^2 - 1} \sinh 2b - L_1 \cos 2a - 2\rho|\eta| \sin(\alpha - \omega') \sin 2a},\\
I_o^{(d)} &= \frac{4 \exp[-(\mu_o/2)(1/\gamma_o + 1/\gamma_h)t]}{L \cosh 2b + \sqrt{L^2 - 1} \sinh 2b - L_1 \cos 2a - 2\rho|\eta| \sin(\alpha - \omega') \sin 2a},
\end{aligned}
\tag{7.12}
$$

where α is the phase angle of the deviation parameter ($\eta = |\eta| \exp i\alpha$) and

$$2a = \frac{2\pi t\rho}{|\Lambda_B|}\cos(\beta+\omega); \qquad 2b = \frac{2\pi t\rho}{|\Lambda_B|}\sin(\beta+\omega),$$

$$L = |\eta|^2 + \rho^2; \qquad L_1 = |\eta|^2 - \rho^2.$$

These expressions are in agreement with those given by Wilkins (1978*b*), who pointed out that there was an unnecessary approximation in the expressions given by Zachariasen (1945).

Figure 7.12 represents the variations of $I_h^{(a)}$ and $I_o^{(d)}$ in the case of a slightly absorbing crystal (Fig. 7.12(a): silicon, Mo$K\alpha$, 220, $t = 2\Lambda_B = 13.65\,\mu$m,

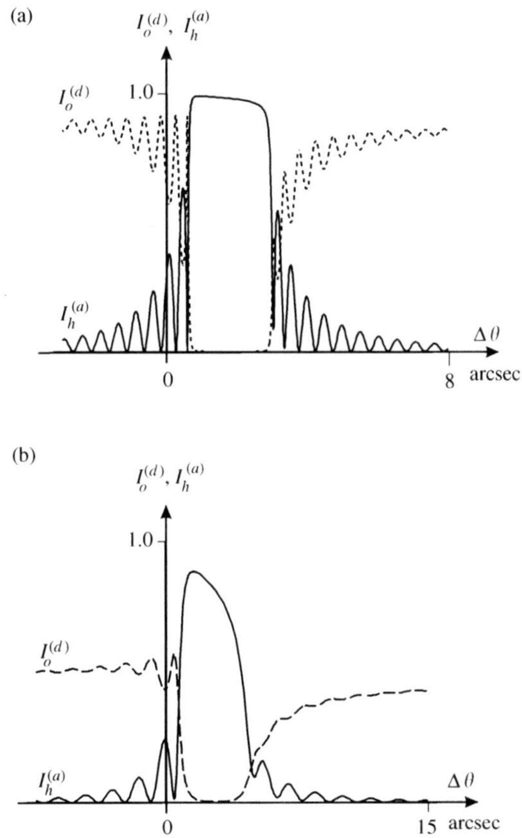

Fig. 7.12. Calculated rocking curves in the reflection geometry, thin crystals, for two different values of the absorption coefficient. Solid curve: reflected beam; dashed curve: refracted beam. (a) Silicon, 220 reflection, Mo$K\alpha$, $t = 2\Lambda_B = 13.65\,\mu$m, $\mu_o t = 0.019$. (b) GaAs, 400, Mo$K\alpha$, $t = 2\Lambda_B = 4.71\,\mu$m, $\mu_o t = 0.15$.

$\mu_o t = 0.019$) and of a more absorbing crystal (Fig. 7.12(b): GaAs, Mo$K\alpha$, 400, $t = 2\Lambda_B = 4.71\,\mu\text{m}$, $\mu_o t = 0.15$). The rocking curves for the reflected and refracted beams have complementary oscillations and, as was shown in Section 4.9.3, if the absorption is zero, $I_h^{(a)} + I_o^{(d)} = 1$ which expresses the conservation of energy. Bragg-geometry rocking curves were recorded experimentally by Batterman and Hildebrandt (1967, 1968), Hashizume *et al.* (1970) and Wilkins (1978*c*).

When the crystal thickness is large with respect to the extinction distance and the absorption and $\mu_o t$ are small, one should take for the expression of the reflectivity the value of (4.46a) averaged over thickness. One finds, in the symmetric case, (Ewald 1917, equation (61′); Zachariasen 1945, equation (3.155)):

$$I_h^{(a)E} = 1 - \frac{\sqrt{\eta^2 - 1}}{|\eta|} \tag{7.13}$$

As pointed out by Ewald (1917), this expression is only valid if there is Pendellösung, namely if there are two wavefields in the crystal, that one created at the entrance surface, and the one partially reflected at the back surface (Fig. 4.20). This situation can be found for instance in the case of neutron diffraction, as was discussed by Takahashi and Hashimoto (1995). These authors showed that expression (7.13), usually known as the *Ewald solution*, is related to (7.1), sometimes called *Darwin solution* by:

$$I_h^{(a)E} = \frac{2I_h^{(a)D}}{1 + I_h^{(a)D}}$$

where $I_h^{(a)D}$ is equal to the value of $I_h^{(a)}$ in (7.1) when $|\chi_h/\chi_{\bar{h}}| = 1$.

The values of $I_h^{(a)E}$ are significantly larger than those of $I_h^{(a)D}$ in the tails of the rocking curve and Villa *et al.* (2003) have described an adaptation of the Bonse-Hart type of channel-cut monochromator to reduce these tails (see Section 15.3.1 for the principle of the Bonse-Hart monochromator).

The variations of the integrated intensity with $\mu_o t$ have been computed from (7.12) by Weiss (1965). He showed that for very small values of $\mu_o t$ it tends towards the value $\pi\delta_{os}$ (equation 4.47b) given both by the geometrical theory and by the dynamical theory using the Ewald solution and that for large values of $\mu_o t$ it tends towards $\frac{8}{3}\delta_{os}$ obtained using the Darwin solution.

8

Dynamical diffraction in highly asymmetric coplanar and non-coplanar geometries

8.1 Introduction

With highly asymmetric reflections it is possible to obtain very narrow rocking curves or very narrow reflected beams. This is the basis of many optical devices (see Chapter 15). It is also possible to reduce drastically the penetration depth, in particular for angles of incidence close to the critical angle, ω_c, for specular reflection (see Section 5.5.5). Highly asymmetric coplanar diffraction and non-coplanar Grazing Incidence Diffraction (GID) have for that reason many applications for the characterization of the structure of surfaces, interfaces, thin films and multilayers (Holý *et al.* 1999).

The approximation made in Chapters 4 and 5 that the spheres of centres O and H and with radii k or nk can be approximated by their tangential planes is, however, no longer valid in those cases; the intersection of the dispersion surface with the diffraction plane cannot be approximated by a hyperbola and the equation of dispersion which is of degree four can no longer be reduced to a second-degree one in the way it was done in Chapter 5; for instance the deviation from Bragg's angle, $\Delta\theta_{os}$, calculated according to (5.16) and the Darwin width, $2\delta_{os}$, calculated according to (4.26) tend towards infinity when the glancing angle tends towards zero, which is physically impossible. Several approaches have been used to solve systems (5.7) or (5.8) in the general case (see references in Sections 8.2, 8.9.1 and Rieutord 1990; Zaus 1993; Fewster 2000). A general solution valid for σ−polarization has been proposed by De Caro *et al.* (1997) and a computation procedure using the analytical solutions of quartic equations has been given by Huang and Dudley (2003); the exact formulation for π-polarization is given in Cho *et al.* (2004). The curvature of the spheres can also be taken into account by solving a third-degree equation instead of a fourth-degree one and that for a qualitative or semi-quantitative analysis it is possible to approximate this third-degree equation by a suitable second-degree equation (Afanas'ev *et al.* 1992; Authier 1998). This is done by using a parabolic approximation instead of a linear one for the spheres of centres O and H and radii k and nk. The deviations from Bragg's incidence and the widths of the rocking curve and of the diffracted beam, the exact shape of the dispersion surface and the values of the specularly reflected and diffracted amplitudes will be calculated using this approach.

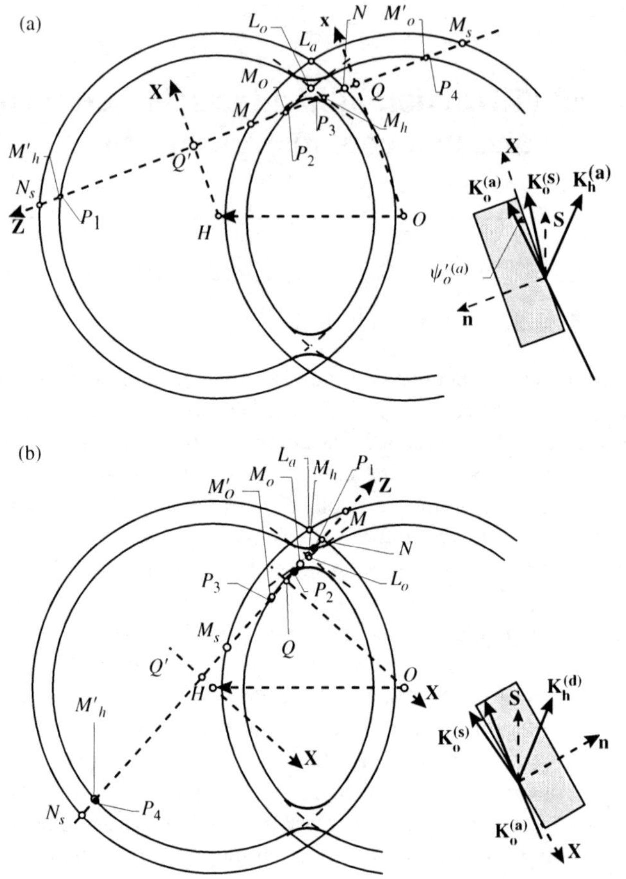

Fig. 8.1.

8.2 Diffraction at grazing incidence or grazing emergence

If the angle of incidence is very close to the critical angle, specular reflection takes place as was discussed in Section 2.10. If the incident beam satisfies at the same time Bragg's condition for a given reflection, there will be both a specularly reflected beam and a Bragg-reflected beam, as shown in Figs 8.1(a) and (b) for the Bragg and Laue geometries, respectively.

In a similar way, if it is not the incident beam, but the Bragg-reflected beam which makes a very small angle with the crystal surface, the latter will be specularly reflected, as shown in Figs 8.1(c) and (d) for the Bragg and Laue geometries, respectively. In these four cases which were reviewed by Härtwig (1978*d*), Afanas'ev *et al.* (1992) and Holý (1996), the normal to the crystal surface is parallel to the diffraction plane, $\mathbf{K_o}$, $\mathbf{K_h}$ (the so-called in-plane or coplanar geometry). They were studied, among

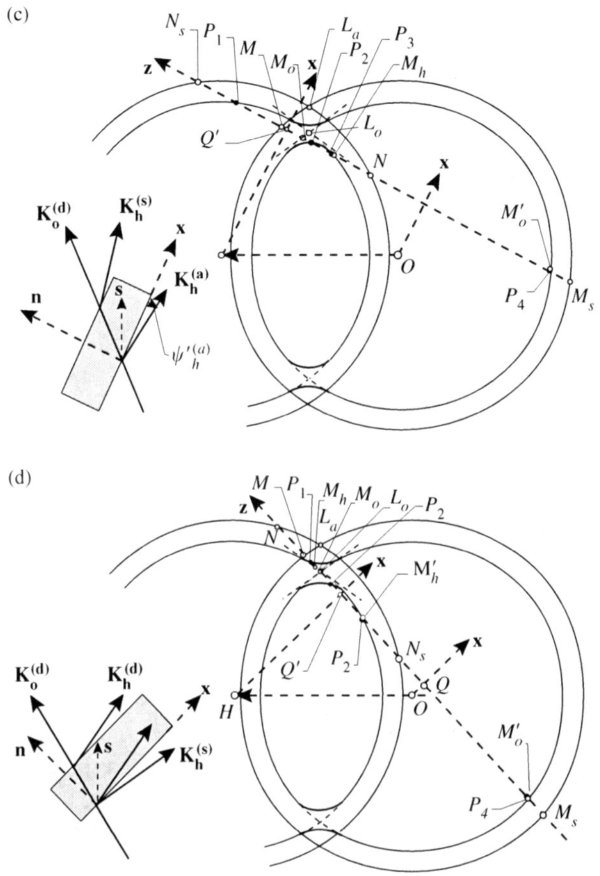

Fig. 8.1. Construction of the tiepoints, P_1, P_2, P_3 and P_4, for a grazing incidence or emergence; L_a: Laue point; L_o: Lorentz point; **Qx**, **Qz**: axes parallel and normal to the crystal surface, respectively; insets: direct space; **n**: normal to the crystal surface; **s**: parallel to the lattice planes. (a) and (b) Grazing incidence, the incident beam, $\mathbf{K}_\mathbf{o}^{(\mathbf{a})}$, is specularly reflected off the crystal surface: $\mathbf{K}_\mathbf{o}^{(\mathbf{s})}$. (a) Reflection geometry. (b) Transmission geometry. (c) and (d) Grazing emergence. (c) Reflection geometry; two wavefields propagate inside the crystal, of tiepoints P_2 or P_3, as in traditional dynamical diffraction, and an additional one, corresponding to tiepoint P_1; if the crystal is not semi-infinite, this gives rise to an emerging beam, $\mathbf{K}_\mathbf{h}^{(\mathbf{s})}$, from the lower surface of the crystal, due to the specular reflection of the Bragg-reflected beam, $\mathbf{K}_\mathbf{h}^{(\mathbf{a})}$. (d) Transmission geometry; the Bragg-reflected beam, $\mathbf{K}_\mathbf{h}^{(\mathbf{d})}$, is specularly reflected off the crystal surface: $\mathbf{K}_\mathbf{h}^{(\mathbf{s})}$.

others, by:

(1) *grazing incidence, Bragg geometry (Fig. 8.1(a))*: Farwig and Schürmann (1967), Kishino and Kohra (1971), Rustichelli (1975), Brümmer *et al.* (1976*a*,*b*, 1982*b*), Härtwig (1978*a*,*c* 1981), Höche and Nieber (1988), Höche *et al.* (1988) and Kimura *et al.* (1994);

(2) *grazing incidence, Laue geometry (Fig. 8.1(b))*: Härtwig (1977), Brümmer *et al.* (1978);

(3) *grazing emergence, Bragg geometry (Fig. 8.1(c))*: Kishino (1971), Bedynska (1973*b*) and Härtwig (1978*b*, 1981);

(4) *grazing emergence, Laue geometry (Fig. 8.1(d))*: Kishino *et al.* (1972), Bedynska (1974) and Härtwig (1976).

The case of grazing incidence or emergence together with three-beam diffraction was discussed, among others, by Andreev *et al.* (1985), Stepanov *et al.* (1991), Stepanov and Ulyanenkov (1994), Stetsko and Chang (1997, 1999*a*) and Chang *et al.* (1998).

With an *out-of-plane or non-coplanar geometry*, so-called GID, when the diffracting planes are perpendicular or nearly perpendicular to the crystal surface, it is possible to have a situation where both the incidence and the emergence angles are grazing. It is described in Section 8.9.

8.3 Deviation from Bragg's incidence of the middle of the reflection domain

8.3.1 *Grazing incidence and Bragg geometry*

It has been shown in Section 5.5.2 that the angular position of the centre of the reflection domain differs from that given by Bragg's law, except in the symmetrical Laue geometry. This is readily seen in Fig. 8.2(a) corresponding to a grazing incidence and Bragg geometry (as in Fig. 8.1(a)): the wavevector corresponding to the wave satisfying Bragg's geometrical condition is $\mathbf{OL_a}$ where L_a is the Laue point and $\mathbf{OI_c}$ is the direction of the wavevector of the incident wave corresponding to the middle of the reflection domain where I_c is at the intersection of the normal $\mathbf{n}$ to the crystal surface, drawn from the Lorentz point, L_o, with the circle of centre O and with radius k (L_o is at the intersection, in the plane of incidence, of the circles of centres O and H, respectively, and of radius nk). If one takes into account the curvature of the spheres, the deviation from Bragg's angle is given by

$$\Delta\theta_{oc} = \frac{\overline{L_a I_c}}{k} = \psi'^{(a)}_{oB} - \psi'^{L}_{oB},$$

where ψ'^{L}_{oB} and $\psi'^{(a)}_{oB}$ are the angles of $\mathbf{OL_a}$ and $\mathbf{OI_c}$ with the crystal surface represented by $\mathbf{Ox}$, respectively. It differs from the value used traditionally, when approximating the spheres by their tangential planes: $\Delta\theta_{os} = \overline{L_a I_s}/k$ (Fig. 8.2(a)), as was first shown by Rustichelli (1975).

Let Q_o be the projection of I_c and L_o on $\mathbf{Ox}$; we have

$$\overline{OQ_o} = k\cos\psi'^{(a)}_{oB} = nk\cos\psi'_{oB}, \tag{8.1}$$

where ψ'_{oB} is the angle made by $\mathbf{OL_o}$ with $\mathbf{Ox}$. At grazing incidence, all the angles considered are very small so that one has, to a very good approximation,

$$\psi'^{(a)}_{oB} = \sqrt{(\psi'_{oB})^2 - \chi_o}.$$

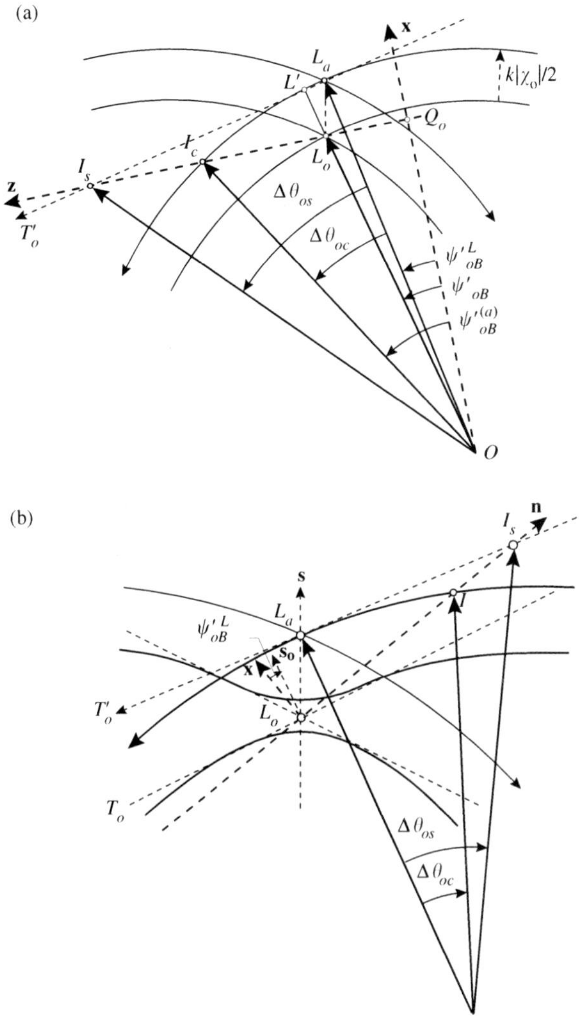

Fig. 8.2. Deviation from Bragg's angle, grazing incidence: $\Delta\theta_{oc} = \overline{L_a I_c}/k$, taking curvature into account, $\Delta\theta_{oc} = \overline{L_a I_s}/k$ neglecting curvature; $\mathbf{OI_s}$: wavevector corresponding to the middle of the reflection domain if one replaces the sphere of centre O by its tangential plane T'_o; $\mathbf{OI_c}$: wavevector corresponding to the middle of the reflection domain if one takes the curvature into account; $\mathbf{Q_o z}$: parallel to the normal $\mathbf{n}$ to the crystal surface. (a) Bragg geometry. (b) Laue geometry.

Let L' be the intersection of OL_o with the circle of radius k. The angles ψ'_{oB} and ψ'^{L}_{oB} are related by

$$\psi'_{oB} - \psi'^{L}_{oB} = \frac{\widehat{L_a L'}}{k}.$$

The arc $\widehat{L_a L}$ can be here approximated by the segment $\overline{L_a L'}$. It makes angle $\pi/2 - \theta_B$ with the segment $L_o L_a$ which is parallel to the lattice planes. It can then easily be seen in triangle $\widehat{L_o L L_a}$ that

$$\psi'_{oB} = \psi'^{L}_{oB} - \frac{\chi_o}{2} \tan\theta_B.$$

The deviation from Bragg's incidence is finally, after substitution of the value of ψ'_{oB} in the preceding equation,

$$\Delta\theta_{oc} = \psi'^{(a)}_{oB} - \psi'^{L}_{oB} = \sqrt{(\psi'^{L}_{oB})^2 - \psi'^{L}_{oB}\chi_o \tan\theta_B - \chi_o} - \psi'^{L}_{oB}. \tag{8.2}$$

This expression was first given in this form by Afanas'ev *et al.* (1992) and in the following form by Rustichelli (1975):

$$\Delta\theta_{oc} = \frac{-\gamma_o + \sqrt{\gamma_o^2 - (\gamma_o - \gamma_h)\sqrt{1-\gamma_o^2}\,\chi_o/\sin 2\theta_B}}{\sqrt{1-\gamma_o^2}}, \tag{8.3}$$

where $\gamma_o = \sin\psi'^{L}_{oB}$ and $\gamma_h = \sin(\psi'^{L}_{oB} - 2\theta_B)$. This expression is quite general. It is easy to show on the one hand that it reduces to (8.2) for very small values of the glancing angle and on the other hand that it reduces to the usual expression (5.16) when the glancing angle is not so small. If ψ'^{L}_{oB} is very small one can

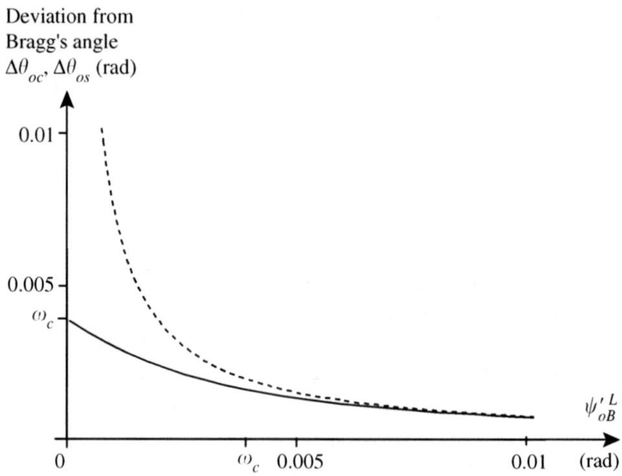

Fig. 8.3. Variations of the deviation from Bragg's angle with ψ'^{L}_{oB}, the angle between the crystal surface and the wavevector satisfying Bragg's condition, grazing incidence, Bragg geometry; solid line: taking curvature into account; dashed line: neglecting curvature; silicon, 400 reflection, Cu$K\alpha$; critical angle: $\omega_c = 0.0389$ rad.

write

$$\gamma_o \approx \psi'^L_{oB}; \qquad \sqrt{1-\gamma_o^2} = \cos\psi'^L_{oB} \approx 1 - \frac{(\psi'^L_{oB})^2}{2}. \tag{8.4}$$

The difference between the direction cosines γ_o and γ_h can be transformed in the following way:

$$\begin{aligned}\gamma_o - \gamma_h &= \sin\psi'^L_{oB} - \sin(\psi'^L_{oB} - 2\theta_B)\\ &= 2\sin^2\theta_B \sin\psi'^L_{oB} + \sin 2\theta_B \cos\psi'^L_{oB}\\ &\approx 2\sin^2\theta_B \psi'^L_{oB} + \sin 2\theta_B.\end{aligned}$$

It follows that

$$\frac{\gamma_o - \gamma_h}{\sin 2\theta_B} = 1 + \psi'^L_{oB}\tan\theta_B. \tag{8.5}$$

Substituting this expression and (8.4) into (8.3) the latter reduces to (8.2).

If $\psi'^L_{oB} \gg \omega_c$ where $\omega_c = \sqrt{-\chi_{ro}}$ is the critical angle, $\gamma_o \gg |\chi_o|$ and it can easily be shown that (8.3) reduces to the usual formula, (5.16):

$$\Delta\theta_{os} = -\frac{\chi_{ro}(1-\gamma)}{2\sin 2\theta_B}, \tag{8.6}$$

where $\gamma = \gamma_h/\gamma_o$.

Figure 8.3 compares the variations of $\Delta\theta_{oc}$ and $\Delta\theta_{os}$ with the angle of incidence. It can be seen that the deviation from Bragg's angle does not increase to infinity when $\psi'^L_{oB} \Rightarrow 0$, but has a finite limit. It is easy to see from (8.2) that this limit is equal to ω_c.

8.3.2 *Grazing incidence, Laue geometry*

The deviation from Bragg's angle of the middle of the reflection domain can be determined in the other geometries using similar derivations to that used for grazing incidence and Bragg geometry. Figure 8.2(b) is an enlargement of the region around the dispersion surface in Fig. 8.1(b). The deviation from Bragg's incidence of the middle of the reflection domain is given by $\Delta\theta_{os} = \overline{L_a I_s}/k$ when the spheres can be approximated by their tangential planes. Its correct expression which must be used when the incidence is grazing is $\Delta\theta_{oc} = \widehat{L_a I_c}/k$. It can be seen in the figure that it is negative and its value is

$$\begin{aligned}\Delta\theta_{oc} &= -\sqrt{(\psi'^L_{oB})^2 - \psi'^L_{oB}\chi_o\tan\theta_B - \chi_o} - \psi'^L_{oB}\\ &= \frac{\gamma_o - \sqrt{\gamma_o^2 + (\gamma_o-\gamma_h)\sqrt{1-\gamma_o^2}\,\chi_o/\sin 2\theta_B}}{\sqrt{1-\gamma_o^2}}.\end{aligned} \tag{8.7}$$

According to Brümmer *et al.* (1978), the peak of the rocking curve does not coincide with this position when the crystal thickness is sufficiently large for branch 2 wavefields to be completely absorbed out. They calculate its position by considering the projection on T'_o (Fig. 8.2(b)) of the apex of branch 1 of the dispersion surface instead of the projection of L_o. They find that the position of the maximum is then given by

$$\Delta\theta_{ocm} = -\sqrt{(\psi'^L_{oB})^2 - \psi'^L_{oB}(\chi_o + C\sqrt{\chi_h\chi_{\bar{h}}})\tan\theta_B - (\chi_o + C\sqrt{\chi_h\chi_{\bar{h}}})} - \psi'^L_{oB},$$

whose limit when $\psi'^L_{oB} \Rightarrow 0$ is $-\sqrt{|\chi_o| - C\sqrt{\chi_h\chi_{\bar{h}}}}$.

8.3.3 *Grazing emergence*

When the emergence angle is grazing, but not the incidence angle, the deviation from Bragg's angle of the middle of the reflection domain and the Darwin width take the usual values; the angular position and the width of the diffracted beam must, however, be determined taking into account the curvatures of the spheres centred at H. The wavevector of the reflected beam corresponding to Bragg's law is $\mathbf{HL_a}$ (Fig. 8.1(c)), while the extremity of the wavevector corresponding to the middle of the diffracted beam is J_s when neglecting the curvature of the spheres and J_c when it is not neglected (Fig. 8.4).

The angle between $\mathbf{HJ_s}$ and $\mathbf{HL_a}$ is

$$\Delta\theta_{hs} = -\frac{\chi_o(1-\gamma^{-1})}{2\sin 2\theta_B},$$

where $\gamma = \gamma_h/\gamma_o$ (here $\gamma_h = \sin\psi'^L_{hB}$ and $\gamma_o = \sin(2\theta_B + \psi'^L_{hB})$, and ψ'^L_{hB} is the angle of $\mathbf{HL_a}$ with the crystal surface). This expression tends towards infinity when $\psi'^L_{hB} \Rightarrow 0$ and is in fact only valid when $\psi'^L_{hB} \gg \omega_c$. If, however, one takes into account the curvature of the Ewald sphere, with a similar derivation to that used for a grazing incidence, one finds:

(1) *grazing emergence, Bragg geometry (Figs 8.1(c) and 8.3)*:

$$\Delta\theta_{hc} = \sqrt{(\psi'^L_{hB})^2 + \psi'^L_{hB}\chi_o\tan\theta_B - \chi_o} + \psi'^L_{hB}$$
$$= \frac{\gamma_h + \sqrt{\gamma_h^2 - (\gamma_o - \gamma_h)\sqrt{1-\gamma_h^2}\,\chi_o/\sin 2\theta_B}}{\sqrt{1-\gamma_h^2}}; \quad (8.8)$$

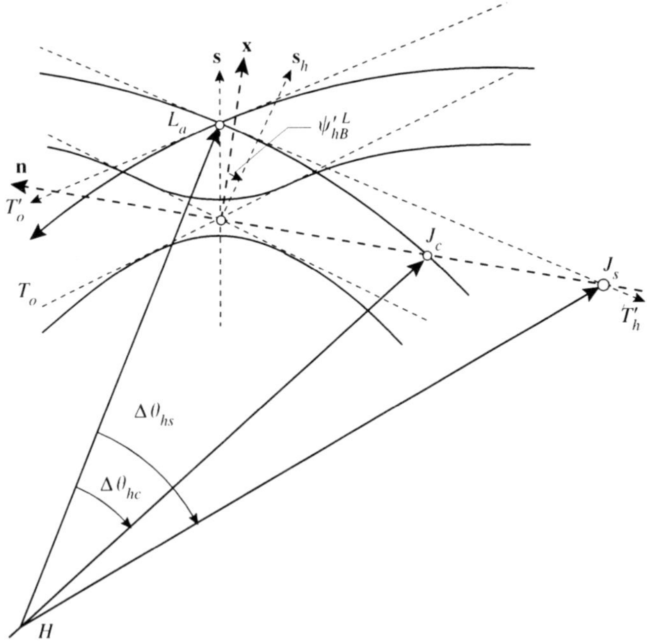

Fig. 8.4. Angular position of the centre of the reflected beam, grazing emergence, Bragg geometry: $\Delta\theta_{hc} = \overline{L_a J_c}/k$, taking the curvature into account; $\Delta\theta_{hs} = \overline{L_a J_s}/k$, neglecting curvature; $\mathbf{HJ_s}$: wavevector corresponding to the centre of the reflected beam if one replaces the sphere of centre H by its tangential plane T'_h; $\mathbf{H_cJ}$: wavevector corresponding to the centre of the reflected beam if one takes the curvature into account.

(2) *grazing emergence, Laue geometry (Fig. 8.1(d))*:

$$\Delta\theta_{hc} = -\sqrt{(\psi'^{L}_{hB})^2 - \psi'^{L}_{hB}\chi_o \tan\theta_B - \chi_o} + \psi'^{L}_{hB}$$

$$= \frac{\gamma_h - \sqrt{\gamma_h^2 - (\gamma_o - \gamma_h)\sqrt{1-\gamma_h^2}\,\chi_o/\sin 2\theta_B}}{\sqrt{1-\gamma_h^2}}. \tag{8.9}$$

8.4 Variation of the Darwin width for a grazing incidence

The angular width of the total reflection domain, or Darwin width, is given by (4.26); it increases towards infinity when $\psi'^{L}_{oB} \Rightarrow 0$. It was shown in Section 5.5.5 (Fig. 5.8b) that it is determined by the angle between the normals to the crystal surface which are tangent to the two branches of the dispersion surface, respectively. This remark was used by Rustichelli (1975) to calculate this width when the two spheres of centres

O and H and with radii nk cannot be approximated by their tangential planes. The principle of the determination is illustrated in Fig. 8.5.

The two tangents to the dispersion surface which are parallel to the normal **n** to the crystal surface intercept segments of width $2\delta_c$ and $2\delta_{os}$ on the sphere of centre O and on its tangent, T_o, respectively; the former is the Darwin width taking the curvature of the sphere into account and the latter the traditional value, neglecting this curvature. Let further 2δ be the segment intercepted by these two tangents on $\overline{L_oL_a}$. One can write

$$2\delta_{os} = 2\delta k \frac{\partial \Delta\theta_{os}}{\partial \overline{L_oL_a}},$$

$$2\delta_{oc} = 2\delta k \frac{\partial \Delta\theta_{oc}}{\partial \overline{L_oL_a}},$$

where $\Delta\theta_{os} = \overline{L_aI_s}/k$ and $\Delta\theta_{oc} = \widehat{L_aI}/k$ (Fig. 8.5). By taking the ratio of these two expressions, it follows that

$$2\delta_{oc} = 2\delta_{os} \frac{\partial \Delta\theta_{oc}}{\partial \overline{L_oL_a}} \Big/ \frac{\partial \Delta\theta_{os}}{\partial \overline{L_oL_a}}.$$

Noting that $\overline{L_oL_a}$ is proportional to χ_o, one can write formally that

$$2\delta_{oc} = 2\delta_{os} \frac{\partial \Delta\theta_{oc}}{\partial \chi_o} \Big/ \frac{\partial \Delta\theta_{os}}{\partial \chi_o}. \tag{8.10}$$

The traditional expression of the Darwin width (4.26) is

$$2\delta_{os} = 2\delta_o\sqrt{|\gamma|}, \tag{8.11}$$

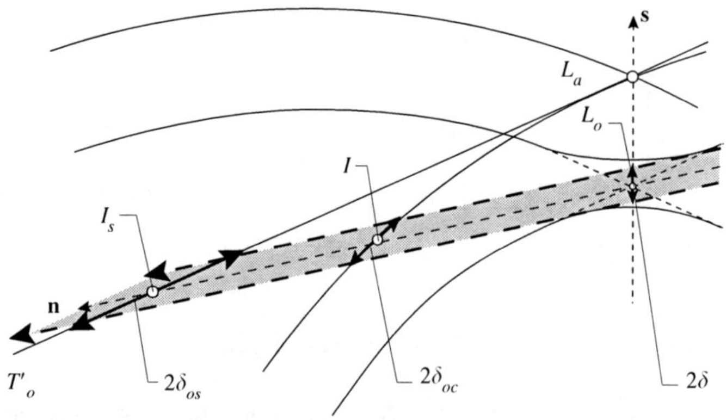

Fig. 8.5. Darwin width: $2\delta_{oc}$, taking into account the curvature of the Ewald sphere; $2\delta_{os}$: neglecting the curvature of the Ewald sphere; L_a: Laue point; L_o: Lorentz point; I: extremity of the wavevector corresponding to the middle of the reflection domain, taking curvature into account; **s**: unit vector parallel to the reflecting planes.

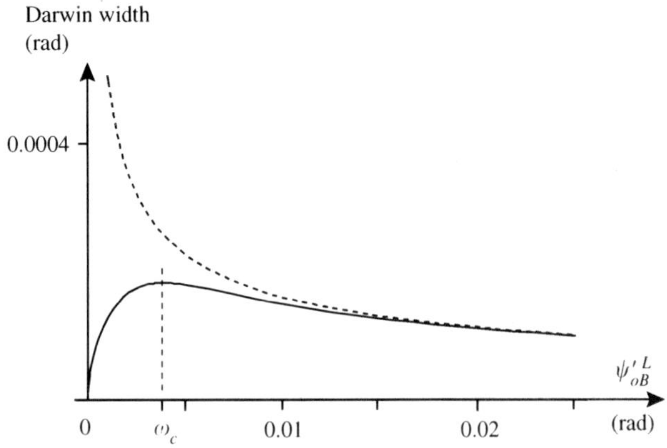

Fig. 8.6. Variations of the Darwin width with ψ'^L_{oB}, the angle between the crystal surface and the wavevector satisfying Bragg's condition. Grazing incidence, Bragg geometry. Solid line: taking curvature into account; dashed line: neglecting curvature. Silicon, 400 reflection, Cu$K\alpha$; critical angle: $\omega_c = 0.0389$ rad.

where $2\delta_o = \sqrt{\chi_h \chi_{\bar{h}}}/\sin 2\theta_B$ is the Darwin width for a symmetrical reflection. By differentiation of (8.2) and (8.6), and using (8.7), it follows that

$$\frac{\partial \Delta\theta_{oc}}{\partial \chi_o} = -\frac{1 + \psi'^L_{oB} \tan\theta_B}{2\sqrt{(\psi'^L_{oB})^2 - \psi'^L_{oB}\chi_o \tan\theta_B - \chi_o}},$$

$$\frac{\partial \Delta\theta_{os}}{\partial \chi_o} = -\frac{1}{2\gamma_o}\frac{\gamma_o - \gamma_h}{\sin 2\theta_B} = -\frac{1 + \psi'^L_{oB} \tan\theta_B}{2\psi'^L_{oB}}.$$

By substitution into (8.10), one finally obtains, using (8.2) and (8.11),

$$2\delta_{oc} = \frac{2\delta_{os}\psi'^{(L)}_{oB}}{\sqrt{(\psi'^L_{oB})^2 - \psi'^L_{oB}\chi_o \tan\theta_B - \chi_o}} = 2\delta_{os}\frac{\psi'^L_{oB}}{\psi'^{(a)}_{oB}}$$

$$= \frac{2\delta_o\sqrt{|\gamma_h|\psi'^L_{oB}}}{\psi'^{(a)}_{oB}}, \qquad (8.12)$$

where $|\gamma_h|$ is practically equal to $\sin 2\theta_B$.

Figure 8.6 represents the variation of $2\delta_{oc}$ with the angle, ψ'^L_{oB}, between the crystal surface and the wavevector satisfying the Bragg condition for a 400 reflection on silicon with Cu$K\alpha$ radiation. It can be seen that it tends towards zero as $\psi'^L_{oB} \Rightarrow 0$ as is obvious from (8.12). It passes through a maximum equal to $2\delta_o\sqrt{|\gamma_h|}/\sqrt{2\omega_c}$ for $\psi'^L_{oB} = \omega_c$. The variation of the Darwin width for a grazing incidence was first checked experimentally by Brümmer *et al.* (1982*b*).

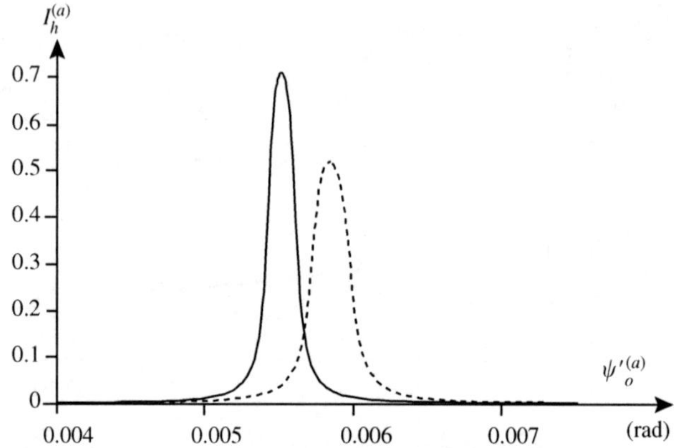

Fig. 8.7. Bragg-reflected intensity, $I_h^{(a)}$. Solid line: taking curvature into account; dashed line: neglecting curvature. Angle between the crystal surface and the wavevector satisfying Bragg's law: $\psi'^L_{oB} = 0.039$ rad. Silicon, 400 reflection, Cu$K\alpha$; critical angle: $\omega_c = 0.00389$ rad.

The difference in angular position and in angular width between the rocking curves calculated taking into account the curvature of the Ewald sphere (see Section 8.7) and neglecting it is illustrated in Fig. 8.7 in the case of a silicon crystal, 400 reflection, Cu$K\alpha$ radiation and an incidence angle of 0.00389 rad, equal to the critical angle for these conditions.

8.5 Variation of the width of the diffracted beam for a grazing emergence

The angular width of the diffracted beam when the crystal is rocked in the incident beam is given by, in the conventional dynamical theory,

$$2\delta_{hs} = 2\delta_o\sqrt{|\gamma|^{-1}} = 2\delta_{os}|\gamma|^{-1}. \tag{8.13}$$

Using a similar derivation to that of Section 8.4, it is possible to show that for a grazing emergence and the Bragg geometry, (8.13) should be replaced by

$$2\delta_{hc} = \frac{2\delta_{os}|\gamma|^{-1}|\psi'^L_{hB}|}{\sqrt{(\psi'^L_{hB})^2 - \psi'^L_{hB}\chi_o \tan\theta_B - \chi_o}}. \tag{8.14}$$

Expression (8.14) shows that the width of the diffracted beam tends towards zero after passing through a maximum when the emergence angle tends towards zero. It is thus possible to obtain a highly collimated beam.

8.6 Equation of the dispersion surface

The equation of the dispersion surface was discussed in Section 5.3 where the curvature of the Ewald sphere was neglected. It is the secular equation of set (5.8):

$$(K_o^2 - n^2k^2)(K_h^2 - n^2k^2) = k^4C^2\chi_h\chi_{\bar{h}}. \tag{8.15}$$

This surface, of degree four, is constituted by the spheres of radius nk and with centres O and H, and a connecting surface. As an example, the case of grazing incidence and Bragg geometry will be discussed here (Fig. 8.1(a)), the other cases can be handled in a similar way. As was seen in Chapters 4 and 5, the positions of the tiepoints are obtained by applying the condition of the continuity of the tangential components of wavevectors on each side of the crystal surface: they lie at the intersection of the parallel, **Mz**, normal to the crystal surface, **n**, drawn from the extremity, M, of the incident wavevector $\mathbf{OM} = \mathbf{K_o^{(a)}}$ with the dispersion surface. The intersections of **Mz** with the two spheres of radius k and with centres O and H are M and M_s, N and N_s, respectively. M_s is the extremity of the specularly reflected wavevector $\mathbf{OM_s} = \mathbf{K_o^{(s)}}$, N the extremity of the Bragg-reflected wavevector $\mathbf{HN} = \mathbf{K_h^{(a)}}$, and N_s the extremity of the wavevector of the wave due to the specular reflection of the Bragg-reflected wave, $\mathbf{HN_s} = \mathbf{K_h^{(s)}}$; this last wave has no physical reality at the entrance surface of the crystal in the case described by Fig 8.1(a) but exists in the cases described by Figs 8.1(b) where its amplitude is negligible, and 8.1(d) and 8.13(b) where it is important. Let Q be the intersection of the parallel to the crystal surface, **Ox**, and the normal **Mz** to the crystal surface. The four tiepoints are denoted by P_1, P_2, P_3 and P_4, sorted by decreasing values of their ordinates $z_j = QP_j$. They are represented by open circles in the figures.

The factors $(K_{oj}^2 - n^2k^2)$ and $(K_{hj}^2 - n^2k^2)$ where $\mathbf{K_{oj}} = \mathbf{OP_j}$ and $\mathbf{K_{hj}} = \mathbf{HP_j}$ intervening in the eqn (8.15) of the dispersion surface can be readily expressed in terms of the ordinates z_j of the tiepoints. Let M_o and M'_o, M_h and M'_h be the intersections of **Mz** with the spheres of radius nk and with centres O and H, respectively (Fig. 8.1(a)). The tiepoints P_1 and P_4 nearly coincide with M'_h and M'_o, respectively. We have $n^2k^2 = \mathbf{OM_o^2}$ and

$$\begin{aligned} K_{oj}^2 - n^2k^2 &= \mathbf{OP_j^2} - \mathbf{OM_o^2} = (\mathbf{OP_j} - \mathbf{OM_o})(\mathbf{OP_j} + \mathbf{OM_o}) \\ &= \mathbf{M_oP_j} \cdot (\mathbf{M_oP_j} + 2\mathbf{OM_o}) \\ &= \overline{M_oP_j}^2 + 2\mathbf{M_oP_j} \cdot \mathbf{QM_o} = \overline{M_oP_j}(\overline{M_oP_j} + 2\overline{QM_o}), \end{aligned}$$

noting that $\mathbf{QM_o}$ is the projection of $\mathbf{OM_o}$ on $\mathbf{M_oP_j}$. In traditional dynamical theory, the term $\overline{M_oP_j}^2$ is neglected. Letting

$$z_o = \overline{QM_o}; \qquad z_{o'} = \overline{QM'_o} = -z_o,$$

it is possible to express $K_{oj}^2 - n^2k^2$ in terms of the coordinates of the tiepoints:

$$K_{oj}^2 - n^2k^2 = (z_j - z_o)(z_j - z_{o'}) = z_j^2 - z_o^2 \tag{8.16a}$$

and, in a similar way,

$$K_{hj}^2 - n^2k^2 = (z_j - z_h)(z_j - z_{h'}), \tag{8.16b}$$

where

$$z_h = \overline{QM_h}; \qquad z_{h'} = \overline{QM'_h}.$$

The equation of the dispersion surface is therefore

$$(z_j^2 - z_o^2)(z_j - z_h)(z_j - z_{h'}) = k^4C^2\chi_h\chi_{\bar{h}}. \tag{8.17}$$

It is an equation of degree four in z_j. However, tiepoint P_1 is always merged into M'_h (Fig. 8.1(a)) and expression (8.16b) can always be approximated by

$$K_{hj}^2 - n^2k^2 = (z_j - z_h)(z_h - z_{h'}).$$

This is equivalent to neglecting $\overline{M_hP_j}^2$ and enables one to reduce the equation of the dispersion surface to a third-degree equation:

$$(z_j^2 - z_o^2)(z_j - z_h)(z_h - z_{h'}) = k^4C^2\chi_h\chi_{\bar{h}} \tag{8.18}$$

($j = 2, 3, 4$). Furthermore, except for angles of incidence such that Q practically lies on the sphere of radius nk, a situation which will be discussed at the end of this section, tiepoint P_4 is merged into M'_o. Using this approximation, expression (8.16a) can be approximated by

$$K_{oj}^2 - n^2k^2 = 2z_o(z_j - z_o) \tag{8.19}$$

and the equation of the dispersion surface is reduced to a second-degree equation:

$$2z_o(z_j - z_o)(z_j - z_h)(z_h - z_{h'}) = k^4C^2\chi_h\chi_{\bar{h}} \quad (j = 2, 3). \tag{8.20}$$

The fact that the dispersion surface can still be approximated by a second-degree equation, even in the case of grazing incidence, was first pointed out by Afanas'ev *et al.* (1992). The roots of the equation of the dispersion surface are, when the approximation of eqn (8.20a) is valid,

$$\begin{aligned} z_1 &= z_{h'}, \\ z_2 &= \frac{1}{2}\left(z_o + z_h + \sqrt{(z_o - z_h)^2 - 2k^4C^2\frac{\chi_h\chi_{\bar{h}}}{z_o(z_{h'} - z_h)}}\right), \\ z_3 &= \frac{1}{2}\left(z_o + z_h - \sqrt{(z_o - z_h)^2 - 2k^4C^2\frac{\chi_h\chi_{\bar{h}}}{z_o(z_{h'} - z_h)}}\right), \\ z_4 &= -z_o. \end{aligned} \tag{8.21}$$

The position of the tiepoints will be determined by their coordinates with respect to the system of coordinates $\mathbf{Q_o x}$, $\mathbf{Q_o z}$ (Fig. 8.8). Their common abscissa is

$$x = \overline{Q_o Q} = \overline{OQ} - \overline{OQ_o},$$

where $\overline{OQ_o}$ is given by (8.1). If $\psi_o'^{(a)}$ and $\psi_{oB}'^{(a)}$ are the angles of the incident wavevector, **OM**, and of the wavevector, **OI**, corresponding to the middle of the reflection domain, respectively, with the crystal surface (Fig. 8.8), we have

$$\begin{aligned} x &= k \cos \psi_o'^{(a)} - k \cos \psi_{oB}'^{(a)} \\ &= \frac{k}{2}\left[\left(\psi_{oB}'^{(a)}\right)^2 - \left(\psi_o'^{(a)}\right)^2\right] \end{aligned}$$

The ordinates z_j of the tiepoints are expressed by (8.21) in terms of the ordinates of M_o, M_h and M_h'. The ordinate of M_o is $z_o = \overline{QM_o} = nk \sin \psi_o' \approx nk\psi_o'$,

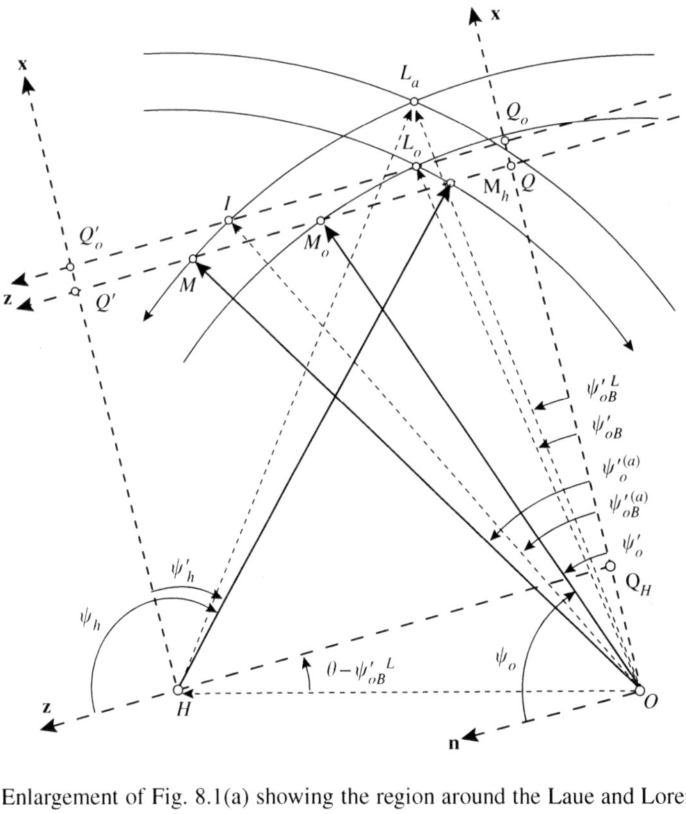

Fig. 8.8. Enlargement of Fig. 8.1(a) showing the region around the Laue and Lorentz points (grazing incidence and Bragg geometry).

where ψ'_o is the angle between OM_o and OQ. This angle is calculated by noting that $OQ = k\cos\psi_o'^{(a)} = nk\cos\psi'_o$. It follows, remembering that these angles are very small, that

$$z_o = \overline{QM_o} = nk\sqrt{\left(\psi_o'^{(a)}\right)^2 + \chi_o}. \tag{8.22}$$

The ordinate of M_h is obtained by expressing that it lies on the sphere of centre H with radius nk. One obtains

$$z_h = \overline{QM_h} = nk\left(\psi_{oB}'^{L} - \frac{\chi_o}{2}\right)\tan\theta_B - x\tan^{-1}2\theta_B. \tag{8.23}$$

Finally, the ordinate of M'_h is

$$\begin{aligned} z'_h &= \overline{QM'_h} = \overline{QM_h} + \overline{M_hM'_h} = 2\overline{QQ'} - \overline{QM_h} \\ &= 4k\sin\theta_B\cos(\theta_B - \psi_{oB}'^{L}) - z_h, \end{aligned} \tag{8.24}$$

where Q' is the projection of reciprocal lattice point H on the normal to the crystal surface drawn from M (Fig. 8.8).

Figure 8.9(a) is a drawing representing the real part of the region around the dispersion surface in Fig. 8.1(a) after a rotation such that the parallel to the crystal surface is horizontal; Fig. 8.9(b) represents the real part of the dispersion surface as *calculated* from (8.21). It is thus possible to have an exact geometrical representation of the dispersion surface, even when taking into account the curvature of the Ewald sphere which is in fact approximated here by a parabola (Authier 1998).

1. *Situations when the angle of incidence, $\psi_o'^{(a)}$, is very close to the critical angle* ω_c: Q is very close to the circle of centre O and radius nk, P_2 is practically merged with M_h (and P_1 is practically merged with M'_h, as before), so that $z_1 = z_{h'}$ and $z_2 = z_h$ are solutions of the fourth-degree equation, but P_4 can no longer be considered to be practically merged with M'_o and the approximation of eqn (8.20) is not valid any longer; the equation of the dispersion surface can in that case be approximated by

$$\left(z_j^2 - z_o^2\right)z_h z_{h'} = k^4C^2\chi_h\chi_{\bar{h}}.$$

The four roots of the equation of the dispersion surface are then

$$\begin{aligned} z_1 &= z_{h'}, \\ z_2 &= z_h, \\ z_3 &= \sqrt{z_o^2 + k^4C^2\frac{\chi_h\chi_{\bar{h}}}{z_{h'}z_h}}, \\ z_4 &= -\sqrt{z_o^2 + k^4C^2\frac{\chi_h\chi_{\bar{h}}}{z_{h'}z_h}}. \end{aligned}$$

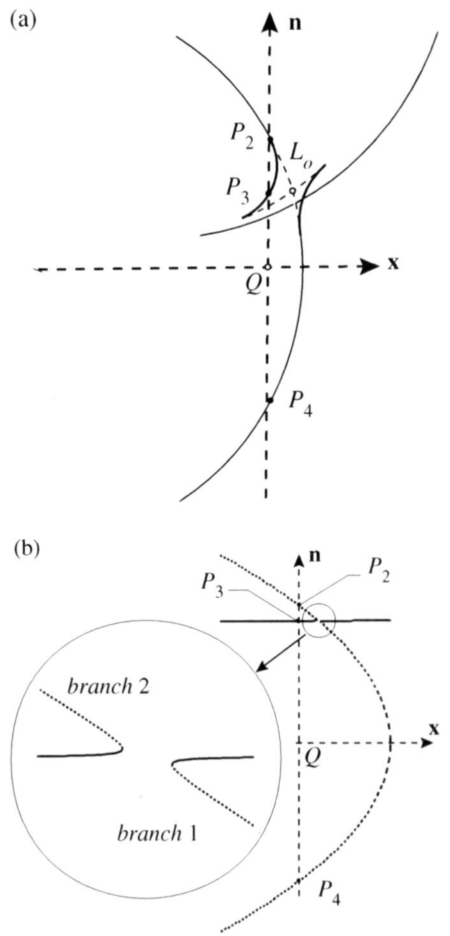

Fig. 8.9. Shape of the dispersion surface for grazing incidence and Bragg geometry. (a) Enlargement of the region around the dispersion surface in Fig. 8.1(a) after a rotation such that the parallel to the crystal surface is horizontal. (b) The dispersion surface as *calculated* from the second-degree equation, with the same orientation as in Fig. 8.9(a). The two circles of centres O and H are approximated, the former by a parabola (dashed line) and the latter by a straight line.

This approximation is only necessary for a small angular range, namely for values of $\psi_o'^{(a)}$ smaller than about $1.25 \times \omega_c$. This can easily be checked by comparing the values of z_j obtained with eqns (8.18) and (8.20) with those obtained using the fourth- or third-degree equation ((8.17) or (8.18)).

2. *Grazing emergence*: In that case, it is more convenient to use the ordinates of M_o, M_h and M_h' with respect to Q'.

8.7 Relation with the traditional dynamical theory

In traditional dynamical theory, the quantities

$$X_{oj} = \frac{K_{oj}^2 - n^2k^2}{2k},$$

$$X_{hj} = \frac{K_{hj}^2 - n^2k^2}{2k}$$

represent the distances of the tiepoint from the asymptotes to the spheres of centres O and H and with radius nk (Section 5.3) and the fact that their product (8.15) is constant shows that the tiepoints lie on a hyperbola. Here they shall be interpreted in terms of the ordinates of the tiepoints. Using (8.16a), it follows that

$$X_{oj} = \frac{z_j^2 - z_o^2}{2k}.$$

Substituting from (8.21) and using the same approximation as in (8.20), we obtain for the tiepoints P_2 and P_3

$$\begin{aligned} X_{oj} &= \frac{z_o(z_j - z_o)}{k} \\ &= \frac{z_o}{2k}\left(z_h - z_o \pm \sqrt{(z_o - z_h)^2 - 2k^4C^2\frac{\chi_h\chi_{\bar{h}}}{z_o(z_{h'} - z_h)}}\right), \end{aligned} \tag{8.25}$$

where the plus sign corresponds to $j = 2$ and the minus sign to $j = 3$.

From the definitions (8.22)–(8.24) of z_o, z_h and z'_h, it follows that

$$\begin{aligned} z_o &= nk \sin\psi'_o = nk\gamma_o, \\ z_o - z_h &= \overline{M_oM_h}, \\ z'_h - z_h &= \overline{M_hM'_h} = -2\overline{Q'M_h} = -2nk\gamma_h, \end{aligned}$$

where $\gamma_o = \cos\psi_o$ with $\psi_o = -\pi/2 + \psi'_o$ and $\gamma_h = \cos\psi_h$ with $\psi_h = -\pi/2 + \psi'_h$ (Fig. 8.8).

Let us set

$$\begin{aligned} \Lambda_B &= \frac{\lambda\sqrt{\gamma_o\,|\gamma_h|}}{|C|\,\sqrt{\chi_h\chi_{\bar{h}}}}, \\ \eta &= \overline{M_oM_h}\,\Lambda_B S(\gamma_h), \end{aligned} \tag{8.26}$$

where $S(\gamma_h)$ means sign of γ_h. These quantities are the extinction distance and the deviation parameter, respectively (see Section 5.5.3, eqns (5.17b) and (5.18)). Substituting from (8.26) into (8.25), we obtain finally

$$X_{oj} = n\frac{\gamma_o}{2\Lambda_B}\left[\eta \pm \sqrt{\eta^2 + S(\gamma_h)}\,\right]. \tag{8.27}$$

Equation (8.27) is identical to the result of traditional dynamical theory (eqn (5.20), Section 5.5.3). Equation (8.25) is valid in all four cases: grazing incidence, Bragg and Laue geometries, grazing emergence, and Bragg and Laue geometries, except for angles of incidence extremely close to the critical angle. Provided the ordinates of z_o, z_h and z'_h are calculated taking into account the curvature of the Ewald sphere, the amplitude ratio ξ can be calculated from (5.10b) and (8.25).

8.8 Specularly and Bragg-reflected intensities

8.8.1 *Boundary conditions for the amplitudes at the entrance surface*

The four boundary conditions (2.7) included in the Maxwell equations are actually reduced to two, as was discussed in Section 6.1. We shall limit ourselves here, for simplicity, to the σ-component of polarization; the same results hold, with a small approximation, for the π-component. The boundary condition for the electric field (or for the electric displacement) must be written separately for the refracted and for the Bragg-reflected wave. In the general case where one has to take into account the four tiepoints and the specular reflection of both the incident and the Bragg-reflected waves, these conditions are written in the following way:

$$
\begin{aligned}
E_o^{(a)} + E_o^{(s)} &= \sum_j E_{oj}, \\
E_h^{(a)} + E_h^{(s)} &= \sum_j E_{hj} = \sum_j \xi_j E_{oj},
\end{aligned}
\tag{8.28}
$$

where $E_o^{(a)}$ is the incident wave, $E_h^{(a)}$ is the Bragg-reflected wave, $E_o^{(s)}$ and $E_h^{(s)}$ are the corresponding specularly reflected waves, E_{oj} and E_{hj} are the diffracted waves inside the crystal and $\xi_j = E_{hj}/E_{oj}$.

The corresponding boundary conditions for the magnetic field are written, using (5.2d), as

$$
\begin{aligned}
K_{oz}^{(a)} E_o^{(a)} - K_{oz}^{(s)} E_o^{(s)} &= \sum_j K_{ojz} E_{oj}, \\
K_{hz}^{(a)} E_h^{(a)} - K_{hz}^{(s)} E_h^{(s)} &= \sum_j K_{hjz} E_{hj} = \sum_j K_{hjz} \xi_j E_{oj},
\end{aligned}
\tag{8.29}
$$

where $K_{oz}^{(a)} = |\overline{QM}|$, $K_{oz}^{(s)} = |\overline{QM_s}|$, $K_{hz}^{(a)} = |\overline{Q'N}|$, $K_{hz}^{(s)} = |\overline{Q'N_s}|$, $K_{oz} = |\overline{QP_j}|$ and $K_{hz} = |\overline{Q'P_j}|$ are the components of the wavevectors parallel to the normal to the crystal surface.

In practice, one takes into account only those terms corresponding to physically significant waves:

1. A wave exists on the outside of the crystal if its wavevector points towards the outside of the crystal.

2. A wave propagates towards the inside of the crystal if the imaginary part of its wavevector, which is normal to the crystal surface, is negative; otherwise it would increase exponentially.
3. A wave has a non-negligible amplitude if the component of its wavevector normal to the crystal surface is small, as can be shown by solving systems (8.28) and (8.29) taking all terms into account. When the incidence and/or the emergence angles are not grazing, the amplitudes associated with M_s and N_s are negligible, there is no specular reflection, and the boundary conditions related to the magnetic field need not be taken into account because they become redundant with those related to the electric field.

Taking these constraints into account, the systems (8.28) and (8.29) can be reduced:

1. *Grazing incidence, Bragg geometry (Fig. 8.1(a)):*

(a) *Semi-infinite crystal*

$$\left.\begin{aligned} E_o^{(a)} + E_o^{(s)} &= E_{oj} \\ E_h^{(a)} &= \xi_j E_{oj} \\ K_{oz}^{(a)} E_o^{(a)} - K_{oz}^{(s)} E_o^{(s)} &= K_{ojz} E_{oj} \\ K_{hz}^{(a)} E_h^{(a)} &= K_{hjz} \xi_j E_{oj} \end{aligned}\right\} \quad (j = 2, 3). \tag{8.30a}$$

The incident beam is specularly reflected. Two roots of the equation of the dispersion surface have a negative imaginary part, z_1 and z_2 or z_3. The length of z_1 is very large and the corresponding amplitude can be neglected, one wavefield only propagates inside the crystal. The fourth equation is in practice redundant with the second one.

(b) *Thin crystal*

(i) *Entrance surface*

$$\begin{aligned} E_o^{(a)} + E_o^{(s)} &= E_{o2} + E_{o3}, \\ E_h^{(a)} &= \xi_2 E_{o2} + \xi_3 E_{o3}, \\ K_{oz}^{(a)} E_o^{(a)} - K_{oz}^{(s)} E_o^{(s)} &= K_{o2z} E_{o2} + K_{o3z} E_{o3}, \\ K_{hz}^{(a)} E_h^{(a)} &= K_{h2z} \xi_2 E_{o2} + K_{h3z} \xi_3 E_{o3}. \end{aligned} \tag{8.30b}$$

One of the two waves, E_{o2} and E_{o3}, is reflected from the back surface—see Section 7.3.

(ii) *Exit surface*

$$\begin{aligned} E_o^{(d)} &= E_{o2}\exp(-2\pi \mathrm{i} K_{o2z}) \\ &\quad + E_{o3}\exp(-2\pi \mathrm{i} K_{o3z}), \\ E_h^{(a)} &= \xi_2 E_{o2}\exp(-2\pi \mathrm{i} K_{h2z}) \\ &\quad + \xi_3 E_{o3}\exp(-2\pi \mathrm{i} K_{h3z}), \\ K_{oz}^{(a)} E_o^{(a)} &= K_{o2z} E_{o2}\exp(-2\pi \mathrm{i} K_{o2z}) \\ &\quad + K_{o3z} E_{o3}\exp(-2\pi \mathrm{i} K_{o3z}), \\ K_{hz}^{(a)} E_h^{(a)} &= K_{h2z}\xi_2 E_{o2}\exp(-2\pi \mathrm{i} K_{h2z}) \\ &\quad + K_{h3z}\xi_3 E_{o3}\exp(-2\pi \mathrm{i} K_{h3z}). \end{aligned} \tag{8.30c}$$

There is no specularly reflected beam to take into account at the exit surface; $E_o^{(d)}$ is the refracted beam coming out the back surface of the thin crystal; the boundary conditions at the exit surface are given by eqn (7.3).

2. *Grazing incidence, Laue geometry (Fig. 8.1(b))*:

$$\begin{aligned} E_o^{(a)} + E_o^{(s)} &= E_{o1} + E_{o2}, \\ 0 &= \xi_1 E_{o1} + \xi_2 E_{o2}, \\ K_{oz}^{(a)} E_o^{(a)} - K_{oz}^{(s)} E_o^{(s)} &= K_{o1z} E_{o1} + K_{o2z} E_{o2}. \end{aligned} \tag{8.31}$$

The incident beam is specularly reflected. Two roots of the equation of the dispersion surface have a negative imaginary part, z_1 and z_2, and two wavefields propagate inside the crystal as in the usual Laue geometry.

3. *Grazing emergence, Bragg geometry (semi-infinite crystal) (Fig. 8.2(a))*:

$$\begin{aligned} E_o^{(a)} &= E_{o1} + E_{oj} \\ E_h^{(a)} &= \xi_1 E_{o1} + \xi_j E_{oj} \qquad (j = 2, 3). \\ K_{hz}^{(a)} E_h^{(a)} &= K_{h1z}\xi_1 E_{o1} + K_{hjz}\xi_j E_{oj} \end{aligned} \tag{8.32}$$

Two roots of the equation of the dispersion surface have a negative imaginary part, z_1 and z_2 or z_3, and two wavefields propagate inside the crystal, one more than in the usual Bragg geometry.

4. *Grazing emergence, Laue geometry* (Fig. 8.2(b)):

$$\begin{aligned} E_o^{(a)} &= E_{o1} + E_{o2}, \\ E_h^{(s)} &= \xi_1 E_1 + \xi_2 E_2, \\ -K_{hz}^{(s)} E_h^{(s)} &= K_{h1z}\xi_1 E_{o1} + K_{h2z}\xi_2 E_{o2}. \end{aligned} \tag{8.33}$$

The Bragg-reflected beam is specularly reflected from the surface of the crystal. Two roots of the equation of the dispersion surface have a negative imaginary part, z_1 and z_2, and two wavefields propagate inside the crystal as in the usual Laue geometry.

5. *Grazing incidence and grazing emergence, GID non-coplanar geometry, Bragg geometry (semi-infinite crystal)*:

$$\begin{aligned} E_o^{(a)} + E_o^{(s)} &= E_{o1} + E_{o2}, \\ E_h^{(a)} &= \xi_1 E_{o1} + \xi_2 E_{o2}, \\ K_{oz}^{(a)} E_o^{(a)} - K_{oz}^{(s)} E_o^{(s)} &= K_{o1z} E_{o1} + K_{o2z} E_{o2}, \\ K_{hz}^{(a)} E_h^{(a)} &= K_{h1z} \xi_1 E_{o1} + K_{h2z} \xi_2 E_{o2}. \end{aligned} \tag{8.34}$$

The incident beam is specularly reflected. Two wavefields propagate inside the crystal.

6. *Grazing incidence and grazing emergence, GID non-coplanar geometry, Laue geometry*:

$$\begin{aligned} E_o^{(a)} + E_o^{(s)} &= E_{o1} + E_{o2}, \\ E_h^{(s)} &= \xi_1 E_{o1} + \xi_2 E_{o2}, \\ K_{oz}^{(a)} E_o^{(a)} - K_{oz}^{(s)} E_o^{(s)} &= K_{o1z} E_{o1} + K_{o2z} E_{o2}, \\ -K_{hz}^{(s)} E_h^{(s)} &= K_{h1z} \xi_1 E_{o1} + K_{h2z} \xi_2 E_{o2}. \end{aligned} \tag{8.35}$$

Both the incident beam and the Bragg-reflected beams are specularly reflected. Two wavefields propagate inside the crystal.

8.8.2 *Specularly and Bragg-reflected intensities for a grazing incidence and the Bragg geometry (semi-infinite crystal)*

The amplitudes of the specularly reflected, $E_o^{(s)}$, and of the Bragg-reflected, $E_h^{(a)}$, waves are obtained by solving system (8.30a):

$$\begin{aligned} E_o^{(s)} &= \frac{K_{oz}^{(a)} - K_{ojz}}{K_{oz}^{(a)} - K_{ojz}} E_o^{(a)} = \frac{k\psi_o'^{(a)} - z_j}{k\psi_o'^{(a)} - z_j} E_o^{(a)}, \\ E_h^{(a)} &= \frac{2\xi_j K_{oz}^{(a)}}{K_{oz}^{(a)} - K_{ojz}} E_o^{(a)} = \frac{2k\psi_o'^{(a)}}{k\psi_o'^{(a)} - z_j} \frac{z_j^2 - z_o^2}{k^2 C \chi_{\bar{h}}} E_o^{(a)}, \end{aligned} \tag{8.36}$$

where z_j is given by (8.21); j is taken to be equal to 2 or 3 according to the sign of the imaginary part of z_j. The specularly reflected and Bragg-reflected intensities are, using (4.38),

$$I_o^{(s)} = \left| \frac{E_o^{(s)}}{E_o^{(a)}} \right|^2,$$

$$I_h^{(a)} = \left| \frac{\gamma_h}{\gamma_o} \right| \left| \frac{E_h^{(a)}}{E_o^{(a)}} \right|^2.$$

The variations of the reflectivity and of the Bragg-reflected intensities have been calculated using the roots of the second-degree equation (8.20) in the case of silicon, 400, Cu$K\alpha$ and an angle of 0.5° between the crystal surface and the incidence satisfying Bragg's condition; these are the same conditions as in Holý (1996) where the roots of the fourth-degree equation (8.17) have been used. They are represented in Fig 8.10. It has been checked that they coincide to a very good approximation with the results obtained using the roots of the fourth-degree equation. In particular, one can see the particular s-shaped features in the curve of the reflectivity at the incidence corresponding to the Bragg peak and in the curve of the Bragg-reflected intensity at the critical angle, which were first observed by Höche *et al.* (1988). The former was also calculated by Rieutord (1990).

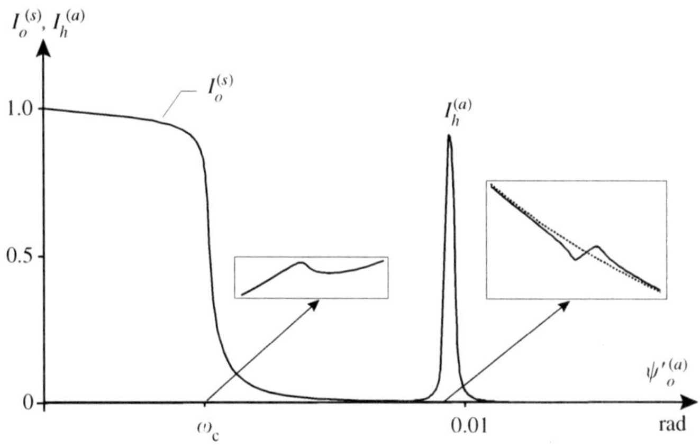

Fig. 8.10. Variation of the reflectivity, $I_o^{(s)}$, and of the Bragg-reflected intensity, $I_h^{(a)}$, with the incidence angle expressed in radians, grazing incidence, Bragg geometry. The insets show the s-shaped features at the incidence angle corresponding to the Bragg peak for the reflectivity and at the critical angle for the Bragg-reflected intensity; silicon, 400 reflection, Cu$K\alpha$; critical angle: $\omega_c = 0.0039$ rad, the angle between Bragg's incidence and the crystal surface is $0.5° = 0.008727$ rad.

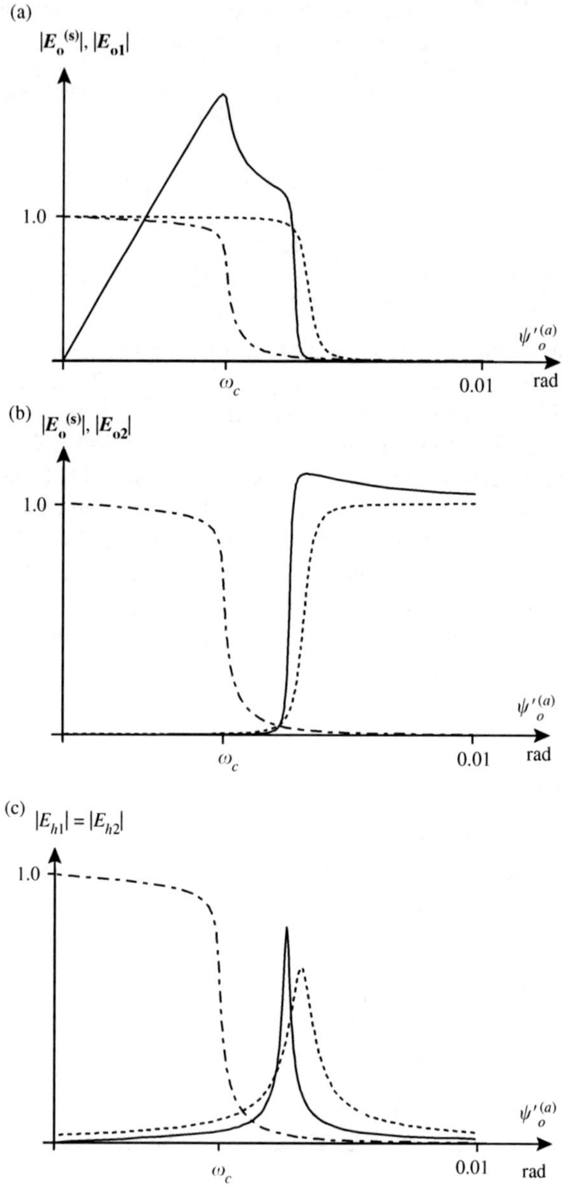

Fig. 8.11. Variations with the incidence angle expressed in radians of the amplitudes of the specularly reflected, $|E_o^{(s)}|$ (dash-dotted line), and of the refracted and reflected waves, $|E_{oj}|$ and $|E_{hj}|$, the latter multiplied by 20 (solid line: taking curvature into account, dashed line: neglecting curvature). Grazing incidence, Laue geometry. Silicon, 400 reflection, Cu$K\alpha$, critical angle: $\omega_c = 0.0039$ rad, the angle between Bragg's incidence and the crystal surface is 0.0039 rad. (a) Refracted wave, branch 1. (b) Refracted wave, branch 2. (c) Reflected waves, branches 1 and 2.

8.8.3 *Specularly and Bragg-reflected intensities for a grazing incidence and the Laue geometry*

The amplitudes of the specularly reflected, $E_o^{(s)}$, and of the diffracted waves E_{oj} are obtained by solving system (8.31):

$$\begin{aligned} E_o^{(s)} &= \frac{\xi_2(K_{oz}^{(a)} - K_{o1z}) - \xi_1(K_{oz}^{(a)} - K_{o2z})}{\xi_2(K_{oz}^{(a)} + K_{o1z}) - \xi_1(K_{oz}^{(a)} + K_{o2z})} E_o^{(a)}, \\ E_{o1} &= \frac{2\xi_2 K_{oz}^{(a)}}{\xi_2(K_{oz}^{(a)} + K_{o1z}) - \xi_1(K_{oz}^{(a)} + K_{o2z})} E_o^{(a)}, \\ E_{o2} &= \frac{-2\xi_1 K_{oz}^{(a)}}{\xi_2(K_{oz}^{(a)} + K_{o1z}) - \xi_1(K_{oz}^{(a)} + K_{o2z})} E_o^{(a)}, \\ E_{hj} &= \xi_j E_{oj}. \end{aligned} \tag{8.37}$$

Figures 8.11(a)–(c) compare the variations of $|E_{o1}|$, $|E_{o2}|$ and $|E_{hj}|$ with angle of incidence calculated using eqn (8.20) and those calculated using the results (6.2a) of traditional dynamical theory, for silicon, 400 and Cu$K\alpha$; they are seen to be very different in the case of $|E_{o1}|$. The variations of the specularly reflected amplitude $|E_o^{(s)}|$ are represented in the same figures as an interrupted line. As in the case of the Bragg geometry, the results calculated using the second-degree equation are perfectly superimposed on the results obtained using the fourth- or third-degree equation (8.17) or (8.18).

The effective absorption coefficients of the two waves penetrating in the crystal are, according to (4.31),

$$\mu_{oj} = -4\pi\gamma_o \mathcal{I}(K_{ojz}) = -4\pi\gamma_o \mathcal{I}(z_j),$$

where j = 1 or 2. Figure 8.12 compares the variations of μ_1 and μ_2 with incidence angle. In this case, the values of the imaginary part of z_1 calculated according to (8.20) are wrong for incidence angles smaller than $\psi_{oB}^{(a)}$ and the corresponding root of the third-degree equation (8.18) must be used. It can be seen that when the incidence is grazing there is an enhancement of the anomalous absorption effect: absorption is strongly reduced for branch 1 and increased for branch 2.

The variations of the diffracted amplitudes for a grazing incidence and the Laue geometry were first calculated by Härtwig (1977).

8.9 Grazing incidence diffraction (non-coplanar geometry)

8.9.1 *Introduction*

Marra *et al.* (1979) introduced a new method combining X-ray specular reflection and conventional X-ray diffraction to study the interface between a molecular beam epitaxially grown aluminium single crystal, 70–2000 Å thick, and a molecular beam

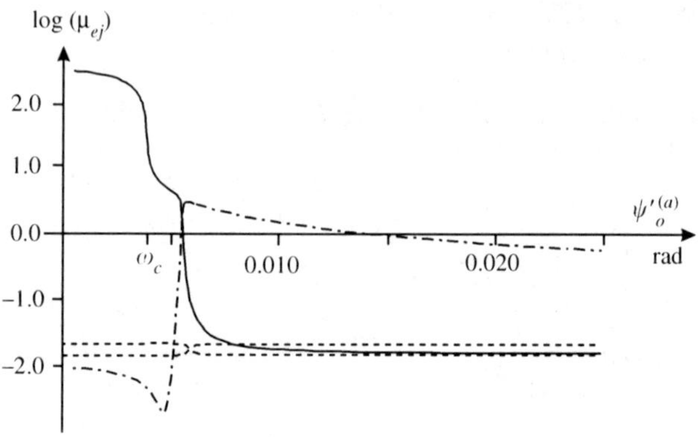

Fig. 8.12. Variations with the incidence angle expressed in radians of the logarithm of the absorption coefficients, μ_{e1} and μ_{e2}, of the two waves penetrating in the crystal (solid line: branch 1, taking curvature into account; dash-dotted line: branch 2, taking curvature into account; dashed line: branches 1 and 2, neglecting curvature); grazing incidence, Laue geometry. Silicon, 400 reflection, Cu$K\alpha$; critical angle: $\omega_c = 0.0039$ rad, the angle between Bragg's incidence and the crystal surface is 0.0039 rad.

epitaxially grown gallium arsenide single-crystal substrate. The same arrangement was used by Eisenberger and Marra (1981) to study the (001) reconstructed surface of a germanium single crystal. In this technique, now called GID, the lattice planes are normal or nearly normal to the crystal surface, and the incident and Bragg-reflected beams make grazing angles, Φ_o and Φ_h respectively, with the crystal surface (Fig. 8.13); if the values of these angles are small enough, specular reflection takes place. Figure 8.13(a) represents the case where the reciprocal lattice vector points towards the outside of the crystal. Depending on the values of the small angle, φ, between the lattice planes and the normal to the crystal surface and of Φ_o, one has a Bragg or a Laue geometry. In the Bragg geometry there is no specular reflection of the Bragg-reflected wave. Figure 8.13(b) represents the case where the reciprocal lattice vector points towards the inside of the crystal; one is then always in the Laue geometry. Afanas'ev and Melikyan (1990) have shown that the extinction distance is given by $\lambda\sqrt{\Phi_o\Phi_h}/\sqrt{\chi_h\chi_{\bar{h}}}$. The penetration depth is therefore very small and GID is now very commonly used for the study of crystal surfaces, in particular for that of multilayers (see, for instance, Baumbach *et al.* 1995; Holý *et al.* 1998).

Two different theoretical approaches have been successfully used: the kinematical one by Vineyard (1982) and the dynamical one by Afanas'ev and Melkonyan (1983). The dynamical approach was further discussed, among others, by Aleksandrov *et al.* (1984*a*,*b*), Afanas'ev *et al.* (1984, 1992), Golovin *et al.* (1984, 1985, 1989), Cowan (1985), Höche *et al.* (1986), Sakata and Hashizume (1987, 1988, 1991), Hashizume and Sakata (1989*a*,*b*), Durbin and Gog (1989), Jach *et al.* (1989), Afanas'ev and Melikyan (1990), Afanas'ev and Esayan (1991) and Holý (1996). Hung and Chang

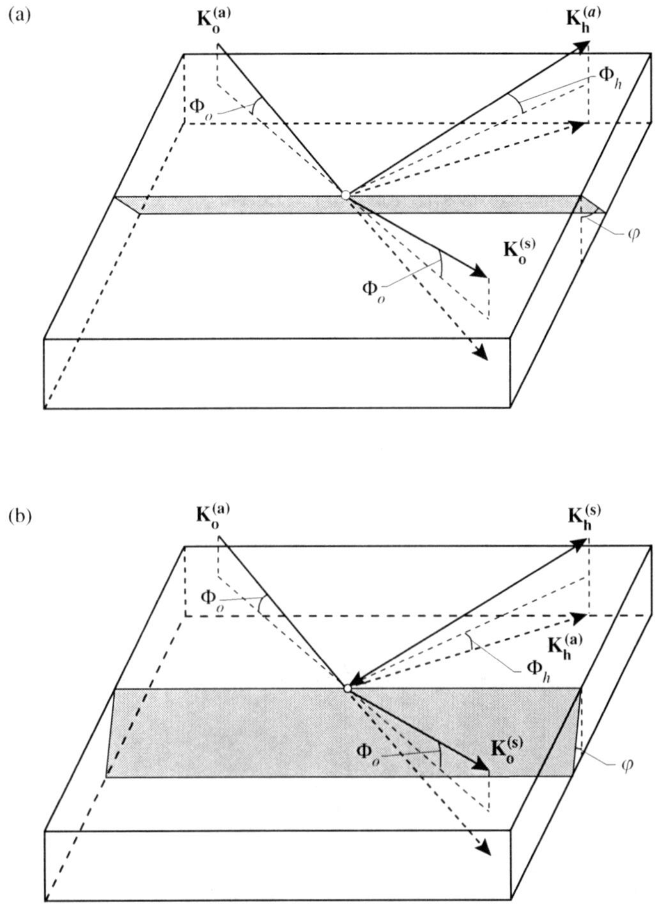

Fig. 8.13. Grazing Incidence Diffraction (non-coplanar geometry). $\mathbf{K}_o^{(s)}$: specularly reflected beam; $\mathbf{K}_h^{(s)}$: specular reflection of the Bragg-reflected beam. The shaded plane is parallel to the lattice planes; φ is their angle with the normal to the crystal surface. (a) The reciprocal lattice vector points towards the outside of the crystal. (b) The reciprocal lattice vector points towards the inside of the crystal.

(1989), Tseng and Chang (1990) and Stetsko and Chang (1997, 1999*a*) considered the case where there are more than two nodes of the reciprocal lattice on the Ewald sphere and three-beam diffraction occurs simultaneously with grazing incidence diffraction. For the study of multilayers, a matrix approach was developed (Andreeva *et al.* 1985; Berreman and Macrander 1988; Rugel *et al.* 1993; Stepanov (1994); Stepanov and Köhler 1994*a*,*b*; Stepanov *et al.* 1995, 1998; Ulyanenkov *et al.* 1995; Baumbach *et al.* 1995; De Caro *et al.* 1997; Holý *et al.* 1999). Cowan *et al.* (1986), Jach *et al.* (1989), Jach and Bedzyk (1990, 1993), Hashizume and Sakata (1989*a*,*b*),

Sakata *et al.* 1993, Sakata and Hashizume (1995, 1997) and Stepanov *et al.* (1998) studied standing waves in the GID geometry and applied them to the analysis of crystal surfaces and epilayers. See also http://sergey.gmca.aps.anl.gov for online calculations.

8.9.2 *Three-dimensional representation of the dispersion surface*

In the GID arrangement the diffraction plane is not parallel to the normal to the crystal surface (Fig. 8.13), contrary to the usual case; the geometry is said to be *off-plane* or *non-coplanar*. Following Höche *et al.* (1986), Jach *et al.* (1989) and Hung and Chang (1989), it is useful to consider a three-dimensional representation of the dispersion surface (Fig. 8.14).

As noticed in Section 4.5, the dispersion surface is a surface of revolution around the reciprocal lattice vector, **OH**. Figure 8.14(a) is a cut by a plane passing through **OH**. It shows the usual shape of the dispersion surface. For simplification we shall assume in the present subsection that this plane is parallel to the crystal surface (in practice, the crystal surface is slightly offset as shown in Figs 8.13(a) and (b)). Figure 8.14(b) is a cut by the median plane, normal to **OH**, and passing through the midpoint, g, of **OH**. It is constituted by two circles passing through the apices of the dispersion surface, a_1 and a_2. Also represented are the circles corresponding to the intersections of the same plane with the spheres of centres O and H and of radii k and nk, respectively. These two circles pass through the Laue and the Lorentz points, l_a and l_o, respectively (Fig. 8.14). The radii of these two circles are, respectively,

$$\begin{aligned} \overline{gl_a} &= k\cos\theta_B, \\ \overline{gl_o} &= k\cos\theta_B + k\frac{\chi_o}{2\cos\theta_B}, \end{aligned} \tag{8.38}$$

where $\cos\theta_B$ is the Bragg angle according to the geometrical theory. The expression of $\overline{gl_o}$ is easily calculated by noticing that $\overline{Og} = k\sin\theta_B = nk\sin\theta_o$ and $\theta_o = \theta_B - (\chi_o/2)\tan\theta_B$, where θ_o is the angle between $\mathbf{O}l_\mathbf{o}$ and the lattice planes.

The radii of the two circles which are the intersections of the dispersion surface with the median plane passing through the apices of the dispersion surface are

$$\begin{aligned} \overline{ga_1} &= k\cos\theta_B + k\frac{\chi_o}{2\cos\theta_B} + k\frac{\sqrt{\chi_h\chi_{\bar{h}}}}{2\cos\theta_B}, \\ \overline{ga_2} &= k\cos\theta_B + k\frac{\chi_o}{2\cos\theta_B} - k\frac{\sqrt{\chi_h\chi_{\bar{h}}}}{2\cos\theta_B}, \end{aligned} \tag{8.39}$$

8.9.3 *Tiepoints excited by the incident wave*

The tiepoints of the wavefields excited by the incident wave are obtained, as usual, by applying the condition of the continuity of the tangential component of wavevectors:

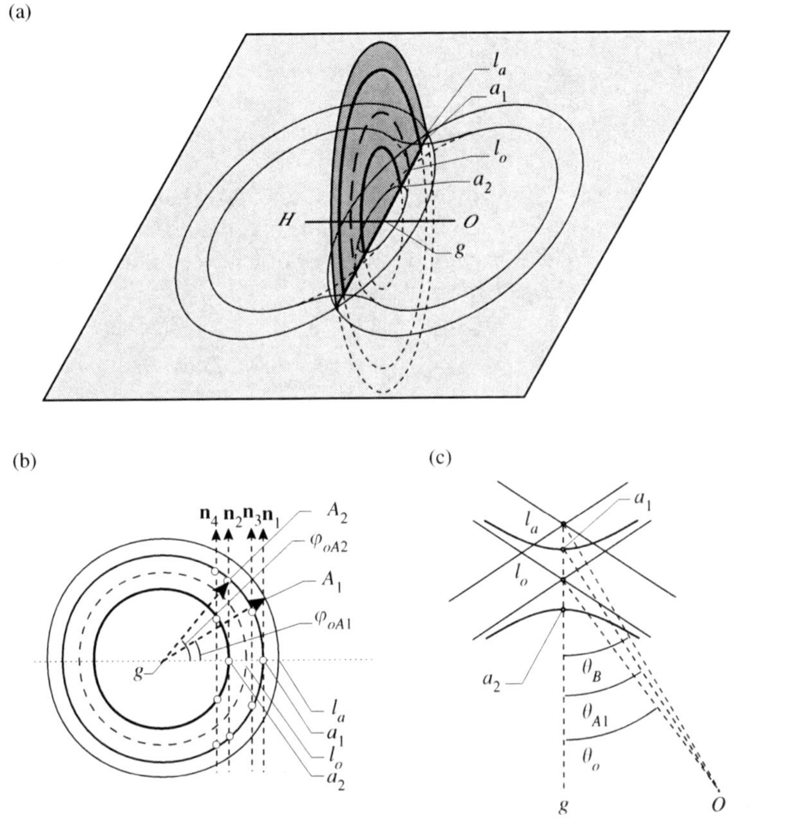

Fig. 8.14. Dispersion surface in the Grazing Incidence Diffraction (non-coplanar) geometry. (a) Tri-dimensional view; the shaded plane is parallel to the crystal surface, assumed here to be parallel to the reciprocal lattice vector; l_a: Laue point; l_o: Lorentz point; a_1, a_2, apices of the dispersion surface. (b) Intersection of the dispersion surface with the median plane; $\mathbf{n_1}$, $\mathbf{n_2}$, $\mathbf{n_3}$, $\mathbf{n_4}$, normals to the crystal surface. (c) Intersection of the dispersion surface with the crystal surface.

they lie at the intersection of the dispersion surface and the normal to the crystal surface drawn from the extremity of the incident wavevector.

8.9.3.1 *The extremity of the incident wavevector lies in the median plane of the dispersion surface—the crystal surface is parallel to the reciprocal lattice vector*

At first, let us consider an incident vector whose extremity lies on the circle of centre g and of radius $\overline{gl_a}$. If the normal to the crystal surface, $\mathbf{n_1}$, passes through the apex of branch 1 the dispersion surface, one tiepoint only is excited, a_1, and the incident wavevector is $\mathbf{OA_1}$ (Figs 8.14(b) and 8.15). If the normal, $\mathbf{n_2}$, passes through the apex a_2 of branch 2, three tiepoints are excited and the incident wavevector is $\mathbf{OA_2}$ (not represented in Fig. 8.15, for simplicity). If the normal, $\mathbf{n_3}$, lies between $\mathbf{n_1}$ and $\mathbf{n_2}$, only two tiepoints are excited (Fig. 8.14(b)). If the normal, $\mathbf{n_4}$, lies between $\mathbf{n_2}$ and g, four tiepoints are excited.

Let Φ_{oA1} (Φ_{oA2}) be the angle between $\mathbf{OA_1}$ ($\mathbf{OA_2}$) and the crystal surface (Fig. 8.15—$\mathbf{OA_2}$ is not represented, to simplify the figure), and let ϕ_{oA1} (ϕ_{oA2}) be the angle between the projection, $\mathbf{gA_1}$ ($\mathbf{gA_2}$), of that vector on the median plane and the crystal surface (Fig. 8.15(b)). One has

$$\cos\phi_{oA1} = \frac{\overline{ga_1}}{\overline{gA_1}} = 1 + \frac{\chi_o}{2\cos^2\theta_B} + \frac{\sqrt{\chi_h\chi_{\bar{h}}}}{2\cos^2\theta_B},$$

$$\cos\phi_{oA2} = \frac{\overline{ga_2}}{\overline{gA_2}} = 1 + \frac{\chi_o}{2\cos^2\theta_B} - \frac{\sqrt{\chi_h\chi_{\bar{h}}}}{2\cos^2\theta_B}.$$

It follows that

$$\begin{aligned} \phi_{oA1} &= \frac{\sqrt{|\chi_o| - \sqrt{\chi_h\chi_{\bar{h}}}}}{\cos\theta_B}, \\ \phi_{oA2} &= \frac{\sqrt{|\chi_o| + \sqrt{\chi_h\chi_{\bar{h}}}}}{\cos\theta_B}. \end{aligned} \tag{8.40}$$

Comparing triangles $\widehat{ga_1A_1}$, $\widehat{Oa_1A_1}$ and $\widehat{gOA_1}$ and noticing that the angle between $\mathbf{gA_1}$ and $\mathbf{OA_1}$ is θ_B, one finds that,

$$\sin\Phi_{oA1} = \cos\theta_B\phi_{oA1} \tag{8.41}$$

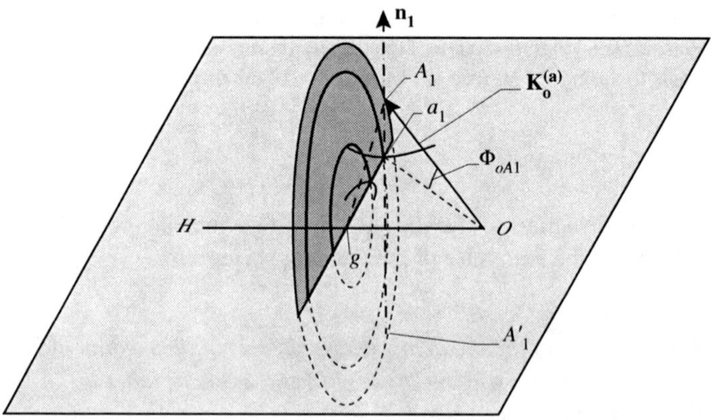

Fig. 8.15. Incident wavevector, $\mathbf{OA_1} = \mathbf{K_0^{(a)}}$, exciting the apex of the dispersion surface, a_1. a_1 and A_1 lie on a normal, $\mathbf{n_1}$, to the crystal surface.

and one obtains, finally,

$$\Phi_{oA1} = \sqrt{|\chi_o| - \sqrt{\chi_h \chi_{\bar{h}}}},$$
$$\Phi_{oA2} = \sqrt{|\chi_o| + \sqrt{\chi_h \chi_{\bar{h}}}}. \tag{8.42}$$

The number of tiepoints excited by the incident wavevector depends on the value of the angle, Φ_o, between that vector and the crystal surface:

- $\Phi_o < \Phi_{oA1}$ no Bragg reflection, specular reflection only
- $\Phi_{oA1} < \Phi_o < \Phi_{oA2}$ two tiepoints
- $\Phi_{oA2} < \Phi_o$ four tiepoints.

In order to specify the orientation of wavevector $\mathbf{OA_1}$ completely, it is necessary to find also the angle, θ_{A1}, between $\mathbf{Oa_1}$ and $\mathbf{ga_1}$ (Fig. 8.14(c)). One has

$$\sin\theta_{A1} = \frac{\overline{Og}}{\overline{Oa_1}} = \frac{k\sin\theta_B}{\cos\Phi_{oA1}}.$$

Let $\theta_{A1} = \theta_B + \Delta\theta_{A1}$. It follows using (8.42) that

$$\Delta\theta_{A1} = \frac{|\chi_o| - \sqrt{\chi_h \chi_{\bar{h}}}}{2}\tan\theta_B,$$
$$\Delta\theta_{A2} = \frac{|\chi_o| + \sqrt{\chi_h \chi_{\bar{h}}}}{2}\tan\theta_B, \tag{8.43}$$

where $\Delta\theta_{A2} = \theta_{A2} - \theta_B$ and θ_{A2} is the angle between $\mathbf{Oa_2}$ and $\mathbf{ga_2}$.

8.9.3.2 *General case—the crystal surface is parallel to the reciprocal lattice vector*

Let M be the extremity of the incident wavevector, $\mathbf{OM} = \mathbf{K_o^{(a)}}$, and let m and N be the intersections of the normal to the crystal surface, $\mathbf{n}$, with the crystal surface and the sphere of centre H and radius k, respectively (Figs 8.16(a) and (b)). It is assumed in this subsection that the crystal surface is exactly perpendicular to the lattice planes. The Bragg-reflected wavevector is $\mathbf{HN} = \mathbf{K_h^{(a)}}$, the specularly reflected wavevector is $\mathbf{OM'} = \mathbf{K_o^{(s)}}$, where M' is the second intersection of $\mathbf{n}$ with the sphere of centre O and radius k; the specular reflection of the Bragg-reflected wavevector is $\mathbf{HN'} = \mathbf{K_h^{(a)}}$, where N' is the second intersection of $\mathbf{n}$ with the sphere of centre H and radius k.

The plane parallel to the normal $\mathbf{n}$ and perpendicular to the reciprocal lattice vector cuts the intersection of the dispersion surface and the crystal surface at p_1 and p_2 (Fig. 8.16(c)). The tiepoints excited by the incident wave lie at the intersection of $\mathbf{n}$ and the small circles which constitute the cuts of the dispersion surface by the plane normal to the reciprocal lattice vector and passing through p_1 and p_2 (Fig. 8.16(b)). As in the preceding case, depending on the angle between $\mathbf{OM}$ and the crystal surface,

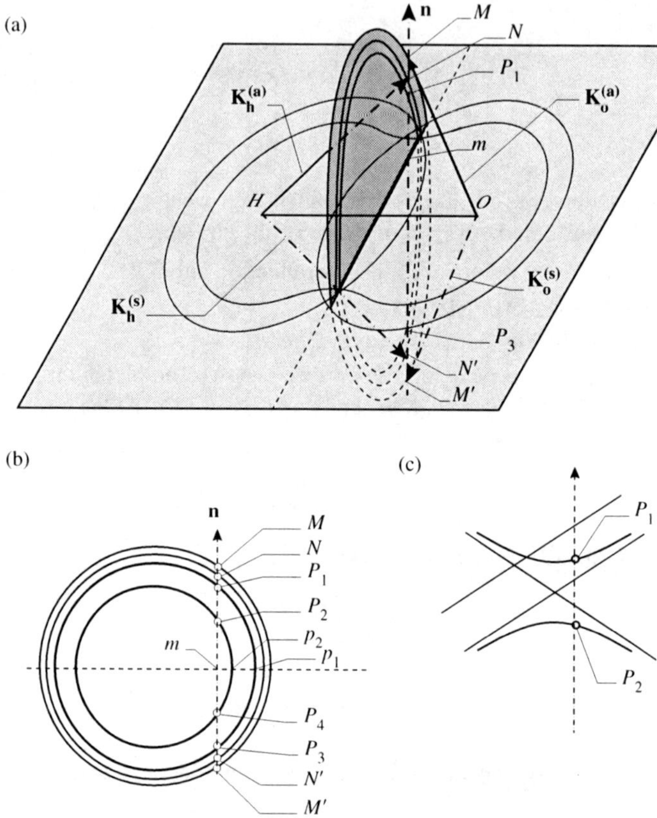

Fig. 8.16. Tiepoints excited by the incident wave, $\mathbf{OM} = \mathbf{K}_o^{(a)}$ (GID geometry, crystal surface perpendicular to the lattice planes). (a) Tri-dimensional view; $\mathbf{OM}' = \mathbf{K}_o^{(s)}$: specularly reflected wave; $\mathbf{ON} = \mathbf{K}_h^{(a)}$: Bragg-reflected wave; $\mathbf{ON}' = \mathbf{K}_h^{(s)}$: specular reflection of the Bragg-reflected wave; branch 2 of the dispersion surface is not represented here for simplification. The shaded plane is parallel to the crystal surface, assumed here to be parallel to the reciprocal lattice vector. (b) Intersection of the dispersion surface with a plane perpendicular to **OH**; **n**: normal to the crystal surface. (c) Intersection of the dispersion surface with the crystal surface.

there are two tiepoints, P_1 and P_3, on the circle passing through p_1 or four tiepoints, P_1, P_3 and P_2 and P_4, on the circle passing through p_2.

8.9.3.3 *General case—the crystal surface is slightly offset with respect to the reciprocal lattice vector*

In practical cases, the reciprocal lattice makes a small angle, φ, with the crystal surface (Fig. 8.17(a)). As usual, the extremities, M and N, of the incident and reflected

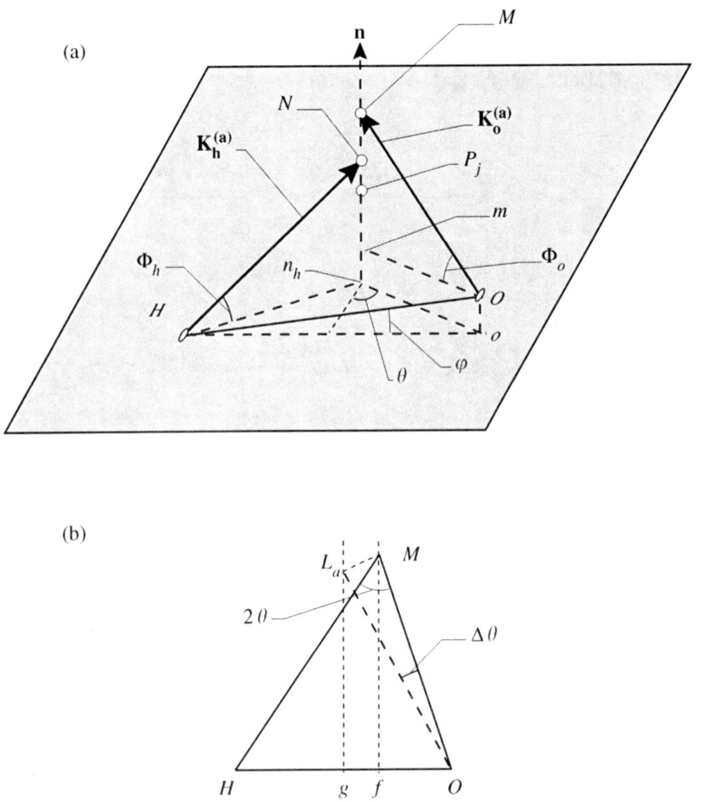

Fig. 8.17. Tiepoints excited by the incident wave, $\mathbf{OM} = \mathbf{K}_o^{(a)}$ (GID geometry, general case). (a) Tri-dimensional view; $\mathbf{ON} = \mathbf{K}_h^{(a)}$: Bragg-reflected wave; **OH**: reciprocal lattice vector; the shaded plane is parallel to the crystal surface; **n**: normal to the crystal surface; P_j: tiepoint of a wavefield excited in the crystal. (b) Departure from Bragg's angle: $\Delta\theta = \theta - \theta_B$; L_a: Laue point.

wavevectors lie on the normal **n** to the crystal surface, as well as the tiepoints, P_j, of the wavefields excited in the crystal. The specularly reflected wavevectors, $\mathbf{OM}' = \mathbf{K}_o^s$ and $\mathbf{ON}' = \mathbf{K}_h^s$, are symmetric of $\mathbf{OM} = \mathbf{K}_o^a$ and $\mathbf{ON} = \mathbf{K}_h^a$ with respect to the crystal surface and are not represented in the figure, for simplification (they are represented in Fig. 8.13 which is drawn in direct space and with a different orientation). The directions of the incident and Bragg-reflected wavevectors are defined by their angles with the crystal surface, Φ_o and Φ_h, respectively (Fig. 8.17(a)), and by the angle between them, 2θ (Fig. 8.17(b)). These three angles are not independent and can be related, as shown below. In a practical experiment, the departure from Bragg's angle, $\Delta\theta = \theta - \theta_B$, can be varied while keeping angle Φ_o constant.

Let n_h be the intersection of $\mathbf{n}$ with a plane parallel to the crystal surface, drawn from reciprocal lattice point H, and let further o and m be the projections of the origin of the reciprocal space, O, on that plane and on $\mathbf{n}$, respectively (Fig. 8.17(a)). Let us consider the quantity

$$\alpha = \frac{\overline{HM}^2 - \overline{HN}^2}{k^2} = \frac{(\mathbf{K_o^a} - \mathbf{h})^2 - k^2}{k^2}.$$

One may write, noticing that $\overline{OH} = 2k \sin \theta_B$ and $\overline{n_h m} = \overline{oO} = 2k \sin \theta_B \sin \varphi \approx 2k\varphi \sin \theta_B$,

$$\begin{aligned} \overline{HM}^2 &= \overline{Hn_h}^2 + \overline{n_h M}^2 = \overline{Hn_h}^2 + (k\Phi_o + 2k\varphi \sin \theta_B)^2, \\ \overline{HN}^2 &= \overline{Hn_h}^2 + \overline{n_h N}^2 = \overline{Hn_h}^2 + k^2 \Phi_h^2. \end{aligned}$$

It follows that

$$\alpha = (\Phi_o + 2\varphi \sin \theta_B)^2 - \Phi_h^2.$$

The quantity α may also be evaluated in triangle $\widehat{HMO}$ (Fig. 8.17(b)). Let f be the projection of M on $\mathbf{OH}$. One has

$$\overline{HM}^2 - \overline{HN}^2 = \overline{Hf}^2 - \overline{Of}^2 = \overline{HO}(\overline{Hf} - \overline{Of}) = -2\overline{OH}\,\overline{gf}$$

and, noticing that

$$\begin{aligned} \overline{OH} &= 2k \sin \theta_B, \\ \overline{gf} &= \overline{L_a M} \cos \theta_B, \end{aligned}$$

where $\overline{L_a M} = k\Delta\theta$, $\Delta\theta$ departure from Bragg's angle of the incident wave, it follows finally that

$$\alpha = \frac{(\mathbf{K_o^a} - \mathbf{h})^2 - k^2}{k^2} = (\Phi_o + 2\varphi \sin \theta_B)^2 - \Phi_h^2 = -2\Delta\theta \sin 2\theta_B. \tag{8.44}$$

This result was first given by Aleksandrov *et al.* (1984).

The angles between the incident and Bragg-reflected wavevectors and the crystal surface, Φ_o and Φ_h, are complementary of the angles between these vectors and the normal to the crystal surface, ψ_o and ψ_h, respectively, defined in Section 4.7.3. One has therefore

$$\begin{aligned} \gamma_o &= \cos \psi_o = \sin \Phi_o \approx \Phi_o, \\ \gamma_h &= \cos \psi_h = \sin \Phi_h \approx \Phi_h \end{aligned} \tag{8.45}$$

(γ_h and Φ_h are negative in the Bragg geometry—Fig. 8.13(a)).

8.9.4 *Equation of the dispersion surface*

The fundamental equations of the dynamical theory can be written, following (5.9), as

$$\begin{aligned} &\left[\frac{K_{oj}^2 - k^2}{k^2} - \chi_o\right] D_{oj} - \chi_{\bar{h}} D_{hj} = 0, \\ -\chi_h D_{oj} + &\left[\frac{K_{hj}^2 - k^2}{k^2} - \chi_o\right] D_{hj} = 0, \end{aligned} \tag{8.46}$$

limiting ourselves to the case of σ-polarization, for simplification.

One may write, using Fig. 8.17(a),

$$K_{oj}^2 = \overline{OP_j}^2 = \overline{Om}^2 + \overline{mP_j}^2 = \overline{Om}^2 + k^2 u_j^2,$$

$$k^2 = \overline{OM}^2 = \overline{Om}^2 + \overline{mM}^2 = \overline{Om}^2 + k^2 \Phi_o^2,$$

where $ku_j = K_{ojz} = \overline{mP_j}$ is the normal component of the refracted wavevector, $\mathbf{K_{oj}}$. In a similar way, one has

$$K_{hj}^2 = \overline{HP_j}^2 = \overline{Hn_h}^2 + \overline{n_h P_j}^2 = \overline{Hn_h}^2 + k^2 (2\varphi \sin\theta_B + u_j)^2,$$

where $K_{hjz} = \overline{n_h P_j} = k(2\varphi \sin\theta_B + u_j)$ is the normal component of the reflected wavevector, $\mathbf{K_{hj}}$.

The fundamental equations (8.46) of the dynamical theory may therefore be written as

$$\begin{aligned} (u_j^2 - \Phi_o^2 - \chi_o) D_{oj} - \chi_{\bar{h}} D_{hj} &= 0, \\ -\chi_h D_{oj} + [(u_j + 2\varphi \sin\theta_B)^2 - \Phi_h^2 - \chi_o] D_{hj} &= 0, \end{aligned} \tag{8.47}$$

and the equation of the dispersion surface is the secular equation of (8.47):

$$(u_j^2 - \Phi_o^2 - \chi_o)[(u_j + 2\varphi \sin\theta_B)^2 - \Phi_h^2 - \chi_o] = \chi_h \chi_{\bar{h}}, \tag{8.48}$$

a result first given by Aleksandrov *et al.* (1984).

Equation (8.48) has four roots; they give the positions of the four tiepoints P_j ($j = 1$ to 4) which lie at the intersection of the normal to the crystal surface and the dispersion surface. Two of the corresponding wavefields propagate towards the inside of the crystal; the two others would be propagating upwards from the bottom of the crystal and are not to be taken into account in the case of a semi-infinite crystal to which we shall limit ourselves here. Equation (8.48) has analytical solutions when $\phi = 0$, namely when the crystal surface is parallel to the reciprocal lattice vector. This particular case was discussed by Afanas'ev and Melkonyan (1983) and Jach *et al.* (1989). The general case was analysed by Aleksandrov *et al.* (1984) and Afanas'ev and Melikyan (1990).

8.9.5 *Amplitudes of the waves*

The amplitudes of the waves are related by the boundary conditions (8.34) and (8.35). Limiting ourselves to the Laue geometry, eqns (8.35) are written, after replacing the normal components K_{ojz} and K_{hjz} by their respective values and noting that $K_{oz}^{(a)} = \Phi_o$ and $K_{hz}^{(a)} = \Phi_h$:

$$
\begin{aligned}
E_o^{(a)} + E_o^{(s)} &= E_{o1} + E_{o2}, \\
E_h^{(s)} &= \xi_1 E_{o1} + \xi_2 E_{o2}, \\
\Phi_o\big(E_o^{(a)} - E_o^{(s)}\big) &= u_1 E_{o1} + u_2 E_{o2}, \\
-\Phi_h E_h^{(s)} &= (2\varphi + u_1)\xi_1 E_{o1} + (2\varphi \sin\theta_B + u_2)\xi_2 E_{o2}.
\end{aligned}
\tag{8.49}
$$

One finds

$$
\begin{aligned}
E_{o1} &= \frac{-2\Phi_o \xi_2 (\Phi_h + 2\varphi \sin\theta_B + u_2) E_o^{(a)}}{\xi_1(\Phi_o + u_2)(\Phi_h + 2\varphi\sin\theta_B + u_1) - \xi_2(\Phi_o + u_1)(\Phi_h + 2\varphi\sin\theta_B + u_2)}, \\
E_{o2} &= \frac{2\Phi_o \xi_1 (\Phi_h + 2\varphi + u_1) E_o^{(a)}}{\xi_1(\Phi_o + u_2)(\Phi_h + 2\varphi\sin\theta_B + u_1) - \xi_2(\Phi_o + u_1)(\Phi_h + 2\varphi\sin\theta_B + u_2)}, \\
E_h^{(s)} &= \frac{2\Phi_o \xi_1 \xi_2 (u_1 - u_2) E_o^{(a)}}{\xi_1(\Phi_o + u_2)(\Phi_h + 2\varphi\sin\theta_B + u_1) - \xi_2(\Phi_o + u_1)(\Phi_h + 2\varphi\sin\theta_B + u_2)}, \\
E_o^{(s)} &= \frac{\xi_1(\Phi_o - u_2)(\Phi_h + 2\varphi\sin\theta_B + u_1) - \xi_2(\Phi_o - u_1)(\Phi_h + 2\varphi\sin\theta_B + u_2)}{\xi_1(\Phi_o + u_2)(\Phi_h + 2\varphi\sin\theta_B + u_1) - \xi_2(\Phi_o + u_1)(\Phi_h + 2\varphi\sin\theta_B + u_2)} E_o^{(a)},
\end{aligned}
\tag{8.50}
$$

where the amplitude ratio ξ_j is given by (8.47):

$$\xi_j = \frac{u_j^2 - \Phi_o^2 - \chi_o}{\chi_{\bar{h}}}.$$

The specularly reflected intensities are, using (4.38) and (8.45):

$$
\begin{aligned}
I_o^{(s)} &= \left|\frac{E_o^{(s)}}{E_o^{(a)}}\right|^2, \\
I_h^{(s)} &= \left|\frac{\gamma_h}{\gamma_o}\right| \left|\frac{E_h^{(a)}}{E_o^{(a)}}\right|^2 = \frac{\Phi_h}{\Phi_o}\left|\frac{E_h^{(a)}}{E_o^{(a)}}\right|^2.
\end{aligned}
\tag{8.51}
$$

9

n-beam dynamical diffraction

9.1 Introduction

In the preceding chapters it has always been assumed that there are only two reciprocal lattice points in the vicinity of the Ewald sphere. The presence of more than two reciprocal lattice points close to the Ewald sphere is accidental for X-rays of usual energies but is quite normal for electron diffraction and for X-rays of very high energies. The early history and the principle of n-beam X-ray diffraction has been recalled in Section 1.9.

The main application of n-beam diffraction is that it provides phase information on the structure factors, which can be used in structure determination and, out of the more than 300 publications devoted to that topic, nearly half concern the phase problem. This application is developed in Section 9.4.

Another interesting property of n-beam diffraction is the enhancement of the Borrmann effect under special circumstances, the so-called super-Borrmann effect (Borrmann and Hartwig 1965; Hildebrandt 1967; Afanas'ev and Kohn 1975). It is described in Section 9.5.

n-beam diffraction has also proved to be an interesting technique to study the influence of an electric field on crystals, through different effects. On the one hand, when an electric field is applied to a crystal such as gallium arsenide where each atom is coordinated to four atoms of the other kind by polar-covalent bonds (see Fig. 7.9) there are modifications of the amplitude and phase of the structure factor. n-beam diffraction is an ideal tool to measure the variations of the phase in terms of the applied electric field (Stahn *et al.* 1999). On the other hand, the applied field induces strains which are related to changes in the lattice parameters that can be revealed by changes in the positions of the peaks excited by *Umweganregung* during a Renninger-scan. This was observed by Avanci *et al.* (1998, 2000) who used the results to measure the piezoelectric constants of several organic non linear optical materials.

The intensity of the primary reflection, $h_1h_2h_3$ (Fig. 9.1) depends on the crystal perfection and provides a means to characterize it (Sasaki *et al.* 1996). In particular, when the direction , $\mathbf{K_g}$, of the reflected beam due to the secondary reflection, $g_1g_2g_3$, is parallel to the crystal surface, the analysis of the intensity of the primary reflection provides information on the perfection of the surface layers. This arrangement was called 'Bragg Surface Diffraction' by Morelhão and Cardoso (1996) and Hayashi *et al.* (1997). The shape of the dispersion surface for that particular geometrical situation was studied by Avanci and Morelhão (2000). Bragg Surface Diffraction was applied

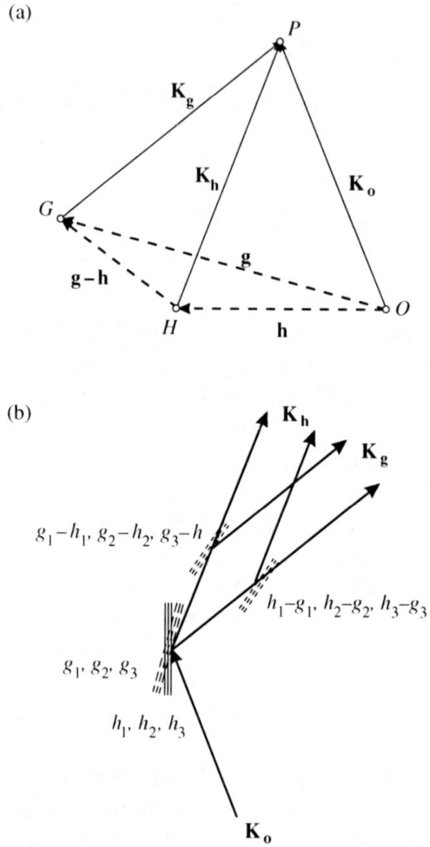

Fig. 9.1. Three-beam diffraction. (a) Reciprocal space. (b) Direct space; h_1, h_2, h_3: primary reflection; g_1, g_2, g_3: secondary reflection; $g_1 - h_1, g_2 - h_2, g_3 - h_3$: coupling reflection.

by Morelhão and Abramof (1998) to the characterization of epilayers on silicon and gallium arsenide substrates. It is also interesting to note that n-beam dynamical diffraction of light has been studied in photonic crystals (Chan *et al.* 1992).

There are several review papers devoted to n-beam diffraction (Chang 1987, 1992, 1998, 2004; Colella 1995; Weckert and Hümmer 1996, 1997); Sections 9.2–9.4 of the present chapter are based essentially on the review by Weckert and Hümmer (1997).

9.2 The general three-beam case

9.2.1 *Renninger-scans*

The number of wavefields excited in the crystal when there are n reciprocal lattice points close to the Ewald sphere is $2n$ in the general case (see Colella 1974, 1980;

Chang 1979; Kohn 1979, 1980). Most of the applications concern the case $n = 3$, a few $n = 4$ or more. The present chapter will be limited to the three-beam case. Let O, H and G be the three reciprocal lattice points (Fig. 9.1(a)) and $\mathbf{K_o}$, $\mathbf{K_h}$ and $\mathbf{K_g}$ the three corresponding wavevectors. The waves of wavevector $\mathbf{K_h}$ can be excited either by direct reflection on the (h_1, h_2, h_3) planes, or by a so-called *detour* (or *Umweg*) excitation: by a reflection first on the (g_1, g_2, g_3) planes and then on the $(h_1 - g_1, h_2 - g_2, h_3 - g_3)$ planes (coupling reflection), as shown in Fig. 9.1(b). In a similar way the waves of wavevector $\mathbf{K_g}$ can be excited either by direct reflection on the (g_1, g_2, g_3) planes, or by a reflection first on the (h_1, h_2, h_3) planes and then on the $(g_1 - h_1, g_2 - h_2, g_3 - h_3)$ planes. The relations between the three wavevectors can be written as

$$\mathbf{K_h} = \mathbf{K_o} - \mathbf{h} = \mathbf{K_g} - (\mathbf{h} - \mathbf{g}),$$
$$\mathbf{K_g} = \mathbf{K_o} - \mathbf{g} = \mathbf{K_h} - (\mathbf{g} - \mathbf{h})$$

with $\mathbf{g} - \mathbf{h} = \mathbf{HG}$. Wavevector $\mathbf{K_o}$ is as usual related to the incident wavevector, $\mathbf{K_o^{(a)}}$, by the condition of the continuity of their tangential components:

$$\mathbf{K_o^{(a)}} - \mathbf{K_o} = \mathbf{OM} - \mathbf{OP} = \mathbf{MP},$$

where $\mathbf{MP}$ is normal to the entrance surface (Fig. 9.2) and P the tiepoint of the wavefield.

In practice, one sets the crystal for the primary reflection, $h_1h_2h_3$, and then one rotates it around the reciprocal lattice vector, $\mathbf{h}$ (Fig. 9.3); this is the so-called Renninger-scan, or Ψ-scan (Renninger 1937*a*). When the reciprocal lattice point G crosses the Ewald sphere, the secondary reflection $g_1g_2g_3$ is also excited. The position of G can be characterized by the angle, β_G, between the projections of $\mathbf{K_o} = \mathbf{OL_o}$ and of $\mathbf{g} = \mathbf{OG}$ on the plane perpendicular to $\mathbf{h}$ and passing through O (Fig. 9.3):

$$\beta_G = (\mathbf{OL'_o}, \mathbf{OG'}).$$

Cole *et al.* (1962) have shown, and it can easily be found in Fig. 9.3, that this angle is given by

$$\cos\beta_G = \frac{\mathbf{g}^2 - \mathbf{h}\cdot\mathbf{g}}{2\sqrt{k^2 - h^2/4}\,\sqrt{\mathbf{g}^2 - (\mathbf{h}\cdot\mathbf{g})^2/\mathbf{h}^2}}.$$

The first Renninger-scan with synchrotron radiation was obtained by Brown *et al.* (1989), but this is now a routine operation. Several algorithms have been developed to simulate Renninger-scans (for instance, Salles da Costa *et al.* 1990; Parente *et al.* 1994; Sasaki *et al.* 1996).

9.2.2 *Fundamental equations of the dynamical theory*

The fundamental equations of dynamical theory, (5.4) for the electric displacement and (5.5) for the electric field, express the coupling between the waves diffracted on

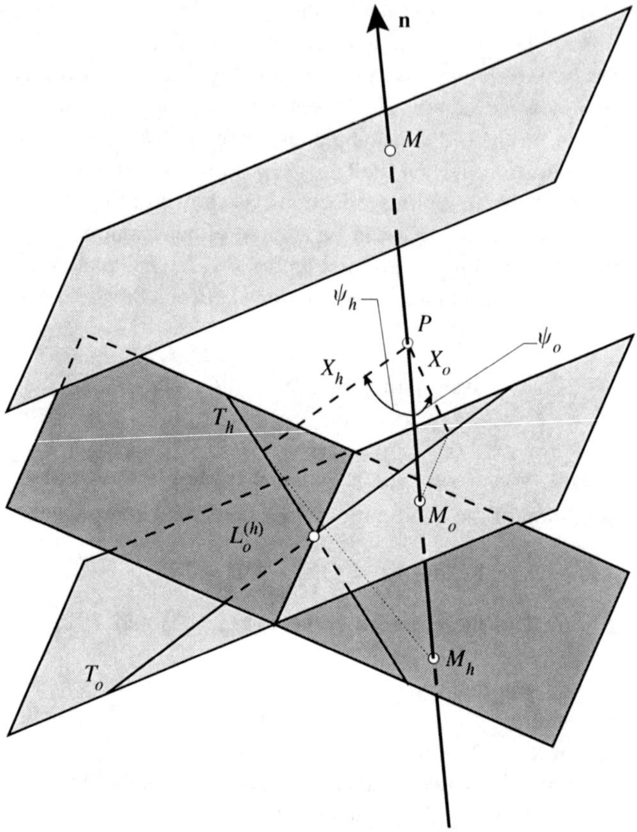

Fig. 9.2. Resonance errors, X_o and X_h; they represent the distances of the tiepoint, P, from the planes asymptotic to the spheres of radius nk and of centres O and H, respectively. T_o, T_h: intersections of these planes with the diffraction plane ($\mathbf{K_o}$, $\mathbf{K_h}$); M: extremity of the incident wavevector; $\mathbf{n}$: normal to the crystal surface; M_o, M_h: intersections of $\mathbf{n}$ with the planes asymptotic to the spheres of radius nk; $L_o(h)$: Lorentz point associated with reciprocal lattice point H.

the three sets of lattice planes:

$$\begin{aligned}
&\left(\frac{K_o^2 - k^2}{K_o^2} - \chi_o\right)\mathbf{D_o} && -\chi_{\bar{h}}\mathbf{D_{h[o]}} && -\chi_{\bar{g}}\mathbf{D_{g[o]}} && = 0\\
&-\chi_h\mathbf{D_{o[h]}} && +\left(\frac{K_h^2 - k^2}{K_h^2} - \chi_o\right)\mathbf{D_h} && -\chi_{h-g}\mathbf{D_{g[h]}} && = 0\\
&-\chi_g\mathbf{D_{o[g]}} && -\chi_{g-h}\mathbf{D_{h[g]}} && +\left(\frac{K_g^2 - k^2}{K_g^2} - \chi_o\right)\mathbf{D_g} && = 0
\end{aligned} \tag{9.1a}$$

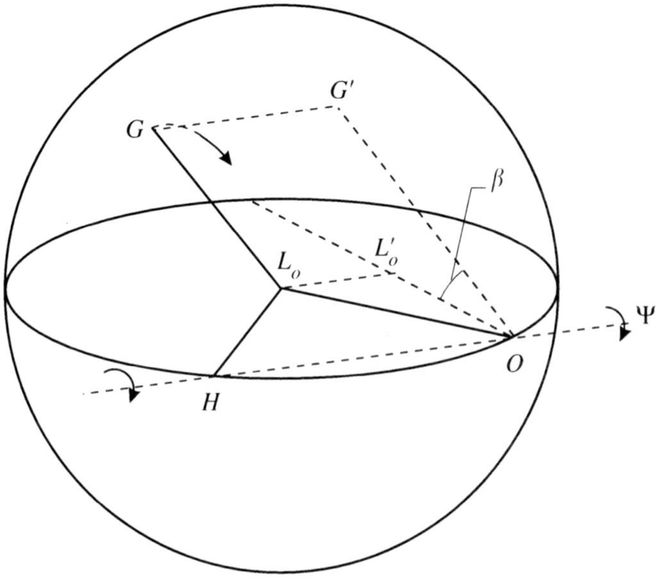

Fig. 9.3. Ψ-scan: as the crystal is rotated around **OH**, reciprocal lattice point G crosses the Ewald sphere; in the present example it moves from inside to outside ('in–out'); L_o: Lorentz point, centre of the Ewald sphere of radius nk; G', L_o': projections of G and L_o on the plane perpendicular to **OH** and passing through O.

$$\begin{array}{llll}
\left(\dfrac{K_o^2-k^2}{k^2}-\chi_o\right)\mathbf{E_{o[o]}} & -\chi_{\bar{h}}\mathbf{E_h} & -\chi_{\bar{g}}\mathbf{E_g} & =0 \\
-\chi_h\mathbf{E_o} & +\left(\dfrac{K_h^2-k^2}{k^2}-\chi_o\right)\mathbf{E_{h[h]}} & -\chi_{h-g}\mathbf{E_g} & =0 \\
-\chi_g\mathbf{E_o} & -\chi_{g-h}\mathbf{E_h} & +\left(\dfrac{K_g^2-k^2}{k^2}-\chi_o\right)\mathbf{E_{g[g]}} & =0
\end{array} \tag{9.1b}$$

where (see Section 5.3)

$$\mathbf{E_{o[o]}} \approx \mathbf{E_o}; \qquad \mathbf{E_{h[h]}} \approx \mathbf{E_h}; \qquad \mathbf{E_{g[g]}} \approx \mathbf{E_g}$$

are related to $\mathbf{D_o}$, $\mathbf{D_h}$, $\mathbf{D_g}$ by (5.3) and, using (2.50),

$$\begin{aligned}
\mathbf{D_{h[o]}} &= -\frac{1}{K_o^2}\mathbf{K_o} \wedge \mathbf{K_o} \wedge \mathbf{D_h} \\
&= \mathbf{D_h} - \frac{1}{K_o^2}\mathbf{K_o}(\mathbf{K_o}\cdot\mathbf{D_h}).
\end{aligned} \tag{9.2}$$

With the same approximations as in Section 5.3, eqns (9.1a,b) take the form

$$\begin{array}{llll} \left(\dfrac{K_o^2 - k^2}{k^2} - \chi_o\right)\mathbf{D_o} & -\chi_{\bar{h}}\mathbf{D_{h[o]}} & -\chi_{\bar{g}}\mathbf{D_{g[o]}} & = 0 \\ -\chi_h \mathbf{D_{o[h]}} & +\left(\dfrac{K_h^2 - k^2}{k^2} - \chi_o\right)\mathbf{D_h} & -\chi_{h-g}\mathbf{D_{g[h]}} & = 0 \\ -\chi_g \mathbf{D_{o[g]}} & -\chi_{g-h}\mathbf{D_{h[g]}} & +\left(\dfrac{K_g^2 - k^2}{k^2} - \chi_o\right)\mathbf{D_g} & = 0 \end{array} \tag{9.3a}$$

$$\begin{array}{llll} \left(\dfrac{K_o^2 - k^2}{k^2} - \chi_o\right)\mathbf{E_o} & -\chi_{\bar{h}}\mathbf{E_h} & -\chi_{\bar{g}}\mathbf{E_g} & = 0 \\ -\chi_h \mathbf{E_o} & +\left(\dfrac{K_h^2 - k^2}{k^2} - \chi_o\right)\mathbf{E_{h[h]}} & -\chi_{h-g}\mathbf{E_g} & = 0 \\ -\chi_g \mathbf{E_g} & -\chi_{g-h}\mathbf{E_h} & +\left(\dfrac{K_g^2 - k^2}{k^2} - \chi_o\right)\mathbf{E_g} & = 0 \end{array} \tag{9.3b}$$

As was shown in Section 5.1, the electric displacement is rigorously normal to the wavevector and the electric field approximately so. The sets (9.3a,b) of three vector equations can be decomposed into sets of six scalar equations by decomposing each electric displacement vector into two components in the plane normal to the corresponding wavevector:

$$\mathbf{D_h} = D_{h\pi}\overrightarrow{\pi_h} + D_{h\sigma}\overrightarrow{\sigma_h}. \tag{9.4}$$

Various authors have chosen various orientations for the unit vectors $\overrightarrow{\pi_h}$ and $\overrightarrow{\sigma_h}$. We shall follow here the conventions chosen by Weckert and Hümmer (1997) after Colella (1974):

$$\overrightarrow{\sigma_h} = \frac{\mathbf{K_h} \wedge \mathbf{K_o}}{|\mathbf{K_h} \wedge \mathbf{K_o}|}; \qquad \overrightarrow{\sigma_g} = \frac{\mathbf{K_g} \wedge \mathbf{K_o}}{|\mathbf{K_h} \wedge \mathbf{K_o}|}; \qquad \overrightarrow{\sigma_o} \perp \mathbf{K_o},$$

$$\overrightarrow{\pi_h} = \frac{\mathbf{K_h} \wedge \overrightarrow{\sigma_h}}{K_h}; \qquad \overrightarrow{\pi_g} = \frac{\mathbf{K_g} \wedge \overrightarrow{\sigma_g}}{K_g}; \qquad \overrightarrow{\pi_o} = \frac{\mathbf{K_o} \wedge \overrightarrow{\sigma_o}}{K_o},$$

so that the vectors $\mathbf{K_h}$, $\overrightarrow{\sigma_h}$ and $\overrightarrow{\pi_h}$, are mutually orthogonal. Unit vector $\overrightarrow{\sigma_o}$ is chosen arbitrarily normal to $\mathbf{K_o}$, in practice normal to the entrance surface.

We have already set (4.12 and 5.10)

$$X_o = \frac{K_o^2 - n^2k^2}{2k},$$

$$X_h = \frac{K_h^2 - n^2k^2}{2k},$$

$$X_g = \frac{K_g^2 - n^2k^2}{2k}.$$

These factors, called resonance errors, are equal to (Section 8.6)

$$X_o = \frac{\overline{M_o P^2} + 2\mathbf{M_o P} \cdot \mathbf{OM_o}}{2k},$$

where M_o is the intersection of the normal to the crystal surface, **MP**, with the sphere of radius nk and centre O (Fig. 9.2), and similar relations hold for X_h and X_g. If neither the incidence angle nor the reflection angles are grazing and if none of the Bragg angles are close to $\pi/2$, $\overline{M_o P}^2$ can be neglected. One has then

$$\begin{aligned}
X_o &\approx \overline{M_o P}\gamma_o = \mathbf{r} \cdot \mathbf{s_o} = \overline{MP}\gamma_o - k\frac{\chi_o}{2}, \\
X_h &\approx \overline{M_h P}\gamma_h = \mathbf{r} \cdot \mathbf{s_h}, \\
X_g &\approx \overline{M_g P}\gamma_g = \mathbf{r} \cdot \mathbf{s_g},
\end{aligned} \tag{9.5}$$

where $\gamma_o = \cos\psi_o$, $\gamma_h = \cos\psi_h$, $\gamma_g = \cos\psi_g$ with $\psi_o = (\mathbf{n}, \mathbf{s_o})$, $\psi_h = (\mathbf{n}, \mathbf{s_h})$ and $\psi_g = (\mathbf{n}, \mathbf{s_g})$; $\mathbf{n}$ is the normal to the crystal surface, $\mathbf{s_o}$, $\mathbf{s_h}$ and $\mathbf{s_g}$ are unit vectors in the directions of $\mathbf{K_o}$, $\mathbf{K_h}$ and $\mathbf{K_g}$, respectively; M_h and M_g are the intersections of the normal to the crystal surface drawn from M with the spheres of radius nk and centres H and G, respectively; $\mathbf{r} = \mathbf{L_o P}$ where L_o is the triple Lorentz point, centre of the sphere of radius nk passing through O, H and G. In two-beam diffraction, one uses the interpretation of the resonance errors as the distances of the tiepoint to the spheres of radius nk and centres O and H. In the treatment of three-beam diffraction one rather uses their expressions in terms of $\mathbf{r}$. The orientation of the incident beam is defined by $\mathbf{L_o M_o}$ (or, alternatively, by $\mathbf{L_a M}$ where L_a is the triple Laue point).

If one now multiplies the first equation of (9.3a) from right successively by $\vec{\sigma_o}$ and $\vec{\pi_o}$, the second one by $\vec{\sigma_h}$ and $\vec{\pi_h}$ and the third one by $\vec{\sigma_g}$ and $\vec{\pi_g}$, one obtains a set of six scalar relations:

$$\begin{aligned}
X_o D_{o\sigma} - \chi_{\bar{h}} D_{h\sigma}\alpha_{\sigma h}^{\sigma o} - \chi_{\bar{h}} D_{h\pi}\alpha_{\pi h}^{\sigma o} - \chi_{\bar{g}} D_{g\sigma}\alpha_{\sigma g}^{\sigma o} - \chi_{\bar{g}} D_{g\pi}\alpha_{\pi g}^{\sigma o} &= 0, \\
X_o D_{o\pi} - \chi_{\bar{h}} D_{h\sigma}\alpha_{\sigma h}^{\pi o} - \chi_{\bar{h}} D_{h\pi}\alpha_{\pi h}^{\pi o} - \chi_{\bar{g}} D_{g\sigma}\alpha_{\sigma g}^{\pi o} - \chi_{\bar{g}} D_{g\pi}\alpha_{\pi g}^{\pi o} &= 0, \\
-\chi_h D_{o\sigma}\alpha_{\sigma o}^{\sigma h} - \chi_h D_{o\pi}\alpha_{\pi o}^{\sigma h} + X_h D_{h\sigma} - \chi_{h-g} D_{g\sigma}\alpha_{\sigma g}^{\sigma h} - \chi_{h-g} D_{g\pi}\alpha_{\pi g}^{\sigma h} &= 0, \\
-\chi_h D_{o\sigma}\alpha_{\sigma o}^{\pi h} - \chi_h D_{o\pi}\alpha_{\pi o}^{\pi h} + X_h D_{h\pi} - \chi_{h-g} D_{g\sigma}\alpha_{\sigma g}^{\pi h} - \chi_{h-g} D_{g\pi}\alpha_{\pi g}^{\pi h} &= 0, \\
-\chi_g D_{o\sigma}\alpha_{\sigma o}^{\sigma g} - \chi_g D_{o\pi}\alpha_{\pi_o}^{\sigma g} - \chi_{g-h} D_{h\sigma}\alpha_{\sigma h}^{\sigma g} - \chi_{g-h} D_{h\pi}\alpha_{\pi h}^{\sigma g} + X_g D_{g\sigma} &= 0, \\
-\chi_g D_{o\sigma}\alpha_{\sigma o}^{\pi g} - \chi_g D_{o\pi}\alpha_{\pi o}^{\pi g} - \chi_{g-h} D_{h\sigma}\alpha_{\sigma h}^{\pi g} - \chi_{g-h} D_{h\pi}\alpha_{\pi h}^{\pi g} + X_g D_{g\pi} &= 0,
\end{aligned} \tag{9.6}$$

where $\alpha^{\pi h}_{\sigma o} = (k/2)\vec{\sigma_o} \cdot \vec{\pi_h}$, and the equivalent relations according to the following table:

	$\vec{\sigma_o}$	$\vec{\pi_o}$	$\vec{\sigma_h}$	$\vec{\pi_h}$	$\vec{\sigma_g}$	$\vec{\pi_g}$
$\vec{\sigma_o}$	1	0	$\alpha^{\sigma h}_{\sigma o}$	$\alpha^{\pi h}_{\sigma o}$	$\alpha^{\sigma g}_{\sigma o}$	$\alpha^{\pi g}_{\sigma o}$
$\vec{\pi_o}$	0	1	$\alpha^{\sigma h}_{\pi o}$	$\alpha^{\pi h}_{\pi o}$	$\alpha^{\sigma g}_{\pi o}$	$\alpha^{\pi g}_{\pi o}$
$\vec{\sigma_h}$	$\alpha^{\sigma o}_{\sigma h}$	$\alpha^{\pi o}_{\sigma h}$	1	0	$\alpha^{\sigma g}_{\sigma h}$	$\alpha^{\pi g}_{\sigma h}$
$\vec{\pi_h}$	$\alpha^{\sigma o}_{\pi h}$	$\alpha^{\pi o}_{\pi h}$	0	1	$\alpha^{\sigma g}_{\pi h}$	$\alpha^{\pi g}_{\pi h}$
$\vec{\sigma_g}$	$\alpha^{\sigma o}_{\sigma g}$	$\alpha^{\pi o}_{\sigma g}$	$\alpha^{\sigma h}_{\sigma g}$	$\alpha^{\pi h}_{\sigma g}$	1	0
$\vec{\pi_g}$	$\alpha^{\sigma o}_{\pi g}$	$\alpha^{\pi o}_{\pi g}$	$\alpha^{\sigma h}_{\pi g}$	$\alpha^{\pi h}_{\pi g}$	0	1

The system (9.6) shows the coupling between the σ- and π-components of polarization. It can also be written in matrix form:

$$AD = 0, \tag{9.7}$$

where

$$A = \begin{pmatrix} X_o & 0 & -\chi_{\bar{h}}\alpha^{\sigma o}_{\sigma h} & -\chi_{\bar{h}}\alpha^{\sigma o}_{\pi h} & -\chi_{\bar{g}}\alpha^{\sigma o}_{\sigma g} & -\chi_{\bar{g}}\alpha^{\sigma o}_{\pi g} \\ 0 & X_o & -\chi_{\bar{h}}\alpha^{\pi o}_{\sigma h} & -\chi_{\bar{h}}\alpha^{\pi o}_{\pi h} & -\chi_{\bar{g}}\alpha^{\pi o}_{\pi g} & -\chi_{\bar{g}}\alpha^{\pi o}_{\pi g} \\ -\chi_h\alpha^{\sigma h}_{\sigma o} & -\chi_h\alpha^{\sigma h}_{\pi o} & X_h & 0 & -\chi_{h-g}\alpha^{\sigma h}_{\sigma g} & -\chi_{h-g}\alpha^{\sigma h}_{\pi g} \\ -\chi_h\alpha^{\pi h}_{\sigma o} & -\chi_h\alpha^{\pi h}_{\pi o} & 0 & X_h & -\chi_{h-g}\alpha^{\pi h}_{\sigma g} & -\chi_{h-g}\alpha^{\pi h}_{\pi g} \\ -\chi_g\alpha^{\sigma g}_{\sigma o} & -\chi_g\alpha^{\sigma g}_{\pi o} & -\chi_{g-h}\alpha^{\sigma g}_{\sigma h} & -\chi_{g-h}\alpha^{\sigma g}_{\pi h} & X_g & 0 \\ -\chi_g\alpha^{\pi g}_{\sigma o} & -\chi_g\alpha^{\pi g}_{\pi o} & -\chi_{g-h}\alpha^{\pi g}_{\sigma h} & -\chi_{g-h}\alpha^{\pi g}_{\pi h} & 0 & X_g \end{pmatrix}$$

and

$$D = \begin{pmatrix} D_{o\sigma} \\ D_{o\pi} \\ D_{h\sigma} \\ D_{h\pi} \\ D_{g\sigma} \\ D_{g\pi} \end{pmatrix}.$$

Equation (9.7) has a trivial solution only if the determinant associated with matrix A is equal to zero:

$$\det A = 0. \tag{9.8}$$

The resonance errors X_o, X_h and X_g define the position of the tiepoint and can all be expressed in terms of K_o^2. The secular equation (9.8) is a 12-degree equation in K_o and there are therefore 12 solutions, the dispersion surface has 12 branches and the incident wave excites a maximum of 12 wavefields. If neither the incidence angle nor one of the reflection angles are grazing and none of the Bragg angles are close to $\pi/2$, the resonance errors can be approximated by (9.5) and the degree of (9.7) is reduced to six: six branches only of the dispersion surface need to be considered and the incident wave excites six wavefields. The polarization factors have been recently discussed by Sheludko (2004).

9.2.3 *Solution in the general case*

A discussion of the solution of (9.7) and (9.8) can be found in Hildebrandt (1967), Ewald and Heno (1968), Heno and Ewald (1968) and Pinsker (1978). In the general case it can only be found numerically. Colella (1974) has given a computational method which has been used also by a number of other authors (see Chang 2004; Weckert and Hümmer 1990, 1997). It was modified by Shen (1993) to include all possible incident and diffracted polarizations. An improved algorithm using matrices and including the cases of grazing incidence and emergence was proposed by Stepanov and Ulyanenkov (1994). Another algorithm has been proposed by Stetsko and Chang (1997) who represent the amplitudes of the diffracted waves directly in a system of coordinates related to the normal to the entrance surface of the crystal and do not decompose them in the plane normal to $\mathbf{K_h}$. Their treatment is valid in the case of a grazing incidence and was tested in the case of an experiment of three-beam diffraction at grazing incidence (Chang *et al.* 1998). A solution of the three-beam diffraction has also been given by Thorkildsen (1987), Larsen and Thorkildsen (1998*a*,*b*), Thorkildsen and Larsen (1998*a*, 2002) and Thorkildsen *et al.* (2001) using the Takagi–Taupin equations which will be described in Chapter 12. It has been extended to the case $n > 3$ by Okitsu (2003). Numerical calculations have also been performed based on the Bethe approximation (Juretschke 1982*a*,*b*, 1984, 1998; Høier and Marthinsen 1983; Hümmer and Billy 1986; Chang *et al.* 1999) and on the second-order Born approximation (Shen 1986, 1998; Shen and Colella 1988; Chang and Tang 1988; Chang *et al.* 1989; Shen and Finkelstein 1990, 1992; Shen *et al.* 1995; Stetsko *et al.* 2000, 2004). A new approach, based on a distorted-wave Born approximation, called "expanded distorted-wave approximation", has been recently developed by Shen (1999*a*, 2000) and Shen and Huang (2001). In that approach, one evaluates the diffraction by the secondary reflection first; it gives rise to a distorted wave inside the crystal, which is used for calculating the diffracted intensity by the primary reflection. The case of three-beam diffraction with a grazing incidence was considered by Andreev *et al.* (1985), Hung and Chang (1989), Tseng and Chang (1990), Stepanov *et al.* (1991, 1995—for online calculations, see also http://sergey.gmca.aps.anl.gov), Stetsko and Chang (1997, 1999*a*) and Chang *et al.* (1998).

There are six solutions to (9.8) and six values of K_{oj} ($j = 1$–6). For each of them there are six amplitudes, $D_{oj\sigma}$, $D_{oj\pi}$, $D_{hj\sigma}$, $D_{hj\pi}$, $D_{gj\sigma}$, $D_{gj\pi}$, and there are therefore 36 amplitudes to be determined. They are related to one another by the 30 ratios $D_{oj\pi}/D_{oj\sigma}$, $D_{hj\pi}/D_{oj\sigma}$, $D_{gj\pi}/D_{oj\sigma}$, $D_{hj\sigma}/D_{oj\sigma}$, $D_{gj\sigma}/D_{oj\sigma}$, deduced from (9.7). The absolute values of the amplitudes are then determined as usual by the boundary conditions. If one or more of the incidence and reflection angles are grazing, or if one of the Bragg angles is close to $\pi/2$, one should use two sets of boundary conditions, one for the electric field and one for the magnetic field as has been described in Chapter 8 in the two-beam case. If this is not the case, one can simply write that the three components of the electric displacement are continuous, as has been done in Chapters 6 and 7:

1. *Laue geometry*: One has, at the entrance surface,

$$\begin{aligned}
\mathbf{D_o^{(a)}} &= D_{o\sigma}^{(a)}\overrightarrow{\sigma_o} + D_{o\pi}^{(a)}\overrightarrow{\pi_o} = \sum_j D_{oj\sigma}\overrightarrow{\sigma_o} + \sum_j D_{oj\pi}\overrightarrow{\pi_o},\\
0 &= \sum_j D_{hj\sigma}\overrightarrow{\sigma_h} + \sum_j D_{hj\pi}\overrightarrow{\pi_h},\\
0 &= \sum_j D_{gj\sigma}\overrightarrow{\sigma_g} + \sum_j D_{gj\pi}\overrightarrow{\pi_g},
\end{aligned} \tag{9.9a}$$

from which one deduces the six relations:

$$\begin{aligned}
D_{o\sigma}^{(a)} &= \sum_j D_{oj\sigma}; & 0 &= \sum_j D_{hj\sigma}; & 0 &= \sum_j D_{gj\sigma},\\
D_{o\pi}^{(a)} &= \sum_j D_{oj\pi}; & 0 &= \sum_j D_{hj\pi}; & 0 &= \sum_j D_{gj\pi},
\end{aligned} \tag{9.9b}$$

which enable to determine fully the amplitudes of the six wavefields.

The boundary conditions at the exit surface are analogous to the relations given in Section 6.4.1 for the two-beam case. Interferences between wavefields associated with different branches of the dispersion surface occur, producing *Pendellösung* effects.

2. *Bragg geometry*: If both the h- and the g-waves are reflected outside the crystal, one has:

(i) *Entrance surface*:

$$\begin{aligned}
\mathbf{D_o^{(a)}} &= D_{o\sigma}^{(a)}\overrightarrow{\sigma_o} + D_{o\pi}^{(a)}\overrightarrow{\pi_o} = \sum_j D_{oj\sigma}\overrightarrow{\sigma_o} + \sum_j D_{oj\pi}\overrightarrow{\pi_o},\\
\mathbf{D_h^{(a)}} &= D_{h\sigma}^{(a)}\overrightarrow{\sigma_h} + D_{h\pi}^{(a)}\overrightarrow{\pi_h} = \sum_j D_{hj\sigma}\overrightarrow{\sigma_h} + \sum_j D_{hj\pi}\overrightarrow{\pi_h},\\
\mathbf{D_g^{(a)}} &= D_{g\sigma}^{(a)}\overrightarrow{\sigma_g} + D_{g\pi}^{(a)}\overrightarrow{\pi_g} = \sum_j D_{gj\sigma}\overrightarrow{\sigma_g} + \sum_j D_{gj\pi}\overrightarrow{\pi_g},
\end{aligned} \tag{9.10a}$$

where $\mathbf{D_h^{(a)}}$ and $\mathbf{D_g^{(a)}}$ are the amplitudes of the h- and g-waves. These three vector equations correspond to six scalar relations.

(ii) *Back surface*: In a similar way to (7.7) in the two-beam case

$$\sum_j D_{oj\sigma} \exp(-2\pi \mathrm{i}\overline{MP}_{j\sigma}\, t)\vec{\sigma_o} + \sum_j D_{oj\pi} \exp(-2\pi \mathrm{i}\overline{MP}_{j\pi}\, t)\vec{\pi_o} = D^{(d)}_{o\sigma}\vec{\sigma_o} + D^{(d)}_{o\pi}\vec{\pi_o},$$

$$\sum_j D_{hj\sigma} \exp(-2\pi \mathrm{i}\overline{MP}_{j\sigma}\, t)\vec{\sigma_h} + \sum_j D_{hj\pi} \exp(-2\pi \mathrm{i}\overline{MP}_{j\pi}\, t)\vec{\pi_h} = 0,$$

$$\sum_j D_{gj\sigma} \exp(-2\pi \mathrm{i}\overline{MP}_{j\sigma}\, t)\vec{\sigma_g} + \sum_j D_{gj\pi} \exp(-2\pi \mathrm{i}\overline{MP}_{j\pi}\, t)\vec{\pi_g} = 0, \tag{9.10b}$$

where $D^{(d)}_{o\sigma}$ and $D^{(d)}_{o\pi}$ are the components of the transmitted wave outside the crystal, M is the extremity of the incident wavevector, $\mathbf{K}^{(\mathbf{a})}_{\mathbf{o}}$, and $P_{j\sigma}$ and $P_{j\pi}$ are the tiepoints of the wavefields excited inside the crystal. Again these vector equations correspond to six scalar equations. In total, one has 42 unknowns: the 36 amplitudes inside the crystal, and the two components of $\mathbf{D}^{(\mathbf{a})}_{\mathbf{h}}$, $\mathbf{D}^{(\mathbf{a})}_{\mathbf{g}}$ and $\mathbf{D}^{(\mathbf{d})}_{\mathbf{o}}$ outside the crystal. They are determined by the 30 ratios between the waves inside the crystal, the 6 boundary conditions at the entrance surface and the 6 boundary conditions at the back surface.

If the crystal is very thick (semi-infinite), the waves propagating towards inside the crystal are absorbed out before reaching the back surface and they are the only waves having a physical meaning. Those for which the imaginary part of the wavevector is positive cannot exist. There are only two possible values of j in this case, 1 and 2 (Pinsker 1978; Chang 1987). There are therefore two wavefields for each direction of polarization and 16 unknown amplitudes, 12 inside the crystal and 4 outside; they are determined using the 10 values of the ratios of the amplitudes inside the crystal and the 6 boundary conditions (9.10a).

9.2.4 *Energy flow*

The direction of the energy flow is given as in the case of two-beam diffraction by the Poynting vector:

$$\langle \mathbf{S_j} \rangle = \frac{c}{\epsilon_0} \exp[4\pi \mathbf{K}^{\mathbf{i}}_{\mathbf{oj}} \cdot \mathbf{r}][\mathbf{s_o}|D_{oj}|^2 + \mathbf{s_h}|D_{hj}|^2 + \mathbf{s_g}|D_{gj}|^2],$$

where $\mathbf{s_o}$, $\mathbf{s_h}$ and $\mathbf{s_g}$ are unit vectors in the directions of $\mathbf{K_o}$, $\mathbf{K_h}$ and $\mathbf{K_g}$, respectively.

Kato (1958) has shown that it is normal to the branch of the dispersion surface passing through P_j, whatever the number of reciprocal lattice points close to the Ewald sphere.

The paths of wavefields have been checked by Campos and Chang (1986) in silicon who observed the Borrmann fan for the four-beam case, 220, $2\bar{2}0$, 400, and by

Heyroth *et al.* (1999, 2001) who studied the distribution of energy flow with pinhole topography in quartz for the $11\bar{2}0$, $2\bar{1}\bar{1}2$. $\bar{1}2\bar{1}2$ case (a topography technique in which the incident beam is limited by a small round aperture).

9.3 The three-beam coplanar case

The system (9.6) is simplified if the three wavevectors, $\mathbf{K_o}$, $\mathbf{K_h}$ and $\mathbf{K_g}$ are coplanar. The three vectors $\vec{\sigma_o}$, $\vec{\sigma_h}$ and $\vec{\sigma_g}$ are parallel and normal to the diffraction plane. They are also perpendicular to $\vec{\pi_o}$, $\vec{\pi_h}$ and $\vec{\pi_g}$ which lie in that plane. The products $\vec{\sigma_o}\cdot\vec{\pi_o}$, etc. are null and σ- and π-polarizations can be completely decoupled. System (9.6) can be divided into two independent subsystems, one for polarization normal to the diffracting plane, and one for polarization in the diffraction plane:

$$\begin{aligned} X_o D_o - \frac{k}{2} C_h \chi_{\bar{h}} D_h - \frac{k}{2} C_g \chi_{\bar{g}} D_g &= 0, \\ -\frac{k}{2} C_h \chi_h D_o + X_h D_h - \frac{k}{2} C_{h-g} \chi_{h-g} D_g &= 0, \\ -\frac{k}{2} C_g \chi_g D_o - \frac{k}{2} C_{g-h} \chi_{g-h} D_h + X_g D_g &= 0, \end{aligned} \tag{9.11}$$

where $C_h = C_g = C_{h-g} = C_{g-h} = 1$ for σ-polarization and $C_h = \cos 2\theta_h$, $C_g = \cos 2\theta_g$, $C_{h-g} = \cos 2\theta_{(h-g)} = C_{g-h}$ for π-polarization.

After elimination of D_g between the second and third equations, the amplitude ratio D_h/D_o can be expressed in terms of the resonance errors X_h and X_g:

$$\frac{D_h}{D_o} = \frac{(k/2) C_h \chi_h X_g + (k^2/4) C_g C_{h-g} \chi_g \chi_{h-g}}{X_h X_g - (k^2/4) C_{h-g}^2 \chi_{h-g} \chi_{g-h}}.$$

This expression, which was already obtained by Laue (1960), is usually written in the following way (Colella 1974; Juretschke 1982*a*; Hümmer and Billy 1982; Høier and Marthinsen 1983; Weckert and Hümmer 1997):

$$\frac{D_h}{D_o} = X_h^{-1} \frac{(k/2) C_h \chi_h + (k^2/4) C_g C_{h-g} \chi_g \chi_{h-g} X_g^{-1}}{1 - (k^2/4) C_{h-g}^2 \chi_{h-g} \chi_{g-h} X_h^{-1} X_g^{-1}}. \tag{9.12}$$

It can easily be checked that when the Bragg condition is far from being satisfied for reflection (g_1,g_2,g_3), that is when $X_g \Rightarrow \infty$, expression (9.12) tends towards expression (5.34) for the two-beam case.

9.4 Determination of phases using n-beam diffraction

The original idea of the determination of phases using three-beam diffraction was given independently by Bijvoet and MacGillavry (1939), Lipscomb (1949) and Miyake and Kambe (1954) and the first experimental evidence is due to Hart and Lang (1961)—see below. As has been said in the introduction, the diffracted amplitude of the wave of wavevector $\mathbf{K_h}$ results from the interference of a wave directly diffracted on the (h_1, h_2, h_3) planes, of amplitude proportional to the structure factor, F_h, and the *Umweg*-wave first diffracted on the (g_1, g_2, g_3) planes and then on the $(h_1 - g_1, h_2 - g_2, h_3 - g_3)$ planes, of amplitude proportional to $F_g F_{h-g}$. It is therefore of the form $\mathcal{A}_1 F_h + \mathcal{A}_2 F_g F_{h-g}$. Let us introduce the phases of the structure factors (2.38):

$$F_h = |F_h| \exp \mathrm{i}\varphi_h = \sum_j f_j \exp(-2\pi \mathrm{i}\mathbf{h} \cdot \mathbf{r_j}),$$

$$F_g = |F_h| \exp \mathrm{i}\varphi_g = \sum_j f_j \exp(-2\pi \mathrm{i}\mathbf{g} \cdot \mathbf{r_j}),$$

$$F_{h-g} = |F_h| \exp \mathrm{i}\varphi_{h-g} = \sum_j f_j \exp[-2\pi \mathrm{i}(\mathbf{h} - \mathbf{g}) \cdot \mathbf{r_j}].$$

The intensity of the h-wave is therefore of the form

$$|\mathcal{A}_1 F_h + \mathcal{A}_2 F_g F_{h-g}|^2 = |\mathcal{A}_1 F_h|^2 + |\mathcal{A}_2 F_g F_{h-g}|^2 + 2|\mathcal{A}_1 F_h \mathcal{A}_2 F_g F_{h-g}| \times \cos(\varphi_g + \varphi_{h-g} - \varphi_h + \Delta\Psi), \quad (9.13)$$

where $\Delta\Psi$ is the phase of $\mathcal{A}_2/\mathcal{A}_1$ (and of X_g^{-1}).

Expression (9.13) shows that the intensity of the h-wave depends on the value of the triplet phase, $\Phi_{3\pm}$, defined by (Weckert and Hümmer 1997)

$$\Phi_{3\pm} = \pm(\varphi_g + \varphi_{h-g} - \varphi_h); \qquad \Phi_{3+} = -\Phi_{3-}.$$

The important property of this triplet phase is its invariance under a change of origin as can easily be understood: if the origin is shifted by $\mathbf{r_o}$, the structure factors are multiplied by $\exp(-2\pi \mathrm{i}\mathbf{h} \cdot \mathbf{r_o})$, $\exp(-2\pi \mathrm{i}\mathbf{g} \cdot \mathbf{r_o})$ and $\exp[-2\pi \mathrm{i}(\mathbf{h} - \mathbf{g}) \cdot \mathbf{r_o}]$, respectively and Φ_3 remains unchanged. The triplet phase is therefore a structure invariant which can be measured by means of a three-beam experiment. The knowledge of a large enough number of such triplets is a great help in the determination of unknown structures. It should be noted, however, that the phase difference thus measured is a triplet phase in the usual sense for direct methods only when Friedel's law is satisfied. The influence of anomalous scattering on the invariant phases was studied experimentally with synchrotron radiation and theoretically by Chang (1986) in the case of gallium arsenide near the absorption edges of Ga and As.

The measurement of the invariant phases can be illustrated easily when the σ- and π-polarizations can be separated. This is the case rigorously in the coplanar case (Section 9.3), but it has been shown numerically (Høier and Aanestad 1981) that the scalar products of the type $\vec{\sigma_h} \cdot \vec{\pi_g}$ can be neglected except in a narrow angular range very close to the exact three-beam Bragg condition. Expression (9.12) of the amplitude ratio D_h/D_o can be written as

$$\frac{D_h}{D_o} = \mathcal{A}_1 F_h + \mathcal{A}_2 F_g F_{h-g} X_g^{-1},$$

where $\mathcal{A}_1$ and $\mathcal{A}_2$ depend on the resonance error X_h. The intensity is indeed of the form (9.13) where $\Delta\Psi$ is the phase of X_g^{-1}. This phase is equal to 0 or π in the zero-absorption case, depending on whether X_g is positive or negative.

The value of the triplet phase Φ_3 can be estimated by recording the variations of the diffracted intensity (9.13) when varying X_g while keeping the other terms constant. This is achieved by means of the Renninger-scan (or Ψ-scan) technique. When one rotates the crystal around the reciprocal lattice vector **OH** of the h_1, h_2, h_3 reflection, the reciprocal lattice point G corresponding to the g_1, g_2, g_3 reflection crosses the Ewald sphere as shown in Fig. 9.3. There are two positions for which G crosses the Ewald sphere: entering ('in') and leaving ('out'). The two can be distinguished using the experimental procedure given by Chang and Caticha-Ellis (1978). In an *in-out* scan, G is at first inside the sphere, wave number K_g is smaller than the radius of the sphere, nk, and X_g is negative; then G is out of the sphere, K_g is larger than the radius of the sphere and X_g is positive. The other terms, $\mathcal{A}_1$ and $\mathcal{A}_2$, are actually not constant but do not vary very much far from the exact three-beam Bragg condition. This so-called two-beam description of a three-beam diffraction has been discussed in detail by Juretschke (1984).

Figure 9.4 shows the variation of $\Delta\Psi$ in the case of an absorbing crystal and Fig. 9.5 shows the variations of the phase-dependent part of the intensity diffracted in the direction of the h-wave, $|X_g^{-1}|\cos(\Phi_3 + \Delta\Psi)$, during this rotation for various values of Φ_3. It can be seen that these variations can easily be distinguished from one another and this provides a way of estimating the triplet phase. These variations obviously depend on the direction of the Ψ-scan (*in-out* or *out-in*), as has been shown by Chang (1982).

As examples, Figs 9.6(a)–(c) show the variations of the calculated rocking curve of the 040 reflection of an arbitrary orthorhombic crystal in the vicinity of the 042 reflection, for three values of the triplet phase Φ_3, 180°, −90° and +90°, respectively,

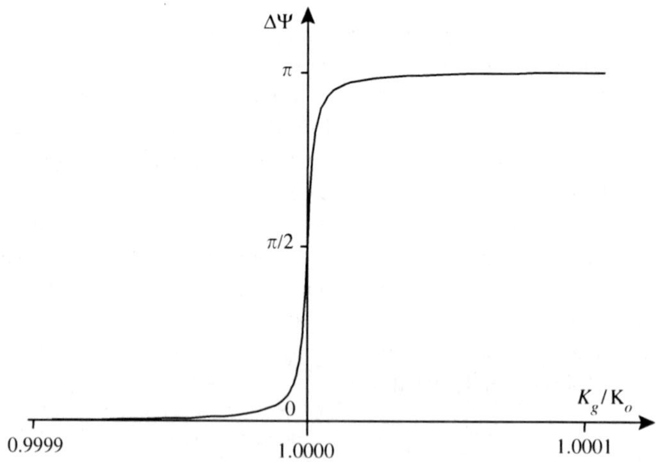

Fig. 9.4. Variations of the phase of the resonance error, X_g^{-1}, in the vicinity of a three-beam point; it is $\pi/2$ for $K_g/K_o = 1$.

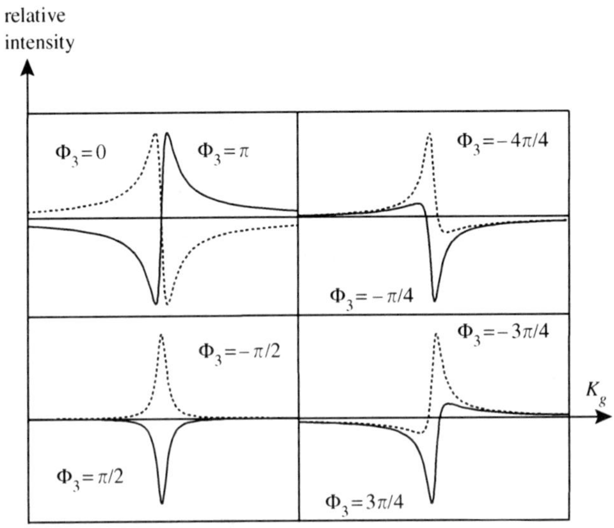

Fig. 9.5. Variations of the relative intensity with K_g for different values of the triplet phase, Φ_3; arbitrary units.

in the Laue geometry (after Weckert and Hümmer 1997). The figures represent three-dimensional plots of the intensity as a function of the departure from Bragg's angle (ω) and of Ψ. It can be seen that the reflection profile depends on the sign of the value and sign of the triplet phase, in accordance with the plots of Fig. 9.5.

In the Laue geometry, very complicated Pendellösung effects occur (Weckert and Hümmer 1997, 1998) for thin crystals (but thicker than one Pendellösung distance). Since the phase varies during a Ψ-scan through a three-beam position, so does the period of the interferences. This results in a variation of the Pendellösung fringe spacing across an *Aufhellungslinie* on an X-ray topograph. An example is given in Fig. 9.7 which represents a 111 projection topograph, taken with Mo$K\alpha$ radiation, of a wedge-shaped silicon crystal. It is crossed by a white line which is an *Aufhellungslinie* corresponding to the simultaneous excitation of the $1\bar{1}\bar{1}$ reflection, resulting in a 111, $1\bar{1}\bar{1}$, 022 three-beam case. The value of Ψ varies along the line. The corresponding phase shift results in a shift of the Pendellösung fringes across the line. Where the three-beam condition is exactly satisfied the phase shift is equal to π and the shift of the fringes is then of a half a fringe spacing as can be seen distinctly on the topograph. It is therefore possible to have information on the reflection phases by observation of the deformation of the fringes. This effect was first observed by Hart and Lang (1961) with a germanium crystal and was simulated by Høier and Aanestad (1981).

The phase-dependent part of the intensity can only be easily detected from the phase-independent part if the structure factors of the two jointly excited reflections

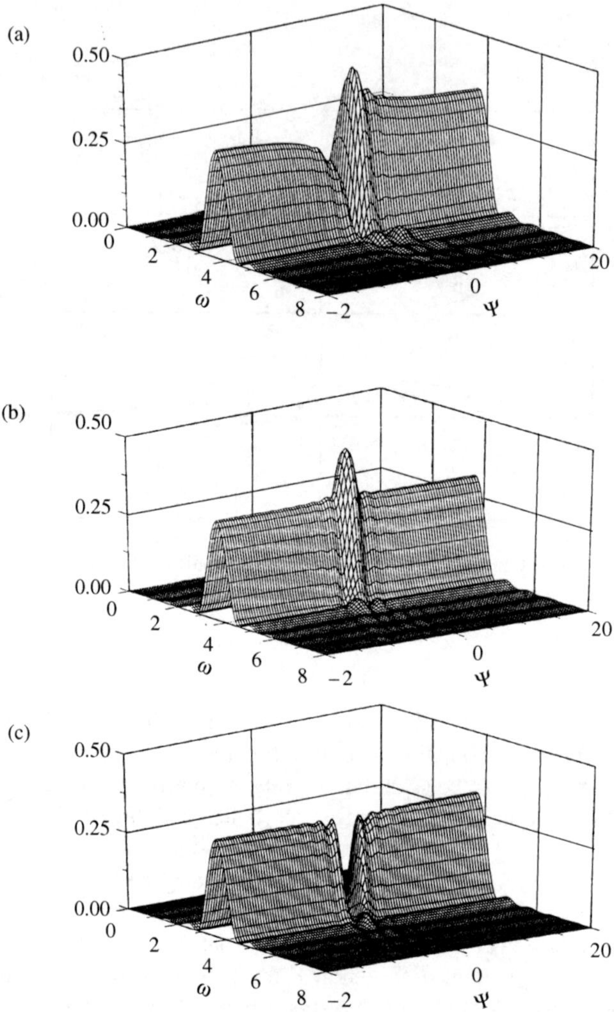

Fig. 9.6. Variations of the reflected intensity in the Laue geometry for the 040, 042, 002 three-beam case. ω: departure from Bragg's angle; Ψ: rotation around the 040 reciprocal lattice vector; arbitrary orthorhombic crystal (a = 5.582 Å, b = 9.812 Å, c = 11.496 Å; $|F_o|$ = 321, $|F_h|$ = 15.5, $|F_g|$ = 20.0, $|F_{h-g}|$ = 30.0, λ = 1.5405 Å, plate thickness 0.05 mm) (Reproduced with permission from Weckert and Hümmer 1997). (a) $\Phi_3 = 180°$. (b) $\Phi_3 = -90°$. (c) $\Phi_3 = 90°$.

are of the same order of magnitude. Otherwise the *Umweganregung* or *Aufhellung* effects predominate and mask the phase-dependent contributions. Different methods have been proposed to separate the phase-dependent part and to determine it quantitatively (Weckert and Hümmer 1997; Chang and Tang 1988; Tang and Chang 1988; Chang 1992; Stetsko and Chang 1999*b*). Stetsko *et al.* (2000, 2001) have shown that

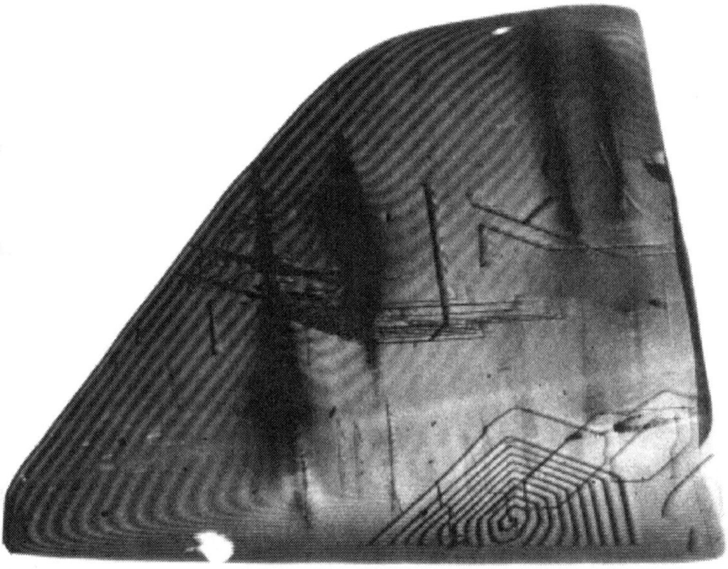

Fig. 9.7. X-ray topograph of a wedge-shaped silicon crystal. 111 reflection, Mo$K\alpha$; the white line is an *Aufhellungslinie* corresponding to the three-beam case 111, $1\bar{1}\bar{1}$, $0\bar{2}\bar{2}$. Note the shift by half a fringe of the Pendellösung fringes across the line as the phase of the reflection varies from 0 to π. The dislocation configuration in this particular crystal is described in Authier and Lang (1964)—see also Figs 17.14 and 17.26.

it can be done by suppressing either the primary or the detour reflection. This occurs for instance when the Bragg angle of the secondary or of the coupling reflection is $\pi/4$ and for appropriate polarization conditions of the incident beam. The relative importance of the phase-dependent part can thus be increased. Such a situation was already discussed by Kshevetskii *et al.* (1985) and is also analysed by Morelhão and Avanci (2001). Morelhão and Kycia (2002) have suggested the accuracy of phase determination can be increased by using a diffraction angle of $\pi/2$.

The determination of triplet phases by means of three-beam diffraction has been applied in practice to the determination of structures of macromolecular compounds (Chang *et al.* 1991; Weckert *et al.* 1993, 2002; Huang *et al.* 1994; Chang *et al.* 1999, 2002; Mo *et al.* 2002; Shen and Wang 2003) and to the determination of the absolute structure of enantiomorphs (Burzlaff *et al.* 1989; Hümmer and Weckert 1995). It has also been extended to the case of quasicrystals by Zhang *et al.* (1998). Usually, the measurements of intensities are performed by means of the Ψ-scan as described above. Recently, another method has been proposed, the reference beam technique (Shen 1999*b*; Shen *et al.* 2000*a*, *b*, 2002), which records the intensities of many spots simultaneously by means of a two-dimensional detector (an image plate or a CCD). The phase information contained in such diagrams is in principle the same as in a Ψ-scan interference profile. Actually, it is less accurate but the large number of interferences that can be measured in a short time counterbalances that drawback. The method has been tested satisfactorily on a known protein, crambin (Weeks *et al.* 2000).

The determination of phases by means of n-beam diffraction has also been applied by Lee *et al.* (1996) to study the deviations from centrosymmetry of the quasicrystal AlPdMn.

9.5 The super-Borrmann effect

9.5.1 *Experimental evidence*

Borrmann and Hartwig (1965) observed that when the Bragg condition is satisfied simultaneously for the 111 and $\bar{1}11$ reflections, the effect of anomalous transmission is enhanced. Figs 9.8(a) and (b) show their results. They represent photographs obtained using the wide-angle transmission technique described in Section 1.10.1 from two dislocation-free germanium crystals, of thicknesses $t = 0.8$ mm and 1.2 mm, respectively, with Cu$K\alpha$ radiation. The values of $\mu_o t$ (μ_o, normal absorption

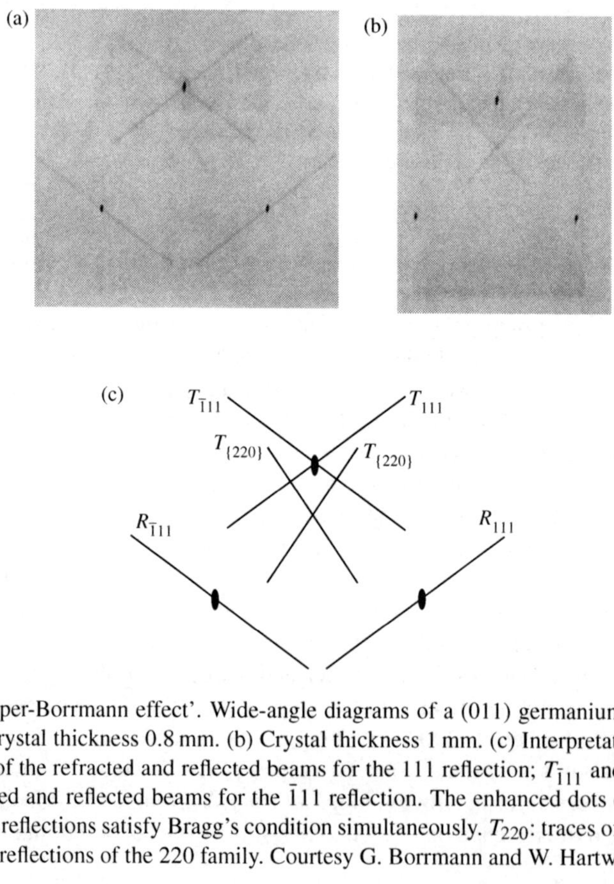

Fig. 9.8. 'Super-Borrmann effect'. Wide-angle diagrams of a (011) germanium crystal with Cu$K\alpha$. (a) Crystal thickness 0.8 mm. (b) Crystal thickness 1 mm. (c) Interpretation: T_{111} and R_{111}: traces of the refracted and reflected beams for the 111 reflection; $T_{\bar{1}11}$ and $R_{\bar{1}11}$: traces of the refracted and reflected beams for the $\bar{1}11$ reflection. The enhanced dots correspond to the case both reflections satisfy Bragg's condition simultaneously. T_{220}: traces of the refracted beam for the reflections of the 220 family. Courtesy G. Borrmann and W. Hartwig.

coefficient) are 27.3 and 41, respectively. The surface of the slabs is $(0\bar{1}1)$. The traces of the transmitted beams in the reflected and refracted directions due to the 111 and $\bar{1}11$ reflections, respectively, can be seen faintly in Fig. 9.8(a) and are represented in the diagram of Fig. 9.8(c). At the intersection of the refracted beams, T_{111} and $T_{\bar{1}11}$, an enhanced spot appears, corresponding to the situation where both reflections are excited, namely to a three-beam case. The third reflection involved, $\mathbf{h}-\mathbf{g}$, is here 200 which is a forbidden reflection for germanium. Similarly, an enhanced spot appears on the reflected beams, R_{111} and $R_{\bar{1}11}$, symmetrically to the intersection point of T_{111} and $T_{\bar{1}11}$ with respect to the respective reflecting planes.

In Fig. 9.8(b), corresponding to a thicker crystal, the faint lines have disappeared, but not the bright spots. This shows that they correspond to rays which have been transmitted through the crystal with an absorption coefficient which is even smaller than the anomalous absorption coefficient of the 111 reflections. In the two figures, two faint lines can also be seen which do not present an enhanced spot at their intersection. They are due to the refracted beams associated with 220 reflections excited by Cu$K\beta$.

9.5.2 *Solution of the 111, $\bar{1}11$ case*

The interpretation of the further decrease of the absorption coefficient when the Bragg condition is satisfied for both 111 and $\bar{1}11$ reflections has been given by Hildebrandt (1967, 1978) who calculated it. A summary of his derivation is given in this section. The experimental observations were repeated more accurately and in other cases (for instance, the $\bar{1}11$, $1\bar{1}1$, $\bar{2}20$ three-beam case) by Uebach and Hildebrandt (1969), Uebach (1973) and Umeno and Hildebrandt (1975). The theoretical explanation was also discussed in detail by Afanas'ev and Kohn (1975, 1976, 1977).

The geometrical conditions of the 111, $\bar{1}11$, 200 diffraction are given in Fig. 9.9. It can be seen that it is characterized by the fact that the (100) plane is a plane of symmetry and this simplifies the calculations. The fact that 200 is a forbidden reflection leads to a further simplification and system (9.3a) can be written here:

$$\begin{aligned}
&2X_o\mathbf{D_o} && -k\chi_{\bar{1}\bar{1}\bar{1}}\mathbf{D_{111[o]}} && -k\chi_{1\bar{1}\bar{1}}\mathbf{D_{\bar{1}11[o]}} && = 0\\
&-k\chi_{111}\mathbf{D_{o[111]}} && +2X_{111}\mathbf{D_{111}} && && = 0\\
&-k\chi_{\bar{1}11}\mathbf{D_{o[\bar{1}11]}} && && +2X_{\bar{1}11}\mathbf{D_{\bar{1}11}} && = 0
\end{aligned} \tag{9.14}$$

From the second and third relations of (9.14) and using (9.2), it is possible to obtain the following relations:

$$\begin{aligned}
\mathbf{D_{111[o]}} &= \frac{k\chi_{111}}{2X_{111}}\left\{\mathbf{D_o} - \frac{\mathbf{K_{111}}\cdot\mathbf{D_o}}{K_{111}^2}\left[\mathbf{K_{111}} - \frac{\mathbf{K_o}}{K_o^2}(\mathbf{K_o}\cdot\mathbf{K_{111}})\right]\right\},\\
\mathbf{D_{\bar{1}11[o]}} &= \frac{k\chi_{\bar{1}11}}{2X_{\bar{1}11}}\left\{\mathbf{D_o} - \frac{\mathbf{K_{\bar{1}11}}\cdot\mathbf{D_o}}{K_{\bar{1}11}^2}\left[\mathbf{K_{\bar{1}11}} - \frac{\mathbf{K_o}}{K_o^2}(\mathbf{K_o}\cdot\mathbf{K_{\bar{1}11}})\right]\right\}.
\end{aligned} \tag{9.15}$$

Substituting these two expressions in the first equation of (9.14), one obtains a relation involving $\mathbf{D_o}$ only. The most convenient here is to decompose it into two components,

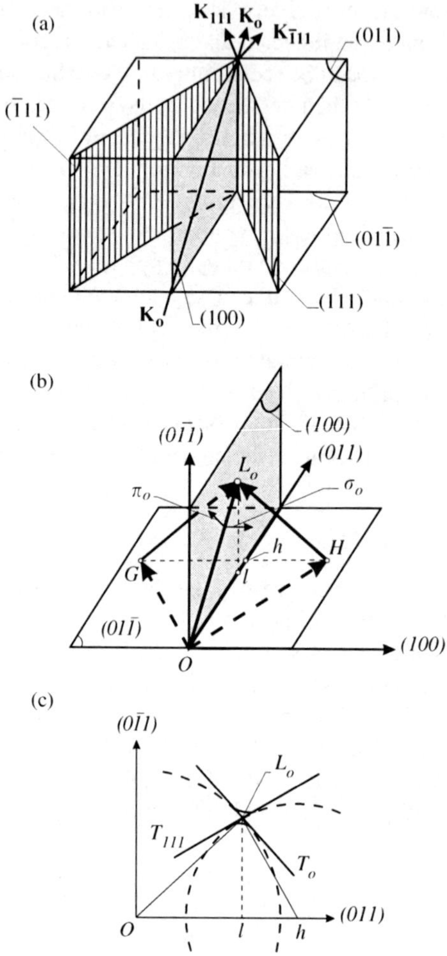

Fig. 9.9. Three-beam case 111, $\bar{1}11$, 200. (a) Direct space; $(01\bar{1})$: entrance surface; K_o: incident wavevector; $\mathbf{K_{111}}$, $\mathbf{K_{\bar{1}11}}$: wavevectors reflected on the (111) and $(\bar{1}11)$ planes; (100): symmetry plane of the figure. (b) Reciprocal space; H and G: 111, and $\bar{1}11$ reciprocal lattice points; L_o: Lorentz point; $\pi_\mathbf{o}$, $\sigma_\mathbf{o}$: polarization directions. (c) Intersection of the dispersion surface with the plane of symmetry, (100). Only one direction of polarization is represented, for simplification; L_o: Lorentz point. The dashed circles represent the intersections of the spheres of radius nk and centres H (or G) and O, respectively, with (100); h: projection of H and G on (100).

one parallel to the plane of symmetry and the other perpendicular to the plane of symmetry:

$$\mathbf{D_o} = D_{o\sigma}\vec{\sigma_o} + D_{o\pi}\vec{\pi_o},$$

where $\vec{\sigma_o}$ is a unit vector along [100] and $\vec{\pi_o}$ is a unit vector parallel to (100) and perpendicular to $\mathbf{K_o}$ (Fig. 9.9(b)). Using this decomposition, the first equation of (9.14), which is a vector equation, can be replaced by two scalar equations which are of the type

$$\begin{aligned} AD_{o\sigma} + BD_{o\pi} &= 0, \\ BD_{o\sigma} + CD_{o\pi} &= 0. \end{aligned} \tag{9.16}$$

The expressions of A, B and C are obtained after noting that the crystal wavevectors $\mathbf{K_o}$, $\mathbf{K_{111}}$ and $\mathbf{K_{\bar{1}11}}$ can here be approximated by $\mathbf{OL_o}$, $\mathbf{HL_o}$ and $\mathbf{GL_o}$, respectively, and that their components in a system of axes respectively parallel to $\mathbf{OL_o}$, $\vec{\sigma_o}$ and $\vec{\pi_o}$ are

$$\mathbf{K_o}\begin{cases} k \\ 0 \\ 0; \end{cases} \qquad \mathbf{K_{111}}\begin{cases} k(1-2\sin^2\theta_{111}) \\ -\frac{2}{\sqrt{3}}k\sin\theta_{111} \\ 2k\sin\theta_{111}\sqrt{\frac{2}{3}-\sin^2\theta_{111}}; \end{cases}$$

$$\mathbf{K_{\bar{1}11}}\begin{cases} k(1-2\sin^2\theta_{111}) \\ +\frac{2}{\sqrt{3}}k\sin\theta_{111} \\ 2k\sin\theta_{111}\sqrt{\frac{2}{3}-\sin^2\theta_{111}}. \end{cases}$$

The expressions of A, B and C are further simplified by noting that the structure factors of the 111 and $\bar{1}11$ reflections are equal. One finally obtains

$$\begin{aligned} A &= 8X_oX_{111}X_{\bar{1}11} - k^2\chi_{111}\chi_{\bar{1}\bar{1}\bar{1}}(1-\tfrac{4}{3}\sin^2\theta_{111})(2X_{111}+2X_{\bar{1}11}), \\ B &= \frac{4}{\sqrt{3}}k^2\chi_{111}\chi_{\bar{1}\bar{1}\bar{1}}\sqrt{\tfrac{2}{3}-\sin^2\theta_{111}}\,\sin^2\theta_{111}(2X_{111}-2X_{\bar{1}11}), \\ C &= 8X_oX_{111}X_{\bar{1}11} - k^2\chi_{111}\chi_{\bar{1}\bar{1}\bar{1}}[1-4(\tfrac{2}{3}-\sin^2\theta_{111})\sin^2\theta_{111}] \\ &\quad\times(2X_{111}+2X_{\bar{1}11}). \end{aligned} \tag{9.17}$$

For eqns (9.16) to have a non-trivial solutions, the determinant must be set equal to zero:

$$AC - B^2 = 0.$$

This is the equation of the dispersion surface. If u, v, w are the coordinates of the tiepoint in the system of axes defined previously, with origin at the triple Lorentz point, one has

$$X_o = u,$$

$$X_{111} = (1 - 2\sin^2\theta_{111})u - \frac{2}{\sqrt{3}}\sin\theta_{111}v + 2\sin\theta_{111}\sqrt{\tfrac{2}{3} - \sin^2\theta_{111}}\,w,$$

$$X_{\bar{1}11} = (1 - 2\sin^2\theta_{111})u + \frac{2}{\sqrt{3}}\sin\theta_{111}v + 2\sin\theta_{111}\sqrt{\tfrac{2}{3} - \sin^2\theta_{111}}\,w.$$

It is very simple to find the shape of the intersection of the dispersion surface with the plane of symmetry since we have, in that plane, $v = 0$ and

$$X_{111} = X_{\bar{1}11}.$$

The coefficient B in (9.16) is null and the equation of the dispersion surface is decomposed in two equations:

$$A = 0 \quad \text{for } \sigma\text{-polarization},$$
$$C = 0 \quad \text{for } \pi\text{-polarization}$$

and, explicitly,

$$\begin{aligned} A &= \left[2X_oX_{111} - k^2\chi_{111}\chi_{\bar{1}\bar{1}\bar{1}}\left(1 - \tfrac{4}{3}\sin^2\theta_{111}\right)\right]X_{111} = 0, \\ C &= \left[2X_oX_{111} - k^2\chi_{111}\chi_{\bar{1}\bar{1}\bar{1}}\left(1 - 4\left(\tfrac{2}{3} - \sin^2\theta_{111}\right)\sin^2\theta_{111}\right)\right]X_{111} = 0. \end{aligned} \tag{9.18}$$

The intersection of the dispersion surface with (100) is therefore composed, for both σ- and π-polarization, of a straight line, of equation $X_{111} = 0$, and of a hyperbola having as asymptotes the straight lines $X_0 = 0$ and $X_{111} = 0$. These lines, called T_o and T_{111} on Fig. 9.9(c), are easily interpreted as the tangents, in the plane (100), to the spheres of radius nk and of centres O and H (or G), respectively.

9.5.3 *Anomalous absorption coefficient*

It has been shown in Section 4.7 that the effective abosrption coefficient is given by

$$\mu_e = \mu_o - 4\pi X_o^i$$

with $\mu_o = -2\pi k\chi_{io}$. The minimum absorption coefficient occurs when X_o is minimum. This is easily deduced from the equation of the dispersion surface. From (9.18) it follows that

$$\sigma\text{-polarization:} \quad X_oX_{111} = k^2\chi_{111}\chi_{\bar{1}\bar{1}\bar{1}}\left(1 - \tfrac{4}{3}\sin^2\theta_{111}\right)/2,$$
$$\pi\text{-polarization:} \quad X_oX_{111} = k^2\chi_{111}\chi_{\bar{1}\bar{1}\bar{1}}\left[1 - 4\left(\tfrac{2}{3} - \sin^2\theta_{111}\right)\sin^2\theta_{111}\right]/2,$$

X_o is minimum when $X_o = X_{111}$ and is

$$\sigma\text{-polarization:}\quad X_o = -k\sqrt{\frac{\chi_{111}\chi_{\bar{1}\bar{1}\bar{1}}}{2}}\sqrt{1 - \tfrac{4}{3}\sin^2\theta_{111}},$$

$$\pi\text{-polarization:}\quad X_o = -k\sqrt{\frac{\chi_{111}\chi_{\bar{1}\bar{1}\bar{1}}}{2}}\sqrt{1 - 4\left(\tfrac{2}{3} - \sin^2\theta_{111}\right)\sin^2\theta_{111}}.$$

The minimum absorption coefficient is given by (Section 4.8.4)

$$\mu_e = \mu_o\left[1 - C\left|\frac{\chi_{i111}}{\chi_{io}}\right|\right].$$

In a symmetrical two-beam diffraction experiment, $C = 1$ or $\cos 2\theta$ for σ- and π-polarizations, respectively. In the present 111, $\bar{1}11$, 200 experiment,

$$\sigma\text{-polarization:}\quad C = \sqrt{2}\sqrt{1 - \tfrac{4}{3}\sin^2\theta_{111}},$$

$$\pi\text{-polarization:}\quad C = \sqrt{2}\sqrt{1 - 4(\tfrac{2}{3} - \sin^2\theta_{111})\sin^2\theta_{111}}.$$

In the example described in Section 9.5.1, the crystal is germanium and Cu$K\alpha$ radiation is used. The values of the minimum absorption coefficient are, respectively, in cm^{-1}, for the two- and three-beam cases, and for π- and σ-polarizations:

μ_o (normal)	μ_e^σ (two-beam)	μ_e^π (two-beam)	μ_e^σ (three-beam)	μ_e^π (three-beam)
341.7	103.2	129.8	17.2	28.2

The minimum attenuation coefficient is $A = \exp(-\mu_e t/\gamma_o)$ (Section 4.8.4) where t is the crystal thickness and $\gamma_o = \mathbf{n}\cdot\mathbf{s_0}$. In a symmetrical two-beam diffraction experiment, $\gamma_o = \cos\theta_{111}$, but in the three-beam diffraction described here, the normal to the entrance surface is $[0\bar{1}1]$ and $\gamma_o = \sqrt{1 - \frac{1}{2}\sin^2\theta_{111}}$.

Table 9.1 compares the values of the minimum attenuation coefficients for π-polarization in the two-beam case and for π- and σ-polarizations in the three-beam cases, for the two crystal thicknesses used in Figs 9.8(a) and (b).

Table 9.1 Comparison of the minimum attenuation coefficient in the two-beam and in the three-beam cases

Crystal thickness t (mm)	$\mu_o t$	$\mu_e^\sigma t$ (two-beam)	$\mu_e^\sigma t$ (three-beam)	$\mu_e^\pi t$ (three-beam)	A^σ (two-beam)	A^σ (three-beam)	A^π (three-beam)
0.8	27.3	8.26	1.38	2.26	2×10^{-4}	0.251	0.103
1.2	41	12.4	2.06	3.39	2.9×10^{-6}	0.126	0.0033

It can be seen that the further decrease of the anomalous absorption coefficient is significant, more so for π-polarization which is perpendicular to the (100) plane than for σ-polarization which is parallel to it. The difference with the two-beam comes from the $\sqrt{2}$ term which appears in the expression of C in the three-beam case.

Hildebrandt (1967) has given a physical interpretation of the 'super'-Borrmann effect using Borrmann's standing-waves approach. Figs 5.15(b) and (c) show the position of the standing waves in the lattice of a silicon or a germanium crystal for a 220 and a 111 reflection, respectively. In the former case, the minimum of electric field occurs on atomic planes, but this is not true for 111, which explains why the Borrmann effect is more important for 220 reflections. Borrmann and Hartwig (1965) had noticed that, in the case of three-beam diffraction, when the Bragg condition is exactly satisfied for both 111 and $\bar{1}11$, one has lines of standing waves instead of planes and Hildebrandt (1967) has shown that these lie exactly on rows of atoms; the absorption coefficient is therefore smaller than for a 111 reflection alone. This effect is not expected for 220 reflections because the standing waves nodes already lie on atomic planes and is not visible in Figs 9.8(a) and (b). An enhancement of the Borrmann effect for four-beam and six-beam cases involving 220 reflections has nevertheless been predicted theoretically by Joko and Fukuhara (1967) and Dalisa *et al.* (1968) and discussed by Afanas'ev and Kohn (1977). It is very difficult to observe and was observed by Koz'mik *et al.* (1976) with the 220, $2\bar{2}0$, 440, four-beam case and a germanium crystal and by Mikhailyuk *et al.* (1978) in the six-beam case 220, 242, 044, 224, 202 of germanium.

10

Spherical-wave dynamical theory: I. Kato's theory

10.1 Extension of the dynamical theory to any kind of incident wave

In the preceding chapters, the incident wave has always been considered to be a plane wave which, by definition, is of infinite lateral extension. Such a wave, however, is never produced in practice in the case of X-rays. The X-rays produced by X-ray tubes in the laboratory and by synchrotrons are emitted by point sources and even if their divergence can be very small for synchrotron sources, they are always divergent. At a large distance in a synchrotron or after successive reflections in a beam conditioner (see Chapter 15), one may approximate the wave locally by an asymptotic plane wave or by a pseudo-plane wave of limited lateral extent which can be described as a wavepacket (see Section 12.1), but one cannot produce a plane wave in the rigorous sense. What is important in practice is the ratio between the divergence of the beam falling on the crystal under study and the width of the rocking curve for the corresponding wavelength (which is measured by G_B/k, where G_B is the Bragg gap defined in Section 4.5). If it is very small, the results of plane-wave dynamical theory can be applied and the propagation of 'individual' wavefields can be discussed (*ray tracing*—see Chapter 12). If this is not the case a new theory must be developed. This has been done in two different ways. In the first one, due to Kato, one expands the incident wave, whatever its nature, in plane waves. In the two-beam case, each one of these plane waves excites two wavefields in the crystal, each of those being constituted by two waves. The total optical field in the crystal is the result of the interference of the waves excited by all the plane waves in the expansion of the incident wave, but it is still possible to interpret it in terms of the corresponding wavefields and to make a *ray tracing* analysis of the intensity distribution within the crystal. Such a treatment was first used by Kato (1961*a*,*b*) to describe the dynamical diffraction of a spherical wave and is explained in the present chapter. One of the major drawbacks of a plane-wave analysis is encountered when extending the dynamical theory to the case of imperfect crystals. It is quite satisfactory for small long-range strains where ray tracing can be done as is done in optics to interpret *mirage* effects (see Chapter 13), but it fails in the case of strong deformations. In this latter case, a plane wave remains a plane wave over a short distance only within the crystal because diffraction effects occur, in the optical sense of diffraction by a slit. This is the reason why a different approach was introduced by Takagi (1962), followed by Taupin (1964). In this approach, which can be called global, one looks directly for the optical field which is simply decomposed into waves associated with the different reciprocal lattice points on the Ewald sphere. Predetermined phases are attributed to these waves and their amplitudes are slowly varying functions of the space coordinates. The fundamental equations of plane-wave dynamical theory are then replaced by partial differential equations. The advantage of such a treatment is that it enables one to handle any kind of incident wave and any kind of deformation within the crystal, but its drawback is that the notion of wavefield and of ray tracing disappears. It will be described in Chapters 11 and 14.

10.2 Fourier expansion of a spherical wave in plane waves

10.2.1 *Principle of Kato's spherical-wave theory*

Kato's spherical-wave dynamical theory is described in his original papers (1961*a*,*b* in the non-absorbing case, 1968*a*,*b* in the absorbing case) and in a review paper (1974). The principle of the theory is as follows:

(1) in a first step, to show, by means of a Fourier transform, that the spherical wave may be expanded as a distribution of plane waves having the same wave number;

(2) in a second step, to apply the plane-wave dynamical theory to each plane-wave component of the spherical wave; and

(3) finally, to obtain the amplitude of the reflected and refracted spherical waves by an inverse Fourier transform performed by means of the stationary phase method.

The principle will first be explained in the case of a scalar spherical wave, then extended to the case of X-ray spherical waves, which are vector waves.

10.2.2 *The incident wave is a scalar wave*

A scalar wave emitted by a point source can be written as

$$\phi(\mathbf{r}) = \phi_o \frac{\exp[2\pi\mathrm{i}(\nu t - kr)]}{4\pi r}, \qquad (10.1)$$

where r is the distance from the source to the point of observation. The total energy emitted per second by the source is $4\pi r^2|\phi(\mathbf{r})|^2 = |\phi_o|^2/(4\pi)$. We shall only consider here waves of constant frequency and we shall assume from now on that the term $\exp(2\pi\mathrm{i}\nu t)$ is included in ϕ_o.

The Fourier transform of $\phi(\mathbf{r})$ is

$$F(\mathbf{K}) = \int_{-\infty}^{+\infty}\int_{-\infty}^{+\infty}\int_{-\infty}^{+\infty} \phi(\mathbf{r}) \exp[2\pi\mathrm{i}\mathbf{K}\cdot\mathbf{r}]\,\mathrm{d}\tau. \qquad (10.2)$$

This integral may be calculated directly. The spherical wave only depends on the magnitude r of the radius vector and, taking into account the properties of the Fourier transforms of functions with spherical symmetry, it follows that

$$F(\mathbf{K}) = \frac{\phi_o}{2\pi K}\int_0^{+\infty} \sin(2\pi K r)\exp(-2\pi \mathrm{i}kr)\,\mathrm{d}r.$$

The integration is easy and

$$F(\mathbf{K}) = \frac{\phi_o}{4\pi^2(K^2 - k^2)}. \qquad (10.3)$$

Taking the inverse transform of (10.2), $\phi(\mathbf{r})$ may be written as

$$\phi(\mathbf{r}) = \int_{-\infty}^{+\infty}\int_{-\infty}^{+\infty}\int_{-\infty}^{+\infty} F(\mathbf{K})\exp[-2\pi \mathrm{i}\mathbf{K}\cdot\mathbf{r}]\,\mathrm{d}^3K. \qquad (10.4)$$

The spherical wave may thus be considered as a distribution of plane waves having **K** as wavevector and $F(\mathbf{K})$ as amplitude. Its propagation in a given direction **Oz** will be studied and the value of its amplitude in the plane $z = z_o$ tangential to the wave front will be evaluated. Taking the directions of reciprocal axes OK_x, OK_y and OK_z to coincide with the direct axes Ox, Oy and Oz, we shall express the spherical wave as a distribution of plane waves whose wavevectors lie within a small solid angle around $\mathbf{OK_z}$. Separating the integrations over K_x, K_y and K_z, (10.4) may be written as

$$\phi(\mathbf{r}) = \int_{-\infty}^{+\infty}\int_{-\infty}^{+\infty} \Phi(\mathbf{K})\exp[-2\pi \mathrm{i}(K_x x + K_y y)]\,\mathrm{d}K_x\,\mathrm{d}K_y, \qquad (10.5)$$

with

$$\Phi(\mathbf{K}) = \int_{-\infty}^{+\infty} F(\mathbf{K})\exp(-2\pi \mathrm{i}K_z z_o)\,\mathrm{d}K_z.$$

From (10.4) and (10.5), it follows that

$$\Phi(K) = \int_{-\infty}^{+\infty} \frac{\phi_o\exp(-2\pi \mathrm{i}K_z z_o)}{4\pi^2(K^2 - k^2)}\,\mathrm{d}K_z.$$

Let us put

$$K'^2 = k^2 - (K_x^2 + K_y^2).$$

Using the properties of Fourier transforms, it follows that

$$\Phi(K) = \mathrm{i}\frac{\phi_o\exp(-2\pi \mathrm{i}K' z_o)}{4\pi K'}$$

with $z > 0$, which indeed corresponds to a wave propagating away from the source and describes a spherical wave propagating in reciprocal space (Fig. 10.1(a)).

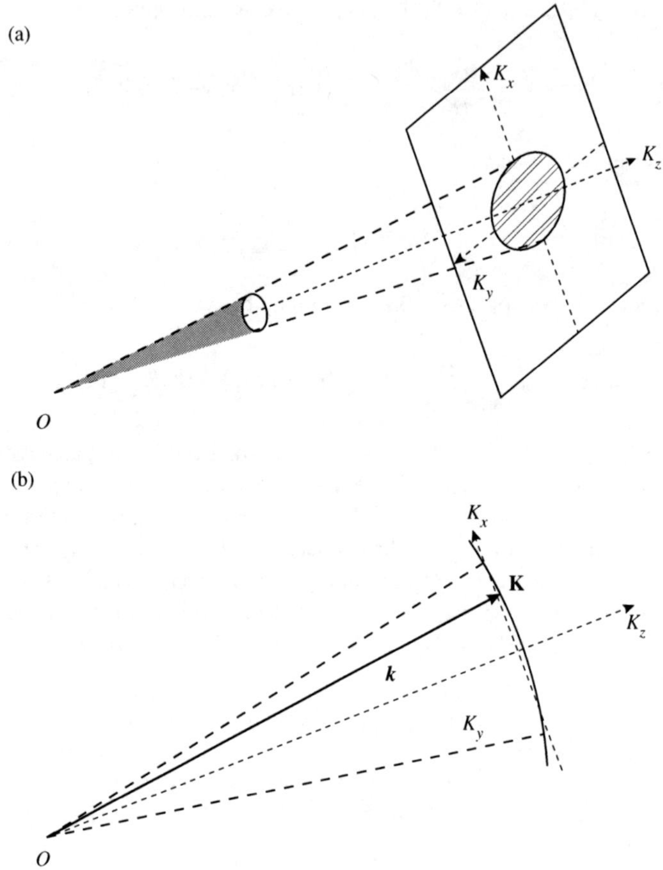

Fig. 10.1. Spherical wave of wave number k in reciprocal space: (a) limited by a small aperture, (b) in the plane K_x, K_z.

Substituting the expression of $\Phi(K)$ in (10.4), we obtain finally the expansion of $\phi(\mathbf{r})$:

$$\phi(\mathbf{r}) = \mathrm{i}\frac{\phi_o}{4\pi}\int_{-\infty}^{+\infty}\int_{-\infty}^{+\infty}\frac{\exp[-2\pi\mathrm{i}\mathbf{K}\cdot\mathbf{r}]}{K_z}\,\mathrm{d}K_x\,\mathrm{d}K_y, \tag{10.6}$$

where **K** is a wavevector of length k and components

$$K_x, K_y, K_z = K' = \sqrt{k^2 - [K_x^2 + K_y^2]}.$$

Expression (10.6) may be reduced to a surface integral on a sphere of radius k in reciprocal space. This result can be shown easily by integrating successively on the

(K_x, K_z) and (K_y, K_z) planes. Let us first consider a distribution of plane waves of wave number k in plane K_x, K_z (Fig. 10.1(b)):

$$\phi(\mathbf{r}) = \int_{-\infty}^{+\infty} F(K)\exp[-2\pi \mathrm{i}(K_x x + K_z z)]\, \mathrm{d}K_x. \tag{10.7}$$

In order to express that the plane waves all have the same wave number k, we may put

$$K_x = k\cos\theta_B; \qquad K_z = k\sin\theta_B;$$
$$\cos\theta_B = C; \qquad \sin\theta_B = S;$$
$$kx = X; \qquad kz = Z.$$

Amplitude $\phi(\mathbf{r})$ may now be written as

$$\phi(\mathbf{r}) = k\int_{-1}^{+1} F(K)\exp[-2\pi \mathrm{i}(CX + SZ)]\, \mathrm{d}C. \tag{10.8}$$

If the integration was performed from $-\infty$ to $+\infty$, as in (10.7), one would have to consider values of $|C|$ larger than 1 and terms such as $F(K)\ \exp(2\pi \mathrm{i}CX)\exp(S'Z)$ would appear in the integral ($S' = \mathrm{i}S = \sqrt{C^2 - 1}$). They correspond to evanescent waves whose amplitudes decrease very rapidly when z increases and which can therefore be neglected. Integrals (10.7) and (10.8) therefore take the same values whether they are performed on the circle of radius k or from $-\infty$ to $+\infty$. A similar integration may be performed in the (K_y, K_z) plane and one may thus show that integral (10.6) is a surface integral on the sphere of radius k. It may therefore be interpreted as an *expansion in plane waves having all the same wave number*, k.

10.2.3 *The incident wave is a vector wave*

Electromagnetic waves are actually vector waves. One may distinguish two situations depending on the nature of the source:

1. The source is an X-ray tube. Kato (1961*a*) has shown that the electric and magnetic fields of the spherical wave may be expanded in the following way:

$$\mathbf{H} = -\mathrm{i}k\frac{H_o}{4\pi}\iint \frac{\mathbf{J}\wedge\mathbf{K}}{K_z}\exp(-2\pi \mathrm{i}\mathbf{K}\cdot\mathbf{r})\, \mathrm{d}K_x\, \mathrm{d}K_y,$$
$$\mathbf{E} = -\mathrm{i}k\frac{E_o}{4\pi}\iint \frac{\mathbf{J}\wedge\mathbf{K}\wedge\mathbf{K}}{K_z}\exp(-2\pi \mathrm{i}\mathbf{K}\cdot\mathbf{r})\, \mathrm{d}K_x\, \mathrm{d}K_y,$$

   where $\mathbf{J}$ is a constant vector proportional to an induced current due to a spontaneous transition within the atom emitting the photon in the anode of the

X-ray tube. If one is interested in the diffraction by a single family of lattice planes, each wave of the vector field may be decomposed into two components respectively parallel and perpendicular to the $\mathbf{K_o}$, $\mathbf{K_h}$ plane. Each of these two components may be treated as a scalar wave and expanded according to (10.5).

2. The source is a synchrotron. It is totally polarized in the plane of the orbit and partially polarized outside. Each wave of the vector field may be decomposed in the same way and each component treated as a scalar wave.

The fundamental assumption of Kato's theory is that each plane wave in expansion (10.5) is diffracted by the crystal as if it were alone. We shall limit ourselves in the present chapter to a spherical wave incident on a perfect crystal with two nodes only, O and H on the Ewald sphere. In the case of a beam emitted by an X-ray tube, its divergence is much larger than the width of the dispersion surface, so that the wave can be considered as a spherical wave and the whole dispersion surface is excited. The useful angular width of the incident beam is of the order of 10^{-5} rad and, for a focus-to-crystal distance of 50 cm, the lateral width of the beam when it reaches the crystal is of the order of 5 μm; the impact of the beam on the crystal in the incidence plane may thus be assimilated to a point.

In the case of the new synchrotron radiation sources, it can still be considered that the wave emitted by the source is a spherical wave, but the source size is very small and the emitted beams have a very high degree of spatial coherence, of the order of several tens of microns (Baruchel *et al.* 1997; Cloetens *et al.* 1997*b*). At ESRF, the source size is estimated to be 50 μm in the vertical direction and 100 μm in the horizontal direction. The corresponding divergences of the wave received by a point of the entrance surface of the crystal are 0.34×10^{-6} and 0.68×10^{-6}, respectively, for a source-to-specimen distance of 145 m or so. They are usually smaller than the width of the rocking curve and part only of the dispersion surface is excited. They may be so small that it can be considered in some cases that a point at the entrance surface of the crystal receives a 'locally plane wave' (Mocella *et al.* 1999, 2000). It can thus be considered that each point of the entrance surface receives a locally plane wave with a different angle of incidence.

The total wave in the crystal is, using (10.6),

$$\mathbf{D} = \frac{\mathrm{i}}{4\pi} \iint K_z^{-1} \sum_j \{\mathbf{D_{oj}} + \mathbf{D_{hj}} \exp[\mathrm{i}\varphi_j(\mathbf{K}, \mathbf{r})]\} \,\mathrm{d}K_x \,\mathrm{d}K_y, \tag{10.9}$$

where the double integration is over the dispersion surface, the summation is over both branches of the dispersion surface ($j = 1, 2$), $\varphi_j(\mathbf{K}, \mathbf{r}) = -2\pi[\mathbf{h} \cdot \mathbf{r} + \mathbf{K} \cdot \mathbf{r_s} + \mathbf{K_{oj}} \cdot (\mathbf{r} - \mathbf{r_s})]$ is a phase factor, $\mathbf{K}$ is a running wavevector in the expansion of the spherical wave, $\mathbf{K_{oj}}$ is the wavevector of the corresponding refracted wavevector inside the crystal, $\mathbf{r}$ is the position vector of the point where the wave is studied and $\mathbf{r_s}$ is the position vector of a point on the crystal surface.

10.3 Direct integration in the transmission geometry

10.3.1 *The geometrical conditions*

The derivations in the present chapter will be done in the transmission or Laue geometry, and the results in the reflection or Bragg geometry will be given in Section 10.9. The geometrical situation is represented in Fig. 10.2. Figure 10.2(a) is represented in direct space; F is a point source, Fz the propagation direction. The distance of the source to the specimen is supposed to be small enough for the impact of the incident beam on the crystal surface to be considered as a point, A. The direct beam intersects the exit surface at C, and the reflected beam generated at A intersects the exit surface at B.

The whole dispersion surface (Fig. 10.2(b)) is excited by the distribution of plane waves which constitutes the spherical wave and the paths of the corresponding

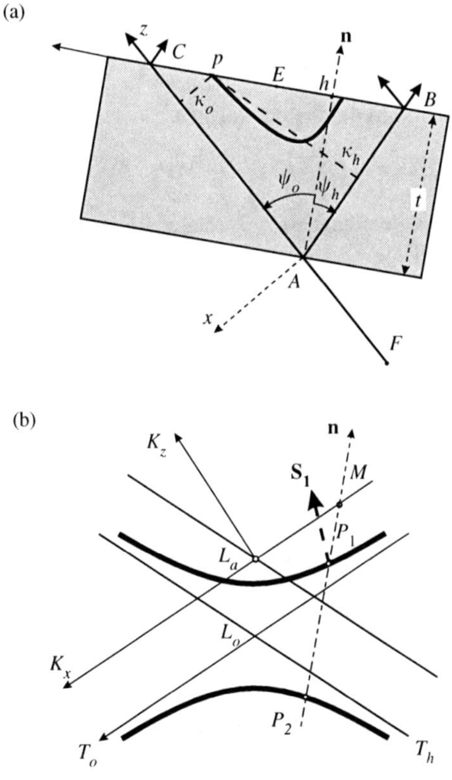

Fig. 10.2. Axes used in spherical wavefield dynamical theory. (a) Direct space, ABC: Borrmann triangle, F: point source. (b) Reciprocal space; $\mathbf{L_a K_x}$ and $\mathbf{L_a K_z}$ are parallel to **Ax** and **Az**, respectively.

wavefields fill out the Borrmann triangle, ABC, introduced in Section 1.10.2. The Ax axis is normal to the incident direction, and Ay is normal to the plane of incidence. Figure 10.2(b) represents the axes in reciprocal space: $L_a K_x$, $L_a K_y$, $L_a K_z$, drawn from the Laue point, are parallel to the axes Ax, Ay, Az in direct space; $L_a K_x$ is along the tangential plane to the Ewald sphere, $L_a K_y$ along the normal to the plane of incidence, $L_a K_z$ along the propagation direction.

Expression (10.9) is reduced to, for the reflected wave,

$$\mathbf{D_h} = \frac{\mathrm{i}}{4\pi}\int_{-\infty}^{+\infty}\int_{-\infty}^{+\infty} K_z^{-1}\sum_j \mathbf{D_{hj}} \exp\{-2\pi\mathrm{i}[\mathbf{K}\cdot\mathbf{r_s} + \mathbf{K_{oj}}\cdot(\mathbf{r}-\mathbf{r_s})]\}$$

$$\times \exp(-2\pi\mathrm{i}\mathbf{h}\cdot\mathbf{r})\,\mathrm{d}K_x\,\mathrm{d}K_y, \qquad (10.10)$$

where $\mathbf{r} = \mathbf{Fp}$ is the position vector of a point, p, on the exit surface of the crystal (Fig. 10.2(a)), $\mathbf{r_s} = \mathbf{FA}$, and $\mathbf{r} - \mathbf{r_s} = \mathbf{Ap}$, $\mathbf{K}$ is a running wavevector in vacuum of length k and of extremity M in reciprocal space (Fig. 10.2(b)), and $\mathbf{K_{o1}}$ and $\mathbf{K_{o2}}$ are the wavevectors of the two waves it generates inside the crystal, of tiepoints P_1 and P_2, respectively. Using (4.22), these wavevectors can be written as

$$\mathbf{K_{oj}} = \mathbf{K} + \mathbf{MP_j}$$

and, since $\mathbf{MP_j}$ is normal to the crystal surface,

$$\mathbf{K}\cdot\mathbf{r_s} + \mathbf{K_{oj}}\cdot(\mathbf{r}-\mathbf{r_s}) = \mathbf{K}\cdot\mathbf{r} + \mathbf{MP_j}\cdot\mathbf{Ap} = \mathbf{K}\cdot\mathbf{r} + \overline{MP_j}\cdot t,$$

where t is the crystal thickness and $\overline{MP_j}$ is related to the coordinate X_{oj} of the tiepoint P_j (eqn (5.23)) by

$$\overline{MP_j} = \frac{1}{\gamma_o}\left(\frac{1}{2}k\chi_o + X_{oj}\right).$$

To a good approximation, the spherical wave can be replaced by a cylindrical wave normal to the plane of incidence. The solutions are therefore independent of K_y and it suffices to study the state of the wave in this plane: the y coordinate of $\mathbf{r}$ can be taken equal to zero. It will furthermore be assumed that the normal $\mathbf{n}$ to the crystal surface lies in the $\mathbf{K_o}$, $\mathbf{K_h}$ plane. The scalar product $\mathbf{K}\cdot\mathbf{r}$ can be written as

$$\mathbf{K}\cdot\mathbf{r} = K_x\cdot x + K_z(z - z_F),$$

where $x, 0, z$ are the coordinates of p and $0, 0, z_F$ those of F and

$$K_z = \sqrt{k^2 - [K_x^2 + K_y^2]}. \qquad (10.11)$$

The double integral (10.10) may be separated into two integrals by first performing the integration over K_y:

$$\mathbf{D_h} = \frac{\mathrm{i}}{4\pi}\mathcal{I}(K_z)\int_{-\infty}^{+\infty}\sum_j\left\{\mathbf{D_{hj}}\exp[-2\pi\mathrm{i}(K_x x + \overline{MP}_j t)]\right\}\mathrm{d}K_x, \qquad (10.12\text{a})$$

with

$$\mathcal{I}(K_z) = \int_{-\infty}^{+\infty} K_z^{-1} \exp[-2\pi \mathrm{i} K_z(z - z_F)] \, \mathrm{d}K_y. \tag{10.12b}$$

The integral $\mathcal{I}(K_z)$ may be calculated using the stationary phase method.

10.3.2 *Stationary phase method*[1]

Let us consider an integral of type

$$I = \int_{-\infty}^{+\infty} f(x) \exp[\mathrm{i}g(x)] \, \mathrm{d}x \tag{10.13}$$

and let x_o be a root of the first derivative of $g(x)$. If $f(x)$ is a slowly varying function in the neighbourhood of x_o, an approximate value of the integral may be obtained by expanding $g(x)$ in Taylor series around x_o:

$$g(x) = g(x_o) + g''(x_o) \frac{[x - x_o]^2}{2} + \cdots .$$

Integral I is then equal to

$$I = f(x_o) \exp[\mathrm{i}g(x_o)] \int_{-\infty}^{+\infty} \exp\left[\mathrm{i}g''(x_o) \frac{(x - x_o)^2}{2}\right] \mathrm{d}x,$$

$$I = f(x_o) \sqrt{\frac{2\pi}{|g''(x_o)|}} \exp\left\{\mathrm{i}\left[g(x_o) \pm \frac{\pi}{4}\right]\right\}, \tag{10.14}$$

where the upper sign is taken when $g''(x_o) > 0$ and the lower sign for $g''(x_o) < 0$.

10.3.3 *Amplitude distribution on the exit surface—reflected wave*

Integral $\mathcal{I}(K_z)$ (eqn (10.12b)) may now be integrated using the stationary phase method around $K_y = 0$ by noticing that, due to (10.11),

$$\partial K_z / \partial K_y = 0 \quad \text{for } K_y = 0.$$

Applying (10.14) to (10.12b), expression (10.12a) of the reflected amplitude becomes

$$\mathbf{D_h} = \frac{\exp(3i\pi/4)}{4\pi\sqrt{z - z_F}} \times \int_{-\infty}^{+\infty} \sum_j \mathbf{D_{hj}} \frac{\exp[-2\pi \mathrm{i}(K_x x + K_z(z - z_F) + \overline{MP_j}\, t)]}{\sqrt{K_z}} \, \mathrm{d}K_x.$$

[1] See, for instance, Born and Wolf 1993.

The various terms in the integrand will now be arranged in order to be able to perform the integration:

1. The component K_x of the wavevector is nothing else but the abscissa $\overline{L_a M} = k\Delta\theta$ of the extremity of the vacuum wavevector (Fig. 10.2(b)). It is related to the deviation parameter η using (5.16) and (5.18):

$$K_x = \overline{L_a M} = k\Delta\theta = \eta \frac{\gamma_h}{\Lambda_L \sin 2\theta_B} - \frac{k\chi_o(1-\gamma)}{2\sin 2\theta_B}. \tag{10.15}$$

The value of K_x is much smaller than that of K_z, which is of the order of k. The quantity $K_z^{-1/2}\exp[-2\pi i K_z(z - z_F)]$ may therefore be considered as constant and taken out of the integral.

2. The abscissa x of p (Fig. 10.2(a)) is more conveniently expressed in terms of a quantity called the reduced coordinate of p along the basis BC of the Borrmann triangle, Y, defined by

$$Y = \frac{\overline{Ep}}{\overline{EC}}, \tag{10.16}$$

where E is the midpoint of BC. They are related by noticing that

$$x = \overline{Cp}\gamma_o,$$

where the basis $\overline{BC}$ of the Borrmann triangle is oriented from B to C (x is negative). The reduced coordinate can also be written as

$$Y = 1 + \frac{\overline{Cp}}{\overline{EC}}.$$

It follows that

$$x = (Y - 1)\gamma_o \overline{EC}.$$

The basis of the Borrmann triangle, BC, can be expressed in terms of the thickness t of the crystal:

$$\overline{BC} = \overline{Bh} + \overline{hC} = -t\tan\psi_h + t\tan\psi_o,$$

where h is the projection of A on BC, ψ_h and ψ_o are the angles between **Ah** and the sides of the Borrmann triangle, **AB**, **AC**, respectively, as defined in Section 4.7.3. From these definitions, $\psi_o - \psi_h = 2\theta_B$, it follows that

$$\overline{BC} = \frac{t\sin 2\theta_B}{\gamma_o\gamma_h}.$$

and, finally,

$$x = (Y - 1)\frac{t \sin 2\theta_B}{2\gamma_h}. \tag{10.17}$$

3. Amplitudes D_{hj} are given by (6.2b) and $\overline{MP_j}$ is given in terms of η by (5.23):

$$\overline{MP_j} = \frac{k\chi_o}{2\gamma_o} + \frac{1}{2\Lambda_L}(\eta \pm \sqrt{1 + \eta^2}). \tag{10.18}$$

By substitution of (10.15), (10.17) and (10.18) in the expression of D_h, and after changing the variable of integration from K_x to η, it follows that

$$D_h = \frac{\mathcal{B}_h}{2} \int_{-\infty}^{+\infty} \frac{\exp(-\pi \mathrm{i} t Y \eta/\Lambda_L)}{\sqrt{1 + \eta^2}} \{\exp[\mathrm{i}\alpha(\eta)] - \exp[-\mathrm{i}\alpha(\eta)]\}\, \mathrm{d}\eta, \tag{10.19a}$$

$$= \mathrm{i}\mathcal{B}_h \int_{-\infty}^{+\infty} \frac{\exp(-\pi \mathrm{i} t Y \eta/\Lambda_L)}{\sqrt{1 + \eta^2}} \sin[\alpha(\eta)]\, \mathrm{d}\eta, \tag{10.19b}$$

where

$$\begin{aligned}
\alpha(\eta) &= \frac{\pi t}{\Lambda_L}\sqrt{1 + \eta^2}, \\
\mathcal{B}_h &= \mathcal{A}_h \exp\left\{- \pi \mathrm{i} k \frac{\chi_o}{2} t \left[\frac{1}{\gamma_o} + \frac{1}{\gamma_h} + Y\left(\frac{1}{\gamma_o} - \frac{1}{\gamma_h}\right)\right]\right\}, \\
\mathcal{A}_h &= \mathcal{A}_o \frac{\sqrt{\chi_h \chi_{\bar{h}}}}{\chi_{\bar{h}}\sqrt{\gamma}}, \\
\mathcal{A}_o &= \frac{\exp(3\mathrm{i}\pi/4) - 2\mathrm{i}\pi K_z(z - z_F)}{4\pi\sqrt{K_z(z - z_F)}} \frac{\gamma_h}{\Lambda_L \sin 2\theta_B} D_o^{(a)}.
\end{aligned} \tag{10.20}$$

Integration (10.19b) can be performed directly, using the properties of Bessel functions (see, for instance, Sneddon 1951), and the reflected amplitude is

$$D_h = -\mathrm{i}\pi \mathcal{B}_h J_0(\zeta), \tag{10.21}$$

where

$$\zeta = \frac{\pi t}{\Lambda_L}\sqrt{1 - Y^2}. \tag{10.22}$$

10.3.4 *Amplitude distribution on the exit surface—refracted wave*

The amplitude of the refracted wave along the exit surface can be written with an expression analogous to (10.19a):

$$D_o = \frac{\mathcal{B}_o}{2}\int_{-\infty}^{+\infty}\sum_j\left[1 \mp \frac{\eta}{\sqrt{1+\eta^2}}\right] \times \exp\left[-\pi \mathrm{i}\frac{t}{\Lambda_L}\left(Y\eta \pm \sqrt{1+\eta^2}\right)\right]\mathrm{d}\eta, \tag{10.23}$$

where the top and bottom signs correspond to $j = 1$ and $j = 2$, respectively, and

$$\mathcal{B}_o = \mathcal{A}_o \exp\left\{-\pi \mathrm{i}k\frac{\chi_o}{2}t\left[\frac{1}{\gamma_o}+\frac{1}{\gamma_h}+Y\left(\frac{1}{\gamma_o}-\frac{1}{\gamma_h}\right)\right]\right\}.$$

The integration is performed using the properties of Bessel functions. One obtains

$$D_o = -\pi\mathcal{B}_o\sqrt{\frac{1+Y}{1-Y}}J_1(\zeta), \tag{10.24}$$

where $J_1(\zeta)$ is the Bessel function of order 1 and ζ is given by (10.22).

Expressions (10.21) and (10.24) of the reflected and refracted waves for an incident spherical wave can also be obtained in a completely different manner by solving Takagi's equations. This will be shown in Chapter 11.

10.4 Intensity distribution on the exit surface

In the case of absorbing crystals, Λ_L and the Bessel functions are complex. One obtains finally, for the intensity distribution of the reflected and refracted beams, respectively, at any point at the basis of the Borrmann triangle,

$$\begin{aligned} I_h(Y) &= \pi^2|\mathcal{A}_h|^2\exp(-\mu_s t)\left|J_0\left[\pi\frac{t}{\Lambda_L}\sqrt{(1-Y^2)}\right]\right|^2, \\ I_o(Y) &= \pi^2|\mathcal{A}_o|^2\exp(-\mu_s t)\frac{1+Y}{1-Y}\left|J_1\left[\pi\frac{t}{\Lambda_L}\sqrt{(1-Y^2)}\right]\right|^2, \end{aligned} \tag{10.25a}$$

where

$$\mu_s = \frac{1}{2}\mu_o\left[\frac{1}{\gamma_o}+\frac{1}{\gamma_h}+Y\left(\frac{1}{\gamma_o}-\frac{1}{\gamma_h}\right)\right]. \tag{10.25b}$$

Tables 10.1 and 10.2 give the position of the maxima of intensity of the reflected and refracted waves for $Y = 0$, namely at the centre of the Borrmann triangle, and for zero absorption.

Table 10.1 Positions and values of the maxima of $J_0(\zeta)^2$ for $Y = 0$ in the zero-absorption case

N	1	2	3	4	5	6	7	8	9	10
t/Λ_L	1.2197	2.2331	3.2383	4.2411	5.2428	6.2439	7.2448	8.2454	9.2459	10.2463
$J_0(\zeta)^2$	0.1622	0.09007	0.06235	0.04768	0.03856	0.03242	0.02795	0.02456	0.021907	0.01977

Table 10.2 Positions and values of the maxima of $J_1(\zeta)^2$ for $Y = 0$ in the zero-absorption case

N	1	2	3	4	5	6	7	8	9	10
t/Λ_L	0.5860	1.6940	2.7172	3.7261	4.7312	5.7345	6.7368	7.7385	8.7399	9.7409
$J_1(\zeta)^2$	0.33857	0.11981	0.07469	0.05443	0.04285	0.03535	0.03004	0.02619	0.02319	0.02081

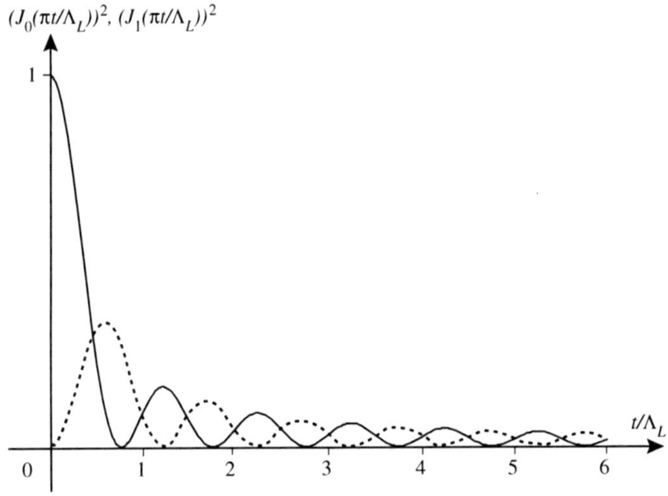

Fig. 10.3. Variations of $J_0(\pi t/\Lambda_L)^2$ (solid line) and $J_1(\pi t/\Lambda_L)^2$ (dashed line) with t/Λ_L.

Figure 10.3 shows the variations of $J_0(\zeta)^2$ and $J_1(\zeta)^2$ with t/Λ_L at the centre of the Borrmann triangle ($Y = 0$), and Fig. 10.4 shows the variations of the intensity of the reflected and refracted beams along the basis of the Borrmann triangle. Figures 10.4(a) and (b) show these variations for two values of t/Λ_L, 10 and 10.5, respectively, and a non-absorbing crystal. It can be seen that for a given value of t/Λ_L the intensity distributions for the reflected and the refracted beams are complementary. At the centre of the Borrmann triangle ($Y = 0$), there is a maximum for the reflected beam and a minimum for the refracted beam if t/Λ_L is an integer and it is the opposite if t/Λ_L is a half integer. Figures 10.4(c) and (d) show the intensity distributions in the case of a real, absorbing crystal (silicon, 220 reflection, Cu$K\alpha$) for $t/\Lambda_L = 10$ and 20, respectively. It can be seen that there is an enhancement

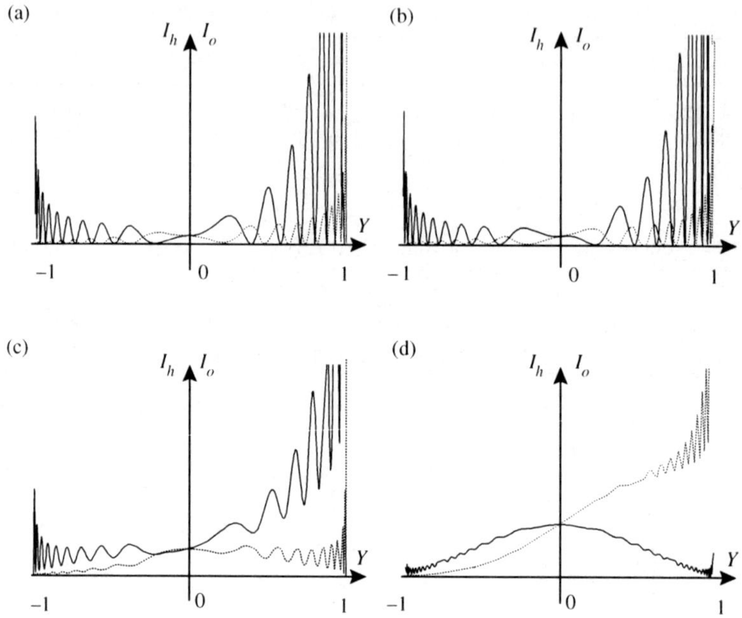

Fig. 10.4. Intensity distribution on the exit surface for an incident spherical wave. solid lines: reflected beam; dashed lines: refracted beam; arbitrary units. (a) No absorption, $t/\Lambda_L = 10.0$. (b) No absorption, $t/\Lambda_L = 10.5$. (c) Silicon, 220 reflection, Cu$K\alpha$, $t/\Lambda_L = 10.0$. (d) Silicon, 220 reflection, Cu$K\alpha$, $t/\Lambda_L = 20.0$.

of the intensities at the centre of the Borrmann triangle, the more so as the crystal thickness increases. This is due to anomalous transmission and will be explained in Chapter 12.

The Bessel functions can be approximated by the following relations:

$$J_0(\zeta) \simeq \sqrt{\frac{2}{\pi\zeta}} \cos\left(\frac{\zeta - \pi}{4}\right),$$
$$J_1(\zeta) \simeq \sqrt{\frac{2}{\pi\zeta}} \cos\left(\frac{\zeta - 3\pi}{4}\right). \tag{10.26}$$

Expressions (10.26) are, of course, only an approximation which is not very good for very small thicknesses. It becomes quite good for values of t/Λ_L larger than 3 in the case of $J_0(\zeta)$ and larger than 5 in the case of $J_1(\zeta)$. From (10.26) as well as from Table 10.1 and Fig. 10.3, it can be seen that the period of the oscillations due to spherical-wave Pendellösung is approximately Λ_L.

10.5 Equal-intensity (Pendellösung) fringes

The argument of the Bessel functions may also be written in a different way by introducing the distances κ_o and κ_h of point p from the sides of the Borrmann triangle (Fig. 10.2(a)). The first one, κ_o, is the opposite of the coordinate x of p whose expression is given by (10.17). The second one is given by a similar expression and we have

$$\kappa_o = |x| = (1 - Y)\frac{t \sin 2\theta_B}{2\gamma_h},$$
$$\kappa_h = (1 + Y)\frac{t \sin 2\theta_B}{2\gamma_o}. \tag{10.27a}$$

It follows that

$$\zeta = \pi \frac{t}{\Lambda_L}\sqrt{(1 - Y^2)} = \frac{2\pi\sqrt{\gamma_o\gamma_h}}{\Lambda_L \sin 2\theta_B}\sqrt{\kappa_o\kappa_h}. \tag{10.27b}$$

The intensity remains constant on lines along which the product of the distances to the sides of the Borrmann triangle is constant, namely hyperbolae having *AB* and *AC* as asymptotes (Fig. 10.5). Figure 10.6 is a section topograph (see Section 17.3.2). It is a photograph of the reflected beam from a wedge-shaped silicon crystal (the thickness of the crystal decreases in the out-of-plane direction in Fig. 10.2(a)). It shows the hyperbolic nature of the fringes. The shape of the fringes is directly related to that, also hyperbolic, of the dispersion surface. The periodic fading of the fringes is due to a beat effect resulting from the superposition of the fringes corresponding

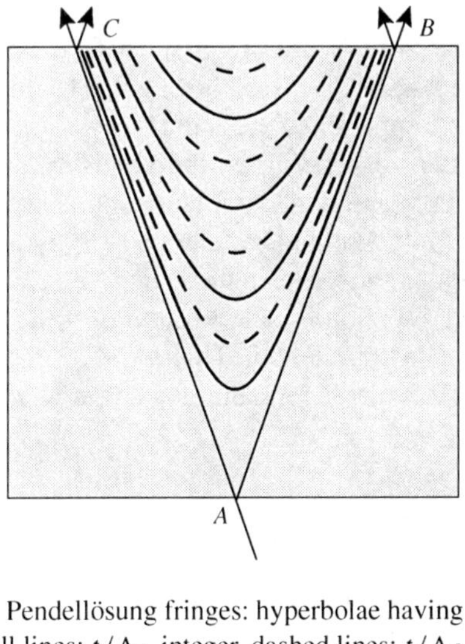

Fig. 10.5. Spherical wave Pendellösung fringes: hyperbolae having the sides of the Borrmann triangle as asymptotes. Full lines: t/Λ_L integer, dashed lines: t/Λ_L half-integer.

Fig. 10.6. X-ray section topograph of a wedge-shaped silicon crystal showing hyperbolic Pendellösung fringes; Ag$K\alpha$, 220 reflection; the periodic fading of the fringes is due to the superposition of the fringes corresponding to the two states of polarization (see Section 10.8). Courtesy N. Kato.

to the two states of polarization, parallel and perpendicular to the plane of incidence (see Section 10.8).

The measurement of the shape of the fringes by Hattori and Kato (1966) constituted a confirmation of the shape of the dispersion surface. For this they needed a perfect wedge and they obtained one by using a twinned silicon crystal and reflecting planes belonging to one individual of the twin only. The exact position of the fringes was measured by Homma *et al.* (1966) and found to be in agreement with the theoretical values given in Table 10.1.

Accurate measurement of the position of Pendellösung fringes provides a means to determine experimentally the structure factors, as was shown by Batterman and Patel (1968), Kato (1969), Hart and Milne (1969*a*, 1970), Aldred and Hart (1973*a*), Takeda and Kato (1978), Saka and Kato (1986, 1987) and Deutsch and Hart (1988) for X-rays and Graf and Schneider (1986) for high-energy γ-rays.

In the preceding discussion it has always been implicitly assumed that the source of the spherical wave is a monochromatic point source at the surface of the crystal. The situation is different if the source is a point source far away from the crystal or if it is an incoherent source limited by a slit before the entrance surface of the crystal. The former case will be discussed in Chapter 11. In the latter case, if the slit is infinitely thin one has a spherical wave, but if the source is finite, one will obtain a more complicated pattern depending on the width of the slit. This situation has been studied experimentally and with computer simulations by Aristov *et al.* (1980, 1982). The influence of polychromaticity can be seen by noticing that parameter ζ given by (10.22) depends on the wavelength through Λ_L. For a given value of Y one therefore expects to have pseudo-periodic variations of the intensity of the reflected and refracted beams with the wavelength. This was indeed observed by Aristov *et al.* (1977*a*,*b*) who studied silicon and germanium wedges with continuous radiation.

10.6 Integration by the stationary phase method

Expressions (10.19a) and (10.23) of the reflected and refracted amplitudes, respectively, may also be integrated using the stationary phase method. The advantage is to provide an interpretation of the fringes described in the preceding section. The two integrals may be written as

$$\begin{aligned} D_h &= \sum_j \pm \frac{\mathcal{B}_h}{2} \int_{-\infty}^{+\infty} \frac{\exp[\mathrm{i} g_j(\eta)]}{\sqrt{1+\eta^2}} \,\mathrm{d}\eta, \\ D_o &= \sum \frac{\mathcal{B}_o}{2} \int_{-\infty}^{+\infty} \left[1 \pm \frac{\eta}{\sqrt{1+\eta^2}}\right] \exp[\mathrm{i} g_j(\eta)] \,\mathrm{d}\eta, \end{aligned} \tag{10.28}$$

where $j = 1$ or 2, the top sign corresponds to $j = 1$ and

$$g_j(\eta) = \frac{-\pi t}{\Lambda_L}\left[\eta Y \pm \sqrt{1+\eta^2}\right].$$

The first and second derivatives of $g_j(\eta)$ are, respectively,

$$\begin{aligned} g_j'(\eta) &= \frac{-\pi t}{\Lambda_L}\left[Y \pm \frac{\eta}{\sqrt{1+\eta^2}}\right], \\ g_j''(\eta) &= \mp \frac{\pi t}{\Lambda_L}(1+\eta^2)^{-3/2}. \end{aligned}$$

The first derivative is zero for a value of η, η_o, such that

$$\mp \frac{\eta_o}{\sqrt{1+\eta_o^2}} = Y \tag{10.29a}$$

$$\Longrightarrow \quad \eta_o = \mp \frac{Y}{\sqrt{1-Y^2}}. \tag{10.29b}$$

The top sign corresponds to $j = 1$, and the bottom one to $j = 2$.

The only non-negligible contributions to D_h and D_o at point p of the exit surface of the crystal come from those waves of expansion (10.9) whose tiepoint is associated with the value η_o of the deviation parameter. The Appendix to the present chapter shows that the left-hand side of (10.29a) is equal to the reduced coordinate Y_o of the point p_o of the exit surface where the two wavefields of deviation parameter $\pm\eta_o$ arrive, while eqn (10.16) has shown that the right-hand side represents the reduced coordinate of p (Fig. 10.7(a)). In other words, p and p_o are confused and the only non-negligible contribution to the amplitude of the reflected and refracted waves at p are the wavefields which have propagated along $\mathbf{Ap_o}$. Indeed, Fig. 10.7 shows that along any path $\mathbf{Ap_o}$ inside the Borrmann triangle two wavefields propagate, one with tiepoint P_1, on branch 1, the other with tiepoint P_2', on branch 2 (Fig. 10.7(b)). These two points lie on the extremities of a diameter of the dispersion surface. The two wavefields interfere, giving rise to the Pendellösung fringes which were first observed by Kato and Lang (1959), and calculated by Kato (1961*b*).

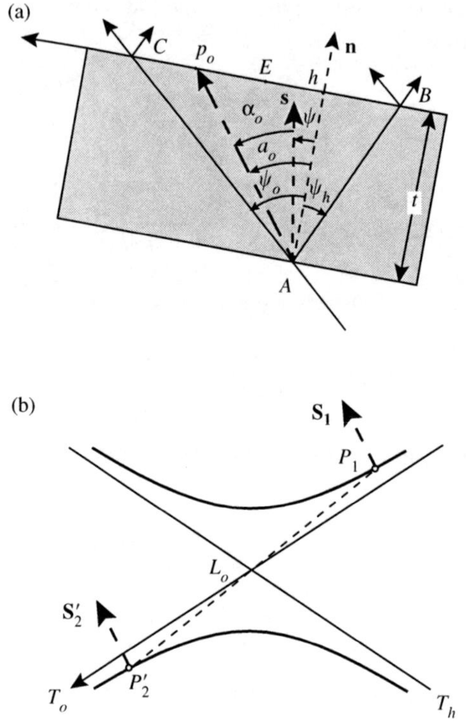

Fig. 10.7. Spherical wave Pendellösung: the two interfering wavefields follow the same path, $\mathbf{Ap_o}$ in direct space. (a) Direct space, **s**: unit vector along the reflecting planes; E: midpoint of $\overline{BC}$. (b) Reciprocal space: tiepoints of the interfering wavefields; they lie at the extremities of a diameter of the dispersion surface and their Poynting vectors, $\mathbf{S_1}$ and $\mathbf{S_2'}$, are parallel.

The total reflected and refracted amplitudes (10.28) may now be derived applying (10.14). Using (10.29) one has

$$g(\eta_o) = \mp\frac{\pi t}{\Lambda_L}\sqrt{1-Y^2},$$

$$g_j''(\eta_o) = \mp\frac{\pi t}{\Lambda_L}(1-Y^2)^{3/2}.$$

Noticing that the second derivative is negative for branch 1 and positive for branch 2, it follows that

$$D_h = \frac{\mathcal{B}_h}{2}\sqrt{\frac{2\Lambda_L}{t}}\frac{\exp[-\mathrm{i}(\zeta+\pi/4)] - \exp[\mathrm{i}(\zeta+a\pi/4)]}{(1-Y^2)^{1/4}},$$

$$D_o = \frac{\mathcal{B}_o}{2}\sqrt{\frac{2\Lambda_L}{t}}\sqrt{\frac{1+Y}{1-Y}}\frac{\exp[-\mathrm{i}(\zeta+\pi/4)] + \exp[\mathrm{i}(\zeta+\pi/4)]}{(1-Y^2)^{1/4}},$$

where ζ is given by (10.22) and, finally,

$$\begin{aligned} D_h &= -\mathrm{i}\pi \mathcal{B}_h \sqrt{\frac{2}{\pi\zeta}} \cos\left(\zeta - \frac{\pi}{4}\right), \\ D_o &= -\pi \mathcal{B}_o \sqrt{\frac{1+Y}{1-Y}} \sqrt{\frac{2}{\pi\zeta}} \cos\left(\zeta - \frac{3\pi}{4}\right). \end{aligned} \tag{10.30}$$

One recognizes in (10.30) the approximate expressions (10.26) of the Bessel functions and if we replace them by the correct ones, the amplitudes of the reflected and refracted waves are identical to (10.21) and (10.24), respectively, which were obtained by direct integration. It is normal that the two methods should lead to the same results and also that the integration by the stationary phase method should lead to the approximate expressions of the Bessel functions.

The corresponding intensities are given, in the case of absorbing crystals, by

$$\begin{aligned} I_h(Y) &= |\mathcal{A}_h|^2 \exp(-\mu_s t) \frac{2|\Lambda_L|}{t} \\ &\times \frac{\sinh^2\left\{\pi\left[(t/\sigma_L)\sqrt{1-Y^2}\right]\right\} + \cos^2\left\{\pi\left[((t/\tau_L)\sqrt{1-Y^2}) - 1/4\right]\right\}}{\sqrt{1-Y^2}}, \end{aligned} \tag{10.31a}$$

$$\begin{aligned} I_o(Y) &= |\mathcal{A}_o|^2 \exp(-\mu_s t) \frac{2|\Lambda_L|}{t} \\ &\times \frac{\sinh^2\left\{\pi\left[(t/\sigma_L)\sqrt{1-Y^2}\right]\right\} + \cos^2\left\{\pi\left[((t/\tau_L)\sqrt{1-Y^2}) - 3/4\right]\right\}}{\sqrt{1-Y^2}}, \end{aligned} \tag{10.31b}$$

where τ_L and σ_L are defined by

$$\frac{1}{\Lambda_L} = \frac{1}{\tau_L} + \mathrm{i}\frac{1}{\sigma_L}$$

and μ_s is given by (10.25b).

As in the case of the amplitudes, the approximate expressions of the intensities agree very well with the rigorous ones for values of t/Λ_L larger than 3 for the reflected beam and 5 for the refracted beam.

The agreement of (10.30) with the results of the direct integration confirms that one may interpret the fringes observed with an incident spherical wave as due to a Pendellösung effect between two wavefields excited by the spherical wave and whose tiepoints lie on a diameter of the dispersion surface. It should be emphasized that this is quite a different effect from the Pendellösung predicted by Ewald in 1917: in his case the two interfering wavefields are excited by a particular plane wave and their tiepoints lie on a given normal to the entrance surface of the crystal.

Table 10.3 Positions of the maxima of $\sqrt{2/\pi\zeta}\cos(\zeta)$ for $Y = 0$ in the zero-absorption case

N	1	2	3	4	5	6	7	8	9
t/Λ_L	0.947	1.974	2.983	3.987	4.99	5.992	6.993	7.994	8.994

It is interesting to note that if one would consider the interferences of two wavefields of tiepoints P_1 and P'_2, respectively (Fig. 10.7(b)), excited by *two different, but coherent plane waves* one would find for the reflected and refracted amplitudes:

$$\begin{aligned} D_h &= -\mathrm{i}\pi\mathcal{B}_h\sqrt{\frac{2}{\pi\zeta}}\cos(\zeta), \\ D_o &= -\pi\mathcal{B}_o\sqrt{\frac{2}{\pi\zeta}}\sqrt{\frac{1+Y}{1-Y}}\cos\left(\zeta-\frac{\pi}{2}\right). \end{aligned} \tag{10.32}$$

Hyperbolic Pendellösung fringes would also appear, which would be similar to the spherical wave ones, but their position would be different because of the phase factor of $\pi/4$ present in the case of spherical waves (10.30) but absent in the case of plane waves (10.32)—see Table 10.3.

The origin of this factor of $\pi/4$ can be tracked down from a mathematical viewpoint by noticing that the sign of the second derivative $g''_j(\eta)$ is opposite for branch 2 and branch 1. This phase shift is a specific property of the spherical waves, and the determination of the exact position of the fringes by Hattori and Kato (1966) and by Homma *et al.* (1966) constituted therefore a direct confirmation of the fact that the fringes were due to spherical-wave Pendellösung and not to some kind of plane-wave Pendellösung.

10.7 Integrated intensity

Kato (1968*c*) has shown that the intensity integrated over the basis BC of the Borrmann triangle is equal to the intensity which would accumulate on a given point of BC if the crystal was translated in the direct beam along a direction parallel to BC (as is done in a *projection topograph*—see Section 17.3.3). This is what he has called the *reciprocity* theorem. He has also shown that it is equal to the integrated intensity obtained in the plane-wave case (Section 6.5). Authier and Simon (1968) have shown that this result is to be expected by noticing that the total amplitude of the refracted wave in the plane-wave case is proportional to the Fourier transform of the total amplitude of the reflected wave at a point of the exit surface for an incident spherical wave:

$$D_h^{PW}(\Delta T) = A\int_{-\infty}^{+\infty} D_h^{SW}(y)\exp[-2\pi\mathrm{i}(y\Delta T)]\,\mathrm{d}y, \tag{10.33}$$

where *PW* and *SW* mean plane wave and spherical wave, respectively, A is a numerical constant, $\Delta T = k\gamma_o \Delta\theta$ ($\Delta\theta$, departure from Bragg's angle) and y, a coordinate along the basis of the Borrmann triangle, are the reciprocal variables in the Fourier transform. Parseval's theorem shows that the integrals of $|D_h^{PW}|^2$ and $|D_h^{SW}|^2$ are equal, which proves that the integrated intensity over the base of the Borrmann triangle for an incident spherical wave is equal to the integrated intensity obtained when rocking the crystal in an incident plane wave. The variations of the integrated intensity with crystal thickness were checked experimentally by Wada and Kato (1977).

10.8 Influence of polarization

The treatment which has been given is valid separately for each of the two components of polarization, the factor C being equal to 1 or to $\cos 2\theta_B$ for the components normal and parallel to the $\mathbf{K_o}$, $\mathbf{K_h}$ plane, respectively.

If the incident wave is not polarized, the observed intensity distribution is equal to the superposition of the intensity distributions corresponding to the two directions of polarization. The periods of the Pendellösung fringes are slightly different in each case, according to equation (5.17b):

$$\Lambda_L^{\perp} = \frac{\pi V \sqrt{\gamma_o \gamma_h}}{R\lambda \sqrt{F_h F_{\bar{h}}}},$$

$$\Lambda_L^{\parallel} = \frac{\Lambda_L^{\perp}}{\cos 2\theta_B}.$$

One therefore observes a beat effect in the contrast of the fringes (Hart and Lang 1965; Hattori *et al.* 1965*a*), of period Λ given by

$$\frac{1}{\Lambda} = \frac{1}{\Lambda_L^{\parallel}} - \frac{1}{\Lambda_L^{\perp}} = \frac{1}{\Lambda_L^{\perp}} \frac{1 - \cos 2\theta_B}{\cos 2\theta_B}. \tag{10.34}$$

Figure 10.6 illustrates this effect.

10.9 Bragg geometry

The expressions of the reflected and refracted waves in the case of an incident spherical wave in the reflection geometry were actually first obtained by solving Takagi's equations (Uragami 1969, 1970; Afanas'ev and Kohn 1971) and a derivation will be given in Chapter 11. Saka *et al.* (1973) have, however, also derived their expression using the plane-wave expansion of the spherical wave. The spherical-wave Pendellösung fringes were observed in the Bragg geometry by Uragami (1969) and by Lang and Mai (1979).

Figure 10.8(a) shows the geometrical conditions in direct space. An incident spherical wave impinges the crystal surface at A and exits the crystal through a lateral face at C; h lies at the intersection of the crystal surface and the lateral face, B lies at the intersection of the reflected beam at A and Ch, and p is a point between C and h. In the zero-absorption case, no intensity penetrates the crystal within the angular range of total reflection, but an evanescent wave propagates along Ah. Outside the domain of total reflection, only part of the intensity is reflected at A and the remainder penetrates inside the crystal.

Figure 10.8(b) shows the corresponding situation in reciprocal space. A given component of the plane-wave expansion of the spherical wave is represented by a point M on the sphere centred at the origin of the reciprocal space and of radius k, here approximated by its tangent T_o'. If the parallel to the normal $\mathbf{n}$ to the crystal surface drawn from M intersects the dispersion surface at imaginary points, there is total reflection. Outside this domain, it intersects the dispersion surface at two tiepoints, P_j' and P_j'', respectively, lying on one of the branches, $j = 1$ or 2, of the dispersion surface (branch 1 for $\eta \leq -1$, branch 2 for $\eta \geq +1$). Of the two wavefields thus excited, only one has a physical reality, that one whose direction of propagation points towards the inside of the crystal. The corresponding tiepoints are P_1' and P_2', respectively (Fig. 10.8(b)), characterized by their distances X_{oj} to T_o:

$$X_{oj} = -\frac{\gamma_o}{2\Lambda_o}\left[\eta \pm \sqrt{\eta^2 - 1}\right],$$

where the upper sign corresponds to P_1' and the lower sign to P_2'.

The Poynting vectors of the wavefields which are excited by the incident spherical wave lie between Ah and AC and their corresponding paths therefore fill out a partial Borrmann triangle, AhC. P_1' and P_2' have been chosen at the extremities of a diameter of the dispersion surface, so that the Poynting vectors of the corresponding wavefields, $\mathbf{S}_1'$ and $\mathbf{S}_2'$, are parallel; their paths in direct space overlap along $\mathbf{Ap}$ (Fig. 10.8(a)). These wavefields interfere giving rise to Pendellösung fringes (represented symbolically in Fig. 10.8(a) by a thick solid line).

Saka *et al.* (1973) have shown that the total respective amplitudes of the reflected and refracted waves at p are

$$\begin{aligned} D_h &= \mathrm{i}\pi \mathcal{A}_B\left[J_0(\zeta) + \frac{|\gamma_h|}{\gamma_o}\frac{\kappa_o}{\kappa_h}J_2(\zeta)\right]D_o^{(a)}, \\ D_o &= \pi \mathcal{B}_B\left[\sqrt{\frac{\kappa_h}{\kappa_o}} - \frac{|\gamma_h|}{\gamma_o}\sqrt{\frac{\kappa_o}{\kappa_h}}\right]J_1(\zeta)D_o^{(a)}, \end{aligned} \tag{10.35}$$

where

$$\zeta = 2\pi k C\frac{\sqrt{\chi_h\chi_{\bar{h}}}}{\sin 2\theta_B}\sqrt{\kappa_o\kappa_h} \tag{10.36a}$$

$$= \pi\frac{t\sqrt{1 - Y^2}}{\Lambda_B\sqrt{|\tan\psi_o|\tan\psi_h}}, \tag{10.36b}$$

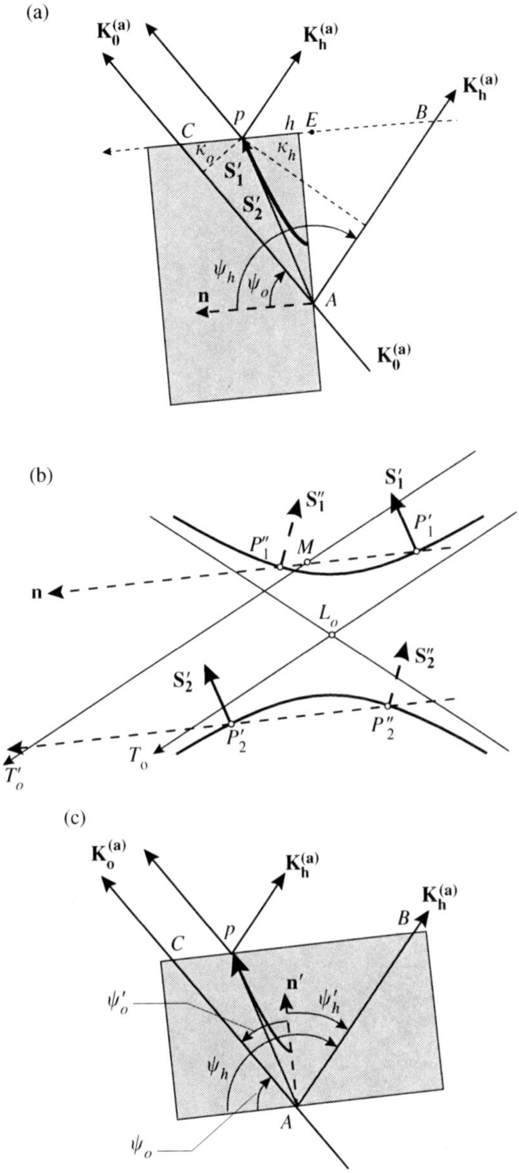

Fig. 10.8. Pendellösung fringes in the reflection geometry: $\mathbf{S}'_1$ and $\mathbf{S}'_2$: Poynting vectors of the wavefields propagating along **Ap**. The wavefields of tiepoints P''_1 and P''_2 do not have physical reality. (a) Direct space, the thick curve represents a Pendellösung fringe. (b) Reciprocal space. (c) Direct space, Pendellösung fringes in the same crystal as in (a), but after it has been translated so as to be in the transmission geometry.

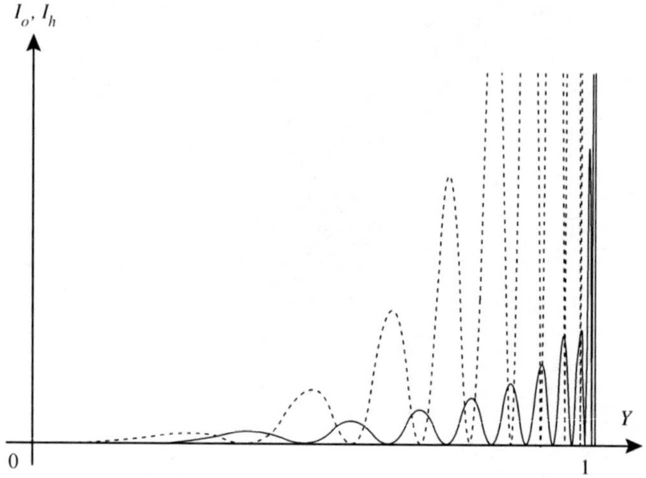

Fig. 10.9. Intensity distribution on the lateral surface for an incident spherical wave in the reflection geometry. Solid line: reflected beam; dashed line: refracted beam. $t/\Lambda_L = 10.0$; no absorption; arbitrary units.

κ_o and κ_h are the distances of p to the incident and reflected beam at A, respectively (Fig. 10.8(a)), $J_2(\zeta)$ is the Bessel function of order 2, t is the distance Ah, $\mathcal{A}_B$ and $\mathcal{B}_B$ are constants, and Y is the reduced coordinate[2] of p along **hC** (oriented from h towards C):

$$Y = \frac{\overline{Ep}}{\overline{EC}},$$

where E is the middle of BC. The distances x_o and x_h can be expressed in terms of Y through the relations

$$x_h = -(1+Y)t\frac{\sin 2\theta_B}{2\sin\psi_o},$$

$$x_o = -(1-Y)t\frac{\sin 2\theta_B}{2\sin\psi_h}.$$

Figure 10.9 represents the variations of the respective intensities of the reflected and the refracted beams along the lateral face of the crystal, hC. It can be noticed that the minima of one of them lie over the maxima of the other one, as expected. Equation (10.36a) shows that the locus of the equal-intensity fringes inside the crystal are hyperbolae having the incident and reflected beams as asymptotes; of course,

[2] The relation between the reduced coordinate Y and the deviation parameter η is much more complicated than in the case of transmission geometry.

there is intensity only along that part of these hyperbolae which lies inside the crystal! It can also be noticed that the intensities of both beams are equal to zero for $Y = 0$, namely along the crystal surface, Ah. This is to be expected on account of the boundary conditions and corresponds to the evanescent wave propagating along Ah. This result can be checked directly in eqns (10.35). On the one hand one has along Ah

$$\frac{x_o}{x_h} = \frac{\gamma_o}{|\gamma_h|}.$$

On the other hand, the second-order Bessel function can be approximated by

$$J_2(\zeta) = -\sqrt{\frac{2}{\pi\zeta}}\cos\left(\zeta - \frac{\pi}{4}\right).$$

Substituting these two expressions in (10.35), it follows that D_h and D_o are indeed zero along Ah.

Equation (10.36b) shows that the period of the fringes in the Bragg geometry is given by $\Lambda_B\sqrt{|\tan\psi_o|\tan\psi_h}$. This can be readily interpreted by noticing that

$$\Lambda_B\sqrt{|\tan\psi_o|\tan\psi_h} = \frac{\lambda\sqrt{|\sin\psi_o\sin\psi_h|}}{|C|\sqrt{\chi_h\chi_{\bar{h}}}}.$$

One considers now a situation where the incident beam in Fig. 10.8(a) falls on the same crystal, with the same orientation of the incident beam with respect to the lattice planes, but in the transmission geometry, after a suitable translation (Fig. 10.8(c)). The Pendellösung fringes have the same position as in the Bragg geometry, but with a different intensity distribution, and they fill up the whole Borrmann triangle. The normal $\mathbf{n}'$ to the entrance surface is now parallel to the entrance surface in the Bragg geometry (Fig. 10.8(a)), and the angles of this normal with the incident and reflected directions are, respectively,

$$\psi_o' = \psi_o + \frac{\pi}{2},$$
$$\psi_h' = \psi_h + \frac{\pi}{2},$$

from which one deduces that the period of the fringes is

$$\frac{\lambda\sqrt{|\sin\psi_o\sin\psi_h|}}{|C|\sqrt{\chi_h\chi_{\bar{h}}}} = \frac{\lambda\sqrt{|\cos\psi_o'\cos\psi_h'|}}{|C|\sqrt{\chi_h\chi_{\bar{h}}}} = \Lambda_L,$$

where Λ_L is the Pendellösung distance associated with the Laue geometry described by Fig. 10.8(c). It follows that eqn (10.36b) can also be written as

$$\zeta = \pi\frac{t\sqrt{1-Y^2}}{\Lambda_L}. \tag{10.37}$$

One concludes that, as could be expected, the effective Pendellösung distance in the Bragg geometry is that of the corresponding Laue geometry where the entrance surface is parallel to the lateral face of the crystal in the reflection geometry.

10.10 Diffraction of ultrashort pulses

An interesting situation occurs when an ultrashort pulse of plane-wave X-rays, such as that produced by a free-electron laser falls on a crystal. Shastri *et al.* (2001*a*, *b*) have suggested that it can be approximated by a Dirac distribution with respect to time, $D_o^{(a)}\delta\,(t - \mathbf{u}\cdot\mathbf{r}/c)$, where $\mathbf{u}$ is a unit vector along the incident wavevector. One can then consider its Fourier expansion in frequency space, written, following Graeff (2002*a*,*b*; 2004) and Malgrange and Graeff (2003):

$$\mathcal{D}_o^{(a)}(\mathbf{r},t) = \int \exp\left[2\pi i\nu\left(t - \frac{\mathbf{u}\cdot\mathbf{r}}{c}\right)\right]d\nu = \int \exp\left[2\pi i\left(\nu t - \mathbf{K}_\mathbf{o}^{(\mathbf{a})}(\nu)\cdot\mathbf{r}\right)\right]d\nu.$$

It can be interpreted as a distribution of plane waves with frequency-dependent wavevectors $\mathbf{K}_\mathbf{o}^{(\mathbf{a})}(\nu) = \mathbf{u}\nu/\mathbf{c}$. They excite in the crystal a distribution of wavefields propagating within the whole Borrmann triangle with an intensity distribution similar to (10.25a). Shastri *et al.* (2001*a*, *b*) and Graeff (2002*a*) have shown that the time duration of the pulse after diffraction is $2T\sin^2\theta_B/c\cos\theta_B$ where T is the crystal thickness. Graeff (2002*b*) discussed the time-dependence of the polarization states after diffraction of a circularly-polarized pulse or a pulse linearly-polarized at 45 ^{o} with respect to the reflection plane of the monochromator and Malgrange and Graeff (2003) considered the general case of asymmetric Laue diffraction.

Appendix: Geometrical interpretation of $\eta/\sqrt{S(\gamma_h)+\eta^2}$ in the transmission geometry

Let p_o be a point of the exit surface and let η_o be the value of the deviation parameter η of a wavefield whose path is $\mathbf{Ap_o}$. Figure 10.7(b) shows that there are actually two wavefields which follow this path within the Borrmann triangle. Their tiepoints are P_1 and P_2' and they belong, one to branch 1 and the other to branch 2; there are therefore two corresponding values of η_o.

The ratio $\eta_o/\sqrt{1+\eta_o^2}$ can be interpreted geometrically by introducing the angle a_o between the path $\mathbf{Ap_o}$ and the normal $\mathbf{Ah}$ to the entrance surface of the crystal. If $\mathbf{s}$ is a unit vector parallel to the diffracting planes, one has

$$\begin{aligned} a_o &= (\mathbf{Ah}, \mathbf{Ap_o}) \\ &= \psi_n + \alpha_o, \end{aligned} \tag{10.A1}$$

where $\psi_n = (\mathbf{Ah}, \mathbf{s})$, the angle between the normal to the crystal surface and the diffracting planes, has been defined in Section 4.7.3 and $\alpha_o = (\mathbf{s}, \mathbf{Ap_o})$ is given by (4.21):

$$\tan \alpha_o = \frac{1 - |\xi_j|^2}{1 + |\xi_j|^2} \tan \theta_B,$$

where j = 1 or 2: for a given path $\mathbf{Ap_o}$, and therefore a given value of the angle α_o, there are two wavefields propagating along this direction, one belonging to branch 1 and one belonging to branch 2.

From the definition of a_o it follows that

$$\frac{\overline{hp_o}}{t} = \tan a_o = \tan(\psi_n + \alpha_o)$$
$$= \frac{\tan \psi_n + \tan \alpha_o}{1 - \tan \psi_n \tan \alpha_o}.$$

After noticing that $2\psi_n = \psi_o + \psi_h$ and $2\theta_B = \psi_o - \psi_h$ and by substitution of (4.21) which relates $\tan \alpha_o$ and the amplitude ratio, ξ_j, it follows that

$$\tan a_o = \frac{\sin \psi_o + |\xi_j|^2 \sin \psi_h}{\cos \psi_o + |\xi_j|^2 \cos \psi_h}.$$

The expression of $|\xi_j|^2$ is given by (4.28):

$$|\xi_j|^2 = \frac{\cos \psi_o}{\cos \psi_h} |\eta_o \pm \sqrt{1 + \eta_o^2}|^2$$

assuming that Friedel's law is satisfied, and considering the transmission geometry. After substitution in the expression of $\tan a_o$, it follows that

$$\tan a_o = \frac{\tan \psi_o + |\eta_o \pm \sqrt{1 + \eta_o^2}|^2 \tan \psi_h}{1 + |\eta_o \pm \sqrt{1 + \eta_o^2}|^2}.$$

Using again the relations $2\psi_n = \psi_o + \psi_h$ and $2\theta_B = \psi_o - \psi_h$, this expression can be transformed into

$$\frac{\overline{hp_o}}{t} = \tan a_o = \frac{\sqrt{1 + \eta_o^2} \sin 2\psi_n \mp \eta_o \sin 2\theta_B}{2\gamma_o \gamma_h \sqrt{1 + \eta_o^2}}, \tag{10.A2}$$

from which one deduces that

$$\mp \frac{\eta_o}{\sqrt{1 + \eta_o^2}} = \frac{2\overline{hp_o}\, \gamma_o \gamma_h}{t \sin 2\theta_B} - \frac{\sin 2\psi_n}{\sin 2\theta_B}.$$

Noticing that

$$\frac{t \sin 2\theta_B}{2\gamma_o \gamma_h} = \overline{EC} \tag{10.A3}$$

and

$$\overline{hE} = \frac{\overline{hB} + \overline{BC}}{2} = \frac{\overline{hB} + \overline{hC}}{2}$$
$$= \frac{t \sin(\psi_o + \psi_h)}{2\gamma_o \gamma_h}$$
$$= \frac{t \sin 2\psi_n}{2\gamma_o \gamma_h},$$

it follows finally that

$$\mp \frac{\eta_o}{\sqrt{1+\eta_o^2}} = \frac{\overline{hp_o} - \overline{hE}}{\overline{EC}} = \frac{\overline{Ep_o}}{\overline{EC}},$$

which gives the geometrical interpretation of $\eta_o/\sqrt{1+\eta_o^2}$. For a given value, η_o, of the deviation parameter, there are two possible paths for the corresponding wavefields, one belonging to branch 1 and one belonging to branch 2, reaching the exit surface at the point p_o. Its reduced coordinate, Y_o, defined by (10.16) is

$$Y_o = \frac{\overline{Ep_o}}{\overline{EC}}$$

and is therefore given in terms of η_o by

$$Y_o = \mp \frac{\eta_o}{\sqrt{1+\eta_o^2}}. \tag{10.A4}$$

Conversely, the deviation parameters of the wavefields reaching the exit surface at p_o are given by

$$\eta_o = \mp \frac{Y_o}{\sqrt{1-Y_o^2}} \tag{10.A5}$$

and the value of the angle a_o is

$$\tan a_o = \frac{\sin 2\psi_n + Y_o \sin 2\theta_B}{2\gamma_o \gamma_h}. \tag{10.A6}$$

In a similar way, it is possible to express the angle α_o in terms of Y_o:

$$\tan \alpha_o = \frac{\tan \psi_n \tan \theta_B + Y_o}{1 + \tan \psi_n \tan \theta_B \; Y_o} \tan \theta_B. \tag{10.A7}$$

11

Spherical-wave dynamical theory: II. Takagi's theory

11.1 Introduction

The theory developed by Takagi (1962, 1969) constitutes a generalization of the dynamical theory for any kind of incident wave, as well as for deformed crystals with any kind of deformation. In the case of perfect crystals, it reduces to the classical dynamical theory or to Kato's theory if the incident wave is a plane wave or a spherical wave. In the case of deformed crystals, it applies even in highly distorted regions. When the column approximation can be applied (Hirsch *et al.* 1960), namely, when the Bragg angle is very small as in the case of electron diffraction, Takagi's theory reduces to Howie and Whelan's two-beam electron diffraction dynamical theory for distorted crystals (Howie and Whelan 1960, 1961, 1962).

Several theories have been derived which are similar to Takagi's (Taupin 1964; Schlangenotto 1967) and equations of the type derived by Takagi and Taupin are often referred to as Takagi–Taupin's equations.[1] Another quite general theory has been derived by Kuriyama which is based on a description of the diffraction phenomena using the formalism of quantum theory (Ashkin and Kuriyama 1966; Kuriyama 1967). The basic equations of this theory which make use of a scattering matrix are derived in Kuriyama (1970, 1972, 1973). Their relation to Takagi's equations is given in Kuriyama (1972) and their application to X-ray topography is described in Kuriyama and Early (1974). They were also derived by Kuriyama (1975) using a classical dynamical theory approach, but they have never been much used in practice.

Plane-wave dynamical theory applies to a semi-infinite perfect crystal and its solutions do not depend on the lateral coordinates. If there is a lateral limitation of the wave, that is when there is diffraction of the incident plane wave by an object, it is obvious that the solutions of the propagation equation must depend on the lateral coordinates. This is the case for a spherical wave which can be generated, for instance, by limiting a plane wave by an infinitely thin slit. One way to solve the problem is to consider a Fourier expansion of the incident wave, as was shown in Chapter 10. The advantage is that this method provides the nature and the distribution of wave-fields inside the crystal as well as the distribution of their propagation directions.

[1] In this book, the expression *Takagi's equations* will refer specifically to the equations derived in Sections 11.2.2 and 14.2.

This is very useful if, as in the case of Kato's spherical-wave dynamical theory, it is possible to integrate the Fourier transform, but, unfortunately, this is not possible in the general case.

In a similar way, plane-wave dynamical theory is not valid any longer when the crystal is deformed, and it will be shown in Chapter 13 that the path of the wavefields can be obtained using ray theory. However, rigorously speaking, it is not correct anymore to speak of wavefields since these are, by definition, characterized by *individual* tiepoints in reciprocal space which are associated with waves of infinite lateral expansion in direct space. One has, therefore, to consider wavebundles having a narrow width both in direct and reciprocal space (Ewald (1958) was the first to show that the propagation of X-rays should be described in terms of wavebundles). Furthermore, this theory breaks down when the strain gradient becomes too large and diffraction (in the optical sense) of the propagating wavebundles becomes predominant. The situation is analogous to the optical case, with the complicating effect that diffraction of wavebundles is accompanied, in the X-ray case, by the generation of new wavefields whose tiepoints belong to the other branch of the dispersion surface than that of the tiepoints of the considered wavebundle. This effect which was intuitively predicted by Penning (1966) and empirically suggested by Authier (1967) to interpret features of defect images on X-ray topographs was demonstrated by Balibar *et al.* (1983).

The two preceding paragraphs have shown that, when diffraction effects occur, either in a perfect or an imperfect crystal, the notion of wavefields which have definite propagation directions in direct space and are represented by a tiepoint in reciprocal space breaks down. Takagi's theory relates the total amplitude and phase of the reflected or refracted wave at any point inside the crystal to the amplitude and phase of another point in the crystal, but it does not provide any analysis of the crystal wave in terms of wavefields or of tiepoints in reciprocal space. The relationship is given through a Green–Riemann function, or *influence* function, which depends on the strain distribution within the crystal. The value of the amplitude and phase at any point of the exit surface is obtained by convolution of this function with the amplitude and phase distribution on the entrance surface, expressed as a distribution of point sources. What is remarkable is that it suffices to take into account this distribution along a segment of the entrance surface which corresponds to the base of an inverted Borrmann triangle whose apex lies at the point within the crystal where the amplitude and phase of the waves are being calculated (Fig. 11.4).

In general, it is not possible to obtain an analytical expression. However, in the case of a perfect crystal and a single point source (the incident wave is a spherical wave), the Green–Riemann function reduces to the Bessel function $J_0(\zeta)$ obtained by Kato's theory (Section 11.8). It is also possible to find an analytical solution in the case of a constant strain gradient (see Section 14.4.1). Except in such favourable situations, it is necessary to integrate Takagi's equations numerically (see Section 14.4.2).

In the present chapter, Takagi's equations will be derived and resolved in the case of a perfect crystal and of two-beam diffraction. They have been resolved using Laplace transforms for three-beam diffraction by Thorkildsen (1987), Thorkildsen and Larsen (1998*a*) and Larsen and Thorkildsen (1998*a*,*b*). The cases for n up to 12 have been treated by computer simulation by Okitsu (2003). Takagi-Taupin

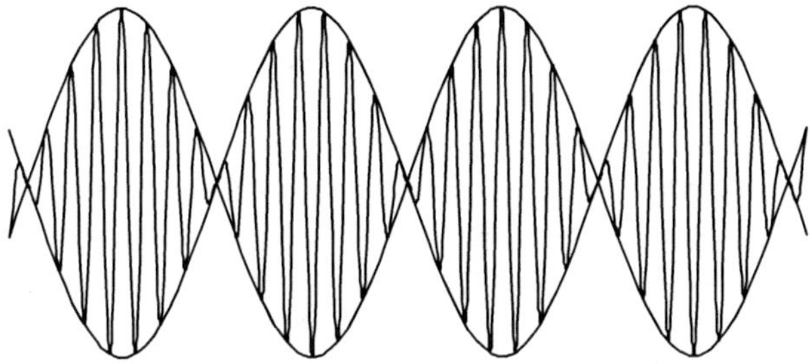

Fig. 11.1. The crystal wave is the product of a fast oscillating microscopic wave by a slowly varying macroscopic modulation.

equations have been extended to time-dependent phenomena by Wark and He (1994); Chukhovskii and Förster (1995); Tomov *et al.* (1998); Wark and Lee (1999); Wark *et al.* (2001); Yamazaki and Ishikawa (2002); Sondhauss and Wark (2003); Adams (2004 *a*, *b*, *c*). The extension to deformed crystals will be discussed in Chapter 14.

11.2 Generalized fundamental equations

11.2.1 *Modulated waves*

Takagi's approach consists in considering the crystal wave to be a modified Ewald wave, which can be developed as a sum of *modulated* waves associated with each reciprocal lattice which lies close to the Ewald sphere:

$$\mathbf{D} = \sum_{\mathbf{h}} \mathbf{D_h}(\mathbf{r}) \exp(-2\pi i \mathbf{K_h} \cdot \mathbf{r}), \tag{11.1}$$

where $\mathbf{K_h} = \mathbf{K_o} - \mathbf{h}$. This development is analogous to (5.1) for plane waves but, here, its coefficients $\mathbf{D_h}(\mathbf{r})$ are slowly varying functions of the position vector $\mathbf{r}$. The wavevector $\mathbf{K_o}$ has an *arbitrary* orientation and is of fixed length nk, where n is the index of refraction of the medium and k the wave number in vacuum. The hypothesis that $\mathbf{D_h}(\mathbf{r})$ is a slowly varying function means that $\Delta[\mathbf{D_h}(\mathbf{r})]$ is negligible.

The idea underlying Takagi's theory is the following: in all the dynamical theories, for perfect or imperfect crystals, making use of the notion of wavefields, the phase of the waves can be considered as being the product of two terms, a fast oscillation and a slow modulation (Fig. 11.1). The fast oscillation, which is a *microscopic* effect, is simply due to the periodicity of the crystal lattice, $1/h$, and represents a carrier wave, while the slow modulation, which is a *macroscopic* effect, is due to interferences between wavefields such as Pendellösung interferences between wavefields associated with both branches of the dispersion surface, or interferences between the wavefields which make up a given wavepacket. The corresponding period is of the order of the Pendellösung distance Λ. If the crystal is deformed, or if there are diffraction effects due to a limitation of the incident wave by a slit, both these periods become space dependent and are no longer constants. The originality of Takagi's theory consists in decomposing the phase factor in a different, phenomenological way, namely as the product of a fast, but constant phase factor, $\exp(-2\pi i\mathbf{K_h}\cdot\mathbf{r})$, by a slowly varying term which can be incorporated into the amplitude $\mathbf{D_h}(\mathbf{r})$. The vectors $\mathbf{K_h}$ are constant and related to one another by the relation $\mathbf{K_h} = \mathbf{K_o} - \mathbf{h}$. $\mathbf{K_o}$ is chosen arbitrarily, but, as indicated in the previous paragraph, in such a way that the phase factors should be very close to the real, varying ones. This has the advantage of simplifying the calculations, but the drawback that these phase factors have no physical meaning.

In a similar way, the incident wave can always be written in the form of a modulated wave:

$$\mathcal{D}_o^{(a)}(\mathbf{r}) = \mathbf{D}_o^{(a)}(\mathbf{r})\exp[-2\pi i\mathbf{k_o}\cdot\mathbf{r}], \tag{11.2}$$

where the length of $\mathbf{k_o}$ is the wave number in vacuum, k. The wavevector in vacuum, $\mathbf{k_o}$, is related to the wavevector inside the crystal, $\mathbf{K_o}$, by the boundary condition on the continuity of the tangential components:

$$\mathbf{K_o}\cdot\mathbf{r_s} = \mathbf{k_o}\cdot\mathbf{r_s}, \tag{11.3}$$

where $\mathbf{r_s}$ is a position vector on the crystal surface. It is therefore actually the couple $\mathbf{k_o}$, $\mathbf{K_o}$ which is chosen arbitrarily.

11.2.2 *Takagi's equations*

In a perfect crystal, the polarizability is expanded in the Fourier series (2.36) where the Fourier coefficients χ_h are constants. This is no longer true in the case of a deformed crystal as will be shown in Chapter 13. Let us now put the developments (11.1) of the modulated wave and (2.36) of the polarizability in the propagation equation (2.55),

$$\Delta\mathbf{D} + \operatorname{curl}\operatorname{curl}\chi\mathbf{D} + 4\pi^2k^2\mathbf{D} = 0.$$

Noting that

$$\operatorname{curl}\chi\mathbf{D} = \vec{\nabla}\chi\wedge\mathbf{D} + \chi\operatorname{curl}\mathbf{D}$$

it follows, after substitution of the developments of χ and $\mathbf{D}$, that

$$\begin{aligned}\operatorname{curl}\chi\mathbf{D} = {} & \sum_{\mathbf{h}}\left(\sum_{\mathbf{h}'}\chi_{h-h'}\operatorname{curl}\mathbf{D_{h'}}\right)\exp(-2\pi i\mathbf{K_h}\cdot\mathbf{r}) \\ & - 2\pi i\mathbf{K_h}\wedge\left(\sum_{\mathbf{h}'}\chi_{h-h'}\operatorname{curl}\mathbf{D_{h'}}\right)\exp(-2\pi i\mathbf{K_h}\cdot\mathbf{r}).\end{aligned}$$

After substitution of this expression in the propagation equation, the left-hand side takes the form of a sum of an infinite number of terms. If one multiplies both sides of the equation by $\exp(+2\pi i\mathbf{K_h}\cdot\mathbf{r})$ and if one then integrates over the unit cell, one finds that each term of this infinite sum should be zero separately. The propagation equation is thus replaced by an infinite series of equations of the type

$$\sum_{\mathbf{h}'} \chi_{\mathbf{h}-\mathbf{h}'} \operatorname{curl} \operatorname{curl} \mathbf{D}_{\mathbf{h}'} - 4\pi \mathrm{i} \mathbf{K}_{\mathbf{h}} \wedge \sum_{\mathbf{h}'} \operatorname{curl} \mathbf{D}_{\mathbf{h}'} + \Delta \mathbf{D}_{\mathbf{h}}$$
$$- 4\pi^2 \mathbf{K}_{\mathbf{h}} \wedge \mathbf{K}_{\mathbf{h}} \wedge \left(\sum_{\mathbf{h}'} \chi_{\mathbf{h}-\mathbf{h}'} \mathbf{D}_{\mathbf{h}'} \right) - 4\pi^2 K_h^2 \mathbf{D}_{\mathbf{h}}$$
$$- 4\pi \mathrm{i} (\mathbf{K}_{\mathbf{h}} \cdot \nabla) \mathbf{D}_{\mathbf{h}} + 4\pi^2 k^2 \mathbf{D}_{\mathbf{h}} = 0.$$

1. *First approximation*: Because of the macroscopic character of the variations of $\mathbf{D}_{\mathbf{h}}$ as compared to the microscopic character of the variations of $\exp[-2\pi \mathrm{i} \mathbf{K}_{\mathbf{h}} \cdot \mathbf{r}]$, the amplitude $\mathbf{D}_{\mathbf{h}}$ can be assumed to be a slowly varying function. The first three terms are therefore second-order terms and will be neglected, following Takagi (1969). After reorganization of the remaining terms and using (2.50), the propagation equation may be written as

$$\pi K_h^2 \sum_{\mathbf{h} \neq \mathbf{h}'} \chi_{\mathbf{h}-\mathbf{h}'} \mathbf{D}_{\mathbf{h}'[\mathbf{h}]} + \pi K_h^2 \chi_o \mathbf{D}_{\mathbf{h}} - \pi (K_h^2 - k^2) \mathbf{D}_{\mathbf{h}} = \mathrm{i} K_h \frac{\partial \mathbf{D}_{\mathbf{h}}}{\partial x_h},$$

where $\mathbf{D}_{\mathbf{h}'[\mathbf{h}]}$ means the projection of $\mathbf{D}_{\mathbf{h}}$ on the plane normal to $\mathbf{K}_{\mathbf{h}}$ (see Fig. 5.2) and x_h is a coordinate in direct space in the reflected direction (Figs 11.2(b) and (c)).

2. *Second approximation*: The length of the reflected wavevector K_h inside the crystal is very close to k and the equation may be rewritten, with a small approximation, as

$$\pi k^2 \sum_{\mathbf{h} \neq \mathbf{h}'} \chi_{\mathbf{h}-\mathbf{h}'} \mathbf{D}_{\mathbf{h}'[\mathbf{h}]} + \pi k^2 \chi_o \mathbf{D}_{\mathbf{h}} - \pi (K_h^2 - k^2) \mathbf{D}_{\mathbf{h}} = \mathrm{i} k \frac{\partial \mathbf{D}_{\mathbf{h}}}{\partial x_h}.$$

This equation is used in the following form:

$$\frac{\partial \mathbf{D}_{\mathbf{h}}}{\partial x_h} = -\mathrm{i} k \pi \left[\left(1 + \chi_o - \frac{K_h^2}{k^2} \right) \mathbf{D}_{\mathbf{h}} + \sum_{\mathbf{h} \neq \mathbf{h}'} \chi_{\mathbf{h}-\mathbf{h}'} \mathbf{D}_{\mathbf{h}'[\mathbf{h}]} \right]. \quad (11.4)$$

This is a set of partial differential equations relating the unknown functions $\mathbf{D}_{\mathbf{h}}(\mathbf{r})$ and their derivatives. In the two-beam case, it reduces to two independent sets of two equations, one set for a polarization perpendicular to the $\mathbf{K}_{\mathbf{o}}$, $\mathbf{K}_{\mathbf{h}}$ plane ($C = 1$) and one set for a polarization parallel to that plane ($C = \cos 2\theta_B$):

$$\frac{\partial D_o(\mathbf{r})}{\partial x_o} = -\mathrm{i} \pi k C \chi_{\bar{h}} D_h(\mathbf{r}), \quad (11.5a)$$

$$\frac{\partial D_h(\mathbf{r})}{\partial x_h} = -\mathrm{i} \pi k [C \chi_h D_o(\mathbf{r}) - 2\beta_h D_h(\mathbf{r})], \quad (11.5b)$$

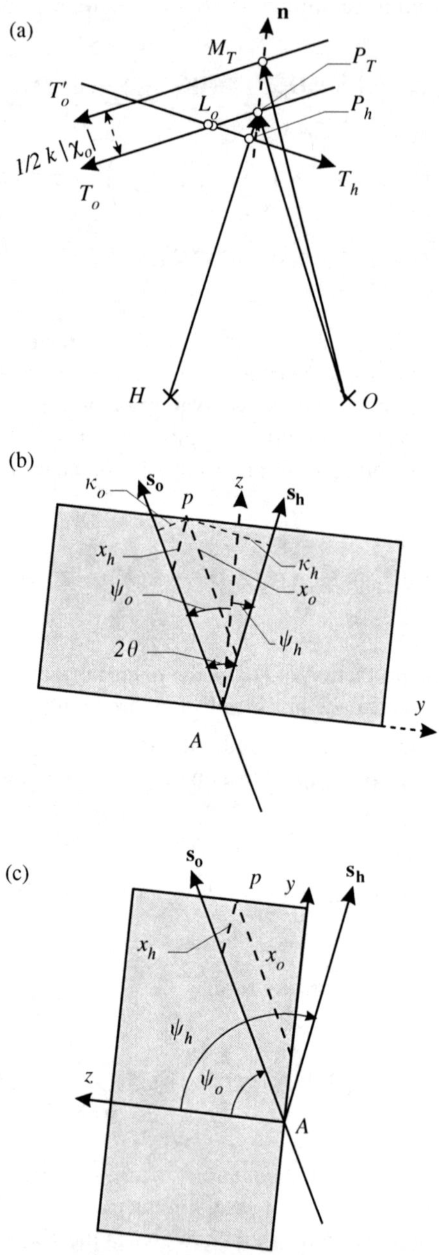

Fig. 11.2. Takagi theory. (a) Reciprocal space (Laue geometry); the tiepoint according to Takagi's theory, P_T, lies on T_o. The extremity, M_T, of the incident wavevector and P_T lie on a normal, **n**, to the crystal surface; L_o is, as usual, the Lorentz point. (b) Direct space (Laue geometry). (c) Direct space (Bragg geometry).

where x_o is a coordinate in direct space along the wavevector, $\mathbf{K_o}$, and the parameter

$$\beta_h = \frac{K_h^2 - k^2(1+\chi_o)}{2k^2} = \frac{K_h^2 - K_o^2}{2k^2} \simeq \frac{K_h - K_o}{k} \tag{11.6}$$

depends on the choice of the couple of vectors $\mathbf{k_o}$, $\mathbf{K_o}$. It is represented geometrically in Fig. 11.2(a). Let, as in the preceding chapters, T_o', T_o and T_h be the traces on the plane of incidence of the tangential planes to the spheres centred at O with radius k, centred at O with radius nk, and centred at H with radius nk, respectively. Let further

$$\mathbf{OM_T} = \mathbf{k_o}; \qquad \mathbf{OP_T} = \mathbf{K_o}; \qquad \mathbf{HP_T} = \mathbf{K_h} = \mathbf{K_o} - \mathbf{OH},$$

where $\mathbf{P_T M_T}$ is along the normal to the crystal surface, in agreement with the boundary condition on the continuity of the tangential components of the wavevectors. The extremity, P_T, of wavevector $\mathbf{K_o}$ lies on T_o, by definition (Fig. 11.2(a)), and is the tiepoint according to Takagi's theory. It can be seen in the figure that

$$\beta_h = \frac{\overline{P_h P_T}}{k} = -\frac{\overline{L_o P_T} \sin 2\theta_B}{k}, \tag{11.7}$$

where P_h is the projection of P_T on T_h. This shows that β_h is proportional to the departure from Bragg's angle of the wavevector $\mathbf{k_o}$. It is often convenient to choose P_T to be confused with L_o; in that case, β_h is zero.

11.2.3 *Boundary conditions for the amplitudes at the entrance surface*

1. *Laue geometry (Fig. 11.2(b))*: The boundary conditions on the entrance surface are

$$\begin{aligned} \mathbf{D_o}(\mathbf{r_s}) &= \mathbf{D_o^{(a)}}(\mathbf{r_s}), \\ \mathbf{D_h}(\mathbf{r_s}) &= 0, \end{aligned} \tag{11.8}$$

where $\mathbf{r_s}$ is a position vector on the surface.

To solve Takagi's equations, it is also necessary to use the boundary conditions for the derivatives. These conditions are

$$\begin{aligned} \left[\frac{\partial D_o}{\partial x_o}\right]_{z=0} &= 0, \qquad \left[\frac{\partial D_o}{\partial x_h}\right]_{z=0} = \frac{\sin 2\theta_B}{\gamma_o}\left[\frac{\partial D_o^{(a)}}{\partial y}\right]_{z=0}, \\ \left[\frac{\partial D_h}{\partial x_o}\right]_{z=0} &= -\mathrm{i}\pi k \frac{\gamma_o}{\gamma_h} C\chi_h D_o^{(a)}(\mathbf{r_s}), \\ \left[\frac{\partial D_h}{\partial x_h}\right]_{z=0} &= -\mathrm{i}\pi k C\chi_h D_o^{(a)}(\mathbf{r_s}), \end{aligned} \tag{11.9}$$

where z is the coordinate perpendicular to the entrance surface. The first and fourth conditions are easily deduced from (11.5) and (11.8). The two others are obtained by derivation of (11.8):

$$\left[\frac{\partial D_h}{\partial y}\right]_{z=0} = 0; \qquad \left[\frac{\partial D_o}{\partial y}\right]_{z=0} = \left[\frac{\partial D_o^{(a)}}{\partial y}\right]_{z=0}, \tag{11.10}$$

where y is a coordinate parallel to the crystal surface (Fig. 11.2(b)). The coordinates, x_o and x_h of a point, p, inside the crystal are related to the orthonormal coordinates y and z by

$$z = x_o \cos\psi_o + x_h \cos\psi_h,$$
$$y = -(x_o \sin\psi_o + x_h \sin\psi_h).$$

One has, along the crystal surface,

$$0 = x_o\gamma_o + x_h\gamma_h,$$
$$y = -x_o\frac{\sin 2\theta_B}{\gamma_h} = x_h\frac{\sin 2\theta_B}{\gamma_o}, \tag{11.11}$$

where γ_o and γ_h have the usual meanings.

The second and third boundary conditions (11.9) are obtained from eqns (11.10) by changing from the y, z to the x_o, x_h coordinate system:

$$\frac{\partial D_o(\mathbf{r_s})}{\partial y} = \frac{\partial D_o(\mathbf{r_s})}{\partial x_o}\frac{\partial x_o}{\partial y} + \frac{\partial D_o(\mathbf{r_s})}{\partial x_h}\frac{\partial x_h}{\partial y} = \frac{\gamma_o}{\sin 2\theta_B}\frac{\partial D_o(\mathbf{r_s})}{\partial x_h}$$

$$\Longrightarrow \quad \frac{\partial D_o(\mathbf{r_s})}{\partial x_h} = \frac{\sin 2\theta_B}{\gamma_o}\frac{\partial D_o^{(a)}(\mathbf{r_s})}{\partial y},$$

$$\frac{\partial D_h(\mathbf{r_s})}{\partial y} = 0 = \frac{\partial D_h(\mathbf{r_s})}{\partial x_o}\frac{\partial x_o}{\partial y} + \frac{\partial D_h(\mathbf{r_s})}{\partial x_h}\frac{\partial x_h}{\partial y}$$

$$\Longrightarrow \quad \frac{\partial D_h(\mathbf{r_s})}{\partial x_o} = -\frac{\partial x_h}{\partial x_o}\frac{\partial D_h(\mathbf{r_s})}{\partial x_h} = -i\pi k\frac{\gamma_o}{\gamma_h}C\chi_h D_o^{(a)}(\mathbf{r_s}).$$

2. *Bragg geometry* (*Fig. 11.2(c)*): The boundary conditions for the amplitudes on the entrance surface are here:

$$\mathbf{D_o}(\mathbf{r_s}) = \mathbf{D_o^{(a)}}(\mathbf{r_s}),$$
$$\mathbf{D_h}(\mathbf{r_s}) = \mathbf{D_h^{(a)}}(\mathbf{r_s}). \tag{11.12}$$

Relation (11.10) still holds and the boundary conditions for the derivatives are obtained in the same way as in the Laue geometry, but are more complicated. It follows that

$$\left[\frac{\partial D_o}{\partial x_o}\right]_{z=0} = -\mathrm{i}\pi k C \chi_{\bar{h}} D_o^{(a)}(\mathbf{r_s}),$$

$$\left[\frac{\partial D_o}{\partial x_h}\right]_{z=0} = -\mathrm{i}\pi k C \chi_{\bar{h}} \frac{\gamma_h}{\gamma_o} D_o^{(a)}(\mathbf{r_s}) + \frac{\sin 2\theta_B}{\gamma_o}\left[\frac{\partial D_o^{(a)}}{\partial y}\right]_{z=0},$$

$$\left[\frac{\partial D_h}{\partial x_o}\right]_{z=0} = -\mathrm{i}\pi k \frac{\gamma_o}{\gamma_h}(C\chi_h D_o^{(a)}(\mathbf{r_s}) - 2\beta_h D_h^{(a)}(\mathbf{r_s})) - \frac{\sin 2\theta_B}{\gamma_h}\left[\frac{\partial D_h^{(a)}}{\partial y}\right]_{z=0}, \quad (11.13)$$

$$\left[\frac{\partial D_h}{\partial x_h}\right]_{z=0} = -\mathrm{i}\pi k(C\chi_h D_o^{(a)}(\mathbf{r_s}) - 2\beta_h D_h^{(a)}(\mathbf{r_s})).$$

11.3 Reduction of Takagi's equations in the plane-wave case

It is interesting to show that system (11.5) reduces to system (5.9) in the case of an incident plane wave. In that case the conditions of the problem, and therefore the solutions, are not space dependent. The Ewald wave must have the form (5.1):

$$\mathbf{D} = \sum_{\mathbf{h}} \mathbf{D_h^{PW}} \exp[-2\pi \mathrm{i} \mathbf{K_h^{PW}} \cdot \mathbf{r}],$$

where $\mathbf{D_h^{PW}}$ is constant and $\mathbf{K_h^{PW}} = \mathbf{OP_{PW}}$ is the real wavevector (Fig. 11.3).

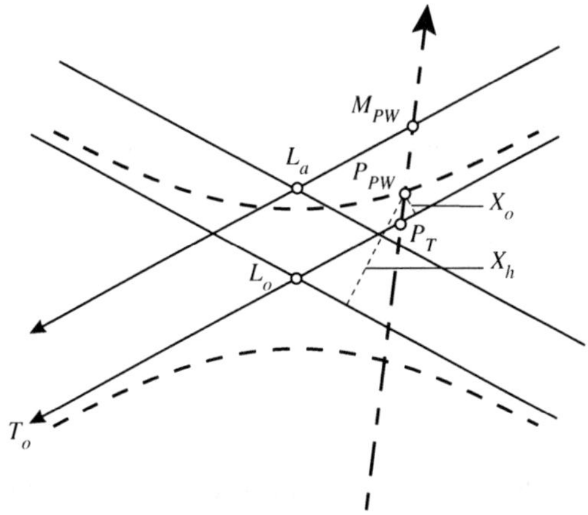

Fig. 11.3. Application of Takagi's theory to the case of an incident plane wave. M_{PW} is the extremity of the incident wavevector; P_{PW} is the tiepoint according to the plane-wave dynamical theory and P_T is the tiepoint according to Takagi's theory.

The expression of **D** can be put in the form (11.1) by writing

$$\mathbf{D} = \sum_{\mathbf{h}} \left\{\mathbf{D}_{\mathbf{h}}^{\mathbf{PW}} \exp\left[-2\pi\mathrm{i}\left(\mathbf{K}_{\mathbf{h}}^{\mathbf{PW}} - \mathbf{K}_{\mathbf{h}}\right)\cdot\mathbf{r}\right]\right\} \exp[-2\pi\mathrm{i}\mathbf{K}_{\mathbf{h}}\cdot\mathbf{r}]$$

$$= \sum_{\mathbf{h}} \mathbf{D}_{\mathbf{h}}(\mathbf{r}) \exp[-2\pi\mathrm{i}\mathbf{K}_{\mathbf{h}}\cdot\mathbf{r}],$$

where $\mathbf{D}_{\mathbf{h}}(\mathbf{r}) = \mathbf{D}_{\mathbf{h}}^{\mathbf{PW}} \exp[-2\pi\mathrm{i}(\mathbf{K}_{\mathbf{h}}^{\mathbf{PW}} - \mathbf{K}_{\mathbf{h}})\cdot\mathbf{r}]$ is the amplitude of the modulated Ewald wave in (11.1). If this expression is substituted in system (11.5), it follows that

$$2\overline{P_T P_{PW}}\,\gamma_o D_o = k\chi_{\bar{h}} C D_h,$$

$$2\overline{P_T P_{PW}}\,\gamma_h D_h = k\chi_h C D_o,$$

where P_T is the tiepoint used in Takagi's theory and γ_o and γ_h have the usual meanings. It can be seen in Fig. 11.3 that $\overline{P_T P_{PW}}\gamma_o = X_o$ and $\overline{P_T P_{PW}}\gamma_h = X_h$. The system reduces thus to

$$2X_o D_o = k\chi_{\bar{h}} C D_h,$$

$$2X_h D_h = k\chi_h C D_o,$$

which is equivalent to system (4.14) for plane waves.

11.4 Absorbing crystals

System (11.5) is still valid in the case of absorbing crystals, but, as in the classical theory, $\mathbf{K}_{\mathbf{o}}$ and $\mathbf{K}_{\mathbf{h}}$ have an imaginary part $\mathbf{K}_{\mathbf{o}}^{\mathbf{i}} = \mathbf{K}_{\mathbf{h}}^{\mathbf{i}}$ which, according to (11.3), is parallel to the normal to the entrance surface. The length of $\mathbf{K}_{\mathbf{o}}$, which has been arbitrarily fixed, can be written as

$$\mathbf{K}_{\mathbf{o}}^{2} = (K_o^r)^2 - (K_o^i)^2 + 2\mathrm{i}\mathbf{K}_{\mathbf{o}}^{\mathbf{r}}\cdot\mathbf{K}_{\mathbf{o}}^{\mathbf{i}} = k^2(1+\chi_o) \simeq (K_o^r)^2 + 2\mathrm{i}\gamma_o K_o^r K_o^i,$$

neglecting K_{oi}^2. The value of the imaginary part of the wavevector is therefore

$$K_o^i \simeq k\frac{\chi_{io}}{2\gamma_o}. \qquad (11.14)$$

The parameter β_h also has an imaginary part which, from this relation, is found to be

$$\beta_{hi} = \frac{\chi_{oi}}{2}\left(\frac{\gamma_h}{\gamma_o} - 1\right). \qquad (11.15)$$

11.5 Analytical resolution of Takagi's equations for perfect crystals

Takagi's equations applied to a spherical wave incident on a perfect crystal lead to the same solution as that derived by Kato using a Fourier expansion. Analytical solutions have also been derived for deformed crystals, but in particular cases only (see Chapter 14). Essentially two methods have been used to solve Takagi's equations.

The first one is due to Takagi (1969). Although only published in 1969, it was already known since 1962 in preprint form. It is based on the transformation of system (11.5) of two coupled first-order partial differential equations into a system of two uncoupled hyperbolic second-order partial differential equations, one for the reflected amplitude, D_h, and the other for the refracted amplitude, D_o. This system is resolved by means of Riemann's functions. This method was used by Authier and Simon (1968), who were acquainted with Takagi's solution, for resolving the equations for an incident spherical wave in the Laue geometry, for a perfect crystal and for a crystal containing a single stacking fault, and by Uragami (1969, 1970, 1971) who resolved the equations in the Bragg geometry. The same results were obtained independently by Afanas'ev and Kohn (1971) for both the Laue and the Bragg geometries. Combinations of Laue and Bragg geometries were treated by Thorkildsen and Larsen who calculated extinction in spheres (Thorkildsen and Larsen 1998*b*,*c*; Larsen and Thorkildsen 1998*c*), in rectangular crystals (Thorkildsen and Larsen 1998*d*, 1999*a*) and in bipyramids (Thorkildsen *et al.* 2001). They also calculated rocking curves from rectangular crystals (Thorkildsen and Larsen 1999*b*).

The other method makes use of integral equations and was developed by Bremer (1984). A very simple derivation based on this concept was recently published (Guigay 1999) and will be described in the next section. It is very interesting in that it does not refer to the results of plane-wave dynamical theory.

11.6 Analytical solution for a point source using the method of integral equations

The derivation due to Guigay is based on the phenomenon of multiple scattering of the incident beam into the reflected direction and *vice versa*. Takagi's equations are transformed into integral equations from which one obtains iteratively the successive terms of the multiple-scattering expansion.

The incident wave is taken to be a point source on the surface and its expression (11.2) is written in terms of a Dirac distribution:

$$\mathcal{D}_o^{(a)}(\mathbf{r}) = \delta(x_h)\exp[-2\pi\mathrm{i}\mathbf{k_o}\cdot\mathbf{r}]D_o^{(a)}. \tag{11.16}$$

The refracted wave *in the absence* of Bragg reflection is

$$\mathcal{D}_o(x_o, x_h) = \delta(x_h)\exp[-2\pi\mathrm{i}\mathbf{K_o}\cdot\mathbf{r}]D_o^{(a)},$$

where the couple $\mathbf{k_o}$, $\mathbf{K_o}$ is chosen in such a way that the term β_h in (11.5) is zero: the refracted wavevector inside the crystal is $\mathbf{K_o} = \mathbf{OL_o}$. System (11.5) reduces to

$$\begin{aligned}\frac{\partial D_o(\mathbf{r})}{\partial x_o} &= -\mathrm{i}\pi kC\chi_{\bar{h}}D_h(\mathbf{r}),\\ \frac{\partial D_h(\mathbf{r})}{\partial x_h} &= -\mathrm{i}\pi kC\chi_h D_o(\mathbf{r}).\end{aligned} \tag{11.17}$$

The next step is to write (11.17) in the form of integral equations but the derivation is different in the transmission and in the reflection geometries and the two cases will be described separately. Integral equations were also used by Werner (1974) to solve the Hamilton–Darwin transfer equations which have a similar mathematical form, but deal with energies and not amplitudes.

11.6.1 *Transmission geometry*

The geometrical situation is described by Fig. 11.2(b). The distribution of non-zero amplitudes is contained within the Borrmann triangle and the coordinates of any point p in that triangle are such that $x_o, x_h \geq 0$. The integral equations equivalent to (11.17) are, here,

$$D_o(x_o, x_h) = \delta(x_h) D_o^{(a)} - \mathrm{i}\pi k C \chi_{\bar{h}} \int_o^{x_o} D_h(s'_o, x_h)\,\mathrm{d}s'_o, \tag{11.18a}$$

$$D_h(x_o, x_h) = -\mathrm{i}\pi k C \chi_h \int_o^{x_h} D_o(x_o, s'_h)\,\mathrm{d}s'_h. \tag{11.18b}$$

At the surface $D_h(x_o, x_h) = 0$ and the zero-order solution of (11.18a) is $D_o^0(x_o, x_h) = \delta(x_h)$. This is put into (11.18b) to obtain the first-order solution, $D_h^1(x_o, x_h)$. The second-order solution, $D_o^2(x_o, x_h)$, is obtained by substituting $D_h^1(x_o, x_h)$ in (11.18a), and so on so that the general expression of the higher-order solutions are

$$D_o^{2n}(x_o, x_h) = -\mathrm{i}\pi k C \chi_{\bar{h}} \int_o^{x_o} D_h^{2n-1}(s'_o, x_h)\,\mathrm{d}s'_o, \tag{11.19a}$$

$$D_h^{2n+1}(x_o, x_h) = -\mathrm{i}\pi k C \chi_h \int_o^{x_h} D_o^{2n}(x_o, s'_h)\,\mathrm{d}s'_h. \tag{11.19b}$$

The successive solutions are

$$D_o^0(x_o, x_h) = \delta(x_h) D_o^{(a)},$$

$$D_h^1(x_o, x_h) = -\mathrm{i}\pi k C \chi_h D_o^{(a)},$$

$$D_o^2(x_o, x_h) = -(\pi k C)^2 \chi_h \chi_{\bar{h}} x_o D_o^{(a)},$$

$$D_h^3(x_o, x_h) = \mathrm{i}(\pi k C)^3 \chi_h^2 \chi_{\bar{h}} x_o x_h D_o^{(a)},$$

$$D_o^4(x_o, x_h) = (\pi k C)^4 \chi_h^2 \chi_{\bar{h}}^2 \frac{x_o^2}{2} x_h D_o^{(a)},$$

$$D_h^5(x_o, x_h) = -\mathrm{i}(\pi k C)^5 \chi_h^3 \chi_{\bar{h}}^2 \frac{x_o^2}{2} \frac{x_h^2}{2} D_o^{(a)},$$

$$\vdots$$

$$D_o^{2n}(x_o, x_h) = -(\pi kC)^{2n}(-\chi_h\chi_{\bar{h}})^n \frac{x_o^n}{n!}\frac{x_h^{n-1}}{(n-1)!}D_o^{(a)}, \tag{11.20a}$$

$$D_h^{2n+1}(x_o, x_h) = -\mathrm{i}(\pi kC)^{2n+1}\chi_h(-\chi_h\chi_{\bar{h}})^n \frac{x_o^n}{n!}\frac{x_h^n}{n!}D_o^{(a)}, \tag{11.20b}$$

$$\vdots$$

The total amplitudes of the refracted and reflected waves are obtained by summing the even and odd terms, respectively:

$$D_o = \sum_0^\infty D_o^{2n}(x_o, x_h), \tag{11.21a}$$

$$D_h = \sum_0^\infty D_h^{2n+1}(x_o, x_h). \tag{11.21b}$$

Now the general terms (11.20a,b) can also be written as

$$D_o^{2n}(x_o, x_h) = \pi kC\sqrt{\chi_h\chi_{\bar{h}}}\sqrt{\frac{x_o}{x_h}}D_o^{(a)} \times \frac{(-1)^n(\zeta/2)^{2n+1}}{n!(n+1)!},$$

$$D_h^{2n+1}(x_o, x_h) = -\mathrm{i}\pi kC\chi_h D_o^{(a)} \times \frac{[-(\zeta/2)^2]^n}{n!n!}$$

with

$$\zeta = 2\pi kC\sqrt{\chi_h\chi_{\bar{h}}}\sqrt{x_o x_h}. \tag{11.22}$$

Substituting into (11.21a,b), it follows that

$$D_o = \delta(x_h)D_o^{(a)} - \pi kC\sqrt{\chi_h\chi_{\bar{h}}}\sqrt{\frac{x_o}{x_h}}D_o^{(a)}$$
$$\times\left[\frac{\zeta}{2} - \left(\frac{\zeta}{2}\right)^3\frac{1}{1!2!} + \cdots + (-1)^n\left(\frac{\zeta}{2}\right)^{2n+1}\frac{1}{n!(n+1)!} + \cdots\right],$$

$$D_h = -\mathrm{i}\pi kC\chi_h D_o^{(a)} \times \left[1 - \left(\frac{\zeta}{2}\right)^2 + \left(-\frac{\zeta^2}{4}\right)^n\frac{1}{n!n!} + \cdots\right].$$

One recognizes in the summations the Taylor expansions of $J_1(\zeta)$ and $J_0(\zeta)$, respectively, and it follows, finally, for the expressions of the refracted and reflected amplitudes that

$$D_o = \delta(x_h)D_o^{(a)} - \pi kC\sqrt{\chi_h\chi_{\bar{h}}}\sqrt{\frac{x_o}{x_h}}D_o^{(a)} \times J_1(\zeta), \tag{11.23a}$$

$$D_h = -\mathrm{i}\pi kC\chi_h D_o^{(a)} \times J_0(\zeta). \tag{11.23b}$$

In order to relate these expressions to the corresponding expressions (10.22) and (10.24) obtained using Kato's theory, one must relate the coordinates x_o and x_h to the distances κ_o and κ_h of p from the sides of the Borrmann triangle. It can be seen in Fig. 11.2(b) that

$$x_h = \frac{\kappa_o}{\sin 2\theta_B}; \qquad x_o = \frac{\kappa_h}{\sin 2\theta_B}.$$

Substituting into (11.22) it follows that ζ is identical to expression (10.27b) and that (11.23a,b) are identical, to a proportionality factor, to Kato's expressions. It should be noted that the term $\delta(x_h)D_o^{(a)}$ represents that part of the direct beam which is outside the domain of reflection of the crystal.

11.6.2 *Reflection geometry*

The geometrical situation is described by Fig. 11.2(c). The equation of the surface of the crystal is

$$x_o = |\gamma| x_h$$

and the coordinates of a point within the partial Borrmann triangle are such that

$$x_o \geq 0; \qquad x_h \geq 0; \qquad x_o \geq |\gamma| x_h.$$

The system of integral equations to resolve is the same as that, (11.18), in the transmission geometry, except that the limits of integration are different:

$$D_o(x_o, x_h) = \delta(x_h)D_o^{(a)} - \mathrm{i}\pi kC\chi_{\bar{h}} \int_{|\gamma|x_h}^{x_o} D_h(s'_o, x_h)\,\mathrm{d}s'_o,$$

$$D_h(x_o, x_h) = -\mathrm{i}\pi kC\chi_h \int_o^{x_h} D_o(x_o, s'_h)\,\mathrm{d}s'_h.$$

The successive solutions are

$$D_o^0(x_o, x_h) = \delta(x_h)D_o^{(a)},$$

$$D_h^1(x_o, x_h) = -\mathrm{i}\pi kC\chi_h D_o^{(a)},$$

$$D_o^2(x_o, x_h) = -(\pi kC)^2\chi_h\chi_{\bar{h}}(x_o - |\gamma|x_h)D_o^{(a)},$$

$$D_h^3(x_o, x_h) = \mathrm{i}(\pi kC)^3\chi_h^2\chi_{\bar{h}}\left(x_o x_h - |\gamma|\frac{x_h^2}{2}\right)D_o^{(a)},$$

$$D_o^4(x_o, x_h) = (\pi kC)^4\chi_h^2\chi_{\bar{h}}^2\frac{x_o - |\gamma|x_h}{2}x_o x_h D_o^{(a)},$$

$$\vdots$$

$$D_o^{2n}(x_o, x_h) = -(\pi kC)^{2n}(-\chi_h\chi_{\bar{h}})^n(x_o - |\gamma|x_h)\frac{x_o^n}{n!}\frac{x_h^{n-1}}{(n-1)!}D_o^{(a)}, \quad (11.24a)$$

$$D_h^{2n+1}(x_o, x_h) = -\mathrm{i}(\pi kC)^{2n+1}\chi_h(-\chi_h\chi_{\bar{h}})^n \times\left[\frac{x_o^n}{n!}\frac{x_h^n}{n!}D_o^{(a)} - |\gamma|\frac{x_o^{n-1}}{(n-1)!}\frac{x_h^{n+1}}{(n+1)!}\right], \quad (11.24b)$$

$$\vdots$$

The total refracted and reflected amplitudes are given as in the case of the transmission geometry, by eqns (11.21a,b). After substitution of (11.24a,b), one may recognize in the summations the Taylor expansions of Bessel functions and it follows, finally, that

$$D_h = -\mathrm{i}\pi kC\chi_h D_o^{(a)} \times \left[J_0(\zeta) + |\gamma|\frac{x_h}{x_o}J_2(\zeta)\right],$$
$$D_o = \delta(x_h)D_o^{(a)} - \pi kC\sqrt{\chi_h\chi_{\bar{h}}}\frac{1}{\sqrt{x_o x_h}}(x_o - |\gamma|x_h)D_o^{(a)} \times J_1(\zeta). \quad (11.25)$$

After replacing x_o and x_h by κ_o and κ_h, the reader will check that expressions (11.25) are identical to (10.35) obtained with Kato's theory. They are also identical with the expressions obtained by Uragami (1969, 1970) and by Afanas'ev and Kohn (1971) using an analytical solution of Takagi's equations.

11.7 Analytical resolution of Takagi's equations using the Riemann function

11.7.1 *Hyperbolic nature of Takagi's equations*

When the crystal is perfect and the coefficients χ_h are constant, the system (11.5) of two first-order partial differential equations can be replaced by a system of two independent second-order partial differential equations. By differentiating (11.5a) with respect to x_h and (11.5b) with respect to x_o, it follows that

$$\frac{\partial^2 D_o}{\partial x_o\,\partial x_h} - 2\pi\mathrm{i}k\beta_h\frac{\partial D_o}{\partial x_o} + \pi^2k^2C^2\chi_h\chi_{\bar{h}}D_o = 0, \quad (11.26a)$$

$$\frac{\partial^2 D_h}{\partial x_o\,\partial x_h} - 2\pi\mathrm{i}k\beta_h\frac{\partial D_h}{\partial x_o} + \pi^2k^2C^2\chi_h\chi_{\bar{h}}D_h = 0. \quad (11.26b)$$

As mentioned in Section 11.2.2, the parameter β_h can always be put equal to zero in the perfect crystal case by an appropriate choice of the couple of wavevectors, $\mathbf{k_o}$,

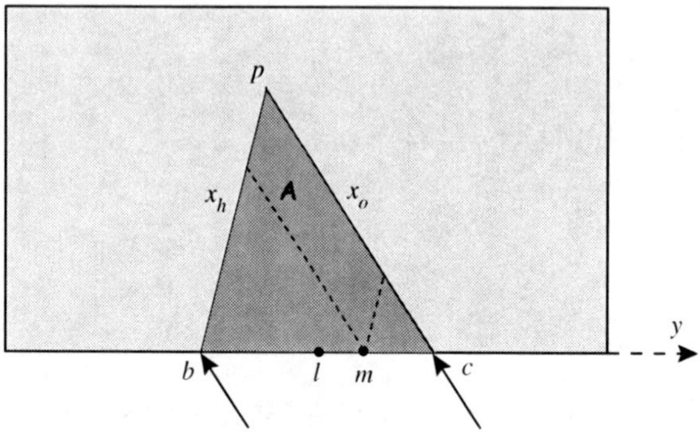

Fig. 11.4. Integration of Takagi's equations using Riemann's method in the inverted Borrmann triangle $\widehat{pbc}$; l is the midpoint of $\overline{bc}$.

$\mathbf{K_o}$. With this simplification, the system (11.26) is further simplified:

$$\frac{\partial^2 D_o}{\partial x_o\,\partial x_h} + \pi^2 k^2 C^2 \chi_h \chi_{\bar{h}} D_o = 0, \qquad (11.27a)$$

$$\frac{\partial^2 D_h}{\partial x_o\,\partial x_h} + \pi^2 k^2 C^2 \chi_h \chi_{\bar{h}} D_h = 0. \qquad (11.27b)$$

These equations are of the form (11.A7)[2] with $D = E = 0$ and are of the hyperbolic type. We shall solve them using Riemann's method described in the Appendix to this chapter. We shall make use here of the x_o, x_h coordinate system, instead of the orthonormal x, y system used in the Appendix and, because of their property of invariance under a change of coordinates, the characteristics are sets of lines parallel to the x_o, x_h directions, namely to the incidence and reflection directions.

Let p be the point inside the crystal where the amplitudes D_h and D_o are being calculated and let b and c be two points of the entrance surface such that cp and bp are parallel to x_o and x_h, respectively (Fig. 11.4). The integration area $\mathcal{A}$ used in the Appendix (Fig. 11.8) is here the inverted Borrmann triangle $\widehat{pbc}$.

11.7.2 *General expression of the reflected and refracted waves*

The general expressions of the amplitude of the reflected and refracted waves, D_h and D_o, at p are given by (11.A10), and, according to that relation, they only depend on the boundary conditions along $\overline{bc}$. Substituting successively in (11.A10) u and its derivatives by D_h and D_o and their derivatives along $\overline{bc}$ which are given in (11.9),

[2] Equation numbers starting with 11.A refer to the Appendix at the end of the chapter.

and noting that

$$\frac{\partial x_h}{\partial y} = \frac{\gamma_o}{\sin 2\theta_B},$$
$$\frac{\partial x_o}{\partial y} = -\frac{\gamma_h}{\sin 2\theta_B},$$

where y is a coordinate along $\overline{bc}$, replacing s in (11.A10), it follows that

$$D_h(p) = \frac{-\mathrm{i}\pi k C \chi_h \gamma_o}{\sin 2\theta_B} \int_{\overline{bc}} D_o^{(a)}(y) v(y)\, \mathrm{d}y, \tag{11.28a}$$

$$D_o(p) = \frac{1}{2}[D_o(b) + D_o(c)] - \frac{1}{2 \sin 2\theta_B} \times \int_{\overline{bc}} \left[D_o^{(a)}(y) \left(\gamma_h \frac{\partial v}{\partial x_o} + \gamma_o \frac{\partial v}{\partial x_h} \right) - v(y) \frac{\partial D_o^{(a)}(y)}{\partial x_h} \right] \mathrm{d}y, \tag{11.28b}$$

where $v(y)$ is the value of the Riemann function at a point source, m, on the entrance surface, between b and c and of abscissa y.

11.7.3 *Determination of the Riemann function*

The Riemann function $v(y)$ satisfies the adjoint differential expression defined by (11.A8) which reduces to

$$M[v] = \frac{\partial^2 v}{\partial x_o\, \partial x_h} + \pi^2 k^2 C^2 \chi_h \chi_{\bar{h}} v = 0. \tag{11.29}$$

According to Section 11.9.2, it satisfies the following conditions:

$$\begin{aligned} \frac{\partial v}{\partial x_o} &= 0 \quad \text{along } \overline{cp}, \\ \frac{\partial v}{\partial x_h} &= 0 \quad \text{along } \overline{bp}, \\ v(p) &= 1. \end{aligned} \tag{11.30}$$

The conditions (11.30) show that $v(p) = 1$ along $\overline{cp}$ and $\overline{bp}$.

The Riemann function will be calculated by first transforming it into a function depending on a single variable. Let x_o and x_h be the coordinates of m, and X_o and X_h be the coordinates of **mp**:

$$X_o = x_o - x_o(p),$$
$$X_h = x_h - x_h(p).$$

Let us now change coordinates from (x_o, x_h) to (X_o, X_h). The Riemann function $v(X_o, X_h)$ satisfies

$$M[v] = \frac{\partial^2 v}{\partial X_o\, \partial X_h} + \pi^2 k^2 C^2 \chi_h \chi_{\bar{h}} v = 0,$$

$$\frac{\partial v}{\partial X_o} = 0 \quad \text{along } \overline{cp},$$

$$\frac{\partial v}{\partial X_h} = 0 \quad \text{along } \overline{bp},$$

$$v(0, 0) = 1.$$

Let us now introduce a new function, $w(X_o, X_h)$, defined by

$$w(X_o, X_h) = v\left(\alpha X_o, \frac{X_h}{\alpha}\right),$$

where α is a constant. It is also a Riemann function since

$$\frac{\partial w}{\partial X_o} = \alpha \frac{\partial v(\alpha X_o, X_h/\alpha)}{\partial X_o} = \frac{\partial v}{\partial X_o}$$

$$\Longrightarrow \quad \frac{\partial w}{\partial X_o} = 0 \quad \text{when } \frac{\partial v}{\partial X_o} = 0,$$

$$\frac{\partial w}{\partial X_h} = \frac{1}{\alpha} \frac{\partial v(\alpha X_o, X_h/\alpha)}{\partial X_h/\alpha} = \frac{\partial v}{\partial X_h},$$

$$\frac{\partial w^2}{\partial X_o\, \partial X_h} = \frac{\partial v^2(\alpha X_o, X_h/\alpha)}{\partial(\alpha X_o)\partial(X_h/\alpha)},$$

$$w(0, 0) = v(0, 0) = 1.$$

From these relations, one may deduce that

$$w(X_o, X_h) = v\left(\alpha X_o, \frac{X_h}{\alpha}\right) = v(X_o, X_h).$$

The Riemann functions v and w do not depend on α and therefore only depend on the product $X_o \times X_h$:

$$w(X_o, X_h) = w(X_o \cdot X_h); \qquad v(X_o, X_h) = v(X_o \cdot X_h).$$

One may thus set

$$v = v(\xi)$$

with $\xi = X_o \cdot X_h = [x_o - x_o(p)][x_h - x_h(p)]$. The Riemann function therefore satisfies the differential equation

$$M[v] \equiv \xi \frac{\mathrm{d}^2 v}{\mathrm{d}\xi^2} + \frac{\mathrm{d}v}{\mathrm{d}\xi} + \pi^2 k^2 C^2 \chi_h \chi_{\bar{h}} v = 0.$$

Finally, let us set

$$\begin{aligned}\zeta &= 2\pi k|C|\sqrt{\xi\chi_h\chi_{\bar{h}}}\\ &= \frac{2\pi k|C|\sqrt{\chi_h\chi_{\bar{h}}}}{\sin 2\theta_B}\sqrt{\gamma_o\gamma_h}\sqrt{L^2 - y^2},\end{aligned}$$

where $2L = \overline{bc}$. It has the same meaning as in (11.23), with the difference that it now refers to the inverted Borrmann triangle $\widehat{pbc}$. It follows that

$$M[v] \equiv \frac{\mathrm{d}^2 v}{\mathrm{d}\zeta^2} + \frac{1}{\zeta}\frac{\mathrm{d}v}{\mathrm{d}\zeta} + v = 0.$$

The solution of this equation is

$$v(\zeta) = J_0(\zeta). \tag{11.31}$$

11.7.4 *General solution of Takagi's equations*

The general solution of Takagi's equations is obtained by substituting expression (11.31) of the Riemann function in eqns (11.28):

$$D_h(p) = \frac{-\mathrm{i}\pi kC\chi_h\gamma_o}{\sin 2\theta_B}\int_{-L}^{+L} D_o^{(a)}(y)J_0(\zeta)\,\mathrm{d}y, \tag{11.32a}$$

$$D_o(p) = D_o^{(a)}(c) - \frac{\pi t(p)}{2L\Lambda_L}\int_{-L}^{+L} D_o^{(a)}(y)\sqrt{\frac{L+y}{L-y}}\,J_1(\zeta)\,\mathrm{d}y. \tag{11.32b}$$

The very important result of (11.32) is that the amplitudes of the reflected and refracted waves at any point p inside the crystal only depend on the distribution of the incident amplitude and its derivatives along that part of the entrance surface which constitutes the base of an inverted Borrmann triangle having p as apex.

11.8 Analytical solution for an incident spherical wave using the method of Riemann functions

11.8.1 *The incident wave is a point source located on the entrance surface*

If the incident wave is a spherical wave, it is emitted by a point source. If that single point source is located on the entrance surface, the incident amplitude can be approximated by a Dirac distribution $D_o^{(a)}(y)\delta(y)$. The expressions (11.32a,b) reduce then to (10.21) and (10.24) given by Kato's theory.

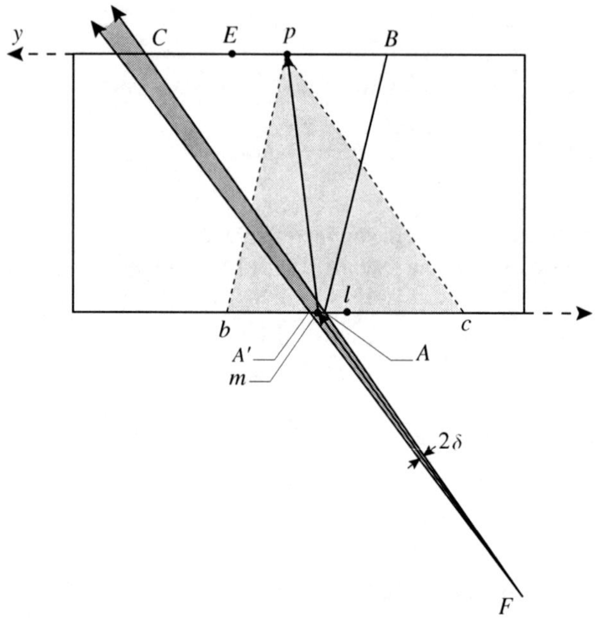

Fig. 11.5. Integration of Takagi's equations for an incident spherical wave. $\overline{AA'}$: trace of the spherical wave on the entrance surface of the crystal for a divergence of $2\delta_{os}$ ($2\delta_{os}$, FWHM of the rocking curve); the amplitude at p is obtained by integration over $\overline{bc}$; $\widehat{pbc}$: inverted Borrmann triangle; l: midpoint of $\overline{bc}$.

11.8.2 *The incident wave is a point source located away from the entrance surface*

When the incident wave is emitted from a point source located away from the entrance surface, one should distinguish different situations depending on the distance *FA* from the point source, F, to the crystal or, rather, depending on the ratio of the trace $AA' = FA \times 2\delta/\cos\theta_B$ of the incident spherical wave on the entrance surface to the base $bc = 2t\,\tan\,\theta_B$ (t, crystal thickness, symmetrical reflection) of an inverted Borrmann triangle $\widehat{pbc}$ where p is a point on the exit surface (Fig. 11.5):

1. $AA' \gg bc$: The wave incident on a point m of the entrance surface can be approximated by a locally plane wave. This is often the case with synchrotron radiation, as has been pointed out in Section 10.2.3. For instance, at the topography beamline ID19 of ESRF, the source-to-specimen distance is about 150 m. For an X-ray energy of 30 keV, silicon and 220, AA' is of the order of 900 μm while bc is 22 μm for a 100 μm thick crystal and 220 μm for a 1 mm thick crystal. The concept of locally plane wave (Mocella *et al.* 1999, 2000) is therefore applicable here.

2. $AA' \approx bc$: The situation is in that case more complicated and requires a special treatment. This was done by Aristov *et al.* (1980) who studied experimentally and

theoretically the fringes obtained with a silicon or germanium wedge-shaped crystal upon which fell the beam emitted by a microfocus rotating anode situated at about 1 m from the crystal. In the very thin part of the crystal, for $t \ll t_s$ (t_s, a certain critical thickness), $AA' \gg bc$ and the observed fringes correspond to the plane-wave case; for the very thick parts of the crystal ($t \gg t_s$), where $AA' \ll bc$, the observed fringes correspond to the spherical-wave case; in the intermediary range, $t \approx t_s$, $AA' \approx bc$ and a new type of fringes appear. For silicon, 220 and AuLα radiation, the critical thickness t_s is 70 μm.

3. $AA' \ll bc$: This is the case in the usual X-ray topography experiments with a conventional X-ray source and a source-to-specimen distance of about half a metre. For instance, for silicon, 220 and Mo$K\alpha$, AA' is of the order of 5 μm while bc is 376 μm for a 1 mm crystal. This case was discussed theoretically by Authier and Simon (1968) and Takagi (1969). As has been shown in the preceding section the total amplitude arriving at a point p of the exit surface (Fig. 11.5) is obtained by integrating eqns (11.32) over the base of the inverted Borrmann triangle pbc. The general expression of a spherical wave is

$$\mathcal{D}_{\mathbf{o}}^{(\mathbf{a})}(\mathbf{r}) = \mathbf{D}_{\mathbf{o}}^{(\mathrm{a})} \frac{\exp(-2\pi\, ikr)}{4\pi r}, \tag{11.33a}$$

where $\mathbf{r}$ is the position vector of m. Expression (11.33a) can also be written, in order to be of the form (11.2), as

$$\mathcal{D}_{\mathbf{o}}^{(\mathbf{a})}(\mathbf{r}) = \mathbf{D}_{\mathbf{o}}^{(\mathrm{a})} \frac{\exp[-2\pi\,\mathrm{i}(kr - \mathbf{k_o}\cdot\mathbf{r})]}{4\pi r} \exp[-2\pi\,\mathrm{i}\mathbf{k_o}\cdot\mathbf{r}]. \tag{11.33b}$$

The angle, $\Delta\theta$, between the running wavevector, $\mathbf{k}$, and $\mathbf{k_o}$ which are respectively parallel to **Fm** and **FA**, is very small since the width of the reflection domain is very small and the scalar product $\mathbf{k_o}\cdot\mathbf{r}$ can be expanded and approximated by

$$\mathbf{k_o}\cdot\mathbf{r} \simeq kr\left(1 - \frac{\Delta\theta^2}{2}\right).$$

It follows that

$$kr - \mathbf{k_o}\cdot\mathbf{r} = kr\frac{\Delta\theta^2}{2}.$$

The angle $\Delta\theta$ is given by

$$\Delta\theta = \frac{\gamma_o\overline{Am}}{r_o} = \frac{\gamma_o(y - y_o)}{r_o},$$

where $r_o = \overline{FA}$, $y = \overline{Im}$ and $y_o = \overline{Ep}$ are the coordinates of m and p along the entrance and exit surfaces, respectively; I is the middle point of $\overline{bc}$ and E that of $\overline{BC}$. After substituting in (11.33b) the expression of the incident wave can be written as

$$\mathcal{D}_{\mathbf{o}}^{(\mathbf{a})}(y) = \mathbf{D}_{\mathbf{o}}^{(\mathbf{a})} \frac{\exp\{-2\pi \mathrm{i}[k\gamma_o^2 (y - y_o)^2/2r_o]\}}{4\pi r} \exp[-2\pi \mathrm{i}\mathbf{k_o}\cdot\mathbf{r}]. \tag{11.33c}$$

This expression can now be substituted in eqns (11.32). The integrations are performed by the stationary phase method and the result is

$$D_h^{SW}(p) = -\mathrm{i}\frac{D_o^{(a)}}{4r_o} \frac{\sqrt{kr_o}C\sqrt{\chi_h\chi_{\bar{h}}}\exp(-\mathrm{i}\pi/4)}{\sin 2\theta_B} J_0(\zeta), \tag{11.34a}$$

$$D_o^{SW}(p) = \frac{D_o^{(a)}}{4\pi r_o}\left\{\exp\left[-\pi\mathrm{i}\frac{k\gamma_o^2(L-y_o)^2}{r_o}\right]\right.$$

$$\left. -\sqrt{\gamma}\frac{\sqrt{kr_o}C\sqrt{\chi_h\chi_{\bar{h}}}\exp(-\mathrm{i}\pi/4)}{\sin 2\theta_B}\sqrt{\frac{L+y_o}{L-y_o}}J_1(\zeta)\right\}, \tag{11.34b}$$

where $y_o = \overline{Ep}$ is the coordinate of p and E is the midpoint of $\overline{BC}$.

The first term in (11.34b) represents the direct beam. The corresponding intensities are

$$I_h^{SW}(p) = I_o^{(a)}\frac{\pi^2 kr_o C^2\chi_h\chi_{\bar{h}}}{\sin^2 2\theta_B} J_0^2(\zeta), \tag{11.35a}$$

$$I_o^{SW}(p) = I_o^{(a)}\left\{1 + \gamma\frac{\pi^2 kr_o C^2\chi_h\chi_{\bar{h}}}{\sin^2 2\theta_B}\frac{L+y_o}{L-y_o}J_1^2(\zeta)\right.$$

$$-2\pi\sqrt{\gamma}\cos\left[\pi\frac{k\gamma_o^2(y-y_o)^2}{r_o} - \frac{\pi}{4}\right]$$

$$\left.\times\frac{\sqrt{kr_o}C\sqrt{\chi_h\chi_{\bar{h}}}\exp(-\mathrm{i}\pi/4)}{\sin 2\theta_B}\sqrt{\frac{1+Y}{1-Y}}J_1(\zeta)\right\} \tag{11.35b}$$

with

$$I_o^{(a)} = \left|\frac{D_o^{(a)}}{4\pi r_o}\right|^2.$$

11.8.3 *Conservation of energy*

Making use of the following properties of Bessel functions,

$$\int_{-L}^{+L} J_0^2\left(B\sqrt{L^2-y_o^2}\right)\mathrm{d}y_o = \frac{1}{B}\int_0^{2BL} J_0(\zeta)\,\mathrm{d}\zeta,$$

$$\int_{-L}^{+L}\frac{L+y_o}{L-y_o}J_1^2\left(B\sqrt{L^2-y_o^2}\right)\mathrm{d}y_o = 2L - \frac{1}{B}\int_0^{2BL} J_0(\zeta)\,\mathrm{d}\zeta,$$

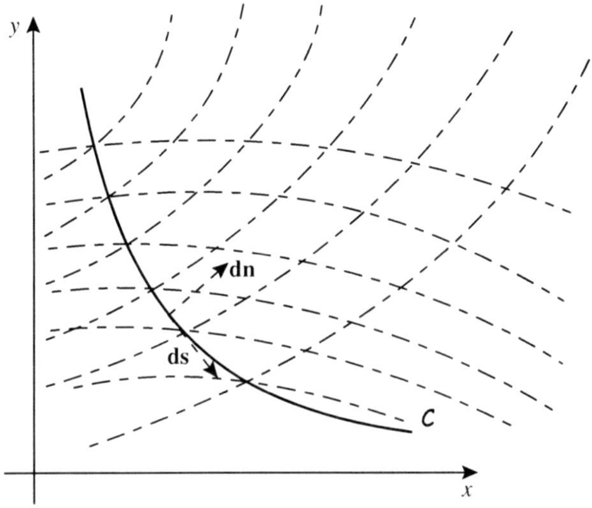

Fig. 11.6. Characteristics of a linear second-order partial differential equation.

it can be shown that

$$2L\gamma_o = \gamma_o \int_{-L}^{+L} I_o \, \mathrm{d}y + \gamma_h \int_{-L}^{+L} I_h \, \mathrm{d}y,$$

which states the conservation of energy in the case of a non-absorbing crystal.

Appendix: Hyperbolic partial differential equations

Characteristics

Riemann's method for integrating hyperbolic differential is described, for instance, in Sommerfeld (1949) or Courant and Hilbert (1962); Sommerfeld's presentation will be followed here. The most general form of a linear partial differential equation of second order is

$$L[u] \equiv A\frac{\partial^2 u}{\partial x^2} + 2B\frac{\partial^2 u}{\partial x\,\partial y} + C\frac{\partial^2 u}{\partial y^2} + D\frac{\partial u}{\partial x} + E\frac{\partial u}{\partial y} + Fu = 0, \qquad (11.A1)$$

which can also be written in the more general way:

$$A\frac{\partial^2 u}{\partial x^2} + 2B\frac{\partial^2 u}{\partial x\,\partial y} + C\frac{\partial^2 u}{\partial y^2} = \Phi\left(u, \frac{\partial u}{\partial x}, \frac{\partial u}{\partial y}, x, y\right). \qquad (11.A2)$$

Let us consider a given curve, C, in the x, y plane along which u and its partial derivative, $\partial u/\partial n$, in the direction of the normal, **n**, to the curve are prescribed

(Fig. 11.6) and the following problem: is it possible to find a solution of (11.A2) satisfying these initial conditions? First of all, if u is known along $\mathcal{C}$, so is $\partial u/\partial s$ and therefore also $\partial u/\partial x$ and $\partial u/\partial y$. Let us set, following Sommerfeld (1949),

$$p = \frac{\partial u}{\partial x}; \qquad q = \frac{\partial u}{\partial y};$$

$$r = \frac{\partial^2 u}{\partial x^2}; \qquad s = \frac{\partial^2 u}{\partial x\,\partial y}; \qquad t = \frac{\partial^2 u}{\partial y^2}.$$

After rewriting (11.A2) using these notations, and after differentiation of p and q, one obtains the following system:

$$\begin{array}{llll} Ar & +2Bs & +Ct & = \Phi, \\ r\,\mathrm{d}x & +s\,\mathrm{d}y & & = \mathrm{d}p, \\ & s\,\mathrm{d}x & +t\,\mathrm{d}y & = \mathrm{d}q. \end{array}$$

One has three linear equations for the determination of r, s and t. This is only possible if the associated determinant

$$\begin{vmatrix} A & 2B & C \\ \mathrm{d}x & \mathrm{d}y & 0 \\ 0 & +\mathrm{d}x & \mathrm{d}y \end{vmatrix} = A\,\mathrm{d}y^2 - 2B\,\mathrm{d}x\,\mathrm{d}y + C\,\mathrm{d}x^2$$

is different from zero. For a given point on $\mathcal{C}$, there are two values of $\mathrm{d}y/\mathrm{d}x$ for which the determinant is zero. They are solutions of

$$A\left(\frac{\mathrm{d}y}{\mathrm{d}x}\right)^2 - 2B\left(\frac{\mathrm{d}y}{\mathrm{d}x}\right) + C = 0. \tag{11.A3}$$

There are therefore two families of curves along which r, s and t cannot be determined. They are called *characteristics* and the solvability of the problem requires that $\mathcal{C}$ is nowhere tangent to the characteristics.

One distinguishes the different types of partial differential equations according to the nature of the solutions of (11.A3):

1. $B^2 - AC > 0$: There are two real families of characteristics. This is the *hyperbolic* type. It is the case of the wave equation, but also of the equation of the vibrating string.
2. $B^2 - AC < 0$: The two families of characteristics are complex conjugate. This is the *elliptic* type. It is the case of the potential equation (Laplace equation).
3. $B^2 - AC = 0$: There is one real family of characteristics. This is the *parabolic* type. It is the case of the heat transfer equation.

One important property of the characteristics that is made use of in the resolution of Takagi's equations is their invariance under a transformation of coordinates. In particular, in the case of the hyperbolic type, if $f(x, y) =$ constant and $g(x, y) =$ constant are the equations of the two families of characteristics, one may make the following change of coordinates:

$$\xi = f(x, y); \qquad \eta = g(x, y).$$

It is possible to show (Sommerfeld 1949) that under this change of coordinates, (11.A2) is transformed into

$$\frac{\partial^2 u}{\partial \xi \, \partial \eta} = \Psi\left[u, \frac{\partial u}{\partial \xi}, \frac{\partial u}{\partial \eta}, \xi, \eta\right]; \tag{11.A4}$$

this is called the *normal* form of the hyperbolic partial differential equation. The characteristics are then simply the lines $\xi =$ constant and $\eta =$ constant in the ξ, η plane.

Adjoint differential expression

The resolution of the linear second-order partial differential equations requires at this point the introduction of an *adjoint differential form*, $M[v]$, such that $vL[u] - uM[v]$ is integrable, that is of the form

$$vL[u] - uM[v] = \frac{\partial X(x, y)}{\partial x} + \frac{\partial Y(x, y)}{\partial y}; \tag{11.A5}$$

it is easily shown that

$$M[v] = \frac{\partial^2 Av}{\partial x^2} + 2\frac{\partial^2 Bv}{\partial x \, \partial y} + \frac{\partial^2 Cv}{\partial y^2} - \frac{\partial Dv}{\partial x} - \frac{\partial Ev}{\partial y} + Fv,$$

$$X(x, y) = A\left(v\frac{\partial u}{\partial x} - u\frac{\partial v}{\partial x}\right) + B\left(v\frac{\partial u}{\partial y} - u\frac{\partial v}{\partial y}\right) + \left(D - \frac{\partial A}{\partial x} - \frac{\partial B}{\partial y}\right)uv,$$

$$Y(x, y) = B\left(v\frac{\partial u}{\partial x} - u\frac{\partial v}{\partial x}\right) + C\left(v\frac{\partial u}{\partial y} - u\frac{\partial v}{\partial y}\right) + \left(E - \frac{\partial B}{\partial x} - \frac{\partial C}{\partial y}\right)uv.$$

It should be noted that the relation between $L[u]$ and $M[v]$ is reciprocal: $L[u]$ is the adjoint of $M[v]$.

Let us now consider a certain area $\mathcal{A}$ bound by a curve $\mathcal{C}$ in the x, y plane (Fig. 11.7) and integrate (11.A5) over $\mathcal{A}$. Applying Gauss' theorem, it follows that

$$\int_{\mathcal{A}} (vL[u] - uM[v])\, \mathrm{d}\sigma = \int_{\mathcal{A}} \left(\frac{\partial X(x, y)}{\partial x} + \frac{\partial Y(x, y)}{\partial y}\right) \mathrm{d}\sigma$$

$$= \int_{\mathcal{C}} [X \cos(\mathbf{n}, \mathbf{x}) + Y \cos(\mathbf{n}, \mathbf{y})]\, \mathrm{d}s, \tag{11.A6}$$

where $\mathbf{n}$ is the normal to curve $\mathcal{C}$ at p (Fig. 11.7).

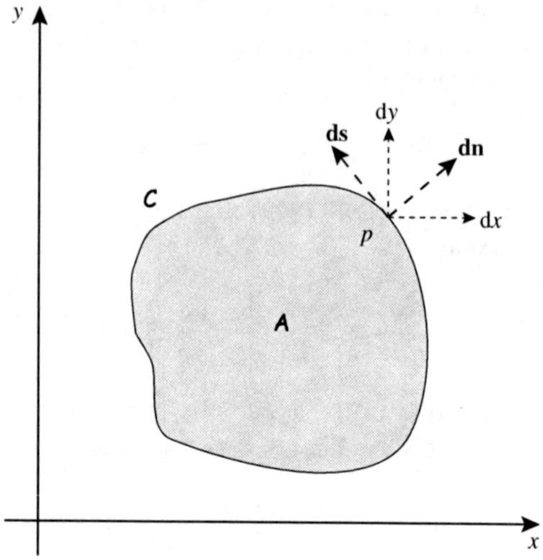

Fig. 11.7. Integration area for Green's theorem.

This result is known as *Green's theorem*, valid for the three types of partial differential equations. In the case of the hyperbolic type, which is the one of interest in our case, one obtains the normal form (11.A4) of the equation by setting $A = C = 0$, $B = \frac{1}{2}$:

$$L[u] = \frac{\partial^2 u}{\partial x\, \partial y} + D\frac{\partial u}{\partial x} + E\frac{\partial u}{\partial y} + Fu = 0 \tag{11.A7}$$

and its adjoint differential expression is

$$M[v] = \frac{\partial^2 v}{\partial x\, \partial y} - D\frac{\partial v}{\partial x} - E\frac{\partial v}{\partial y} + Fv = 0 \tag{11.A8}$$

and the two functions $X(x, y)$ and $Y(x, y)$ are

$$\begin{aligned} X(x, y) &= \frac{1}{2}\left(v\frac{\partial u}{\partial y} - u\frac{\partial v}{\partial y}\right) + Duv, \\ Y(x, y) &= \frac{1}{2}\left(v\frac{\partial u}{\partial x} - u\frac{\partial v}{\partial x}\right) + Euv. \end{aligned} \tag{11.A9}$$

When the hyperbolic equation is written in its normal form, the characteristics are the lines $x = \xi =$ constant and $y = \eta =$ constant. The function v is then determined according to Riemann's method with the following conditions:

- the area $\mathcal{A}$ is limited by two straight segments pb and pc respectively parallel to the characteristics and a curve $\mathcal{C}$ (Fig. 11.8)

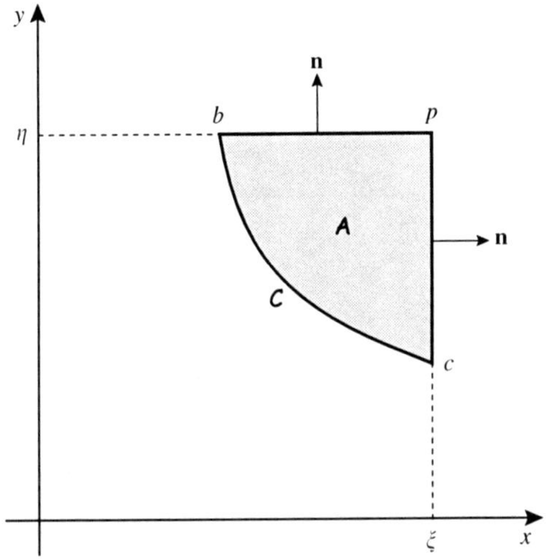

Fig. 11.8. Integration of a hyperbolic partial differential equation using Riemann's method. $\overline{bp}$ and $\overline{cp}$ are along the characteristics η = constant and ξ = constant, respectively.

- $M[v] = 0$ within the area $\mathcal{A}$
- $\partial v/\partial y - Dv = 0$ along the characteristic pb ($\cos(\mathbf{n}, \mathbf{x}) = 0$; $\cos(\mathbf{n}, \mathbf{y}) = 1$)
- $\partial v/\partial x - Ev = 0$ along the characteristic pc ($\cos(\mathbf{n}, \mathbf{x}) = 1$; $\cos(\mathbf{n}, \mathbf{y}) = 0$)
- $v(p) = 1$
- the curve $\widehat{bc}$ is never parallel to Ox or Oy.

Let us now integrate (11.A6) along the path $bcpb$. Remembering that $L[u] = 0$ and $M[v] = 0$, it follows that

$$0 = \int_{\widehat{bc}} \cdots \, \mathrm{d}s + \int_{\overline{cp}} \cdots \, \mathrm{d}s + \int_{\overline{pb}} \cdots \, \mathrm{d}s,$$

where ... means $\left(\partial X(x, y)/\partial x + \partial Y(x, y)/\partial y\right)$. The integration is performed by parts and one finds, using the conditions which have been set, that

$$u(p) = \frac{1}{2}u(c)v(c) + \frac{1}{2}u(b)v(b) - \frac{1}{2}\int_{\widehat{bc}} \left\{ \left[v\frac{\partial u}{\partial y} - u\frac{\partial v}{\partial y} + 2Duv \right] \frac{\partial y}{\partial s} \right.$$
$$\left. - \left[v\frac{\partial u}{\partial x} - u\frac{\partial v}{\partial x} + 2Euv \right] \frac{\partial x}{\partial s} \right\} \mathrm{d}s. \qquad (11.\mathrm{A}10)$$

The very important result is that the value of u at p is simply obtained by an integration along curve C along which initial values are known, once the Riemann function v has been determined. The determination of the Riemann function depends on the boundary conditions and an example has been given in Section 11.7 in the case of the Takagi equations.

12

Ray tracing in perfect crystals

12.1 Ray tracing

The dynamical theory of diffraction has been developed in the preceding chapters for two types of incident waves: a plane wave and a spherical wave. It has been seen that the basic properties of the dynamical diffraction of waves by matter are interpreted in terms of wavefields. The importance of this notion, introduced by Ewald (1913) as a mathematical tool, comes from the fact that it also covers a physical reality. This is revealed, for instance, by the Borrmann effect or by the spherical-wave Pendellösung fringes. Real waves, whether they are emitted by an X-ray tube or in a synchrotron and have been conditioned by various optical devices, are, however, neither ideal plane waves nor spherical waves. They have a more complicated nature. The real crystals in which their propagation is studied in practice are never ideally perfect. The diffraction properties of real waves in real crystals can nevertheless also be interpreted in terms of generalized wavefields. The propagation of X-rays in optical devices and in deformed crystals is in many ways analogous to that of visible light: one may observe focusing effects, bending of trajectories in regions of variable index of refraction, diffraction by slits, etc. These effects can, however, only be observed close to a Bragg reflection. The fundamental difference with usual optics comes from the angular amplification effect which takes place in the crystal within the reflection domain and which is due to the curvature of the dispersion surface. Far from a Bragg reflection, the index of refraction of matter for X-rays is so close to one that these effects are very difficult to observe.

The notion of *light rays* used in geometrical optics applies, strictly speaking, to plane waves but may be generalized to wavepackets or wavebundles (Ewald 1958). It will be extended to packets of wavefields propagating in a crystal. These wavepackets (outside the crystal) or packets of wavefields (within a crystal under conditions of Bragg diffraction) have a certain angular width in reciprocal space and a certain lateral extension, or spatial coherence, in direct space. If they are limited by a slit or if they propagate in a region where there is a strong gradient of index of refraction, Fresnel or Fraunhofer diffraction effects occur; if this is not the case one will speak of *ray tracing*, which simply means determining wavefield trajectories in perfect or slightly deformed crystals. This will be discussed in the case of perfect crystals in the present chapter and in the case of slightly deformed crystals in Chapter 13. Diffraction effects by a slit will be discussed in Section 12.11 and diffraction effects in a highly deformed crystal will be discussed in Chapter 14.

12.2 The structure of real waves

Any wave is represented by a certain distribution of the electric displacement which may be expanded as a Fourier integral:

$$\mathcal{D}_{\mathbf{o}}^{(\mathrm{a})}(\mathbf{r},t) = \iiint \mathbf{D}_{\mathbf{o}}^{(\mathrm{a})}(\mathbf{K}) \exp\{2\pi \mathrm{i}[\nu(\mathbf{K})t - \mathbf{K}\cdot\mathbf{r}]\} \, \mathrm{d}^3 K, \tag{12.1}$$

where $\nu(\mathbf{K})$ shall be assumed to be a constant.

The incident wave can thus always be interpreted as a superposition of plane waves, called *wavepacket*. It is characterized by its *spatial extension* in real space (Δx, Δy, Δz) or by its *spread* in reciprocal space (ΔK_x, ΔK_y, ΔK_z) which are related through the basic relations:

$$\begin{aligned} \Delta x \cdot \Delta K_x &\sim 1, \\ \Delta y \cdot \Delta K_y &\sim 1, \\ \Delta z \cdot \Delta K_z &\sim 1. \end{aligned} \tag{12.2}$$

Ox, **Oy**, **Oz** and $\mathbf{OK_x}$, $\mathbf{OK_y}$, $\mathbf{OK_z}$ are coordinate axes in direct and reciprocal space, respectively: $\mathbf{OK_x}$ is parallel to the reciprocal lattice vector, $\mathbf{h} = \mathbf{OH}$ (Fig. 12.1), $\mathbf{OK_y}$ is perpendicular to the diffraction plane, $\mathbf{OK_z}$ is parallel to the intersection of

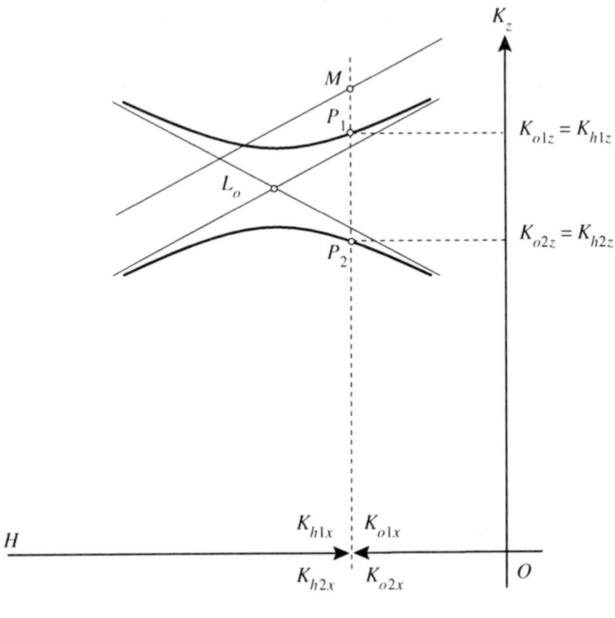

Fig. 12.1. Components of the wavevectors of the waves excited by a single plane wave in the symmetric Laue geometry.

the diffraction plane and the lattice planes, and **Ox**, **Oy**, **Oz** are respectively parallel to $\mathbf{OK_x}$, $\mathbf{OK_y}$, $\mathbf{OK_z}$.

The physical meaning of these relations will be discussed, following Balibar and Malgrange (1975), in two cases:

1. The wavepacket is a superposition of separate plane waves. This will simply serve as a means for reformulating some of the classical results of the dynamical theory.
2. The wavepacket is represented by a *continuous* distribution of wavevectors.

12.3 Wavepackets made of the superposition of separate plane waves

Let us first consider a wavepacket made of *two* plane waves. ΔK_x is in this case simply $|K_{1x} - K_{2x}|$ and the interference between the two plane waves results in *beats*, that is a special type of interference effect characterized by the existence of a double periodicity in each direction: a *fast* oscillation corresponding to the mean periodicity of the waves and a *modulation* of the amplitude of this oscillation whose spatial frequency is the difference of the spatial frequencies of the two components (see Fig. 11.1). The superposition of *two*, and only two, sinusoidal phenomena is therefore again a sinusoidal phenomenon extending from $-\infty$ to $+\infty$ with an amplitude modulation spread over $\Delta K_x^{-1} = |K_{1x} - K_{2x}|^{-1}$ in the **Ox** direction.

Let us now consider two wavefields generated inside a crystal by a plane wave whose incidence is close to the Bragg incidence. There are therefore *four* waves inside the crystal which constitute a wavepacket (we consider here one direction of polarization only). For the simplicity of the argument, we shall limit ourselves to the symmetrical Laue geometry; the figure represents the K_x and K_z components of the four waves associated with the two tiepoints, P_1 and P_2, excited by the incident plane wave. Figure 12.2 is a three-dimensional representation where the horizontal plane is the K_x, K_z plane in which the vectors $\mathbf{K_o}$ and $\mathbf{K_h} = \mathbf{K_o} - \mathbf{h}$ are drawn from the same origin and the coordinates in the third direction are proportional to the amplitudes of the four waves. It can be seen in Fig. 12.1 that

$$\begin{aligned} \Delta K_x &= K_{o1x} - K_{h1x} = K_{o2x} - K_{h2x} = |\mathbf{OH}| = \frac{1}{d_{hkl}}, \\ \Delta K_z &= K_{o1z} - K_{o2z} = K_{h1z} - K_{h2z} = \overline{P_1 P_2} = \frac{1}{\Lambda}, \end{aligned} \tag{12.3}$$

where Λ is the Pendellösung distance defined in Section 4.8.6 and d_{hkl} is the lattice spacing divided by the order of the reflection.

The corresponding modulations in direct space are, respectively,

$$\Delta x = d_{hkl}; \qquad \Delta z = \Lambda.$$

The modulation along the **Ox** direction describes the standing wave pattern discussed in Chapter 7 while the modulation along the **Oz** direction describes the Pendellösung

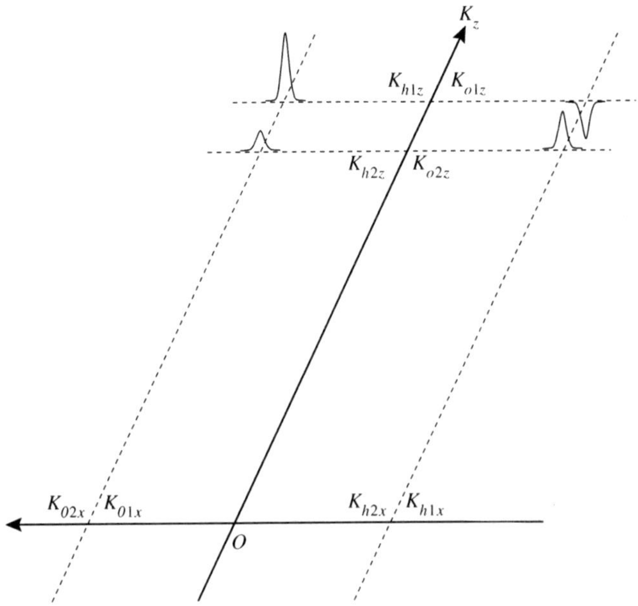

Fig. 12.2. Amplitudes of the waves excited by a single incident plane wave in the symmetric Laue geometry.

effect. Since D_{o1} and D_{o2} are of the same sign while D_{h1} and D_{h2} are of opposite signs, the corresponding Pendellösung fringe systems are shifted by $\Lambda/2$ with respect to one another. This modulation and the existence of wavefields are characteristic of perfect or nearly perfect crystals, but disappear in highly distorted regions.

It has been noticed in Section 4.4 that the general solution of the propagation equation is a linear combination (4.5a),

$$\mathbf{D} = \sum_j A_j \mathbf{D_j},$$

of particular solutions $\mathbf{D_j}$ which are Ewald waves (5.1). The modulation described above may be exhibited by writing the general solution in the following way:

$$\mathbf{D} = \sum_{\mathbf{h}} \mathbf{D_h} \exp(-2\pi \mathrm{i} \boldsymbol{\Gamma}_{\mathbf{h}} \cdot \mathbf{r}) \exp(-2\pi \mathrm{i} \mathbf{K_h} \cdot \mathbf{r}), \tag{12.4}$$

where $|\boldsymbol{\Gamma}_{\mathbf{o}}|^{-1}$ and $|\boldsymbol{\Gamma}_{\mathbf{h}}|^{-1}$ are of the order of Λ, and K_o^{-1} and K_h^{-1} of the order of λ.

Expansion (12.4) which exhibits both the Λ and the λ modulations shows that the general solution may be interpreted as a modulated Ewald wave. The spread of this wavepacket in the x and z directions is infinite.

12.4 Wavepackets made of a continuous distribution of wavevectors

Let us now consider a continuous distribution of wavevectors with a finite bandwidth ΔK_x in reciprocal space. Its spread in direct space is also finite. This is due to the fact that interference between waves of wavevectors very close to one another will give rise to destructive interference everywhere except in a very localized region. At a given point of abscissa x in direct space the resultant complex amplitude is the sum of all the plane-wave components. It depends on the phase relationships between these complex components. If they are in phase at a point of abscissa x_o, the phase difference between the two points is $2\pi\,\Delta K_x \cdot (x - x_o)$ for a given component. For a bandwidth ΔK_x, the phase interval covered by these components is therefore $2\pi\,\Delta K_x \cdot (x - x_o)$ and the resultant amplitude is non-zero only if

$$2\pi\,\Delta K_x \cdot (x - x_o) \leq 2\pi.$$

The amplitude distribution in direct space is therefore zero *outside* a *wavefront* of lateral width $\Delta x \sim \Delta K_x^{-1}$. The propagation velocity of the wavepacket, that is its group velocity $\mathbf{V_g}$, is the gradient $\overrightarrow{\nabla}_K[\nu(\mathbf{K})]$ of the frequency with respect to the coordinates in reciprocal space.

Let us now assume that this wavepacket is incident on a crystal. Each of its plane-wave components induces a modulated Ewald wave such as (12.4). Since they all have similar characteristics, their resultant can be written as a packet of wavefields:

$$\mathbf{D}(\mathbf{r}) = \sum_j \int \Big[\sum_{\mathbf{h}} \mathbf{D_{hj}}(\mathbf{K_o}) \exp(2\pi i \mathbf{h} \cdot \mathbf{r})\Big] \exp\{2\pi i[\nu(\mathbf{K})t - \mathbf{K_o} \cdot \mathbf{r}]\}\, \mathrm{d}^3K, \tag{12.5}$$

where $\mathbf{h}$ is the reciprocal lattice vector.

Ewald (1958) and Wagner (1959*a*,*b*) have shown that the group velocity is here also given by $\overrightarrow{\nabla}_K[\nu(\mathbf{K})]$. Their argument is as follows. The expression between brackets in (12.5) varies much more slowly than the exponential outside the brackets. The integral therefore takes non-negligible values only when the argument of this exponential is stationary:

$$\overrightarrow{\nabla}_K[\nu(\mathbf{K})t - \mathbf{K_o} \cdot \mathbf{r}] = 0,$$

that is, at points such that

$$\mathbf{r} = t\,\overrightarrow{\nabla}_K[\nu(\mathbf{K})]. \tag{12.6}$$

This expression gives the energy distribution in direct space of the wavepacket. Its time derivative is the group velocity

$$\mathbf{V_g} = \overrightarrow{\nabla}_K[\nu(\mathbf{K})]. \tag{12.7}$$

The direction of propagation is therefore along the normal to the surface

$$\nu(\mathbf{K}) = \text{constant}$$

for each one of the j packets of wavefields, namely along the normal to the dispersion surface. It will be shown directly in the next section that this direction coincides with that of the Poynting vector. Far from Bragg's incidence, the group velocity reduces to the phase velocity.

As in the case of an incident plane wave, (12.5) can be written as a modulated Ewald wavepacket:

$$\mathbf{D}(\mathbf{r}) = \sum_{\mathbf{h}} \int \mathbf{D_h}(\eta) \exp(-2\pi \mathrm{i}\boldsymbol{\Gamma}_{\mathbf{h}} \cdot \mathbf{r}) \exp(-2\pi \mathrm{i}\mathbf{K_h} \cdot \mathbf{r})\, \mathrm{d}\eta, \tag{12.8}$$

where $|\boldsymbol{\Gamma}_{\mathbf{h}}|^{-1} \sim \Lambda$, η is the usual deviation parameter (5.18a) and the integration in K space has been reduced to a one-dimensional integral by limiting our study to a given plane of incidence. All the components of the incident wavepacket have the same wave number, $k = \nu c^{-1}$. The integration is performed along the intersection of the sphere of radius k centred in O and the plane of incidence, approximated by its tangent, T_o' (Fig. 12.3).

The integration is carried over the width $\Delta\eta = \Delta k/(k\delta_{os})$ of the wavepacket where δ_{os} is the half-width of the rocking curve (5.18c). If this width is sufficiently narrow,

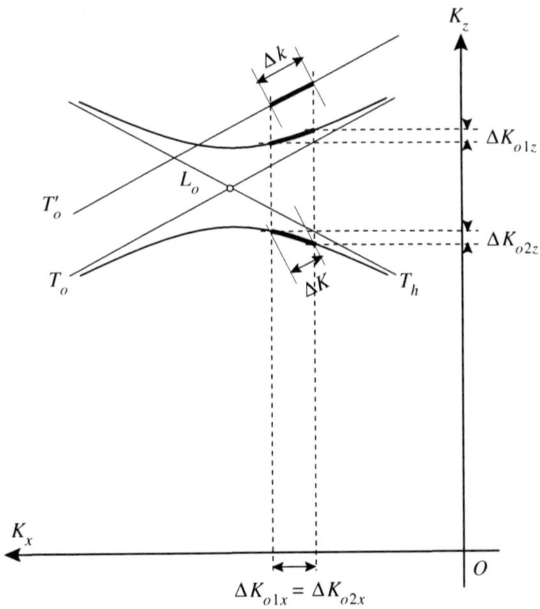

Fig. 12.3. Packets of the wavefields excited by an incident wavepacket. Δk: width of the incident wavepacket; ΔK: width of the packet of excited wavefields.

that is if $\Delta\eta$ is much smaller than 1, $\mathbf{\Gamma_h}$ and $\mathbf{K_h}$ may be considered to be constant; it follows that

$$\mathbf{D(r)} = \sum_{\mathbf{h}} \mathbf{B_h}(\eta, \Delta\eta) \exp(-2\pi i \mathbf{\Gamma_h} \cdot \mathbf{r}) \exp(-2\pi i \mathbf{K_h} \cdot \mathbf{r}), \qquad (12.9)$$

where $\mathbf{B_h}(\eta, \Delta\eta)$ depends both on the mean direction and the structure of the incident wavepacket.

It may thus be concluded that the modulated Ewald wavepackets behave roughly like the Ewald wavepackets excited by an incident plane wave and that they have the same propagation properties.

Figure 12.3 shows the widths ΔK_{ojx} and ΔK_{ojz} ($j = 1$ or 2) of the packets of wavefields excited in the crystal by the incident wavepacket of width Δk. The extension of these packets of wavefields is also limited in direct space, through the relations

$$\Delta x \cdot \Delta K_x \sim 1; \qquad \Delta z \cdot \Delta K_z \sim 1.$$

What is of special interest here is the lateral extension of the beam in direct space in the direction normal to the direction of propagation, namely its lateral spatial coherence. It can be seen in Fig. 12.3 that the width ΔK of the packet of wavefields varies along the dispersion surface: due to the curvature, it is much smaller than Δk near the apices of the hyperbola, while it is equal to it near the asymptotes. The divergence of the beam in the crystal will therefore be much larger for wavefields propagating along the lattice planes than for wavefields propagating near the sides of the Borrmann triangle. This effect, called *angular amplification* (Authier 1970), will be discussed in Section 12.6.

12.5 Group velocity and Poynting vector

In order to calculate the group velocity (12.7), one must first find the expression of $\nu(\mathbf{K})$. This can be done using the fundamental equations of dynamical theory (5.9):

$$\begin{aligned} [K_o^2 - k^2(1+\chi_o)]D_o - k^2 C \chi_{\bar{h}} D_h &= 0, \\ -k^2 C \chi_h D_o + [K_h^2 - k^2(1+\chi_o)]D_h &= 0. \end{aligned} \qquad (12.10)$$

Let us set, using (2.37),

$$F_o' = k^2 \chi_o = -\frac{R F_o}{\pi V},$$

$$F_h' = k^2 \chi_h = -\frac{R F_h}{\pi V}.$$

These quantities do not depend on the wavelength or on the frequency. For the set of fundamental equations to have a solution different from the trivial one, its determinant should be zero. It follows that

$$f(\nu, K_o) \equiv k^4 - k^2(K_h^2 + K_o^2 - 2F_o') + (K_h^2 - F_o')(K_o^2 - F_o') \\ - C^2 F_h' F_{\bar{h}}' = 0. \tag{12.11}$$

This is a relation between the frequency, $\nu = kc$ and K_o. The group velocity is obtained by taking the gradient in K space of (12.11). It follows that

$$\vec{\nabla}_K(f) \equiv -\frac{k}{c}\vec{\nabla}_K(\nu)[(K_o^2 - k^2 - F_o') + (K_h^2 - k^2 - F_o')] \\ + \mathbf{K_o}(K_h^2 - k^2 - F_o') + \mathbf{K_h}(K_o^2 - k^2 - F_o') = 0,$$

and, using (12.10),

$$\mathbf{V_g} = \vec{\nabla}_K(\nu) = c\frac{\mathbf{s_o}|D_o|^2 + \mathbf{s_h}|D_h|^2}{|D_o|^2 + |D_h|^2},$$

where $\mathbf{s_o}$ and $\mathbf{s_h}$ are, as usual, unit vectors in the incident and reflected directions, respectively. This expression confirms that the group velocity is indeed parallel to the Poynting vector, which is given by (5.11).

Equation (12.11) is biquadratic and its solutions are

$$k^2 = \frac{\nu^2}{c^2} = \frac{1}{2}\left[K_h^2 + K_o^2 - 2F_o' \pm \sqrt{(K_h^2 - K_o^2)^2 + 4C^2 F_h' F_{\bar{h}}'}\right],$$

and, after replacing $\mathbf{K_h}$ by $\mathbf{K_o} - \mathbf{h}$,

$$\nu^2 = c^2\left[K_o^2 - \mathbf{h}\cdot\left(\mathbf{K_o} - \frac{\mathbf{h}}{2}\right) - F_o' \pm \sqrt{\left[\mathbf{h}\cdot\left(\mathbf{K_o} - \frac{\mathbf{h}}{2}\right)\right]^2 + C^2 F_h' F_{\bar{h}}'}\right]. \tag{12.12}$$

At constant energy, ν is constant and (12.12) is the equation of the dispersion surface. It was first given in this form by Penning and Polder (1961).

12.6 Angular amplification

An incident wavepacket of width $\Delta k = k\Delta\theta$ excites two packets of wavefields inside the crystal, one on each branch of the dispersion surface (Fig. 12.4(a)). Let $\Delta\alpha_j$ be the angular spread of their paths in direct space (Fig. 12.4(b)). It varies quite strongly depending on the average value of the deviation parameter of the incident wavepacket, and the ratio

$$\mathcal{A} = \left|\frac{\mathrm{d}\alpha}{\mathrm{d}(\Delta\theta)}\right| \tag{12.13}$$

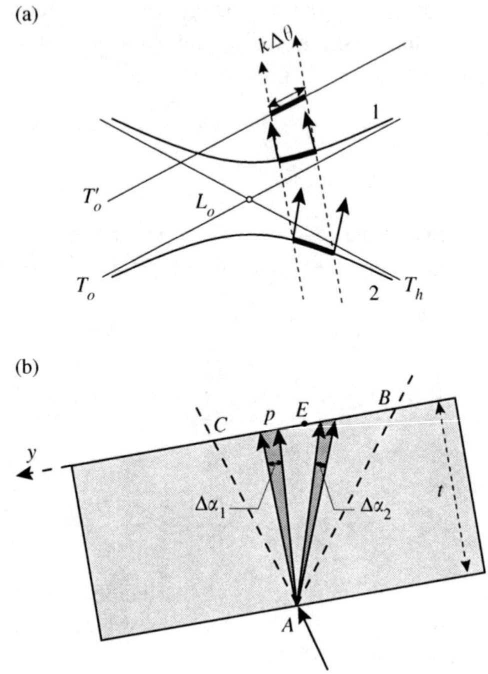

Fig. 12.4. Packets of wavefields excited by an incident wavepacket of angular width $\Delta\theta$. (a) Reciprocal space. (b) Direct space: $\Delta\alpha_1$, $\Delta\alpha_2$ are the divergences of the paths of the wavepackets excited on branches 1 and 2, respectively, by the incident wavepacket.

called *angular amplification* is much larger in the centre of the reflection domain than on the edges.

In a similar way, the mean deviation parameter of the wavepacket arriving at a point, p, of the exit surface (Fig. 12.4(b)) does not vary uniformly with the abscissa, $y = \overline{Ep}$, of p (E is the midpoint of the base, BC, of the Borrmann triangle). Its variations in terms of the reduced coordinate $Y = y/y_{\mathrm{Max}}$, where $y_{\mathrm{Max}} = \overline{EC}$ (10.A3), are given by (10.A5):

$$\eta = \frac{\mp Y}{\sqrt{1 - Y^2}}$$

They are represented in Fig. 12.5 for branch 1. It can be seen that they are very slow in the centre and very rapid at the edges.

The amplification ratio is calculated by writing it in the form

$$\mathcal{A} = \left|\frac{\mathrm{d}\alpha}{\mathrm{d}Y}\right|\left|\frac{\mathrm{d}Y}{\mathrm{d}\eta}\right|\left|\frac{\mathrm{d}\eta}{\mathrm{d}(\Delta\theta)}\right|. \tag{12.14}$$

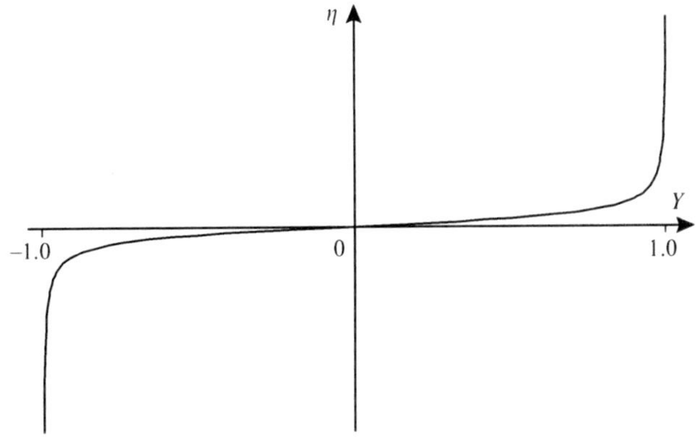

Fig. 12.5. Deviation parameter, η, of the wavefield arriving at point p of the exit surface (branch 1), as a function of its relative coordinate, Y; arbitrary units.

Using (10.A4), (10.A6) and (5.18), it follows, in the transmission geometry, that

$$\mathcal{A} = \frac{\Lambda_o}{\lambda}\frac{\sin^2 2\theta_B}{2\gamma_o\gamma_h^2}\frac{\cos^2(\alpha+\psi_n)}{(\eta^2+1)^{3/2}} = \frac{\Lambda_o}{\lambda}\frac{\sin^2 2\theta_B}{2\gamma_o\gamma_h^2}\frac{\cos^2 a}{(\eta^2+1)^{3/2}}. \tag{12.15a}$$

In the case of reflection geometry, (10.A6) is replaced with

$$\tan a_o = \frac{\sin 2\psi_n}{2\gamma_o|\gamma_h|} + \frac{\sin 2\theta_B}{2\gamma_o|\gamma_h|}\frac{\eta}{\sqrt{\eta^2-1}}$$

and (12.15a) is replaced with

$$\mathcal{A} = \frac{\Lambda_o}{\lambda}\frac{\sin^2 2\theta_B}{2\gamma_o|\gamma_h|^2}\frac{\cos^2(\alpha+\psi_n)}{(\eta^2-1)^{3/2}}. \tag{12.15b}$$

In the case of a symmetrical reflection, this expression reduces to

$$\mathcal{A} = \frac{\Lambda_o}{\lambda}\frac{2\sin^2\theta_B\cos\theta_B}{(\eta^2+\cos^2\theta_B)\sqrt{1+\eta^2}} \tag{12.16a}$$

for transmission geometry and to

$$\mathcal{A} = \frac{\Lambda_o}{\lambda}\frac{2\sin^2\theta_B\cos\theta_B}{(\eta^2-\sin^2\theta_B)\sqrt{\eta^2-1}} \tag{12.16b}$$

for reflection geometry ($|\eta| > 1$).

The variations of the amplification ratio with the deviation parameter are represented in Fig. 12.6 in the case of silicon, 220 and Mo$K\alpha$ and a symmetrical Laue

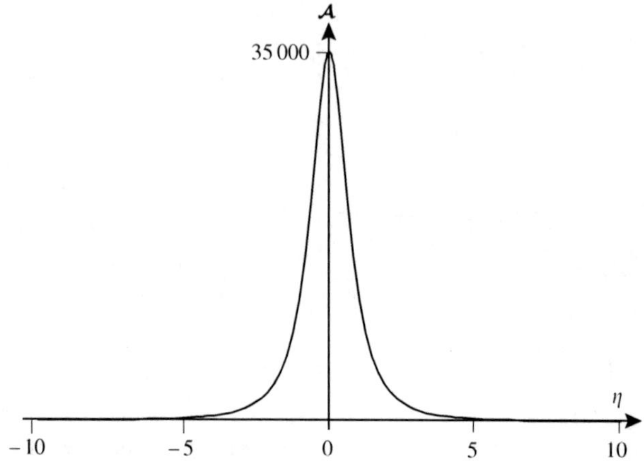

Fig. 12.6. Variations of the amplification ratio, $\mathcal{A}$, with the deviation parameter. Symmetric Laue geometry; silicon, 220 reflection, Mo$K\alpha$.

geometry. It is maximum at the centre of the reflection domain, for $\eta = 0$ ($Y = 0$), and is then

$$\mathcal{A}_{\text{Max}} = \frac{\Lambda_o}{\lambda} \frac{2\sin^2\theta_B}{\cos\theta_B} = \frac{\pi V}{2d_{hkl}^2 R|C|\sqrt{F_h F_{\bar{h}}}}. \tag{12.16c}$$

This maximum is of the order of 10^4 (35 000 in the example of Fig. 12.6). It does not depend on the wavelength and is larger for low-order reflections and for reflections with a weak structure factor. Table 12.1 shows the values of $\mathcal{A}_{\text{Max}}$ for different reflections of silicon.

The value of $\mathcal{A}$ falls off very rapidly on each side of the maximum. In that same example, it is 35 per cent of the maximum value for $|\eta| = 1.0$, that is half of the rocking curve width, and 10 per cent of the maximum value for $|\eta| = 1.89$. The amplification ratio is thus very large only within a narrow angular range of the order of the rocking curve width. This has very important consequences for the distribution of the wavefield paths within the Borrmann triangle, as will be shown in the following sections.

According to (12.15) and (12.16), far from the reflection domain, for $|\eta| \to \infty$, the amplification ratio is seen to tend towards 0. This is of course not possible as it should tend towards 1! The reason is that the derivation has been made with the assumption that the asymptotes to the dispersion surface are planes while they are actually spheres, as has been discussed in Chapter 8. In order to overcome this difficulty, the amplification ratio can be expressed in terms of the radius of curvature, $\mathcal{R}$, of the dispersion surface. For an incident wavepacket of width in reciprocal space $MM' = \mathrm{d}k$, exciting a packet of wavefields of width $PP' = \mathrm{d}K$ on the dispersion

Table 12.1 Maxima of the amplification ratio for different reflections of silicon

hkl	$\sin\theta_B/\lambda$	F_{hkl}	$\mathcal{A}_{\mathrm{Max}}$
111	0.159	60.55	1.5×10^4
220	0.260	67.84	3.57×10^4
311	0.305	44.0	7.57×10^4
400	0.368	56.6	8.55×10^4
331	0.401	38.5	1.49×10^5
422	0.451	49.7	1.46×10^5
511	0.478	33.2	2.46×10^5
440	0.52	43.3	2.24×10^5
444	0.638	33.61	4.32×10^5
642	0.689	29.86	5.68×10^5
660	0.781	23.9	9.12×10^5
555	0.797	16.20	1.40×10^6
844	0.902	17.68	1.64×10^6
880	1.042	14.3	2.71×10^6

surface, one has (Fig. 12.7)

$$\begin{aligned} k &= \frac{\mathrm{d}k}{\mathrm{d}(\Delta\theta)} \quad \text{(radius of the Ewald sphere)}, \\ \mathcal{R} &= \frac{\mathrm{d}K}{\mathrm{d}\alpha} \quad \text{(radius of the dispersion surface)}. \end{aligned} \tag{12.17}$$

Let **n** be the direction of the normal to the dispersion surface, and M'' and P'' the projections of M and P on the parallel to **n** drawn from M', respectively. From the figure it is readily seen that

$$\begin{aligned} \overline{MM''} &= \overline{MM'}\cos\psi_o = \gamma_o\,\mathrm{d}k \\ &= \overline{PP''} = \overline{PP'}\cos(\alpha+\psi_n) = \mathrm{d}K\cos(\alpha+\psi_n), \end{aligned} \tag{12.18}$$

where ψ_n is, as usual, the angle between **n** and the reflecting planes. From (12.13), (12.17) and (12.18), it follows that

$$\begin{aligned} \mathcal{R} &= k\frac{\mathrm{d}K}{\mathrm{d}k}\frac{\mathrm{d}(\Delta\theta)}{\mathrm{d}\alpha} \\ &= \frac{k\gamma_o}{\mathcal{A}\cos(\alpha+\psi_n)}, \end{aligned}$$

from which one deduces that

$$\mathcal{A} = \frac{k\gamma_o}{\mathcal{R}\cos(\alpha+\psi_n)} \tag{12.19a}$$

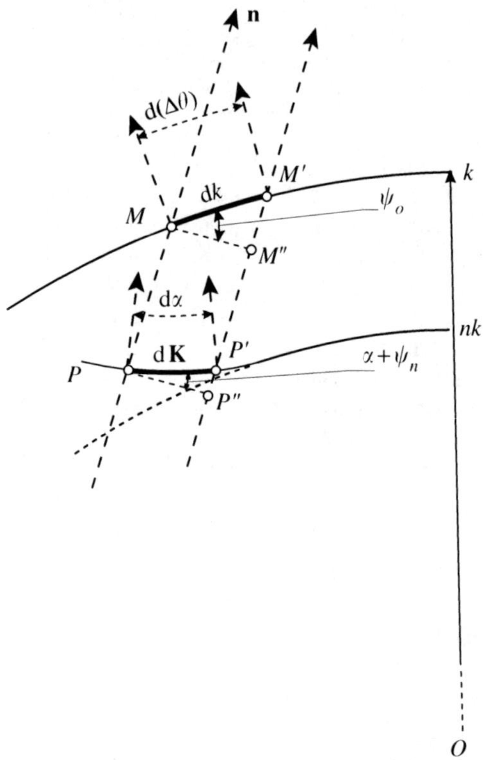

Fig. 12.7. Representation of the dispersion surface taking into account the curvature of its asymptotes. **n**: normal to the crystal surface; ψ_n: angle between **n** and the reflecting planes. MM': incident wavepacket; PP': excited wavepacket on the dispersion surface. $d(\Delta\theta)$: divergence of the incident wavepacket; $d\alpha$: divergence of the excited wavepacket.

in the case of an asymmetric reflection, and

$$\mathcal{A} = \frac{k\cos\theta_B}{\mathcal{R}\cos\alpha} \tag{12.19b}$$

in the case of a symmetric reflection.

Near the apices of the dispersion surface, the radius of curvature of the dispersion surface is very small. Far from the centre of the reflection domain,

$$|\alpha| \longrightarrow \theta_B; \qquad \cos(\alpha + \psi_n) \longrightarrow \gamma_o; \qquad \mathcal{R} \longrightarrow k$$

and the amplification ratio $\mathcal{A}$ tends towards 1, as expected.

12.7 Intensity distribution along the base of the Borrmann triangle (transmission geometry)

In this section, we shall consider an incident beam limited by a fine vertical slit normal to the plane of incidence, in the transmission geometry (Fig. 12.8). The structure of the beam is given in the general case by (12.1), but we shall limit ourselves here to the case of a uniform distribution, $\mathbf{D}_\mathbf{o}^{(\mathbf{a})}(\mathbf{K})\exp[-2\pi \mathrm{i}\mathbf{K}\cdot\mathbf{r}]$, in reciprocal space. We shall neglect the vertical divergence of the beam and the curvature of the dispersion surface in the direction normal to the plane of incidence. The incident beam can therefore be approximated by a cylindrical wave and the intensity distribution calculated in a plane of incidence is independent of the position of that plane.

The intensity, $I_o(\Delta\theta)$, of the incident beam is defined as the energy of the incident beam per unit time, per unit angle and per unit height, and the energy passing through the entrance slit per unit time is

$$I_o(\Delta\theta)h\,\mathrm{d}(\Delta\theta),$$

where h is the height of the vertical entrance slit (in the direction normal to the plane of incidence, that is the out-of-plane direction in Fig. 12.8) and $\Delta\theta$ is its mean departure from Bragg's incidence.

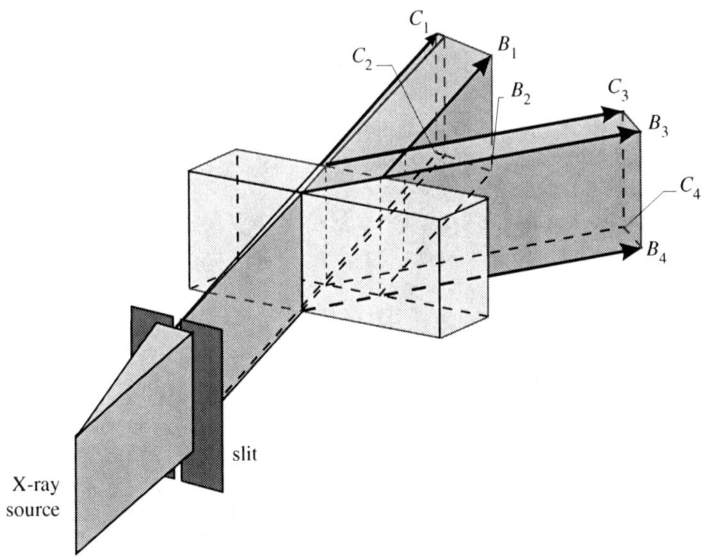

Fig. 12.8. Experimental setup showing the incident, reflected and refracted beams. The intensity distributions in the planes normal to the reflected and refracted beams are discussed in Section 12.7.

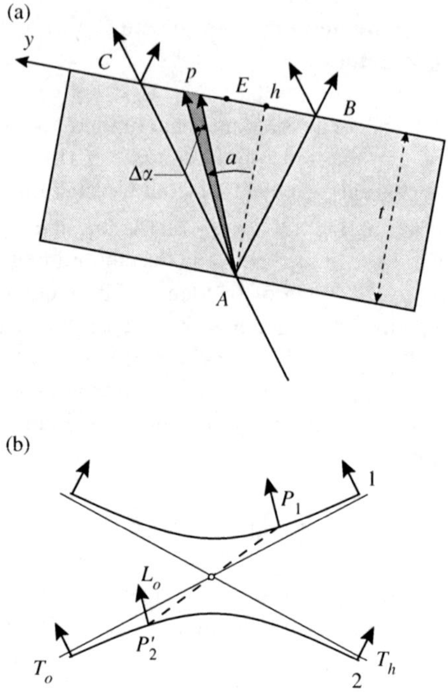

Fig. 12.9. Wavefields propagating along a given direction within the Borrmann triangle. (a) Direct space; $\Delta\alpha$: divergence of the wavepacket. (b) Reciprocal space; the tiepoints of the wavefields lie on a diameter of the dispersion surface.

Depending on its angular width, the incident beam excites all or part of the dispersion surface and generates a refracted and a reflected beam outside the crystal. The intensity of these beams is the flux of energy per unit surface and unit time passing through their cross-sections, $C_1B_1B_2C_2$ and $C_3B_3B_4C_4$, respectively (Fig. 12.8). Let ABC be the Borrmann triangle in a given plane of incidence and p a point on the exit surface (Fig. 12.9(a)), whose position is defined by $y = \overline{Ep}$, where E is the midpoint of $\overline{BC}$ and **Ey** is an oriented axis parallel to **BC**. As has been seen in Chapter 10, two wavefields propagate along Ap, of tiepoints P_1 and P'_2, respectively (Fig. 12.9(b)). The interference of these wavefields gives rise to the Kato Pendellösung fringes discussed in Chapter 10. This effect will be neglected in the present chapter and the intensity distributions will be calculated separately for the wavefields belonging to branches 1 and 2, respectively.

A narrow incident wavepacket of angular width $\delta(\Delta\theta)$ generates within the crystal a packet of wavefields of angular width $\Delta\alpha$ (Fig. 12.9(a)). The energy of the wavepacket of each component of the wavefields arriving at p on the exit surface is

$$I_{mj}(y)h\,\mathrm{d}y,$$

where $m = o$ or h and $j = 1$ or 2.

The reflecting power of the crystal is the ratio of the reflected or refracted energy to the incident energy. It is, for each branch of the dispersion surface,

$$P_{mj} = \frac{I_{mj}(y)h\,\mathrm{d}y}{I_o(\Delta\theta)h\,\mathrm{d}(\Delta\theta)} = \frac{I_{mj}(y)\,\mathrm{d}y}{I_o(\Delta\theta)\,\mathrm{d}(\Delta\theta)}. \tag{12.20}$$

A narrow wavepacket behaves like the plane wave whose wavevector coincides with the mean wavevector of the wavepacket. The reflecting power of a plane wave in the transmission geometry is given by (6.6) and

$$P_{mj} = \frac{\gamma_m |D_{mj}^{(d)}|^2}{\gamma_o |D_o^{(a)}|^2}. \tag{12.21}$$

The intensity distribution along the exit surface is deduced from (12.20) and (12.21):

$$I_{mj}(y) = \frac{\gamma_m}{\gamma_o}\frac{|D_{mj}^{(d)}|^2}{|D_o^{(a)}|^2}\frac{\mathrm{d}(\Delta\theta)}{\mathrm{d}y}\, I_o(\Delta\theta), \tag{12.22}$$

where

$$\frac{\mathrm{d}(\Delta\theta)}{\mathrm{d}y} = \frac{\mathrm{d}(\Delta\theta)}{\mathrm{d}\alpha}\frac{\mathrm{d}\alpha}{\mathrm{d}y} = \frac{1}{\mathcal{A}}\frac{\mathrm{d}\alpha}{\mathrm{d}y}.$$

One has, in Fig. 12.9(a),

$$y = \overline{Ep} = \overline{hp} - \overline{hE}$$

and

$$\mathrm{d}y = \mathrm{d}(\overline{hp}) = t\,\mathrm{d}(\tan a) = \frac{t}{\cos^2 a}\,\mathrm{d}a,$$

where t is the crystal thickness. The angles a and α are related through (10.A1) and $\mathrm{d}a = \mathrm{d}\alpha$. It follows that

$$\frac{\mathrm{d}\alpha}{\mathrm{d}y} = \frac{\cos^2 a}{t}$$

and, finally, using the reduced coordinate, $Y = y/y_{\mathrm{Max}}$:

$$\frac{\mathrm{d}(\Delta\theta)}{\mathrm{d}Y} = \frac{2\gamma_o\gamma_h^2}{\sin^2 2\theta_B}(1 - Y^2)^{-3/2}\frac{\lambda}{t\Lambda_o}. \tag{12.23}$$

This quantity represents the density of wavefields reaching the base, BC, of the Borrmann triangle. Its variations along BC are represented in Fig. 12.10 along with those of the reflecting power of the wavefields reaching each point of the exit surface.

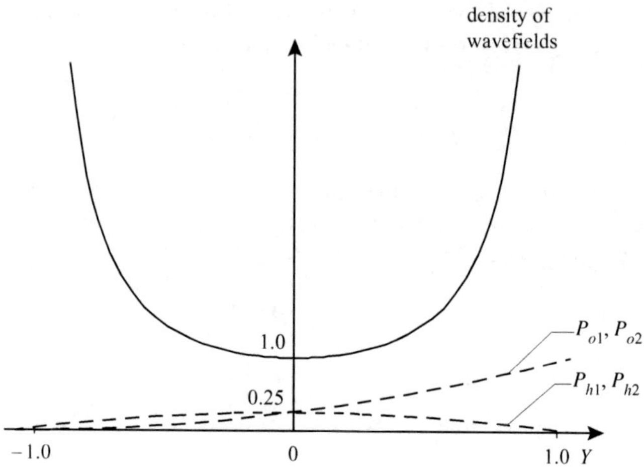

Fig. 12.10. Variations of the reflecting power, P_{mj}, of the wavefields reaching the exit surface (dashed lines) and of the density of wavefields, d(Δθ)/dy (solid line), along the base of the Borrmann triangle, arbitrary units; symmetric reflection, no absorption; Y is the reduced coordinate.

The distribution of wavefield trajectories within the Borrmann triangle is schematically represented in Fig. 1.10. The final expressions of the intensity distribution along BC are given by the product of the density of wavefields by the relevant expressions of the reflecting power; it follows, using (4.33), that

$$\begin{aligned} I_{hj}(Y) &= \frac{\gamma_o\sqrt{\gamma_o\gamma_h}}{2\sin^2 2\theta_B}|C|\sqrt{\chi_h\chi_{\bar{h}}}\,\frac{\exp(-\mu_{ej}t/\gamma_o)}{t(1-Y_j^2)^{1/2}}I_o(\Delta\theta),\\ I_{oj}(Y) &= \frac{\gamma_o\sqrt{\gamma_o\gamma_h}}{2\sin^2 2\theta_B}|C|\sqrt{\chi_h\chi_{\bar{h}}}\,\frac{1+Y_j}{1-Y_j}\,\frac{\exp(-\mu_{ej}t/\gamma_o)}{t(1-Y_j^2)^{1/2}}I_o(\Delta\theta), \end{aligned} \tag{12.24}$$

where the effective absorption coefficient, μ_{ej}, is given by (5.40) and the upper sign corresponds to $j = 1$ (wavefields with tiepoints on branch 1 of the dispersion surface) and the lower one to $j = 2$ (wavefields with tiepoints on branch 2). It should be noted that, in agreement with (12.22), the intensity distributions, $I_{mj}(y)$, have the dimensions of the inverse of a length.

The intensity distribution on the exit surface is represented in Fig. 12.11 for the reflected and refracted beams and for both wavefields in the case of zero absorption (Fig. 12.11(a)) and for $\mu_o t = 1$ where μ_o is the normal absorption coefficient (Fig. 12.11(b)). The total intensity distributions corresponding to the sum of the wavefields belonging to the two branches of the dispersion surface (neglecting Pendellösung) are represented in Fig. 12.12 for different values of $\mu_o t$.

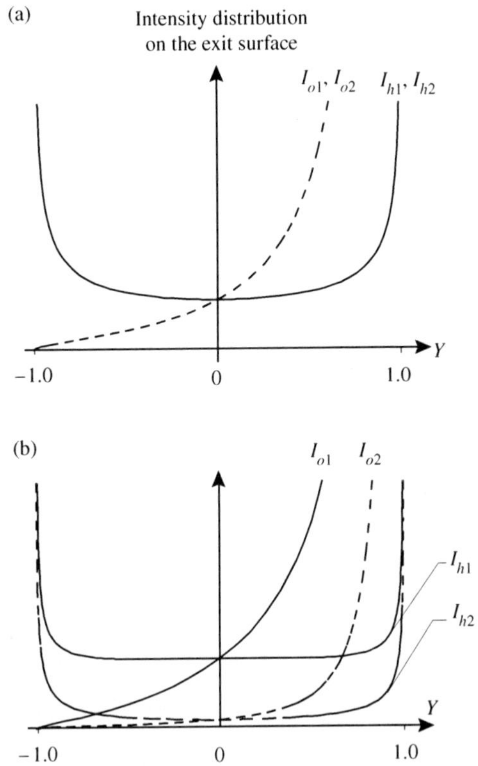

Fig. 12.11. Intensity distribution along the base of the Borrmann triangle; symmetric reflection, arbitrary units; Y is the reduced coordinate. (a) No absorption. (b) $\mu_o t = 1$.

1. *Small values of $\mu_o t$ (less than 1) and zero absorption*: It can be seen in Figs 12.11 and 12.12 that the intensity distribution for the reflected beam presents a wide minimum in the centre and increases very sharply at the edges of the Borrmann triangle, while it is the contrary for the reflecting power, P_h: it is maximum at the centre and decreases to zero on the edges where Bragg's condition is not fulfilled. The reason for this paradox is simple: it is given by Fig. 12.10. The deviation parameter of wavefields reaching the base of the Borrmann triangle close to B ($Y = -1$, Fig. 12.9(b)) corresponds to $\eta \to +\infty$; the reflecting powers $|P_{h1}|^2$ and $|P_{h2}|^2$ are very small, but the density of wavefields which is proportional to $(1 - Y^2)^{-3/2}$ is very large so that the intensity distribution is also very large. At the centre, E, of the Borrmann triangle, $Y = 0$, $\eta = 0$, the density of wavefields is minimum and so is the intensity distribution. At the other edge of the Borrmann triangle, C ($Y = +1$, $\eta \to -\infty$), the reflected intensities are also very small, the density of wavefields is very large and the intensity distribution is also very large. This effect, called the margin effect,

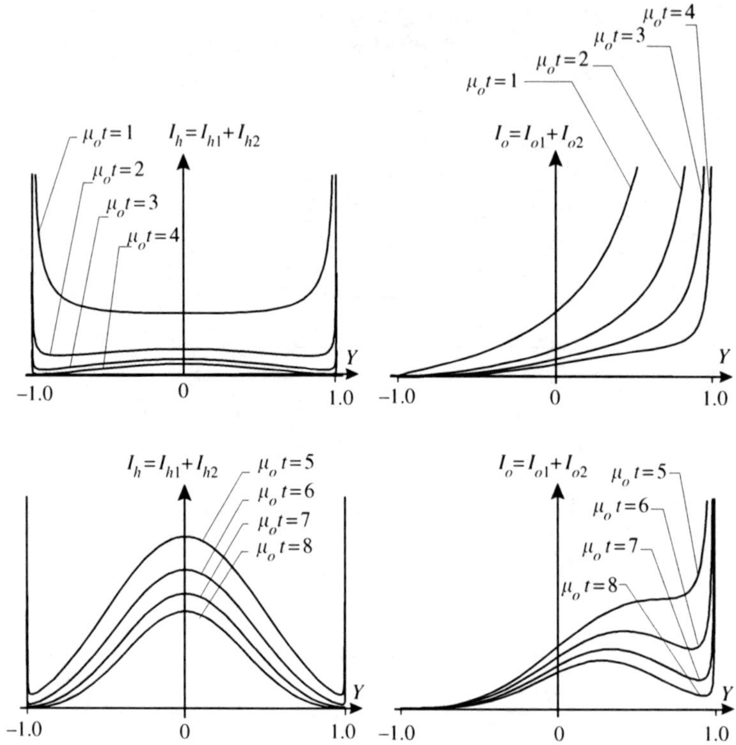

Fig. 12.12. Intensity distribution along the base of the Borrmann triangle for increasing values of $\mu_o t$; symmetric reflection; arbitrary units, the scale is different for each figure; I_h: reflected beam; I_o: refracted beam.

was predicted qualitatively by Borrmann (1959*a*) and Laue (1960), demonstrated experimentally by Kato and Lang (1959), and calculated by Kato (1960*a*). In the case of the refracted intensities, $|D_{o1}|^2$ and $|D_{o2}|^2$ are very small for $Y = -1$ and decrease towards zero much more rapidly than the density of wavefields increases towards $+\infty$ so that the resultant intensity distribution is zero; for $Y = +1$, both are very large and the intensity distribution tends towards $+\infty$. As $\mu_o t$ increases, a maximum appears in the centre of the distribution, which is due to anomalous absorption, as described in the next paragraph.

2. *Large values of* $\mu_o t$ *(of the order of* 6 *or more)*: The predominant factor is now anomalous absorption, expressed by the factor $\exp(-\mu_{ej}t/\gamma_o)$. The wavefields propagating along the edges of the Borrmann triangle ($Y = \pm 1$, $\eta \rightarrow \mp\infty$) undergo normal absorption, while those propagating parallel, or nearly parallel, to the lattice planes ($Y \approx 0$) correspond to tiepoints in the centre of the dispersion surface ($\eta \approx 0$) and undergo anomalously low absorption. The intensity distribution has now

a maximum in the centre, both for the reflected and the refracted beams. Wavefields whose tiepoints belong to branch 1 may be neglected. For values of $\mu_o t$ larger than 10 or so, only the wavefields whose Poynting vector is parallel to the lattice planes propagate through the crystal which acts as a waveguide: this is the Borrmann effect.

3. *Experimental determination of the imaginary part of the polarizability when the absorption is very large*: The value of the coefficient $|\chi_{ih}|$ of the Fourier expansion of the imaginary part of the polarizability may be deduced from the width of the intensity distribution of the reflected beam along the base of the Borrmann triangle when $\mu_o t \geq 6$. At the end of Section 6.4 the FWHM of the rocking curve was calculated for such large values. The same argument can be used to find the width of the intensity distribution along the exit surface. Let Y_o be the value of the reduced coordinate Y when the intensity $I_{h1}(y)$ is half the maximum value. It follows, for a symmetric reflection ($\gamma = 1$), that

$$|\chi_{ih}| = \frac{\lambda \cos\theta_B \ln 2}{\pi t Y_o^2},$$

an expression which was first given by Borrmann (1971).

12.8 Geometrical properties of wavefield trajectories within the Borrmann triangle

12.8.1 *Wavefields propagating along the median, AE, of the Borrmann triangle*

For the wavefields propagating along the median, AE, of the Borrmann triangle (Fig. 12.13(a)), the reduced coordinate, Y and the deviation parameter, η, are zero. Their tiepoints, I_1 and I_2, therefore lie on the normal to the entrance surface drawn from the Lorentz point, L_o (Fig. 12.13(b)), and are excited by the same incident plane wave. Conversely, it can be shown that the wavefields created at I_1 and I_2 propagate along the median AE. The asymptotes to the hyperbola, T_o and T_h, a diameter $I_1 I_2$ and the tangents to the hyperbola at I_1 and I_2 constitute what is called in geometry a harmonic bundle. This is also true of their normals, namely the incident and reflected directions, AC and AB, the entrance surface and the propagation direction of the wavefields of tiepoints I_1 and I_2, respectively. The crystal is assumed here to be plane parallel. The exit surface is parallel to the entrance surface and it is a property of the harmonic bundles that the wavefield trajectory intersects $\overline{BC}$ at its midpoint, E (Fig. 12.13(a)).

12.8.2 *Properties of the trajectories of the two wavefields excited by a plane wave*

Let P_1 and P_2 be the tiepoints of the two wavefields excited by an incident wave, of deviation parameter η. The trajectories of these wavefields within the Borrmann

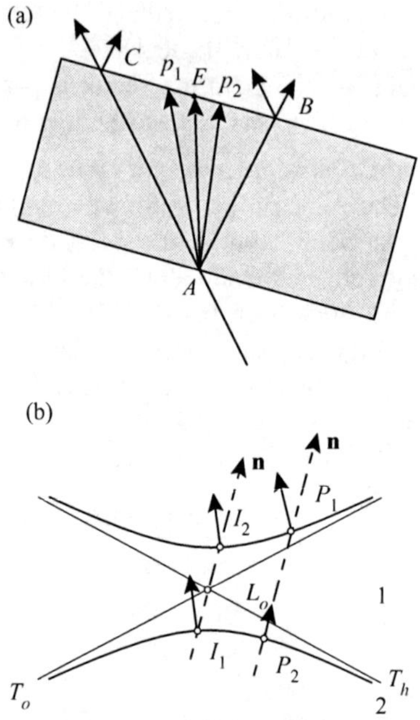

Fig. 12.13. Geometrical properties of wavefields propagating within the Borrmann triangle: the tiepoints of the wavefields propagating along AE lie on a diameter $I_2 I_1$ of the dispersion surface parallel to the normal, **n**, to the entrance surface. (a) Direct space. (b) Reciprocal space.

triangle are Ap_1 and Ap_2, respectively. According to (10.A4), the values of the reduced coordinates of p_1 and p_2 are opposite, which shows that p_1 and p_2 are symmetric with respect to the midpoint of BC, even in the case of a non-symmetric reflection. It is interesting to note that this result can be shown directly using the properties of conics and of harmonic bundles: the tangents to the hyperbola at P_1 and P_2, $P_1 P_2$, and the tangents to the hyperbola at the extremities I_1 and I_2 of a diameter parallel to $P_1 P_2$ constitute a harmonic bundle. The same argument as in the preceding section holds: the normals to these lines also constitute a harmonic bundle, namely the paths, Ap_1 and Ap_2 of the two wavefields, the entrance surface and, as noted in Section 12.8.1, the median AE. This shows that E is the midpoint of $p_1 p_2$.

12.9 Experimental proof of double refraction

The refraction of X-rays by matter was discussed theoretically in Chapter 2. Far from a Bragg reflection, one has simple refraction; it was first observed, by Larsson *et al.* (1924) and by Davis and Slack (1926), using a prism; the latter measured the deviation by inserting a prism of aluminium between the two calcite crystals of a double-crystal diffractometer. The basic property of the dynamical diffraction of X-rays by crystals is, however, the phenomenon of dispersion, namely the variation of the index of refraction with the direction of wavevectors. This is described by the dispersion surface and, as we have seen, the result is that, in the two-beam case, a non-polarized incident plane wave excites two waves in the crystal for each direction of propagation: this is the double refraction effect predicted by Ewald (1938) and by Borrmann (1955). This effect has the same consequences as in visible optics: the paths of the two waves inside the crystal are different (birefringence) and, if the crystal is in the shape of a prism, the directions of the wavevectors of the two waves emergent from the crystal are different. The separation of the paths of the two waves inside the crystal was observed by Authier (1960*a*, 1961*a*), that of the wavevectors of the two waves emerging from a prism was observed by Kohra *et al.* (1965),[1] while the variation of the direction of the wavevectors inside the crystal was displayed directly using a prism by Lang *et al.* (1986*a*). These experiments required a highly collimated incident beam; Authier's and Lang's required in addition that this beam should also be very parallel, in other words, that the incident beam be a wavepacket of limited extension in both direct and reciprocal space. These two conditions are contradictory and it is therefore difficult to fulfil them simultaneously; (12.2) shows that a compromise has to be found. Several techniques exist now, in particular taking advantage of the properties of synchrotron radiation; they are described in Section 15.3.3. The principles of the three experiments by Authier, by Kohra and coworkers and by Lang and coworkers are as follows.

1. *Separation of the wavefields paths inside the crystal (Authier 1960a, 1961a)*: The incident beam is obtained by using a relatively thick first crystal in the transmission geometry and a slit between that crystal and a second crystal, identical but thinner, also in the transmission geometry (Fig. 12.14). Due to the angular amplification which takes place in the first crystal, the beam passing through the slit is very narrow both in direct and in reciprocal space. The paths of the wavefields excited in the second

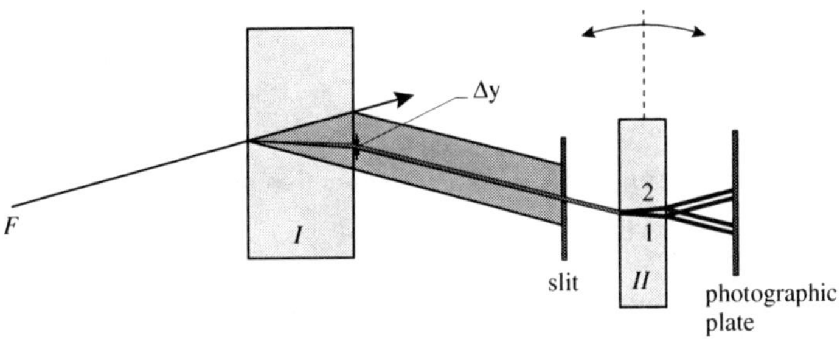

Fig. 12.14. Experimental proof of the separation of the paths of wavefields 1 and 2 (after Authier 1961*a*). *I*: first crystal; *II*: second crystal.

[1] This effect was first observed in the case of electron diffraction, by Cowley and Rees (1947) and Sturkey (1948) and discussed theoretically by Kato (1949, 1952*a*, *b*).

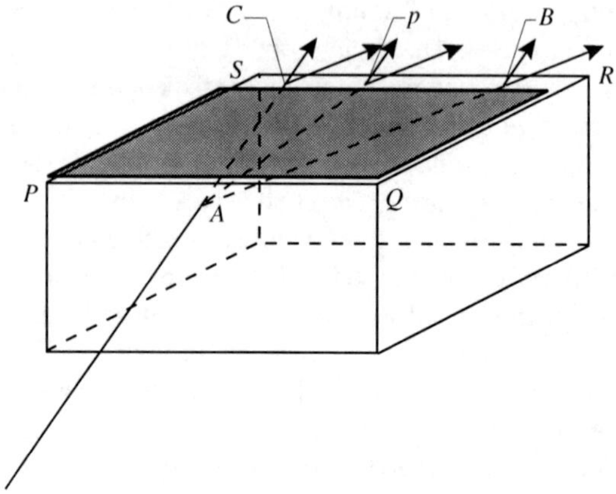

Fig. 12.15. Measurement of double refraction using a prism (adapted from Lang *et al.* 1986*a*).

crystal may thus be separated. The detailed explanation is given in Section 12.10. The separation of wavefields due to the double refraction effect was also simulated by Authier *et al.*(1968*a*) by solving Takagi's equations numerically for an incident plane wave limited by a slit (see Fig. 12.24 and Section 12.11).

2. *Measurement of the refraction effect using a prism* (*Kohra* et al. *1965*): The setup is shown in Fig. 6.6 and was discussed in Section 6.3.2.

3. *Direct determination of the path of wavefields* (*Lang* et al. *1986a*): The very parallel incident beam from a synchrotron is highly collimated by slits. It falls on a crystal set for the transmission geometry and oriented in such a way that the diffraction plane, $\mathbf{K_o}$, $\mathbf{K_h}$ is slightly inclined with respect to the top lateral face, *PQRS* of the crystal (Fig. 12.15): the base, *BC*, of the Borrmann triangle is located on that lateral face. That face is masked nearly entirely, leaving only a thin strip around BC. It is therefore easy to locate the point of emergence, p, of a given wavefield and to correlate that position with the exact angle of incidence of the beam, deduced from the rocking curve.

12.10 Experimental observation of the separation of the wavefield paths

12.10.1 *Experimental setup*

The experimental setup is shown in Fig. 12.14. The X-ray source is a linear fine focus (20 μm × 8 mm), of cross-section F, normal to the mean plane of incidence. The primary beam is incident on a first crystal, I, thick enough for type 2 wavefields to be absorbed out, for instance a silicon crystal 3 or 4 mm thick ($\mu t \approx 4.4$ or 5.8 with Mo$K\alpha$ radiation). Its divergence is much larger than that of the rocking curve and the whole dispersion surface is excited. The intensity distribution on the exit surface is that given by (12.24) and is similar to that corresponding to $\mu_o t = 5$ in Fig. 12.12.

Let Δy be a small segment on the exit surface. Due to the angular amplification, the angular divergence of the corresponding wavepacket emerging from the crystal is very small. Its value, given by (12.23), is, for a 3 mm thick silicon crystal, a 220 reflection and Mo$K\alpha$:

$$\mathrm{d}(\Delta\theta) = 0.8 \times 10^{-8}\,\mathrm{d}y,$$

that is 0.1 arcsec for $\mathrm{d}y = 50\,\mu\mathrm{m}$.

A slit 50 μm wide put in the reflected beam from the first crystal selects therefore a pencil of rays of angular width 0.1 arcsec. It is incident on a second crystal of the same nature set for the same reflection, but thinner, about 1 mm thick. The angular divergence of the pencil of rays is roughly ≤5 per cent of the FWHM of the second crystal; its spread in reciprocal space is therefore sufficiently small to be considered as a *pseudo*-plane wave; on the other hand, its width in direct space is small enough for the two wavefields excited in the second crystal to be separated: a wavepacket which is narrow both in direct and in reciprocal space has thus been isolated!

There are two important limiting factors to the experiment:

1. The width of the slit limiting the pencil of rays cannot be too narrow, otherwise Fraunhofer diffraction becomes important and the angular width of the wavepacket increases. This point will be discussed in Section 12.11.
2. Up to now it has been assumed implicitly that the incident beam is monochromatic. This is however not true in practice and the paths due to different wavelengths overlap inside the crystal (Fig. 12.16); the intensity distributions corresponding to the various wavelengths are mixed up on the exit surface and it is not possible to make the selection described above if the slit is put at an arbitrary distance of the first crystal. The solution, as shown in Section 12.10.2,

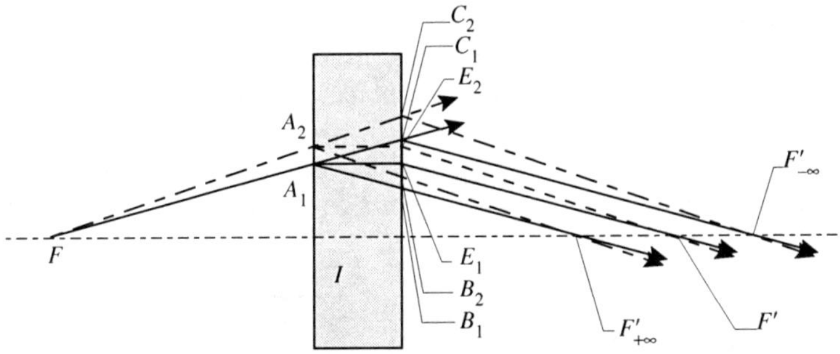

Fig. 12.16. Focalization effect in the transmission geometry for different wavelengths. F: X-ray source; $A_1B_1C_1$: Borrmann triangle for $K\alpha_1$ radiation; $A_2B_2C_2$: Borrmann triangle for $K\alpha_2$ radiation; F': focalization point for $\eta = 0$; $F'_{+\infty}$: focalization point for $\eta \Rightarrow +\infty$; $F'_{-\infty}$: focalization point for $\eta \Rightarrow -\infty$.

is to put the slit at a focalization point, F', located symmetrically to the X-ray focus, with respect to the crystal.

12.10.2 *Focalization of the various wavelengths*

Figure 12.16 shows the paths of the reflected beam inside the crystal for two wavelengths, $\lambda_{K\alpha 1} = 0.709$ Å and $\lambda_{K\alpha 2} = 0.713$ Å. The corresponding difference in Bragg angle is, for the silicon 220 reflection,

$$\Delta\theta = \frac{\delta\lambda}{\lambda}\tan\theta_B \approx 220\,\text{s}.$$

The positions of the respective Borrmann triangles, $A_1B_1C_1$ and $A_2B_2C_2$, differ therefore in a non-negligible way: for a distance X-ray focus to crystal of about half a metre, the distance A_1A_2 is of the order of 0.5 mm, which is much larger than the width of the slit ($\Delta y = 50$ μm).

Let E_1 and E_2 be the midpoints of B_1C_1 and B_2C_2, respectively. The reflected beams at E_1 and E_2 converge at a point F', symmetric of F with respect to the median plane of crystal I. It is a focalization effect which is a property of settings in the Laue geometry and which was first mentioned by Bragg (1913) and made use of by Guinier and Tennevin (1949). In a similar way, the reflected beams at B_1 and B_2 and C_1 and C_2 converge at points $F'_{+\infty}$ and $F'_{-\infty}$, respectively. More generally, the reflected beams corresponding to wavefields having the *same* departure from Bragg's angle for all the wavelengths converge to a given point, F_α, of the line FF' defined by

$$\overline{F'F_\alpha} = t\frac{\tan\alpha}{\tan\theta_B},$$

where t is the thickness of crystal I and $\tan\alpha$ the slope of the path of a given wavefield with respect to the lattice planes. When η varies from $-\infty$ to $+\infty$, the position of F_α moves from $F'_{-\infty}$ to $F'_{+\infty}$. A slit of width Δy put at F_α thus selects a beam of X-rays for which there is a *one-to-one correlation between direction and wavelength*. The accuracy with which this correlation is established depends on the width Δy. It is highest when the slit is put at F', but does not increase infinitely when Δy decreases because of the diffraction effects.

The intensity distribution (12.24) does not depend on the wavelength if the spectral width is not too large and is of the order of the $K\alpha_1 - K\alpha_2$ doublet. One has therefore

$$\frac{I_{h1}(\lambda_1)}{I_o(\lambda_1)} = \frac{I_{h1}(\lambda_2)}{I_o(\lambda_2)} = \frac{\int I_{h1}(\lambda)\,\mathrm{d}\lambda}{\int I_o(\lambda)\,\mathrm{d}\lambda}$$

and (12.24) is a good expression for the total intensity distribution.

The focalization effect which occurs for the reflected beams does not apply to the refracted beams. These diverge from a virtual focus located at a distance from the real focus, F, equal to $0.5t(1 + \tan\alpha/\tan\theta_B)$.

12.10.3 *Separation of wavefield paths in the transmission case*

The experimental setup is as described in Fig. 12.14, with the slit located between $F'_{-\infty}$ and $F'_{+\infty}$ (Fig. 12.16). It isolates a pencil of rays whose mean departure from Bragg's angle, $(\Delta\theta)_1$, depends on the position of the slit with respect to F'. The angle of the corresponding incident wavevector with the lattice planes of the first crystal is $\theta_B + (\Delta\theta)_1$, where θ_B is the Bragg angle. The pencil of rays selected by the slit is a *reflected* beam for the first crystal and it makes an angle $\theta_B - (\Delta\theta)_1$ with the lattice planes (see Section 6.3.2 and Fig. 6.4). It is an incident beam for the second crystal. This second crystal is identical to the first one, but thinner. The same reflection is used and the two-crystal arrangement is in the non-dispersive mode (see Section 15.2.3). If $(\Delta\theta)_o$ is the angle between the reflecting planes of the second and the first crystal, the angle of incidence of the pencil of rays selected by the slit on the second crystal is

$$\theta_B + (\Delta\theta)_2 = \theta_B - (\Delta\theta)_1 + (\Delta\theta)_o.$$

To these departures from Bragg's law are associated corresponding values of the deviation parameter, η_o, η_1 and η_2, and, according to (5.18),

$$\eta_2 = \eta_o - \eta_1. \tag{12.25}$$

The width of the slit and the thickness of the first crystal have been chosen in such a way that the angular width of the wavepacket incident on the second crystal is much smaller than the width of the rocking curve and that this wavepacket may be approximated by a pseudo-plane wave. It excites two wavefields in the second crystal as shown in Fig. 12.4. The paths of the two wavefields separate if the second crystal is thick enough; when they reach the exit surface, they break up into their two wave components and there are therefore four beams coming out of the crystal (Fig. 12.17).

The slit is linear, as is the X-ray focus, and the four beams are recorded as four lines on a photographic plate placed symmetrically to the slit with respect to the second crystal because of the focalization effect mentioned in Section 12.10.2. As the

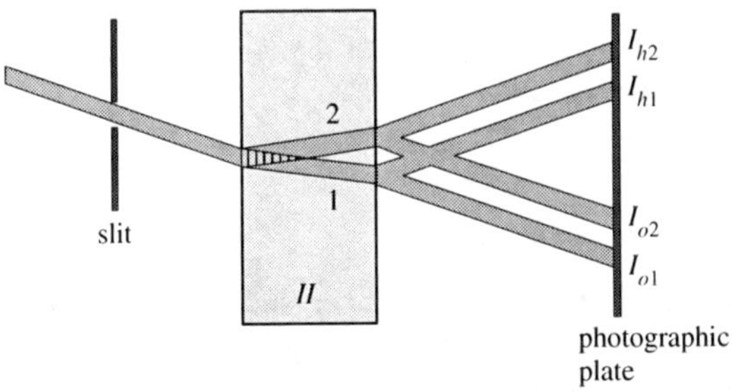

Fig. 12.17. Separation of wavefields in the second crystal when the incident beam is a pseudo-plane wave. I_{o1}, I_{o2}: traces of the refracted beams for wavefields 1 and 2, respectively; I_{h1}, I_{h2}: traces of the reflected beams for wavefields 1 and 2, respectively.

focalization effect does not hold for the refracted beams, the distances between the slit and the second crystal, and the second crystal and the photographic plate should be small, of the order of 5–10 mm.

1. *The slit is parallel to the focus line*, the second crystal makes a small angle, 0.44 arcsec, with respect to the first crystal. One observes four parallel lines on the photographic plate (Fig. 12.18(a)), which proves that the wavefields have indeed followed separate paths within the second crystal.
2. *The slit is not parallel to the focus line*. If the slit inclined, as shown in Fig. 12.19, the departure from Bragg's angle of the beam selected by the slit varies along the slit: at one extremity it selects a pencil of rays which has followed path Ap in the first crystal and corresponds to a negative value of η_1, and at the other extremity it selects a pencil of rays which has followed path $A'p'$ and corresponds to a positive value of η_1. The deviation parameter for the *second* crystal, given by (12.25), varies therefore along the trace of the beam. The separation of the two wavefields in the second crystal varies accordingly and the traces of the emerging beams on the photographic plate are no longer two parallel lines. Two examples are given in Figs 12.18(b) and (c).

In the case of Fig. 12.18(b), the two crystals are parallel and the departure from Bragg's angle is zero in the centre of the picture. In the case of Fig. 12.18(c), the two crystals are inclined by 0.44 arcsec with respect to one another. The departure from Bragg's angle is then zero for one extremity of the slit and very large by positive values at the other end. The intensity variations of the reflected and refracted beams for branches 1 and 2 observed in the two pictures are in agreement with the calculated values represented in Fig. 6.2(b). It can be noted that the reflected beam is always more intense for branch 1 than for branch 2, which is normal because it is less absorbed, due to the Borrmann effect. In contrast, if the refracted beam is more intense for branch 1 than for branch 2 when η_2 is negative, it is the contrary when η_2 is positive and larger than a certain value; the reason is that, on the one hand, the anomalous absorption effect decreases far from the centre of the reflection domain and, on the other hand, it is then branch 2 wavefields which merge into the direct beam; this last point may be checked in Fig. 12.9. It can be seen that the direction of the normal to the dispersion surface tends towards that of the incident beam for $\eta \to -\infty$ (branch 1) and for $\eta \to +\infty$ (branch 2).

12.10.4 *Plane-wave Pendellösung*

In the experiments of Fig. 12.18, the second crystal is plane parallel and thick enough for the paths of the two wavefields to be separated inside the crystal, but if the second crystal is wedge shaped (Fig. 12.20) the wavefields overlap in the thinner part (corresponding to the hacked area in Fig. 12.17) and interfere, giving rise to Pendellösung fringes which are *the* plane-wave Pendellösung fringes predicted by Ewald; they can be seen on the left-hand side of Fig. 12.20. In the thicker part of the

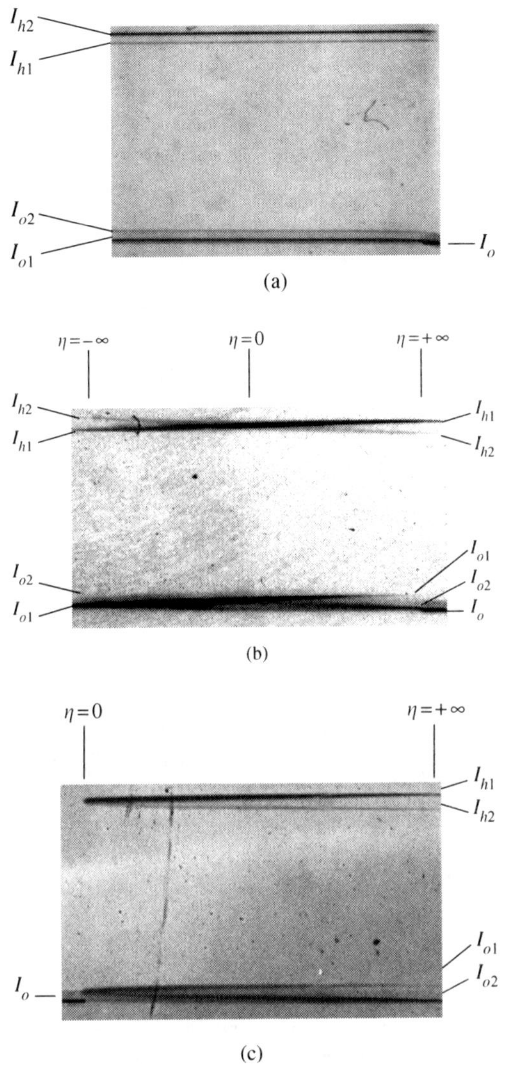

Fig. 12.18. Ray tracing in a perfect crystal; I_{h1}, I_{h2}, I_{o1}, I_{o2}, traces of the reflected and refracted components of the two wavefields; I_o, trace of the primary beam; the crystal is plane parallel; silicon, 220 reflection, Mo$K\alpha$. (a) The slit located at F' (Fig. 12.16) is parallel to the X-ray focus; the second crystal is inclined with respect to the first one by 0.44 arcsec; after Authier (1961*a*). (b) The slit located at F' (Fig 12.16) is not parallel to the X-ray focus; the second crystal is parallel to the first one; the departure from Bragg's angle is exactly zero at the centre of the picture; after Authier (1961*a*). (c) The slit located at F' (Fig. 12.16) is not parallel to the X-ray focus; the second crystal is inclined with respect to the first one by 0.44 arcsec; after Authier (1961*a*).

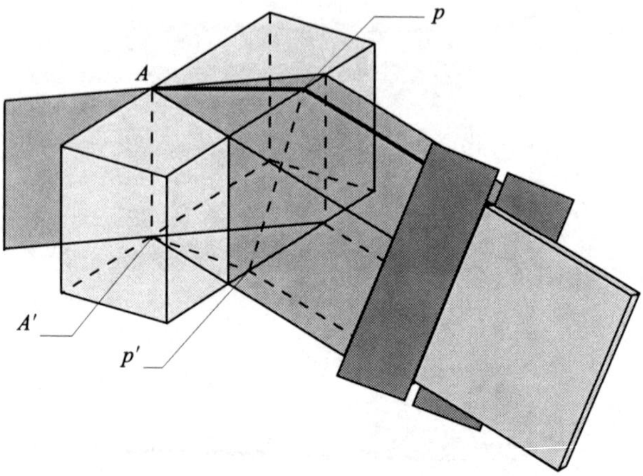

Fig. 12.19. Variation of the paths of the selected wavefields in the first crystal when the slit is inclined: Ap at one extremity of the slit, $A'p'$ at the other extremity.

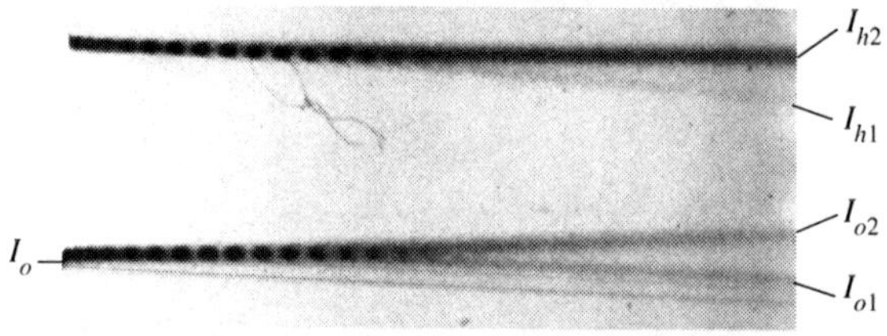

Fig. 12.20. Plane-wave Pendellösung; I_{h1}, I_{h2}, I_{o1}, I_{o2}, traces of the reflected and refracted components of the two wavefields; I_o, trace of the primary beam; wedge-shaped crystal. Silicon, 220 reflection, Mo$K\alpha$. The slit located at F' (Fig. 12.16) is parallel to the X-ray focus; the second crystal is inclined with respect to the first one by 0.3 arcsec; after Malgrange and Authier (1965).

crystal the wavefields do not overlap and their traces are separated as can be seen on the right-hand side of Fig. 12.20; this separation is proportional to the thickness of the crystal.

These plane-wave Pendellösung fringes were first observed by Malgrange and Authier (1965); they have since also been observed by other techniques which have been summarized in Section 1.10.3.

12.10.5 *Application to the measurement of the index of refraction*

The index of refraction of matter for X-rays being so close to one is very difficult to measure. Two of the most widespread methods nowadays are the determination of the critical angle for specular reflection (see Section 2.10) and the use of the X-ray and neutron interferometers developed by Bonse and Hart (1965*b*) which will be described in Section 15.5.

It is also possible to use a two-crystal setup and a locally plane wave. This can be done in a variety of ways:

1. Using the setup described in Fig. 12.14: If one puts a prism in the beam between the slit and the second crystal, the deviation from Bragg's angle of that beam on the second crystal is modified and so is the separation of the paths of the wavefields. The index of refraction of the prism is easily deduced from the variation of this separation.

The deviation angle D due to the prism is

$$D = 0.5\chi_o[\tan i + \tan(\Omega - i)],$$

where $\chi_o = -R\lambda^2\mathcal{N}/\pi$ is the average polarizability of the prism, R the classical radius of the electron, $\mathcal{N}$ the total number of electrons per unit volume of the prism, λ the wavelength, i the angle of incidence and Ω the angle of the prism. The maximum accuracy is obtained at the minimum deviation angle, D_{Min}, which occurs for $i = \Omega/2$ ($D_{\text{Min}} = \chi_o \tan(\Omega/2)$). The method is possible if D_{Min} is comparable with the width of the rocking curve of the second crystal.

Figure 12.21 shows an example of such a measurement performed on an aluminium crystal with Mo$K\alpha$ radiation by Malgrange *et al.* (1968). The slit is parallel to the X-ray focus, the second crystal makes a small angle with the first one and the aluminium prism is placed along half the length of the slit in Fig. 12.14: the right-hand side of the figure shows the separation of the wavefields without the prism and the left-hand side the separation with the prism. The experimental result is

$$|\chi_o|_{\text{exp}} = (3.53 \pm 0.05) \times 10^{-6}$$

while the theoretical value is

$$|\chi_o|_{\text{theor}} = 3.55 \times 10^{-6}.$$

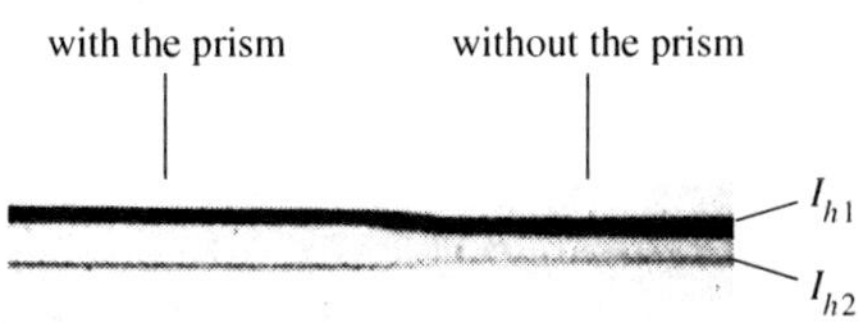

Fig. 12.21. Measurement of the index of refraction of an aluminium prism; the separation of the paths of the wavefields is different with and without the prism; I_{h1}, I_{h2}, traces of the reflected components of the two wavefields; the second crystal is plane parallel. Silicon, 220 reflection, Mo$K\alpha$. Reproduced with permission from Malgrange *et al.* (1968).

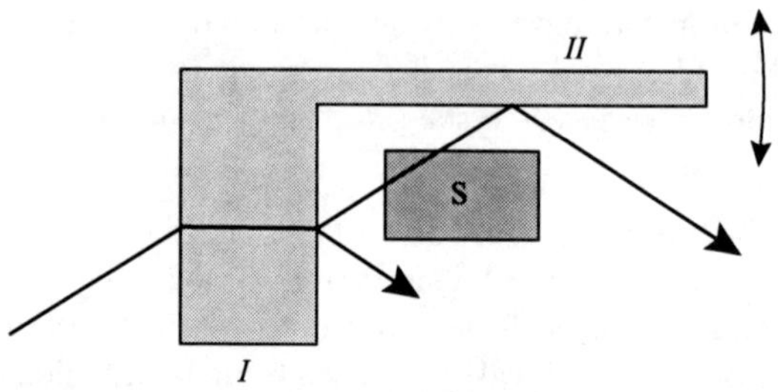

Fig. 12.22. Measurement of the index of refraction of a sample, S. It is deduced from the change of the angular position of the Bragg peak on part II of the crystal with and without the sample. Its value is obtained by elastically rotating part II of the crystal with respect to part I. Adapted from Okkerse (1963).

Another experiment performed on a zirconium crystal gave the following results:

$$|\chi_o|_{\text{exp}} = (7.21 \pm 0.12) \times 10^{-6},$$

$$|\chi_o|_{\text{theor}} = 7.78 \times 10^{-6} \text{ without taking the anomalous dispersion correction into account,}$$

$$|\chi_o|_{\text{theor}} = 7.24 \times 10^{-6} \text{ taking the anomalous dispersion correction into account.}$$

The agreement between experimental and theoretical values is therefore quite satisfactory.

2. Using a very thick first crystal so that the locally plane wave is selected by means of the Borrmann effect. This has been done by Okkerse (1963). He used an L-shaped germanium crystal oriented in such a way that the incident beam is set at the Bragg angle for transmission through on one branch of the 'L' and for reflection on the other branch (Fig. 12.22).

The first part is very thick so that, due to the Borrmann effect, only the rays very close to Bragg's angle cross the crystal. These rays also satisfy Bragg's condition on the second part of the crystal and undergo total reflection. If a prism of some material, S, is inserted between the two parts of the crystal, the angle of incidence of the beam on the second crystal is modified. Part II of the crystal is brought back to Bragg's incidence by deforming slightly one branch of the 'L' with respect to the other one. The deviation due to the prism can thus be measured.

3. Using an extended locally plane wave produced by synchrotron radiation, with a large source-to-crystal distance, L, as shown by Mocella *et al.* (1999, 2000). The insertion of a prism, near the entrance surface of a perfect crystal diffracting in the transmission geometry, produces a displacement $\Delta x = DL/\cos\theta_B$ of the diffracted intensity pattern on a photographic plate, where D is the angular deviation of the

prism. Since L is large, Δx is sufficiently large to allow a straightforward and accurate measurement of the angular deviation, D.

12.11 Fresnel diffraction near the Bragg incidence

Fresnel and Fraunhofer diffraction occur in the same way for hard X-rays as for visible optics. Diffraction of X-rays by a slit was first observed by Walter and Pohl (1908, 1909) using white radiation and gave an indication as to their wave nature. But the first very distinct diffraction fringes were observed by Kellström (1932) and by Bäcklin (1935) using characteristic lines.[2]

Fraunhofer diffraction occurs when the distance r between the observation plane and the slit is large enough for the width of the diffraction pattern to be much bigger than that, e, of the slit. Its angular width, β, is of the order of λ/e and the condition is

$$\beta r \gg e \implies r \gg \frac{e^2}{\lambda}. \tag{12.26}$$

For instance, for a 50 μm slit and Mo$K\alpha$ ($\lambda = 0.709$ Å), r should be larger than 35 m. In their respective experiments, Kellström used long wavelengths, Al$K\alpha$ (8 Å), CuLα (13 Å) and CLα (44 Å), a 6.6 μm slit and a 42 cm slit to film distance, and Bäcklin also used a long wavelength, CLα (44 Å), a 3.3 μm slit and a 51 cm slit to film distance. In recent years, it has been possible to obtain diffraction patterns with much shorter wavelengths, using synchrotron radiation: Mancini and Bilderback (1983), at CHESS, US, used 1.54 Å and 0.77 Å wavelengths, a 10 μm slit and a 1.3 m slit to film distance, and Lang *et al.* (1987), at Daresbury, UK, used a 1 Å wavelength, slit widths down to 8.8 μm and 2.8 μm and a 16 m slit to film distance.

Fresnel diffraction of hard X-rays has been used by different authors for phase contrast imaging of small objects or small variations of the index of refraction in an object: Pogany *et al.* (1997) using a microfocus X-ray source, Snigirev *et al.* (1995*b*) and Cloetens *et al.* (1996) using synchrotron radiation.

The application of (12.26) is different if the beam which has passed through the slit is incident on a crystal close to Bragg's incidence. Due to the angular amplification effect, the angular width of the diffraction pattern is now $\mathcal{A}\beta$, the lateral width of the diffraction pattern is $\mathcal{A}\beta r$ and (12.26) is replaced by

$$r \gg \frac{e^2}{\mathcal{A}\lambda}. \tag{12.27}$$

For a 50 μm slit it is satisfied at $r \gg 1$ mm if $\mathcal{A} = 3.5 \times 10^4$ (silicon, 220, Mo$K\alpha$). This effect was predicted by Kato (1961*a*,*b*) and calculated and checked experimentally by Authier and Malgrange (1966). The influence of a slit on rocking curves was studied by Teworte and Bonse (1987).

[2] A beautiful diffraction pattern by Bäcklin is reproduced in Laue (1960).

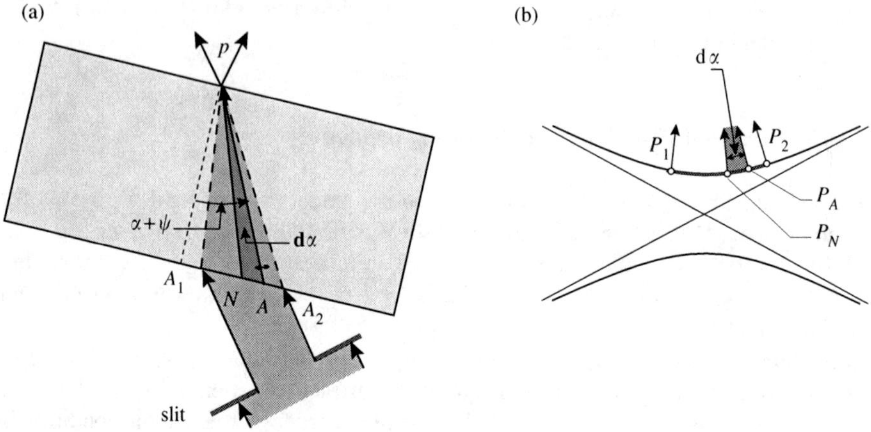

Fig. 12.23. Wavefields excited in a crystal by the diffraction pattern due to a plane wave limited by a slit. (a) Direct space. (b) Reciprocal space.

Let us consider a plane wave of wavevector $\mathbf{K_o^{(a)}}$ incident on a perfect crystal close to the Bragg incidence and limited by a narrow slit (Fig. 12.23(a)). Let $A_1A_2 = e$ be the trace of the X-ray beam on the entrance surface of the crystal and p a point on the exit surface. It receives wavefields whose paths $\mathbf{Np}$ lie between $\mathbf{A_1p}$ and $\mathbf{A_2p}$ and whose tiepoints P_N lie between P_1 and P_2 on the dispersion surface (Fig. 12.23(b)). Let $\overrightarrow{Ap}$ be the path of the wavefield excited by the incident plane wave in the *absence* of any diffraction and P_A its tiepoint. The phase factors of the wavefields of tiepoints P_N and P_A are, respectively,

$$\text{for } \overrightarrow{Ap}: \quad \exp(-2\pi \mathrm{i}\mathbf{K_{oA}} \cdot \mathbf{r}),$$

$$\text{for } \overrightarrow{Np}: \quad \exp\{-2\pi \mathrm{i}[\mathbf{K_o^{(a)}} \cdot \mathbf{r_s} + \mathbf{K_{oN}}\cdot(\mathbf{r} - \mathbf{r_s})]\},$$

where $\mathbf{r} = \mathbf{Ap}$, $\mathbf{r_s} = \mathbf{AN}$ and $\mathbf{K_{oA}}$ and $\mathbf{K_{oN}}$ are the wavevectors of P_A and P_N. The phase difference between the two wavefields is

$$\begin{aligned}\Delta\phi &= 2\pi[\mathbf{K_o^{(a)}} \cdot \mathbf{r_s} + \mathbf{K_{oN}}\cdot(\mathbf{r} - \mathbf{r_s}) - \mathbf{K_{oA}} \cdot \mathbf{r}] \\ &= 2\pi[(\mathbf{K_o^{(a)}} - \mathbf{K_{oA}})\cdot\mathbf{r_s} + (\mathbf{K_{oN}} - \mathbf{K_{oA}})\cdot(\mathbf{r} - \mathbf{r_s})].\end{aligned}$$

The first term is equal to zero because of the continuity of the tangential components of wavevectors ($\mathbf{K_{oA}}$ is the wavevector excited by the incident wavevector, $\mathbf{K_o^{(a)}}$). It follows that

$$\begin{aligned}\Delta\phi &= 2\pi(\mathbf{K_{oN}} - \mathbf{K_{oA}})\cdot(\mathbf{r} - \mathbf{r_s}) = 2\pi\mathbf{P_AP_N} \cdot \mathbf{Np} \\ &\approx 2\pi\, P_A P_N \frac{r\, \mathrm{d}\alpha}{2},\end{aligned}$$

where $\mathrm{d}\alpha$ is the angle between paths **Ap** and **Np** and segment $P_A P_N$ of the dispersion surface is assumed to be small. Let $\mathrm{d}(\Delta\theta)$ be the variation in departure from Bragg's angle between P_A and P_N. Using Fig. 12.7 and (12.18), it follows, in the general case of an asymmetric reflection, that

$$P_A P_N = k\,\mathrm{d}(\Delta\theta)\frac{\gamma_o}{\cos(\alpha+\psi_n)},$$

where α and ψ_n are the angles between the lattice planes on the one hand and the wavefield trajectory and the normal to the entrance surface, respectively, on the other hand. Introducing the angular amplification ratio (12.13), the expression of the phase shift becomes

$$\Delta\phi = \pi k r(\mathrm{d}\alpha)^2\frac{\gamma_o}{\mathcal{A}\cos(\alpha+\psi_n)}.$$

Let us set $y = \overline{AN} = r\,\mathrm{d}\alpha/\cos(\alpha+\psi_n)$. It follows that

$$\Delta\phi = \pi k y^2\frac{\cos(\alpha+\psi_n)\gamma_o}{\mathcal{A}r}.$$

The total amplitude received in p is obtained by summing over all the positions of N along A_1A_2. The intensity of the reflected beam in p is therefore proportional to

$$I_h(p) \propto \left|\int_{y(A_1)}^{y(A_2)} D_h(y)\exp\Big[-\pi\mathrm{i}k\frac{\cos(\alpha+\psi_n)\gamma_o}{\mathcal{A}r}y^2\Big]\,\mathrm{d}y\right|^2.$$

The amplitude $D_h(y)$ diffracted by the segment $\mathrm{d}y$ may be considered as constant if $\mathrm{d}(\Delta\theta)$ is small and may be taken out of the integral. The calculation of the intensity then reduces, as in visible optics, to that of a Fresnel integral of type

$$\int_{v_1}^{v_2}\exp\Big(-\mathrm{i}\frac{\pi v^2}{2}\Big)\,\mathrm{d}v,$$

with

$$\begin{aligned} v &= y\sqrt{\frac{2k\gamma_o\cos(\alpha+\psi_n)}{\mathcal{A}r}},\\ v_2 - v_1 = w &= e\sqrt{\frac{2k\gamma_o\cos(\alpha+\psi_n)}{\mathcal{A}r}} = e'\sqrt{\frac{2k\gamma_o}{\cos(\alpha+\psi_n)\mathcal{A}r}} \qquad (12.28)\\ &= e'\sqrt{\frac{2\mathcal{R}}{r}}, \end{aligned}$$

where $e' = e\cos(\alpha+\psi_n)$ is the width of the slit in the direction normal to **Np** and $\mathcal{R}$ is the radius of curvature of the dispersion surface.

Outside the reflection domain ($\alpha = \theta_B$, $\mathcal{A} = 1$, $\mathcal{R} = k$), the expression of w is identical to the usual expression in visible optics. The diffraction pattern depends only on the value of w; one distinguishes three domains in optics:

1. $w \ll 0.5$: Fraunhofer region (this is the case of the experiments by Kellström ($w = 0.51$, 0.40 and 0.22, respectively, for the three wavelengths he used), Bäcklin ($w = 0.1$), and Lang *et al.* ($w = 0.3$ and 0.08, respectively, for the two slit widths they used)).
2. $0.5 < w < 2.5$: pseudo-Fraunhofer region; the diffraction pattern presents a single main maximum but its width is of the same order of magnitude as that of the slit; this is the case of the experiments by Mancini and Bilderback ($w = 1.0$ and 1.4 for the two wavelengths they used).
3. $w \gg 2.5$: Fresnel region.

Within the reflection domain, the diffraction pattern is identical to what it would be in vacuum at a distance $\mathcal{A}$ times larger than the crystal thickness. This was checked experimentally by Authier and Malgrange (1966), who studied the fine structure of the reflected beam after the second crystal in the setup of Fig. 12.14 for various slit widths and for various values of the deviation parameter of the beam for the second crystal, hence of $\mathcal{A}$: the diffraction patterns that they observed were in

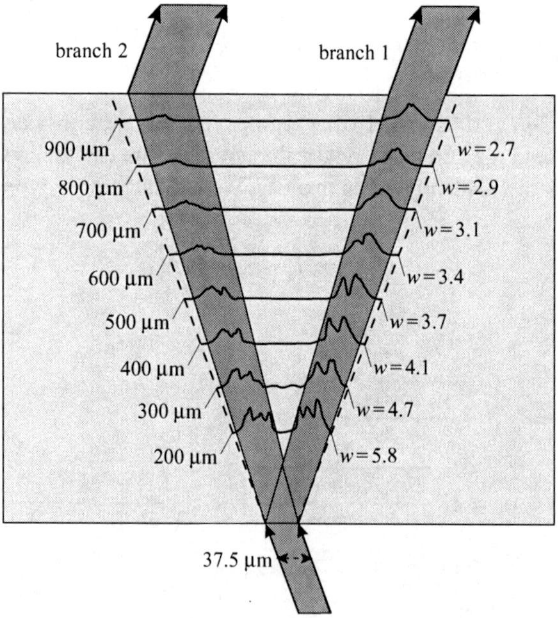

Fig. 12.24. Computer simulation of double refraction and Fresnel diffraction by a slit (adapted from Authier *et al.* 1968*a*).

good agreement with those predicted from the values of w calculated with (12.28). The physical reason for this effect lies in the dispersive nature of the crystalline medium near a Bragg reflection. It can also be studied by calculating the variation of the intensity distribution of the beam as it propagates through the crystal. This can be done by solving numerically Takagi's equations (Authier *et al.* 1968*a*—see Section 14.4.2).

Figure 12.24 shows the intensity distribution in a plane parallel to the entrance surface at various depths in a silicon crystal, for a 220 reflection, Mo$K\alpha$ radiation, and an incident plane wave with a departure from Bragg's angle $\Delta\theta = 1.7$ arcsec. The simulation has been performed for a slit of width 37.5 μm in the direction parallel to the entrance surface. One can observe two peaks corresponding to branches 1 and 2, respectively. The fine structure of these peaks is due to the Fresnel diffraction by the slit; the corresponding values of w, taking into account the angular amplification, $\mathcal{A} = 6 \times 10^3$ for the conditions of the simulation, are indicated in the figure. They may be compared to the theoretical profiles calculated for the same values w at the depths of 200 μm, 400 μm, 600 μm and 800 μm: $w = 5.8$, 4.1, 3.4 and 2.9, respectively (Fig. 12.25). The agreement is qualitatively good, taking into account the fact that the simulated intensity profile is the result of the convolution of the profile due to Fresnel diffraction by the intensity distribution due to Bragg diffraction.

12.12 Ray tracing in finite crystals

12.12.1 *Introduction*

In all the examples discussed so far, the crystal has been assumed to be semi-infinite. In practice, however, crystals are limited by lateral faces. Different geometrical situations were classified and discussed by Saka, Katagawa and Kato who distinguished *Laue–Bragg* geometries where the incident beam enters the crystal in the transmission geometry and is then partially reflected once or several times by the lateral faces (Saka *et al.* 1972*a*,*b*), as illustrated in Fig. 12.26, and the *Bragg–Bragg* geometry where the beam is incident on the crystal in the reflection geometry, partially transmitted and then partially reflected at the back surface (Saka *et al.* 1973), as illustrated in Fig. 12.27(c). In all these cases they considered an incident spherical wave and calculated the resultant reflected and refracted amplitudes with Kato's spherical-wave dynamical theory. Similar calculations have been repeated and extended recently by Thorkildsen and Larsen (1998*d*, 1999*a–c*) using Takagi's theory.

In Sections 12.12.2 and 12.12.3 the Bragg–Laue and Bragg–Bragg geometries will be discussed in the simple case where the incident beam is a collimated pseudo-plane wave.

In the case of the Laue–Bragg geometries very interesting interference effects occur which have first been observed by Borrmann and Lehmann (1962, 1963) and give rise to what is now called *Borrmann–Lehmann fringes*. They will be discussed in Section 12.12.4.

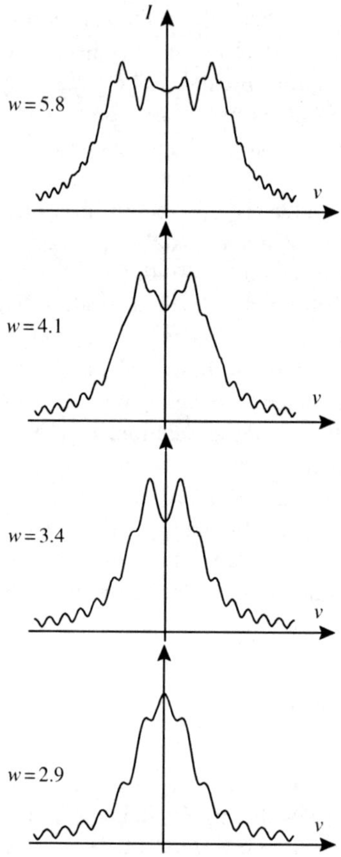

Fig. 12.25. Fresnel diffraction patterns as a function of parameter v for different values of w.

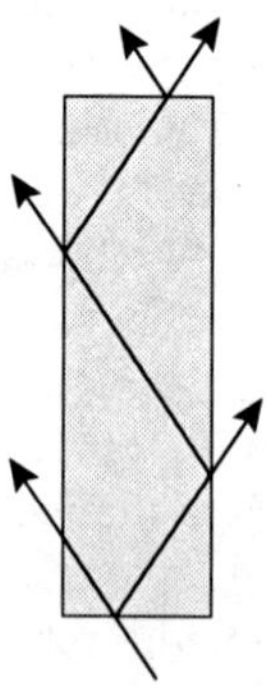

Fig. 12.26. Laue–Bragg geometry.

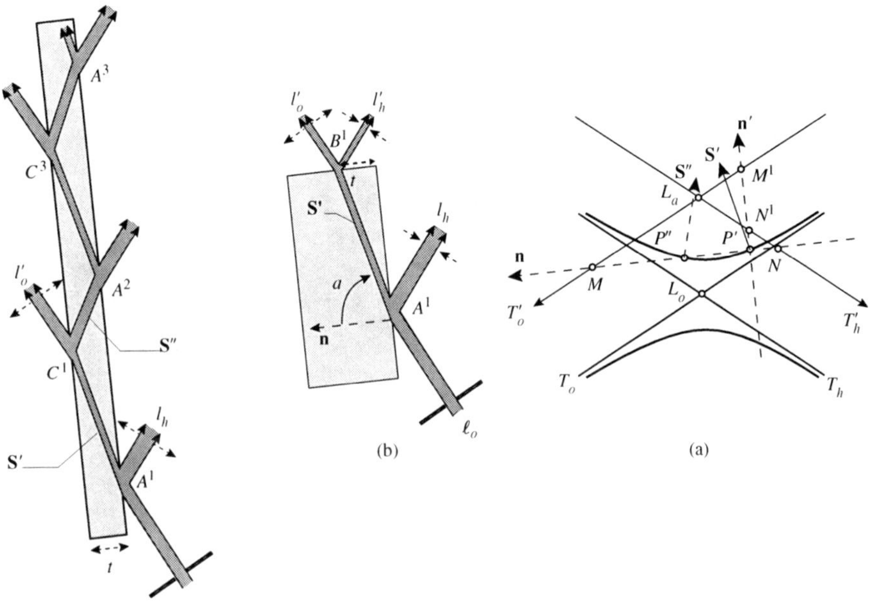

Fig. 12.27. Ray tracing in the reflection geometry. (a) Reciprocal space; **n**: normal to the entrance surface; **S**$'$ and **S**$''$: Poynting vectors. (b) Direct space: Bragg–Laue case; **n**$'$: normal to the lateral face. (c) Direct space: Bragg–Bragg case.

12.12.2 *Bragg–Laue geometry—pseudo-plane waves*

In this section and the next we shall consider the case of a beam limited in direct and in reciprocal space which is incident at A^1 on a crystal set for the reflection geometry (Fig. 12.27).

The boundary conditions for the wavevectors and the amplitudes have been described in Sections 4.7 and 4.9. The normal **n** to the crystal surface drawn from the excitation point, M, cuts the dispersion surface at two points, P' and P'' (Fig. 12.27(a)). The propagation direction of the wavefield with tiepoint P' points towards the inside of the crystal while that of the wavefield with tiepoint P'' points from the inside of the crystal towards the entrance surface. This latter wavefield is not excited at A^1 because there is no place where it could have been generated. The boundary conditions for the amplitudes at A^1 therefore reduce to (Section 4.9.2)

$$D_o'^{(1)} = D_o^{(a)}; \quad D_h'^{(1)} = D_h^{(1)(a)}, \tag{12.29}$$

where $D_o^{(a)}$ is the incident amplitude, $D_o'^{(1)}$ and $D_h'^{(1)}$ are the amplitudes of the two waves generated inside the crystal and $D_h^{(1)(a)}$ is the amplitude of the wave reflected outside the crystal at A^1; the superscript '(1)' corresponds to the position A^1.

Two cases are to be considered, depending on whether the wavefield created at A^1 emerges at a lateral surface or reaches the bottom surface of the crystal. The geometry of the former case is called *Bragg–Laue* geometry; it is discussed in the present subsection; the latter one is called *Bragg–Bragg* geometry and will be discussed in the next subsection.

The wavefield created at A^1 emerges at B^1 on a lateral face (Fig. 12.27(b)). There it separates into its two wave components. A photographic plate located behind the crystal will then record three beams: the reflected beam at A^1 and the refracted and reflected beams emerging at B^1; within the domain of total reflection, however, no energy penetrates into the crystal and the beams emerging at B^1 disappear. This has been observed experimentally by Authier (1962) using the setup of Fig. 12.14 with the second crystal in Bragg geometry.

The boundary conditions for the amplitudes at the lateral face are

$$D_o^{(d)} \exp(-2\pi i \mathbf{K_o^{(d)}} \cdot \mathbf{r}) = D_o^{\prime(1)} \exp(-2\pi i \mathbf{K}'_\mathbf{o} \cdot \mathbf{r}),$$

$$D_h^{1(d)} \exp(-2\pi i \mathbf{K_h^{(d)}} \cdot \mathbf{r}) = D_h^{\prime(1)} \exp(-2\pi i \mathbf{K}'_\mathbf{h} \cdot \mathbf{r}),$$

where $\mathbf{K}'_\mathbf{o}$ and $\mathbf{K}'_\mathbf{h}$ are the wavevectors of the waves of tiepoint P', $D_o^{(d)}$ and $D_h^{(d)}$ the amplitudes of the refracted and reflected waves emerging from the lateral face. If t is the distance of A^1 from the lateral face, it follows that

$$D_o^{(d)} = D_o^{\prime(1)} \exp(-2\pi i \overline{M^1 P'} t),$$

$$D_h^{1(d)} = D_h^{\prime(1)} \exp(-2\pi i \overline{N^1 P'} t),$$

where M^1 and N^1 are the intersections of the normal to the exit surface with T_o and T_h (Fig. 12.27(a)). The corresponding intensities are

$$|D_o^{(d)}|^2 = |D_o^{(a)}|^2 \exp\left(-\frac{\mu_e t}{\sin\psi_o}\right),$$

$$|D_h^{(d)}|^2 = |\xi'_h|^2 |D_o^{(d)}|^2 \exp\left(-\frac{\mu_e t}{\sin\psi_o}\right),$$

where $\xi'_h = D_h^{\prime(1)}/D_o^{\prime(1)}$ and μ_e is given by (5.46).

The reflecting powers are the ratios of the energies of the reflected or refracted beams to that of the incident beam:

$$I_h^{1(a)} = \frac{l_h}{l_o}\left|\frac{D_h^{1(a)}}{D_o^{(a)}}\right|^2 = \frac{l_h}{l_o}|\xi'_h|^2,$$

$$I_h^{1(d)} = \frac{l'_h}{l_o}\left|\frac{D_h^{1(d)}}{D_o^{(a)}}\right|^2 = \frac{l'_h}{l_o}|\xi'_h|^2 \exp\left(-\frac{\mu_e t}{\sin\psi_o}\right),$$

$$I_o^{1(d)} = \frac{l'_o}{l_o}\left|\frac{D_o^{1(d)}}{D_o^{(a)}}\right|^2 = \frac{l'_h}{l_o} \exp\left(-\frac{\mu_e t}{\sin\psi_o}\right),$$

where l_o, l_h, l'_o and l'_h are the cross-sections of the incident beam, the reflected beam at A^1, the refracted and reflected beams at B^1, respectively (Fig. 12.27). They are evaluated in terms of the angle a between the normal to the entrance surface and the path of the wavefield (see Appendix of Chapter 10). It follows, assuming that the lateral face is perpendicular to the entrance surface, that

$$I_h^{1(a)} = \frac{|\gamma_h|}{\gamma_o}|\xi'_h|^2,$$

$$I_h^{1(d)} = \tan^{-1}(a)\frac{\sin\psi_h}{\gamma_o}|\xi'_h|^2 \exp\left(-\frac{\mu_e t}{\sin\psi_o}\right),$$

$$I_o^{1(d)} = \tan^{-1}(a)\tan\psi_o \exp\left(-\frac{\mu_e t}{\sin\psi_o}\right).$$

In the case of a symmetric reflection these expressions reduce to

$$I_h^{1(a)} = |\xi'_h|^2,$$

$$I_h^{1(d)} = \frac{1-|\xi'_h|^2}{1+|\xi'_h|^2}|\xi'_h|^2 \exp\left(-\frac{\mu_e t}{\sin\psi_o}\right),$$

$$I_o^{1(d)} = \frac{1-|\xi'_h|^2}{1+|\xi'_h|^2}\exp\left(-\frac{\mu_e t}{\sin\psi_o}\right).$$

It is easy to check that, when there is no absorption, $I_h^{1(a)} + I_h^{1(d)} + I_o^{1(d)} = 1$, which states the conservation of energy.

12.12.3 *Bragg–Bragg geometry; multiple reflections of a pseudo-plane wave in thin crystals*

When the wavefield created at A^1 reaches the bottom surface of the crystal, at C^1, partial reflection takes place (Fig. 12.27(c)): the wavefield of tiepoint P'' is now excited. P'' lies at the intersection of the dispersion surface with the outward normal to the back surface. It is here assumed that the crystal is plane parallel. The amplitudes of the wavefield components are $D_o''^{(1)}$ and $D_h''^{(1)}$, respectively. The wavefield reaches the top surface at A^2 where it is partially transmitted, giving rise to a reflected wave, $D_h^{2(a)}$. It is also partially reflected, giving rise to a wavefield penetrating in the crystal, of tiepoint P' and amplitudes $D_o'^{(2)}$ and $D_h'^{(2)}$, respectively. This wavefield reaches the back surface at C^2, where it is partially reflected, giving rise to a wavefield of tiepoint P''; this wavefield reaches the top surface at A^3 where it gives rise to a reflected wave, $D_h^{3(a)}$, and so on. The process will be repeated as long as the lateral extension of the crystal is large enough and there will be a succession of reflected beams: $D_h^{1(a)}$, $D_h^{2(a)}$, $D_h^{3(a)}$, etc. This situation was discussed by Wagner (1956) for a pseudo-plane wave. The case of an incident spherical wave was studied by Saka *et al.* (1973).

The boundary conditions at A^1 are given by (12.29). The boundary conditions at C^1 are given in Section 7.3.1. It follows that

$$D_h''^{(1)} = -\xi_h' \exp(-2\pi \mathrm{i}\overline{P''P'}t)D_o^{(a)},$$

$$D_o''^{(1)} = -\frac{\xi_h'}{\xi_h''} \exp(-2\pi \mathrm{i}\overline{P''P'}t)D_o^{(a)},$$

$$D_o^{1(d)} = \left(1 - \frac{\xi_h'}{\xi_h''}\right) \exp(-2\pi \mathrm{i}\overline{MP'}t)D_o^{(a)},$$

where $\xi_h'' = D_h''^{(1)}/D_o''^{(1)}$. The boundary conditions at A^2 are

$$D_o''^{(1)} + D_o'^{(2)} = 0,$$

$$D_h^{1''(1)} + D_h'^{(2)} = D_o^{2(a)},$$

and the corresponding amplitudes are

$$D_o'^{(2)} = \frac{\xi_h'}{\xi_h''} \exp(-2\pi \mathrm{i}\overline{P''P'}t)D_o^{(a)},$$

$$D_h'^{(2)} = \frac{(\xi_h')^2}{\xi_h''} \exp(-2\pi \mathrm{i}\overline{P''P'}t)D_o^{(a)},$$

$$D_h^{2(a)} = \xi_h'\left(\frac{\xi_h'}{\xi_h''} - 1\right) \exp(-2\pi \mathrm{i}\overline{P''P'}t)D_o^{(a)}.$$

The corresponding intensities reflected at A^1, A^2, A^3, etc. are rapidly decreasing. Since all the beams propagating through the crystal belong to the same branch of the dispersion surface, they all undergo either anomalously low or anomalously high absorption.

12.12.4 *Laue–Bragg geometry—Borrmann–Lehmann fringes*

Figure 12.26 illustrates the Laue–Bragg geometry. We shall consider here the simple case where there is a single partial reflection on a lateral face (Fig. 12.28) which is sometimes called Laue–Bragg–Laue. The incident beam is a spherical wave impinging at A on the entrance surface and the reflection is symmetrical.

The Borrmann triangle ABC is partially intercepted by a lateral face and one distinguishes two cases according to whether it is the reflected beam side, AB (Fig. 12.28(a)), or the direct beam side, AC (Fig. 12.28(b)), of the Borrmann triangle which cuts a lateral face. They are called *type I* according to Saka *et al.* (1972*a*) or *negative* according to Lang *et al.* (1986*b*), and *type II* according to Saka *et al.* (1972*a*) or *positive* according to Lang *et al.* (1986*b*), respectively.

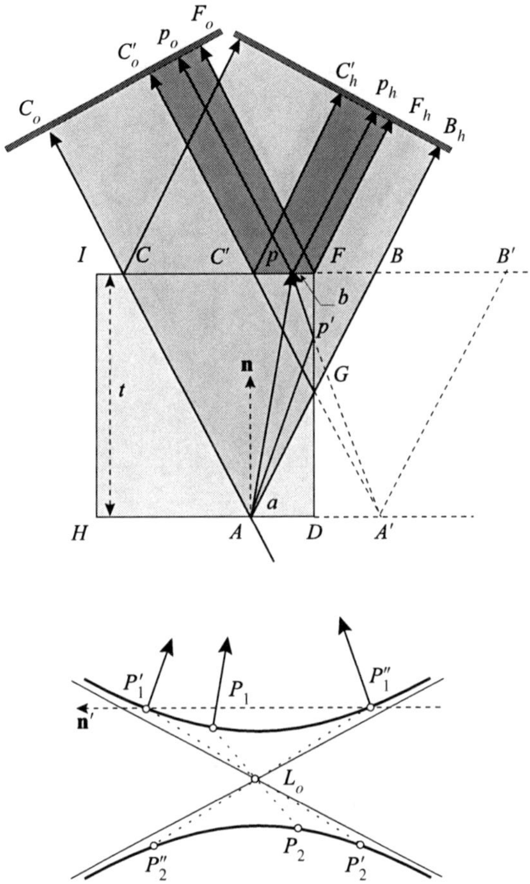

Fig. 12.28. (a)

Let G be the intersection of AB with the lateral face, DF (type I, Fig. 12.28(a)), C' the intersection with the exit surface of the beam partially reflected at G and p a point of the exit surface between F and C'. At that point arrive four wavefields: two have propagated directly along $\mathbf{Ap}$, of tiepoints P_1 and P_2, respectively; the two others, of tiepoints P_1'' and P_2'', have propagated along $\mathbf{p'p}$ and are due to the partial reflection of two wavefields, of tiepoints P_1' and P_2', which have propagated along $\mathbf{Ap}'$. The path $\mathbf{p'p}$ proceeds from the summit, A', of a virtual Borrmann triangle, $A'B'C'$, whose side $A'C'$ passes through G. A similar situation arises for type II geometry and is represented in Fig. 12.28(b).

Two types of interferences occur at p: on the one hand, the interference of the wavefields of tiepoints P_1 and P_2 and that of the wavefields of tiepoints P_1'' and P_2'', which give rise to the usual spherical-wave Pendellösung fringes; on the other hand, the interference of the wavefields of tiepoints P_1 and P_1'' and that of the wavefields of tiepoints P_2 and P_2'' which give rise to the fringes first observed by Borrmann and

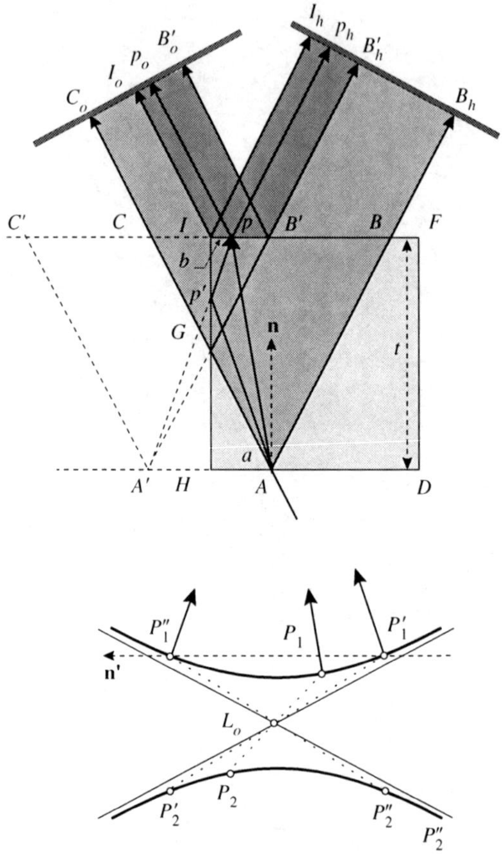

Fig. 12.28. (b)

Fig. 12.28. Borrmann–Lehmann fringes, Laue–Bragg–Laue geometry; right-hand side: reciprocal space; $\mathbf{n}'$: normal to the lateral face; left-hand side: direct space; the region where the Borrmann–Lehmann fringes occur is in dark grey. (a) Type *I*, negative. (b) Type *II*, positive.

Lehmann (1962) and first interpreted by Borrmann and Lehmann (1963) and Lehmann and Borrmann (1967); an example of their observations is shown in Fig. 12.29. Their studies were actually limited to the high-absorption case where branch 2 wavefields are absorbed out and Pendellösung fringes are not visible. It should be noted that this new type of interference is between tiepoints which belong to the *same* branch of the dispersion surface. These fringes were studied in the low-absorption case by Lang *et al.* (1986*b*, 1990) using synchrotron radiation and by Snigirev and Suvorov (1993) in the case of a periodic relief on silicon crystals.

The intensity distribution on the exit surface was calculated by Mai and Zhao (1989) and by Lang *et al.* (1990) using Kato's spherical-wave dynamical theory and the results of Saka *et al.* (1972*a*), with a good agreement between experimental results

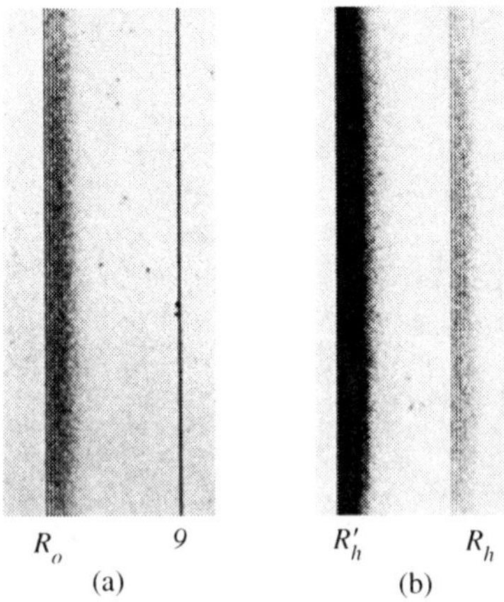

Fig. 12.29. Borrmann–Lehmann fringes; silicon, 220 reflection, Mo$K\alpha$; crystal thickness $t = 5.25$ mm; $a = 0.2$ mm; type I geometry. Courtesy K. Lehmann. (a) Refracted beam; P: primary beam; R_o: fringes due to the partially reflected beam. (b) Reflected beam; R'_h: reflected beam through the lateral face; R_h: fringes due to the partially reflected beam.

and computer simulations. It was also calculated by Thorkildsen and Larsen (1999*c*) using Takagi's theory.

The amplitude distribution along the base of the Borrmann triangle of the wavefields having propagated along **Ap** is given by (10.21):

$$D_h = -\mathrm{i}\pi \mathcal{B}_h J_o\Big(\frac{\pi t}{\Lambda_L}\sqrt{1 - Y^2}\Big),$$

where the reduced coordinate, Y, of p with respect to the base BC defined by (10.16) is given by

$$Y = \frac{\mp(a - b)}{t \tan \theta_B},$$

where the top sign corresponds to type I and the bottom sign to type II, a and b are the distances of A and B, respectively, from the lateral face where the partial reflection takes place (Fig. 12.28) and t is the crystal thickness.

The amplitude distribution of the wavefields having propagated along $\mathbf{Ap'p}$ was shown by Saka *et al.* (1972*a*) to be

$$D'_h = -\mathrm{i}\pi \mathcal{B}_h \frac{1 - Y'}{1 + Y'} J_2\Big(\frac{\pi t}{\Lambda_L}\sqrt{1 - Y'^2}\Big) \quad \text{(type } I\text{)},$$

$$D'_h = \mathrm{i}\pi \mathcal{B}_h J_o\Big(\frac{\pi t}{\Lambda_L}\sqrt{1 - Y'^2}\Big) \quad \text{(type } II\text{)},$$

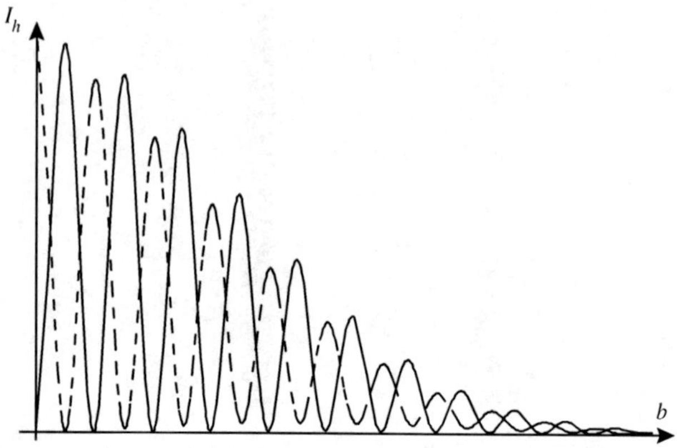

Fig. 12.30. Borrmann–Lehmann fringes, reflected beam. Dashed line: type *I*; solid line: type *II*. Experimental conditions: the same as in Fig. 12.29; arbitrary units.

where the reduced coordinate, Y', of p with respect to the base $B'C'$ of the virtual Borrmann triangle $A'B'C'$ is given by

$$Y' = \frac{\mp(a+b)}{t\tan\theta_B},$$

where the minus and plus signs correspond to type *I* and *II*, respectively.

The intensity distribution along $C'E$ (type *I*) or $B'I$ (type *II*) is given simply by

$$I_h = |D_h + D_h'|^2.$$

Figure 12.30 represents the variation of I_h with b for the same experimental conditions as in Fig. 12.29. One can see that the positions of the fringes in type *I* and in type *II* geometries are shifted with respect to one another by half a fringe.

The Bessel functions can be replaced in practice by their asymptotic expressions and Lang *et al.* (1990) have shown that, for b small and for high absorption, this expression reduces to

$$I_h \simeq \sin^2\left[\frac{\pi t}{2\Lambda_L}\left(\sqrt{1-Y^2}-\sqrt{1-Y'^2}\right)\right]$$

for type *II*.

With the approximation that a and b are small, the expressions $\sqrt{1-Y^2}$ and $\sqrt{1-Y'^2}$ can be expanded. It follows simply that

$$I_h \simeq \sin^2\left(\frac{\pi}{\Lambda_L}\frac{ab}{t\tan^2\theta_B}\right).$$

This shows that the period of the Borrmann–Lehmann fringes along the base of the Borrmann triangle, close to the lateral face of the crystal, is

$$\Delta = \frac{t}{a}\Lambda_L \tan^2\theta_B.$$

This expression was first derived by Borrmann and Lehmann (1963) by calculating the path difference for the waves propagating along **Ap** and **Ap′p**.

12.13 Coherence of extended, non-strictly monochromatic sources

Real sources, whether an X-ray tube or a synchrotron, are never point sources, as has been assumed up to now, but are extended and they are never strictly monochromatic. This results in a decrease of the coherence of the beam. This can be understood by considering the interference of two beams having followed two different paths, as in Young's slits' experiment in visible optics or in the Pendellösung interference of two wavefields described by Fig. 4.15. Consider first the effect of non-monochromaticity:

1. *Temporal or longitudinal coherence length*: If the beam emitted by the source is not monochromatic, it can be considered as a sum of plane waves each having a different wavelength which can be described by a Fourier expansion:

$$F(t) = \int_{-\infty}^{+\infty} f(\nu)\exp[-2\pi i\nu t]\mathrm{d}t,$$

where the amplitude distribution can be expressed using the inverse Fourier transform:

$$f(\nu) = \int_{-\infty}^{+\infty} F(t)\exp[2\pi i\nu t]\mathrm{d}\nu.$$

It can also be considered as made up of wave trains of finite length passing at random time intervals (Born and Wolf 1993). In the experiment this train is divided into two parts following different paths to arrive at the point of observation, p. If the difference in optical path is large enough, the train having followed the second path arrives at p after the first one has left and there is no interference. For interferences to take place, the length of the wave trains should be as large as possible. This length is called the temporal or longitudinal coherence length. It can be evaluated in the following way. Let Δt be the duration of a wave train and $\Delta\nu$ the frequency range over which the intensity of the Fourier components, $|f(\nu)|^2$, has an appreciable intensity. It is to be expected that the wave trains are longer, the narrower the frequency range $\Delta\nu$ and it can be shown (Born and Wolf 1993 p. 319) that:

$$\Delta t\Delta\nu \sim 1.$$

The time Δt is called the *coherence time* of the beam. If λ_o is the mean wavelength, one has, by the derivation of the relation $\nu = c/\lambda$:

$$\Delta\nu = \frac{c\Delta\lambda}{\lambda_o^2}.$$

It follows that:

$$\Delta\ell_T = c\Delta t \sim \frac{\lambda_o^2}{\Delta\lambda}$$

where $\Delta\ell_T$ is the *temporal coherence length.*

2. *Partial coherence – Degree of coherence – Transverse or spatial coherence:* Let us now consider two points in the beam coming from an extended source. These points lie for instance on each of the two slits of Young's experiment, or are the points p_1 and p_2 of Fig. 4.15. The degree of correlation between the vibrations of electric field at these two points may be estimated from the sharpness of the fringes observed at point p and due to the interference of these two vibrations: if the fringes are very sharp, the vibrations are said to be coherent and if there are no fringes they are incoherent. In between these two situations, they are *partially coherent.* In order to describe these intermediary cases, Zernike (1938) has introduced the notion of *degree of coherence* which is analysed in detail in Born and Wolf (1993). Following these authors, one may write the intensity of the interference field at p as:

$$I(p) = I_1(p) + I_2(p) + 2\sqrt{I_1(p)I_2(p)}\,|\gamma_{12}(\tau)|\cos[\alpha_{12}(\tau) - \delta],$$

where $I_1(p)$, $I_2(p)$ are the intensities of the waves having followed the paths p_1p and p_2p, respectively, $\tau = (p_2p - p_1p)/c$, $\delta = 2\pi(p_2p - p_1p)/\lambda_o$ and $|\gamma_{12}(\tau)|$ is, by definition, the *degree of coherence.*

Let I_{max} and I_{min} be the maxima and minima of intensity of the interference pattern in the vicinity of p. The visibility of the fringes at p is

$$\mathcal{V}(p) = \frac{I_{max} - I_{min}}{I_{max} + I_{min}} = \frac{2\sqrt{I_1(p)I_2(p)}}{I_1(p) + I_2(p)}|\gamma_{12}(\tau)|$$

If $I_1(p) = I_2(p)$ which is usually the case in optics (and which would be the case in the experiment described by Fig. 4.15 in the absence of absorption), the degree of coherence is directly given by the visibility of the fringes.

If the source size and the distance p_1p_2 are small with respect to the distance from the source to these points, the degree of coherence is equal to the absolute value of the normalized Fourier transform of the intensity distribution of the source. This is the van Cittert–Zernike theorem (Born and Wolf 1993). It is also shown that the degree of coherence decreases towards very small values when

$$p_1p_2 \leq \Delta\ell_S = \frac{L\lambda_o}{a}$$

where L is the distance of the source from p, a the size of the source and $\Delta \ell_S$ is called the spatial or transverse coherence length (it is measured in the direction perpendicular to the propagation direction and is usually different in the horizontal and vertical directions).

From the definition of the degree of coherence it is seen that it can be deduced from the visibility of interference fringes. This has been applied to determine the spatial coherence length of synchrotron radiation:

(i) by Snigirev (1996) who observed the Fresnel diffraction fringes of boron fibers;

(ii) by Baruchel *et al.* (1997) and Cloetens *et al.* (1997*b*) who observed Talbot fringes (under coherent illumination a periodic object will produce through Fresnel diffraction an image which is not only periodic in the direction perpendicular to the direction of propagation but also along the direction of propagation – this is the Talbot effect);

(iii) by Ishikawa (1988) and Tamasaku and Ishikawa (2001) who observed the equal-thickness Pendellösung fringes due to a silicon wedge; and

(iv) by Mocella *et al.* (2000) who observed the equal-orientation Pendellösung fringes in a silicon crystal.

(v) by Lin *et al.* (1996) who studied the propagation of the spatial coherence from the diffraction patterns of a pair of slits.

(vi) by Panzner *et al.* (2003) who studied Fraunhofer patterns of a 5 μm entrance pin hole.

(vii) by Lang and Makepeace (1999*c*) who observed Fresnel biprism interference fringes from two diamond plates;

(viii) by Kohn *et al.* (2000, 2001) and Leitenberger *et al.* (2001) using the Fresnel diffraction by a slit.

(ix) by Yabashi *et al.* (2001) using the Hanbury Brown-Twiss experiment and an intensity interferometer.

The propagation of the coherence of an X-ray beam during dynamical diffraction has been studied recently by Yamazaki and Ishikawa (2002) using time-dependent Takagi equations. They have shown in particular that the temporal and spatial components of the coherence cannot in fact be handled separately. Yamazaki and Ishikawa (2004) have further shown that rocking curves are in fact convolutions of the intrinsic profile with the Fourier transform of the mutual coherence function of the incident wave.

III

Extension of the dynamical theory to deformed crystals

13

Ray tracing in slightly deformed crystals

13.1 X-ray propagation in deformed materials

13.1.1 *The different degrees of deformation*

Since the early days when the calculated diffracted intensities were first compared with experimental values, it was realized that most crystals contain imperfections, some being nearly perfect, such as calcite and quartz, the others more or less highly imperfect. It was not until silicon and germanium were grown for their applications as semiconductors that crystals of a high degree of perfection were obtained. For many applications of crystals as high-technology materials, good-quality crystals must be used and it is usually necessary to characterize the nature and the distribution of defects. For this purpose one must understand the propagation of X-rays in deformed crystals, and it is for this reason that extensions of the diffraction theory for imperfect crystals were developed. One may distinguish several cases (see also Härtwig 2001):

1. The crystal is only slightly deformed. The lateral width, e, or coherence of wavepackets remains roughly constant over a distance of the order of $e^2/\mathcal{A}\lambda \approx e^2/\Lambda_o$, where $\mathcal{A}$ is the angular amplification ratio defined in Section 12.6 (Fresnel regime—see eqn (12.27)); if this width is of the order of the Pendellösung distance, Λ_o, this distance is also of the order of Λ_o. The notions of dispersion surface and of wavefields as they were defined in the first two parts of this book are still valid; the paths of the wavefields in the deformed materials are bent as are rays of light in a region of varying index of refraction (the mirage effect, Fig. 13.1); these paths and the variations of the reflected intensity are calculated using 'ray theories' based on modifications of the traditional dynamical theory and Fermat's principle. They will be discussed in the present chapter.

2. The crystal is strongly deformed, but there is a model to describe the distribution of strain; when the strain gradient becomes very large the divergence of wavepackets increases, as in visible optics, because of Fraunhofer diffraction effects; this is accompanied for X-rays by 'interbranch scattering' (creation of new wavefields on the opposite branch of the dispersion surface) and the ray theories are no longer valid; new forms of the dynamical theories, such as those developed by Takagi (1962, 1969) and Taupin (1964), must be used; their principle was given in Chapter 11 which was limited to their application to perfect crystals. Their use for the study of highly deformed materials will be discussed in Chapter 14. If one applies a strain gradient to a perfect crystal and increases it, passing progressively through regimes I and II,

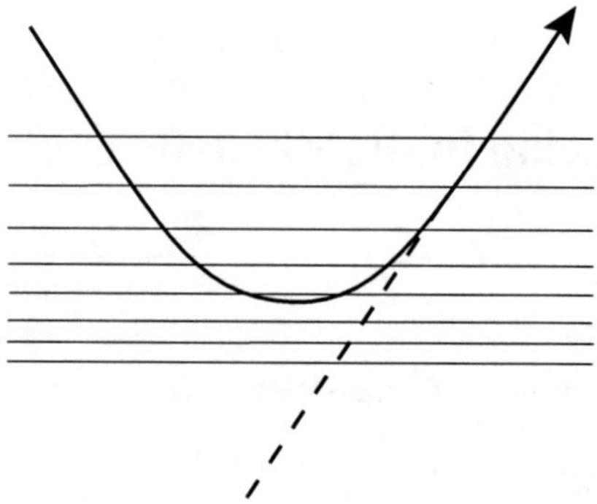

Fig. 13.1. Propagation of a wave in a medium with a variable index of refraction—mirage effect.

one may span the whole domain of variations of intensities, from the values given by the perfect crystal dynamical theory to the values given by geometrical theory for the 'ideally imperfect' crystal.

3. The crystal is strongly deformed, but the distribution of defects is very complicated and there is no model to describe the distribution of strain; extensions of the dynamical theory called statistical dynamical theories, which constitute the modern versions of extinction theory, must be used. They will be briefly discussed in Section 14.5.

13.1.2 *Principle of ray theories for weak deformations*

It is wellknown in light optics that if a wavepacket propagates in a region where the index of refraction, n, varies, the direction of the main vector of the wavepacket varies accordingly. The medium can be divided into small nearly homogeneous regions where the index of refraction can be considered as constant. The boundaries between these regions are the surfaces n = constant. There is continuity of the tangential component of the wavevectors across these boundaries and the variation of the wavevectors is

$$\delta \mathbf{k} \sim \vec{\nabla}(n). \tag{13.1}$$

It is along the direction of maximum variation of the index of refraction. The direction of propagation of the wavepacket is along the normal to the surface of indices and can be found by applying Fermat's principle.

Penning and Polder (1961) and Kato (1963*d*, 1964*a*,*b*) have applied this principle to X-rays propagating in a crystal under conditions of Bragg diffraction. When the deformation of the crystal is small enough, it is possible to consider at each point a local perfect crystal, *asymptotic* to the real deformed one to which dynamical theory applies. The local variation of the departure from Bragg's angle of the incident beam during the deformation is called the *effective misorientation*. It is due to the local rotation of the lattice planes and to the local variation of the lattice parameter. It will be shown in Section 13.2 how it can be related to the components of the strain tensor, and in Section 13.3 how the expression of $\delta\mathbf{k}$ can be expressed in terms of the *effective misorientation*. The resultant wavefield trajectories will be discussed in Section 13.5.

13.2 Effective misorientation

13.2.1 *Local reciprocal lattice vector*

If the crystal undergoes a deformation, any point p of position vector $\mathbf{r}$ is transformed in a point p' of position vector $\mathbf{r}'$ (Fig. 13.2):

$$\mathbf{r}' = \mathbf{r} + \mathbf{u}(\mathbf{r}).$$

This equation can also be written as

$$\mathbf{r} = \mathbf{r}' - \mathbf{u}(\mathbf{r})$$

and with a small approximation, valid if the coefficients of the strain tensor, $\partial u_i/\partial x_j$, are small with regard to unity, which is the usual approximation in elasticity:

$$\mathbf{r} \simeq \mathbf{r}' - \mathbf{u}(\mathbf{r}'). \tag{13.2}$$

Let us consider a perfect infinite crystal asymptotic at p to the local, real crystal, and let $\mathbf{h}'$ be its reciprocal lattice vector. It can be related to the reciprocal lattice vector

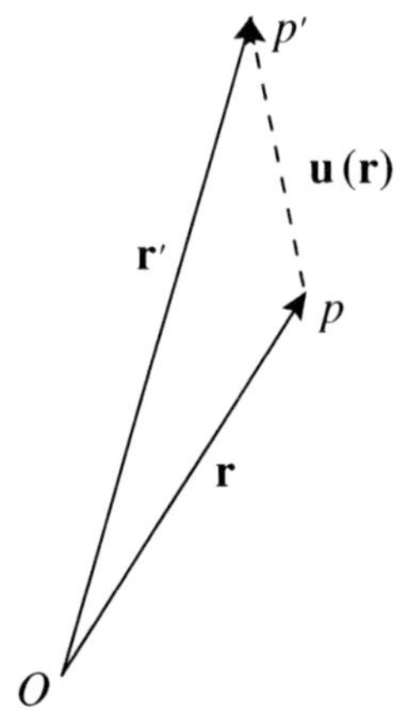

Fig. 13.2. Deformation of a material. $\mathbf{u}(\mathbf{r})$: deformation vector; $\mathbf{r}'$: position vector after deformation.

before deformation, **h**, by means of the displacement vector, $\mathbf{u}(\mathbf{r})$. The equation of a plane of the family of direct lattice planes (hkl) is

$$H \equiv \mathbf{h} \cdot \mathbf{r} = \mathcal{N}. \tag{13.3a}$$

The reciprocal lattice vector **h** can be defined as the gradient of H:

$$\mathbf{h} = \vec{\nabla}(H).$$

If one replaces in (13.3a) **r** by its expression in function of the position vector *after* deformation, $\mathbf{r}'$, the equation of the lattice planes becomes

$$H' \equiv \mathbf{h} \cdot \mathbf{r}' - \mathbf{h} \cdot \mathbf{u}(\mathbf{r}') = \mathcal{N}. \tag{13.3b}$$

This is the equation of the surfaces into which the lattice planes have been transformed by the deformation. The reciprocal lattice vector, $\mathbf{h}'$, of the asymptotic perfect crystal is given by

$$\mathbf{h}' = \vec{\nabla} H' = \mathbf{h} - \vec{\nabla}(\mathbf{h} \cdot \mathbf{u}). \tag{13.4a}$$

The local change $\delta\mathbf{h}$ of the reciprocal lattice is therefore

$$\delta\mathbf{h} = \mathbf{h}' - \mathbf{h} = -\vec{\nabla}(\mathbf{h} \cdot \mathbf{u}). \tag{13.4b}$$

It is a function of the direct space coordinates in the deformed region and its variations are given by

$$\mathrm{d}(\delta\mathbf{h}) = (\vec{\nabla} \otimes \delta\mathbf{h})\cdot \mathrm{d}\mathbf{r} = -\vec{\nabla} \otimes \vec{\nabla}(\mathbf{h} \cdot \mathbf{u})\cdot \mathrm{d}\mathbf{r},$$

where $\otimes$ denotes a tensor product. Its components are

$$\mathrm{d}(\delta\mathbf{h_j}) = -\frac{\partial^2(\mathbf{h} \cdot \mathbf{u})}{\partial x_i\, \partial x_j}\, \mathrm{d}x_i.$$

The ray theories are only valid if this quantity is small with respect to the width of the dispersion surface, that is smaller than Λ_o^{-1}. If the lateral width of the wavepacket is of the order of Λ_o, the condition of validity of the ray theories can be written as

$$\frac{\partial^2(\mathbf{h} \cdot \mathbf{u})}{\partial x_i\, \partial x_j} \ll \Lambda_o^{-2}. \tag{13.5}$$

It will be shown in Section 13.2.3 how $\delta\mathbf{h}$ can be expressed in terms of the local variation of the departure from Bragg's angle of the incident wavevector.

13.2.2 *Effective misorientation in direct space*

In general, during a deformation, the lattice planes are rotated and the lattice spacing varies. The angle, θ, between the incident wavevector and the lattice planes and the Bragg angle, θ_B, are both modified. The departure from Bragg's angle of the incident wavevector is, before deformation,

$$\Delta\theta = \theta - \theta_B.$$

The value of the Bragg angle after deformation is obtained through derivation of Bragg's law:

$$\theta'_B = \theta_B - \frac{\delta d}{d_{hkl}} \tan\theta_B,$$

where δd is the variation of the reflecting plane spacing. The *new* departure from Bragg's angle of the incident beam, *after* deformation, is

$$\Delta\theta' = \theta' - \theta'_B,$$

where θ' is the angle between the incident wavevector and the lattice planes, after deformation (Fig. 13.3(a)).

The value of θ' depends on the angle of rotation of the lattice planes. Let $\mathbf{n}'$ be the direction of the normal to the lattice planes after deformation, α the angle between $\mathbf{n}$ and $\mathbf{n}'$, and φ the angle between the normal to the $(\mathbf{n}, \mathbf{n}')$ plane and the normal to the diffraction plane (Fig. 13.3(b)). It is obtained by calculating $\mathbf{n}' \cdot \mathbf{s_o}$ where $\mathbf{s_o}$ is a unit vector parallel to the incident wavevector. It follows that

$$\sin\theta' = \sin\alpha\cos\varphi\cos\theta + \cos\alpha\sin\theta.$$

1. If the rotation axis is normal to the diffraction plane ($\varphi = 0$), one has

$$\theta' - \theta = \alpha.$$

2. If the rotation axis is parallel to the diffraction plane ($\varphi = \pi/2$), one has

$$\theta' - \theta = -\frac{\alpha^2}{2}\tan\theta.$$

3. In the general case one has (to the first order)

$$\theta' - \theta = \alpha\cos\varphi.$$

The local variation of the departure from Bragg's angle, or *effective misorientation*, is in the general case

$$\delta\theta = \Delta\theta' - \Delta\theta = \theta' - \theta'_B - (\theta - \theta_B) = \alpha\cos\varphi + \frac{\delta d}{d}\tan\theta_B. \tag{13.6}$$

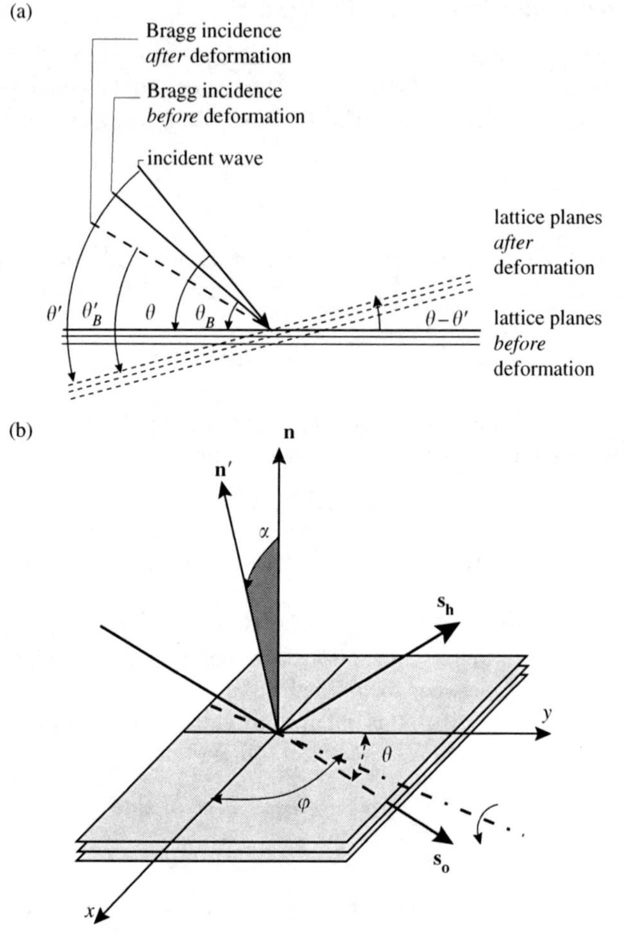

Fig. 13.3. Effective misorientation. (a) Variation of the departure from Bragg's angle with deformation: it is $\theta - \theta_B$ before deformation and $\theta' - \theta'_B$ after deformation. (b) Variation of the angle, θ, between the incident wavevector, $\mathbf{s_o}$, and the lattice planes when they are rotated; $\mathbf{n}'$: normal to the lattice planes after deformation; α: rotation angle; φ: angle between the rotation axis and the normal to the diffraction plane.

13.2.3 *Effective misorientation in reciprocal space*

The effective misorientation can also be expressed in reciprocal space (Authier 1966) in terms of the local variation of the reciprocal lattice vector, $\delta\mathbf{h}$, given by (13.4). Let $\mathbf{HH}' = \delta\mathbf{h}'$ be the projection of $\delta\mathbf{h}$ on the diffraction plane (Fig. 13.4). The Laue point, L_a, lies, by definition, on the sphere of radius $k = 1/\lambda$ and centre O, and on the mediatrix of $\mathbf{OH}$. Let G be the middle point of $\mathbf{OH}$. The new, local, Laue point,

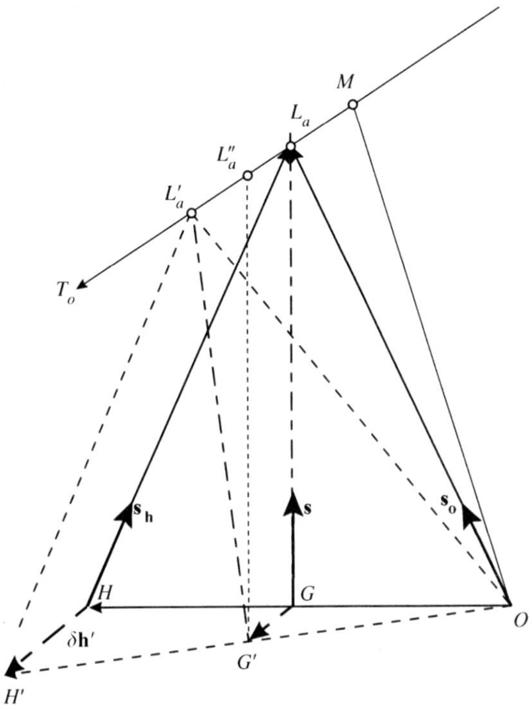

Fig. 13.4. Reciprocal space: variation of the departure from Bragg's angle with deformation. It is $L_a M/k$ before deformation and $L'_a M/k$ after deformation (L_a: Laue point before deformation, L'_a: Laue point after deformation); $\delta\mathbf{h}'$: variation of the reciprocal lattice vector.

L'_a, after deformation, lies on the same sphere and on the mediatrix, $\mathbf{G'L'_a}$, of the new reciprocal lattice vector, $\mathbf{OH'}$. Let $\mathbf{OM} = \mathbf{K_o^{(a)}}$ be an incident wavevector.

If the deformation is small enough, the angle ($\mathbf{OL_a}$, $\mathbf{OL'_a}$) is small and the arc $\widehat{ML_aL_a}$ can be merged with the segment $\overline{ML_aL'}$ on the tangent, T_o, to the sphere. The effective misorientation is

$$\delta\theta = \Delta\theta' - \Delta\theta = \frac{\overline{L'_a M} - \overline{L_a M}}{k} = \frac{\overline{L'_a L_a}}{k}.$$

The value of $\overline{L'_a L_a}$ can be evaluated in the following way. One has

$$\mathbf{L_a L'_a} = \mathbf{L_a H} + \mathbf{HH'} + \mathbf{H'L'_a} = \mathbf{K'_h} - \mathbf{K_h} + \delta\mathbf{h}'.$$

Let us take the scalar product with $\mathbf{K_h}$:

$$\begin{aligned} \mathbf{L_a L'_a} \cdot \mathbf{K_h} = \overline{L'_a L_a}\, k \sin 2\theta_B &= (\mathbf{K'_h} - \mathbf{K_h})\cdot\mathbf{K_h} + \delta\mathbf{h}' \cdot \mathbf{K_h} \\ &= \delta\mathbf{h} \cdot \mathbf{K_h} \end{aligned} \tag{13.7a}$$

since $K_h' = K_h = k$ and the scalar product of $\mathbf{K_h}$ with the component of $\delta\mathbf{h}$ in the direction normal to the diffracting plane is zero. It follows that

$$\delta\theta = \frac{\delta\mathbf{h}\cdot\mathbf{K_h}}{k^2\sin 2\theta_B} = -\frac{\mathbf{K_h}\cdot\vec{\nabla}(\mathbf{h}\cdot\mathbf{u})}{k^2\sin 2\theta_B} = -\frac{1}{k\sin 2\theta_B}\frac{\partial(\mathbf{h}\cdot\mathbf{u})}{\partial x_h} \tag{13.7b}$$

after replacing $\delta\mathbf{h}$ by its expression (13.4); x_h is a coordinate in the reflected direction.

The effective misorientation is zero if $\delta\mathbf{h}\cdot\mathbf{K_h} = 0$, that is if the local variation of the reciprocal lattice vector is normal to the reflected direction. This means that any deformation of the crystal such that

$$\mathbf{h}'\cdot\mathbf{K_h} = \text{constant}$$

at every point has no influence on the propagation of X-rays.

The effective misorientation is a linear combination of components of the strain tensor $\partial u_i/\partial x_j$. It is also interesting to decompose it as the sum of two terms, one representing the rotation of the lattice planes and the other representing the variation of the lattice parameter. Let L_a'' be the intersection with T_o of the parallel $\overline{G'L_a''}$ to $\overline{GL_a}$ drawn from the midpoint, G', of $\mathbf{OH}'$. It can be seen in the figure that

$$\overline{L_a'L_a} = \overline{L_a'L_a''} + \overline{L_a''L_a}. \tag{13.8}$$

The first term corresponds to the component of the rotation of the lattice planes around the normal to the lattice planes. Noticing that the angle $(\mathbf{G'L_a''}, \mathbf{G'L_a'})$ is equal to the angle $(\mathbf{OH}, \mathbf{OH}')$, it follows that

$$\overline{L_a'L_a''} = \frac{\delta\mathbf{h}'\cdot\mathbf{s}}{2\sin\theta_B}, \tag{13.9a}$$

where $\mathbf{s}$ is a unit vector along $\mathbf{GL_a}$.

The second term corresponds to the variation of the lattice parameter:

$$\overline{L_a''L_a} = -\frac{\delta\mathbf{h}'\cdot\mathbf{h}}{2k\sin 2\theta_B}. \tag{13.9b}$$

Substituting (13.9*a*) and (13.9*b*) in (13.8), and noticing that $\mathbf{K_h} = k\cos\theta_B\mathbf{s} - \mathbf{h}/2$, one gets (13.7*b*).

13.2.4 *Strain gradient*

The derivative of the effective misorientation along the incident direction is one of the components of the strain gradient and plays an important role in the properties of the propagation of X-rays in deformed crystals. It is usually introduced by means of a coefficient β proportional to its ratio over the half-width δ_{os} (4.26) of the reflection domain and which is, in the symmetrical Laue geometry,

$$\beta = \frac{1}{\delta_{os}\cos\theta_B}\frac{\partial(\delta\theta)}{\partial x_o} = -\frac{\Lambda_o}{\cos^2\theta_B}\frac{\partial^2(\mathbf{h}\cdot\mathbf{u})}{\partial x_o\,\partial x_h}, \tag{13.10a}$$

where x_o is a coordinate in the incident direction. Using this parameter, the condition of validity (13.5) of the ray theories is simply

$$|\beta| \ll \left|\Lambda_o\right|^{-1}. \tag{13.10b}$$

The case of a uniform strain gradient is of particular interest because the ray trajectories can be obtained analytically. It corresponds, for instance, to a homogeneous bending (see Section 13.6.2), to a constant thermal gradient (see Section 13.6.3) or to a constant concentration gradient (see Section 14.4.2).

13.3 Polarizability of a deformed crystal

The polarizability is a triply periodic function of space coordinates in a perfect crystal. This is no longer the case in a deformed crystal. If one replaces **r** in the development (2.36) of the dielectric susceptibility by its expression (13.2), there comes

$$\chi(\mathbf{r}') = \sum_h \chi_h \exp\{2\pi \mathrm{i}\mathbf{h} \cdot [\mathbf{r}' - \mathbf{u}(\mathbf{r}')]\}. \tag{13.11a}$$

This expression can be rewritten, using (13.3b), as

$$\chi(\mathbf{r}) = \sum_h \chi_h \exp\{2\pi \mathrm{i} H'(\mathbf{r})\} \tag{13.11b}$$

or

$$\chi(\mathbf{r}) = \sum_h \chi_h \exp\{2\pi \mathrm{i}\mathbf{h}' \cdot \mathbf{r}\} \tag{13.11c}$$

remembering that the local reciprocal lattice vector $\mathbf{h}'$ is given by (13.4a):

$$\mathbf{h}' = \vec{\nabla} H'(\mathbf{r}).$$

13.4 The Eikonal approximation

13.4.1 *Justification of the concept of local dispersion surface*

In an isotropic, homogeneous, medium a propagating plane wave is of the form

$$\mathbf{D} \exp(-2\pi \mathrm{i}\mathbf{K} \cdot \mathbf{r}).$$

It is well known in general optics that, if the medium is inhomogeneous but the inhomogeneities are very small, it takes the form

$$\mathbf{D} \exp[-2\pi \mathrm{i} S(\mathbf{r})],$$

where $S(\mathbf{r})$ is called the Eikonal. As has been recalled in Section 13.1.2, the waves propagate along the direction normal to the surfaces n = constant, namely the surfaces $S(\mathbf{r})$ = constant. Their local wavevectors are given by

$$\mathbf{K}' = \vec{\nabla} S(\mathbf{r}).$$

By analogy, a wavefield propagating in an isotropic, slightly deformed crystal will be of the form

$$\mathbf{D} = \sum_{\mathbf{h}} \mathbf{D_h}(\mathbf{r}) \exp[-2\pi \mathrm{i} S_h(\mathbf{r})], \tag{13.12a}$$

where

$$S_h(\mathbf{r}) = S_o(\mathbf{r}) - H(\mathbf{r}). \tag{13.12b}$$

and the local wavevectors are given by

$$\mathbf{K_h}(\mathbf{r}) = \vec{\nabla} S_h(\mathbf{r}). \tag{13.12c}$$

These relations show that the wavevectors (as well as the local reciprocal lattice vector) are rotation free.

If, following Kato (1963*d*), one replaces $\mathbf{D}$ and χ by their modified expressions in the propagation equation (2.55), one finds that this propagation equation can be written as[1]

$$K_1 + K_2 + K_3 = 0,$$

with

$$K_1 = -4\pi^2 [\mathbf{K_h} \wedge \mathbf{K_h} \wedge \sum_g \chi_{h-g} \mathbf{D_g} + K_h^2 \mathbf{D_h} - k^2 \mathbf{D_h}]$$
$$\times \exp(-2\pi \mathrm{i} \mathbf{K_h} \cdot \mathbf{r}),$$
$$K_2 = -2\pi \mathrm{i}[2\mathbf{K_h} \wedge \sum_g \chi_{h-g} \mathrm{curl}\, \mathbf{D_g} + 2(\mathbf{K_h} \cdot \vec{\nabla})\mathbf{D_h} + \mathbf{D_h} \vec{\nabla} \cdot \mathbf{K_h}]$$
$$\times \exp(-2\pi \mathrm{i} \mathbf{K_h} \cdot \mathbf{r}),$$
$$K_3 = \left[\sum_g \chi_{h-g} \mathrm{curl}\,\mathrm{curl}\, \mathbf{D_g} + \Delta \mathbf{D_h} \right] \exp(-2\pi \mathrm{i} \mathbf{K_h} \cdot \mathbf{r}).$$

[1] The calculation is analogous to that used in Section 11.2.2 to derive Takagi's equations. The difference is that here the local wavevectors $\mathbf{K_h}$ are not constant, which is at the origin of extra terms with respect to the expressions in Section 11.2.2. The reason is that in the Eikonal approximation one is assuming that the concept of wavefield still holds despite the fact that the crystal is deformed; one can therefore apply the results of geometrical optics (ray theory). This is only possible for weak deformations. In contrast, as will be seen in Chapter 14, Takagi's equations are valid for strong deformations, but they do not provide an interpretation of the total field in terms of wavefields.

K_1 is proportional to the square of the first derivative of the Eikonal, K_2 includes terms proportional to the first derivative of the amplitudes and the first derivative of the Eikonal or to the second derivative of the Eikonal, and K_3 is proportional to the second derivative of the amplitude.

If one considers the amplitudes and the Eikonal as independent functions, there are too many equations and one may add one further constraint. One takes $K_1 = 0$ which will give a good approximation for the wavefield if the wavevectors are large and the amplitudes and the Eikonal are slowly varying functions (small deformations). One notices immediately that this equation is identical to the fundamental dispersion equations (5.4) in a perfect crystal. One can therefore indeed define a local dispersion surface whose equation is deduced from $K_1 = 0$. As in the case of perfect crystals, the local values of the amplitude ratios $\xi = D_h/D_o$ are derived from this equation.

The second derivatives of the amplitudes can be neglected if the amplitudes are slowly varying functions and one therefore has $K_3 = 0$ and $K_2 = 0$. Kato (1963*d*) has shown that through multiplying K_2 by the complex conjugate of the wavefield, $\mathbf{D}^*$, and neglecting the oscillating terms, equation $K_2 = 0$ reduces to

$$\vec{\nabla} \sum_h \mathbf{K_h} |D_h|^2 = 0.$$

Knowing the amplitude ratios from $K_1 = 0$, the absolute value of $|D_0|$ is deduced from this equation and the local wavefield can be completely determined.

13.4.2 *Fermat's principle*

Kato (1963*d*) has shown that the paths followed by the wavefields in the deformed crystal can be determined using Fermat's principle. Let us consider two points within the crystal, A and B, and two paths from A to B; one is the real one, Ap_1p_2B, and the other is arbitrary, Aq_1q_2B (Fig. 13.5(a)); p_1 and p_2, q_1 and q_2 are the intersections of these two paths with two neighbouring Eikonal surfaces (S_o = constant), S_{o1} and S_{o2}. Let us now consider a vector field, $\mathbf{k_o}$, defined by

$$\mathbf{k_o} = \vec{\nabla} S_o.$$

At any point, q, for instance along Aq_1q_2B, $\mathbf{k_o}$ is normal to the Eikonal surface, S_o, at that point. This vector satisfies

$$\operatorname{curl} \mathbf{k_o} = 0$$

and, therefore, the path integral, $\int_A^B \mathbf{k_o} \cdot \mathrm{d}\vec{l} = S_o(B) - S_o(A)$, where $\mathrm{d}\vec{l}$ is a path element, is independent of the path from A to B. For instance,

$$\int_{q_1}^{q_2} \mathbf{k_o} \cdot \mathrm{d}\vec{l} = \int_{p_1}^{p_2} \mathbf{k_o} \cdot \mathrm{d}\vec{l} = S_{o2} - S_{o1}. \tag{13.13a}$$

Let us now define at any point, for instance q, of the path Aq_1q_2B, a vector, $\mathbf{K_o}$, satisfying the two following conditions:

1. It is parallel to a wavevector **OQ** of the local dispersion surface in q (Fig. 13.5(b)).
2. The direction of propagation of the corresponding wavefield (that of the normal to the local dispersion surface at Q, of unit vector $\vec{\nu}$) is tangent in q to the path Aq_1qq_2B.

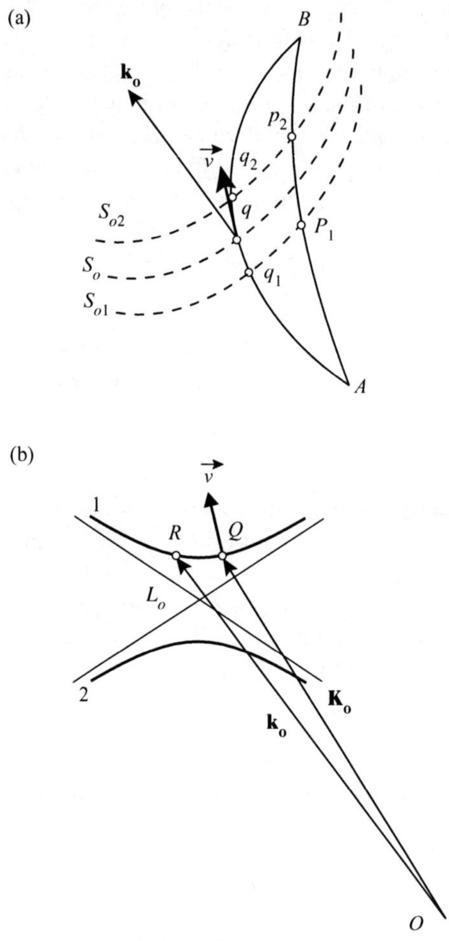

Fig. 13.5. Application of Fermat's principle to the determination of ray trajectories. (a) Direct space; Ap_1p_2B is a real trajectory, Aq_1q_2B is an arbitrary trajectory, the wavevector $\mathbf{k_o}$ at q is normal to the Eikonal surface at that point. (b) Reciprocal space; the normal ν to the dispersion surface at Q is parallel to the tangent to the ray trajectory at q, R extremity of $\mathbf{k_o} = \mathbf{OR}$.

The vector $\mathbf{K_o}$ is in general different from the vector $\mathbf{k_o}$ normal to the Eikonal surface, S_o in q. One may define the integral

$$S = \int_A^B \mathbf{K_o} \cdot \mathrm{d}\vec{l} = \int_A^B \mathbf{K_o} \cdot \vec{\nu}\, \mathrm{d}l, \tag{13.13b}$$

which depends of course on the path from A to B. Let us now compare the path integrals

$$\int_{q_1}^{q_2} \mathbf{K_o} \cdot \vec{\nu}\, \mathrm{d}l \quad \text{and} \quad \int_{q_1}^{q_2} \mathbf{k_o} \cdot \vec{\nu}\, \mathrm{d}l.$$

In Fig. 13.5(b), Q lies on branch 1 of the dispersion surface. One has

$$\mathbf{k_o} = \mathbf{OR} = \mathbf{OQ} + \mathbf{QR} = \mathbf{K_o} + \mathbf{QR},$$
$$\mathbf{k_o} \cdot \vec{\nu} = \mathbf{K_o} \cdot \vec{\nu} + \mathbf{QR} \cdot \vec{\nu}.$$

It is obvious that, whatever the position of R, $\mathbf{QR} \cdot \vec{\nu} \geq 0$ and, therefore, that $\mathbf{k_o} \cdot \vec{\nu} \geq \mathbf{K_o} \cdot \vec{\nu}$. If Q lies on branch 2, whose concavity is of opposite sign to that of branch 1, one has $\mathbf{k_o} \cdot \vec{\nu} \leq \mathbf{K_o} \cdot \vec{\nu}$. One concludes that

$$\int_{q_1}^{q_2} \mathbf{K_o} \cdot \vec{\nu}\, \mathrm{d}l \geq (\leq) \int_{q_1}^{q_2} \mathbf{k_o} \cdot \vec{\nu}\, \mathrm{d}l,$$

where the first sign corresponds to branch 1 and the second one, $(\leq)$, to branch 2.

Along the real trajectory, Ap_1p_2B, the points R and Q of the dispersion surface coincide and one has

$$\int_{p_1}^{p_2} \mathbf{K_o} \cdot \vec{\nu}\, \mathrm{d}l = \int_{p_1}^{p_2} \mathbf{k_o} \cdot \vec{\nu}\, \mathrm{d}l$$

and, therefore, using (13.13a),

$$\int_{q_1}^{q_2} \mathbf{K_o} \cdot \vec{\nu}\, \mathrm{d}l \geq (\leq) \int_{p_1}^{p_2} \mathbf{K_o} \cdot \vec{\nu}\, \mathrm{d}l.$$

This shows that the integral $\int_A^B \mathbf{K_o} \cdot \vec{\nu}\, \mathrm{d}l$ is either minimum (branch 1) or maximum (branch 2) along the real trajectory. In other words, the real trajectory is defined by

$$\delta \left[\int_A^B \mathbf{K_o} \cdot \vec{\nu}\, \mathrm{d}l \right] = 0.$$

This is Fermat's principle.

13.5 Ray trajectories

13.5.1 *Local dispersion surface*

In this section for the ray trajectories and in Section 13.6 for the intensities of the waves, we shall follow the treatment of Penning and Polder (1961) which is equivalent to that of Kato (1964*a*).

The reciprocal lattice vector of the perfect crystal asymptotic to the real crystal has been determined in Section 13.1.3. We have seen that the result of the deformation is a shift $\overline{L_a L_a'} = k\delta\theta$ of the Laue point along the tangent T_o' to the sphere of centre O and radius k (Fig. 13.6) where the effective misorientation, $\delta\theta$, is given by (13.6). The Lorentz point is shifted by the same amount, $\overline{L_a L_a'} = \overline{L_o L_o'}$ along the tangent T_o to the sphere of centre O and radius nk (Fig. 13.6).

As a first approximation it can be assumed that the structure factor has not been modified during the deformation. The local dispersion surface associated with this

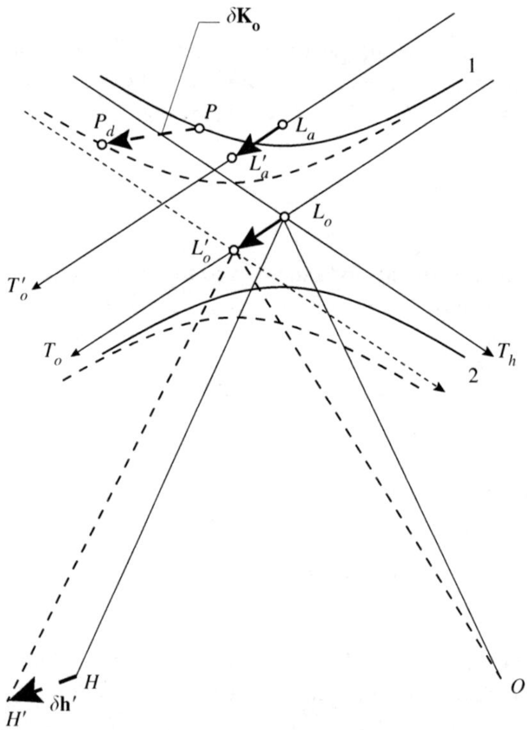

Fig. 13.6. Ray theory. The dispersion surface after a deformation is represented by dashed lines. P: tiepoint before deformation; P_d: tiepoint after deformation; L_o': Lorentz point after deformation; $\delta\mathbf{h}$: variation of the reciprocal lattice vector.

asymptotic crystal is identical with that defined for the perfect crystal; it has simply been translated by the same amount, $k\delta\theta$, along T_o.

13.5.2 *Local wavevectors*

Let us consider a certain wavefield passing through a point p in direct space (Fig. 13.2). Its tiepoint is a point P on the dispersion surface *before* deformation (Fig. 13.6). During the deformation, point p will be transformed into point p', and the same wavefield, now passing through p', will have as tiepoint, *after* deformation, a certain point P_d on the translated dispersion surface. The wavevector, $\mathbf{K_o}$, of the refracted wave has changed. It is now $\mathbf{K'_o}$ and

$$\delta\mathbf{K_o} = \mathbf{K'_o} - \mathbf{K_o} = \mathbf{OP_d} - \mathbf{OP} = \mathbf{PP_d}.$$

It has been shown at the end of Section 13.2.3 that the deformation does not affect the Bragg diffraction if $\delta\mathbf{h} \cdot \mathbf{s_h} = 0$. The dispersion surface is in that case invariant and equation

$$\mathbf{h} \cdot \mathbf{K_h} = \text{constant}$$

can therefore be interpreted as the equation of constant index of refraction and, by analogy with the optical case (eqn (13.1)), the variation of the local wavevector is given by (Penning and Polder 1961; Kambe 1968)

$$\delta\mathbf{K_o} \propto \vec{\nabla}.(\mathbf{h} \cdot \mathbf{K_h}). \tag{13.14}$$

The reflected wavevector, $\mathbf{K_h}$, changes accordingly and

$$\begin{aligned}\delta\mathbf{K_h} &= \mathbf{K'_h} - \mathbf{K_h} = \mathbf{H'P_d} - \mathbf{HP} = \mathbf{H'H} + \mathbf{HP} + \mathbf{PP_d} - \mathbf{HP} \\ &= \delta\mathbf{K_o} - \delta\mathbf{h}.\end{aligned}$$

13.5.3 *Differential equation of the wavefield trajectories*

There is an alternate way of presenting the displacement of the tiepoint. Instead of considering that the dispersion surface glides along T_o, one may consider the dispersion surface as fixed and the tiepoint as gliding along the dispersion surface (Fig. 13.7).

The position P' of the tiepoint after deformation is obtained as follows: let Q be a point deduced from the tiepoint P before deformation by the translation

$$\mathbf{PQ} = -\mathbf{L_o L'_o}$$

which is minus the shift described in Section 13.5.1. P' is deduced from Q by the translation

$$\mathbf{QP'} = \mathbf{PP_d} = \delta\mathbf{K_o}$$

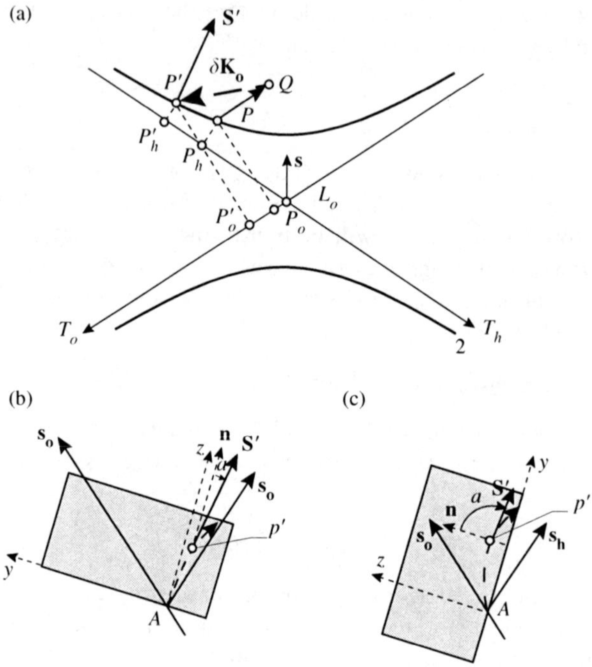

Fig. 13.7. Ray theory: alternate presentation. (a) Reciprocal space: the dispersion surface is considered to be fixed but the tiepoint glides along its branches during the deformation. P: tiepoint before deformation; P': tiepoint after deformation; $\mathbf{PQ} = -\mathbf{L_o L'_o}$, where L'_o, not represented here, has been defined in Fig. 13.6; **s**: unit vector parallel to the lattice planes. (b), (c) Direct space; the ray trajectory is represented by a thick dashed line; **n**: normal to the entrance surface; $\mathbf{S}'$: Poynting vector of the wavefield at p'. (b) Transmission geometry. (c) Reflection geometry.

so that

$$\mathbf{PP'} = \delta\mathbf{K_o} - \mathbf{L_o L'_o}.$$

It is important to note that according to the basic assumption of ray theories the deformation is small enough for no Fraunhofer diffraction effects to appear. There is no interbranch scattering and the tiepoint remains on the *same* branch of the dispersion surface during the deformation: a type 1 wavefield remains a type 1 wavefield, and the same thing holds for a type 2 wavefield.

Let P_o, P'_o, P_h, P'_h be the projections of P and P' on T_o and T_h, respectively (Fig. 13.7). The coordinates of P and P' are, respectively,

$$P \quad \begin{cases} X_o = \overline{P_o P}, \\ X_h = \overline{P_h P}, \end{cases} \qquad P' \quad \begin{cases} X'_o = \overline{P'_o P'}, \\ X'_h = \overline{P'_h P'}. \end{cases}$$

Noting that **PQ** is parallel to T_o and using (13.7a), it follows, after projecting $\mathbf{PP}'$ on $\mathbf{s_o}$ (unit vector normal to T_o) and $\mathbf{s_h}$ (unit vector normal to T_h), respectively, that

$$\mathrm{d}X_o = X'_o - X_o = \mathbf{PP}' \cdot \mathbf{s_o} = \delta\mathbf{K_o} \cdot \mathbf{s_o}, \tag{13.15a}$$

$$\mathrm{d}X_h = X'_h - X_h = \mathbf{PP}' \cdot \mathbf{s_h} = \delta\mathbf{K_o} \cdot \mathbf{s_h} - \delta\mathbf{h} \cdot \mathbf{s_h}. \tag{13.15b}$$

$\mathrm{d}X_o$ and $\mathrm{d}X_h$ are related by differentiation of eqn (4.15) of the dispersion surface:

$$X_o\,\mathrm{d}X_h + X_h\,\mathrm{d}X_o = 0.$$

Substituting $\mathrm{d}X_o$ and $\mathrm{d}X_h$ from (13.15a,b), one obtains

$$\delta\mathbf{h} \cdot \mathbf{s_h} = \delta\mathbf{K_o} \cdot \left(\frac{X_h}{X_o}\mathbf{s_o} + \mathbf{s_h}\right). \tag{13.16a}$$

Let us set

$$\mathbf{S}' = \frac{X_h}{X_o}\mathbf{s_o} + \mathbf{s_h}.$$

Remembering that $X_h/X_o = \xi^{-2} = (D_o/D_h)^2$ (eqn (4.17)), if one neglects anomalous dispersion (in other words, assuming that $|\chi_h/\chi_{\bar{h}}| = 1$), it can be seen that $\mathbf{S}'$ can also be written,

$$\mathbf{S}' = \xi^{-2}\,\mathbf{s_o} + \mathbf{s_h},$$

and that it is proportional to the Poynting vector (4.20), and therefore that it is parallel to the local trajectory of the wavefield. Noting that $\delta\mathbf{h} = (\mathbf{dr} \cdot \vec{\nabla})\mathbf{h}$, (13.16a) may be written as

$$\delta\mathbf{K_o} \cdot \left(\frac{X_h}{X_o}\mathbf{s_o} + \mathbf{s_h}\right) = \vec{\nabla}(\mathbf{h} \cdot \mathbf{s_h}) \cdot \mathbf{dr}, \tag{13.16b}$$

$$\delta\mathbf{K_o} \cdot (\xi^{-2}\,\mathbf{s_o} + \mathbf{s_h}) = \vec{\nabla}(\mathbf{h} \cdot \mathbf{s_h}) \cdot \mathbf{dr}. \tag{13.16c}$$

The equation of the trajectories can be represented either with respect to axes related to the lattice planes or with respect to axes related to the geometry of the crystal, namely to the crystal surface. Both representations are useful and will be discussed successively. In the former case the tiepoint of the propagating wavefield will be characterized by the value of the amplitude ratio, ξ, and in the latter by the deviation parameter, η.

1. *Representation of the ray trajectory with respect to the lattice planes*
The ray trajectory will be represented with respect to axes $A\zeta$ and $A\upsilon$, respectively parallel and perpendicular to the lattice planes (Fig. 13.8).

Let **s** be a unit vector parallel to $A\zeta$. The projection of $\mathbf{S}'$ on the lattice planes is

$$\mathbf{S}' \cdot \mathbf{s} = S' \cos\alpha,$$

$$\mathbf{S}' \cdot \mathbf{s} = (\xi^{-2} + 1)\cos\theta_B.$$

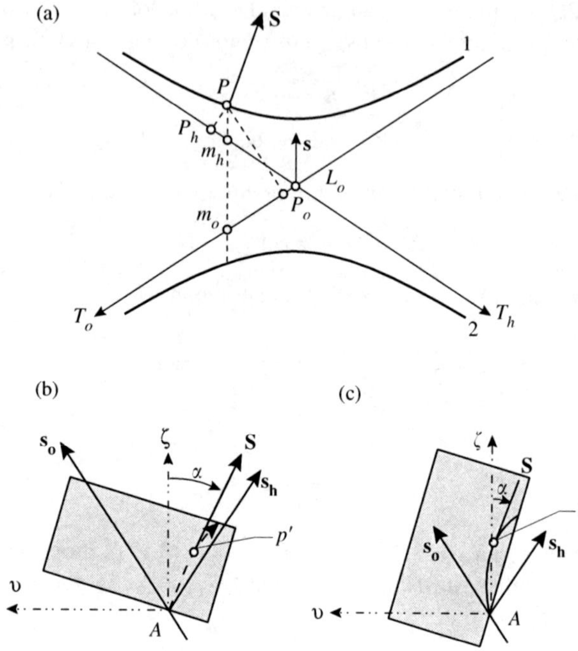

Fig. 13.8. Representation of the ray trajectory with respect to axes related to the lattice planes. (a) Reciprocal space. (b), (c) Direct space; the ray trajectory is represented by a thick dashed line; **S**: Poynting vector of the wavefield at p'; $A\zeta$, $A\upsilon$: axes parallel and perpendicular to the lattice planes, respectively. (b) Transmission geometry. (c) Reflection geometry.

It follows that

$$\mathbf{S}' = (\xi^{-2} + 1)\frac{\cos\theta_B}{\cos\alpha}\frac{\mathrm{d}\mathbf{r}}{\mathrm{d}r},$$

where $\mathrm{d}\mathbf{r}$ is an element parallel to the trajectory of the wavefield. Equation (13.16c) becomes

$$(\xi^{-2} + 1)\frac{\cos\theta_B}{\cos\alpha}\frac{\delta\mathbf{K_o}}{\mathrm{d}r}\cdot \mathrm{d}\mathbf{r} = \vec{\nabla}(\mathbf{h}\cdot\mathbf{s_h})\cdot \mathrm{d}\mathbf{r}.$$

As has been discussed above, $\delta\mathbf{K_o}$ is proportional to $\vec{\nabla}(\mathbf{h}\cdot\mathbf{s_h})$ (eqn (13.14)). It follows therefore that

$$\delta\mathbf{K_o} = (\xi^{-2} + 1)^{-1}\frac{\cos\alpha}{\cos\theta_B}\vec{\nabla}(\mathbf{h}\cdot\mathbf{s_h})\,\mathrm{d}r$$

and, since the local reciprocal lattice vector $\mathbf{h}$ is given in terms of the undeformed reciprocal lattice vector, $\mathbf{h_o}$ by (13.4), $\mathbf{h} = \mathbf{h_o} - \overrightarrow{\nabla}(\mathbf{h} \cdot \mathbf{u})$, one may write

$$\delta\mathbf{K_o} \cdot \mathbf{s_o} = -(\xi^{-2} + 1)^{-1} \frac{\cos\alpha}{\cos\theta_B} \frac{\partial^2(\mathbf{h} \cdot \mathbf{u})}{\partial x_o \partial x_h} \, \mathrm{d}r. \tag{13.17}$$

The variation $\delta\mathbf{K_o}$ of the position of the tiepoint corresponds to a variation $\mathrm{d}\xi$ of the amplitude ratio. Their relation is obtained by differentiating the first of the fundamental equations of the dispersion surface, (5.9), which can be written as

$$K_o^2 = k^2(1 + \chi_o) + k^2 C \chi_{\bar{h}} \xi.$$

It follows, after differentiation and division by $K_o \approx k$, that

$$\delta\mathbf{K_o} \cdot \mathbf{s_o} = \frac{kC\chi_{\bar{h}}}{2} \, \mathrm{d}\xi$$

and, after substitution in (13.17),

$$\begin{aligned} (\xi^{-2} + 1)\,\mathrm{d}\xi = \mathrm{d}\left(\xi - \frac{1}{\xi}\right) &= -\frac{2}{kC\chi_{\bar{h}} \cos\theta_B} \frac{\partial^2(\mathbf{h} \cdot \mathbf{u})}{\partial x_o \partial x_h} \, \mathrm{d}\zeta \\ &= -2\beta \, \mathrm{d}\zeta \end{aligned} \tag{13.18}$$

where $\mathrm{d}\zeta = \mathrm{d}r \cos\alpha$. The strain gradient

$$\beta = \frac{1}{kC\chi_{\bar{h}} \cos\theta_B} \frac{\partial^2(\mathbf{h} \cdot \mathbf{u})}{\partial x_o \, \partial x_h} = \frac{-1}{S(C) G_B \cos^2\theta_B} \frac{\partial^2(\mathbf{h} \cdot \mathbf{u})}{\partial x_o \, \partial x_h} \tag{13.19}$$

is independent of the orientation of the crystal surface with respect to the lattice planes. The Bragg gap, G_B, is the diameter of the dispersion surface (eqn (4.16)).

Equation (13.18) can also be expressed in terms of the slope (5.13),

$$p = \frac{\tan\alpha}{\tan\theta_B} = \frac{1}{\tan\theta_B} \frac{\mathrm{d}\upsilon}{\mathrm{d}\zeta}, \tag{13.20}$$

of the trajectory. One obtains, using (5.12b),

$$\xi - \frac{1}{\xi} = \pm \frac{2p}{\sqrt{1 - p^2}},$$

where the top sign corresponds to branch 1 of the dispersion surface and the bottom sign to branch 2. Equation (13.18) becomes

$$\mathrm{d}\left(\frac{\mp p}{\sqrt{1 - p^2}}\right) = \beta \, \mathrm{d}\zeta. \tag{13.21}$$

Equation (13.18) was first obtained by Penning and Polder (1961) and eqn (13.21) by Kato (1964*a*). Gronkowski and Malgrange (1984*a*) introduced parameter

$$a_M = \frac{1}{2}\left(\frac{1}{\xi} - \xi\right) = \mp\frac{p}{\sqrt{1-p^2}} \tag{13.22}$$

and gave its geometrical interpretation. One has, according to eqns (5.9),

$$\xi = -\frac{2}{S(C)G_B\cos\theta_B}X_o,$$
$$\frac{1}{\xi} = -\frac{2}{S(C)G_B\cos\theta_B}X_h.$$

It follows that

$$a_M = \frac{X_o - X_h}{S(C)G_B\cos\theta_B}.$$

Let m_o and m_h be the intersections of the parallel to the lattice planes drawn from the tiepoint with the asymptotes to the dispersion surface, respectively (Fig. 13.8). It can be seen in the figure that

$$X_o - X_h = \overline{m_o m_h}\cos\theta_B$$
$$= \overline{L_o m_o}\sin 2\theta_B.$$

One has, therefore,

$$a_M = \frac{2\overline{L_o m_o}\sin\theta_B}{G_B}.$$

This shows that parameter a_M characterizes very simply the position of the tiepoint on the dispersion surface. It reduces to the deviation parameter η in the symmetrical transmission geometry.

2. *Representation of the ray trajectory with respect to the crystal surface*
The vector $\mathbf{S}'$ is now evaluated by projecting it on $\mathbf{n}$, unit vector along the normal to the crystal surface (Fig. 13.7). One has

$$\mathbf{S}'\cdot\mathbf{n} = S'\cos a,$$
$$\mathbf{S}'\cdot\mathbf{n} = \left(\frac{X_h}{X_o}\gamma_o + \gamma_h\right),$$

where a is the angle between the Poynting vector and $\mathbf{n}$ (as defined by eqn (10.A1)). It follows that

$$\mathbf{S}' = \left(\frac{X_h}{X_o}\gamma_o + \gamma_h\right)\frac{1}{\cos a}\frac{\mathrm{d}\mathbf{r}}{\mathrm{d}r}.$$

After a similar derivation, eqn (13.17) is replaced by, using (13.15a),

$$\delta \mathbf{K_o} \cdot \mathbf{s_o} = \mathrm{d}X_o = -\left(\gamma_h + \frac{X_h}{X_o}\gamma_o\right)^{-1} \cos a \frac{\partial^2(\mathbf{h} \cdot \mathbf{u})}{\partial x_o\, \partial x_h}\, \mathrm{d}r. \tag{13.23}$$

The element $\mathrm{d}X_o$ can be expressed in terms of the usual deviation parameter, η, using (5.22a). Substituting into (13.23), it follows that

$$\mathrm{d}\eta = -\frac{\Lambda_o}{\gamma_o|\gamma_h|}\frac{\partial^2(\mathbf{h} \cdot \mathbf{u})}{\partial x_o\, \partial x_h} \cos a\, \mathrm{d}r = \beta^* \cos a\, \mathrm{d}r, \tag{13.24}$$

where β^* is given by

$$\beta^* = -\frac{\Lambda_o}{\gamma_o\,|\gamma_h|}\frac{\partial^2(\mathbf{h} \cdot \mathbf{u})}{\partial x_o\, \partial x_h} = \frac{\cos\theta_B}{\sqrt{\gamma_o\,|\gamma_h|}}\beta. \tag{13.25}$$

For a symmetric reflection in transmission geometry, $\beta^* = \beta$, and for a symmetric reflection in reflection geometry, $\beta^* = \beta/\tan\theta_B$.

13.6 The case of a constant strain gradient

The integration of (13.21) or (13.24) enables one in principle to determine the wave-field trajectories but there is no analytical solution in general and a numerical solution must be obtained using a computer. There is, however, an analytical solution in the case of a constant strain gradient, which will be described in the present section.

13.6.1 *Equation of the ray trajectory with respect to the lattice planes*

Following Gronkowski and Malgrange (1984*a*), the equation of the ray trajectories is derived by expressing the coordinates of a point on the trajectory in terms of parameter a_M. The expression of ζ is obtained by integrating eqn (13.21), using (13.22):

$$a_M = \beta\zeta + a_M^{(i)}, \tag{13.26}$$

where $a_M^{(i)}$ is the initial value of a_M at the entrance surface. In the transmission geometry, the incident wave generates two wavefields, of tiepoints $P_1^{(i)}$ and $P_2^{(i)}$, when it impinges on the crystal (Fig. 13.9).

It should be noted that the value of $a_M^{(i)}$ is different for $P_1^{(i)}$ and $P_2^{(i)}$ if the reflection is not symmetrical.

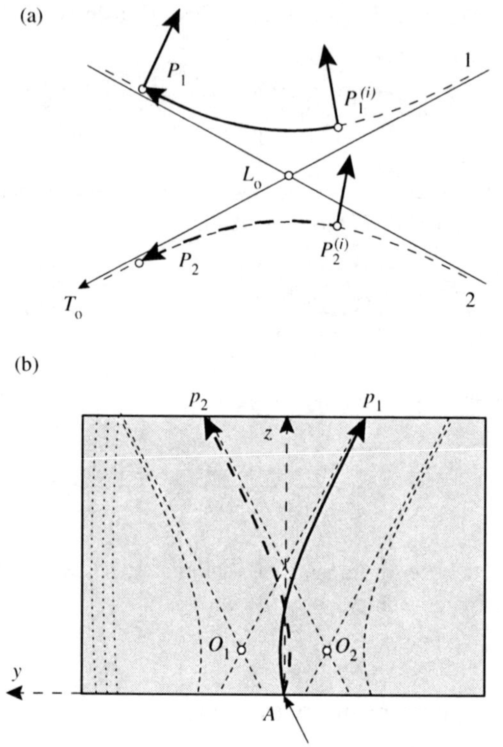

Fig. 13.9. Ray trajectories in a crystal with a constant strain gradient β and for an incident plane wave. (a) Reciprocal space: as the wavefield propagates in the strained crystal, its tiepoint is displaced from $P_1^{(i)}$ to P_1 (branch 1) and from $P_2^{(i)}$ to P_2 (branch 2). (b) Direct space: the paths of the wavefields are branches of hyperbolae (thick solid line for branch 1, thick dashed line for branch 2); their asymptotes are thin broken lines; the lattice planes are represented as curved thin dotted lines.

The expression of υ is obtained by integrating eqn (13.22) which can be written as

$$\mathrm{d}\upsilon = p \tan\theta_B \, \mathrm{d}\zeta.$$

Parameter p and $\mathrm{d}\zeta$ can be expressed in terms of a_M using (13.22) and (13.21). It follows that

$$\begin{aligned}\mathrm{d}\upsilon &= \frac{\mp a_M \tan\theta_B \, \mathrm{d}\zeta}{\sqrt{1+a_M^2}} \\ &= \frac{\mp \tan\theta_B}{\beta} \frac{a_M \, \mathrm{d}a_M}{\sqrt{1+a_M^2}} \\ &= \frac{\mp \tan\theta_B}{\beta} \mathrm{d}\left(\sqrt{1+a_M^2}\right).\end{aligned}$$

This equation is readily integrated and

$$\sqrt{1+a_M^2} = \sqrt{1+\left(a_M^{(i)}\right)^2} \mp \frac{\beta}{\tan\theta_B}\upsilon. \tag{13.27}$$

The equation of the trajectories is obtained by eliminating a_M between (13.26) and (13.27):

$$\left(\frac{\beta}{\tan\theta_B}\upsilon \mp \sqrt{1+\left(a_M^{(i)}\right)^2}\right)^2 - \left(\beta\zeta + a_M^{(i)}\right)^2 = 1, \tag{13.28}$$

where the $-$ sign corresponds to branch 1 and the $+$ sign to branch 2 of the dispersion surface. It can be rearranged by changing coordinates and setting

$$X = \upsilon - \upsilon_{O_j}, \quad \text{with } \upsilon_{O_j} = \pm\frac{\tan\theta_B}{\beta}\sqrt{1+\left(a_M^{(i)}\right)^2},$$

$$Z = \zeta - \zeta_O, \quad \text{with } \zeta_O = -\frac{a_M^{(i)}}{\beta}.$$

Equation (13.28) becomes

$$\frac{X^2}{\sin^2\theta_B} - \frac{Z^2}{\cos^2\theta_B} = \frac{1}{\beta^2\cos^2\theta_B}. \tag{13.29}$$

It is the equation of a hyperbola whose asymptotes are parallel to the incident and reflected directions and whose diameter is $2\tan\theta_B/\beta$. There are two hyperbolae, one for branch 1 and one for branch 2. Their shape is independent of the geometrical orientation of the crystal surface and in particular of whether one is in the transmission or the reflection geometry; the ray trajectory only depends on the strain gradient β, for a given reflection. As the wavefield propagates inside the crystal, its tiepoint glides along the branch of the dispersion surface on which it was generated. It does not jump to the other branch when the strain gradient is small enough for the assumptions of the ray theories to be valid (the case of strong gradients will be considered in Chapter 14). In the same way, the trajectory in direct space of the wavefield follows a given branch of the corresponding hyperbola. The position of the hyperbolae with respect to the crystal surface will be discussed in Sections 13.6.2 and 13.6.5 for the transmission and reflection geometries, respectively.

13.6.2 *Ray trajectories in the transmission geometry*

The derivation is very similar to that of the preceding section. The equation of the ray trajectories will be represented with respect to the axes Ay and Az, respectively perpendicular and parallel to the crystal surface both in transmission and reflection geometry (Fig. 13.7). One has

$$dz = dr \cos a,$$
$$dy = dr \sin a.$$

It follows, using (13.24), that

$$d\eta = \beta^* dz, \tag{13.30a}$$
$$d\eta = \cot a \, \beta^* dy. \tag{13.30b}$$

Equation (13.30a) is readily integrated and

$$\eta^e = \beta^* z + \eta^i, \tag{13.31}$$

where η^i is the initial value of the deviation parameter η at the entrance surface and η^e is its value at depth z. The expression of $\tan a$ is given by (10.A6) after replacing η_o by η and a_o by a:

$$\tan a = \frac{Y \sin 2\theta_B + \sin 2\psi_n}{2\gamma_o \gamma_h},$$

where Y is given by (10.A4),

$$Y = \mp \frac{\eta}{\sqrt{1+\eta^2}},$$

the top sign corresponds to branch 1 and the bottom sign to branch 2. It reduces to $Y = p$, in the case of a symmetric reflection.

Substituting in (13.30b), it follows that

$$\begin{aligned} dy &= \left(\mp \frac{\eta}{\sqrt{1+\eta^2}} \frac{\sin 2\theta_B}{2\gamma_o \gamma_h} + \frac{\sin 2\psi_n}{2\gamma_o \gamma_h} \right) \frac{d\eta}{\beta^*} \\ &= \mp \frac{\sin 2\theta_B}{2\beta^* \gamma_o \gamma_h} d\left(\sqrt{1+\eta^2}\right) + \frac{\sin 2\psi_n}{2\gamma_o \gamma_h} \frac{d\eta}{\beta^*}. \end{aligned}$$

After integration and substitution of (13.31), it follows that

$$\left[\frac{2\beta^* \gamma_o \gamma_h}{\sin 2\theta_B} (y - z \sin 2\psi_n) \mp \sqrt{1 + (\eta^{(i)})^2} \right]^2 - (\beta^* z + \eta^i)^2 = 1, \tag{13.32a}$$

which, like (13.28), is the equation of a hyperbola. For simplicity, we shall restrict ourselves to the case of a symmetric reflection ($\psi_n = 0, \gamma_o = \gamma_h = \cos\theta_B, \beta = \beta^*$). Equation (13.32a) reduces to

$$\left(\frac{\beta}{\tan\theta_B} y \mp \sqrt{1 + (\eta^{(i)})^2} \right)^2 - (\beta z + \eta^i)^2 = 1. \tag{13.32b}$$

It is identical to (13.28) since axes Av, $A\zeta$ and Ay, Az, respectively, now coincide. Equation (13.29) holds with respect to axes Ay, Az and with

$$X = y - y_{O_j}, \quad \text{with } y_{O_j} = \pm\frac{\tan\theta_B}{\beta}\sqrt{1 + (\eta^{(i)})^2},$$

$$Z = z - z_O, \quad \text{with } z_O = -\frac{\eta^i}{\beta}.$$

The two hyperbolae corresponding to branches 1 and 2, respectively, are represented in Fig. 13.9. Their centres are O_1, located at the point of coordinates $y = y_{O_1}, z = z_O$ for branch 1, and O_2, located at the point of coordinates $y = y_{O_2}, z = z_O$ for branch 2 (Fig. 13.9(b)).

Figures 13.9(a) and (b) illustrate the gliding of the tiepoints along the branch of the dispersion surface and the corresponding trajectory the wavefield propagates inside the crystal. The sense of the movement of the tiepoint on the dispersion surface and the branch of the hyperbola followed in direct space depends on the sign of β. Figure 13.9 corresponds to $\beta > 0$ and one has

- branch 1: $y_{O1} > 0$, $X < 0$
- branch 2: $y_{O2} < 0$, $X > 0$.

For $\beta < 0$ it would be the contrary, the sense of the movement of the tiepoints in Fig. 13.9(a) would be opposite; the positions of O_1 and O_2 in Fig. 13.9(b) and the trajectories of the two wavefields would be exchanged (the corresponding values of η^i would be opposite of those of the wavefields represented for the case $\beta > 0$).

Figure 13.10 summarizes the shape of the wavefield trajectories for the two branches as a function of the sign of β and of the value of η^i, that is of the position of the tiepoint on the dispersion surface at the entrance surface.

The minimum radius of curvature of the trajectory occurs for $\eta = 0$ and is given by

$$\mathcal{R}_t = \frac{\text{diameter of the hyperbola}}{2\tan^2\theta_B}.$$

Since the diameter is $2\tan\theta_B/\beta$, the minimum radius of curvature is

$$\mathcal{R}_t = \frac{1}{\beta\tan\theta_B}.$$

13.6.3 *Pure bending*

Pure bending is an example of deformation where the strain gradient β is constant. Let O be the centre of curvature and $\mathcal{R}$ the radius of curvature of the lattice planes (Fig. 13.11). For simplicity we shall assume that the reflection is symmetrical and that the radius of curvature is much larger than the dimensions of the crystal; we shall also neglect the radial variation of the lattice parameter. During the bending a point, p, of the crystal is transformed in p' and the lattice plane Ap is bent along the cylinder

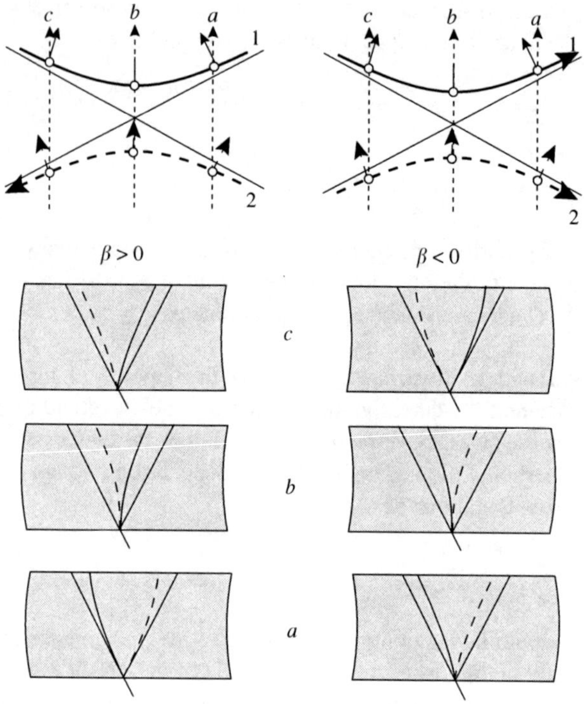

Fig. 13.10. Ray trajectories in a crystal with a constant strain gradient β as a function of the sign of β and of the position of the tiepoint at the entrance surface; a: $\eta^i \ll 0$; b: $\eta^i = 0$; c: $\eta^i \gg 0$. Solid lines: branch 1; dashed lines: branch 2. The sense of the displacement of the tiepoints on the dispersion surface is indicated by an arrow.

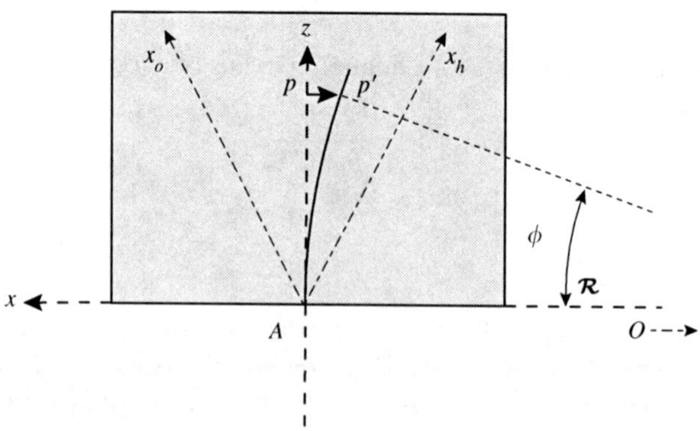

Fig. 13.11. Deformation of the lattice planes through pure bending; O is the centre of curvature, $\mathcal{R}$ the radius of curvature.

of cross-section Ap'. Let us express the displacement $\mathbf{u} = \mathbf{pp}'$ with respect to axes **Ax**, **Az**, where **Ax** is parallel to the entrance surface and **Az** is normal to the entrance surface.

The coordinates of p and p' are, respectively,

$$p \quad \begin{cases} x = 0, \\ z = \mathcal{R}\sin\phi \approx \mathcal{R}\phi, \end{cases} \qquad p' \quad \begin{cases} x' = \pm\mathcal{R}(\cos\phi - 1) \approx \mp\mathcal{R}\dfrac{\phi^2}{2}, \\ z' \approx z = \mathcal{R}\sin\phi \approx \mathcal{R}\phi, \end{cases}$$

where ϕ is the angle between Ox and Op', and the top sign corresponds to a position of the centre of curvature on the negative side of the **Ax** axis and the bottom sign to a position of the centre of curvature on the positive side of the **Ax** axis; it follows that

$$u = \overline{pp'} \approx \mp\mathcal{R}\frac{\phi^2}{2} = \mp\frac{z^2}{2\mathcal{R}}.$$

The reciprocal lattice vector **h** is parallel to **Ox** and the scalar product $\mathbf{u}\cdot\mathbf{h}$ in expression (13.10a) of β is simply

$$\mathbf{u}\cdot\mathbf{h} = \mp\frac{hz^2}{2\mathcal{R}}, \tag{13.33}$$

with $h = 2k\sin\theta_B$.

Let us now introduce the oblique system of axes, Ax_o, Ax_h, parallel to the incident and reflected directions, respectively. The coordinate, z, of p is $z = (x_o + x_h)\cos\theta_B$ and, after substitution in (13.33), it follows that

$$\frac{\partial^2(\mathbf{h}\cdot\mathbf{u})}{\partial x_o\,\partial x_h} = \mp\frac{h\cos^2\theta_B}{\mathcal{R}}$$

and, using (5.17b) and (13.10a),

$$\beta = \pm\frac{h\Lambda_o}{\mathcal{R}} = \pm\frac{\sin 2\theta_B}{\mathcal{R}C\sqrt{\chi_h\chi_{\bar{h}}}}. \tag{13.34}$$

A positive sign of β therefore corresponds to a position of the centre of curvature on the negative side of the **Ax** axis. The results of the preceding subection show that the curvature of the trajectory is of the same sign as that of the lattice planes for wavefields belonging to branch 1. It is of the opposite sign for branch 2. This situation was studied theoretically and experimentally by Okkerse and Penning (1963). The integrated intensity in bent silicon crystals was studied systematically as a function of strain and crystal thickness by Datsenko *et al.* (1981). They obtained good agreement with the theoretical values calculated according to the expressions given by Chukhovskii and Petrashen (1977).

The criterion (13.10b) for the validity of ray theory becomes, using (13.34),

$$\mathcal{R} \gg h\,|\Lambda_o|^2.$$

For instance, in the case of silicon, Mo$K\alpha$ and the 220 reflection, $|\Lambda_o| = 36.3\,\mu$m, $|\beta|$ should be much smaller than $270\,\text{cm}^{-1}$ and $\mathcal{R}$ should be much larger than 7 m.

13.6.4 *Temperature gradient*

If a temperature gradient $\vec{\nabla}(T)$ is applied to a crystal, the lattice planes perpendicular to the gradient are curved with a radius of curvature $\mathcal{R}$ given by $\mathcal{R}^{-1} = \nabla(\alpha T)$ where α is the coefficient of linear thermal expansion. The local displacement is (Penning and Polder 1961)

$$\mathbf{u}(\mathbf{r}) = \mathbf{r}\left[\mathbf{r}\cdot\vec{\nabla}(\alpha T)\right] - \tfrac{1}{2}\mathbf{r}\cdot\mathbf{r}\vec{\nabla}(\alpha T).$$

This expression is valid only if $\vec{\nabla}(\alpha T)$ is constant in direction and in magnitude. It follows that

$$\frac{\partial^2(\mathbf{h}\cdot\mathbf{u})}{\partial x_o\,\partial x_h} = \mathbf{h}\cdot\vec{\nabla}(\alpha T)$$

and, using (13.10a), in the symmetrical Laue geometry,

$$\beta = -\frac{\Lambda_o}{\gamma_o\gamma_h}\mathbf{h}\cdot\vec{\nabla}(\alpha T). \qquad (13.35)$$

The strain gradient is constant for an isotropic thermal expansion coefficient. It results in a bending of the lattice planes and the trajectories of the wavefields are analogous to those observed in the case of a pure bending. The criterion (13.10b) for the validity of ray theory becomes, using (13.34),

$$\frac{\mathrm{d}T}{\mathrm{d}x} \ll \frac{\cos^2\theta_B}{\Lambda_o^2 h\alpha}.$$

For instance, in the case of silicon, the coefficient of linear thermal expansion is $\alpha \approx 2.7 \times 10^{-6}/^\circ\mathrm{C}$, and for Mo$K\alpha$ and 220, the temperature gradient $\mathrm{d}T/\mathrm{d}x$ should be much smaller than 500 °C/cm.

The influence of a temperature gradient was studied theoretically and experimentally by Hart (1966) in the case of an incident spherical wave and by Malgrange (1969) and Hart and Milne (1971) in the case of an incident pseudo-plane wave generated with a two-crystal setting analogous to that described in Section 12.10. They determined the trajectories of the wavefields and checked that they were in agreement with those predicted by the ray theory. In the experiment by Malgrange, the two crystals were mounted on different goniometer axes, while in the experiment by Hart and Milne the two crystals were actually cut as parts of the same monolithic silicon block. In that case the rotation of the second crystal with respect to the first one was obtained by elastic body rotation around the region which connects it to the common base of the two wafers.

13.6.5 *Ray trajectories in the reflection geometry*

The first study of X-ray propagation in deformed crystals in the Bragg geometry is due to Bonse (1964*a*). This theory, based on a variational principle, allowed for wavevectors with large imaginary parts in order to be applicable to the situations where the incidence angle is within the total reflection range. As shown by Penning (1966) and Chukhovskii (1981), it is however not possible to describe wavefields with complex wavevectors in the framework of ray theory. Bonse's theory is nevertheless valid in the region outside the range of total reflection and was applied to the case of a bent crystal (Bonse 1964*b*; Bonse and Graeff 1973) and to the case of a crystal containing a dislocation line (Bonse 1964*c*). We shall follow here the treatment by Gronkowski and Malgrange (1984*a*), who showed that Penning and Polder's theory can be extended to the Bragg geometry, outside the domain of total reflection. They also showed that the results are in agreement with those obtained in the same situation by numerical resolution of Takagi's equations.

The derivation will be limited to the symmetric Bragg geometry (Fig. 13.12). Equation (13.30a) is still valid and is integrated as in the Laue geometry; eqn (13.31) holds. Axis **Ay** is here parallel to the lattice planes and the slope of the ray trajectory is

$$\frac{\mathrm{d}z}{\mathrm{d}y} = \tan\alpha.$$

Using (13.25), (13.30a) and (5.22b), it follows that

$$\mathrm{d}\eta = -\tan\alpha\,\beta^*\,\mathrm{d}y = \mp\frac{\sqrt{\eta^2-1}}{\eta}\beta\,\mathrm{d}y.$$

It can be rewritten in the form

$$\mp\frac{\eta}{\sqrt{\eta^2-1}}\mathrm{d}\eta = \mp\mathrm{d}\left(\sqrt{\eta^2-1}\right) = \beta\,\mathrm{d}y.$$

This equation is readily integrated:

$$\mp\sqrt{\eta^2-1} = \beta y + \text{constant}.$$

For $y = 0$ one should only consider the wavefield propagating towards the inside of the crystal; it follows that

$$\mp\sqrt{\eta^2-1} = \beta y + S(\eta)\sqrt{(\eta^i)^2-1}, \tag{13.36}$$

where η^i is the value of the deviation parameter at the entrance surface and $S(\eta)$ means sign of η.

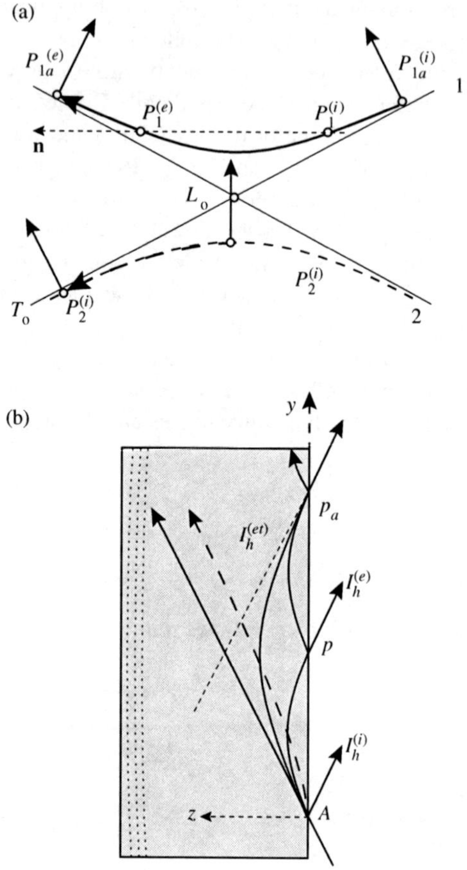

Fig. 13.12. Ray trajectories in a crystal with a constant strain gradient $\beta > 0$ and for an incident plane wave, reflection geometry. (a) Reciprocal space. (b) Direct space: the paths of the wavefields are branches of hyperbolae (thick solid line for branch 1, thick dashed line for branch 2).

The equation of the trajectory is obtained by eliminating η between (13.31) and (13.36):

$$\left(\frac{\beta}{\tan\theta_B}z + \eta^i\right)^2 - \left[\beta y + S(\eta^i)\sqrt{(\eta^i)^2 - 1}\right]^2 = 1. \qquad (13.37a)$$

This equation can be written as

$$\frac{Z^2}{\sin^2\theta_B} - \frac{X^2}{\cos^2\theta_B} = \frac{1}{\beta^2\cos^2\theta_B}, \qquad (13.37b)$$

with

$$X = y + S(\eta^i)\frac{\sqrt{(\eta^i)^2 - 1}}{\beta},$$

$$Z = z + \frac{\eta^i \tan\theta_B}{\beta}.$$

As in the Laue geometry, it is a branch of a hyperbola whose asymptotes are parallel to the incident and parallel directions, respectively (Fig. 13.12(b)). This branch intersects the crystal surface at an abscissa,

$$y_e = -\frac{2S(\eta^i)\sqrt{(\eta^i)^2 - 1}}{\beta},$$

and gives rise to a beam of intensity $I_h^{(e)}$ reflected from the surface (Fig. 13.12(b)). This 'extra-reflected' beam was first predicted and observed by Bonse (1964*b*) and was studied experimentally and theoretically by Bonse and Graeff (1973).

The abscissa, y_e, of the intersection only has a physical sense if it is positive. This is the case if $S(\eta^i)\beta$ is negative. The ray trajectory corresponds to tiepoints on branch 1 if β is positive and to branch 2 if β is negative. This is illustrated in Fig. 13.12 drawn in the case $\beta > 0$.

1. *Branch 1*: Two trajectories are shown for wavefields belonging to branch 1, $\widehat{AP}$ whose tiepoint moves from $P_1^{(i)}$ to $P_1^{(e)}$ (Fig. 13.12(a)) and $\widehat{Ap_a}$ whose tiepoint moves from $P_{1a}^{(i)}$ to $P_{1a}^{(e)}$. It can be seen that the tiepoint ($P_1^{(e)}$ or $P_{1a}^{(e)}$) at the point where the ray trajectories intersect the crystal surface (p or p_a, respectively) is symmetric of the tiepoint of the wavefield when it is created at the crystal surface ($P_{1a}^{(i)}$ or $P_1^{(i)}$), with respect to the crystal surface. The amplitude ratio ξ_e at the end of the trajectory is therefore the inverse of its value at the beginning of the trajectory ($\xi_e = 1/\xi_i$). When the wavefield of initial tiepoint $P_1^{(i)}$ reaches the crystal surface, at p, it is partially reflected. The tiepoint of the wavefield thus created is obtained as usual by applying the condition of the tangential continuity of wavevectors. It lies at the intersection of the normal **n** to the crystal surface and of the dispersion surface (Fig. 13.12(a)) and it is again $P_1^{(i)}$. The partially reflected wavefield propagates therefore along a similar trajectory, which is simply translated: it reaches the crystal surface at p_a. It can be seen in the figure that there is also a wavefield having propagated directly along $\widehat{Ap_a}$. Both wavefields are simultaneously excited if the incident wave is spherical; they interfere and give rise to Pendellösung fringes which were first observed experimentally by Shul'pina *et al.* (1984) and discussed theoretically by Chukhovskii and Petrashen (1988). In a similar way, the rocking curve for a bent crystal in the reflection geometry presents oscillations which were first calculated by Taupin (1964) by solving numerically equations similar to Takagi's and observed experimentally by Uschmann *et al.* (1993, 1997).

2. *Branch 2*: The trajectory of the wavefields belonging to branch 2 penetrates towards the inside of the crystal (represented as a thick dashed line in Fig. 13.12(b)). It only exists for values of $\eta > 1$. For $\eta < -1$ no wavefield is created on branch 2.

The roles of branches 1 and 2 are interchanged if β is negative.

13.7 Diffracted intensities—plane-wave case

13.7.1 *Zero absorption*

1. *Transmission geometry*

We shall first consider the propagation in a deformed crystal of a wavepacket which can be represented by a pseudo-plane wave (Fig. 13.13). We shall assume that the incident wavepacket is limited by a slit of height unity and width $\delta l_o^{(a)}$. Its cross-section is therefore given by its width. The widths $\delta l^{(i)}$, $\delta l^{(e)}$, $\delta l_o^{(d)}$ and $\delta l_h^{(d)}$ will define in a similar way the cross-sections of the diffracted wavepacket inside the crystal at the entrance surface and at the exit surface, the refracted and reflected wavepackets outside the crystal, respectively. Further, let $\mathbf{S_o^{(a)}}$, $\mathbf{S_j^{(i)}}$, $\mathbf{S_j^{(e)}}$, $\mathbf{S_{oj}^{(d)}}$ and $\mathbf{S_{hj}^{(d)}}$ be the corresponding Poynting vectors (the subscript j indicates whether the tiepoint of the wavefield belongs to branch 1 or 2 and the superscript i or e indicates that the quantity is taken at the entrance or exit surface, respectively); to a proportionality factor, they are, using eqn (5.11a),

$$\begin{aligned}
\mathbf{S_o^{(a)}} &= \left|D_o^{(a)}\right|^2 \mathbf{s_o},\\
\mathbf{S_j^{(i)}} &= \left|D_{oj}^{(i)}\right|^2 \mathbf{s_o} + \left|D_{hj}^{(i)}\right|^2 \mathbf{s_h},\\
\mathbf{S_j^{(e)}} &= \left|D_{oj}^{(e)}\right|^2 \mathbf{s_o} + \left|D_{hj}^{(e)}\right|^2 \mathbf{s_h},\\
\mathbf{S_{oj}^{(d)}} &= \left|D_o^{(d)}\right|^2 \mathbf{s_o},\\
\mathbf{S_{hj}^{(d)}} &= \left|D_h^{(d)}\right|^2 \mathbf{s_h}.
\end{aligned} \tag{13.38}$$

The conservation of energy requires that, in the absence of absorption,

$$\begin{aligned}
S_j^{(i)} \delta l^{(i)} &= P_j^{(i)} S_o^{(a)} \delta l_o^{(a)},\\
S_j^{(e)} \delta l^{(e)} &= S_j^{(i)} \delta l^{(i)},\\
S_{oj}^{(d)} \delta l_o^{(d)} &= P_{oj}^{(e)} S_j^{(e)} \delta l^{(e)},\\
S_{hj}^{(d)} \delta l_h^{(d)} &= P_{hj}^{(e)} S_j^{(e)} \delta l^{(e)},
\end{aligned}$$

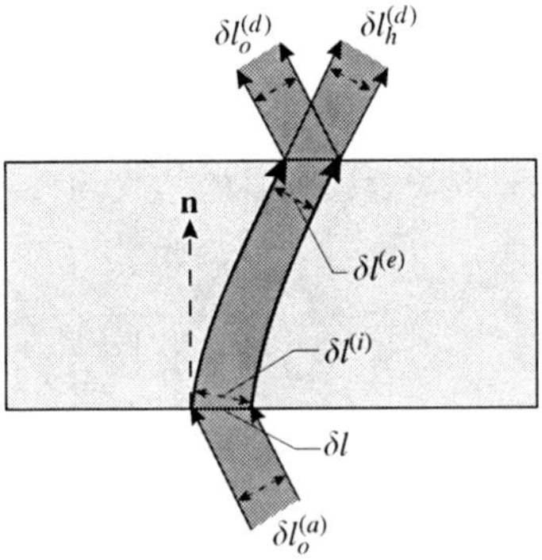

Fig. 13.13. Propagation of a wavepacket in a deformed crystal.

where $P_j^{(i)}$, $P_{oj}^{(e)}$ and $P_{hj}^{(e)}$ are reflecting powers. According to these relations, it follows that

$$\begin{aligned}\frac{S_{oj}^{(d)}}{S_o^{(a)}} &= \frac{|D_{oj}^{(d)}|^2}{|D_o^{(a)}|^2} = \frac{|D_{oj}^{(e)}|^2}{|D_o^{(a)}|^2} = P_j^{(i)} P_{oj}^{(e)},\\ \frac{S_{hj}^{(d)}}{S_o^{(a)}} &= \frac{|D_{hj}^{(d)}|^2}{|D_o^{(a)}|^2} = \frac{|D_{hj}^{(e)}|^2}{|D_o^{(a)}|^2} = P_j^{(i)} P_{hj}^{(e)} \gamma^{-1}\end{aligned} \tag{13.39}$$

for a plane-parallel crystal ($\delta l_o^{(d)} = \delta l_o^{(a)}$, $\delta l_h^{(d)} = \gamma^{-1} \delta l_o^{(a)}$). The reflecting power P_j^i is obtained using the first two equations (13.38):

$$P_j^i = \frac{S_j^{(i)} \delta l^{(i)}}{S_o^{(a)} \delta l_o^{(a)}} = \frac{S_j^{(i)} \delta l \cos(\mathbf{n}, \mathbf{S_j^{(i)}})}{S_o^{(a)} \delta l \cos(\mathbf{n}, \mathbf{S_o^{(a)}})} = \frac{\mathbf{S_j^{(i)}} \cdot \mathbf{n}}{\mathbf{S_o^{(a)}} \cdot \mathbf{n}},$$

where δl is the intersection of the incident wavepacket with the entrance surface (Fig. 13.13). In a similar way, the reflecting powers $P_o^{(e)}$ and $P_h^{(e)}$ are

$$\begin{aligned}P_{oj}^{(e)} &= \frac{S_{oj}^{(d)} \delta l_o^{(d)}}{S_j^{(e)} \delta l^{(e)}} = \frac{\mathbf{S_{oj}^{(d)}} \cdot \mathbf{n}}{\mathbf{S_j^{(e)}} \cdot \mathbf{n}},\\ P_{hj}^{(e)} &= \frac{S_{hj}^{(d)} \delta l_h^{(d)}}{S_j^{(e)} \delta \ell^{(e)}} = \frac{\mathbf{S_{hj}^{(d)}} \cdot \mathbf{n}}{\mathbf{S_j^{(e)}} \cdot \mathbf{n}}.\end{aligned}$$

It follows, using (13.38), that

$$P_j^i = \frac{\gamma_o|D_{oj}^{(i)}|^2 + \gamma_h|D_{hj}^{(i)}|^2}{\gamma_o|D_o^{(a)}|^2},$$

$$P_{oj}^{(e)} = \frac{\gamma_o|D_{oj}^{(d)}|^2}{\gamma_o|D_{oj}^{(e)}|^2 + \gamma_h|D_{hj}^{(e)}|^2}, \tag{13.40}$$

$$P_{hj}^{(e)} = \frac{\gamma_h|D_{hj}^{(d)}|^2}{\gamma_o|D_{oj}^{(e)}|^2 + \gamma_h|D_{hj}^{(e)}|^2}.$$

Expression (6.2a) of the refracted amplitude can be written, using (5.34) and in the absence of absorption:

$$\frac{D_{oj}}{D_o^{(a)}} = \frac{1}{1+\gamma\xi_j^2}, \tag{13.41}$$

where $\xi_j = D_{hj}/D_{oj}$ is the amplitude ratio. Substituting from (13.40) and (13.41) in (13.39), we obtain finally

$$I_{oj}^{(d)} = \frac{|D_{oj}^{(d)}|^2}{|D_o^{(a)}|^2} = \frac{1}{1+\gamma|\xi_j^{(i)}|^2}\,\frac{1}{1+\gamma|\xi_j^{(e)}|^2},$$
$$I_{hj}^{(d)} = \frac{|D_{hj}^{(d)}|^2}{|D_o^{(a)}|^2} = \frac{1}{1+\gamma|\xi_j^{(i)}|^2}\,\frac{|\xi_j^{(e)}|^2}{1+\gamma|\xi_j^{(e)}|^2}, \tag{13.42}$$

where $\xi_j^{(i)}$ and $\xi_j^{(e)}$ are the amplitude ratios at the entrance and exit surface, respectively.

2. *Reflection geometry*

The boundary conditions are simpler in the reflection geometry since one wavefield only penetrates inside the crystal. The incident beam at A (Fig. 13.12(b)) is split in a reflected beam of intensity $I_h^{(i)}$ and a partially transmitted beam of intensity $I_h^{(it)}$. According to (7.1), one has, for a symmetric reflection,

$$I_h^{(i)} = |\xi_i|^2,$$
$$I_h^{(it)} = 1 - |\xi_i|^2. \tag{13.43a}$$

When the wavefield reaches the crystal surface again, at p, after having followed a curved trajectory, its intensity is split in the same way in a beam reflected outside the crystal, of intensity $I_h^{(e)}$, and a beam propagating inside the crystal, of intensity

$I_h^{(et)}$ (Fig. 13.12(b)). After applying the boundary conditions, as was done in the transmission geometry, one finds (Gronkowski and Malgrange 1984*a*)

$$I_h^{(e)} = \left(1 - |\xi_i|^2\right)^2, \qquad (13.43b)$$
$$I_h^{(et)} = \left(1 - |\xi_i|^2\right)|\xi_i|^2.$$

13.7.2 *Absorbing crystals (transmission geometry)*

If the crystal is absorbing, which is the normal situation, the intensities (13.42) should be multiplied by the attenuation coefficient of the beam along the propagation direction. One has

$$\frac{\mathrm{d}I}{I} = -\mu_S \, \mathrm{d}r,$$

where μ_S is given by (5.41b). Noting that in the case of a symmetrical reflection

$$\cos\alpha \, \mathrm{d}r = \mathrm{d}z = \frac{\mathrm{d}\eta}{\beta},$$

$$\frac{\mathrm{d}I}{I} = -\frac{\mu_o}{\cos\theta_B}\mathrm{d}z \pm \frac{\mu_o}{\cos\theta_B} C\cos\varphi \left|\frac{\chi_{ih}}{\chi_{io}}\right| \frac{\mathrm{d}\eta}{\beta\sqrt{1+\eta^2}}. \qquad (13.44)$$

The integration is easy in the case of a constant strain gradient as is shown in the next subsection.

13.7.3 *Expression of the diffracted intensities for a constant strain gradient*

If we set

$$\eta = \sinh v \qquad (13.45)$$

and noticing that

$$\frac{\mathrm{d}\eta}{\sqrt{1+\eta^2}} = \mathrm{d}v,$$

the integration of (13.44) gives

$$B_j = \frac{I}{I_o} = \exp\left\{-\left[\frac{\mu_o}{\cos\theta_B}\left(z \mp C\cos\varphi\left|\frac{\chi_{ih}}{\chi_{io}}\right|\frac{v - v_i}{\beta}\right)\right]\right\}.$$

A convenient expression for v is obtained by writing

$$\exp v = \sinh v + \cosh v$$

and, using (13.45),

$$v = \log\left(\eta + \sqrt{1+\eta^2}\right),$$

it follows that

$$B_j = \exp\left\{-\frac{\mu_o}{\cos\theta_B}\left[z \mp \frac{C\cos\varphi}{\beta}\left|\frac{\chi_{ih}}{\chi_{io}}\right|\log\left(\frac{\eta^e + \sqrt{1+(\eta^e)^2}}{\eta^i + \sqrt{1+(\eta^i)^2}}\right)\right]\right\}, \qquad (13.46a)$$

where the value of the deviation parameter of the wavefield at depth z, η^e, is given in terms of its value at the entrance surface, η^i, by (13.31). Using (5.34) the attenuation ratio B can be expressed in terms of the amplitude ratio:

$$B_j = \exp\left(-\left\{\frac{\mu_o}{\cos\theta_B}\left[z \mp \frac{C\cos\varphi}{\beta}\left|\frac{\chi_{ih}}{\chi_{io}}\right|\log\left(\frac{\xi_j^{(e)}}{\xi_j^{(i)}}\right)\right]\right\}\right). \qquad (13.46b)$$

Using (13.42), we obtain the final expressions of the intensities of the components of a wavefield having propagated through a crystal deformed by a constant gradient, in the case of a symmetrical reflection:

$$\begin{aligned} \frac{|D_{oj}^{(d)}|^2}{|D_o^{(a)}|^2} &= A_j A'_j B_j, \\ \frac{|D_{hj}^{(d)}|^2}{|D_o^{(a)}|^2} &= A_j A''_j B_j, \end{aligned} \qquad (13.47)$$

with

$$\begin{aligned} A_j &= \frac{1}{1+|\xi_j^{(i)}|^2}, \\ A'_j &= \frac{1}{1+|\xi_j^{(e)}|^2}, \\ A''_j &= \frac{|\xi_j^{(e)}|^2}{1+|\xi_j^{(e)}|^2} \end{aligned}$$

or, in terms of the initial and final values of the deviation parameter:

$$\begin{aligned} A_1 &= \frac{1}{2\sqrt{1+(\eta^i)^2}(\eta^i + \sqrt{1+(\eta^i)^2})}; \qquad & A_2 &= \frac{\eta^i + \sqrt{1+(\eta^i)^2}}{2\sqrt{1+(\eta^i)^2}}, \\ A'_1 &= \frac{1}{2\sqrt{1+(\eta^e)^2}(\eta^e + \sqrt{1+(\eta^e)^2})}; \qquad & A'_2 &= \frac{\eta^e + \sqrt{1+(\eta^e)^2}}{2\sqrt{1+(\eta^e)^2}}, \\ A''_1 &= \frac{\eta^e + \sqrt{1+(\eta^e)^2}}{2\sqrt{1+(\eta^e)^2}}; \qquad & A''_2 &= \frac{1}{2\sqrt{1+(\eta^e)^2}(\eta^e + \sqrt{1+(\eta^e)^2})}. \end{aligned} \qquad (13.48)$$

Figures 13.14(a)–(d) show the variations of the reflected and refracted intensities for branches 1 and 2 in terms of the initial value of the deviation parameter, η^i, and for different values, positive and negative, of the strain gradient, β, in the case of a 1 mm thick silicon crystal, Mo$K\alpha$ ($\mu_o t = 1.4$), 220 and the symmetric Laue geometry. As an indication of the physical interpretation of the strain gradient, the value of the corresponding temperature gradient which should be applied normal to the lattice planes is given for each value of β.

One may observe that the curves are shifted towards the increasing values of η^i as β decreases. The position of the maximum of the reflected intensity corresponds to $\eta^i = -\beta z/2$. These variations are discussed in further detail in the next subsection.

13.7.4 *Discussion of the intensity distribution for a constant strain gradient*

1. *Exponential factor*: The variations of B are easily interpreted by remembering that $\eta \pm \sqrt{1+\eta^2}$ is proportional to the distance X_{oj} of the tiepoint corresponding to branch j from the asymptote to the dispersion surface (eqn (5.20)). One finds

$$B_j = \exp\left[-\frac{\mu_o}{\cos\theta_B}\left(z \mp \frac{C\cos\varphi}{\beta}\left|\frac{\chi_{ih}}{\chi_{io}}\right|\right)\log\left(\frac{X_{oj}^{(e)}}{X_{oj}^{(i)}}\right)\right].$$

Figure 13.15 shows that when $\beta < 0$, $\log(X_{oj}^{(e)}/X_{oj}^{(i)})$ is negative for branch 1 and positive for branch 2 and that it is the opposite for $\beta > 0$. *The absorption is therefore always anomalously low for branch 1 and anomalously high for branch 2, independent of the sign of β.*

The value of B_j is extremum for

$$\eta^i = -\eta^e = -\tfrac{1}{2}\beta z,$$

and anomalous absorption is maximum at the centre of the rocking curves for the reflected beam.

2. *Distribution factor*: The intensity variations observed in Fig. 13.14 can be interpreted by noticing that there are transfers of intensity between wavefields belonging to branches 1 and 2 and transfers of intensity between the reflected and the refracted beam:

(i) *Distribution of intensity between branches 1 and 2.* The ratio between the reflected intensities for branches 1 and 2 is, using (13.47), (13.48) and (13.46a),

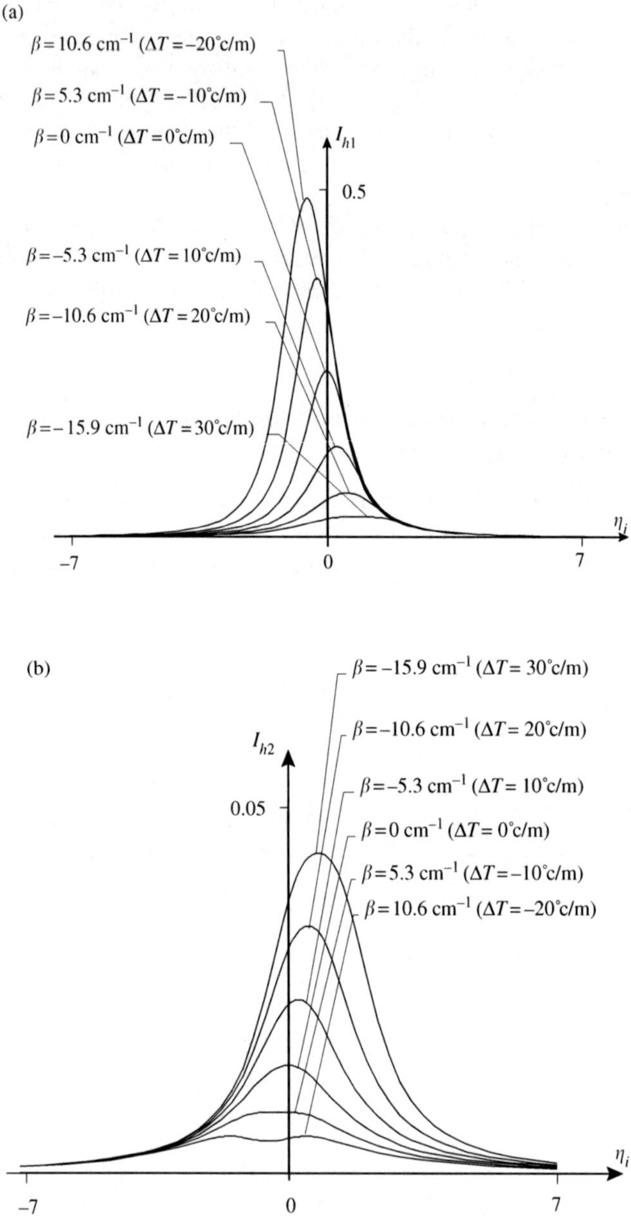
(a)
β = 10.6 cm⁻¹ (ΔT = –20°c/m)
β = 5.3 cm⁻¹ (ΔT = –10°c/m)
β = 0 cm⁻¹ (ΔT = 0°c/m)
β = –5.3 cm⁻¹ (ΔT = 10°c/m)
β = –10.6 cm⁻¹ (ΔT = 20°c/m)
β = – 15.9 cm⁻¹ (ΔT = 30°c/m)
I_h1
0.5
η_i
–7
0
7
(b)
β = –15.9 cm⁻¹ (ΔT = 30°c/m)
β = –10.6 cm⁻¹ (ΔT = 20°c/m)
β = –5.3 cm⁻¹ (ΔT = 10°c/m)
β = 0 cm⁻¹ (ΔT = 0°c/m)
β = 5.3 cm⁻¹ (ΔT = –10°c/m)
β = 10.6 cm⁻¹ (ΔT = –20°c/m)
I_h2
0.05
η_i
–7
0
7

Fig. 13.14 (a,b)

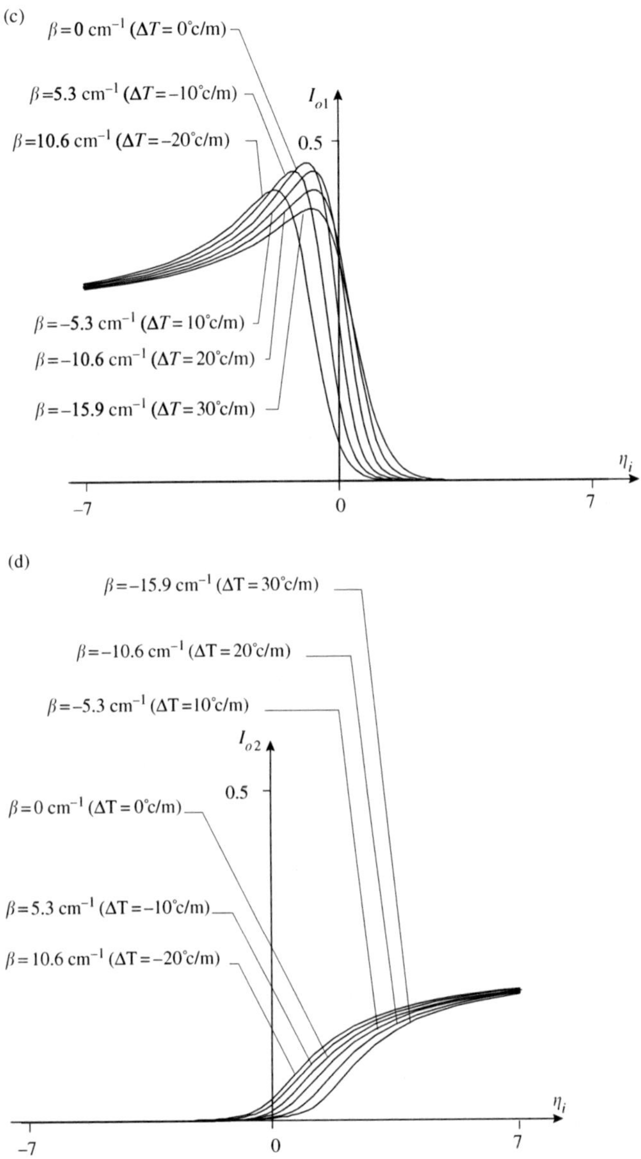

Fig. 13.14 (c,d)

Fig. 13.14. Variations of the reflected and refracted intensities of a wavepacket diffracted by a 1 mm thick silicon crystal submitted to a uniform strain gradient, for different values of β; symmetrical Laue geometry. As an example the corresponding values of the temperature gradient which should be applied in the direction perpendicular to the lattice planes are shown; transmission geometry; 220 reflection, Mo$K\alpha$. (a) Reflected beam, branch 1. (b) Reflected beam, branch 2; note that the scale is different from that of Figs 13.14(a), (c) and (d). (c) Refracted beam, branch 1. (d) Refracted beam, branch 2.

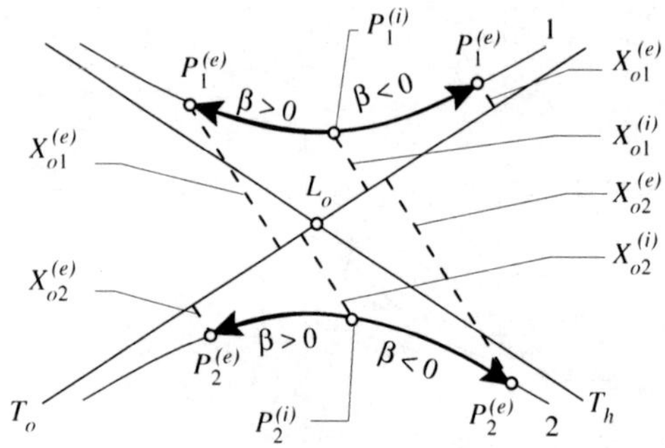

Fig. 13.15. Variation of the position of the tiepoint as the wavefield propagates inside the crystal, according to the sign of β. $P_j^{(i)}$: position at the entrance surface of the crystal; $P_j^{(e)}$: position at the exit surface of the crystal; $j = 1$, branch 1; $j = 2$, branch 2.

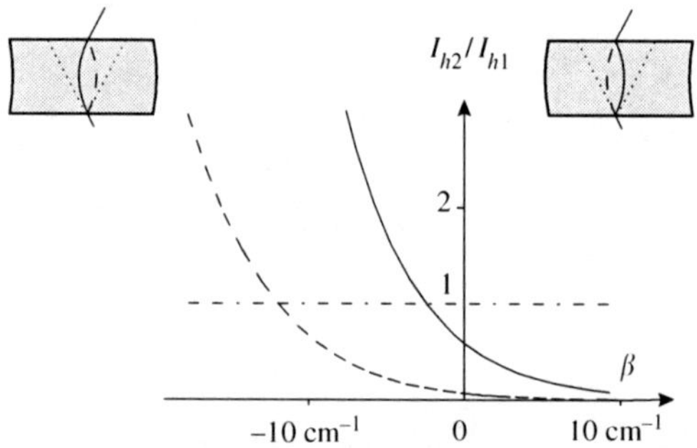

Fig. 13.16. Variation with β of the distribution factor between the reflected beams for branches 1 and 2, respectively. Solid line: Mo$K\alpha$; dashed line: $\lambda = 0.413$ Å. Silicon, 220 reflection, thickness 1 mm; in the inserts, the path of type 1 wavefields is represented as a solid line.

at the maximum of reflected intensities ($\eta^i = -\eta^e = -\frac{1}{2}\beta z$)

$$\frac{|D_{h1}^{(d)}|^2}{|D_{h2}^{(d)}|^2} = \frac{A_2 A_2'' B_2}{A_1 A_1'' B_1}$$

$$= (p + \sqrt{1+p^2})^4$$

$$\times \exp\left[-2\frac{\mu_o}{\cos\theta_B}\frac{C\cos\varphi}{\beta}\left|\frac{\chi_{ih}}{\chi_{io}}\right| \log(p - \sqrt{1+p^2})^2\right], \tag{13.49a}$$

where $p = -\beta z/2$. It is represented in Fig. 13.16 in terms of β for two wavelengths, $\lambda = 0.413$ Å and Mo$K\alpha$. It can be seen that, for β smaller than a certain value which depends on the absorption and therefore on the wavelength, this ratio is larger than 1 and the intensity is increased for wavefield 2 and decreased for wavefield 1. It is the opposite when β is larger than that same value.

(ii) *Distribution of intensity between the reflected and refracted beams.* The ratio between the reflected and refracted intensities is, using (13.47) and (13.48),

$$\frac{|D_{hj}^{(d)}|^2}{|D_{oj}^{(d)}|^2} = \frac{A_j''}{A_j'} = \left(p \mp \sqrt{1+p^2}\right)^2, \tag{13.49b}$$

where the top sign corresponds to $j = 1$ and the bottom sign to $j = 2$. It is independent of the wavelength. It can be seen that when β is negative this ratio is smaller than 1 for branch 1 and there is a transfer of intensity from the reflected beam to the refracted beam. When β is positive the ratio is larger than 1 and the transfer of energy is from the refracted to the reflected beam. It is the opposite for branch 2.

The relative variations of the refracted and reflected intensities for branches 1 and 2 according to the sign of β are represented in Fig. 17.3. The ratio of the total reflected intensity to the total refracted intensity,

$$\frac{I_h^{(d)}}{I_o^{(d)}} = \frac{|D_{h1}^{(d)}|^2 + |D_{h2}^{(d)}|^2}{|D_{o1}^{(d)}|^2 + |D_{o2}^{(d)}|^2}$$

(neglecting the interferences between the two branches) is represented in Fig. 13.17(a) in the low-absorption case ($\lambda = 0.413$ Å and Mo$K\alpha$) and in Fig. 13.17(b) in the high-absorption case. In the latter case the influence of branch 2 is negligible and the transfers of intensity between the reflected and refracted beams are like for branch 1 only. They are symbolized by + and − signs in the insert. According to its definition 13.10(a), the sign of β can be changed by changing the sign of the reciprocal lattice vector **h**. A contrast reversal is therefore expected for the images of distorted areas in *hkl* and $\bar{h}\bar{k}\bar{l}$ projection topographs as can be seen in Figs. 17.14(a),(b). This effect is very useful to detect at first glance the sense of the deformation.

13.8 Diffracted intensities—spherical-wave case

13.8.1 *Pendellösung in slightly deformed crystals*

We shall now consider the case of an incident spherical wave in the symmetric transmission geometry. As has been shown in Chapter 10, every point of both branches of the dispersion surface is excited. It can be seen in Fig. 13.18 that, at any point p of the base of the Borrmann triangle, BC, arrive two wavefields, one excited on branch 1 of the dispersion surface (solid line) and one excited on branch 2 (dotted line).

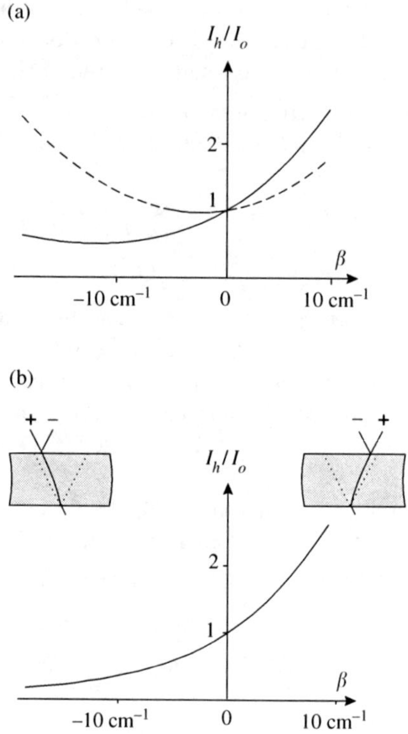

Fig. 13.17. Variation with β of the distribution factor between the reflected and refracted beams. Silicon, 220 reflection, thickness 1 mm. (a) Low-absorption case: solid line, Mo$K\alpha$; dashed line: $\lambda = 0.413$ Å. (b) High-absorption case: Cu$K\alpha$.

This intuitive observation is confirmed by determining the paths of the wavefields travelling from A to p. Their equations are given by (13.29) in the symmetric case:

$$\frac{(y - y_O)^2}{\sin^2 \theta_B} - \frac{(z - z_O)^2}{\cos^2 \theta_B} = \frac{1}{\beta^2 \cos^2 \theta_B}, \tag{13.50a}$$

where y_O and z_O represent the coordinates of the centre, O, of the hyperbola with respect to the set of coordinates Ay, Az. The hyperbola passes through A ($y = 0$, $z = 0$) so that one also has

$$\frac{y_O^2}{\sin^2 \theta_B} - \frac{z_O^2}{\cos^2 \theta_B} = \frac{1}{\beta^2 \cos^2 \theta_B}. \tag{13.50b}$$

The coordinates of the centre of the hyperbola are solutions of eqns (13.50a,b). There are two solutions with opposite signs for y_O, one corresponding to a branch 1 wavefield and the other to a branch 2 wavefield.

The two wavefields arriving in p interfere, giving rise to equal-strain Pendellösung fringes visible on X-ray diffraction topographs (Kato 1964*b*; Ando and Kato 1966, 1970; Hart 1966; Patel and Kato 1973)—see Section 13.8.3. The intensities of the refracted and reflected waves at the exit surface are of the form

$$\begin{aligned} |D_h|^2 &= |D_{h1}^{(d)} + D_{h2}^{(d)}|^2 = |D_{h1}^{(d)}|^2 + |D_{h2}^{(d)}|^2 + 2|D_{h1}^{(d)} D_{h2}^{(d)}| \cos\varphi, \\ |D_o|^2 &= |D_{o1}^{(d)} + D_{o2}^{(d)}|^2 = |D_{o1}^{(d)}|^2 + |D_{o2}^{(d)}|^2 + 2|D_{o1^{(d)}} D_{o2}^{(d)}| \cos\varphi, \end{aligned} \tag{13.51}$$

where $\varphi = \varphi_{h2} - \varphi_{h1} = \varphi_{o2} - \varphi_{o1}$ is the phase difference between the two wavefields and $|D_{hj}^{(d)}|$ and $|D_{oj}^{(d)}|$ are given by (13.47) ($j = 1$ or 2).

13.8.2 *Phase of the refracted wave in a deformed crystal*

According to (13.13b), the phase of the refracted wave at p (Fig. 13.18) is

$$\varphi_{oj} = 2\pi \int_A^p \mathbf{K_{oj}} \cdot \mathrm{d}\vec{l}. \tag{13.52}$$

The local variation of the wavevector is, using (13.23),

$$\delta\mathbf{K_{oj}} = -\frac{1}{1 + X_{hj}/X_{oj}} \vec{\nabla}\left(\frac{\partial \mathbf{h} \cdot \mathbf{u}}{\partial x_h}\right)\frac{\mathrm{d}z}{\cos\theta_B}$$

assuming the reflection to be symmetric (lattice planes normal to the crystal surface, the deformation being negligible).

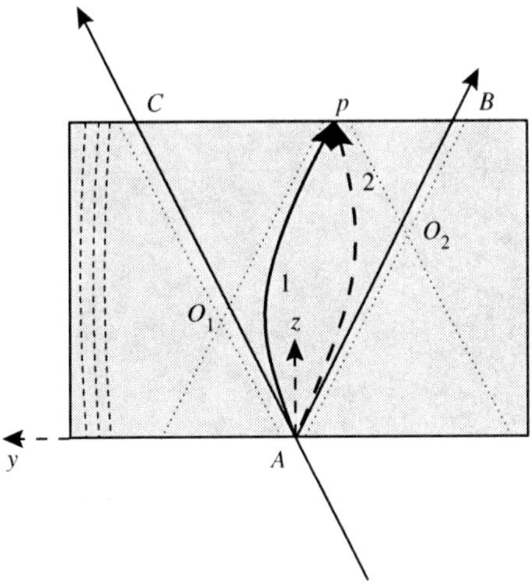

Fig. 13.18. Spherical-wave Pendellösung in a deformed crystal with a constant strain gradient.

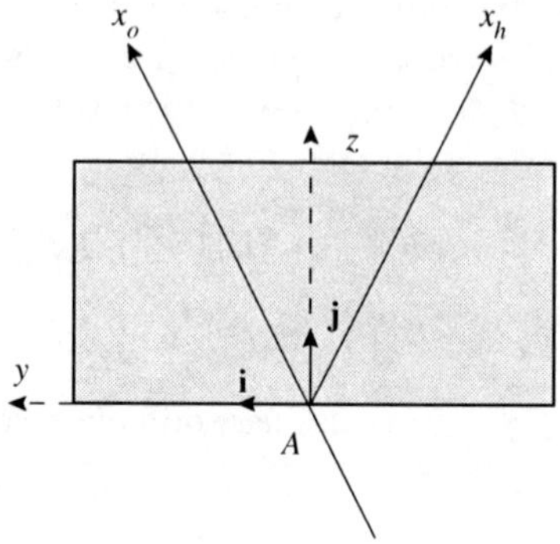

Fig. 13.19. Coordinate systems used in the calculation of the phase.

This expression is calculated by changing from the coordinates x_o, x_h along the incident and refracted directions, respectively, to the coordinates y, z along the directions parallel and normal to the crystal surface, respectively (Fig. 13.19). Let us set

$$f = \frac{\partial^2(\mathbf{h}\cdot\mathbf{u})}{\partial x_o\, \partial x_h},$$

$$g = \frac{\partial^2(\mathbf{h}\cdot\mathbf{u})}{\partial x_h^2}.$$

In order to simplify the integrations, we shall limit ourselves to the case where quantities f and g are constants (constant strain gradient); the scalar product $\mathbf{h}\cdot\mathbf{u}$ will therefore be assumed to be a quadratic function of x_o and x_h. It follows, for the expression of $\delta\mathbf{K_{oj}}$, that

$$\delta\mathbf{K_{oj}} = -\frac{f\mathbf{m}}{1 + X_{hj}/X_{oj}} \frac{\mathrm{d}z}{\cos\theta_B \sin 2\theta_B}$$

with

$$\mathbf{m} = \cos\theta_B\left(1 - \frac{g}{f}\right)\mathbf{i} + \sin\theta_B(1 + \frac{g}{f}\Big)\mathbf{j},$$

where $\mathbf{i}$ and $\mathbf{j}$ are unit vectors in the directions parallel and normal to the entrance surface of the crystal, respectively (Fig. 13.19).

It can be expressed in terms of its projection along the incident direction, $\mathrm{d}X_{oj}$, using (13.23):

$$\delta\mathbf{K_{oj}} = \frac{\mathbf{m}}{\sin 2\theta_B}\,\mathrm{d}X_{oj}.$$

This relation may be integrated and

$$\mathbf{K_{oj}} = \frac{\mathbf{m}}{\sin 2\theta_B}\left[X_{oj} - X_{oj}^{(i)}\right] + \mathbf{K_{oj}^{(i)}},$$

where the superscript i means that the quantity is taken at point A on the entrance surface (Fig. 13.18), following the notation introduced in Section 13.7.1. Substituting into (13.52), it follows that

$$\varphi_{oj} = 2\pi \int_A^P \left[X_{oj} - X_{oj}^{(i)}\right]\frac{\mathbf{m}\cdot \mathrm{d}\vec{l}}{\sin 2\theta_B} + 2\pi \int_A^P \mathbf{K_{oj}^{(i)}} \cdot \mathrm{d}\vec{l}. \tag{13.53}$$

The path element is parallel to the Poynting vector of the wavefield and can be written as

$$\mathrm{d}\vec{l} = \frac{\mathbf{S}}{\mathbf{S}\cdot\mathbf{s}}\,\mathrm{d}z,$$

where, following the notations of Chapters 4 and 5, $\mathbf{S}$ is the Poynting vector and $\mathbf{s}$ a unit vector in the direction of the lattice planes. Using (4.20) and (4.17), it can be expressed in terms of the ratio X_{hj}/X_{oj}:

$$\mathrm{d}\vec{l} = \frac{\mathbf{s_h} + (X_{hj}/X_{oj})\mathbf{s_o}}{\cos\theta_B(1 + X_{hj}/X_{oj})}\,\mathrm{d}z.$$

The coordinate X_{hj} can be expressed in terms of X_{oj} using the equation of the dispersion surface (4.15) and it follows, using (13.23), that

$$\mathrm{d}\vec{l} = -\left(\mathbf{s_h} + \frac{\cos^2\theta_B}{4\Lambda_o^2}\frac{\mathbf{s_o}}{X_{oj}^2}\right)\frac{\Lambda_o\,\mathrm{d}X_{oj}}{\beta\cos^2\theta_B}. \tag{13.54}$$

After substitution in (13.53), φ_{oj} can be put in the form of a sum of three terms:

$$\varphi_{oj} = \varphi_{oj}^{(1)} + \varphi_{oj}^{(2)} + \varphi_{oj}^{(3)},$$

with

$$\varphi_{oj}^{(1)} = \frac{-2\pi\Lambda_o}{\beta\cos^2\theta_B \sin 2\theta_B}\int_A^P \left[X_{oj} - X_{oj}^{(i)}\right]\mathbf{s_h}\cdot\mathbf{m}\,\mathrm{d}X_{oj},$$

$$\varphi_{oj}^{(2)} = \frac{-2\pi}{4\beta\Lambda_o}\int_A^P \frac{\left[X_{oj} - X_{oj}^{(i)}\right]}{X_{oj}^2}\,\mathrm{d}X_{oj},$$

$$\varphi_{oj}^{(3)} = 2\pi \int_A^P \mathbf{K_{oj}^{(i)}} \cdot \mathrm{d}\vec{l}.$$

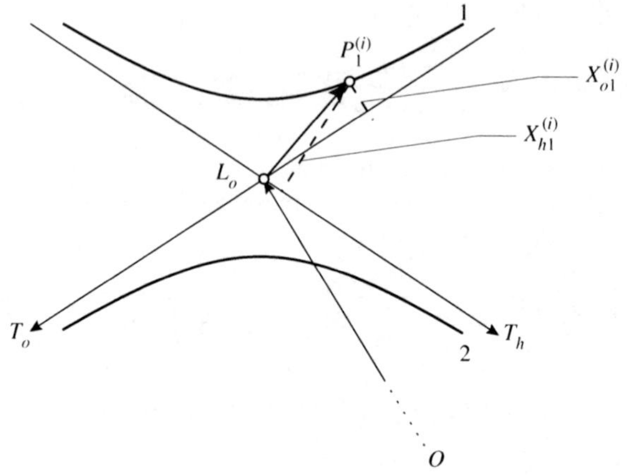

Fig. 13.20. Decomposition of the wavevector: $\mathbf{K_{oj}^{(i)}} = \mathbf{OP_j^{(i)}} = \mathbf{OL_o} + \mathbf{L_oP_j^{(i)}}$ where $P_j^{(i)}$ is the position of the tiepoint at the entrance surface of the crystal.

The first two integrals are readily integrated:

$$\begin{aligned} \varphi_{oj}^{(1)} &= \frac{-2\pi\Lambda_o}{\beta\cos^2\theta_B}\frac{g}{f}\left[X_{oj}^{(e)} - X_{oj}^{(i)}\right]^2, \\ \varphi_{oj}^{(2)} &= \frac{-2\pi}{4\beta\Lambda_o}\left[\log\left(\frac{X_{oj}^{(e)}}{X_{oj}^{(i)}}\right) + \frac{X_{oj}^{(i)}}{X_{oj}^{(e)}} - 1\right], \end{aligned} \tag{13.55}$$

where the superscript e means that the quantity is taken at point p on the exit surface (Fig. 13.18).

The third integral is calculated by decomposing the wavevector at the entrance surface (Fig. 13.20):

$$\mathbf{K_{oj}^{(i)}} = \mathbf{OL_o} + \mathbf{L_oP_j^{(i)}},$$

where $P_j^{(i)}$ is the tiepoint at the entrance surface for wavefield j and L_o is the Lorentz point. The third integral is therefore decomposed into a sum of two terms, of which the first one, $2\pi\int_A^p \mathbf{OL_o}\cdot \mathrm{d}\vec{l}$, oscillates in the crystal with a period of the order of the wavelength and can be neglected (in any case this term does not depend on the branch of the dispersion surface and disappears when calculating $\varphi = \varphi_{o2} - \varphi_{o1}$). The second one is readily evaluated using (13.54) and noticing that

$$\begin{aligned} \mathbf{L_oP_j^{(i)}}\cdot\mathbf{s_o} &= X_{oj}^{(i)}, \\ \mathbf{L_oP_j^{(i)}}\cdot\mathbf{s_h} &= X_{hj}^{(i)}, \end{aligned}$$

it follows that

$$\mathbf{K}_{\mathbf{oj}}^{(\mathbf{i})} \cdot \mathrm{d}\vec{l} = -\frac{X_{oj}^{(i)}}{4\beta\Lambda_o}\left[\frac{1}{(X_{oj}^{(i)})^2} + \frac{1}{(X_{oj})^2}\right]$$

$$\Longrightarrow \quad \varphi_{oj}^{(3)} = -\frac{2\pi}{4\beta\Lambda_o}\left[\frac{X_{oj}^{(e)}}{X_{oj}^{(i)}} - \frac{X_{oj}^{(i)}}{X_{oj}^{(e)}}\right]$$

and, finally, using (13.55),

$$\begin{aligned}\varphi_{oj} &= \varphi_{oj}^{(1)} + \varphi_{oj}^{(2)} + \varphi_{oj}^{(3)} \\ &= \frac{-2\pi}{4\beta\Lambda_o}\left[\log\left(\frac{X_{oj}^{(e)}}{X_{oj}^{(i)}}\right) + \frac{X_{oj}^{(e)}}{X_{oj}^{(i)}} - 1\right] - \frac{2\pi\Lambda_o}{\beta\cos^2\theta_B}\frac{g}{f}[X_{oj}^{(e)} - X_{oj}^{(i)}]^2. \quad (13.56)\end{aligned}$$

13.8.3 *Expression of the phase in terms of the coordinates in direct space*

In order to be able to use expression (13.56) in practice, the reciprocal coordinates $X_{oj}^{(e)}$ and $X_{oj}^{(i)}$ must be expressed in terms of the coordinates of the extremity, p, of the path of the wavefield. This is done by first expressing the coordinates, y and z, of p in terms of the local deviation parameter, η_j. These coordinates are related by the differential relations

$$\frac{\mathrm{d}y}{\mathrm{d}z} = \tan\alpha = \frac{\mp\eta_j}{\sqrt{1+\eta_j^2}}\tan\theta_B \quad \text{(local slope of the path)},$$

$$-\beta\,\mathrm{d}z = \mathrm{d}\eta_j \quad \text{(from (13.21))}.$$

Let us set

$$\left.\begin{aligned} y' &= \beta\frac{y}{\tan\theta_B} \\ z' &= \beta z \end{aligned}\right\} \quad \Longrightarrow \quad \left.\begin{aligned} \mathrm{d}y' &= \frac{\mp\eta_j\,\mathrm{d}\eta_j}{\sqrt{1+\eta_j^2}} = \pm\mathrm{d}\left(\sqrt{1+\eta_j^2}\right) \\ \mathrm{d}z' &= -\mathrm{d}\eta_j \end{aligned}\right\}.$$

It follows, by integration, that

$$\begin{aligned} y' &= \pm\left[\sqrt{1+\eta_j^2}\right]_A^p, \\ z' &= -[\eta_j]_A^p. \end{aligned}$$

Using (5.21a), which gives X_{oj} in terms of η_j, it is possible to relate the ratio $X_{oj}^{(e)}/X_{oj}^{(i)}$ to the modified coordinates y' and z' of point p of the exit surface. It follows, in the symmetric case, that

$$\frac{X_{oj}^{(e)}}{X_{oj}^{(i)}} = \frac{1}{4}\left(\sqrt{z'^2 - y'^2 + 4} \mp \sqrt{z'^2 - y'^2}\right). \tag{13.57}$$

The sign can be attributed to branches 1 and 2 using Fig. 13.15:

1. $\beta > 0$: $X_{oj}^{(e)}/X_{oj}^{(i)} > 1$ for branch 1 and < 1 for branch 2; the upper sign corresponds to branch 1 and the lower sign to branch 2.
2. $\beta < 0$: $X_{oj}^{(e)}/X_{oj}^{(i)} < 1$ for branch 1 and > 1 for branch 2; the lower sign corresponds to branch 1 and the upper sign to branch 2.

By setting

$$\left.\begin{aligned}\sinh v &= \frac{\sqrt{z'^2 - y'^2}}{2}\\ \cosh v &= \frac{\sqrt{z'^2 - y'^2 + 4}}{2}\end{aligned}\right\}$$

$$\text{hence}\quad \exp(\pm y) = \frac{\sqrt{z'^2 - y'^2 + 4} \pm \sqrt{z'^2 - y'^2}}{2},$$

it is easy to calculate the logarithm in (13.56):

$$\log\left(\frac{X_{oj}^{(e)}}{X_{oj}^{(i)}}\right) = \mp 2 \arg\sinh\left(\frac{\sqrt{z'^2 - y'^2}}{2}\right). \tag{13.58}$$

Using (13.57) and (13.58), the phase (13.56) can be expressed in terms of the coordinates of p:

$$\begin{aligned}\varphi_{oj} = {} & \frac{\pm\pi}{2\beta\Lambda_o}\left[2 \arg\sinh\left(\frac{\sqrt{z'^2 - y'^2}}{2}\right) + \frac{\sqrt{z'^2 - y'^2}\sqrt{z'^2 - y'^2 + 4}}{2}\right]\\ & \frac{-\pi}{4\beta\Lambda_o}\left[(z'^2 - y'^2) + \frac{g}{f}(z' - y')^2\right].\end{aligned}$$

It is the phase difference between the wavefields belonging to branches 1 and 2 which comes in the expressions (13.51) of the reflected and refracted intensities:

$$\begin{aligned}\varphi &= \varphi_{o2} - \varphi_{o1}\\ &= \frac{2\pi}{\beta\Lambda_o}\left[\arg\sinh\left(\frac{\sqrt{z'^2 - y'^2}}{2}\right) + \frac{\sqrt{z'^2 - y'^2}\sqrt{z'^2 - y'^2 + 4}}{4}\right].\end{aligned} \tag{13.59}$$

This result was first obtained by Kato (1964*b*) and Hart (1966). The following remarks can be made:

1. The phase difference does not depend on $g = \partial^2(\mathbf{h}\cdot\mathbf{u})/\partial x_h^2$.
2. The same result would have been obtained for the phase difference if it had been calculated from $\int_A^p \mathbf{K_{hj}} \cdot d\overrightarrow{l}$.

13.8.4 *Shape of the Pendellösung fringes in a deformed crystal*

The loci of constant phase, and therefore of constant intensity, are given by

$$z'^2 - y'^2 = \text{constant},$$

that is, replacing y' and z' by their expression in terms of the coordinates y and z of p:

$$\frac{z^2}{\sin^2\theta_B} - \frac{y^2}{\cos^2\theta_B} = \text{constant}.$$

They are, therefore, hyperbolae as in the case of perfect crystals. The interference order, given by $\varphi/2\pi$, is always larger than its value in the case of a perfect crystal,

$$\frac{\varphi_{\text{perfect}}}{2\pi} = \frac{1}{\Lambda_o}\sqrt{z^2 - \frac{y^2}{\tan^2\theta_B}},$$

as can be seen readily by replacing y' and z' by their expression in terms of y and z in (13.59) and taking the limit of φ when $\beta \to 0$.

Figure 13.21 represents the variation with crystal thickness of the difference in interference order between the deformed and the perfect crystal, at the centre of the

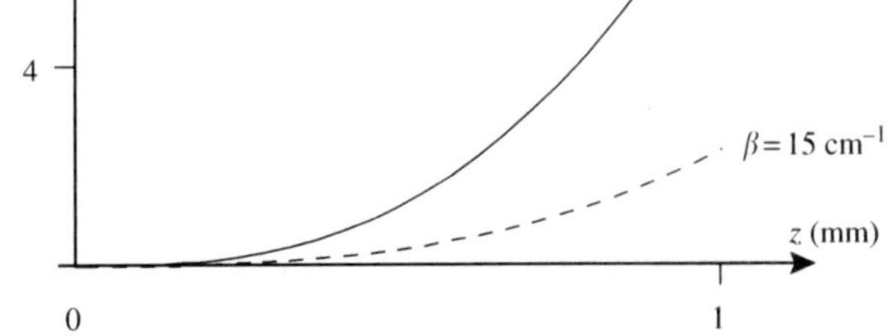

Fig. 13.21. Variation with depth of the difference in interference order between the deformed and the perfect crystal for two values of β, at the centre of the Borrmann triangle, for an incident spherical wave. Silicon, 220 reflection, Mo$K\alpha$ ($\Lambda_B = 36.3\,\mu\text{m}$).

(a)

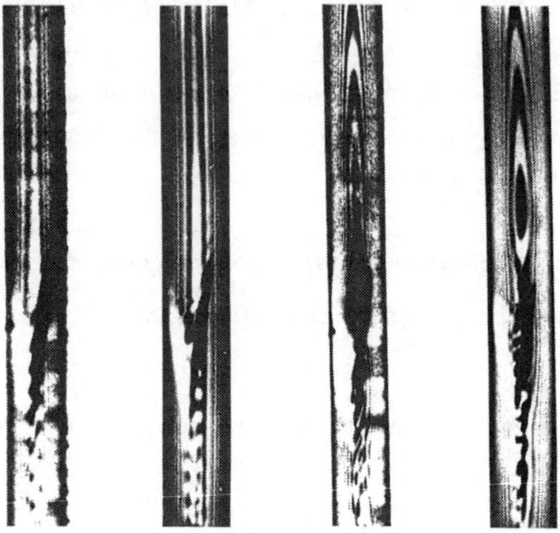

1 2 3 4

(b)

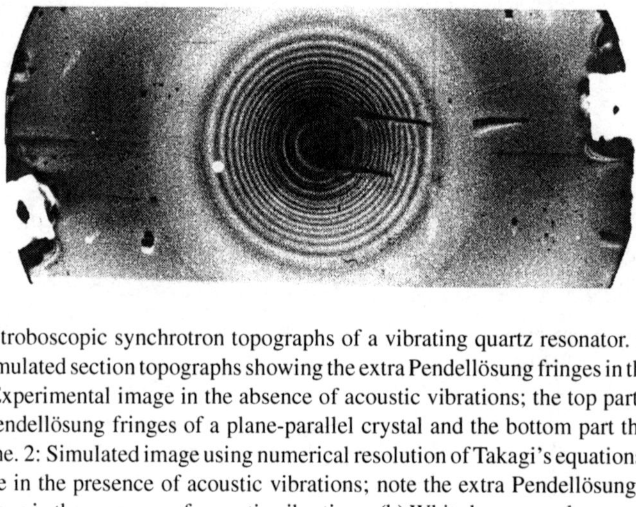

Fig. 13.22. Stroboscopic synchrotron topographs of a vibrating quartz resonator. (a) Experimental and simulated section topographs showing the extra Pendellösung fringes in the presence of strain. 1: Experimental image in the absence of acoustic vibrations; the top part represents the normal Pendellösung fringes of a plane-parallel crystal and the bottom part the image of dislocation line. 2: Simulated image using numerical resolution of Takagi's equations. 3: Experimental image in the presence of acoustic vibrations; note the extra Pendellösung fringes. 4: Simulated image in the presence of acoustic vibrations. (b) White beam synchrotron topograph. The fringes represent equal strain contours.

Borrmann triangle ($y = 0$) and for different values of β. This difference represents the number of extra Pendellösung fringes which appear in the deformed crystal. This phenomenon of 'contraction' of the fringes was first observed by Hart (1965) and Ando and Kato (1966) and discussed by Kato and Ando (1966). It was systematically studied by Ando and Kato (1970) in a dislocation-free silicon crystal deformed by an applied compression and by Hart (1966) in a silicon crystal submitted to a thermal gradient. It was used by Patel and Kato (1968, 1973), Kato and Patel (1973) and Ando *et al.* (1973) to study the deformations induced in a silicon crystal by a discontinuous oxide film deposited on its surface and by Zheng *et al.* (1989) to interpret the Pendellösung fringes in a weakly excited quartz resonator.

On a projection topograph (see Section 17.3.3) the observed intensity is an integrated intensity and if the crystal thickness is constant no fringe appears, even if the incident wave is spherical. However, as a result of the presence of the extra Pendellösung fringes in a deformed crystal, fringes will appear in the projection topograph of a deformed crystal of constant thickness. These fringes will be along the contours of constant phase (and therefore of constant strain) and their number is given by the difference in interference order between the deformed and the perfect crystal (Ando and Kato 1966, 1970).

Figure 13.22 gives an example in the case of a vibrating quartz resonator. The acoustic vibrations are made visible using synchrotron stroboscopic topography (Zarka *et al.* 1988*a*,*b*). The principle of the experiment is the exact synchronization between the acoustic vibration excited in the resonator by an applied a.c. field and the time structure of the pulsed synchrotron radiation. Figure 13.22(a) presents experimental and simulated section topographs (see Section 17.3.2) (the incident beam is limited by a slit and can be assimilated to a spherical wave) illustrating the extra Pendellösung fringes (after Capelle *et al.* 1989, reproduced in Authier and Capelle 1992). The simulations have been performed using numerical resolution of Takagi's equations (see Section 14.3.3). Figure 13.22(b) is a white beam synchrotron topograph (see Section 17.3.3.2) of the quartz resonator showing the equal strain fringes (after Zarka *et al.* 1988*b*, reproduced in Authier and Capelle 1992); in a white beam synchrotron topograph, the intensity is integrated over wavelength (see Section 17.3.3.3). Zolotoyabko *et al.* (1998, 1999) have also studied in a similar way the strain associated with acoustic waves in $LiNbO_3$.

As a practical application of the influence of defects on Pendellösung fringes, the shape of the fringes will be calculated in Section 17.3.4.5 for the section topograph of a crystal containing a screw dislocation.

13.9 Lamellar model

In order to describe a deformed crystal, one may consider it as a stack of perfect lamellae. This was first done by Authier (1961*b*) and, in a more elaborate way, by Kato (1963*b*,*c*). This lamellar model was abandoned by Kato in favour of the Eikonal model, which provides a more precise description of the strain and allows to calculate the ray trajectories inside the crystal as well as the contraction of Pendellösung fringes. By simply considering the variation of the orientation of lattice from one lamella to the next, this simple model has, however, proved useful to evaluate the rocking curve widths of bent crystal monochromators (Egert and Dachs 1970; Albertini *et al.* 1976; Erola *et al.* 1990). It was sagittally extended, taking into account elastic anisotropy, to the evaluation of rocking curve widths (Zhong *et al.* 2002) and integrated intensities of sagittally bent crystals (Zhong *et al.* 2003), and applied by Schevchenko (2003) to the interpretation of interbranch scattering.

14

Propagation of X-rays in highly deformed crystals

14.1 Introduction

Ray theory which has been developed in Chapter 13 does not apply when the strain gradient is too large and, furthermore, the analytical determination of the trajectories and of the diffracted intensities is only possible in particular cases. The calculation of the diffracted intensities must therefore be done using Takagi's equations. In particular cases they have an analytical solution (see Section 14.3.2) but not in the general case. It is then necessary to integrate Takagi's equations numerically. Suitable algorithms have been derived for this purpose and are described in Section 14.3.3. Whenever a suitable model exists to describe the distribution of strain within the crystal, Takagi's equations have proved extremely powerful to solve a great variety of problems ranging from the determination of the contrast of defect images on X-ray and neutron topographs to the calculation of the rocking curve profiles of crystals containing many kinds of defects, epilayers, heterostructures, implanted layers, etc. In some cases, it has even been possible to solve the inverse problem, in other words to extract the parameters of the deformation from the experimental rocking curve profiles. Some examples of the applications of the numerical integration of Takagi's equations are listed in Section 14.3.4.

One drawback of Takagi's treatment is that it does not provide a direct analytical description of the crystal wave in terms of wavefields and therefore does not allow one to determine their energy flow. It is nevertheless possible to generalize the notion of wavefield and to show that, when the strain gradient is larger than a certain critical value, new wavefields are created by interbranch scattering (Section 14.4.1). By combining usual Eikonal theory (as described in Chapter 13) and interbranch scattering, it is thus possible to extend the ray concept to highly distorted crystals. An example in the case of a crystal with a concentration gradient is given in Section 14.4.2. Another example will be found in the interpretation of the contrast of the dislocation images on X-ray topographs (Section 17.3.4.3).

When the distribution of defects cannot be modelled, or is too complicated as in the case of microdefects, a different approach must be developed as is done in the so-called statistical dynamical theories which also use Takagi's equations as a basis. Their principle is briefly described in Section 14.5.

14.2 Takagi's equations in a deformed crystal

The principle of Takagi's derivation (Takagi 1969) is the same for a deformed crystal as that for a perfect crystal (Section 11.2). The wave propagating within the crystal is assumed to be a modulated wave analogous to (11.1):

$$\mathbf{D} = \sum_{\mathbf{h}} \mathbf{D_h}(\mathbf{r}) \exp(-2\pi i \mathbf{K_h} \cdot \mathbf{r}),$$

where, as in Section 11.2.1, the wavevectors are constant. They are related by $\mathbf{K_h} = \mathbf{K_o} - \mathbf{h}$ and $\mathbf{K_o}$ is chosen arbitrarily.

In this way the phase modulation is included in $\mathbf{D_h}(\mathbf{r})$. However, it is more suitable to use a form analogous to (13.12a):

$$\mathbf{D} = \sum_{\mathbf{h}} \mathbf{D'_h}(\mathbf{r}) \exp\{-2\pi i [\mathbf{K_h} \cdot \mathbf{r} - \mathbf{h} \cdot \mathbf{u}(\mathbf{r})]\}. \tag{14.1a}$$

The amplitudes $\mathbf{D'_h}(\mathbf{r})$ represent the *local* values of the modulated amplitudes (which include the local variations of the phase). They are related to $\mathbf{D_h}(\mathbf{r})$ by

$$\mathbf{D'_h}(\mathbf{r}) = \mathbf{D_h}(\mathbf{r}) \exp[-2\pi i \mathbf{h} \cdot \mathbf{u}(\mathbf{r})]. \tag{14.1b}$$

The intensities, which are the final result needed, are the same:

$$|D'_h(\mathbf{r})|^2 = |D_h(\mathbf{r})|^2$$

and there is no drawback in using the local modulated amplitudes. The expression of the polarizability of a deformed crystal is given by (13.11a):

$$\chi(\mathbf{r}') = \sum_{h} \chi_h \exp\{2\pi i \mathbf{h} \cdot [\mathbf{r}' - \mathbf{u}(\mathbf{r}')]\}.$$

After substitution of the expressions (14.1a) of the modulated waves and (13.11a) of the polarizability, the propagation equation (2.55) is replaced, as in the case of perfect crystals (see Section 11.2.2), by an infinite series of equations of the type

$$\begin{aligned}
&\sum_{\mathbf{h}'} \chi_{\mathbf{h}-\mathbf{h}'} \,\mathrm{curl}\,\mathrm{curl}\, \mathbf{D'_{h'}} - 4\pi i \big[\mathbf{K_h} - \vec{\nabla}(\mathbf{h}\cdot\mathbf{u})\big] \wedge \sum_{\mathbf{h}'} \mathrm{curl}\, \mathbf{D'_{h'}} \\
&\quad - 4\pi^2 \mathbf{K_h} \wedge \mathbf{K_h} \wedge \Big(\sum_{\mathbf{h}'} \chi_{\mathbf{h}-\mathbf{h}'} \mathbf{D'_{h'}}\Big) \\
&\quad - 4\pi^2 \big[\mathbf{K_h} - \vec{\nabla}(\mathbf{h}\cdot\mathbf{u})\big] \wedge \big[\mathbf{K_h} - \vec{\nabla}(\mathbf{h}\cdot\mathbf{u})\big] \wedge \Big(\sum_{\mathbf{h}'} \chi_{\mathbf{h}-\mathbf{h}'} \mathbf{D'_{h'}}\Big) \\
&\quad + \Delta \mathbf{D'_h} - 4\pi^2 \big[\mathbf{K_h} - \vec{\nabla}(\mathbf{h}\cdot\mathbf{u})\big]^2 \mathbf{D'_h} \\
&\quad - 4\pi i \big(\big[\mathbf{K_h} - \vec{\nabla}(\mathbf{h}\cdot\mathbf{u})\big] \cdot \vec{\nabla}\big) \mathbf{D'_h} + 4\pi^2 k^2 \mathbf{D'_h} = 0.
\end{aligned}$$

Following Takagi (1969), it will be assumed that $\mathbf{D}'_\mathbf{h}$ is a slowly varying function, as was done in Section 11.2.2, and that the same is true of $\vec{\nabla}(\mathbf{h}\cdot\mathbf{u})$. The second derivatives of these two quantities or the product of their first derivatives can thus be neglected. The equation is further simplified by noting, as in Section 11.2.2, that the length of the reflected wavevector K_h inside the crystal is very close to k. The set of equations (11.5) is then replaced by

$$\begin{aligned}\frac{\partial D'_o(\mathbf{r})}{\partial x_o} &= -\mathrm{i}\pi k C\chi_{\bar{h}} D'_h(\mathbf{r}),\\ \frac{\partial D'_h(\mathbf{r})}{\partial x_h} &= -\mathrm{i}\pi k[C\chi_h D'_o(\mathbf{r}) - 2\beta'_h D'_h(\mathbf{r})],\end{aligned} \tag{14.2a}$$

where x_o and x_h are as usual the coordinates of the position vector $\mathbf{r}$ along directions parallel to the incident and reflected beams, respectively, and

$$\beta'_h = \beta_h + \frac{1}{k}\mathbf{s_h}\cdot\vec{\nabla}(\mathbf{h}\cdot\mathbf{u}). \tag{14.2b}$$

The parameter β_h has the same meaning (11.7) as in the perfect crystal case and is in general taken to be zero.

If one uses the modulated amplitudes $\mathbf{D_h}(\mathbf{r})$ instead of local modulated amplitudes $\mathbf{D}'_\mathbf{h}(\mathbf{r})$, the set of equations (14.2a) is replaced by the set

$$\begin{aligned}\frac{\partial D_o(\mathbf{r})}{\partial x_o} &= -\mathrm{i}\pi k C\chi_{\bar{h}} D_h(\mathbf{r})\exp[2\pi\mathrm{i}\mathbf{h}\cdot\mathbf{u}(\mathbf{r})],\\ \frac{\partial D_h(\mathbf{r})}{\partial x_h} &= -\mathrm{i}\pi k[C\chi_h D_o(\mathbf{r}) - 2\beta_h D_h(\mathbf{r})]\exp[-2\pi\mathrm{i}\mathbf{h}\cdot\mathbf{u}(\mathbf{r})],\end{aligned} \tag{14.2c}$$

where β_h is a constant. Depending on circumstances it may be more convenient to use either of the two sets of equations, (14.2a) or (14.2c).

The boundary conditions are also the same as in the perfect crystal case (see Section 11.2.3).

Using the same modified waves, the second-order differential equations (11.26) are replaced by

$$\frac{\partial^2 D'_o}{\partial x_o\,\partial x_h} - 2\pi\mathrm{i}k\beta'_h\frac{\partial D'_o}{\partial x_o} + \pi^2k^2C^2\chi_h\chi_{\bar{h}}D'_o = 0, \tag{14.3a}$$

$$\frac{\partial^2 D'_h}{\partial x_o\,\partial x_h} - 2\pi\mathrm{i}k\beta'_h\frac{\partial D'_h}{\partial x_o} + \left[\pi^2k^2C^2\chi_h\chi_{\bar{h}} - 2\pi\mathrm{i}k\frac{\partial\beta'_h}{\partial x_o}\right]D'_h = 0. \tag{14.3b}$$

14.3 Resolution of Takagi's equations in the deformed crystal case

14.3.1 *Small deformations, limit of the validity of the Eikonal approximation*

If the deformation is small enough for $|2\pi k\, \partial\beta_h'/\partial x_o|$ to be much smaller than the quantity $\pi^2 k^2 C^2 \chi_h \chi_{\bar{h}}$, in other words if, using (14.3b),

$$\left| 2\pi \frac{\partial^2(\mathbf{h}\cdot\mathbf{u})}{\partial x_o\, \partial x_h} \right| \ll \pi^2 k^2 C^2 \chi_h \chi_{\bar{h}}, \tag{14.4a}$$

the set of equations (14.3) reduces to the set (11.26) where the amplitudes D_h have been replaced by the local amplitudes D_h'. This is equivalent to the concept of local dispersion surface defined in Section 13.5.1. Using (14.2b) and (13.10a), it is possible to relate parameter β_h' to the parameter β defined in Chapter 13:

$$\beta = -\frac{\Lambda_o}{\lambda \cos^2\theta_B} \frac{\partial \beta_h'}{\partial x_o}.$$

The condition (14.4a) therefore reduces to condition (13.10b) for the validity of Kato's and Penning and Polder's ray theories. The parameter β_h', as was shown by (11.7) for parameter β_h in the perfect crystal case, is proportional to the local value of the deviation parameter, η. Differentiation of (14.2b) therefore gives

$$\frac{\mathrm{d}\eta}{\mathrm{d}z} = -\frac{\Lambda_o}{\lambda \cos^2\theta_B} \frac{\partial \beta_h'}{\partial x_o} = \beta,$$

which is equivalent to eqn (13.21) of the ray trajectories. Takagi's equations therefore reduce to those of the ray theories when the deformations are small enough, as was shown by Katagawa and Kato (1974) and Chukhovskii and Petrashen (1977).

Condition (14.4a) can also be written as

$$\left| \frac{\partial^2(\mathbf{h}\cdot\mathbf{u})}{\partial x_o\, \partial x_h} \right| \ll \frac{1}{\Lambda_o}^2. \tag{14.4b}$$

Authier and Balibar (1970) have given a physical interpretation of this condition. It is equivalent to saying that the deformation of the lattice over the unit distance is much smaller than an effective misorientation of the lattice planes of the order of the width of the rocking curve over a distance equal to the Pendellösung distance Λ_o:

$$\frac{\partial \Delta\theta_B}{\partial l} \ll \frac{\delta_{os}}{\Lambda_o}. \tag{14.4c}$$

Authier and Balibar (1970) have shown that the condition can also be expressed in terms of the curvature $1/\mathcal{R}$ of the path of the X-rays in the deformed regions of the crystal. Let α be, as usual, the angle of the direction of propagation of the wavefields with the lattice planes. It is related to the local departure from Bragg's angle, $\Delta\theta$, by means of the amplification ratio (12.13):

$$\mathcal{A} = \left| \frac{\mathrm{d}\alpha}{\mathrm{d}(\Delta\theta)} \right|.$$

As $1/\mathcal{R} \simeq \mathrm{d}\alpha/\mathrm{d}x_o$, condition (14.4b) can be written, in the symmetric case:

$$\frac{1}{\mathcal{A}R} \ll \frac{\pi\gamma_o}{4\sin\theta_B}\frac{\lambda}{\Lambda_o^2}.$$

Using (12.16c), the condition becomes

$$\frac{\Lambda_o}{\mathcal{R}} \ll \frac{\pi}{2}\sin\theta_B.$$

The quantity $\pi\sin\theta_B/2$ is of the order of one, and it follows, finally, that

$$\frac{\Lambda_o}{\mathcal{R}} \ll 1. \tag{14.4d}$$

The condition of validity of the Eikonal or ray theory for small deformations is then the same as for visible optics, provided one considers the Pendellösung distance Λ_o as the characteristic wavelength for the propagation of X-rays.

14.3.2 *Analytical resolution of Takagi's equations*

1. *Introduction*

Takagi's equations have been solved analytically in the case of a constant strain gradient and for the case of a crystal with a transition layer. In the case of a constant strain gradient, the scalar product $\mathbf{h}\cdot\mathbf{u}$ is a quadratic function of x_o and x_h, and parameter β_h' (14.2b) is of the form

$$\beta_h' = \frac{\partial(\mathbf{h}\cdot\mathbf{u})}{\partial x_h} = \frac{-1}{2\pi ik}[Px_o + Qx_h], \tag{14.5}$$

where β_h has been chosen equal to zero. The quantities $P = -2\pi ik\,\partial^2(\mathbf{h}\cdot\mathbf{u})/\partial x_o\,\partial x_h$ which is proportional to the strain gradient (13.10a) and Q are constants and the phase factor $2\pi i\mathbf{h}\cdot\mathbf{u}$ is of the form

$$-2\pi ik\mathbf{h}\cdot\mathbf{u} = Px_ox_h + Qx_h^2 + R(x_o) + \text{constant},$$

where $R(x_o)$ is any function of x_o.

Litzman and Janacek (1974) and Petrashen (1973), using Riemann's function, Chukhovskii (1974), using Green–Riemann's function, and Katagawa and Kato (1974), using Laplace transforms, have shown that the solution of (14.3) can be expressed in terms of confluent hypergeometric functions. The derivation according to Litzman and Janacek follows, that according to Katagawa and Kato is given in the Appendix to this chapter.

2. *Resolution in the case of constant strain gradient using Riemann's function—Laue geometry (Litzman and Janacek 1974)*
Riemann's function has been defined in Sections 11.7.3 and 11.9.1. It satisfies the adjoint differential expression defined by (11.A8) which reduces to

$$M[v] = \frac{\partial^2 v}{\partial x_o\,\partial x_h} - [Px_o + Qx_h]\frac{\partial v}{\partial x_o} + \pi^2 k^2 C^2 \chi_h \chi_{\bar{h}} v = 0. \tag{14.6}$$

The boundary conditions have been defined in Section 11.9.2:

$$\begin{aligned} &v(p) = 1, \\ &\frac{\partial v}{\partial x_h} = [Px_o + Qx_h]v \quad \text{along } bp, \\ &\frac{\partial v}{\partial x_o} = 0 \quad \text{along } cp, \end{aligned} \tag{14.7}$$

(see Fig. 14.1).

The condition along bp ($x_o = x_o(p)$) is readily integrated:

$$\log\left(\frac{v}{v_o}\right) = Px_o(p)x_h + \frac{1}{2}Qx_h^2.$$

The Riemann function, v, is therefore of the form

$$v(x_o, x_h) = w(x_o, x_h)\exp\left[Px_o(p)x_h + \tfrac{1}{2}Qx_h^2\right]. \tag{14.8}$$

If one calculates the derivatives of v from (14.8), and substitutes them into (14.6), it can easily be shown that the function w satisfies

$$\frac{\partial^2 w(x_o, x_h)}{\partial x_o\,\partial x_h} - P[x_o - x_o(p)]\frac{\partial w(x_o, x_h)}{\partial x_o} + \pi^2 k^2 C^2 \chi_h \chi_{\bar{h}} w = 0.$$

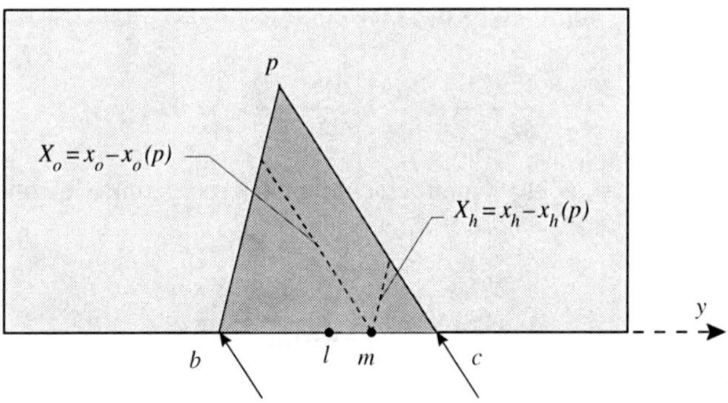

Fig. 14.1. Integration of the Riemann function in the inverted Borrmann triangle $\widehat{pbc}$; l is the midpoint of $\overline{bc}$.

The function $w(x_o, x_h)$ is therefore also a Riemann function which will be calculated, as was done in Section 11.7.3 for the perfect crystal case, by first transforming it in a function depending on a single variable. Let us set

$$X_o = x_o - x_o(p),$$
$$X_h = x_h - x_h(p).$$

It follows, after changing from coordinates (x_o, x_h) to coordinates (X_o, X_h), that

$$\frac{\partial^2 w(X_o, X_h)}{\partial X_o \partial X_h} - PX_o \frac{\partial w(X_o, X_h)}{\partial X_o} + \pi^2 k^2 C^2 \chi_h \chi_{\bar{h}}\, w(X_o, X_h) = 0. \quad (14.9)$$

It is easy to check from (14.9) that

$$w(X_o, X_h) = w\left(\alpha X_o, \frac{X_h}{\alpha}\right) = w(X_o X_h),$$

where α is any constant. Let us set

$$\lambda = PX_o X_h = P[x_o - x_o(p)][x_h - x_h(p)].$$

Equation (14.9) can now be written as

$$\lambda \frac{\mathrm{d}^2 w}{\mathrm{d}\lambda^2} + (1 - \lambda)\frac{\mathrm{d}w}{\mathrm{d}\lambda} - \sigma w = 0, \quad (14.10)$$

with

$$\sigma = -\frac{\pi^2 k^2 C^2 \chi_h \chi_{\bar{h}}}{P} = -\frac{\pi k \mathrm{i}}{2} \frac{C^2 \chi_h \chi_{\bar{h}}}{\partial \beta'_h / \partial x_o}. \quad (14.11)$$

This equation is a simplified form of the confluent hypergeometric equation,

$$\lambda \frac{\mathrm{d}^2 w}{\mathrm{d}\lambda^2} + (\gamma - \lambda)\frac{\mathrm{d}w}{\mathrm{d}\lambda} - \sigma w = 0,$$

of which the solution is the Kummer confluent hypergeometric function of the first kind:[1]

$${}_1F_1(\sigma; \gamma; \lambda) = \sum_{n=0}^{\infty} \frac{(\sigma)_n}{(\gamma)_n} \frac{\lambda^n}{n!},$$

[1] See, for instance, H. Hochstadt (1971). *The functions of mathematical physics*. John Wiley & Sons, New York , or D. Zwillinger (1997). *Handbook of differential equation*. Academic Press.

where $(\alpha)_n$, or the Pochhammer symbol, is given by

$$(\alpha)_n = \frac{\Gamma(\alpha+n)}{\Gamma(\alpha)} = \alpha(\alpha+1)\cdots(\alpha+n-1). \tag{14.12}$$

The Gamma function is defined by

$$\Gamma(\alpha) = \int_0^\infty x^{\alpha-1} \exp(-x)\,\mathrm{d}x$$

and has the following properties:

$$\Gamma(1) = 1; \qquad \Gamma(\alpha+1) = \alpha\Gamma(\alpha); \qquad \Gamma(n+1) = n! \quad (n \text{ integer}).$$

Using (14.12) the solution of the confluent hypergeometric equation is

$$_1F_1(\sigma;\gamma;\lambda) = \sum_{n=0}^{\infty} \frac{\Gamma(\sigma+n)}{\Gamma(\sigma)} \frac{\Gamma(\gamma)}{\Gamma(\gamma+n)} \frac{\lambda^n}{n!}.$$

Here $\gamma = 1$ and one has, finally,

$$_1F_1(\sigma;1;\lambda) = \frac{1}{\Gamma(\sigma)} \sum_{n=0}^{\infty} \frac{\Gamma(\sigma+n)\lambda^n}{(n!)^2}.$$

The expression of the amplitude at p is therefore, according to (11.28a),

$$D'_h(p) = \frac{-\mathrm{i}\pi k C \chi_h \gamma_o}{\sin 2\theta_B} \int_{\overline{bc}} D_o^{(a)}(y)\, {}_1F_1(\sigma;1;P X_o X_h)\,\mathrm{d}y, \tag{14.13}$$

where $y = \overline{lm}$ is the coordinate along $\overline{bc}$ (Fig. 14.1) of a point, m, of the entrance surface (l is the midpoint of $\overline{bc}$).
Two particular cases are of interest:

1. If the incident wave is of the form $D_o^{(a)}(y) = \mathcal{A}\delta(y)$, in other words if the incident wave is a spherical wave originating at the entrance surface, the amplitude at p is

$$D'_h(p) = \frac{-\mathrm{i}\pi k C \chi_h \gamma_o}{\sin 2\theta_B} \mathcal{A}\, {}_1F_1[\sigma;1;P x_o(p) x_h(p)]. \tag{14.14}$$

   The Riemann function can therefore be interpreted as representing the amplitude at p due to a point source at the crystal surface.

2. When $\partial\beta'_h/\partial x_o \Rightarrow 0$, $P \Rightarrow 0$, $\sigma \Rightarrow \infty$, but $\sigma\lambda$ is finite and $_1F_1(\sigma;1;PX_oX_h)$ reduces to the Bessel function, $J_0\sqrt{-4\sigma\lambda} = J_0(2\pi k C\sqrt{\chi_h\chi_{\bar{h}}}\sqrt{X_oX_h})$. The solution is then identical, as it should, to that obtained in the perfect crystal case (11.23b).

Riemann's method of resolution shows that, as in the perfect crystal case, the amplitude at p only depends on the value of the incident wave along bc, and therefore *depends only on the strain distribution within the inverted Borrmann triangle* $\widehat{bcp}$. The incident wave is described by a distribution of point sources on the surface and the solution for a given point source is expressed in terms of a confluent hypergeometric function (14.14).

This notion generalizes Huygens–Fresnel's principle. Following this idea, Petrashen and Chukhovskii (1976) and Chukhovskii and Petrashen (1977) have expressed in the general case the solution at point p of the exit surface in terms of the amplitude distribution of the point sources along bc and of the Green function describing the influence of the point source m at p:

$$D_h(p) = \int_{\overline{bc}} \mathcal{G}_{ho}(m, p) D_o(p)\,\mathrm{d}x_o + \int_{\overline{bc}} \mathcal{G}_{hh}(m, p) D_h(p)\,\mathrm{d}x_h,$$

$$D_o(p) = \int_{\overline{bc}} \mathcal{G}_{oo}(m, p) D_o(p)\,\mathrm{d}x_o + \int_{\overline{bc}} \mathcal{G}_{oh}(m, p) D_h(p)\,\mathrm{d}x_h.$$

3. *Resolution in the case of constant strain gradient using Riemann's function—Bragg geometry*

Using Riemann's method and Laplace transforms, Chukhovskii *et al.* (1978) found the exact form of the Green, or influence, function in the case of a constant strain gradient. They also determined analytically the X-ray trajectories in a thick crystal where the lattice spacing is a linear function of the crystal depth, a problem which had been solved numerically by Taupin (1964) in the Laue geometry. Chukhovskii and Petrashen (1988) used the Green function in the case of a bent crystal in the reflection geometry to interpret the interferences between wavefields having followed curved trajectories (see Section 13.6.3).

4. *Resolution in the case of a crystal with a transition layer*

Takagi's equations were also solved analytically in the Bragg geometry for a crystal with a transition layer by Chukhovskii and Khapachev (1985). They assumed that the strain in the layer is of the form

$$\epsilon(\tau) = \frac{\exp[M(\tau - \tau_o)]}{1 + \exp[M(\tau - \tau_o)]},$$

where $\tau = z/t$ (z, depth inside the crystal; t, crystal thickness), τ_o defines the relative depth of the middle of the transition layer inside the crystal and M is the strain gradient in the transition layer.

Takagi's equation (14.3b) is reduced in that case to a one-dimensional differential equation. Chukhovskii and Khapachev (1985) obtained solutions in terms of hypergeometric functions and gave expressions for the diffracted intensity. They discussed in particular the case of a sharp gradient ($M \gg 1$) and showed that these expressions can be used to solve the inverse problems, namely to determine the characteristics of the transition layer from the rocking curve.

5. *Diffraction by a doubly curved crystal*

Chukhovskii *et al.* (1995*a*,*b*) calculated the distribution of the intensity diffracted by a doubly curved crystal in a focusing geometry for a point source and for an extended incident source. For this they solved the propagation equation by means of perturbation theory and calculated the solutions numerically.

14.3.3 *Numerical integration*

When the strain distribution is more complicated, the only way to obtain the solution of Takagi's equations is by numerical integration. Authier *et al.* (1968*a*) have given an algorithm enabling one to perform the integration by means of a grid illustrated in Fig. 14.2(a). The principle of the calculation is as follows.

Let us divide the inverted Borrmann triangle $\widehat{pbc}$ by a grid along its sides pb and pc which are parallel to the incident and reflected directions, respectively (Fig. 14.2). The amplitudes D_o and D_h at point p can be expressed in terms of their values at

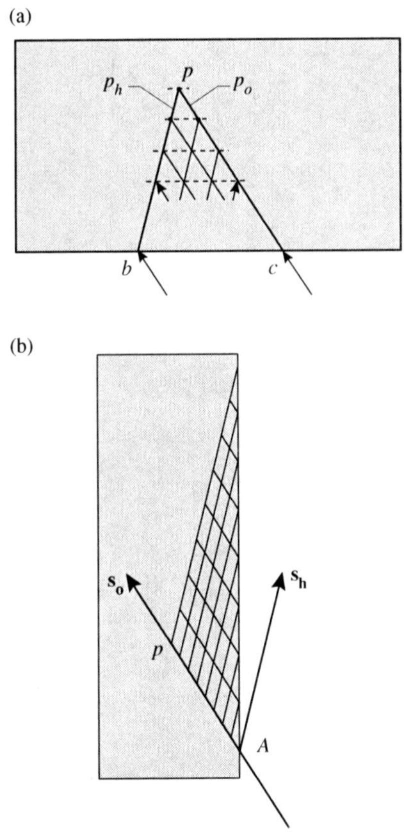

Fig. 14.2. Grid for the numerical integration of Takagi's equations. (a) Laue, or transmission geometry; p_o: integration step in the incident direction; p_h: integration step in the reflected direction. (b) Bragg, or reflection geometry, adapted from Epelboin and Riglet (1979).

neighbouring points q and r on the grid by means of a Taylor series expansion. The latter involves their partial derivatives which can in turn be expressed in terms of the amplitudes by means of the set of equations (14.2a). This provides a linear set of equations which can be calculated by the computer step by step down from the surface to point p. The value of the step is of the order of the micron and the number of steps is very large. It is therefore of paramount importance to reduce as much as possible the error committed at each step. The approximation chosen for that purpose is the so-called 'half-step' derivative approximation. Its principle can be explained using an ordinary function $f(x)$. Its first-order derivative can be written as

$$\frac{\mathrm{d}f(x)}{\mathrm{d}x} = F(f, x).$$

The Taylor expansion of $f(x + \Delta x)$ can best be approximated by considering the derivative at $x + \Delta x/2$:

$$f(x + \Delta x) \simeq f(x) + \Delta x \left(\frac{\mathrm{d}f}{\mathrm{d}x}\right)_{x+\Delta x/2},$$

which, in turn, can be approximated by

$$f(x + \Delta x) \simeq f(x) + \Delta x F\left[\frac{1}{2}\{f(x) + f(x + \Delta x)\}, x + \frac{\Delta x}{2}\right].$$

Using this approximation, the amplitudes at p can be written:

$$\begin{aligned} D'_o(p) &= D'_o(q) + \frac{\overline{qp}}{2}\left\{\left[\frac{\partial D'_o}{\partial s_o}\right]_q + \left[\frac{\partial D'_o}{\partial s_o}\right]_p\right\}, \\ D'_h(p) &= D'_h(r) + \frac{\overline{rp}}{2}\left\{\left[\frac{\partial D'_h}{\partial s_h}\right]_r + \left[\frac{\partial D'_h}{\partial s_h}\right]_p\right\}. \end{aligned} \tag{14.15}$$

These expressions can be calculated using the set of equations (14.2a) which, for that purpose, can be written, using simplified notations:

$$\begin{aligned} p_o \frac{\partial D'_o(p)}{\partial s_o} &= 2AD'_h(p), \\ p_h \frac{\partial D'_h(p)}{\partial s_h} &= 2BD'_o(p) + 2WD'_h(p), \end{aligned} \tag{14.16}$$

where

$$2A = -\mathrm{i}\pi kC\chi_{\bar{h}}p_o; \qquad 2B = -\mathrm{i}\pi kC\chi_h p_h; \qquad 2W = 2\mathrm{i}\pi k\beta'_h p_h$$

and $p_o = \overline{qp}$ and $p_h = \overline{rp}$ are the integration steps. Using (14.15) and (14.16), one can relate the values of the amplitudes at p in terms of the values at q and r by means of a matrix product:

$$\begin{pmatrix} D'_o(p) \\ D'_h(p) \end{pmatrix} = \frac{1}{1 - W - AB} \begin{pmatrix} 1 - W & A(1 - W) & AB & A(1 + W) \\ B & AB & B & 1 + W \end{pmatrix} \times \begin{pmatrix} D'_o(q) \\ D'_h(q) \\ D'_o(r) \\ D'_h(r) \end{pmatrix}. \tag{14.17}$$

This method was used for the first simulations of a dislocation image on a section topograph[2] (Balibar and Authier 1967; Epelboin 1974; Chukhovskii 1974), of planar defects (Epelboin 1979*b*) and of ferromagnetic domain walls (Nourtier *et al.* 1977, 1979) in the Laue geometry, and of dislocations (Bedynska 1973*a*; Bedynska *et al.* 1976; Bubáková and Šourek 1976; Gronkowski 1980; Riglet *et al.* 1980) in the Bragg geometry. Fig. 14.2(b) shows the integration grid in the reflection geometry, after Epelboin and Riglet 1979.

The choice of the integration step is very important. It must be small enough to be adapted to the variations of the strain in the deformed areas. However, since the computing time is proportional to the square of the number of steps, there is a conflict between computing time and resolution, and a compromise must be made. In the case of an incident plane wave, the constant step algorithm is satisfactory, as shown by Petrashen *et al.* (1980), but if the incident wave is a spherical wave, the intensity varies very rapidly along the edges of the Borrmann triangle. In order to solve this difficulty, Epelboin (1983) has introduced a *varying step algorithm* which uses a very small step along the edges of the Borrmann triangle and in the regions where the strain varies rapidly, for instance close to a dislocation line, and a larger one where the strain varies more slowly. Along the edges of the Borrmann triangle, the steps p_o and p_h are adapted to the period of the oscillations of the spherical wave Pendellösung fringes which are known from the analytical solution (see Chapters 10 and 11). In the deformed areas, the change in the steps is done automatically according to the value of the strain gradient. Figure 14.3 shows the integration grid proposed by Epelboin in order to simulate the image of a dislocation line. Reviews of the simulations performed with this method are given in Epelboin (1985, 1987).

The algorithm proposed by Authier *et al.* (1968*a*) has the advantage of being simple. Its precision is, however, limited and this constitutes a severe limitation, in particular for the simulation of traverse topographs or synchrotron white beam topographs. Carvalho and Epelboin (1993*a*) have proposed a new method, more

[2] Intensity distributions in a crystal containing a dislocation line were also calculated numerically by Taupin (1967) using equations similar to Takagi's (Taupin 1964).

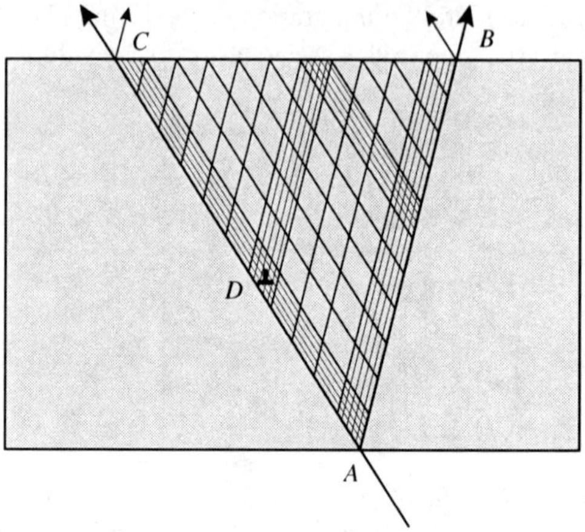

Fig. 14.3. Grid for the integration of Takagi's equations using the varying step algorithm (adapted from Epelboin 1983). D represents the position of a dislocation line.

rapid and more accurate. As, in general, one is only interested in the reflected wave, they use eqn (14.3b) only and do not calculate D'_o, which saves computing time. The boundary conditions along the edges of the Borrmann triangle are determined by reference to the Riemann function: the amplitude of the reflected wave is given in terms of the Riemann function by (11.28a) and the boundary conditions for the Riemann function are specified in Section 11.9.2. It follows, for a point source located at m on the entrance surface (Fig. 14.1), that

$$\begin{aligned} D'_h(p) &= -\frac{\mathrm{i}\pi kC\chi_h}{\sin 2\theta_B} \quad \text{along } \mathbf{s_o}, \\ D'_h(p) &= -\frac{\mathrm{i}\pi kC\chi_h}{\sin 2\theta_B} \exp\{-2\pi\mathrm{i}\mathbf{h}\cdot[\mathbf{u}(p)-\mathbf{u}(m)]\} \quad \text{along } \mathbf{s_h}. \end{aligned} \tag{14.18}$$

Carvalho and Epelboin (1993*a*) introduce a new amplitude, Ψ_h, defined by

$$\Psi_h(p) = -\frac{\sin 2\theta_B}{\mathrm{i}\pi kC\chi_h} \exp\{-2\pi\mathrm{i}\mathbf{h}\cdot[\mathbf{u}(x_o, x_h) - \mathbf{u}(x_o, 0)]\} D'_h. \tag{14.19}$$

From (14.19), the boundary condition (14.18) becomes $\Psi_h = 1$ along $\mathbf{s_o}$ and $\mathbf{s_h}$. By substitution of (14.19) into (14.3b), it follows that

$$\frac{\partial^2 \Psi_h}{\partial x_o\, \partial x_h} - W_\Psi \frac{\partial \Psi_h}{\partial x_o} + \pi^2 k^2 C^2 \chi_h \chi_{\bar{h}} \Psi_h = 0, \tag{14.20}$$

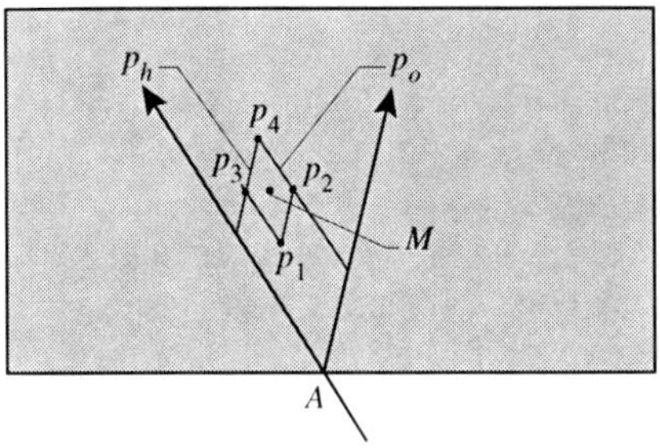

Fig. 14.4. Grid for the numerical integration of Takagi's equations according to Carvalho and Epelboin (1993*a*). p_o: integration step in the incident direction; p_h: integration step in the reflected direction.

with

$$W_\Psi = 2\pi \mathrm{i}\left[\frac{\partial \mathbf{h}\cdot\mathbf{u}(x_o, x_h)}{\partial x_o} - \frac{\partial \mathbf{h}\cdot\mathbf{u}(x_o, 0)}{\partial x_o}\right].$$

For the numerical integration, Carvalho and Epelboin (1993*a*) use the grid represented in Fig. 14.4. They have shown that the amplitude at point p_4 is given by

$$\Psi_h(p_4) = \frac{(Q-G)\Psi_h(p_3) - (Q+G)[\Psi_h(p_2)+\Psi_h(p_1)] + 2\Psi_h(p_2)}{Q^* + G}, \tag{14.21}$$

with

$$Q = 1 - \tfrac{1}{2} p_o W_\Psi(M),$$
$$G = \frac{p_o p_h}{4}\pi^2 k^2 C^2 \chi_h \chi_{\bar{h}},$$

where M is a point at the centre of the rhombus $p_1 p_2 p_3 p_4$ (Fig. 14.4); Q^* is the complex conjugate of Q.

Carvalho and Epelboin (1993*a*) have shown that the number of arithmetic operations involved in (14.21) is significantly smaller than that involved in (14.17) derived from the half-step derivative algorithm proposed by Authier *et al.* (1968*a*). Carvalho and Epelboin's method is therefore much faster. It allowed them, combined with the varying step algorithm, to simulate for the first time traverse and white beam synchrotron topographs (Carvalho and Epelboin 1993*b*).

Further improvement has been introduced by Epelboin (1996) with algorithms for multiprocessor machines and massively parallel computers. These algorithms separate the computation of the derivatives of the deformation inside the crystal from the actual simulation of the image. This allows one to change the elastic model of the defect easily and makes the program very versatile. It is particularly well adapted for the simulation of synchrotron white beam images.

14.3.4 *Applications*

The numerical integration of Takagi's equations was used for theoretical studies such as, for instance, by Gronkowski and Malgrange (1984*a*,*b*) to follow the propagation of X-rays in strongly deformed crystals in the Bragg geometry or by Malgrange and Gronkowski (1984) to investigate Fraunhofer diffraction of X-ray beams in the Bragg geometry (see also Gronkowski 1991). But it was also used to study the influence of defects in many practical cases, for instance:

1. *Simulation of the images of defects on X-ray topographs*:
   - *dislocations (Laue case)* (Authier *et al.* 1985; Epelboin and Patel 1982; Epelboin and Soyer 1984, 1985; Chukhovskii and Petrashen 1988; Green and Tanner 1987; Green *et al.* 1991; Ishida 1980)
   - *dislocations, including misfit dislocations (Bragg case)* (Pétroff *et al.* 1980; Riglet *et al.* 1980; Wierzchowski 1985; Kaganer and Möhling 1991; Cottrell *et al.* 1993; Spirkl *et al.* 1994*a*,*b*; Wierzchowski *et al.* 1995, 1997)
   - *precipitates* (Cui *et al.* 1989; Green *et al.* 1990; Okitsu *et al.* 1992; Stephenson 1992; Holland and Tanner 1995)
   - *microdefects* (Holý 1982*a*,*b*; Indenbom and Kaganer 1985; Petrashen 1986)
   - *stacking faults* (Epelboin 1985, 1987; Kowalski and Lang 1986; Kowalski *et al.* 1989)
   - *ferromagnetic domains* (Nourtier *et al.* 1977, 1979)
   - *antiphase domain boundaries* (Capelle *et al.* 1982; Capelle and Malgrange 1982)
   - *growth striations* (Härtwig *et al.* 1987)
   - *bent crystal in the Bragg case* (Bąk-Misiuk *et al.* 1987).
   - *propagation of acoustic waves in piezoelectric media* (Zheng *et al.* 1989; Epelboin *et al.* 1998; Schelovkov *et al.* 2004)
2. *Calculation of the standing wave pattern in a crystal with a strained overlayer* (Authier *et al.* 1989) or in a *crystal with a uniform strain gradient* (Vartanyants *et al.* 1993*a*,*b*, 1994; Chukhovskii *et al.* 1996*a*).
3. *Calculation of rocking curve profiles*, using an appropriate model for the strain distribution, and sometimes to restore the strain profile by best-fit techniques:
   - *crystals with surface layers or epilayers* (Fukuhara and Takano 1977*a*, *b*; Miltat 1984*b*; Halliwell *et al.* 1984; Bartels *et al.* 1986; Bensoussan *et al.* 1987*a*,*b*; Baumbach *et al.* 1988; Faleev *et al.* 1990; Cui and Mai 1990; Aristov *et al.* 1991; Fewster 1992; Servidori *et al.* 1992)
   - *crystals with implanted layers* (Takeuchi *et al.* 1983; Cembali *et al.* 1985; Servidori *et al.* 1987; Brühl *et al.* 1993; Klappe and Fewster 1994; Servidori *et al.* 1996; Wieteska *et al.* 2000)
   - *crystals with surface acoustic waves* (Gabrielyan and Aslanian 1988)

- *bent crystals* (Taupin 1964; Klar and Rustichelli 1973; Gabrielyan *et al.* 1986; Uschmann *et al.* 1993, 1997; Yamaoka *et al.* 1998; Hölzer *et al.* 1998; Podorov and Förster 2000) and *crystals with a constant strain gradient* (Mana *et al.* 2004*a*, *b*).
- thermally deformed monochromators (Hoszowska *et al.* 2001; Mocella *et al.* 2003)

4. *Restoration of the value of the strain gradient* (Chukhovskii *et al.* 1978; Voronkov *et al.* 1984; Darbinyan *et al.* 1991; Fedorov *et al.* 1993).
5. *Investigation of the diffraction by periodic distortions at the surface of crystals designed for X-ray optics* (Aristov *et al.* 1991, 1992, 1993).

14.4 Ray concept applied to highly distorted crystals

14.4.1 *Generalization of the notion of wavefields, interbranch scattering*

The ray concept supported by the Eikonal theory is very useful in understanding the X-ray trajectories in deformed crystals and in interpreting them physically as the reaction of the electromagnetic wave to the strain gradient field. This is a very familiar procedure in visible optics. As shown in Section 14.3.1, however, this approach is no longer valid when the distortions of the crystal lattice are too large. It is well known in visible optics that geometrical optics does not hold any longer in that case and that Fraunhofer diffraction occurs then. This means that the angular width of wavepackets increases (Fig. 14.5).

A similar effect is expected in the case of X-rays and has been discussed theoretically by Balibar and Malgrange (1975). It is, however, accompanied by another effect, the generation of new wavefields whose tiepoints belong to the other branch of the dispersion surface than those which are propagating, this is the so-called '*interbranch scattering*' effect. This phenomenon was first predicted by Penning in his thesis (Penning 1966) when discussing intuitively the limits of the ray theory and was empirically suggested by Authier in 1966 at the 15th Denver Conference on X-ray analysis to explain qualitatively some features of the images of dislocations on X-ray diffraction topographs (Authier 1967). Intuitively, one expects this effect to occur

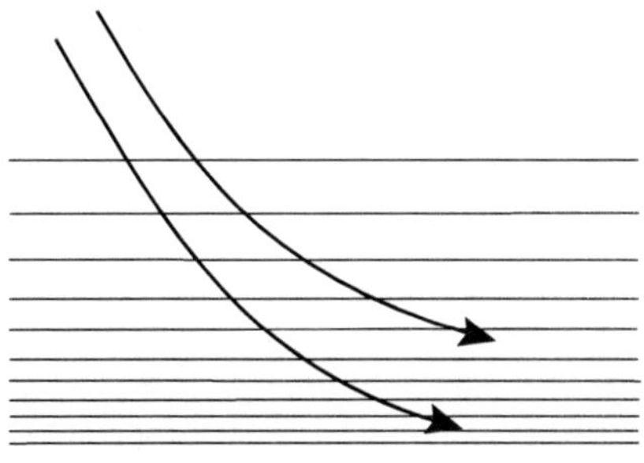

Fig. 14.5. Fraunhofer diffraction of a wavepacket when the variation of the index of refraction is very rapid.

when there is diffraction by a slit, as shown by Balibar (1968), or when there is an isolated diffracting object in the crystal (Authier and Balibar 1970). It was quantitatively confirmed by means of a computer experiment by Balibar *et al.* (1975), who solved Takagi's equations numerically for a crystal with a large temperature gradient. It was observed directly by Zolotoyabko and Panov (1992), who studied intensity diffracted by a single crystal of silicon submitted both to an elastic strain due to a simple bending device and to ultrasonic vibrations. Due to the bending, the tiepoints moved along the dispersion surface according to ray theory; when the magnitude of the wavevector of the ultrasonic waves was of the order of the Bragg gap between the branches of the dispersion surface, interbranch scattering occurred, giving rise to additional Pendellösung oscillations.

The theoretical and physical understanding of the phenomenon of interbranch scattering is difficult because Takagi's treatment does not provide a direct analytical expression of the crystal wave. After transforming the hyperbolical differential equation (14.3b) into an elliptical equation by means of a suitable change of coordinates, Balibar (1969) showed that the Green function can be expressed as the sum of two terms involving Hankel functions, $H_o^1(\zeta)$ and $H_o^2(\zeta)$ (where ζ is given by (10.22)). These terms constitute generalized wavefields and Balibar (1969) proved that two new generalized wavefields appear when the X-ray beams cross a region with a high strain gradient. This was indeed observed, theoretically and experimentally, by Authier *et al.* (1970) in the case of a dislocation, both for spherical and plane waves. Chukhovskii and Petrashen (1977) extended that procedure to the case of a crystal with a uniform strain gradient and could also express the Green function as a sum of two functions which are the equivalent of Balibar's Hankel functions.

Balibar *et al.* (1983) have remarked that these mathematical presentations of the solutions have no real physical meaning and do not provide an understanding of the structure of the crystal wave in reciprocal space. According to these authors, the reason is that the Green function which represents the influence of a point source operates in direct space, while the wavefields are expressed in reciprocal space (they are actually represented by a single point in reciprocal space). In order to describe physically the phenomenon of interbranch scattering by means of the solutions of Takagi's equation, Balibar *et al.* (1983) have expanded the Green function in reciprocal space by means of the Laplace transform. They were thus able to show that the generation of new wavefields occurs when the modulus of the strain gradient β is larger than a critical value $\beta_c = \pi/(2\Lambda_o)$ and when the curvature of the wavefield trajectories is maximum. This occurs when they are parallel to the lattice planes (Fig. 14.6), a result which had already been obtained empirically by Balibar *et al.* (1975), by solving Takagi–Taupin equations numerically.

Balibar *et al.* (1983) showed also that the fraction of the intensity which is transferred to the new wavefield is $\exp(-2\pi\beta_c/|\beta|)$. It is interesting to note that the quantity $\beta_c/|\beta|$ is proportional to Λ_o^{-2} and therefore to $\lambda^2 F_h F_{\bar{h}}$, which shows that the probability of the generation of new wavefields increases for short wavelengths (high X-ray energies). This derivation had been performed in the Laue geometry. The same result was shown analytically by Chukhovskii and Malgrange (1989) in the Bragg geometry.

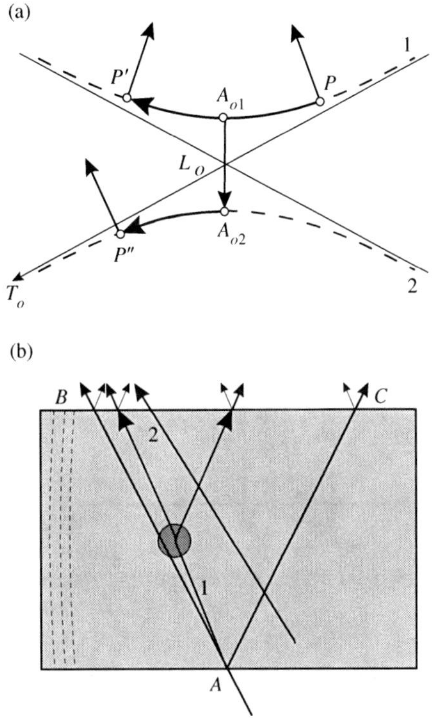

Fig. 14.6. Interbranch scattering when the strain gradient β is very large (reproduced with permission from Authier and Malgrange 1998). (a) Reciprocal space: a new wavefield is excited on the other branch of the dispersion surface when the tiepoint reaches A_{o1}. (b) Direct space: the path of the new wavefield is labelled 2.

It was also shown numerically to be true by Gronkowski and Malgrange (1984*b*) in the case of a variable strain gradient (see also Gronkowski 1991).

These results enable one to extend the ray theory to highly deformed crystals and to give in some cases a quantitative interpretation of their rocking curves, as is shown in the example described in Section 14.4.2.

14.4.2 *Example: X-ray propagation in a crystal with a concentration gradient (Keitel et al. 1999)*

The properties of the propagation of wavefields in a deformed crystal will be illustrated by a practical example taken from a recent study by Keitel *et al.* (1998). Their aim is to design a crystal with a wide FWHM and a high peak reflectivity for use in crystal optics for high-energy synchrotron X-ray radiation (100–200 keV). For this purpose they have tested a large Czochralski-grown silicon crystal with a grown-in concentration gradient of germanium (so-called graded crystal—see Section 15.4.1)

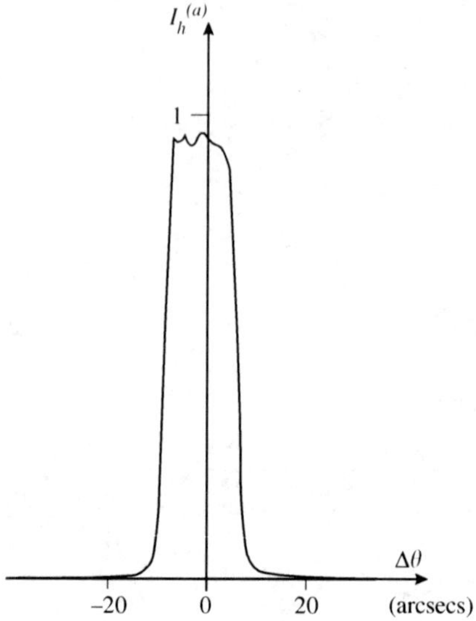

Fig. 14.7. Intrinsic reflectivity of a $Si_{1-x}Ge_x$ crystal; 111 reflection, 100 keV (adapted from Keitel *et al.* 1999).

and a composition $Si_{1-x}Ge_x$ ($0.02 \leq x \leq 0.07$). They recorded the rocking curve for several positions in the crystal corresponding to increasing concentrations of germanium, the 111 reflection and 100 keV photons. They observed a FWHM of 14.6 seconds of arc and a reflectivity of 97 per cent for the lower concentration of germanium (3.5 at.%), neglecting absorption (Fig. 14.7), and a FWHM of 70.9 s and a 74 per cent reflectivity for the higher concentration (5.3 at.%), neglecting absorption, to be compared with a FWHM of 0.36 seconds of arc for a perfect silicon crystal. These results have been explained quantitatively by Keitel *et al.* (1999), using ray theory, but taking into account the creation of new wavefields for strain gradients larger than the critical value for interbranch scattering.

In a crystal with a concentration gradient, the lattice planes are bent as they would be by a temperature gradient and the paths of the wavefields in the crystal are as described in Section 13.6. Their curvature increases with increasing strain gradient and when the strain gradient β is very high as in the example illustrated here, mainly because at high energies Λ_L is large, it takes place in a very small region (Fig. 14.8). The hyperbolae are then constituted by two nearly straight segments respectively parallel to the incident and the reflected directions and joined by a highly curved region. The kink occurs when the tiepoint passes through the apex A_{o1} of the dispersion surface, namely when η is equal to zero. The corresponding depth z inside the crystal is given by (13.31) in terms of the value of the deviation parameter at the entrance surface, η^i, and of parameter β describing the strain gradient ($\eta^i = -\beta z$). The trajectory will therefore present such a kink if the corresponding value of z is smaller than the crystal thickness t ($0 < -\eta^i/\beta < t$). When the wavefield reaches the exit surface, it splits into its two

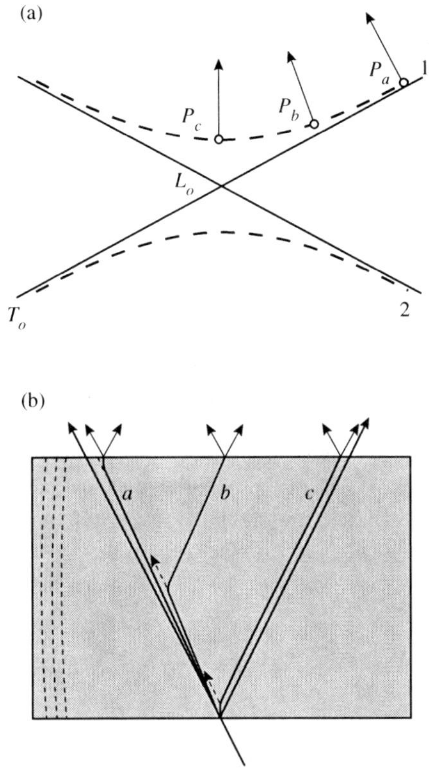

Fig. 14.8. Wavefield trajectories in a crystal with a high strain gradient. (a) Reciprocal space: P_a, P_b, P_c are the tiepoints of wavefields a, b, c at the entrance surface of the crystal. (b) Direct space: full line, trajectories of wavefields excited on branch 1; dashed line, trajectories of wavefields excited by interbranch scattering, their intensity depends on the value of the strain gradient.

wave components. The ratio between their intensities depends on the orientation of the wavefield path at the exit surface, as shown by expression (4.20) of the Poynting vector: the closer the path to the reflected direction, the higher the fraction of the intensity which goes into the reflected beam. The fraction of the intensity which goes into a wavefield from the *incident* beam depends on the orientation of its path at the entrance surface: the closer the path to the incident direction, the higher the intensity.

The crystal was mounted on a triple crystal diffractometer at the Hasylab BW5 beamline (used as a double crystal diffractometer), the monochromator and analyser crystals being perfect silicon crystals also diffracting from (111) planes in transmission. As the sample crystal was rotated and the rocking curve was recorded, successive points such as P_a, P_b, P_c were excited on the dispersion surface at the entrance surface of the crystal. For all the paths whose apex lies inside the crystal, such as b, the incident intensity is nearly fully transferred to the reflected beam if one does not take absorption into account. For the wavefields whose path is parallel to the lattice planes either at the entrance surface (like c) or at the exit surface (like a), the final transfer is 50 per cent (50 per cent at the entrance surface and 100 per cent at the exit surface for c, 100 per cent at the entrance surface and 50 per cent at the exit surface for a). The effect of the concentration gradient is, therefore, to transfer nearly all the intensity of the incident beam into the reflected beam and this for all the wavefields reaching a large proportion of the base of the Borrmann triangle as shown in Fig. 14.8. One also has to take into account the generation of wavefields when the strain gradient is large enough. As is shown in Fig. 14.6, its effect is to take away intensity from the wavefield travelling in the reflected direction, the more so, the higher is the strain gradient. The results of the experiment by Keitel *et al.* (1998) show that this fraction is 4 per cent for the lower concentration of germanium and 28 per cent for the higher concentration, at 100 keV. From the variations of the lattice parameter along the surface of the silicon–germanium crystal, it is easy to deduce the concentration gradient at each point, and Keitel *et al.* (1999) checked that the fraction of the intensity transferred away from a wavefield by interbranch scattering, as calculated at the end of Section 14.4.1, is in very good agreement with the experimental values. All the diffraction properties of this new material can be explained satisfactorily in this way.

14.5 Statistical dynamical theories

14.5.1 *Introduction*

Real crystals are usually neither perfect nor 'ideally imperfect' (in the sense of Section 3.6). If an appropriate model exists to describe the strain distribution inside the crystal and if coherence relations hold between the crystal waves throughout the crystal, the crystal wave may be determined at every point of the crystal using a treatment such as those which have been described in Chapter 13 and the present one. Such coherence relations break down, for instance, if the defect structure of the crystal is of the type described by Darwin's mosaic model and the crystal is made up by a stacking of more or less small and irregular blocks. As noticed by Kato (1976), within each block coherence relations hold and amplitudes can be related by Takagi's equations; between blocks there is no coherence and intensities must be combined. They are related by equations such as the Hamilton–Darwin transfer equations (1.2). This model is however too crude and the description of the crystal as a mosaic of physically separated blocks is usually unrealistic. It is more probable that there is locally, at every point of the crystal, a certain coherence length over which coherence between amplitudes of the crystal wave is lost.

When the density of defects and the local strain gradient become very large, the approximations of Takagi's theory are no longer valid. In the regions of very large deformations where coherence breaks down, diffuse scattering takes place which is incoherent with the coherent scattering from the bulk of the crystal. Dederichs (1967, 1970) was the first to assume that the intensity diffracted by a crystal containing statistically distributed point defects equals the intensity averaged over all possible microscopic defect configurations. In order to calculate the coherent part he extended the dynamical theory to include a generalized Debye–Waller factor taking into account the correlation of static and thermal atomic displacements. The diffuse scattering from crystals containing defects with displacement fields was considered by various authors, including Huang (1947), Cochran (1956), Borie (1957, 1959), Krivoglaz (1961, 1969) and Dederichs (1971, 1973), and observed experimentally, among others, by Austerman and Miller (1965) in BeO, Colella and Merlini (1966) in germanium and silicon, Baldwin *et al.* (1968) and Thomas *et al.* (1971) in copper, Patel (1975) and Lomov *et al.* (1985) in silicon. The line broadening due to a dislocation distribution was studied theoretically and experimentally by Wilkens (1970, 1976) in the cases of copper and sodium chloride. The kinematical theory of the scattering by statistically distributed microdefects and dislocations is reviewed in Krivoglaz (1996). A lot of work has been devoted to the study of the scattering by crystals containing dislocations. As recent examples, Kaganer *et al.* (1997*a*,*b*) studied the scattering by arrays of misfit dislocations in heteroepitaxial microstructures and Breuer *et al.* (2000) have determined the dislocation density in plastically deformed copper.

Kato (1980*a*,*b*) introduced a statistical dynamical theory of diffraction which enables one to express the coherent and the incoherent (diffuse) intensities. Its principle is summarized in Section 14.5.2. It was later modified by Al Haddad and Becker (1988) and Becker and Al Haddad (1989, 1990, 1992), and reformulated by Kato (1991, 1994, 1996), Guigay (1989), Guigay and Chukhovskii (1992, 1995) and Guigay and Vartanyants (1999). It was applied to the study of the thickness dependence of the integrated intensity and diffuse scattering of crystals with microdefects (Molodkin *et al.* 1985; Kochelab *et al.* 1988).

Various authors proposed alternative approaches: the Random Elastic Deformation theory (Kulda 1987, 1988, 1991) in which the imperfect crystal is assumed to consist of a stochastic ensemble of elastically deformed domains; Davis' theory (1991, 1994) in which the X-ray phases in a crystal containing a random distribution of defects are treated indirectly by describing the behaviour of the complex reflectance as a function of position and in which the state of the X-rays, considered as part of a statistical ensemble, is represented by a point in the complex reflectance plane; Holý (1980, 1982*a*,*b*, 1984) and Holý and Gabrielyan (1987) developed a theory based on the mutual coherence function and applied it to the case of crystals containing microdefects (*e.g.* Holý and Kuběna 1987, 1989*a*, *b*, 1992; Holý, Kuběna and Bochnicek 1991).

Bushuev (1989*a*,*b*) extended Kato's statistical dynamical theory to the case of plane waves, taking into account the angular distribution of scattered intensity. This approach was applied by Punegov (1990, 1991, 1993) to heteroepiaxial systems and by Pavlov and Punegov (1998, 2000) to microdefects. Using the same approach, Punegov *et al.* (1996) could solve the inverse problem of diffraction for non-uniform epitaxial layers and Pavlov *et al.* (1995) for laser heterostructures; Faleev *et al.* (1999*a*,*b*) and Pavlov *et al.* (1999) determined the parameters of statistically disturbed quantum dots in InAs–GaAs/GaAs heterostructures used as laser diodes and Nesterets *et al.* (2000) studied the reflection and transmission of factors of rows of pyrolitic graphite plates used as echelon monochromators.

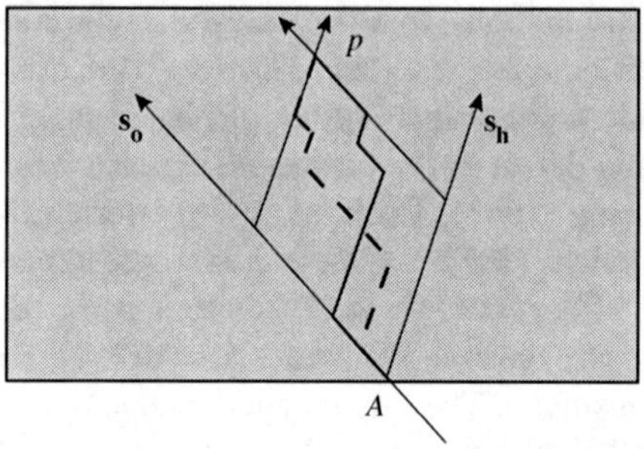

Fig. 14.9. Statistical dynamical theory; multiple reflections (after Kato 1976). The wave travelling from A to p suffers many diffraction events from the incident direction, $\mathbf{s_o}$, to the reflected direction, $\mathbf{s_h}$, and back. Two such paths are represented as examples.

14.5.2 *Principle of Kato's statistical dynamical theory*

Let us consider a point source, at A on the entrance surface, and a point, p, inside the crystal. The crystal wave may have followed any of an infinity of zig-zag routes, two of which are pictured in Fig. 14.9. These waves have been successively diffracted a number of times from the incident direction, $\mathbf{s_o}$, to the reflected direction, $\mathbf{s_h}$, or backdiffracted from the reflected direction, $\mathbf{s_h}$, to the incident direction, $\mathbf{s_o}$ (Kato 1976, 1980*a*). The observed amplitudes at p are therefore average amplitudes, $\langle D_o \rangle$ and $\langle D_h \rangle$.

The principle of the statistical dynamical theory is to consider the total diffracted intensity as a sum of two terms:

$$I_h = I_h^c + I_h^i,$$

where the coherent part of the intensity

$$I_h^c = |\langle D_h \rangle|^2$$

is related to the primary extinction and the incoherent part

$$I_h^i = \langle |\delta D_h|^2 \rangle$$

is related to the secondary extinction. The quantity δD_h represents the fluctuation:

$$\delta D_h = D_h - \langle D_h \rangle.$$

The amplitudes of a wave having travelled along one of the zig-zag routes pictured in Fig. 14.9 satisfy Takagi's equations (14.2c) which Kato rewrites as

$$\begin{aligned} \frac{\partial D_o(x_o, x_h)}{\partial x_o} &= -\mathrm{i}\pi k C \chi_{\bar{h}} \Phi(x_o, x_h) D_h(x_o, x_h), \\ \frac{\partial D_h(x_o, x_h)}{\partial x_h} &= -\mathrm{i}\pi k C \chi_h \Phi(x_o, x_h)^* D_o(x_o, x_h), \end{aligned} \tag{14.22a}$$

where β_h has been taken as zero and

$$\Phi = \exp[2\pi \mathrm{i}\mathbf{h} \cdot \mathbf{u}(\mathbf{r})]$$

is called the lattice phase by Kato.

Equations (14.22a) may also be written in the form of integral equations:

$$D_o(x_o, x_h) = \delta(x_h) - \mathrm{i}\pi k C \chi_{\bar{h}} \int_0^{x_o} \Phi(x_o - u, x_h) D_h(x_o - u, x_h)\, \mathrm{d}u,$$
$$D_h(x_o, x_h) = -\mathrm{i}\pi k C \chi_h \int_0^{x_h} \Phi^*(x_o, x_h - v) D_o(x_o, x_h - v)\, \mathrm{d}v, \tag{14.22b}$$

where the Dirac distribution $\delta(x_h)$ represents a point source on the surface of the crystal.[3]

The average values of the amplitudes, $\langle D_o(\mathbf{r})\rangle$ and $\langle D_h(\mathbf{r})\rangle$, are solutions of the set of equations obtained by taking the average of eqns (14.22a):

$$\frac{\partial \langle D_o(\mathbf{r})\rangle}{\partial x_o} = -\mathrm{i}\pi k C \chi_{\bar{h}} \langle \Phi(x_o, x_h) D_h(\mathbf{r})\rangle,$$
$$\frac{\partial \langle D_h(\mathbf{r})\rangle}{\partial x_h} = -\mathrm{i}\pi k C \chi_h \langle \Phi^*(x_o, x_h) D_o(\mathbf{r})\rangle. \tag{14.23}$$

The correlation term appearing in the first of equations (14.23) can be written as

$$\langle \Phi(x_o, x_h) D_h(\mathbf{r})\rangle = \langle \Phi(x_o, x_h)\rangle \langle D_h(\mathbf{r})\rangle + \langle \delta\Phi(x_o, x_h) D_h(\mathbf{r})\rangle,$$

where $\delta\Phi(x_o, x_h) = \Phi(x_o, x_h) - \langle \Phi(x_o, x_h)\rangle$. By definition $\langle \delta\Phi(x_o, x_h)\rangle = 0$.

Considering only the intrinsic correlation of the second order, Kato (1980*a*) shows that, using (14.22b), this correlation term may be put in the form

$$\langle \Phi(x_o, x_h) D_h(\mathbf{r})\rangle = \langle \Phi(x_o, x_h)\rangle \langle D_h(\mathbf{r})\rangle - \mathrm{i}\pi k C \chi_h \times \int_0^{x_h} \langle \delta\Phi(x_o, x_h) \delta\Phi^*(x_o, x_h - v)\rangle \langle D_o(x_o, x_h - v)\rangle\, \mathrm{d}v. \tag{14.24}$$

Kato sets

$$E = \langle \Phi \rangle = \langle \Phi^* \rangle. \tag{14.25}$$

It is the average of the lattice phase and characterizes long-range order. It can be considered as a static Debye–Waller factor. It ranges from $E = 1$ for the perfect crystal to $E \to 0$ for the mosaic crystal (ideally imperfect).[4]

[3] A solution of the integral equations has been proposed recently by Guigay and Vartanyants (1999) using Laplace transforms.

[4] The special case $E = 0$ is interesting because it allows a detailed analytical development. This was done, with different results, by Becker and Al Haddad (1990) and by Guigay and Chukhovskii (1995).

Let us now consider the spatial correlation function defined by

$$\begin{aligned} G(\mathbf{z}) &= \langle \Phi(\mathbf{r}+\mathbf{z})\Phi^*(\mathbf{r})\rangle \\ &= \langle \Phi(\mathbf{r})\rangle\langle \Phi^*(\mathbf{r})\rangle + \langle \delta\Phi(\mathbf{r}+\mathbf{z})\delta\Phi^*(\mathbf{r})\rangle \\ &= E^2 + (1-E^2)g(\mathbf{z}), \end{aligned} \tag{14.26}$$

where it has been noticed that $\langle \Phi(\mathbf{r})\Phi^*(\mathbf{r})\rangle$ is always unity and $g(\mathbf{z})$ is called the intrinsic correlation function. If homogeneity and isotropy of the medium is assumed, $g(\mathbf{z})$ is simply a function of the separation distance z, $g(z)$.

In addition, Kato (1980*a*) introduces a short-range order correlation length:

$$\tau = \int_0^\infty g(z)\,\mathrm{d}z. \tag{14.27}$$

Using (14.24)–(14.27), the set of equations (14.23) becomes

$$\begin{aligned} \frac{\partial\langle D_o\rangle}{\partial x_o} &= -\mathrm{i}\pi k C\chi_{\bar{h}} E\langle D_h\rangle - \pi^2 k^2 C^2 \chi_h \chi_{\bar{h}}(1-E^2)\tau\langle D_o\rangle, \\ \frac{\partial\langle D_h\rangle}{\partial x_h} &= -\mathrm{i}\pi k C\chi_h E\langle D_o\rangle - \pi^2 k^2 C^2 \chi_h \chi_{\bar{h}}(1-E^2)\tau\langle D_h\rangle, \end{aligned} \tag{14.28}$$

assuming p to be sufficiently far away from the edges of the Borrmann triangle (by more than τ by an order of magnitude) and τ to be much smaller than Λ_o.

Similar differential equations are derived for the average incoherent intensities (Kato 1980*a*; Al Haddad and Becker 1988):

$$\begin{aligned} \frac{\partial I_o^i}{\partial x_o} &= -2\pi^2 k^2 C^2\big[\mathcal{R}e\big(\chi_h\chi_{\bar{h}}\big)\tau_e I_o^i + |\chi_{\bar{h}}|^2\big(-\tau_e I_h^i + \tau(1-E^2)I_h^c\big)\big], \\ \frac{\partial I_h^i}{\partial x_h} &= -2\pi^2 k^2 C^2\big[\mathcal{R}e\big(\chi_h\chi_{\bar{h}}\big)\tau_e I_h^i + |\chi_h|^2\big(-\tau_e I_o^i + \tau(1-E^2)I_h^c\big)\big], \end{aligned} \tag{14.29}$$

where

$$\tau_e = (1-E^2)\tau + E^2\Gamma$$

is called the effective short-range order correlation length. The correlation length for the fluctuations of the intensities, Γ, is given by (Al Haddad and Becker 1988)

$$\Gamma I_h^i = \int_0^{x_o} \langle \delta D_h^*(x_o-u, x_h)\delta D_h(x_o, x_h)\rangle\,\mathrm{d}u.$$

One of the important practical difficulties of the statistical dynamical theory is that it is impossible to calculate Γ in a direct way from the distribution of distortions in the crystal. It is estimated by Kato (1980*a*) to be of the order of Λ_o/E, but Becker and Al Haddad (1990, 1992) suggest that it varies within the crystal and fluctuates around τ.

The integrated coherent and incoherent intensities have been calculated by Kato (1980*a*,*b*) and by Becker and Al Haddad (1990). Their results are the same for the coherent intensity:

$$I_h^c = I_h^{\text{dyn}}\left(\frac{\Lambda_o}{E}\right)\exp\left[-2(1-E^2)\frac{\tau t}{\Lambda_o^2}\right],$$

where $I_h^{\text{dyn}}(\Lambda_o/E)$ is the expression (4.40) of the integrated intensity given by the usual dynamical theory in which Λ_o has been replaced by Λ_o/E: the consequences of the crystal distortions are to multiply the coherent intensity by the static Debye–Waller factor, E, and to increase the Pendellösung distance by $1/E$.

The two parameters, E, and the short-range correlation length, τ, cannot in general be calculated from the defect and strain distribution in the crystal. They have to be determined by a fitting process with the experimental data. It can be noticed that, when $E \Rightarrow 1$, namely when the crystal is perfect, I_h^c tends to the expression for the perfect crystal, I_h^{dyn}. The exponential factor may be considered as a damping factor which becomes equal to 1 when the short-range correlation length tends towards 0; in that case only the increase in the Pendellösung distance can be observed. As E decreases, line broadening increases, and when E is very small, the coherent part of the intensity becomes negligible; this occurs for what Krivoglaz (1996) calls '*second class defects*', such as dislocation lines.

The expressions for the incoherent intensity are rather complicated and need not be given here. They are slightly different in Kato's and Becker and Al Haddad's derivations (see also Guigay 1989) and are discussed in detail in Schneider *et al.* (1992).

14.5.3 *Experimental tests of the statistical dynamical theory*

The first experimental test of the statistical theory is due to Olekhnovich *et al.* (1983), who studied silicon single crystals containing dislocation densities in the range 3×10^5 to 5×10^7 cm^{-2} but their experimental data could not be interpreted satisfactorily using Kato's theoretical expressions. Good agreement with the data of Olekhnovich *et al.* (1983) was, however, obtained by Al Haddad and Becker (1988) using the expressions they had derived for the incoherent intensity.

Schneider *et al.* (1992) observed the variations with crystal thickness of the intensities of 0.0392 Å γ-radiation diffracted by nearly perfect annealed Czochralski-grown silicon crystals containing microprecipitates of silicon oxide, SiO_2. They found that they could be adequately described by the expression of $I_h^c + I_h^i$ given by Becker and Al Haddad (1990), but not by that given by Kato. The values of the static Debye–Waller factor they obtained by the fitting process were around $E = 0.9595$ to $E = 0.9976$ for the 220 reflection, depending on the samples, and around $E = 0.856$ to $E = 0.915$ for the 440 reflection. The short-range order parameter, τ, was very close to zero, which means that the intensities diffracted by these samples could be characterized by the static Debye–Waller factor alone. The values of the correlation length for the fluctuations of the intensities, Γ, were found to be of the order of τ ($\Gamma \simeq 0.5 \times \tau$).

The two theories (Kato's and Becker and Al Haddad's) were also tested by Takama and Harima (1994) on annealed Czochralski-grown silicon crystals containing micro-precipitates of silicon oxide. They found that the experimental results could not be described by Kato's theory; in particular, Kato's assumption that the correlation length for the intensities, Γ, is of the order of Λ_o/E is not correct; like Schneider *et al.* (1992) they found that it is comparable with τ ($\Gamma \simeq 1$ to $6 \times \tau$). They obtained a better agreement with Becker and Al Haddad's theory, but for some of their samples only. The annealing temperature they used was different from that used by Schneider *et al.* (1992) and the microdefects produced were probably different, which might explain why the properties of the intensities diffracted by their samples were also different. The values they obtained for the short-range order parameter, τ, were also very close to zero.

More recently, Kurbakov and Sokolov (1999) applied the results of the statistical dynamical theory to dislocation-free silicon crystals containing microdefects and to quartz single crystals containing a small density of randomly distributed dislocations which they studied with γ-radiation. They also found that Becker and Al Haddad's expression for the diffracted intensities accounted quite satisfactorily for the observed intensities and better than Kato's.

An *et al.* (1995) and Li *et al.* (1995) generalized Kato's statistical theory to the diffraction of epitaxial layers. Analysing the rocking curve profiles, they characterized the perfection and the parameters, such as thickness, strain relaxation, of epilayers of $In_xGa_{1-x}As$ on InP. They showed that the statistical theory is necessary to take into account the contribution of diffuse scattering to the tails of the rocking curves, which cannot be done with a treatment based on Takagi's equations alone.

Appendix: Resolution of Takagi's equations in the case of a constant strain gradient using Laplace transforms (Katagawa and Kato 1974)

Katagawa and Kato (1974) consider the simpler case where

$$-2\pi \mathrm{i}\mathbf{h} \cdot \mathbf{u} = P x_o x_h$$

($Q = R = 0$) and the set of equations (14.2c) with $\beta_h = 0$:

$$\frac{\partial D_o^1(\mathbf{r})}{\partial x_o} = -\mathrm{i}\pi k C \chi_{\bar{h}}(\mathbf{r}) D_h^1(\mathbf{r}) \exp[2\pi \mathrm{i}\mathbf{h} \cdot \mathbf{u}(\mathbf{r})], \tag{14.A1}$$

$$\frac{\partial D_h^1(\mathbf{r})}{\partial x_h} = -\mathrm{i}\pi k C \chi_h(\mathbf{r}) D_o^1(\mathbf{r}) \exp[-2\pi \mathrm{i}\mathbf{h} \cdot \mathbf{u}(\mathbf{r})], \tag{14.A2}$$

where D_o^1 and D_h^1 denote the solution in this simple case.

Let us further consider that the incident wave is limited to a point source, A, on the entrance surface (Fig. 14.10), so that one has

$$\left.\begin{aligned} D_o(x_o, x_h) &= \mathcal{A}\delta(x_h) \\ D_h(x_o, x_h) &= 0 \end{aligned}\right\} \quad \text{along the entrance surface,}$$

where $\mathcal{A}$ is a constant.

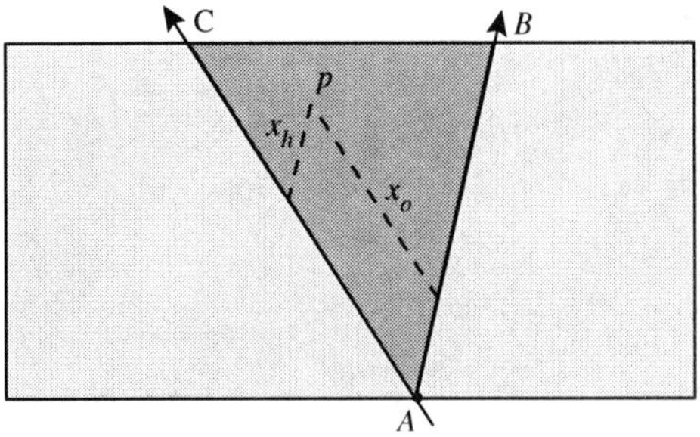

Fig. 14.10. Boundary conditions for the integration of Takagi's equations.

The boundary conditions are

$$D_o(0, x_h) = \mathcal{A}\delta(x_h) \quad \text{along } AB,$$
$$D_h(x_o, -\epsilon) = 0 \quad \text{along } AC,$$

where ϵ is a quantity as small as one wants.

Let us take the Laplace transform of $D_o^1(\mathbf{r})$:

$$V(x_o, u) = \int_{-\epsilon}^{\infty} D_o^1(x_o, x_h)\exp(-ux_h)\,\mathrm{d}x_h. \tag{14.A3}$$

We revert to $D_o^1(x_o, x_h)$ using the complex inversion theorem:

$$D_o^1(x_o, x_h) = \frac{1}{2\pi\mathrm{i}}\int_{\gamma-\mathrm{i}\infty}^{\gamma+\mathrm{i}\infty} V(x_o, u)\exp(ux_h)\,\mathrm{d}u, \tag{14.A4}$$

where γ is a constant such that $0 \leq \gamma \leq \mathcal{R}(u)$ ($\mathcal{R}(u)$ means real part of u).

By substitution in (14.A1), one obtains

$$D_h^1(x_o, x_h) = \frac{\exp(Px_o x_h)}{2\pi^2 kC\chi_{\bar{h}}}\int_{\gamma-\mathrm{i}\infty}^{\gamma+\mathrm{i}\infty} \frac{\partial V(x_o, u)}{\partial x_o}\exp(ux_h)\,\mathrm{d}u. \tag{14.A5}$$

Let us now substitute this expression in (14.A2); it follows that

$$(Px_o + u)D_h^1(x_o, x_h) = -\mathrm{i}\pi kC\chi_h \exp(Px_o x_h)D_o^1(x_o, x_h). \tag{14.A6}$$

Finally, substituting expressions (14.A4) of $D_o^1(x_o, x_h)$ and (14.A5) of $D_h^1(x_o, x_h)$ into (14.A6), and using the boundary conditions for $D_h^1(x_o, x_h)$, we get a differential equation for the Laplace transform $V(x_o, u)$:

$$(Px_o + u)\frac{\partial V(x_o, u)}{\partial x_o} + \pi^2 k^2 C^2 \chi_h \chi_{\bar{h}} V(x_o, u) = 0$$

of which the solution is

$$V(x_o, u) = V(0, u)\Big(\frac{Px_o + u}{u}\Big)^{\sigma} \tag{14.A7}$$

with

$$\sigma = -\frac{\pi^2 k^2 C^2 \chi_h \chi_{\bar{h}}}{P}. \tag{14.30}$$

The value of $V(0, u)$ is easily obtained using (14.A3) and the boundary condition for $D_o^1(x_o, x_h)$:

$$V(0, u) = \int_{-\epsilon}^{\infty} \mathcal{A}\delta(x_h)\exp[-ux_h]\,\mathrm{d}x_h = \mathcal{A}.$$

Substituting expression (14.A7) of the Laplace transform into (14.A4) and (14.A5), it follows that

$$D_o^1(x_o, x_h) = \frac{\mathcal{A}}{2\pi\mathrm{i}}\int_{\gamma-\mathrm{i}\infty}^{\gamma+\mathrm{i}\infty}\Big(1 + \frac{Px_o}{u}\Big)^{\sigma}\exp[ux_h]\,\mathrm{d}u,$$

$$D_h^1(x_o, x_h) = \frac{\mathcal{A}kC\chi_h}{2}\exp[Px_o x_h]\int_{\gamma-\mathrm{i}\infty}^{\gamma+\mathrm{i}\infty}\Big(1 + \frac{Px_o}{u}\Big)^{\sigma}\frac{\exp[ux_h]}{u + Px_o}\,\mathrm{d}u.$$

Using the boundary conditions and an expansion in Laurent series, one obtains the values of the refracted and reflected amplitudes for an incident wave limited to a point source on the entrance surface:

$$\begin{aligned} D_o^1(x_o, x_h) &= -\mathcal{A}Px_o\sum_{n=o}^{\infty}\frac{\Gamma(n+1-\sigma)}{\Gamma(-\sigma)n!(n+1)!}(Px_o x_h)^n \\ &= \mathcal{A}\pi^2 k^2 C^2 \chi_h \chi_{\bar{h}}\, x_o\, {}_1F_1(1-\sigma; 2; -Px_o x_h), \end{aligned}$$

$$\begin{aligned} D_h^1(x_o, x_h) &= \mathrm{i}\mathcal{A}kC\chi_h\exp(Px_o x_h)\sum_{n=o}^{\infty}\frac{\Gamma(n+1-\sigma)}{\Gamma(-\sigma)n!(n+1)!}(Px_o x_h)^n \\ &= \mathrm{i}\mathcal{A}\pi kC\chi_h\, {}_1F_1(\sigma; 1; Px_o x_h). \end{aligned}$$

The result is identical to that obtained with the Riemann function (14.14).

Katagawa and Kato (1974) have shown that the solutions for the general case,

$$-2\pi\mathrm{i}k\mathbf{h}\cdot\mathbf{u} = Px_o x_h + Q(x_h) + R(x_o),$$

where $Q(x_h)$ and $R(x_o)$ are arbitrary functions, are given by

$$\begin{aligned} D_o &= D_o^1\exp(Q(0) - Q(x_h)), \\ D_h &= D_h^1\exp(Q(0) + R(x_o)). \end{aligned}$$

IV

Applications

15

X-ray optics

15.1 X-ray sources

X-rays used for diffraction studies have a high energy (short wavelength). They are produced either by electronic processes, namely X-ray tubes, synchrotron radiation, free electron lasers, or by decay of natural or artificial radioactive isotopes (γ-rays). X-ray tubes and synchrotron radiation will be briefly described in the next two subsections.

The relation between the energy, E, and the wavelength, λ, of photons is

$$E = h\nu = h\frac{c}{\lambda},$$

where the Planck constant $h = 4.13566727 \times 10^{-15}$ eV s and the velocity of light $c = 2.99792458 \times 10^{8}$ m s^{-1}.[1] The conversion between the energy, in keV, and the wavelength, in angstroms, is therefore

$$E \times \lambda = 12.3984186 \ (\text{keV} \times \text{Å}).$$

The approximation $E \times \lambda \approx 12.4$ (keV$\times$Å) is often sufficient.

15.1.1 *X-ray tubes*

The most common source of X-rays in the laboratory is the high-vacuum, or Coolidge, tube. The emitted spectrum is the superposition of a continuous ('white') spectrum, and of discrete characteristic lines.

Electromagnetic radiation is generated by acceleration of charged particles. In particular, the continuous spectrum emitted by an X-ray tube is due to the slowing down in the target of the electrons from the filament of the tube (*Bremsstrahlung*). The maximum frequency, ν_o, of the photons corresponds to the situation where all the energy, eV (e, charge of the electrons; V, applied voltage), of the bombarding electrons is dissipated in the anode in a single process:

$$\nu_o = \frac{eV}{h}.$$

[1] CODATA recommended values of the fundamental Physical Constants: 1998 (Mohr and Taylor 2000).

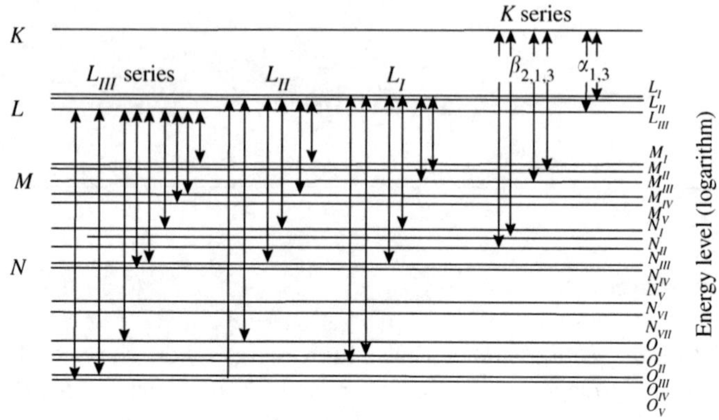

Fig. 15.1. Structure of permitted electron transitions between energy levels in atoms and associated X-ray characteristic lines (reproduced with permission from Aslanov *et al.* 1998).

Table 15.1a Wavelengths of the K emission lines and K absorption edges; numbers in parentheses are standard uncertainties in the least significant figures (reproduced with permission from Deslattes *et al.* 1999*a*)

Element	$K\alpha_1$ energy in keV	$K\alpha_1$ wavelength in Å	$K\alpha_2$ in Å	K abs. edge in Å
W	59.319	0.20901313 (18)	0.2138312 (20)	0.178373 (15)
Ag	22.163	0.55942177 (76)	0.5638131 (26)	0.4859155 (57)
Mo	17.479	0.70931713 (41)	0.713607 (12)	0.61991006 (62)
Cu	8.048	1.54059292 (45)	1.544414 (19)	1.3805971 (31)
Co	6.930	1.789001 (14)	1.792886 (14)	1.6083510 (42)
Fe	6.404	1.936081 (14)	1.940019 (14)	1.7436170 (49)
Cr	5.415	2.289746 (30)	2.2936517 (32)	2.070193 (14)

The radiation produced by an X-ray tube is unpolarized and can be considered as a sum of incoherent spherical waves emitted by the different points of the anode.

The characteristic lines result from the decay of excited states of the element constituting the anode of the tube (see Fig. 15.1, after Aslanov *et al.* 1998). Table 15.1a gives the wavelengths of the K emission lines and of the K absorption edges of the most usual elements, and Table 15.1b the wavelengths of the L emission lines and of the L_I absorption edge of tungsten.

There are two kinds of tubes:

1. The *sealed tube* where the anode is fixed and water cooled. The usual power is of the order of a kilowatt (applied voltage of the order of 30–50 kV, current in the filament of the order of 10–30 mA). It is limited by the necessity to keep the temperature in the anode below the melting point.

Table 15.1b Wavelengths of the L emission lines and L_I absorption edge of tungsten; numbers in parentheses are standard uncertainties in the least significant figures (reproduced with permission from Deslattes *et al.* 1999*a*)

Element	$L\alpha_1$ energy in keV	$L\alpha_1$ wavelength in Å	$L\alpha_2$ in Å	L_I abs. edge in Å
W	8.398	1.476412 (30)	1.487452 (30)	1.024685 (74)

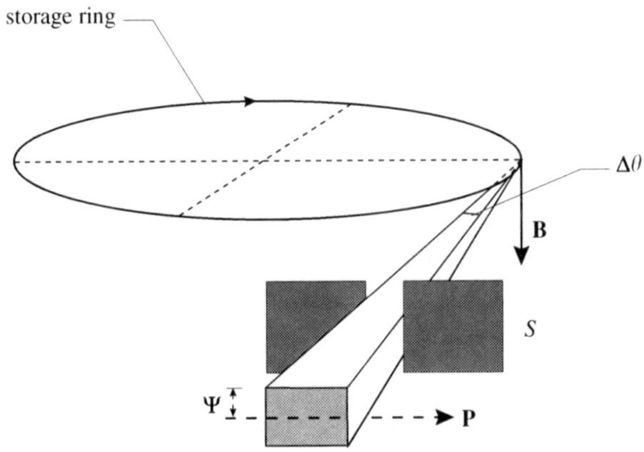

Fig. 15.2. Synchrotron radiation emitted by a relativistic electron travelling along a circular orbit. **B**: magnetic field perpendicular to the plane of the orbit; **P**: direction of the polarization; S: slit limiting the angular width, $\Delta\theta$, of the beam; Ψ: natural vertical divergence of the beam.

2. The *rotating anode*, introduced around 1960, which enables to improve significantly the dissipation of heat and which is continuously pumped. Typical rotations speeds are of the order of 3000 rpm. Maximum values in extreme cases are up to 12 000 rpm. The power is typically of the order of 15 kW and is up to 120 kW in the best case, with a relatively large focus.

For high-resolution diffractometry or topography a fine focus is recommended, either a line focus or a square focus, depending on the experiment. The beam is emitted at a low angle with respect to the surface of the anode and the apparent size of the smallest dimension of the focus is of the order of 10–100 μm.

15.1.2 *Synchrotron radiation*

15.1.2.1 *Principles of synchrotron radiation*

The electromagnetic radiation emitted when an electric charge moves in a circular orbit, thus undergoing centripetal acceleration, at velocities close to the velocity of light where relativistic effects are important, is called synchrotron radiation (Fig. 15.2). The charged particles are maintained in their orbit by a magnetic field,

B, perpendicular to the plane of the orbit. Its value is deduced from the expression of the Lorentz force and is

$$B = \frac{m\gamma c}{Re},$$

where $\gamma = E/mc^2$, E is the energy of the particles, e their charge and m their rest mass, R is the radius of the orbit and c is the velocity of light. The necessary energy is supplied by an oscillating radiofrequency electric field at one or more places along the orbit. The natural vertical divergence of the beam is of the order of (Winick 1988)

$$2\Psi \approx 2\gamma^{-1}.$$

Since $\gamma = 1957E$ where E is in GeV for electrons and positrons, Ψ is of the order of 80 μrad for $E = 6$ GeV, but it actually also depends on the wavelength of the photons.

Synchrotron radiation was first directly observed in 1947 at a General Electric laboratory, in the USA (Elder *et al.* 1947, 1948—for an account of the early history, see Blewett 1998). Its potentialities for diffraction studies were soon realized and the first-generation synchrotrons, designed for high-energy physics, were used parasitically in the seventies by crystallographers. For instance, the first white beam synchrotron X-ray topographs were taken in 1973 (Tuomi *et al.* 1973, 1974). The first dedicated sources (second-generation synchrotrons) appeared in the early eighties. They were followed in the nineties by the third-generation synchrotrons, such as Elettra in Italy, ESRF in France, APS in the USA, SPring-8 in Japan, etc., characterized by highly improved performances. As an indication, the energies of the Elettra, ESRF, APS and SPring-8 synchrotrons are 2 GeV, 6 GeV, 7 GeV and 8 GeV, respectively.

The bases of the theory of the emission of electromagnetic radiation by a charged particle travelling along a circular orbit at relativistic velocities were laid down at the end of the nineteenth century. The theory was developed by Schwinger (1949) and the principles can be found in many papers (see, for instance, Kim 1989; Raoux 1993; Margaritondo 1995; Jackson 1999). For a review of the properties of synchrotron radiation and its applications, see, for instance, Winick (1980, 1988, 1998), Helliwell (1998), Arndt (1999*a*) Duke (2000); Als-Nielsen and McMorrow (2001); Margaritondo (2002).

The power, P, radiated by a charged particle of constant speed v in a solid angle 4π is

$$P = \frac{2}{3}\frac{e^2 c}{R^2}\left(\frac{v}{c}\right)^4 \gamma^4.$$

It decreases very rapidly with the mass of the particles, and for this reason only electrons or positrons are used.

A very important parameter for the users of synchrotron radiation is the *brilliance*, sometimes also called *spectral brightness*. It represents the flux of photons emitted by the source per solid angle, per unit transverse area of the source and per bandwidth.

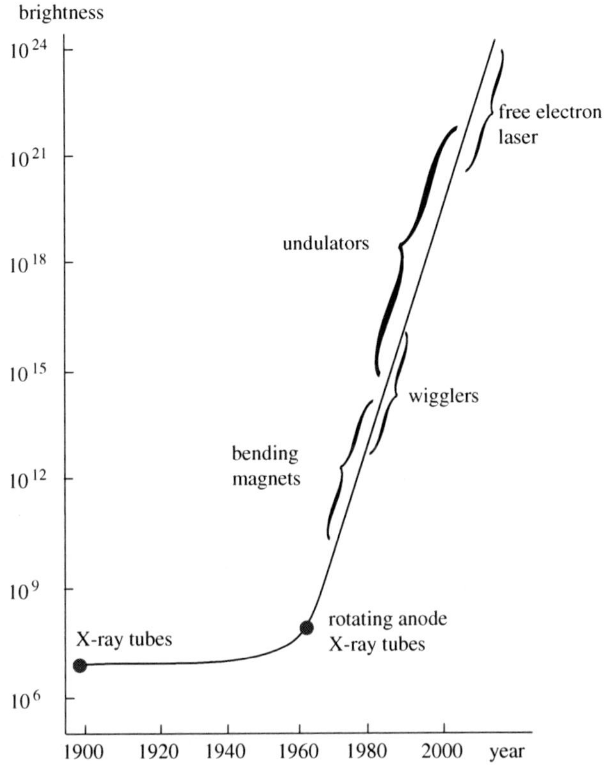

Fig. 15.3. X-ray source brilliance in photons $\times$ s^{-1} $\times$ mm^{-2} $\times$ mrad^{-2} $\times$ (0.1% bandwidth)$^{-1}$, as a function of time; reproduced with permission from Winick (1998).

It is expressed in photons s^{-1} mrad^{-2}mm^{-2} $(0.1\%\delta\ \lambda/\lambda)^{-1}$. Fig. 15.3 shows the increase of the brilliance of X-ray sources with time during past century.

Another important parameter of the synchrotron radiation sources is the *emittance*, ϵ, which is defined as the product of the transverse size, σ, of the beam and its divergence, σ'. One distinguishes the horizontal emittance, $\epsilon_x = \sigma_x \sigma'_x$, which is determined by the electron energy and the ring design, and the vertical emittance, $\epsilon_y = \sigma_y \sigma'_y$, which is due to the coupling with the horizontal magnetic lattice. As machine design improves, lower and lower emittances are achieved. It was of the order of one hundred to several hundred nm rad in second-generation synchrotrons and goes down to 5–20 nm rad with the insertion devices of third-generation synchrotrons.

Synchrotron radiation covers a very wide spectral range, from the infrared to the hard X-rays. It is centred around a broad maximum and is characterized by a critical wavelength

$$\lambda_c = \tfrac{4}{3}\pi R\gamma^{-3},$$

it is defined in such a way that half the total power emitted by the synchrotron is at wavelengths $\lambda < \lambda_c$. The critical wavelength and the corresponding critical energy may also be written, after replacing the constants by their numerical values:

$$\lambda_c = \frac{18.64}{BE^2},$$
$$E_c = 0.665BE^2,$$

where λ_c is in Å, E_c in keV, $B = 3.34E/R$ in tesla (T) , E in GeV and R in metres. For instance, for the ESRF storage ring, $E = 6$ GeV, $R = 24.95$ m and $B = 0.803$ T, $\lambda_c = 0.6448$ Å and $E_c = 19.22$ keV. At higher energies ($\lambda < \lambda_c$) the intensity falls rapidly, to about 10^{-3} of the maximum for $\lambda \approx \lambda_c/10$. At lower energies the intensity is almost constant; it decreases by a factor of 10 over a wavelength range up to $1000 \times \lambda_c$.

The brightness is expressed in terms of the wavelength of the photons through

$$b = 1.327 \times 10^{13} E^2 I H_2\left(\frac{\lambda_c}{\lambda}\right),$$

where E is in GeV, the intensity I of the current stored in the machine is in amperes (A) and $H_2(\lambda_c/\lambda)$ is a dimensionless function which has a maximum close to when λ_c/λ is close to 1 and decreases rapidly for wavelengths larger or smaller than the critical wavelength.

15.1.2.2 *Insertion devices: wigglers and undulators*

The storage ring in which the electrons or positrons are maintained is constituted by an alternation of bending magnets and straight sections. So-called *insertion devices* are inserted in the straight sections and beamlines are installed at the bending magnets and at the insertion devices. These devices (*wigglers* and *undulators*) consist of several dipole magnets of alternating polarity, the orientation of the magnetic fields being perpendicular to the trajectory of the particles. The particles therefore 'wiggle' around their mean path. The radius of curvature of the path in the wiggles is usually smaller than that of the orbit in the bending magnets and the critical wavelength is shifted towards the shorter wavelengths. For instance, at ESRF, the field of the wiggler of the topography beamline is 1.5 T, $R = 13.4$ m and $\lambda_c = 0.345$ Å. For an introduction to the properties of wigglers and undulators, see Winick *et al.* (1981); for a more detailed development see, for instance, Raoux (1993) or Kim (1989).

In a *wiggler*, the total divergence, α, of the beam emitted by the device is larger than the divergence, $\Psi = 1/\gamma$, of the radiation emitted by a point on the orbit. The total beam emitted by the device is therefore the incoherent sum of the beams emitted by each pole and, if the number of periods is N, the intensity is increased by $2N$ (there are two poles per period, one of each sign).

In an *undulator*, the angle α is smaller than Ψ and the beams emitted by the individual poles interfere. The radiation emitted along the undulator axis is therefore quasi-monochromatic; it is constituted by the sum of a fundamental line and its harmonics. Their frequency, λ_n, is

$$\lambda_n = \lambda_u \frac{1 + K^2/2 + \gamma^2\theta^2}{2n\gamma^2},$$

where λ_u is the period of the magnetic field, n the harmonic number, θ the observation angle and $K = 0.934 B_o \lambda_u$ with B_o in T and λ_u in cm. Typically, λ_u is of the order of a few centimetres.

The frequency of the radiation emitted by an undulator is therefore tunable. If $K \ll 1$, the harmonics are very weak and there is essentially one peak, the fundamental ($n = 1$). If $K \approx 1$, the wavelength of the fundamental is shifted towards the longer wavelengths, the power in the fundamental increases (it is maximum at $K = 1.2$) and more harmonics appear. When $K \gg 1$, there are many harmonics and, as they have a certain width, they blend in a broad band when K becomes very large, as is the case for wigglers and bending magnets.

Because of their interference, the amplitudes of the beams emitted by each pole in an undulator add up and the intensity increases with N^2.

Figure 15.4 compares the brilliances of the X-ray beams emitted at the ESRF by bending magnets, wigglers and undulators (energy and current in the storage ring, 6 GeV and 100 mA, respectively).

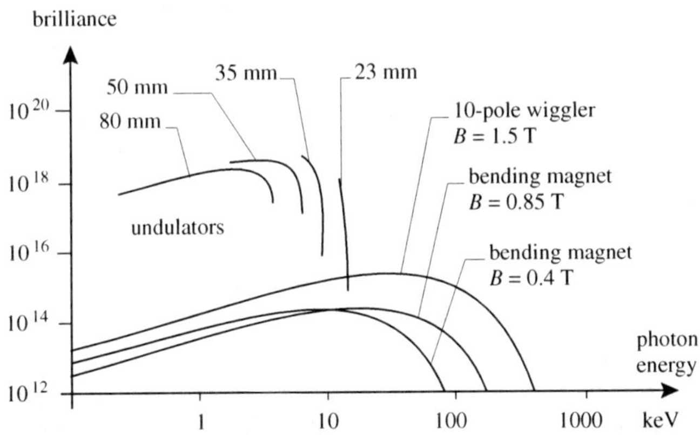

Fig. 15.4. Comparison of the brilliances of the X-ray beams emitted at the ESRF by bending magnets, wigglers and undulators, in photons $\times$ s^{-1} $\times$ mm^{-2} $\times$ mrad^{-2} $\times$ (0.1% bandwith)$^{-1}$, as a function of the energy of the photons. The magnetic fields B are given in teslas and the lengths of the periods of the undulators in mm (gap: 20 mm); adapted from Raoux (1993).

15.1.2.3 *Summary of the properties of third-generation synchrotrons*

The synchrotron radiation sources are characterized by:

1. A *very high flux* of photons and *a high brilliance*, ranging from 10^{14} photons s^{-1} mrad^{-2} mm^{-2} $(0.1\%\, \delta\lambda/\lambda)^{-1}$ for bending magnets to 10^{19} to 10^{21} photons s^{-1} mrad^{-2} mm^{-2} $(0.1\%\, \delta\lambda/\lambda)^{-1}$ for undulators.
2. A *very broad continuous spectrum* from bending magnets and wigglers, extending in the X-ray range from a few keV to a few hundred keV, and monochromatic peaks from undulators.
3. A *highly collimated beam*, with a divergence of the order of 10^{-4} to 10^{-5} rad.
4. A *very small source size*, of the order of a few tens of micrometres vertically and one or two hundred micrometres horizontally.
5. A *very low emittance*, as a consequence of the small source size and the small divergence of the beam. The horizontal emittance is of the order of a few nm rad for third-generation synchrotrons and the vertical emittance can be as low as 0.5 or 1 per cent of the horizontal emittance; for instance, the horizontal and vertical emittances of ESRF are 4 and 0.03 nm rad, respectively.
6. *Long source-to-sample distances*, ranging from several tens of metres to more than one hundred metres. One beamline at SPring-8 (Japan) is 1000 m long.
7. A *high degree of polarization*; the radiation emitted by particles travelling along a curved orbit is plane polarized in the plane of the orbit (the electric field **E** is parallel to the plane of the orbit). As the direction of observation deviates from the plane of the orbit, the radiation becomes elliptically polarized. Special multipole magnet devices can rotate the plane of polarization (planar undulators) or produce circularly polarized X-rays (helical undulators).
8. *High coherence*; due to the very small source size and the very long source-to-sample distances in third-generation synchrotrons, the spatial coherence is very high. It has been discussed by Coïsson (1995), who considered the spatial coherence properties of a monochromatic component from an insertion device in the Fraunhofer region. It has been estimated to be of the order of 100 μm at ESRF (Snigirev 1996; Cloetens *et al.* 1997*a*; Baruchel *et al.* 1997; Espeso *et al.* 1998—see Section 12.13). The synchrotron source of radiation can be described as a distribution of elementary quasi-monochromatic sources of infinitesimal size. These sources emit spherical waves, but Mocella *et al.* (1999, 2000) have shown that they can be approximated locally by a plane wave (see Section 10.2.3).
9. A *regularly pulsed time structure*; electrons in storage rings are grouped in bunches whose length is determined by the radiofrequency system that replenishes the energy, so that an observer will see a succession of flashes of X-rays. The duration of each flash is less than 1 ns and the time separating two successive flashes is of the order of a few nanoseconds or a few tens of nanoseconds. This property has been used, for instance, for the stroboscopic observation of acoustic vibrations in piezoelectric materials (see Section 17.3.6).

15.1.2.4 *Fourth-generation X-ray sources*

A new type of X-ray sources, so-called fourth-generation X-ray sources, is being developed at present. It is an application of X-ray free-electron lasers (FELs) to produce highly collimated (about 1 μrad) pulsed X-ray beams with a brilliance larger than that of the third-generation synchrotron sources (see Fig. 15.3) and a high bunch density. One way to achieve this is based on 'self-amplified spontaneous emission in which the spontaneously emitted X-ray radiation from an electron bunch in an undulator becomes strong enough to modulate the electron density in the bunch in such a way as to further amplify the spontaneous emission in phase with the originating X-ray radiation field' (Doniach 2000). For an introduction to the concept and possibilities of fourth-generation X-ray sources, see, for instance, Winick (1998), Doniach (2000) and Margaritondo (2002); for projects, see, for instance, SLAC (1998) or DESY (2001); for first results see, for instance, Ayvazyan *et al.* (2002) and Wabnitz *et al.* (2002); for the diffraction of ultrashort pulses, see Sections 10.10 and 11.1.

15.2 Flat monochromators

15.2.1 *Introduction*

Some applications such as synchrotron white beam topography or collection of data using Laue diffraction for structure determination purposes take advantage of the white spectrum of X-ray sources, but in most cases, such as diffractometry, topography, small-angle scattering, X-ray optics, magnetic scattering, etc., monochromatic radiation with a more or less narrow bandwidth is required (from 0.1 to 10^{-7} depending on the applications). In the case of EXAFS or XANES, the possibility of scanning the white spectrum with a high-energy resolution must be provided. More generally, the beam impinging on the sample must be conditioned in order to have maximum intensity, well-defined spectral and angular bandwidths, a wide tunable energy range, no unwanted harmonics, specific polarization properties and a high peak-to-background ratio. Furthermore, a good thermal stability and a good resistance to radiation damage are required. The conditioning of the beam is achieved, in general, by using a combination of flat monochromator crystals and focusing optics. The former will be described in the present section and the latter in Section 15.4. Reviews of monochromators and focusing optics are given, for instance, in Matsushita and Hashizume (1983), Freund (1989, 1993), Batterman and Bilderback (1991), Malgrange (1992), Hart (1996), Holý and Mikulik (1996), Hart and Berman (1998) and Treimer (1998).

Monochromatization of the incident beam from the source was done in the early days of X-ray studies using appropriate absorbing filters. It was, however, recognized very early on that a much better way is to make use of the Bragg reflection of the incident radiation on a first crystal. This was done, for instance, by Compton (1917) and by Bragg *et al.* (1921) in order to measure integrated intensities, and by Davis and Stempel (1921, 1922) to obtain the width of the rocking curve from freshly cleaved calcite crystals.

15.2.2 *Monochromator crystals*

15.2.2.1 *Resolution*

Bragg's law, $2d_{hkl} \sin\theta_B = \lambda$, relates the angle of reflection and the wavelength of the radiation which is reflected. By differentiation, one finds that, if $\Delta\theta$ is the angular divergence of the beam diffracted by the crystal, the range, $\Delta\lambda$, of the reflected wavelengths and the corresponding bandwidth are

$$\Delta\lambda = 2d_{hkl} \cos\theta_B \Delta\theta,$$

$$\frac{\Delta\lambda}{\lambda} = \frac{\Delta E}{E} = \cot\theta_B \Delta\theta. \tag{15.1}$$

The divergence of the beam diffracted by the crystal depends, on the one hand, on the natural divergence, Ω, of the incident beam and, on the other hand, on the width $2\delta_{os}$ (FWHM) of the rocking curve of the crystal. The energy resolution can be obtained by combining the two with $\Delta\theta = \sqrt{\Omega^2 + (2\delta_{os})^2}$. In the non-dispersive $(+n, -n)$ setting described in Section 15.2.3.1, the energy resolution is independent of the divergence of the source and depends only on the width of the rocking curve of the monochromator crystal.

It is then obvious that good monochromators require highly perfect crystals since the FWHM of imperfect crystals is much larger than that of ideally perfect crystals, as has been shown in Chapter 3. Such perfect crystals have only become routinely available in the early sixties when very good silicon and germanium single crystals have been commercially produced.

Monochromators are most generally used in the reflection geometry, and the Darwin width, $2\delta_{os}$, is given by (4.48b):

$$2\delta_{os} = \frac{2R\lambda^2}{\pi V \sin 2\theta_B} \sqrt{|\gamma|}\,|C| \sqrt{F_h F_{\bar{h}}}. \tag{15.2}$$

It depends on the wavelength, the structure factor, the polarization and the asymmetry and the influence of these various parameters has been discussed in Sections 4.10 and 7.1.2. Using (15.1) and (15.2), it follows, for the bandwidth, that

$$\frac{\Delta\lambda}{\lambda} = \frac{\Delta E}{E} = \frac{4Rd_{hkl}^2}{\pi V} \sqrt{|\gamma|}\,|C| \sqrt{F_h F_{\bar{h}}}. \tag{15.3}$$

It is independent of the wavelength. Table 15.2 gives the value of the Darwin width, in μrad, for selected crystals, reflections and wavelengths and of the corresponding bandwidth. The narrowest bandwidths are obtained for the high-order reflections, but this is at the expense of the reflected intensity.

Table 15.2 Darwin widths, in μrad, for selected crystals and reflections (symmetrical Bragg geometry)

Radiation	Si 111	Si 220	Si 440	Ge 111	Ge 220	Ge 440	C 111	C 220
100 keV (0.124 Å)	1.3	0.90	0.29	3.04	2.22	0.67	0.94	0.47
30 keV (0.413 Å)	4.4	3.0	0.98	10.3	7.5	2.29	3.1	1.60
22.16 keV (Ag$K\alpha_1$)	6.0	4.1	1.36	13.9	10.2	3.2	4.3	2.19
17.48 keV (Mo$K\alpha_1$)	7.6	5.3	1.8	17.8	13.3	4.3	5.4	2.8
8.05 keV (Cu$K\alpha_1$)	16.9	12.3	6.0	37.7	29.0	12.6	12.6	7.4
Bandwidth $\Delta E/E \times 10^6$	133.3	56.0	8.9	319.8	143.2	21.4	62.2	19.2

It is in general very convenient to represent graphically the bandwidth selected by a monochromator by means of the DuMond diagram (DuMond 1937[2]). It is illustrated in Fig. 15.5, where Bragg's law is represented by a sine curve and the wavelength range $\Delta\lambda$ associated to an angular divergence $\Delta\theta$ is shown. The incident beam such as that emitted by bending magnets or wigglers is constituted by white radiation and several harmonics may be simultaneously reflected by the monochromator. For instance, if a silicon crystal is set for the 220 reflection of 7.5 keV X-rays ($\lambda = 1.653$ Å, $\theta_B = 25.5°$), it will automatically also be set for the 440 reflection of 15 keV X-rays ($\lambda = 0.8265$ Å) and the 660 reflection of 22.55 keV X-rays ($\lambda = 0.551$ Å). More generally, Bragg's law may be written as

$$2\frac{d_{hkl}}{n}\sin\theta_B = \frac{\lambda}{n}, \tag{15.4}$$

where n is the order of the reflection and of the harmonic reflected at θ_B. The sine curves corresponding to several harmonics are represented in Fig. 15.5.

In the Bragg geometry, the middle of the reflection domain is slightly shifted with respect to the geometric Bragg angle by the quantity $\Delta\theta_{os}$ given by (4.48a):

$$\Delta\theta_{os} = \frac{1-\gamma}{2\sin 2\theta_B}\frac{RF_o\lambda^2}{\pi V}. \tag{15.5}$$

The angular range of total reflection is comprised between the angles $\theta_B + \Delta\theta_{os} - \delta_{os}$ and $\theta_B + \Delta\theta_{os} + \delta_{os}$ and is delimited by two portions of sine curves in the top-left insert of Fig. 15.5 for a symmetric reflection.

If the reflection is asymmetric, the total reflection domain and the angular distribution of the reflected beam have different widths, as seen in Section 4.10, and are represented by different areas on the DuMond diagram (top-right insert of Fig. 15.5).

[2] Another graphic representation, called *phase-space analysis*, has been introduced by Matsushita and Hashizume (1983) for the description of synchrotron radiation optics (see also Suortti and Freund 1989).

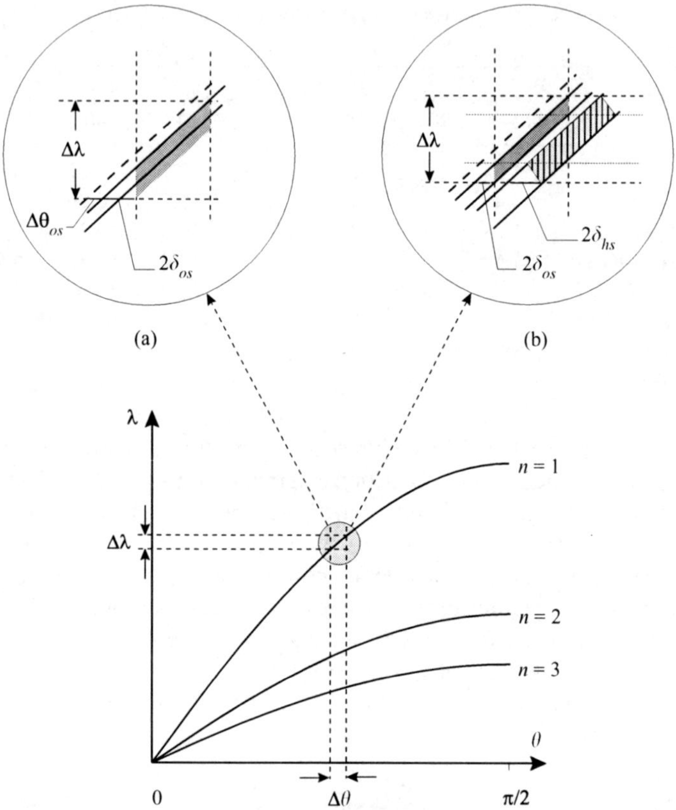

Fig. 15.5. DuMond diagram; the curves denoted $n = 1, 2, 3$ correspond to different harmonics. (a) Enlargement of the circled zone in the case of a symmetric reflection; the sine curve corresponding to the geometric Bragg condition is represented by a dashed line; the total reflection domain, of width $2\delta_{os}$, is represented in grey. (b) Enlargement of the circled zone in the case of an asymmetric reflection; same conventions as in (a), the angular range of the reflected beam, of width $2\delta_{hs}$, is hacked; the shifts of the domain of total reflection and of the angular distribution of the reflected beam are exaggerated.

The angular distribution of the reflected beam is comprised between the angles $\Delta\theta_{hs} - \delta_{hs}$ and $\Delta\theta_{hs} + \delta_{hs}$ where the position, $\Delta\theta_{hs}$, and the width, $2\delta_{hs}$, are given by (4.49):

$$2\delta_{hs} = \frac{2R\lambda^2}{\pi V \sin 2\theta_B} \frac{1}{\sqrt{|\gamma|}} |C| \sqrt{F_h F_{\bar{h}}}, \tag{15.6a}$$

$$\Delta\theta_{hs} = -\frac{R\lambda^2 F_o}{2\pi V \sin 2\theta_B}\left(1 - \frac{1}{\gamma}\right). \tag{15.6b}$$

15.2.2.2 *Choice of the monochromator crystal*

Two parameters are important for the choice of a monochromator crystal: one is the resolution and the other is the resistance to the heat load. The intensity of synchrotron radiation is very high and the dissipation of heat in the monochromator causes serious problems, in particular in the first of a multicrystal arrangement. One problem is the thermal gradient which may cause the crystal to behave as an imperfect crystal for the diffraction; another problem is the variation of the lattice parameter due to thermal expansion. If α is the coefficient of thermal expansion, the relative variation of lattice parameter induced by a temperature change ΔT is

$$\frac{\Delta d_{hkl}}{d_{hkl}} = \alpha \Delta T$$

and the corresponding variation of the Bragg angle is

$$\Delta\theta = -\alpha \Delta T \tan\theta_B.$$

There are several ways to solve these problems. One is to adapt the shape of the monochromator so as to compensate for the thermal deformation (Hart 1990; Berman and Hart 1991*a*, *b*, 1993; Berman *et al.* 1992; Quintana *et al.* 1995; Quintana and Hart 1995; Schulte–Schrepping *et al.* 1995, 1998). Another one is to choose a material with a high thermal conductivity, κ, and a low expansion coefficient, α; the ratio κ/α is a good figure of merit for that purpose (Hart 1996; Hart and Berman 1998). Two materials have a strong advantage in this respect, silicon at 120 K for which the thermal expansion coefficient is zero (Bilderback 1986) and diamond at room temperature (Freund 1995). Cryogenically cooled monochromators have been developed by Joksch *et al.* (1991), Marot *et al.* (1992) and Rogers *et al.* (1995); for a review, see Marot (1995) and Bilderback *et al.* (2000). The performance of monochromators under heat load have been discussed by Zhang *et al.* (2001), Mocella *et al.* (2001, 2003), Hoszowska *et al.* (2001).

15.2.3 ***Multiple-reflection monochromators***

15.2.3.1 *Double-crystal settings*

Wagner and Kulenkampf (1922) were the first to distinguish the two types of double-crystal settings, antiparallel and parallel, represented in Figs 15.6, 15.7 and 15.9, 15.10, respectively. The optical and geometrical principles of the double-crystal diffractometer were described by Schwarzschild (1928) and a detailed discussion can be found, for instance, in Compton and Allison (1935) and in James (1950). The notation $(+n, +m)$ and $(+n, -m)$ to characterize these settings was introduced by Allison and Williams (1930). The notations $(+n, +n)$, $(+n, -n)$ mean that the reflections are identical, namely from identical crystals and from the same set of lattice planes, and of the nth order on both crystals (Figs 15.6 and 15.8). The notation $(+, +)$ means that the incident beam on the first crystal and the reflected from the second crystal are on the same side of the beam going from the first to the second crystal (Figs 15.6 and 15.7), while they are on different sides in the $(+, -)$ settings (Figs 15.8 and 15.10). The notations $(+n, \pm m)$ mean that the orders of the reflections

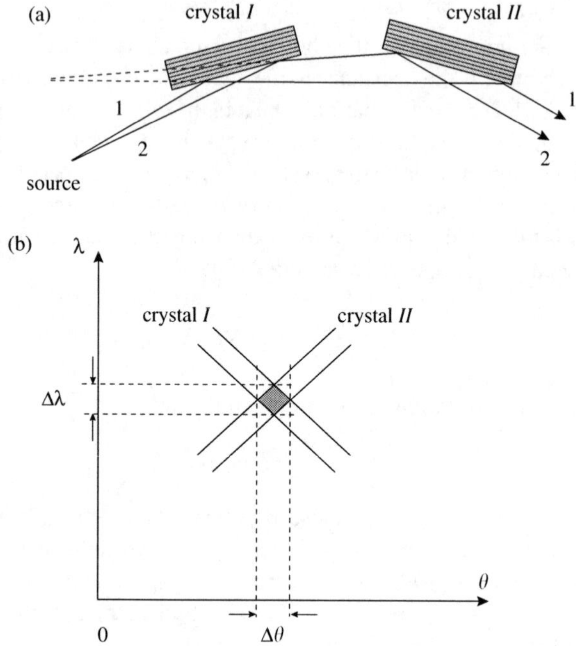

Fig. 15.6. Antiparallel, $(+n, +n)$ dispersive, two-crystal setting, symmetric reflections. (a) Experimental arrangement. (b) DuMond diagram.

on the two crystals are different and, by extension, that the reflecting-plane spacings are different for the two crystals (either because the indices of the reflections are different or because the crystals themselves are different). In the $(+n, -n)$ setting (Fig. 15.8), the crystals and the lattice planes are parallel, but this is not the case in the $(+n, -m)$ setting. Recent development of double-crystal diffractometry are reviewed in Treimer (1998).

1. $(+n, +n)$ *dispersive, antiparallel* setting, symmetric reflections (Fig. 15.6). In this setting the angle between the lattice planes of the two crystals is $2\theta_B$. Let us first consider a monochromatic incident beam and two rays incident on the first crystal. If the angle of incidence of ray 1 is Bragg's angle, θ_B, and the angle of incidence of ray 2 is $\theta_B + \delta\theta$, the angle of incidence of ray 1 on the second crystal is also θ_B but the angle of incidence of ray 2 is $\theta_B - \delta\theta$. If one now rotates the second crystal by $\Delta\theta$, the angle of incidence becomes $\theta_B + \Delta\theta - \delta\theta$ and the reflectivity of the second crystal when it is rocked is

$$I_h(\Delta\theta) = \frac{\int_{-\infty}^{+\infty} I_h^I(\delta\theta) I_h^{II}(\Delta\theta - \delta\theta)\,\mathrm{d}(\delta\theta)}{\int_{-\infty}^{+\infty} I_h^I(\delta\theta)\,\mathrm{d}(\delta\theta)},$$

$$I_h(\Delta\theta) = \frac{I_h^I * I_h^{II}}{I_h^{iI}}, \tag{15.7}$$

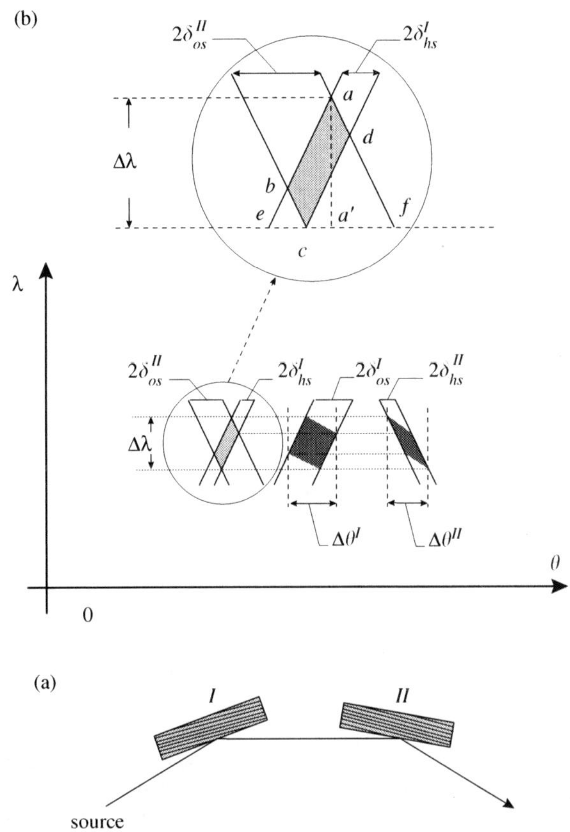

Fig. 15.7. Antiparallel, $(+n, +m)$ dispersive, two-crystal setting, asymmetric reflections, the angle of incidence on both crystals is glancing. (a) Experimental arrangement. (b) DuMond diagram.

where I_h^I and I_h^{II} are the intrinsic reflectivities of the first and second crystals, respectively, I_h^{iI} is the integrated intensity of the first crystal and $*$ means convolution. If the reflections on the two crystals are identical ($I_h^I = I_h^{II}$), the numerator of (15.7) is an autoconvolution.

Let us now consider a polychromatic incident beam and let λ_1 and λ_2 be the wavelengths for which rays 1 and 2 satisfy Bragg's condition. Ray 1 still satisfies Bragg's condition on the second crystal, but this is not the case for ray 2: the setting is *dispersive*. The sine curves corresponding to the two crystals are shifted by $2\theta_B$ with respect to one another and cross each other. The combinations angle of incidence–wavelength for which reflection is possible are limited to the shaded area in the diagram. For a fixed position of the two crystals, the wavelength bandwidth of the reflected beam is determined by the Darwin width.

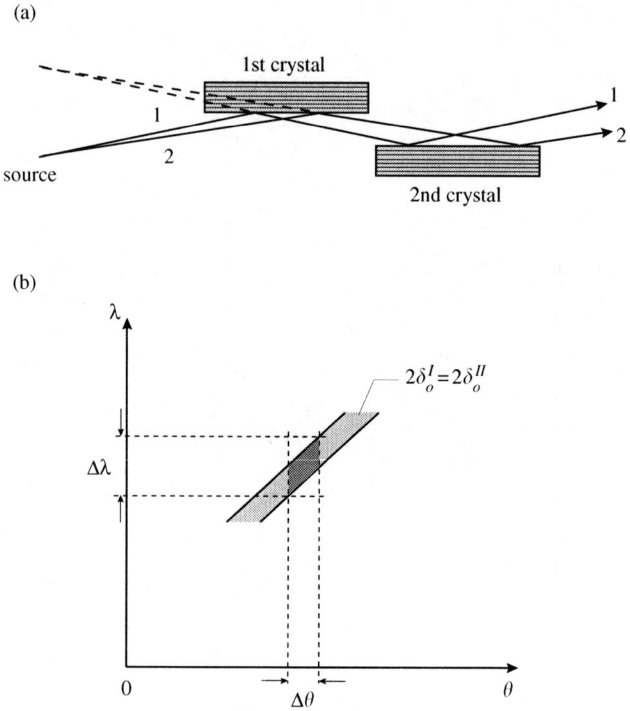

Fig. 15.8. Parallel, $(+n, -n)$ non-dispersive, two-crystal setting, symmetric reflections. (a) Experimental arrangement. (b) DuMond diagram. The reflection domains of the two crystals overlap. The wavelength bandwidth $\Delta\lambda/\lambda$ is determined by the divergence $\Delta\theta$ of the beam.

2. $(+n, +m)$ *dispersive, antiparallel* setting, asymmetric reflections on both crystals (Fig. 15.7). This case has been discussed by Nakayama *et al.* (1973) and Matsushita and Hashizume (1983). Let δ_{os}^I, δ_{hs}^I, δ_{os}^{II} and δ_{hs}^{II} be the half-widths of the total reflection domains and of the angular distributions of the reflected beams for the two crystals, respectively (Fig. 15.7(b)). For a given angular position of the two crystals with respect to one another, the rays which have the proper angle of incidence on the second crystal and the proper wavelength to be reflected by the arrangement correspond to the intersection of the angular distribution domain of the first crystal, of width $2\delta_{hs}^I$, and of the total reflection domain of the second crystal, $2\delta_{os}^{II}$; it is shaded in light grey in Fig. 15.7(b) and delimited by the parallelogram *abcd* in the inset of Fig. 15.7(b).

The energy bandwidth reflected by the arrangement can be calculated using the geometrical construction of the inset of Fig. 15.7(b). Let a' be the projection of a on the parallel to the θ axis drawn from c, and let e and f be the intersections of ab and

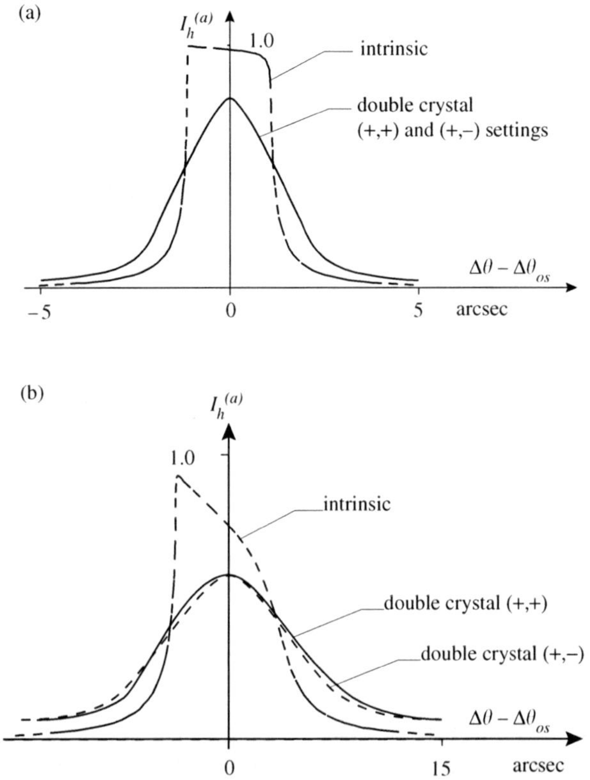

Fig. 15.9. Calculated intrinsic and double-crystal rocking curves. (a) Silicon, 220 reflection, Mo$K\alpha$. (b) Germanium, 220 reflection, $\lambda = 1$ Å.

ad with the same line. One has, using (15.1),

$$\frac{\overline{a'a}}{\overline{ea'}} = \lambda \cot \theta_B^I,$$

$$\frac{\overline{a'a}}{\overline{a'f}} = \lambda \cot \theta_B^{II},$$

where θ_B^I and θ_B^{II} are the Bragg angles of the two reflections. It follows, noting that $\overline{a'a} = \Delta\lambda$, that

$$\overline{ea'} = \frac{\Delta\lambda}{\lambda} \tan \theta_B^I,$$

$$\overline{a'f} = \frac{\Delta\lambda}{\lambda} \tan \theta_B^{II}.$$

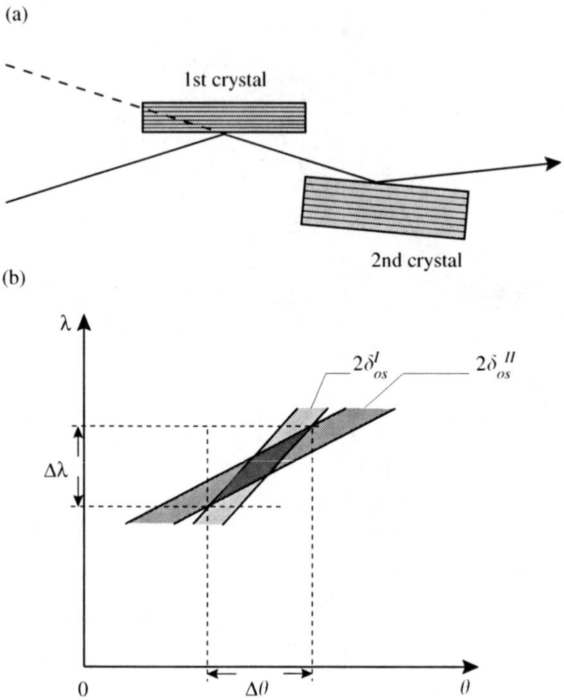

Fig. 15.10. $(+n, -m)$ slightly dispersive, two-crystal setting, symmetric reflections; the lattice parameters of the two crystals are different. (a) Experimental arrangement. (b) DuMond diagram. The reflection domains of the two crystals overlap partially.

Since

$$\overline{ea'} + \overline{a'f} = \overline{ec} + \overline{cf} = 2\delta^I_{hs} + 2\delta^{II}_{os},$$

it follows, finally, that

$$\frac{\Delta\lambda}{\lambda} = \frac{2\delta^I_{hs} + 2\delta^{II}_{os}}{\tan\theta^I_B + \tan\theta^{II}_B}.$$

The range of angles of incidence on the first crystal, or *acceptance*, $\Delta\theta^I$, is obtained by projecting the parallelogram *abcd* on the domain of total reflection of the first crystal and the width, $\Delta\theta^{II}$, of the beam reflected by the second crystal is obtained by projection on the angular distribution of the reflected beam. They are deduced following a similar derivation. One finds (Matsushita and Hashizume 1983)

$$\Delta\theta^I = 2\delta^{II}_{os}\frac{\tan\theta^I_B}{\tan\theta^I_B + \tan\theta^{II}_B} + \left|2\delta^I_{os} - 2\delta^I_{hs}\frac{\tan\theta^I_B}{\tan\theta^I_B + \tan\theta^{II}_B}\right|,$$

$$\Delta\theta^{II} = 2\delta^{I}_{hs}\frac{\tan\theta^{II}_{B}}{\tan\theta^{I}_{B} + \tan\theta^{II}_{B}} + \left|2\delta^{II}_{hs} - 2\delta^{II}_{os}\frac{\tan\theta^{II}_{B}}{\tan\theta^{I}_{B} + \tan\theta^{II}_{B}}\right|.$$

This arrangement enables one to produce very narrow beams and has been used in many different ways, with the two crystals being separate or cut in the same monolithic block.

3. $(+n, -n)$ *non-dispersive, parallel* setting (Fig. 15.8). In this setting, the reflecting planes of the two crystals are parallel. If ray 1 satisfies Bragg's condition on the first crystal for a given wavelength, λ_1, and ray 2 for another wavelength, λ_2, they both satisfy Bragg's condition on the second crystal: the setting is *non-dispersive*. On DuMond's diagram the domains of total reflection of both crystals overlap (Fig. 15.8(b)).

Let us now consider a single wavelength. If ray 1 satisfies Bragg's condition for the first crystal and the angle of incidence of ray 2 is $\theta_B + \delta\theta$, the angle of incidence of ray 1 on the second crystal is also θ_B and the angle of incidence of ray 2 is also $\theta_B + \delta\theta$. If one now rotates the second crystal by $\Delta\theta$, the angle of incidence becomes $\theta_B + \delta\theta - \Delta\theta$ and the reflectivity of the second crystal when it is rocked is

$$\begin{aligned} I_h(\Delta\theta) &= \frac{\int_{-\infty}^{+\infty} I^{I}_{h}(\delta\theta) I^{II}_{h}(\delta\theta - \Delta\theta)\,\mathrm{d}(\delta\theta)}{\int_{-\infty}^{+\infty} I^{I}_{h}(\delta\theta)\,\mathrm{d}(\delta\theta)}, \\ &= \frac{I^{I}_{h}(\delta\theta) * I^{II}_{h}(-\delta\theta)}{I^{iI}_{h}}, \end{aligned} \tag{15.8}$$

where I^{I}_{h} and I^{II}_{h} are the intrinsic reflectivities of the first and second crystals, respectively. In contrast to the result for the $(+n, +n)$ setting, the convolution is now with $I^{II}_{h}(-\delta\theta)$. Since in the $(+n, -n)$ setting the two reflections are identical, (15.8) is actually

$$I_h(\Delta\theta) = \frac{I^{I}_{h}(\delta\theta) * I^{I}_{h}(-\delta\theta)}{I^{iI}_{h}}.$$

The convolution is commutative and the rocking curve is symmetric for the $(+n, -n)$ setting even if the intrinsic rocking curve is asymmetric, which is not the case for the $(+n, +n)$ setting. However, if the asymmetry is small, the difference of the rocking curves obtained with the two settings is negligible. As an example, the rocking curves for the $(+n, -n)$ and $(+n, +n)$ settings are compared in Fig. 15.9, once for a slightly asymmetric intrinsic rocking curve and once for a very asymmetric intrinsic rocking curve.

4. $(+n, -m)$ *parallel* setting (Fig. 15.10). In this case the overlap of the total reflection domains of the two crystals is only partial (Fig. 15.10(b)) and the setting is said to be *slightly dispersive*.

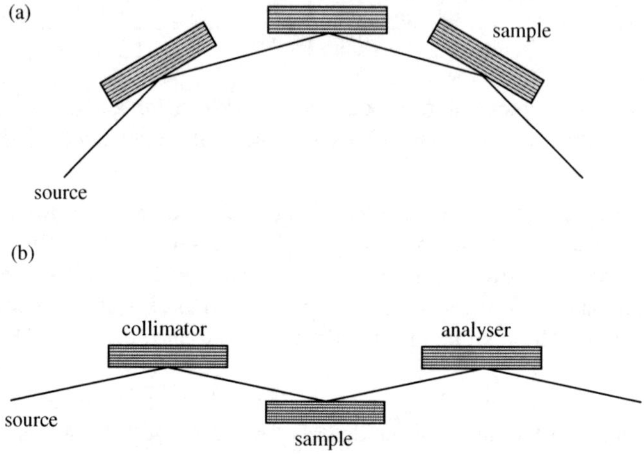

Fig. 15.11. Triple-crystal diffractometers. (a) $(+n, +n, +n)$. (b) $(+n, -n, +n)$.

15.2.3.2 *Triple-crystal settings*

Multiple-crystal settings were first described by DuMond (1937) and very many papers have been devoted to their applications. One type of arrangement uses a $(+, +, \pm)$ setting (Fig. 15.11(a)), the sample being the third crystal. This ensures a narrow spread in wavelength and in angle. It was used by Renninger (1955) for the first experimental observation of the Darwin–Prins rocking curve. The theory of the setting was discussed in detail by Bubáková *et al.* (1961)—see also Matsushita (1974). This setting is used to study rocking curves and for plane-wave topography.

Another type of arrangement uses a $(+, -, +)$ setting (Fig. 15.11(b)) in which the sample is the second crystal and the third crystal acts as an analyser (Larson and Schmatz 1974; Iida and Kohra 1979; Zaumseil and Winter 1982, 1990). This type of setting is used to analyse in reciprocal space the distribution of intensity scattered by the sample. Dynamically diffracted and kinematically scattered intensity can thus be separated. For recent applications, see, for instance, Zaumseil *et al.* (1987), Kazimirov *et al.* (1987), Stangl *et al.* (1997), Zaumseil (1997), Liss *et al.* (1998) and Bouchard *et al.* (1998). The resolution of triple-crystal diffractometers in the Laue dispersive and non-dispersive geometries is discussed in Neumann *et al.* (1994) and Rütt *et al.* (1995).

15.3 Applications of multiple-crystal arrangements to beam conditioning

15.3.1 *Suppression of tails*

It was first shown by Bonse and Hart (1965*c*) that multiple reflections in the $(+n, -n)$ non-dispersive setting from the two sides of a groove sawn or etched in a single crystalline block (so-called *channel-cut crystal*, Fig. 15.12) result in a reduction of

(a)

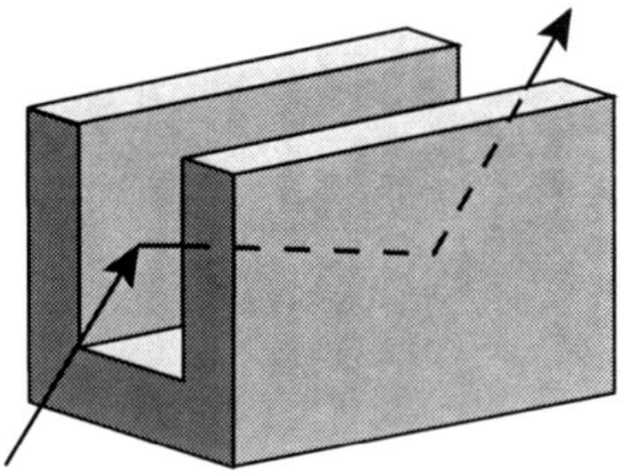

(b)

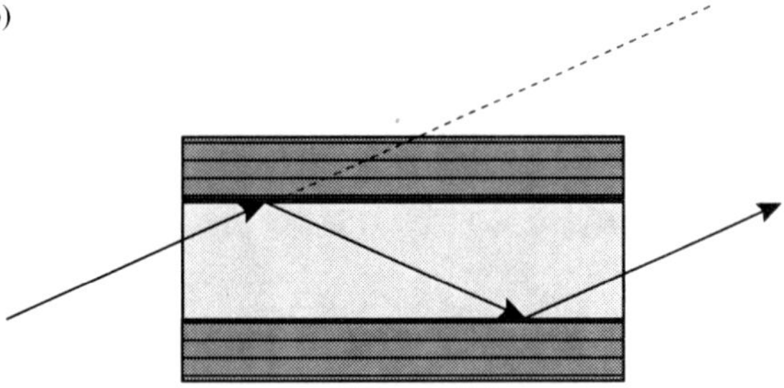

Fig. 15.12. Channel-cut monochromator for the suppression of tails. (a) Side view. (b) Top view.

the tails of the rocking curve. If $I_h(\Delta\theta)$ is the reflectivity of a single crystal, the reflectivity after n reflections is $I_h^n(\Delta\theta)$.

Figure 15.13 compares the rocking curves after one, two, three and four reflections, respectively, for a low-absorption case, silicon, 111, 25 keV ($\lambda = 0.496$ Å) and a high-absorption case, silicon, 111, Cu$K\alpha$. The tails of the rocking curve decrease very rapidly with the number of reflections, as can easily be understood since it has been noticed in Section 4.9.2 that the tails decrease as $1/\eta^2$; after n reflections, they decrease as $1/\eta^{2n}$. The width of the curve remains however constant and equal to the Darwin width.

This arrangement is very simple and does not have the drawback of instability which exists when the two crystals are separate. It is widely used at synchrotron radiation beamlines as an element of many types of multiple reflection systems such as those for harmonic suppression (Section 15.3.4) or tunable polarizers (Section 15.3.5); it is very useful, for instance, for small-angle scattering experiments (Bonse and Hart 1966*a*; Deutsch 1980*a*,*b*, who improved it using asymmetric reflections).

The single-groove channel-cut crystal has, however, the drawback that the exiting beam is translated with respect to the incident beam (see Fig. 15.12(b)), the shift depending on the Bragg angle. One way to counter this is to use two single-groove channel-cut crystals in the (+, +) setting (Fig. 15.14); one thus obtains a

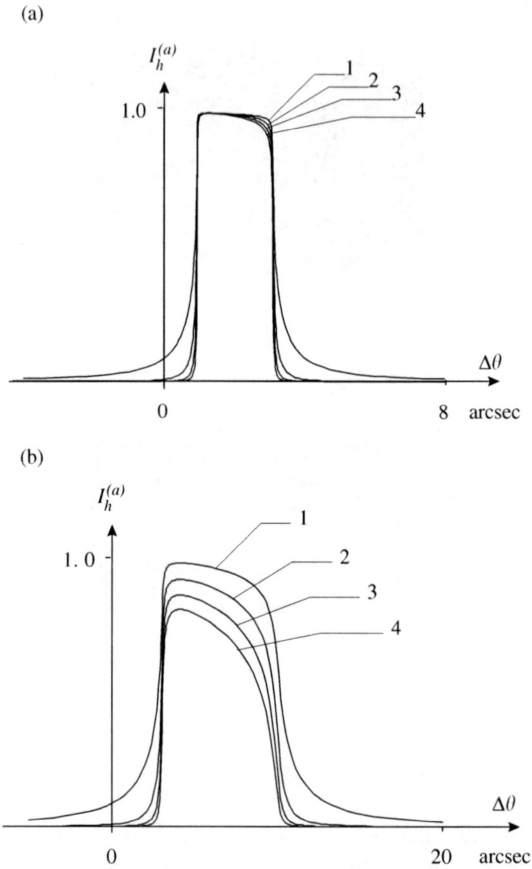

Fig. 15.13. Channel-cut monochromator: rocking curves; 1: one bounce; 2: two bounces; 3: three bounces; 4: four bounces. Silicon 111 reflection. (a) $\lambda = 0.496$ Å (25 keV). (b) 111 reflection, Cu$K\alpha$ (8.05 keV).

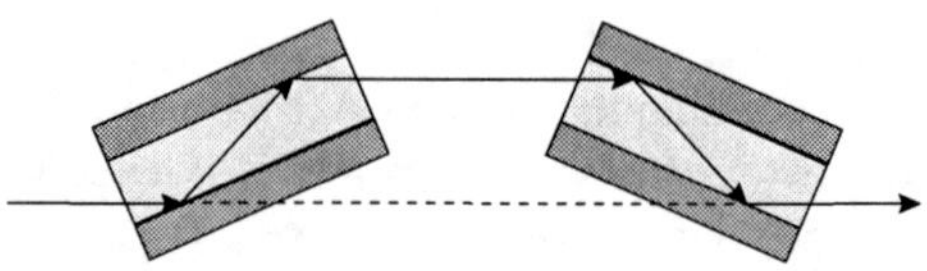

Fig. 15.14. Fixed-exit four-crystal monochromator.

fixed-exit beam. This arrangement, or four-crystal monochromator, was proposed by Beaumont and Hart (1974) to make a wavelength scanner and by Bartels (1983) for high-resolution diffractometry. It was improved by Fewster (1989, 1991) who used it in combination with a single or channel-cut crystal analyser after the sample, the analyser serving to explore the distribution of the intensity diffracted by the sample in reciprocal space (Fewster and Andrew 1995*b*). For a review, see Fewster

(2000). A fixed-exit channel-cut monochromator with profiled surfaces designed for microscopy purposes was proposed by Oestreich *et al.* (1999).

15.3.2 *Wavelength scanner*

When used in order to scan the wavelength, the two blocks of the four-crystal monochromator represented in Fig. 15.14 are counter-rotated with the same angular speed. The four crystals can also be cut in a monolithic block to produce a very narrow spectral window and, thus, a wavelength calibrator (Hart 1979).

15.3.3 *Production of beams with a very narrow angular spread*

15.3.3.1 *Use of asymmetric reflections*

It has been shown in Section 4.10 that asymmetric reflections reduce drastically the width of the angular distribution of the beam reflected by a monochromator. This procedure was pioneered by Renninger (1961). It was developed by Kohra and Kikuta (1968), Kikuta and Kohra (1970), Kikuta (1971) and Kohra (1972) using successive asymmetric reflections from two or three crystals cut from a single silicon monolith and in the (+, −) setting. Figure 15.15 shows the principle in the case of two crystals; the angle of incidence on the surface is the same for the two crystals and very small.

In the case shown by Kikuta (1971) the reflection was 422, the Bragg angle for the radiation used, Cu$K\alpha$, was 44.1° and the angle of incidence on the surface was 6° (asymmetry ratio $\gamma = -9.45$). At each reflection the angular spread of the reflected beam is decreased and its lateral extension increased. However, because of the refraction effect, the angular reflection domain of the second crystal has only a partial overlap with the angular distribution of the beam reflected by the second crystal. This can be understood by means of the DuMond diagram. As noticed in Fig. 15.8, the total reflection domains of the two crystals overlap perfectly in the parallel (+, −) setting, when the reflections are symmetric. However, when the reflections are asymmetric the shift of the angular distribution of the reflected beam is different from that of the total reflection domain, as can be seen in the top-right insert of Fig. 15.5. If there is only partial overlap between the angular distribution of the beam reflected by the first crystal, which lies in the interval $\Delta\theta_h^I \pm \delta_{hs}^I$, and the acceptance of the second crystal,

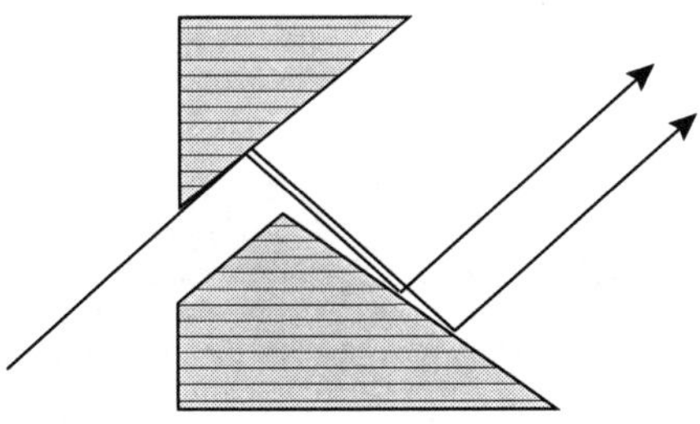

Fig. 15.15. Monochromator for the production of a beam with a very narrow angular spread.

which lies in the interval $\Delta\theta_o^{II} \pm \delta_{os}^{II}$, the intensity reflected by the monochromator is reduced. To overcome this difficulty, Matsushita *et al.* (1971) have added the possibility to tilt slightly the second crystal in order to have a better overlap of the domain of total reflection of the second crystal and of the angular distribution of the beam reflected by the second crystal. This is achieved by introducing a weak link between the two crystals and pushing the second one with a screw. An asymmetric/symmetric version has been proposed by Materlik and Kostroun (1980) and Kostroun and Materlik (1980).

15.3.3.2 *Offsetting successive reflections*

Consider a single-groove channel-cut crystal and symmetric reflections as in Fig. 15.12. If one slightly offsets from parallel alignment one of the two crystals by rotating it with respect to the other one by means of a spring, and if one then rocks the channel-cut crystal, the reflectivity becomes $I_h(\Delta\theta)I_h(\Delta\theta - \epsilon)$, where ϵ is the offset.

Figure 15.16 shows that the rocking curve is narrowed down. This manner of decreasing the angular spread was first proposed by Hart and Rodrigues (1978). The system has been improved by using a system where two single-groove crystals linked together are tilted against one another as shown in Fig. 15.17(a) (Bonse *et al.* 1983; Hart *et al.* 1984; Berman *et al.* 1985; Ferrari and Korytár 2001). Figure 15.17(b) shows the profile obtained when rocking one of the channel-cut crystals.[3] The reflectivity

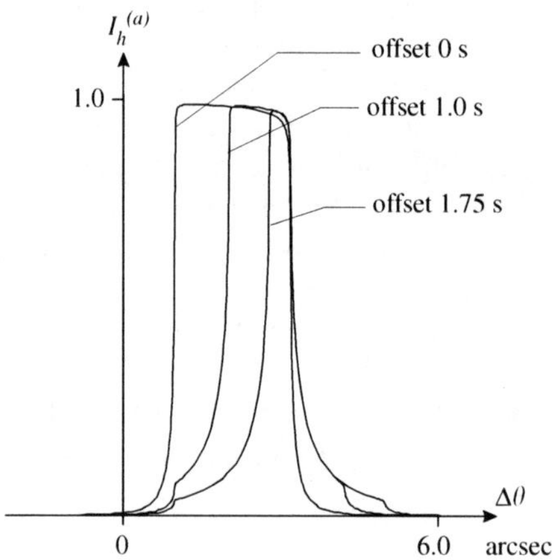

Fig. 15.16. Offset narrowing of the rocking curve. Silicon, 111 reflection, 25 keV ($\lambda = 0.496$ Å).

[3] This profile is nearly an isosceles triangle, as confirmed experimentally by Hart *et al.* (1984). The approximation is excellent if the intensity reflected by the channel-cut crystals is exactly 1 inside the total reflection domain and 0 outside, which is practically the case for a low-absorbing crystal and a high number of reflections. It is interesting to note that James (1950) had given the example of an isosceles triangle as the convolution of two such rocking curves, adding that they '*correspond to nothing that actually happens in nature but are nevertheless interesting*'!

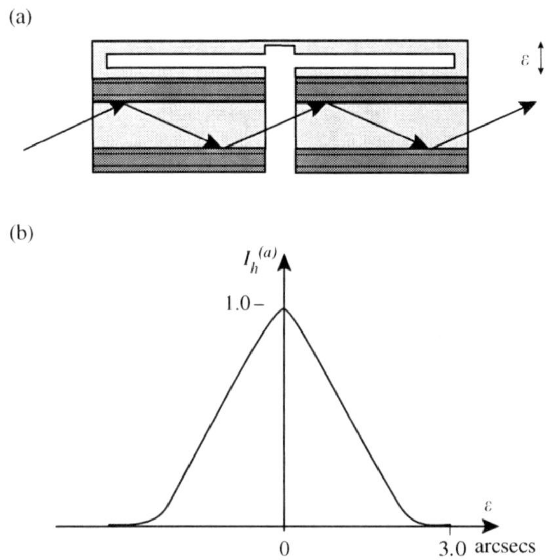

Fig. 15.17. Monochromator constituting in two channel-cut silicon crystals linked by a spring for adjustable resolution. (a) Schematic experimental arrangement (ϵ: offset). (b) Theoretical profile obtained when rocking one of the crystals against the other one (silicon, 111 reflection, 25 keV).

of the system is $I_h^2(\Delta\theta) \times I_h^2(\Delta\theta - \epsilon)$ where ϵ is the offset and the rocking curve profile is given by the convolution $I_h^2(\Delta\theta) * I_h^2(-\Delta\theta)$. Such systems are very useful to achieve adjustable resolution.

15.3.4 *Harmonic suppression*

When the incident beam is constituted by white radiation such as that emitted by bending magnets or wigglers, harmonics are present as shown by (15.4). Figure 15.18 shows the intrinsic rocking curves for the 220, 440 and 660 reflections of silicon in the Bragg geometry, corresponding to energies of 7.5 keV ($\lambda = 1.653$ Å), 15 keV ($\lambda = 0.8265$ Å) and 22.5 keV ($\lambda = 0.551$ Å) X-rays, respectively.

There are many ways to separate the first harmonic ($n = 1$) from the others. Since they correspond to photons of different energies, one may use absorbing foils with well-chosen absorption edges or a photon detector with a pulse height analyser, but the usual way is to use a multiple-crystal monochromator which takes advantage of the differences between the shifts and widths of the total reflection domain and of the angular distribution of the reflected beam for the various harmonics. The corresponding expressions (15.2), (15.5) and (15.6a,b) are generalized as follows in terms of the

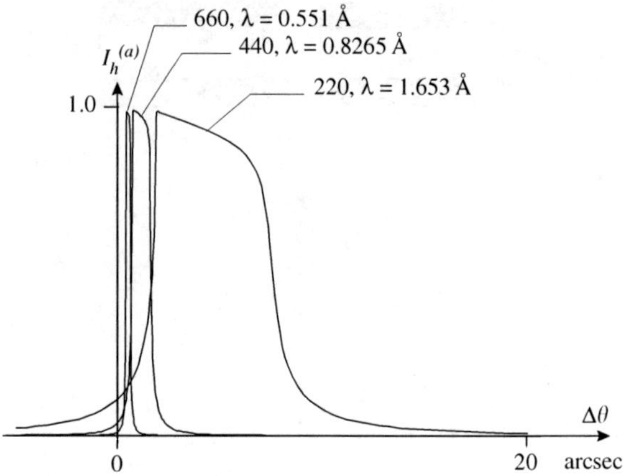

Fig. 15.18. Intrinsic rocking curves of the 220 reflection and the following harmonics, 440 and 660. Silicon single crystal, symmetric reflections.

rank n of the harmonic:

$$\begin{aligned} \Delta_{os}^{(n)} &= (1-\gamma)\frac{RF_o}{2\pi V}\frac{1}{\sin 2\theta_B}\left(\frac{\lambda}{n}\right)^2, \\ 2\delta_{os}^{(n)} &= 2\sqrt{|\gamma|}\frac{RF_{nh,nk,nl}}{\pi V}\frac{1}{\sin 2\theta_B}\left(\frac{\lambda}{n}\right)^2, \end{aligned} \tag{15.9a}$$

$$\begin{aligned} \Delta_{hs}^{(n)} &= \left(1-\frac{1}{\gamma}\right)\frac{RF_o}{2\pi V}\frac{1}{\sin 2\theta_B}\left(\frac{\lambda}{n}\right)^2, \\ 2\delta_{hs}^{(n)} &= 2\sqrt{\frac{1}{|\gamma|}}\frac{RF_{nh,nk,nl}}{\pi V}\frac{1}{\sin 2\theta_B}\left(\frac{\lambda}{n}\right)^2, \end{aligned} \tag{15.9b}$$

where $F_{nh,nk,nl}$ is the structure factor of the reflection of nth order.

Several schemes are possible:

1. A double-crystal arrangement in the dispersive (+, +) setting; this was used by Beaumont and Hart (1974) in the multiple-reflection version.

2. A double-crystal arrangement in the slightly dispersive (+, −) setting where the reflecting-plane spacings of the two crystals are slightly different; this was used by Bonse *et al.* (1976) with a 220 reflection from a silicon channel-cut crystal and a 220 reflection from a germanium crystal.

3. A double-crystal arrangement with the same crystal and the same reflection where one crystal is set in the Laue geometry and the other crystal in the Bragg

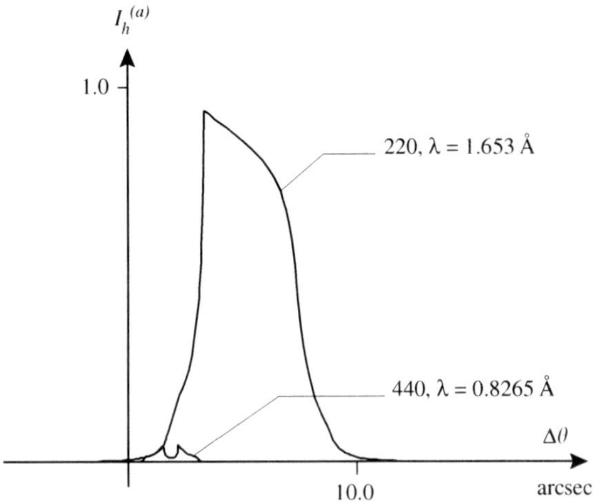

Fig. 15.19. Intrinsic rocking curves of the 220 and 440 reflections from a channel-cut silicon crystal with an offset of 1.5 arcsec between the two crystals.

geometry; this was done by Bonse and Materlik (1974) and by Bonse *et al.* (1976); the latter used separate crystals (one a channel-cut crystal in the $(+,-)$ setting, the second one an *LLL* interferometer) and by Kostroun and Materlik (1980) and Materlik and Kostroun (1980) who built a tunable Laue–Bragg combination from a single silicon crystal with the 220 reflection.

4. A multiple-crystal arrangement cut in a single block in which one of the crystals is slightly offset with respect to the others. The offset angle is chosen such that the rocking curves of the two crystals overlap for the fundamental and not for the harmonics. As an example, consider the 220 and 440 reflections at the same energies as in Fig. 15.18. The width of the 220 reflection is 5.6 arcsec and the angular difference between the middle of the total reflection domains of these two reflections is 3.5 arcsec. Figure 15.19 shows the rocking curves for 220 and 440 reflections from a channel-cut silicon crystal where the second crystal is offset with respect to the first one by 1.5 arcsec. It can be seen that the intensity of the 440 reflection is very small; that of the 660 reflection is negligible. The offset of the second crystal is obtained by linking the two crystals by a silicon spring and rotating the second crystal by means of an electrostatic force generator or an electromagnetic system (Hart and Rodrigues 1978). Bonse *et al.* (1983) used a monolithic silicon double-groove monochromator constituted by two linked channel-cut crystals with three reflections in the first one and two in the second one to achieve tunable harmonic suppression as well as suppression of unwanted Laue spots (reflections from other sets of lattice planes); the odd number of reflections is advantageous for the elimination of these extra Laue spots.

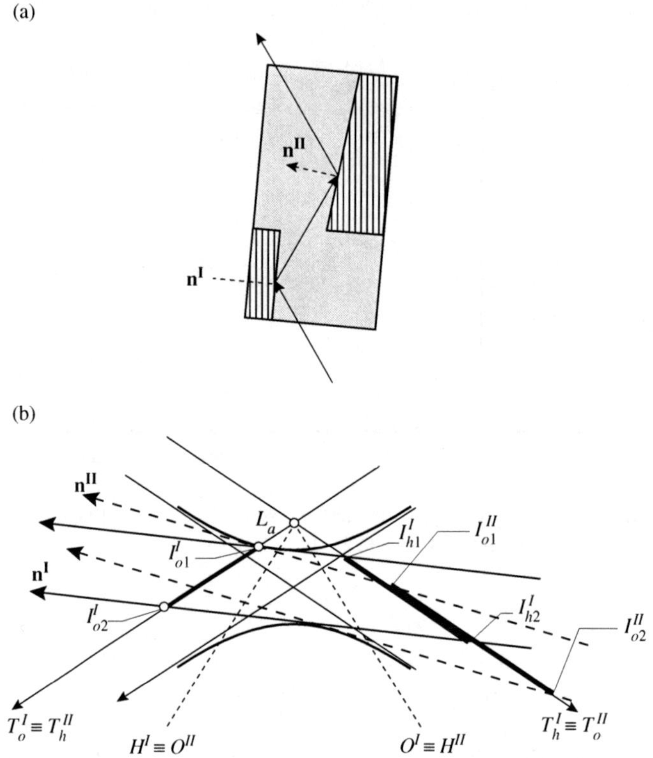

Fig. 15.20. Monochromator for harmonic suppression. $\mathbf{n}^{\mathbf{I}}$: normal to the surface of crystal ***I***; $\mathbf{n}^{\mathbf{II}}$: normal to the surface of crystal ***II***. (a) Direct space. (b) Reciprocal space.

5. Another scheme makes use of two successive asymmetric reflections from two crystals in the $(+n, -n)$ setting cut in the same block, but with different asymmetries (Hashizume 1983; Treimer and Hildebrandt 1989). The experimental setup is shown schematically in Fig. 15.20(a). The principle of the action of the monochromator can be easily understood using the DuMond diagram but, as an exercise, we shall use here the dispersion surface (Fig. 15.20(b)), which is of course equivalent.

The incident beam is reflected by a first crystal whose normal to the surface is indicated by $\mathbf{n}^{\mathbf{I}}$. The range of total reflection is indicated by $I^I_{o1} I^I_{o2}$ on the tangent T^I_o to the sphere of centre O^I and radius k (see Section 4.9.3 for detailed explanations). The extremities of the reflected wavevectors lie on the tangent T^I_h to the sphere of centre H^I and radius k, and the width of the reflected beam corresponding to the total reflection range is indicated by the segment $\overline{I_{h1}^{I} I_{h2}^{I}}$ where

$$\overline{L_a I_{h1}^{I}} = \Delta_{hs}^{(n)I} - \delta_{hs}^{(n)I},$$
$$\overline{L_a I_{h2}^{I}} = \Delta_{hs}^{(n)I} + \delta_{hs}^{(n)I}.$$

$\Delta_{hs}^{(n)I}$ and $\delta_{hs}^{(n)I}$ are given by (15.9b) in which γ has been replaced by γ^{I}, asymmetry ratio of the first crystal. The beam reflected by the first crystal is incident on the second crystal whose normal to the surface is indicated by $\mathbf{n}^{\mathbf{II}}$. The origin, O^{II}, of the reciprocal space for this second crystal is blended into H^{I} and the associated tangent T_{o}^{II} to the sphere of radius k is blended into T_{h}^{I}. The range of total reflection on the second crystal is indicated by $I_{o1}^{II} I_{o2}^{II}$ where

$$\overline{L_a I_{o1}^{II}} = \Delta_{os}^{(n)II} - \delta_{os}^{(n)II},$$
$$\overline{L_a I_{o2}^{II}} = \Delta_{os}^{(n)II} + \delta_{os}^{(n)II}.$$

$\Delta_{os}^{(n)II}$ and $\delta_{os}^{(n)II}$ are given by (15.9a) in which γ is replaced by γ^{II}, asymmetry ratio of the second crystal.

The beam reflected by the first crystal is also reflected by the second one if its angular width, $I_{h1}^{I} I_{h2}^{I}$, overlaps at least partially the reflection range of the second crystal, $I_{o1}^{II} I_{o2}^{II}$. Since $\Delta_{os}^{(n)}$, $\delta_{os}^{(n)}$, $\Delta_{hs}^{(n)}$ and $\delta_{hs}^{(n)}$ depend both on the order of the harmonic and on the asymmetry ratio of the reflection, it is possible to adjust γ^{I} and γ^{II} so that the degree of overlap of these ranges is larger for $n = 1$ than for the following harmonics. For instance, let us assume that the surface of the first crystal is cut parallel to the (110) planes ($\gamma^{I} = -1$) and that the surface of the second crystal is offset by 10° with respect to the (110) planes ($\gamma^{II} = -2.17$). Figure 15.21 shows, for silicon the 220 reflection at 7.5 keV ($\lambda = 1.65312$Å) and the 440 and 660 harmonics, respectively, the angular distribution, $I_{2n,2n,0}^{I}$, of the beam reflected by the first crystal ($n = 1, 2, 3$), the acceptance, $I_{2n,2n,0}^{II}$, of the second crystal and the angular distribution of the beam reflected by the monochromator, $I_{2n,2n,0}^{I-II}$:

$$I_{2n,2n,0}^{I-II}(\Delta\theta) = I_{2n,2n,0}^{I}(\Delta\theta) \times I_{2n,2n,0}^{II}\left[\Delta\theta - \left(\Delta_{hs}^{(n)I} - \Delta_{os}^{(n)II}\right)\right].$$

It can be seen in the figure that the overlap is quite large for 220, small for 440 and very small for 660, and that the contributions of the 440 and 660 harmonics to the intensity reflected by the monochromator are decreasingly small and negligible.

This scheme was used for instance by Pétroff *et al.* (1980) to study epilayers of GaAlAsP on a gallium arsenide substrate using synchrotron radiation plane-wave topography. In that case the monochromator used three reflections, the first two in the $(+n, +n)$ setting to reduce the angular and spectral bandwidths (333 and $\bar{1}31$) and the last two ($\bar{1}31$ and $1\bar{3}\bar{1}$) to suppress the harmonics (Fig. 15.22).

6. Recently it has been proposed to use a perfect quasicrystal as an analyser after a channel-cut silicon monochromator (Capitan and Alvarez 2000). Thanks to the

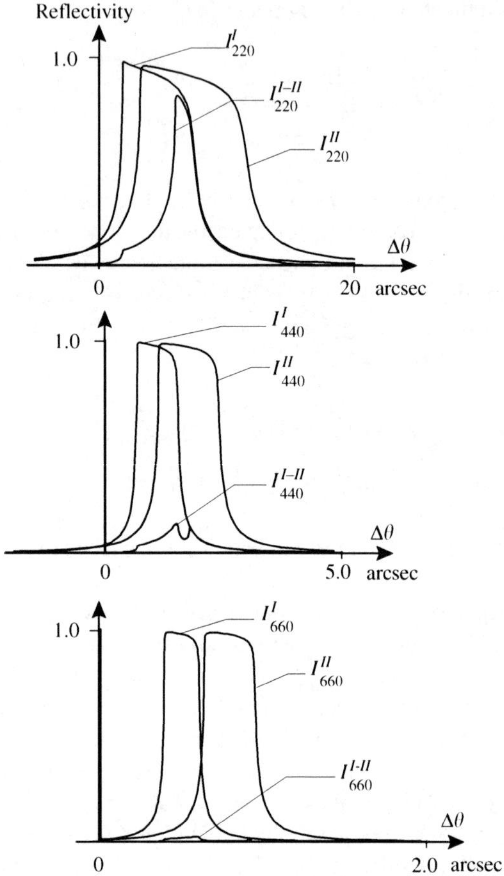

Fig. 15.21. Harmonic suppression with the monochromator shown in Fig. 15.20. $I^{I}_{2n,2n,0}$; ($n = 1, 2, 3$): angular distribution of the intensity diffracted by the first crystal; $I^{II}_{2n,2n,0}$: rocking curve of the second crystal; $I^{I-II}_{2n,2n,0}$: angular distribution of the intensity diffracted by the monochromator. Silicon, $\lambda = 1.65312$Å for the 220 reflection.

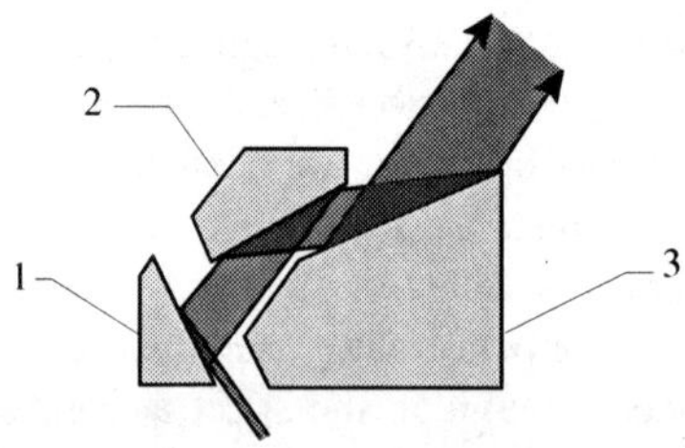

Fig. 15.22. Silicon triple-reflection monochromator designed for $\lambda = 1.2378$ Å. 1: 333 reflection; 2: $\bar{1}31$ reflection; 3: $1\bar{3}\bar{1}$ reflection. Adapted from Pétroff *et al.* (1980).

quasiperiodicity, the harmonics reflected by the silicon crystal are not reflected by the quasicrystal. The authors have announced a 5×10^{-4} harmonic rejection for an AlPdMn quasicrystal.

15.3.5 *Production of polarized radiation*

Radiation from an X-ray tube is unpolarized, synchrotron radiation is plane polarized in the plane of the orbit and elliptically polarized away from this plane. Many experiments require linear or circular polarization and the following subsections will show how this can be achieved.

15.3.5.1 *Linear polarization*

As has been seen in Chapter 5, in all expressions of the dynamical theory, in the two-beam case, the structure factor should be multiplied by C equal to 1 for the σ-component and to $\cos 2\theta_B$ for the π-component. There are therefore four branches to the dispersion surface (Fig. 15.23) and a different anomalous absorption coefficient for each one (eqn (5.42)). If the incident wave is unpolarized, as in the case of an X-ray tube, the two components are incoherent; if the incident wave is polarized, the two components are coherent.

Linear polarization is produced from unpolarized radiation by eliminating the π- component. Different types of polarizers have been suggested:

1. *A crystal and a reflection for which the Bragg angle θ_B is close to $\pi/4$.* The π-component is then zero since C is then close to zero (Chandrasekaran 1959).

2. *A Borrmann effect polarizer* operating in the transmission geometry and taking advantage of the difference in absorption coefficient for the four branches of the dispersion surface (Cole *et al.* 1961). Cole *et al.* used the 220 reflection of germanium, Cu$K\alpha$ and a 1 mm thick crystal. In that case, one has

$$\mu_o t = 34.17 \quad \Longrightarrow \quad \exp(-\mu_o t) = 1.4 \times 10^{-15},$$
$$\mu_e^\sigma t = 1.3 \quad \Longrightarrow \quad \exp(-\mu_e^\sigma t) = 0.272,$$
$$\mu_e^\pi t = 11.9 \quad \Longrightarrow \quad \exp(-\mu_e^\pi t) = 6.5 \times 10^{-6},$$

where μ_o is the normal absorption coefficient and μ_e^σ and μ_e^π are the effective absorption coefficients for branch 1 at the centre of the dispersion surface for the σ- and the π-components, respectively. The coefficients for branch 2 are of course very much larger. A Borrmann polarizer was used, for instance, by Staudenmann *et al.* (1985), Templeton and Templeton (1988) and Staudenmann and Chapman (1989) to determine the polarization of X-ray beams.

3. *A channel-cut multiple-crystal polarizer* operating in the reflection geometry (Hart and Rodrigues 1979). The ratio of the integrated intensities for the σ- and π-components is $\cos 2\theta_B$, hence the idea that this ratio will be decreased by using a multiple-reflection channel-cut monochromator as that represented in Fig. 15.11. For instance, Fig. 15.24 compares the intrinsic rocking curves of the σ- and π-components for germanium, 440 and Cu$K\alpha$ ($\theta_B = 50.5°$) with the corresponding rocking

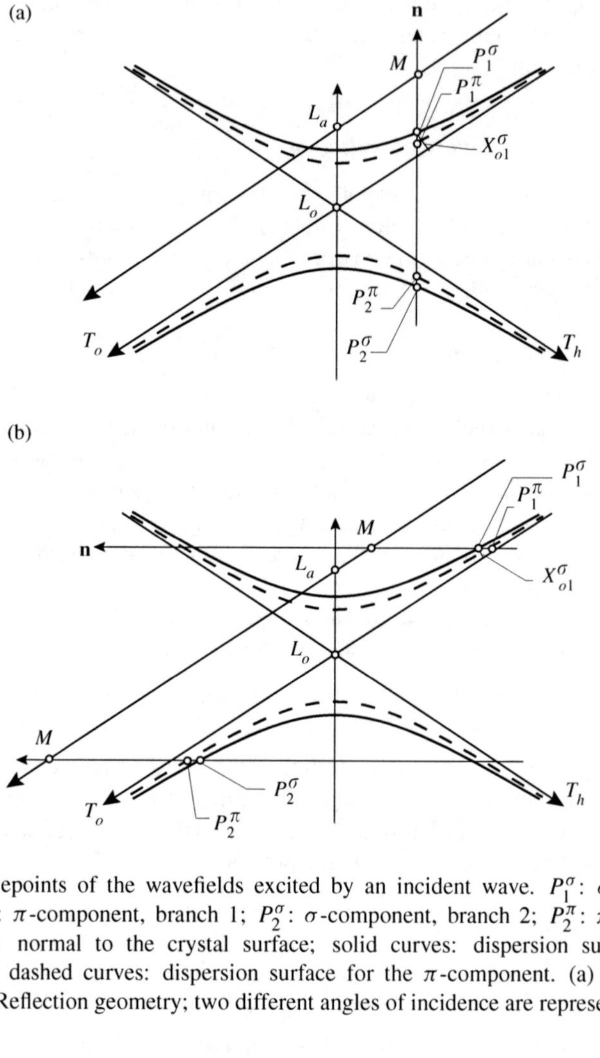

Fig. 15.23. Tiepoints of the wavefields excited by an incident wave. P_1^σ: σ-component, branch 1; P_1^π: π-component, branch 1; P_2^σ: σ-component, branch 2; P_2^π: π-component, branch 2; (**n**: normal to the crystal surface; solid curves: dispersion surface for the σ-component; dashed curves: dispersion surface for the π-component. (a) Transmission geometry. (b) Reflection geometry; two different angles of incidence are represented.

curves after two reflections. The reduction of the ratio of the σ-component to the π-component can be further amplified by slightly offsetting the second crystal with respect to the first one by an angle smaller than the Darwin width for the σ-component but larger than that for the π-component. This is also shown in Fig. 15.24 with an offset of 1 arcsec (the Darwin width is 0.99 arcsec for the π-component and 5.19 arcsec for the σ-component). By increasing the number of bounces Hart and Rodrigues (1979) have achieved very high polarization ratios, as high as 10^9 in the best case. This type of polarizer has the advantage of being tunable.

As an example of the use of channel-cut offset polorizers, Sato *et al.* (2000) have recently demonstrated the possibility of linear dichroism imaging with hard X-rays.

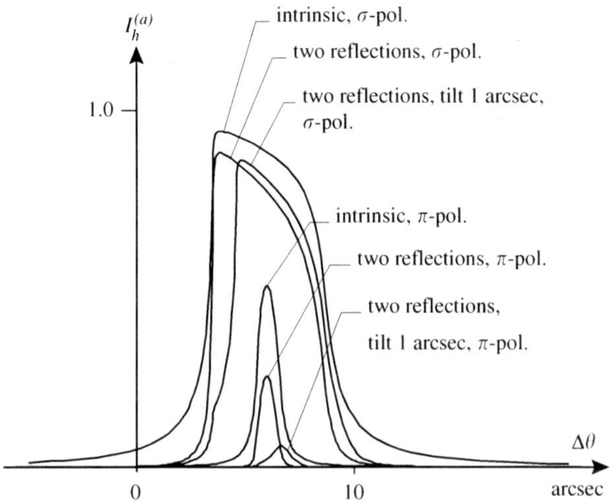

Fig. 15.24. Multi-reflection linear polarizer. Theoretical rocking curves for σ- and π-polarization. Germanium, 440 reflection, Cu$K\alpha$.

4. *A polarizer set for the reflection geometry, but using the transmitted beam* (Hasegawa *et al.* 1998, 1999). As has been seen in Section 7.1.2, the width of the total reflection domain of the σ-component is wider than that of the π-component (see Fig. 7.6). If the crystal is set at an angle for which the σ-component is totally reflected, but not the π-component, and if the incident beam is sufficiently collimated, the transmitted beam is then nearly 100 per cent polarized. Hasegawa *et al.* tested both silicon and diamond crystals and achieved 10^5 polarization ratios in the best case. This polarizer is tunable and has the advantage of not modifying the position of the beam, whatever the wavelength used.

15.3.5.2 *X-ray phase plates*

It is possible to produce circularly polarized X-rays with special insertion devices (see Section 15.1.2.3), but also by taking advantage of the phase difference between the σ- and the π-components. The phase of the refracted wave of a wavefield propagating through a crystal is $2\pi \mathbf{OP_j^p} \cdot \mathbf{r}$ where P_j^p is the tiepoint of the wavefield (Fig. 15.23), $j = 1$ or 2 and $p = \sigma$ or π.

For an incident linearly polarized wave, the two components are coherent and their phase difference after propagating through a crystal thickness t is

$$\Psi_j^{\sigma-\pi} = 2\pi \mathbf{P_j}^\pi \mathbf{P_j}^\sigma \cdot \mathbf{r} = 2\pi (X_{oj}^\sigma - X_{oj}^\pi) t,$$

where X_{oj}^p is the coordinate of P_j^p in the direction of the incident wave (Fig. 15.23). The value of this phase difference is of course the same for the reflected wave. The expression of X_{oj}^{p} in terms of the deviation parameter, η, is given by (5.21a). It should, however, be remembered that the deviation parameter itself depends on the value of C (see eqn (5.18)) and it is better to express X_{oj}^{p} in terms of the departure from Bragg's angle, $\Delta\theta$, of the incident wave. It follows that

$$X_{oj}^{p} = \frac{\gamma_o}{2}\left[\frac{\delta\theta}{L} \pm \sqrt{\left(\frac{1}{\Lambda_o^p}\right)^2 + \left(\frac{\delta\theta}{L}\right)^2}\right],$$

where $L = \lambda|\gamma_h|/\sin 2\theta_B$, the $+$ corresponds to branch 1 and the $-$ sign to branch 2, and $\delta\theta = \Delta\theta - \Delta\theta_{os}$ is the departure of the incident wave from the middle of the reflection domain.

It follows, for the phase difference between the σ- and π-components, that

$$\Psi_j^{\sigma-\pi} = \pm\pi\gamma_o\left[\sqrt{\left(\frac{1}{\Lambda_o^\sigma}\right)^2 + \left(\frac{\delta\theta}{L}\right)^2} - \sqrt{\left(\frac{1}{\Lambda_o^\pi}\right)^2 + \left(\frac{\delta\theta}{L}\right)^2}\right]t. \tag{15.10}$$

Two types of setups are to be distinguished, those which operate in the region where the reflected beam is important and those which operate far from the reflection domain.

1. *Phase plates operating within the reflection domain*

(i) *Setups in the transmission geometry*
If the crystal thickness is such that $|\Psi_j^{\sigma-\pi}|$ is equal to $\pi/2 + 2k\pi$, the emerging wave which is the coherent superposition of its two components, σ and π, is circularly polarized, actually elliptically polarized since, due to the difference in absorption, the amplitudes of the two components are slightly different. If the wavefields corresponding to branch 1 and 2 do not overlap, one has two elliptically polarized emerging waves (Skalicky and Malgrange 1972). If they overlap, the two elliptically polarized waves which have opposite signs of helicity ($\Psi_1^{\sigma-\pi} = -\Psi_2^{\sigma-\pi}$) but different amplitudes due to anomalous absorption interfere as experimentally shown by Skalicky and Malgrange (1972) with an X-ray tube. Hart (1978) used this phenomenon to prepare elliptically polarized X-rays with a three-quarter wave plate silicon crystal, Cu$K\alpha$ and at Bragg's incidence. The same scheme was applied by Annaka *et al.* (1981) and Annaka (1982). It was extended to synchrotron radiation, among others, by Golovchenko *et al.* (1986), Mills (1987), Hirano *et al.* (1993, 1995) Goulon *et al.* (1996), Malgrange *et al.* (1999), Okitsu *et al.* (2001, 2002), Berman *et al.* (2002) to produce phase retarders.

(ii) *Setups in the reflection geometry*
Brümmer *et al.* (1982*a*, 1984) and Batterman (1992) proposed an entirely different scheme based on the properties of phases in the reflection geometry. It has been seen in Chapter 7 that the phase of the reflected beam is π on the low-angle side of the reflection domain and 0 on the high-angle side, and that it varies from π to 0 inside the total reflection domain. The Darwin width is proportional to the structure factor, and also to C. There are therefore angles of incidence for

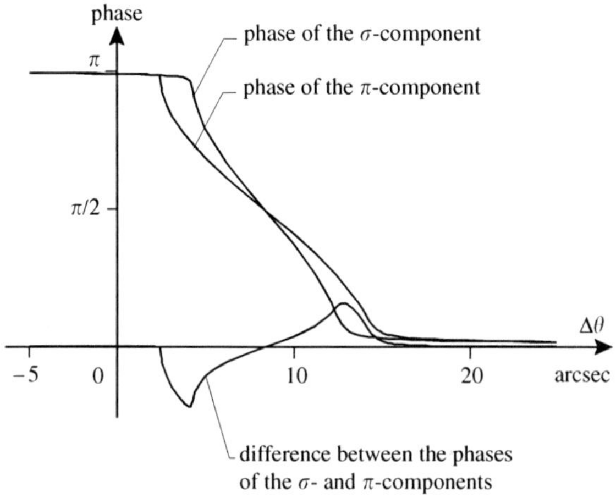

Fig. 15.25. Phases of the σ- and π-components (reflection geometry). Germanium, 220 reflection, Cu$K\alpha$.

which these two phases differ as shown in Fig. 15.25. The difference is never as large as $\pi/2$, but if one uses a channel-cut crystal of the type represented in Fig. 15.12, the phase difference between the σ- and the π-components is multiplied by the number of bounces.

Batterman (1992) used germanium, the 220 reflection, 8 keV radiation (1.54 Å) and a channel-cut crystal with five bounces. It suffices then to choose an incidence angle such that the phase difference between the σ- and the π-components is $\pi/10$ for a single reflection to have a $\pi/2$ phase difference after five reflections. In his experimental setup, Batterman used a 333 reflection on a first germanium crystal in the Bragg geometry ($\theta_B \approx 45°$) to produce linearly polarized radiation and to collimate the beam, and a silicon 111 Borrmann effect polarizer to analyse the circularly polarized radiation.

2. *Phase plates operating far from the reflection domain*

One of the drawbacks of the schemes described in paragraph (1) is that they require a very precise angular and spectral adjustment. Another solution is to operate far from the reflection domain. When $\Delta\theta$ is very large, the square roots in (15.10) can be expanded and it follows, after replacing Λ_o by its expression (5.17b), that

$$\Psi_j^{\sigma-\pi} = \pm\frac{\pi}{2}\left(\frac{R}{\pi V}\right)^2 \lambda^3 \sin 2\theta_B \frac{t}{|\delta\theta|},$$

where, as was mentioned before, the + sign corresponds to branch 1 and the − sign to branch 2. Far from the reflection domain only one type of wavefields has a non-negligible intensity: those whose direction of propagation is parallel to the incident direction, namely branch 1 wavefields for $\delta\theta \ll 0$ and branch 2 wavefields for $\delta\theta \gg 0$

(Fig. 15.23(b)). One has therefore, finally,

$$\Psi^{\sigma-\pi} = -\frac{\pi}{2}\left(\frac{R}{\pi V}\right)^2 \lambda^3 \sin 2\theta_B \frac{t}{\delta\theta} \tag{15.11}$$

dropping the index j.

As opposed to the phase plates operating in the reflection domain, the phase plates operating far from the reflection domain have the following properties:

(i) Only one wavefield is present.

(ii) The amplitudes of the σ- and π-components are equal so that it is possible to have exactly circularly polarized radiation.

(iii) The sensitivity to angular divergence of the beam is smaller.

(iv) The sensitivity to crystal imperfections is smaller; it is enough that $|\Delta\theta|$ be larger than the width of the rocking curve of the imperfect crystal.

Equation (15.11) shows that the circular polarization is right or left handed, depending on the sign of $\delta\theta$, and that the value of the phase shift is proportional to the crystal thickness and inversely proportional to $\delta\theta$. The thickness for which the phase shift is equal to $\pm\pi/2$ (quarter-wave plate) is

$$t = \left(\frac{\pi V}{R}\right)^2 \frac{|\delta\theta|}{\lambda^3 \sin 2\theta_B}.$$

The use of phase plates far from the reflection domain was first suggested theoretically by Dmitrienko and Belyakov (1980*b*) and Belyakov and Dmitrienko (1989) and, independently, by Baranova and Zel'dovich (1980); it was first applied experimentally by Hirano *et al.* (1991, 1992*a*,*b*) with thin silicon plates and a well-collimated beam. They operate in the reflection geometry and they use the transmitted beam.

The circular polarization rate

$$\tau = \frac{I_R - I_L}{I_R + I_L},$$

where I_R and I_L are the right-handed and left-handed polarized intensities, respectively, obviously depends on the divergence of the beam. The sensitivity is the smallest if $\delta\theta$ is large but then the necessary crystal thickness is also large and one has to worry about the absorption through the phase plate. For this reason Giles *et al.* (1994*a*,*b*, 1995*a*,*b*) have used diamond crystals. For a diamond crystal of 1.3 mm and the 220 reflection, a $\pi/2$ phase shift is obtained for $\delta\theta = 118$ arcsec, while for a silicon crystal and the 220 reflection, one would need a crystal thickness of 1.02 mm for the same value of $\delta\theta$, but the absorption would be very large or a thickness of 0.14 mm for the same absorption, but $\delta\theta$ would only be 16 arcsec (Malgrange 1992).

To test the efficiency of a phase plate one may either use a physical property sensitive to the circular polarization, such as magnetic Compton scattering (Mills 1988) or X-ray magnetic circular dichroism (Giles *et al.* 1994*a*,*b*, 1995*a*,*b*), or analyse the vibration after the phase plate with a linear polarizer and a quarter-wave plate (Ishikawa *et al.* 1991; Giles *et al.* 1995*c*).

15.3.5.3 *Optical activity*

The availability of highly efficient polarizers such as the channel-cut multiple-reflection polarizers has been an incitement to look for evidence of optical activity in the X-ray range. After preliminary attempts by Hart (1978) and Hart and Rodrigues (1981) with quartz and nickel sulphate hexahydrate ($NiSO_4{\cdot}6H_2O$) and Cu$K\alpha$ radiation, the first conclusive measurements were made using synchrotron radiation, which enables one to use wavelengths very close to an absorption edge where an effect is expected (see Sections 1.8 and 2.7.1), by Sauvage *et al.* (1983) with sodium bromate ($NaBrO_3$) and a wavelength very close to the bromine K absorption edge, and by Siddons *et al.* (1990) who observed optical activity and the Faraday effect in crystalline and non-crystalline cobalt compounds with a wavelength very close to the cobalt K absorption edge. Hart *et al.* (1991) developed a tunable polarimeter for the measurement of optical rotary power at synchrotron sources.

15.4 Focusing optics

15.4.1 *Introduction*

X-ray sources are always divergent. The divergence of synchrotron radiation is very small but the source-to-sample distances are very large. Perfect single-crystal flat monochromators have a very high reflectivity but only within a very narrow angular range. There are several ways to increase this angular range.

One is to introduce a degree of mosaicity in the crystal; this can be done through heat treatment (Schneider *et al.* 1988, 1989; Manninen *et al.* 1990), by using composite crystals such as Si/$TaSi_2$ (Neumann *et al.* 1996) or graded crystals in which the lattice parameter varies with position (Smither 1982; Smither and Fernandez 1992; Bradaczek *et al.* 1993). This last concept has known recent applications with the development by Abrosimov *et al.* (1996) of growth techniques for good quality $Si_{1-x}Ge_x$ crystals (Erko *et al.* 1996*a*,*b*; Keitel *et al.* 1998); the diffraction properties of these crystals have been discussed by Keitel *et al.* 1999 – see Subsection 14.4.2).

The other way to lose as little as possible of the intensity coming from the source, it is in general necessary to focus the beam both horizontally and vertically so as to concentrate it on a small area of the sample. Two types of focusing are thus to be considered: *meridional* focusing, where the curvature is parallel to the scattering plane (Fig. 15.26), and *sagittal* focusing, where the curvature is perpendicular to the scattering plane (Fig. 15.27). In usual applications, three types of optical elements are used for geometric focusing: mirrors, multilayers and curved crystals. Reflection by crystals and multilayers follows Bragg's law and these optical elements are dispersive. Reflection by mirrors is specular and is not dispersive. For

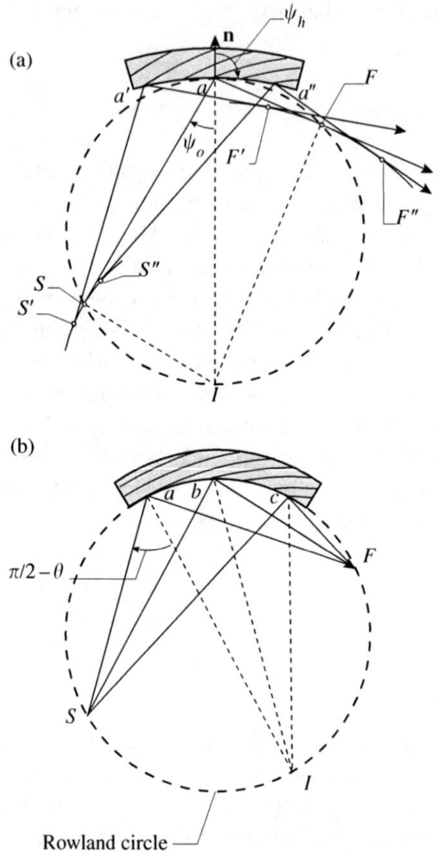

Fig. 15.26. Curved crystal monochromator. (a) Johann geometry; I: centre of cur‘vature of the crystal; S: source; F: focus; $\widehat{SS''}$, $\widehat{FF''}$: source and focus caustics. (b) Johansson geometry; S: source; F: focus; I: centre of curvature of the lattice planes.

other applications such as microdiffraction or microimaging requesting very small focal sizes, in particular submicron focal sizes, special types of optical elements are used such as those described in Sections 15.4.4–15.4.8. A recent review is given in Hildebrandt and Bradaczek (2000).

15.4.2 *Mirrors*

X-ray mirrors (Kirkpatrick and Baez 1948) are important optical elements which can focus efficiently a wide beam of X-rays to a small spot size and can be used to match the divergence of the white beam to the acceptance of perfect crystal monochromators. They are also filters which reflect X-rays below a certain cutoff energy and stop X-rays

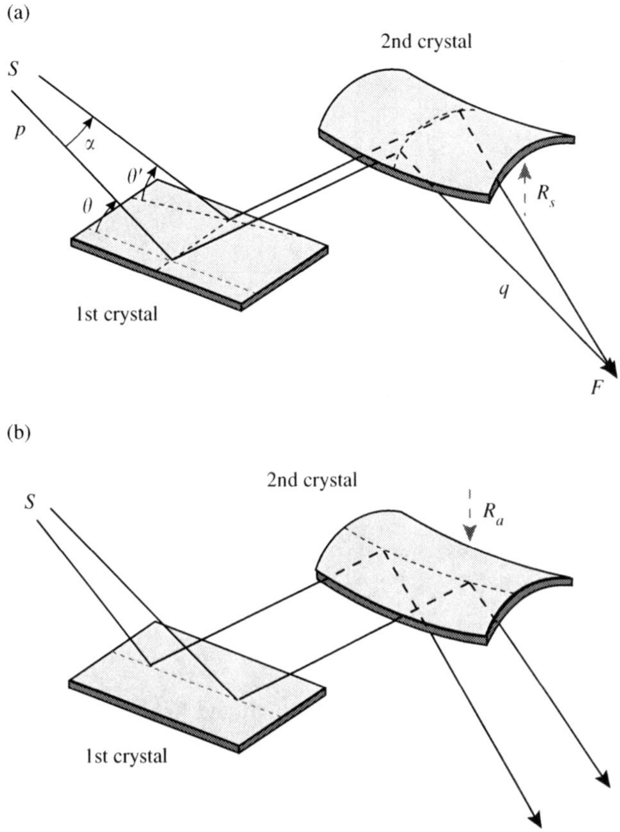

Fig. 15.27. Sagittal focusing. (a) Principle of the method showing how a ray out of the median diffraction plane is brought into focus; $\mathcal{R}_s$: radius of sagittal curvature; S: source; F: focus. (b) Anticlastic curvature, $\mathcal{R}_a$.

above this energy; for instance, they can be used to eliminate higher-order harmonics of Bragg reflections from crystal monochromators. Aberrations can be reduced by using two cylindrically bent mirrors at 90° (so-called Kirkpatrick-Baez arrangement – see for instance Yang *et al.* 1993, Underwood *et al.* 1996, Eng *et al.* 1995, 1998). Mirrors make use of the specular reflection of X-rays at glancing angles. Figs 2.7 and 2.8 show their cut–off properties with glancing angle and with energy. A detailed discussion of the theory of specular reflection and of its application to optical elements for synchrotron radiation is given in Matsushita and Hashizume (1983).

The maximum aperture of mirrors does not exceed the value of the critical angle and this is a limitation for the use of mirrors. The value of the critical angle is given

Table 15.3 Critical angles, in mrad, of a few materials at 10 keV (after Hart 1996)

Material	Silicon	Copper	Gold	Platinum
ω_c (mrad)	3.17	5.50	7.69	8.07

by (Section 2.10)

$$\omega_c = \lambda\sqrt{\frac{RF_o}{\pi V}}.$$

As this angle increases with the atomic number, it is preferable to use a heavy element; unfortunately, it is difficult to fabricate mirrors out of such materials. The usual solution is to coat a material such as silicon with a thin layer of gold or platinum. Table 15.3 compares the values of critical angle for a few materials at 10 keV energy.

One of the consequences of the smallness of the critical angles is the importance of surface roughness which scatters photons out of the focus and which must be kept as low as possible.

15.4.3 *Multilayers*

Multilayers consist of alternating layers of high and low atomic number materials, such as tungsten or molybdenum and silicon or boron carbide, with thicknesses of the order of 20–30 Å, deposited on appropriate substrates, the number of bilayers being of the order of 100 (Underwood and Barbee 1981; Underwood *et al.* 1986). They act as interference reflectors and have a high reflectivity. Their period is about 10 times larger than that of the crystals used as monochromators which results, as shown by (15.3), in a much broader bandwidth ($10^{-2} < \Delta E/E < 10^{-1}$) and a higher intensity which makes them suitable for certain applications. They provide controllable band pass characteristics and a high reflectivity (Schuster *et al.* 1988; Schuster 1991). By varying the period of the layers with depth (graded multilayers), it is possible to increase the energy or the angular range of reflection (Nagel *et al.* 1982; Joensen *et al.* 1995; Erko *et al.* 1998). Focal sizes of the order of a few micrometres have been obtained which make multilayers useful for microdiffraction or fluorescence microprobes (Underwood *et al.* 1998). Parabolically curved graded multilayers have been proposed by Schuster and Göbel (1995). Multilayers have also been used in the design of Bragg–Fresnel lenses for a scanning X-ray microscope (Chevallier *et al.* 1995). For reviews of the applications of multilayers at synchrotron sources, see Ziegler (1995), Andronova *et al.* (1995*a*, *b*), Underwood (1995) and Morawe *et al.* (2001).

As for all optical elements used with synchrotron radiation, the problem of thermal stability under the heat-load of the beam arises. For a recent study, see, for instance, Jergel *et al.* (2000) and the references therein.

15.4.4 *Curved crystals*

15.4.4.1 *Meridional focusing*

Curved crystal monochromators have been common practice with X-ray tubes and are also used for neutron diffraction (see, for instance Albertini *et al.* 1977; Mikula *et al.* 1984, 1988; and Šaroun *et al.* 1994) and for synchrotron radiation (see, for instance, Boeuf *et al.* 1978; Caciuffo *et al.* 1987; Wittry and Sun 1990; Suortti *et al.* 1990, 1992, 2000; Suortti 1992; Suortti and Schulze 1995; Chang and Förster 1997; Schulze *et al.* 1998; and Lienert *et al.* 1998).

The two most common schemes are (see, for instance, Aslanov *et al.* 1998; Parrish and Langford 1999):

1. The *Johann geometry* in which the crystal is simply curved (Johann 1931). Figure 15.26(a) represents the case of an asymmetric reflection. I is the centre of curvature of the crystal and $Ia = R$ the radius of curvature. Let ψ_o and ψ_h be the angles between the normal to crystal surface, **n**, at point a, the incident **Sa** and reflected **aF** directions, respectively, following the notations of Section 4.7.3. It can be seen in the figure that

$$p = \overline{Sa} = R \cos \psi_o,$$

$$q = \overline{aF} = R |\cos \psi_h|.$$

The ratio p/q of the distances mirror to sample over source to mirror is therefore the asymmetry ratio, $|\gamma|$. The Johann geometry has the drawback of presenting spherical aberrations which are discussed, for instance, in Matsushita and Kaminaga (1980*a*,*b*).

2. The *Johansson geometry* (Johansson 1933) which has no spherical aberration and which is represented in Fig. 15.26(b). In that setup, the crystal is first ground so as to present a cylindrical surface of radius R. It is then elastically bent so as to present a curvature of radius $R/2$. The curvature of the lattice planes is the difference between the curvatures of the surface after bending, $2/R$, and before bending, $1/R$, and is therefore $1/R$. The various incident rays, **Sa**, **Sb**, **Sc**, all satisfy simultaneously the Bragg condition. The source, the focus and the centre of curvature of the lattice planes, I, lie on a circle, called the Rowland circle. Such a setup is not easily tunable.

Another way of eliminating the spherical aberrations is to use an ellipsoidal surface, the source and the focus being located at the focal points of the ellipse. Other systems use crystals curved to a logarithmic spiral (Wittry *et al.* 1993) or doubly curved crystals (Wittry 1990; Wittry and Sun 1993, Wittry and Chang 1993; Chukhovskii *et al.* 1995*a*, *b*).

15.4.4.2 *Sagittal focusing*

The bases of sagittal focusing were laid down by Sparks *et al.* (1980, 1982), who showed that for energies larger than 10 keV curved crystals are more efficient than mirrors. The principle is shown in Fig. 15.27(a). Consider a two-crystal setup in which the first crystal is flat and consider two rays:

1. One for which the diffraction plane containing the incident and the reflected rays coincides with the plane of incidence containing the incident ray and the normal to the reflecting planes; its angle of incidence on the lattice planes is θ.
2. One for which the reflecting plane and the incidence plane do not coincide. It makes a small angle, α, with the first one and its angle of incidence on the lattice planes is θ'.

The angles of incidence of the two rays are related by

$$\theta' = \theta - \frac{\alpha^2}{2} \tan\theta$$

as shown in Section 13.2.2. The two rays satisfy Bragg's condition for slightly different wavelengths.

The two rays are further reflected by the second crystal and, if this second crystal is also flat, they will continue to diverge; they will hit the sample at two points, F and F', separated by the distance $FF' = \alpha(p+q)$ where p is the distance from source to monochromator and q the distance from monochromator to sample. If this distance is larger than the sample, the energy carried by the second ray is lost. In order to bring that ray back into the focus F, the second crystal should be bent as shown in Fig. 15.27(a), with a radius of curvature $R_s = 2pq \sin\theta/(p+q)$. Once the crystal is curved, the angle of incidence of the second ray is modified. Sparks *et al.* (1980) have shown that for it to be within the reflection domain, the ratio q/p should be of the order of $\frac{1}{3}$. This constraint can be reduced and the aberrations decreased if one uses a conically shaped surface (Habenschuss *et al.* 1988; Ice and Sparks 1994).

There are several technical problems with bent crystals for sagittal focusing. One is related to the heat-load from the synchrotron beam as for every monochromator; another one is the transverse curvature which occurs when bending the crystal, called anticlastic curvature (Fig. 15.27(b)). In order to avoid it several methods have been proposed, such as putting stiffening ribs at the backside of the crystal, the crystal and the ribs being cut in the same monolithic block and bent with a four-point bending setup (Sparks *et al.* 1982), or various devices whose principle is to have a slotted crystal cut from a single crystal of triangular or more complicated shape (Batterman and Berman 1983; Matsushita *et al.* 1986). Kushnir *et al.* (1993) have shown theoretically, using finite element analysis, that anticlastic curvature can be avoided with an appropriate bending device. Another design in which both heat-load and anticlastic curvature are taken into account and which has also been calculated by finite element analysis has been proposed by Zontone and Comin (1992); in that design both crystals of a non-dispersive Bragg–Bragg pair are bent, the first one in the meridional direction and the second one in the sagittal geometry. The dynamical diffraction of a spherical wave by two bent crystals in the Kirkpatrick-Baez arrangement is discussed by Tchen (2002).

15.4.5 *Fresnel zone plates*

A Fresnel zone plate is constituted by a succession of circular zones such that the optical path between the source and image focal point varies by $\lambda/2$ from one zone to the next (Fig. 15.28).

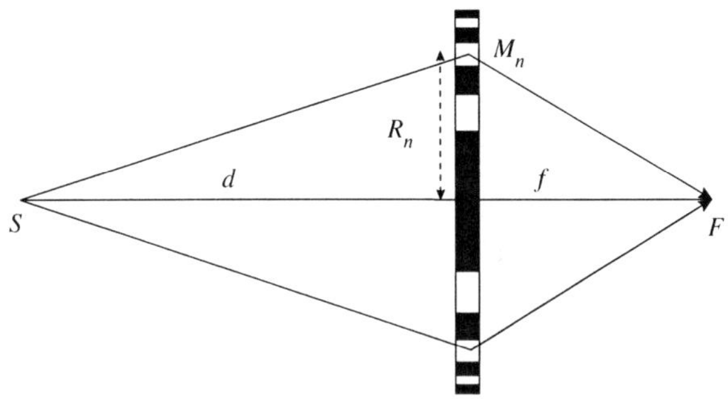

Fig. 15.28. Fresnel zone plate. S: source; F: focus.

If S represents the source, F the focal point and M_n is a point on the nth zone, the difference between the optical paths SM_nF and SF should be $n\lambda/2$:

$$\sqrt{d^2 + R_n^2} + \sqrt{f^2 + R_n^2} - (d + f) = n\frac{\lambda}{2},$$

where R_n is the radius of the nth zone. If d and f are much larger than R_n, one has, to a first approximation,

$$R_n = \sqrt{n\lambda\frac{df}{d + f}}.$$

Typically, at a synchrotron, the distance from the source to the zone plate is of the order of a few tens of metres or more and the distance between the zone plate and the focal point a few tens of centimetres. The expression of the radius of the nth zone reduces to

$$R_n = \sqrt{n\lambda f}.$$

Zone plates were first used for hard X-rays by Baez (1952). They operate in transmission and one distinguishes *amplitude zone plates* where the zones are alternatively opaque and transparent for X-rays and *phase zone plates* where the zones are transparent for X-rays, but the phase shift through the plate is alternatively 0 and π for successive plates (Lai *et al.* 1992, 1995; Yun *et al.* 1992). The zones may be prepared by lithography for energies below 10 keV (Lai *et al.* 1992; Di Fabrizio *et al.* 1994, 1999; Yun *et al.* 1999) or sputtering of opaque or transparent multilayers for energies larger than 8 keV (Saitoh *et al.* 1989; Tamura *et al.* 1994; Kamijo *et al.* 1995, 2002; Suzuki *et al.* 1997, 2001). If the thickness of the layer is larger than 10 μm, the zone plate operates as an amplitude plate; if it is smaller, it operates as a phase plate (Bionta *et al.* 1989, 1990, 1994). Another possibility is to use gold electroplating to prepare the zones. The efficiency of the plate is better for phase plates but, in practice, the zone plates always operate in a mixed regime, partially phase and partially amplitude.

The size of the focal spot is in the submicrometre region. The applicability of zone plates to the design of an X-ray microscope has been demonstrated, for instance, by Schmahl *et al.* (1993, 1994) for soft X-rays, Kaulich *et al.* (1999), Lai *et al.* (1995), Yokosuka *et al.* (2002), Kagishima *et al.* (2002), Kamijo *et al.* (2002), Awaji *et al.* (2002), Neuhäusler *et al.* (2003) and Kilcoyne *et al.* (2003) or hard X-rays.

Indenbom (1979) proposed to use the intensity distribution at the base of the Borrmann triangle for an incident spherical wave (see Section 10.4) to produce a Fresnel zone plate analog and to apply it to the case of neutrons.

15.4.6 *Bragg–Fresnel lenses*

The idea of combining the principle of Fresnel zone plates and Bragg diffraction from a perfect silicon crystal in the reflection geometry was suggested by Aristov *et al.* (1986, 1987, 1989). The Fresnel zones are obtained by grooving the surface of the crystal (Fig. 15.29); the wave reflected by the lower surface has a phase shift of π with respect to that reflected by the upper surface and the zone plate acts as a phase zone plate. The zones can also be obtained by ion implanting (Souvorov *et al.* 1996) or by acoustic excitation (Souvorov *et al.* 1997).

There are two types of Bragg–Fresnel lenses: the circular one, as represented in Fig. 15.29, and the linear one. In the circular lens, the focus is a point focus determined by Bragg's law (Basov *et al.* 1991; Snigirev *et al.* 1995*a*). The order of magnitude of the size of the focus is 0.7–0.8 μm (Kuznetsov *et al.* 1994). The linear lens (Aristov *et al.* 1992; Bonse *et al.* 1992; Snigirev and Kohn 1995) is a planar plate and is used in sagittal geometry; it produces a linear focus. It can also be fabricated on a cylindrically bent plate (see, for instance, Chang *et al.* 1995).

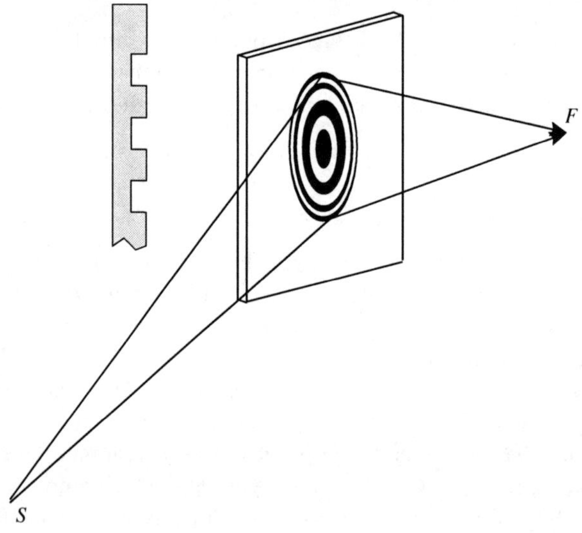

Fig. 15.29. Bragg–Fresnel lens. S: source; F: focus. The insert shows the cross-section of the lens.

Bragg–Fresnel lenses take full advantage of the coherence properties of third-generation synchrotron sources and preserve this coherence (Basov *et al.* 1991; Aristov *et al.* 1992). Their focusing properties are applied to microprobe analysis (Bonse *et al.* 1992; Kuznetsov *et al.* 1994 and Snigirev *et al.* 1995*a*), high-resolution microdiffraction, small-angle scattering and microimaging. Reviews can be found in Snigirev (1995) and Snigireva *et al.* (1998).

15.4.7 *Refractive lenses*

Because the index of refraction of matter for X-rays is so close to 1, it is not possible to make a refractive lens in the usual way. Recently, Snigirev *et al.* (1996, 1998) have proposed a lens based on the principle described schematically in Fig. 15.30.[4] A series of cylindrical holes are drilled in an aluminium block. The focal length for a lens made with a single hole is

$$f = \frac{R}{2(1-n)},$$

where R is the radius of the hole and n the index of refraction. For a lens with N holes, the focal length is

$$f = \frac{R}{2N(1-n)}.$$

For aluminium, $1 - n = 2.8 \times 10^{-6}$ for 14 keV X-rays; Snigirev *et al.* (1998) use $N = 30$ and $R = 0.3$ mm; the focal length is therefore $f = 1.8$ m. The focal size they have obtained is about 8 μm. With the scheme represented in Fig. 15.30, the focus is linear, but if one uses two series of holes, at 90°, a point focus can be obtained (Snigereva *et al.* 1998; Lengeler *et al.* 1998). Recently, the refractive lenses have been improved by using a different technique. They consist of a stacking of single lenses with a parabolic profile (Lengeler *et al.* 1999*a*,*b*, 2002), thus reducing the spherical aberrations. They were applied to the construction of an X-ray microscope by the same authors. Aristov *et al.* (2000, 2001) and Nöhammer *et al.* (2003) have also improved the design using planar parabolic lenses with minimized absorption.

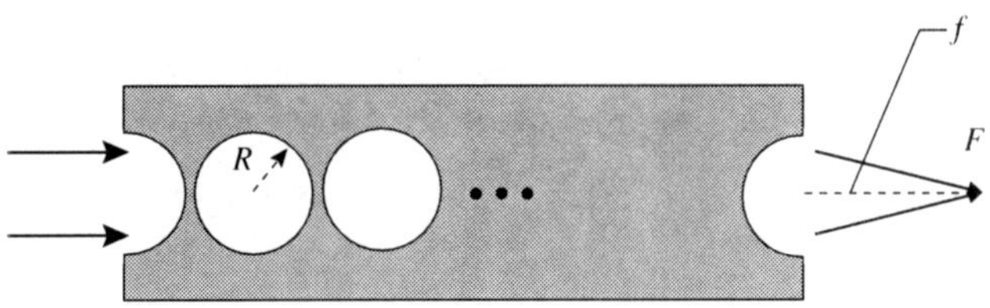

Fig. 15.30. Compound refractive lens. F: focus; f: focal length.

[4] A similar idea was proposed by Tomie (1997)—see Lund (1997).

15.4.8 *X-ray wave-guides*

Another way of producing submicrometric X-ray beams of medium to high energy is offered by the X-ray waveguides. The first attempt to fabricate hard X-ray waveguides is due to Spiller and Segmüller (1974) and the first observations were made by Feng *et al.* (1995) and by Lagomarsino *et al.* (1996). The waveguides are made of a multilayer structure shown schematically in Fig. 15.31.[5] It consists in the stacking of:

(1) a very flat substrate made of silicon oxide;

(2) a metal layer of high or relatively high atomic number with a thickness of a ten of nanometres, such as chromium or nickel;

(3) a layer of low-density material such as carbon, about one hundred nanometres thick;

(4) a thin metal layer a few nanometres thick.

A collimated, highly coherent, beam from an undulator impinges on the structure at a grazing angle. The index of refraction of the low-density material is closer to 1 than that of the high-density metal; the critical angle for specular reflection ω_c is therefore smaller for the low-density material (Sections 2.7.3 and 2.10). If the incidence angle is chosen in between these two values, the beam will undergo total reflection in the carbon film at the interface between carbon and metal (Fig. 15.31). As the source is highly coherent, the beams reflected at the upper and lower surfaces of the carbon layer interfere and a set of standing waves are produced. If the thickness of the layer corresponds to an integer number of nodes, resonance appears with the resulting effect that the X-ray beam is narrower and has a higher peak intensity (Jark *et al.* 1996). The beam exiting from the waveguide is highly focused in the direction normal to the plane of the layers (about 0.15 μm, the thickness of the carbon layer) and can be used for phase contrast (Lagomarsino *et al.* 1997; Di Fonzo *et al.* 1998, 2000), strain analysis (Giannini *et al.* 2001), microscopy (Lagomarsino *et al.* 2000; Jark *et al.* 2001; Jark and Di Fonzo 2004) or microdiffraction. For a review, see Cedola *et al.* (1999) and Müller *et al.* (2000)

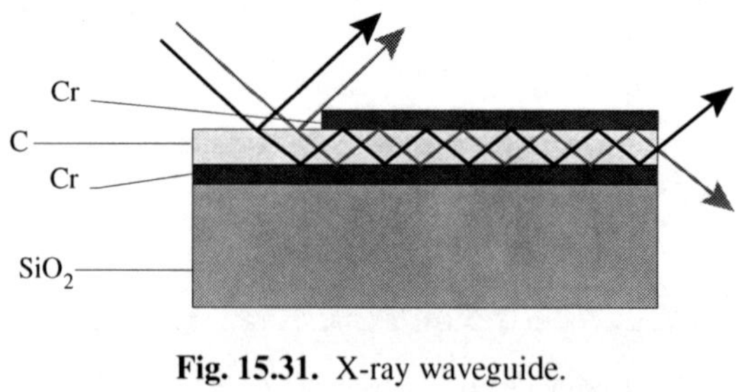

Fig. 15.31. X-ray waveguide.

[5] New designs of X-ray waveguides have also been proposed recently by Zwanenburg *et al.* (1999, 2000), Pfeiffer *et al.* (2000, 2002), Bongaerts *et al.* (2002).

15.5 X-ray interferometers

15.5.1 *Principle*

The principle of optical interferometry is to have a beam split into two separated coherent beams which can be recombined and then interfere. If a phase object is placed on the path of one of the beams, a shift of the fringes can be observed from which the phase of the object can be measured. The X-ray interferometer invented by Bonse and Hart (1965*a*–*c*, 1966*c*) follows exactly the same principle and the measurement of phases is one of its main applications. The simplest form, the so-called *LLL* interferometer, is represented in Fig. 15.32. It consists of three crystals which have been cut and etched in a monolithic crystal, usually silicon, and which are all of them used in the transmission, or Laue geometry (hence the name). It has also been fabricated in other materials, such as quartz and germanium (Hart 1968*b*). Interferometers can also be operated in the reflection geometry (Bonse and Hart 1966*c*; Graeff and Bonse 1977) or in a combined geometry (Bonse and Hart 1968); in this latter case the shift due to refraction in the Bragg case must be taken into account. The same principle is also applied to neutron interferometry (Bauspiess *et al.* 1974, 1977; Rauch *et al.* 1974; Rauch and Werner 2000) where germanium crystals are often used.

The extension to X-rays of the principle of visible optics Michelson interfermetry was proposed by Deslattes (1968) and Hart (1975*b*) with beam geometries involving special four-beam cases. A Michelson interferometer operating successfully with X-rays and neutrons mixing Laue and Bragg geometries was constructed by Appel and Bonse (1991) and improved by Bartscher and Bonse (1998) and Nusshardt and Bonse (2003). A four-Bragg reflection Michelson interferometer with a large and variable path length difference was recently built and tested by Fezzaa and Lee (2001). Other designs have been proposed by Tamasaku *et al.* (2002, 2003) and by Sutter *et al.* (2003, 2004). An interferometer of the Fabry–Perot type consisting of two thin parallel crystals operating at a Bragg angle close to $\pi/2$ has also been proposed, both for X-rays (Caticha and Caticha-Ellis 1990*b*; Caticha *et al.* 1990) and for neutrons (Caticha and Caticha-Ellis 1996). The theory of Fabry–Perot interferometers is discussed in Kohn *et al.* (2000).

The simplest case to consider is that of an incident plane wave, and the theory has been developed by Bonse and Hart (1965*c*) and Bonse and Graeff (1977)—see also Colella (1988). It was extended to the case of an incident spherical wave by Bauspiess *et al.* (1976). We shall limit ourselves here to the case of an incident plane wave. It is further simplified if wavefields belonging to branch 2 of the dispersion surface can be neglected, which is possible for silicon and crystal thicknesses of about 1 mm, but not in the case of neutrons for which there is no absorption (Rauch and Suda 1974; Bauspiess *et al.* 1976; Staudenmann *et al.* 1980). The coherence properties and the geometrical and stability requirements have been studied theoretically and experimentally by Bonse and Hart (1965*d*), Kuriyama (1971), Bonse and te Kaat (1968, 1971) and Becker and Bonse (1974). The case of absorbing crystals with arbitrary absorption (from zero to strong) was discussed by Petrascheck and Folk (1976), both for plane and spherical incident waves.

The first crystal of the interferometer is the splitter, S, which splits the incident beam into two beams, R and T, which are transmitted in the reflected and incident

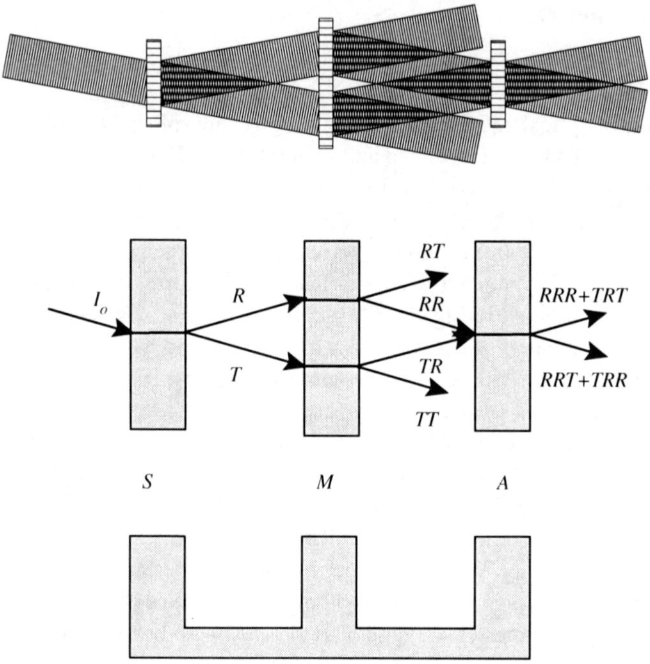

Fig. 15.32. Principle of the *LLL* interferometer. *S*: splitter; *M*: mirror; *A*: analyser; from bottom to top: side view, top view, beam paths.

directions, respectively. They are incident on the second crystal, the mirror, M. The beams R and T give rise each to two beams, *RT* and *RR*, *TT* and *TR*, respectively. The beams *RR* and *T R* which have been both Bragg-reflected by the mirror, with reciprocal lattice vectors **h** and −**h** respectively, converge on the analyser, A. They interfere in the region where they overlap, giving rise to standing waves which, as has been seen in Section 5.8, have a period equal to that of the reflecting planes, d_{hkl}. The beams *RR* and *TR* are then diffracted by the analyser, giving rise to two beams each, *RRR* and *RRT*, *TRR* and *TRT*, respectively. The beams *RRR* and *TRT* are in the reflected directions; they are coherent and interfere. The same thing holds for the beams *RRT* and *TRR* which are in the incident direction. The intensity of the beams depends on the position of the lattice planes of the analyser with respect to the nodes of the standing waves generated by the *RR* and *TR* beams: if the third crystal is translated in the direction of the reciprocal lattice vector the intensities of the two beams emerging from the interferometer will vary periodically, with the period of the reflecting planes. This can be achieved by introducing a spring flexure between the second and the third crystal; it is represented schematically in Fig. 15.33. The drive mechanism for translating the crystal is usually magnetic or piezoelectric. This system is the base of many of the most important applications of interferometers, as will be described in Section 15.5.2.

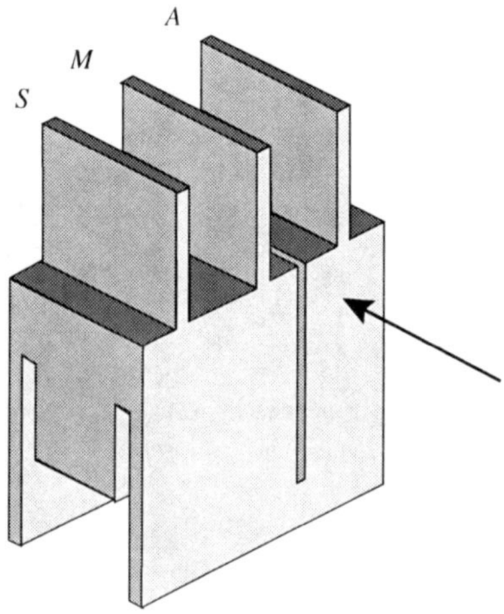

Fig. 15.33. Design of an interferometer. S: splitter; M: mirror; A: analyser; the analyser can be translated by pushing along the arrow.

Some of the main advantages of having the three crystals as part of the same monolithic block are the automatic alignment of their lattice planes, the stability, and the relative insensitivity to vibrations. But, in some cases, it is useful to have a two-crystal interferometer, for instance for wavelength-independent measurements of interplanar spacings (Bonse and te Kaat 1968, 1971), for the measurement of tiny rotations or to have long path lengths, particularly in the case of neutrons. In the first design by Bonse and te Kaat, the splitter and the analyser were part of one crystal and the analyser was a separate crystal. Another design, the so-called *skew-symmetric* two-crystal interferometer (Fig. 15.34), was shown by Becker and Bonse (1974) to be much more stable. Such a design allows small rotations of the two parts with respect to one another, and the angular dependence of the reflectivity was investigated theoretically and experimentally by Windish and Becker (1992). In the so-called wavefront-dividing interferometers (Izumi *et al.* 1995; Yamazuki and Ishikawa 1996, 2003) the splitter-mirror and mirror-analyzer spacings are different.

For reviews on X-ray and neutron interferometers and their applications, see, for instance, Bonse (1969, 1988), Bonse and Graeff (1977), Bonse and Rauch (1979), Bowen (1996), Colella (1996), Hart (1975*b*, 1980), Treimer (1978), Werner (1980) and Rauch and Werner (2000).

15.5.2 *Applications*

For most applications, interferometers require a very high precision in their fabrication, surfaces smooth to a few microns, mechanical precision down to 10–20 μm, perfect, dislocation-free, homogeneous crystals, with as few impurities as possible. The design of interferometers and the problems encountered in their machining

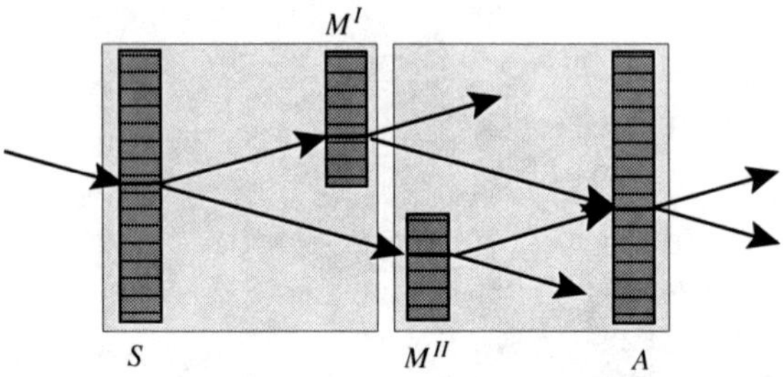

Fig. 15.34. Principle of the skew-symmetric two-crystal interferometer. S: splitter; M^I and M^{II}: mirrors; A: analyser.

are described in Smith and Chetwynd (1992). The best crystal is silicon which can be obtained with a high degree of perfection. Float-zone silicon is preferred to Czochralski-grown silicon as it contains less oxygen. The influence of common impurities, in particular carbon, on the lattice parameter of silicon has been studied by Windish and Becker (1988). A reflection on '*how perfect a crystal can be*' is given in Deslattes *et al.* (1999*b*).

15.5.2.1 *Moiré topographs and the study of defects*

If the three crystals of an interferometer are absolutely perfect, with no distortions and no lattice parameter variations, there would be uniform contrast on a photographic plate placed in one of the beams after the analyser. If there is a slight deviation from perfectness, fringes will appear. They were interpreted by Bonse and Hart (1966*b*) as a simple optical moiré pattern formed by the superposition of the two grids constituted by the lattice planes of the analyser on the one hand and the system of standing waves formed by the interference of the *RR* and *TR* beams on the other hand (Fig. 15.32). A moiré topograph of the interferometer can thus be obtained (Bonse *et al.* 1976*a*). Moiré images of single dislocations were first observed in this manner by Bonse and Hart (1966*b*). Hart (1972) could determine completely the Burgers vectors of dislocations by interferometry and Christiansen *et al.* (1971) analysed the strain field of an array of 60 grown-in dislocations in silicon from their moiré pattern, also by interferometry. A recent example is given in Fodchuk and Raransky (2003). Moiré topographs (see Section 17.3.5.1), as shown also by Lang and Miuscov (1965), by Brádler and Lang (1968) and Lang (1968) with usual topography, are a very sensitive way to determine strains, but with, of course, a poor spatial resolution.

15.5.2.2 *Measurement of indices of refraction, X-ray scattering factors and anomalous dispersion corrections*

The principle of the measurement of the index of refraction and, hence, of the scattering factor and anomalous dispersion correction is to introduce a thin sheet of the material in one of the beams and to measure the phase shift, either with a compensating plate (Bonse and Hellköter 1969; Creagh and Hart 1970; Creagh 1970; and Bonse and Materlik (1972, 1974, 1976) or by translating the analyser (Hart 1968*a*; Cusatis and Hart 1974). It was applied to many materials, first with X-ray tubes, then with synchrotron radiation (see, for instance, Bonse *et al.* 1980; Templeton *et al.* 1980, 1982). The method was improved by Cusatis and Hart (1976) who developed a two-wavelength method which increased the accuracy. Further improvements to the scanning interferometer have been implemented by Hart and Siddons (1981), Siddons and Hart (1983) and Begum *et al.* (1986) who have achieved a very high precision and could measure the variations of the dispersion corrections close to the K and L absorption edges of several elements. The quality of the measurements was excellent and so accurate that the values could be compared with those computed by Cromer and Liberman (1981); the agreement is very good.

15.5.2.3 *Phase contrast microscopy*

See Section 15.6.2.1.

15.5.2.4 *Absolute lattice parameter measurements and metrology at the nanometre scale*

X-ray and neutron interferometers are remarkable instruments in that they can be used to measure with a very high absolute accuracy distances with a nanometre precision and angles with a nanosecond precision. Although the technology is very delicate, the principle is very simple. If one translates the analyser by a number n of fringes, the displacement is simply $x = nd_{hkl}$, where h, k, l are the indices of the reflection and d_{hkl} is the reflecting plane spacing. Therefore it is enough to count a number of fringes to measure accurately a distance of a few nanometres. If the crystal is perfect and homogeneous enough, there are no problems of non-linearity and the method is better than the usual ones. For this reason, Hart, as early as 1968, called the X-ray interferometer an *angstrom ruler* (Hart 1968*a*). The first step, however, is to have an absolute measurement of the lattice parameter of the crystal.

Two types of techniques have been used to measure lattice parameters; they are reviewed in Hart (1981) and Fewster (1999). In the first one, an absolute value of the lattice parameter is obtained which is related to the X-ray wavelengths by means of Bragg's law, corrected for refraction. Before the time of the interferometer the X-ray wavelengths were measured using gratings. The techniques comprise the Bond method (Bond 1960) improved by Fewster and Andrew (1995*a*) by using a channel-cut analyser, the Kossel method, the high-angle double-crystal diffraction method (Okazaki and Ohama 1979; Ohama *et al.* 1979; Okazaki and Soejima 2001) and the energy dispersive methods (Fukumachi *et al.* 1973; Bordas *et al.* 1977) which are low resolution but fast. In the second type of technique, the parameter of the sample is compared to that of a reference standard with a double or multiple-crystal diffractometer in a non-dispersive or pseudo-non-dispersive arrangement. The first such comparator was a three-crystal arrangement with which Buschert (1965) measured the lattice parameter of silicon to 1 part in 10^7; Hart (1969) introduced the simultaneous recording of the rocking curves obtained with several beams; in that case he used a double-crystal arrangement and two X-ray sources, the key feature being that the precision is unlimited by X-ray optics and determined only by counting statistics. The accuracy was successively improved by Ando *et al.* (1978) who achieved an accuracy of 3 parts in 10^8 with two crystals, one X-ray source, but three detectors, and by Buschert *et al.* (1983) who achieved an accuracy of 1 part in 10^9 with a very sophisticated arrangement requiring two crystals, three X-ray sources and four detectors. Häusermann and Hart (1990) proposed a method using one X-ray source, a triple-axis diffractometer where two of the axes are double-leaf silicon springs and two detectors. The accuracy is of 1 part in 10^8, but the method has the advantage of being very fast. Other methods use a multiple-crystal arrangement and two detectors (Cembali *et al.* 1992; Zhang *et al.* 2003) or backscattering (Shvyd'ko *et al.* 2000; Sozontov *et al.* 2001; Hu *et al.* 2003–in these last two the effect of isotope composition on lattice parameters was measured).

With the X–ray interferometer, it is possible to measure the distance between several interference fringes with an optical Fabry–Perot interferometer. The absolute value of the lattice parameter of the crystal can thus be evaluated directly in terms of a standard optical wavelength and is independent of the X-ray wavelength. The lattice parameter of another crystal can then be deduced from that one using a comparator as was done for instance by Baker and Hart (1975) who deduced the lattice parameter of a germanium crystal from that of a silicon crystal which had been measured by interferometry and by Baker *et al.* (1976) for gallium arsenide.

The lattice parameter obtained by interferometry is given in terms of the International Metre (itself determined using time measurements) with an accuracy which is of the order of ± 2.5 parts in 10^8 for the most recent measurements. This work was pioneered by several groups (Deslattes 1969; Hart and Morgan 1971; Deslattes and Henins 1969, 1973; Becker *et al.* 1981; Deslattes *et al.* 1987; Basile *et al.* 1989; Windish and Becker 1990, etc.). The most recent values are those of Basile *et al.* (1995*a*) and Fujimoto *et al.* (1997). The value of the lattice parameter of silicon given by Basile *et al.* (1995*a*) is $a = 0.543102047(14)$ nm and that given by CODATA (1998)[6] is $a = 0.543102088(16)$ nm. These values are understood for a very highly pure silicon crystal and depend on the residual impurities. The influence of strain in the analyzer has been studied by Mana *et al.* 2004*b* using Takagi's equations.

The next step was the determination of X-ray wavelengths from the optically calibrated value of the lattice parameter of silicon. The values obtained by Deslattes and Henins (1973), Deutsch and Hart (1982) and Maskil and Deutsch (1988) were very accurate but suffered from a slight systematic error in the lattice parameter determined by Deslattes and Henins (1973), as shown by Deslattes *et al.* (1987). The values obtained by Härtwig *et al.* (1991) do not have this drawback and the analysis of the profile of the $K\alpha$ emission line of copper (Härtwig *et al.* 1993) and of other emission lines (Härtwig *et al.* 1994) is the most complete.

Once an absolute value is obtained for the lattice parameter of silicon, other physical parameters and universal constants can be determined such as, for instance, Avogadro's number, N_A. It is given by

$$N_A = \frac{M}{\rho\sqrt{8}d_{220}^3},$$

[6] See footnote 1.

where M is the molar mass and ρ is the specific mass. The latter is difficult to measure and requires the fabrication of perfect optically polished spheres of silicon and the determination of the amount of residual impurities in the silicon.

The determination of Avogadro's number has been performed by the same groups which measured the absolute value of the lattice parameter. The first value determined using interferometry was that given by Deslattes *et al.* (1974) based on the measurements by Deslattes and Henins (1973). It was improved successively by Basile *et al.* (1995*b*) and Becker *et al.* (2003). The latest value is $N_A = 6.0221353 \times 10^{23}\,\text{mol}^{-1}$. It is slightly smaller than the value obtained by the electrical method and which depends on the value of Planck's constant, $N_A = 6.02214199(47) \times 10^{23}\,\text{mol}^{-1}$.

The X-ray interferometer can also be used to calibrate transducers and other engineering systems (Bowen *et al.* 1985, 1990*a*,*b*; Chetwynd *et al.* 1991) and to measure angular rotations to a precision of a few nanoradians (Windish and Becker 1992).

For an overview of the possibilities of interferometers in nanometre metrology, see Bowen (1996).

15.6 Imaging with X-rays

15.6.1 *Introduction*

The oldest imaging technique with X-rays dates back to Röntgen's discovery: *radiography*, based on the differences in absorption of the various elements which constitute the probe. It has, however, weak contrast for low-Z elements and a poor resolution.

A completely different technique is *X-ray diffraction topography* which is based on the variations in the local contributions of the different volume elements of the sample to the intensity diffracted by the crystal when it is bathed in an incident spherical or plane wave. It is the subject of Chapter 17.

Cathodo- and *photoluminescence* topography provides a map of the distribution of certain types of colour centres related to impurities. It has often been used in conjunction with diffraction topography, in minerals such as natural or synthetic diamond (Ramachandran 1946*a*,*b*; Kiflawi and Lang 1974, 1976, 1977; Lang and Moore 1991; Wierzchowski *et al.* 1991; Frank *et al.* 1994; Lang *et al.* 2004) and in semiconductors, for instance in gallium arsenide (Fornari *et al.* 1985; Feng *et al.* 1988), in indium phosphide (Holt *et al.* 1989) or indium arsenide (Matveev *et al.* 1988), etc.

A method which is much more sensitive than absorption imaging for low-Z elements is *phase imaging*, despite the fact that, as shown in Section 2.7.3, the index of refraction is very close to one. It is briefly described in the next subsections.

15.6.2 *Phase contrast imaging*

The amplitude of a plane wave propagating in a medium of index of refraction n is

$$E = E_0 \exp[-2\pi i n k t],$$

where (Section 2.7.3)

$$n \approx 1 + \frac{\chi_{ro}}{2} + \mathrm{i}\frac{\chi_{io}}{2}$$

and t is the length travelled by the wave.

If the sample is thin and is constituted by light elements, one can neglect absorption as a first approximation. Of course, for a quantitative analysis it should be taken into account. The phase shift of the beam with respect to a beam having travelled through the same thickness in air is

$$\varphi = 2\pi k(n-1)t,$$

where n is considered as real when one neglects absorption.

As an indication, the values of t corresponding to a phase shift of 2π are given in Table 15.4 for a few materials and a few wavelengths. The difference between the values for carbon and water is of particular interest for biological samples.

The intensity $|E|^2$ is independent of the phase and, in order to visualize and to record phase differences, it is necessary to be able to convert them into differences in amplitude. The two main methods used in optics for this purpose are Michelson interferometry and phase contrast (Zernike 1935; see also Françon 1950; Born and Wolf 1993). Both have their counterpart for X-rays which are described in the following subsections.

Phase imaging is nowadays an important topic of research and has been applied to many types of objects: biological samples such as bone tissue or insects (Davis *et al.* 1995*a*; Gao *et al.* 1998; Raven *et al.* 1997; Mori *et al.* 2002), polymer fibres (Gao *et al.* 1995; Snigirev *et al.* 1995*b*) or boron fibres (Raven *et al.* 1996), silicon crystals with cracks (Cloetens *et al.* 1997*c*), aluminium alloys with SiC inclusions (Buffière *et al.* 1999), quasicrystals with holes (Mancini *et al.* 1998*a*,*b*; Gastaldi *et al.* 1999), ferroelectric domains (Hu *et al.* 1998; Rejmánková *et al.* 1998), *in situ*, time-resolved dendritic growth of binary alloys (Mathiesen *et al.* 1999), etc.

Table 15.4 Thicknesses corresponding to a phase advance of 2π, in μm, for a few materials (after Bowen 1996); the values of $n-1$ can be worked out from the wavelengths which are given in Table 15.1

Material	W$K\alpha_1$	Ag$K\alpha_1$	Mo$K\alpha_1$	Cu$K\alpha_1$	Cr$K\alpha_1$
C	186.6	69.73	54.98	25.25	16.94
H_2O	319.1	119.2	93.96	43.09	28.87
Si	152.7	56.86	44.72	20.37	13.62
Ge	75.10	27.97	22.21	10.67	7.179
LiF	144.7	54.05	42.58	19.53	13.08
NaCl	170.8	63.53	50.02	22.9	15.33

15.6.2.1 *Phase imaging with interferometry*

The possibility to use interferometry to detect phase changes was recognized immediately by Bonse and Hart (1965*b*), who imaged the constant thickness contours of a phase plate, and is discussed in further detail in the review by Hart (1975*b*)—see also Bowen (1996). It was applied by Ando and Hosoya (1971, 1972*a*) to obtain images of bone tissue and other inorganic and biological material. The sample is put inside any of the beams inside an *LLL* interferometer, for instance in *RR* or *TR* (Fig. 15.32). It introduces a phase shift and if it is not homogeneous, the shifts will vary across its area. A phase shifter constituted by a wedge is placed in the T beam. This external phase shift is used to reconstruct the phase modulation introduced by the sample. The analyser, A, acts as a phase-sensitive detector and the object is made visible. The method is very sensitive as intensity changes of less than a per cent can be detected and the corresponding depth of resolution is of the order of 1 μm (Bowen 1996). In practice, a scanning interferometer is used. The phase shifter is then a plane-parallel phase plate which can be rotated, thus varying its effective thickness and the phase shift it introduces. The phase modulation is measured by scanning the analyser. One records a set of interference patterns for different values of the phase introduced by the phase shifter. This phase is changed step by step. For each step an interference pattern is recorded with and without the sample. As absorption contrast always exists, it may be separated from the phase contrast by taking two images, with the analyser A translated parallel to the diffraction vector by $\pm d_{hkl}/4$ relative to the positions of intensity maxima; the mean intensity is then absorption contrast and the difference is a pure phase contrast image (Hart 1975*a*).

In recent years, the method has been applied to computer-assisted microtomography by Graeff and Engelke (1991), Momose (1995, 2002), Momose and Fukuda (1995), Momose *et al.* (1995, 1996*a*,*b*, 1998), Bonse and Busch (1996), Takeda *et al.* (1998), Beckman *et al.* (1997, 1999) and Bonse and Beckman (2001). In order to increase the field of view, Momose *et al.* (1997) and Yoneyama *et al.* (2002) have used a two-crystal Laue interferometer which has, however, the drawback of a stringent requirement to achieve very good mechanical stability. For the same purpose Takeda *et al.* (2000) have recently been able to fabricate a large Laue interferometer with a 25 × 25 mm field of view from a 10 cm diameter silicon ingot. Bonse and Beckman (2001) doubled the field of view by using non-planar three-beam diffraction, with the $hk\ell$ and $hk\bar{\ell}$ reflections (e.g. 553 and $55\bar{3}$).

15.6.2.2 *Diffraction-enhanced phase contrast imaging*

This method was proposed by Förster *et al.* (1980) and was called by them *Schlieren* method by analogy to a phase imaging technique used in optics (see, for instance, Born and Wolf 1993). The sample is put between the first crystal, called collimator, and a one- or two-crystal analyser, of a two-crystal diffractometer in the $(+, -)$ setting, either in the Bragg–Bragg geometry (Fig. 15.35(a)) or in the Laue–Laue or Bragg–Laue geometry. Without the sample put in, the rocking curve is the convolution of the intrinsic rocking curves of the collimator and the analyser (Section 15.2.3.1).

Let us introduce the phase object in the path of the beam. Local variations of the thickness or of the index of refraction in the sample will create phase gradients due to a prism effect and small deviations in the directions of the X-ray wavevectors will occur (Fig. 15.35(a)). Let us assume the second crystal is set so that the mean intensity corresponds to that of the position marked 1 on the rocking curve (Fig. 15.35(b)). For a small variation of the direction of the X-rays due to the local phase gradients from the sample, the angle of incidence on the analyser will correspond to position 2 or 3 on the rocking curve. This will result in a variation of the diffracted intensity which can be recorded on a photographic film or plate. A phase contrast image of the sample can thus be obtained. It is therefore the diffraction in the analyser which enables one to detect the variations of the phase. Strong contrast is expected where sharp phase gradients are present, for instance at the edges of the sample or of holes or inclusions within the sample and at various boundaries. The refractions at the opposite edges of a sample are opposite and the image of a symmetric object becomes asymmetric; to compensate for this effect it is useful to compare the images taken at positions 1 and 4 on the rocking curve.

Several groups have developed and expanded this technique in recent years: Somenkov *et al.* (1991), Ingal and Beliaevskaya (1995, 1996, 1997*a*,*b*), Bushuev *et al.* (1996, 1997, 1998), Davis *et al.* (1995*a*,*b*), Gao *et al.* (1995), Davis and Stevenson (1996), Nikulin (1997), Chapman *et al.* (1997) and Rigon *et al.* (2003). For a recent review of the state of the art, see Bravin (2003).

In practice, the experimental setup is more complicated than the one schematically represented in Fig. 15.35(a). Double-crystal collimators or analysers are used to improve the geometrical conditions. The rocking curve of the collimator should be much narrower than that of the analyser so as to achieve better resolution and, for this, asymmetric reflections are necessary, as shown in Section 15.3.3.1.

The interpretation of the contrast is also much more complicated than the primitive geometric explanation given above. Absorption, scattering and refraction of the beam occur at the same time inside the sample, resulting in a complex modified wavefront incident on the analyser. Gureyev and Wilkins (1997) have shown that two regimes should be distinguished: *phase contrast imaging* when the angular rotations due to the phase gradients are smaller than the rocking curve width of the analyser, and *refractometric imaging* when they are larger. The nature of the modified wave may thus be anything between a plane wave and a spherical wave. The formalism of Takagi's equations is well adapted to treat the propagation of such a wave (Davis 1996; Gureyev and Wilkins 1997; Bushuev *et al.* 1997).

The method has also been extended to microtomography, both experimentally (Bushuev *et al.* 1998; Hartman *et al.* 1997) and from the viewpoint of image formation (Pavlov *et al.* 2000).

15.6.2.3 *Direct phase imaging without an analyser, or propagation technique*

Direct imaging of phase contrast can be observed due to Fresnel diffraction at edges or sharp phase variations. This is a very simple method which is possible with highly coherent monochromatic radiation from a synchrotron as shown by two groups at

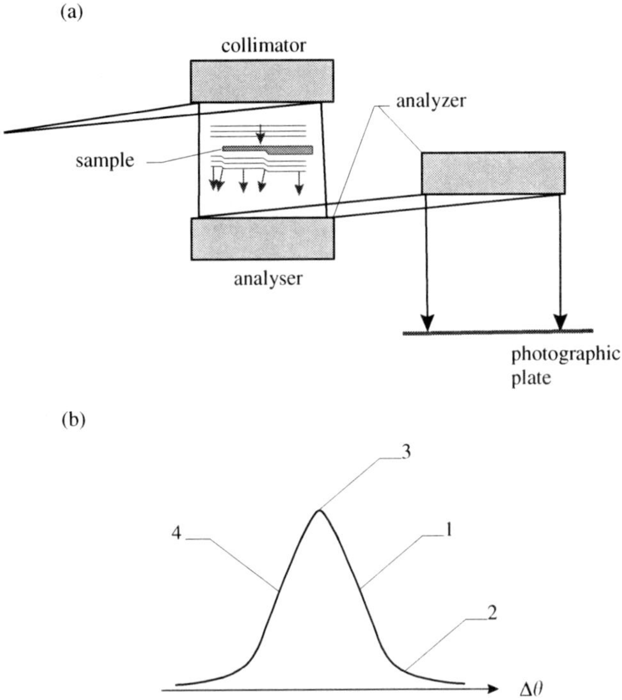

Fig. 15.35. Phase contrast imaging using a double-crystal analyser. The angular position of the analyser is given by $\Delta\theta$. (a) Experimental setup; the phase gradients due to the sample are schematically illustrated. (b) Rocking curve. Adapted from Davis *et al.* (1995*b*).

ESRF: Snigirev *et al.* (1995*b*), Snigirev (1996), Snigirev and Raven (1997), Hu *et al.* (1998) and Cloetens *et al.* (1996, 1999*a*), Baruchel *et al.* (1997), Rejmánková *et al.* (1998) and by a group at the Australian beamline at the Photon factory in Japan (Nugent *et al.* 1996). Direct phase contrast imaging is also possible in the laboratory with a microfocus (Wilkins *et al.* 1996; Stevenson *et al.* 1997; Gao *et al.* 1998; Gureyev and Wilkins 1998).

The resolution of phase contrast microscopy is improved by using a focusing device such as a Bragg–Fresnel lens (Snigirev *et al.* 1997) or an X-ray waveguide (Lagomarsino *et al.* 1997).

The propagation method is not restricted to imaging of phase edges. Continuous phase profiles also produce contrast with relatively large object-to-detector distances which depend on the spatial frequency range of the object phase profile, in agreement with the image formation by Fresnel diffraction discussed by Pogany *et al.* (1997). The recorded images are actually in-line holograms. A quantitative determination of the phase profile requires processing which may be based on intensity-transport equations (Gureyev *et al.* 1995, 1999; Gureyev and Wilkins 1998; Gureyev *et al.* 2001, 2002).

A quantitative holographic phase retrieval method, using simultaneously an arbitrary number of images, usually four, recorded at different distances has been demonstrated by Cloetens *et al.* (1997*b*).

15.6.2.4 *Microtomography*

The method has been extended to microtomography by taking several hundred projections from different angles and reconstructing three-dimensional images with the usual algorithms for absorption tomography (Raven *et al.* 1996, 1997; Cloetens *et al.* 1997*c*). In a more complete approach, for a quantitative three-dimensional reconstruction of the distribution of the index of refraction of the object, each projection is obtained by recording images at different distances and using the holographic phase retrieval method cited in the preceding subsection (*holotomography*, Cloetens *et al.* 1999*b*). Hard X-ray microtomography has provided new information on many different objects such as biological samples (Spanne *et al.* 1999), fractures of silicon carbide-reinforced aluminium alloys (Cloetens *et al.* 1997*c*; Buffière *et al.* 1999) and quasicrystals (Mancini *et al.* 1998*b*; Gastaldi *et al.* 1999). A review of the principles and applications is given in Baruchel *et al.* (2000).

16

Location of atoms at surfaces and interfaces using X-ray standing waves

16.1 Principle

When two coherent progressive waves overlap, they interfere, generating a set of standing waves. This is the case, for instance, when the second wave is due to the reflection, either Bragg or specular, of the first one from a surface; the standing wave pattern extends in the overlapping region, as well in air as inside the material where the reflection takes place (Fig. 16.1). How far the pattern extends inside the material depends on the penetration depth, which can be very small. The absorption of photons by matter is proportional to the intensity of the electric field and if an atom, either an impurity or an atom of the reflecting material, is located at an antinode of electric field, secondary emission of fluorescence radiation, photoelectrons, Auger electrons, etc. is excited. This emission can be recorded with a detector and, by analysing its energy, the nature of the atom can be found. By moving the position of the nodes and antinodes in a known way, it is possible to find where the atom is located with respect to the set of standing waves and, therefore, with respect to the crystal lattice or to its surface. One characterizes that way the type of site occupied by an adsorbed atom or the position of the surface or near-surface atoms with respect to those in the bulk.

It has been shown in Section 7.2 that in the case of Bragg reflection from a surface, the position of the nodes and antinodes is shifted by $d_{hkl}/2$ as the crystal is rocked through the reflection domain (Figs 16.2 and 7.11). When $-\infty < \eta_r < -1$, wavefields belonging to branch 1 of the dispersion surface are excited; the phase difference between the reflected and refracted beams is π for a non-absorbing crystal, and the nodes lie on the atomic planes. The phase changes progressively through the total reflection domain and is 0 when $+1 < \eta_r < +\infty$. The wavefields belonging to branch 2 are then excited and the antinodes lie on the atomic planes. The exact position of the nodes and antinodes has been discussed in Section 7.2 and their relation to the crystal structure in Section 5.8.3. Let an impurity atom lie at a certain position in between the lattice planes. For a certain angle of incidence, as the crystal is rocked through the reflection domain, the antinodes will pass through that position (Fig. 16.1). Secondary emission will then take place; if it is detected in synchronism with the recording of the rocking curve, that angle of incidence will be determined. From its value the phase of the reflected beam is measured and the position of the atom within the unit cell is deduced from it.

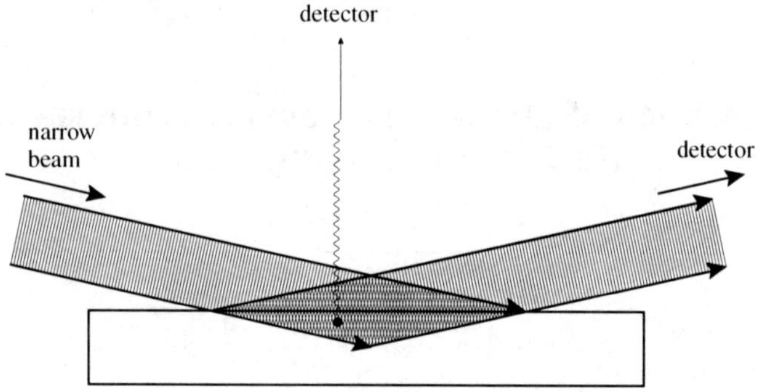

Fig. 16.1. Principle of the location of atoms with the standing wave method. Standing waves occur in the region where the incident and reflected beams overlap. If an atom is located at an antinode secondary emission of photons and electrons is excited.

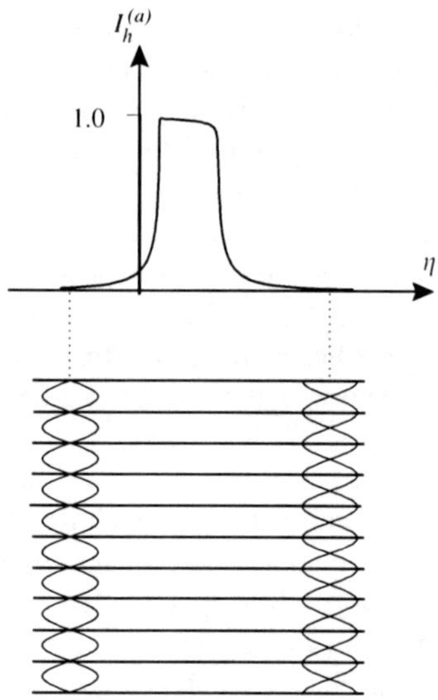

Fig. 16.2. Position of the nodes and antinodes of the standing waves in the Bragg geometry on each side of the rocking curve. The nodes lie on the atomic planes when $\eta \to -\infty$ (left-hand side), and it is the antinodes when $\eta \to \infty$ (right-hand side).

It is very important to note that the synchronous detection of the secondary emission with a separate detector provides thus a means to measure the phase of that reflected beam, which is impossible to do from the reflected intensity alone.

The variation of the intensity of the wavefields through the total reflection domain was first detected by Knowles (1956) for neutron diffraction, using γ-ray emission and by Batterman (1964*a*) for X-rays, using fluorescence scattering. The associated variation of thermal and Compton scattering were first observed by Annaka, Kikuta and Kohra (1965, 1966). The possibility of detecting the position and nature of foreign atoms in this manner was first suggested by Batterman (1969). A feasibility study was performed by Golovchenko *et al.* (1974) and the first accurate measurement of the position of surface and near-surface impurities was made by Andersen *et al.* (1976), who were also the first to realize that the standing wave pattern extends above the surface. Cowan *et al.* (1980) opened the way to the examination of adsorbed surface layers.

The first studies of the photoelectron yield due to the standing wave pattern were performed by Kikuta *et al.* (1975), Kruglov *et al.* (1977), Afanas'ev *et al.* (1977) and Takahashi and Kikuta (1977). Kruglov *et al.* (1978) used this method to detect the displacements of atomic planes in near-surface layers. As a recent example, Mukhamedzhanov and Bocchi (2000) have analysed in this manner a strained single layer of $Ga_{1-x}Al_xSb$ on gallium antimonide.

Auger electrons were used for instance by Durbin *et al.* (1985) and by Woodruff *et al.* (1987, 1988).

Trucano (1976), using fluorescence radiation, and Takahashi and Kikuta (1979), using photoelectrons, showed that the (111) and $(\bar{1}\bar{1}\bar{1})$ surfaces of a GaP crystal can be distinguished by means of the secondary emission associated with standing waves and that the polarity of that crystal can therefore be determined.

Standing waves can also be formed by a Bragg reflection from a non-crystalline layered structure as was shown by Barbee and Warburton (1984), Barbee (1985), Iida *et al.* (1985), Bedzyk *et al.* (1986) and Matsushita *et al.* (1986). Bedzyk *et al.* (1988) located in this manner a layer of zinc ions in a Langmuir–Blodgett thin film.

The first studies of X-ray standing waves at a reflecting mirror surface were done by Bedzyk (1988) and by Bedzyk *et al.* (1989) who analysed a Langmuir–Blodgett film deposited on a gold mirror.

The determination of atomic positions by X-ray standing wave analysis is accurate to a few per cents of a lattice spacing and the detection of foreign atoms is sensitive to a fraction of a monolayer. That technique is applied, essentially in semiconductors, to study surface reconstructed structures, the position of adsorbates, chemisorbed and implanted ions, ultrathin buried layers such as quantum wells, heterointerfaces, epilayers, bulk crystals, etc., but it has also been used for locating trace elements in minerals such as calcite (Qian *et al.* 1994), to evaluate the effect of isotope composition on the lattice parameters of Si and Ge (Sozontov *et al.* 2001) and to study the valence electronic structure of materials (Woicic *et al.* 2001; Kim *et al.* 2002). Reviews of the theory and applications can be found in Koval'chuk and Kohn (1986), Authier (1989), Malgrange and Ferret (1992), Zegenhagen (1993), Patel (1996), Patel and Fontes (1996), Lagomarsino (1996*a*), Zegenhagen *et al.* (1997), Vartanyants and Zegenhagen (1999) and Vartanyants and Koval'chuk (2001). For the applications of long-period standing waves from layered synthetic microstructures, see, for instance, Abruña *et al.* (1990) and Lagomarsino (1996*b*).

16.2 Theory

16.2.1 *Fluorescence yield*

The normalized intensity of the standing wavefield at a position $z = \mathbf{h} \cdot \mathbf{r}$ in the unit cell is given by (5.43) and (5.45):

$$\begin{aligned} I_{SW}(\Delta\theta, z) = \frac{|D|^2}{|D_o|^2} &= 1 + |\xi|^2 + 2C|\xi| \cos(2\pi \mathbf{h} \cdot \mathbf{r} + \psi) \\ &= 1 + |\xi|^2 + 2|\xi| \cos(2\pi z + \psi), \end{aligned} \tag{16.1}$$

where ψ is the phase of ξ and $\xi = D_h/D_o$ is given by (5.34). It is important to note that z is a numerical coordinate representing the position of a point in the unit cell as a fraction of the reflecting-plane spacing, d_{hkl} (see Fig. 5.14):

$$z = \frac{\Delta d}{d_{hkl}} \quad \text{modulo} \quad d_{hkl}.$$

Figure 16.3 represents the variations of I_{SW} with incidence angle for different values of z.

The probability of the secondary emission is proportional to the intensity of the standing wavefield. Following Hertel *et al.* (1985), Bedzyk and Materlik (1985) and Vlieg *et al.* (1991), the fluorescence yield normalized to the yield far from the reflection domain is

$$Y(\Delta\theta, z) = c \int_{\text{unit cell}} I_{SW}(\Delta\theta, z)\rho_u \, dz, \tag{16.2}$$

where c is a normalization constant and ρ_u is the distribution of the electron density of the fluorescent atom in the unit cell. Assuming that the secondary emission comes from the core electrons, one can write

$$\rho_u = \sum_{j=1}^{N} \delta(z - z_j), \tag{16.3a}$$

where N is the number of atoms per unit cell. If the fluorescent atoms are periodically distributed in the lattice,[1] the corresponding electronic density can be expanded in Fourier series. One has, by analogy with (2.36),

$$\rho_u(\mathbf{r}) = \sum_h A_h \exp[2\pi i \mathbf{h} \cdot \mathbf{r}]. \tag{16.3b}$$

[1] This is obviously true if the fluorescent atoms are the atoms of the crystal itself; it is also the case if they are substitutional atoms or atoms adsorbed near the surface and occupying specific sites with respect to the lattice; however, part of them are usually also randomly distributed and their contribution will not be coherent with that of the others.

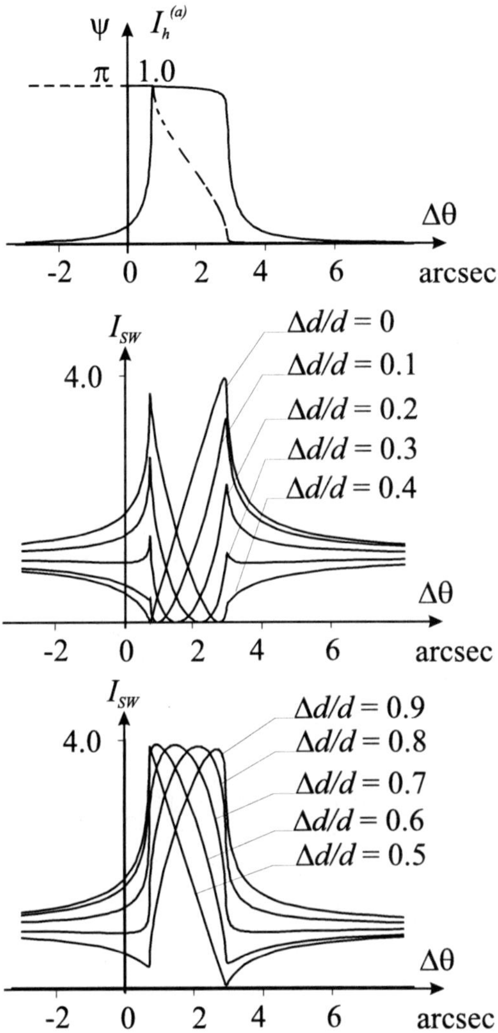

Fig. 16.3. Variations of the intensity of the wavefield at different positions in the unit cell. Silicon, 220 reflection, Mo$K\alpha$. The variations of the phase (dashed curve) and of the intensity of the reflected wave (solid curve) are represented in the top drawing.

It follows for the Fourier coefficients, using (16.3), that

$$
\begin{aligned}
A_h &= \int_{\text{unit cell}} \rho_u(\mathbf{r}) \exp -[2\pi \mathrm{i}\mathbf{h}\cdot\mathbf{r}]\,\mathrm{d}\mathbf{r} \\
&= \frac{1}{N}\sum_{j=1}^{N} \exp -[2\pi \mathrm{i}\mathbf{h}\cdot\mathbf{r_j}] = \frac{1}{N}\sum_{j=1}^{N} \exp -[2\pi \mathrm{i} z_j].
\end{aligned}
$$

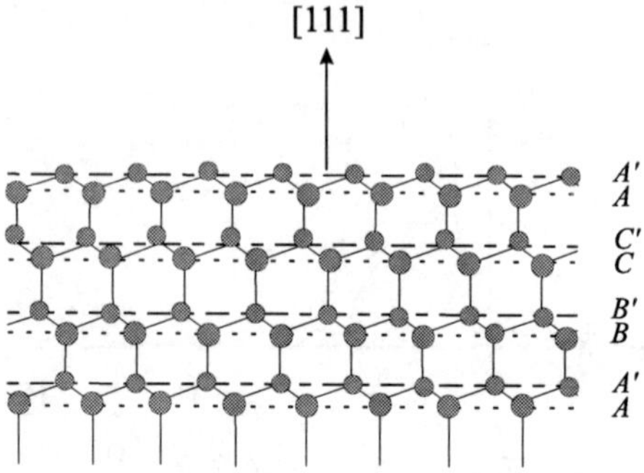

Fig. 16.4. Structure of silicon, projected on $(1\bar{1}0)$ plane.

For instance, let us consider the fluorescence signal from the silicon atoms in a silicon crystal and the 111 reflection. The structure of silicon is face-centred cubic with one atom at the nodes of the fcc lattice and one at half the tetrahedral sites; it can be described as a compact packing $AA'BB'CC'$ of $\{111\}$ planes, the distance between the A and A' layers being one third of that between the A' and B layers (Fig. 16.4).

The reflecting-plane spacing, d_{111}, corresponds to the distance between an A and a B layer, and the distance between an A and an A' layer is therefore $d_{111}/4$. Let us take the origin at mid-distance between the A and A' layers; the numerical coordinates of the two layers are $z_j = \pm\frac{1}{8}$ and the Fourier coefficient is

$$A_h = \frac{1}{2}\left[\exp\left(-\mathrm{i}\frac{\pi}{4}\right) + \exp\left(\mathrm{i}\frac{\pi}{4}\right)\right] = 0.707.$$

The fluorescence yield normalized per atom is, from (16.1), (16.2) and (16.3),

$$\frac{Y(\Delta\theta, z)}{Nc} = 1 + |\xi|^2 + 2|\xi|\frac{1}{N}\sum_{j=1}^{N}\cos(2\pi z_j + \psi). \tag{16.4}$$

It is important to note that it is given in terms of a coordinate in the direction normal to the lattice planes. A standing wave experiment with one set of lattice planes only will therefore only give the position of the fluorescent atom in that direction.

In order to analyse the experimental results, it is usual to express the normalized yield (16.4) in the following way:

$$\frac{Y(\Delta\theta, z)}{Nc} = 1 + |\xi|^2 + 2|\xi|F\cos(2\pi P + \psi), \tag{16.5}$$

where F and P are two adjustable parameters to fit the experimental curve by comparison with the theoretical curves (16.1); F is called the coherent fraction, namely the fraction of the fluorescent atoms localized at a given site in the unit cell, and P represents the position of that site. From (16.3) and (16.4) it follows, after replacing

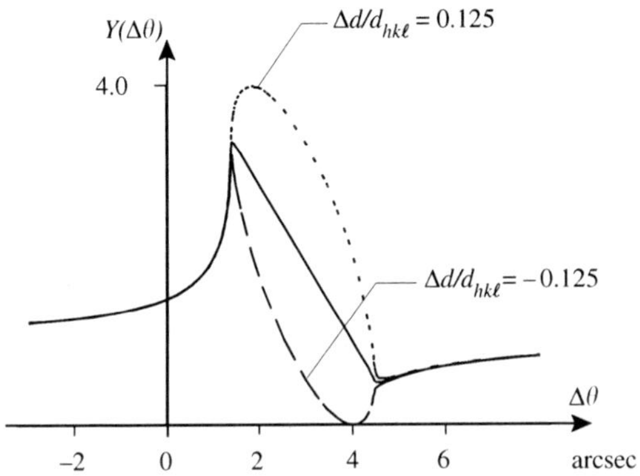

Fig. 16.5. Variations of the fluorescence yield of germanium atoms adsorbed on silicon (solid curve). 111 reflection, Mo$K\alpha$. The dashed curve corresponds to the fluorescence yield of atoms located on the A layer of the silicon structure (Fig. 16.4) and the dotted curve to that of atoms located on the A' layer.

the cosine term by a sum of exponentials, that

$$F \exp(2\pi \mathrm{i} P) = \frac{1}{N} \sum_{j=1}^{N} \exp(2\pi \mathrm{i} z_j)$$

$$= \frac{1}{N} \sum_{j=1}^{N} \exp(2\pi \mathrm{i} \mathbf{h} \cdot \mathbf{r_j}) = A_h = F_{SW}, \tag{16.6}$$

where F_{SW} is called the standing wavefield structure factor. A detailed development of the analysis of the standing wave pattern is given in Zegenhagen (1993).

As an example, let us consider the case of germanium atoms deposited on silicon (Patel *et al.* 1986). The adsorbed germanium atoms occupy the same positions as those of the silicon crystal, in an A and A' double layer above the silicon surface described in Fig. 16.4. The experimental curve can be fitted to the solid curve in Fig. 16.5 with $F = 0.67$ and $P = 1$. This result can be interpreted as follows: the fluorescence signal from the germanium atoms is the sum of the fluorescence signals from the atoms on the A layer located at $\Delta d/d_{hkl} = -\frac{1}{8} = -0.125$ (dashed curve) and of that from the atoms on the A' layer located at $\Delta d/d_{hkl} = +\frac{1}{8} = 0.125$ (dotted curve). This sum corresponds to the signal which would come from the position $\Delta d/d_{hkl} = 0$ and is in agreement with the experimental value $P = 1$. The fact that the coherent fraction F is slightly different from the theoretical value of 0.707 is due to the thermal vibrations (see Section 16.2.2); after correction the agreement is good to ± 0.01. Since the atoms oscillate around their mean position, it is to be expected that the occupancy rate of the sites should be less than 100 per cent. If one measures a sufficient number of structure factors (16.6), it is possible to reconstruct the electronic density distribution (16.3b) by inverse Fourier transform (*Standing waves imaging* — Zhang *et al.* 2004; Zegenhagen 2004).

16.2.2 *Influence of thermal vibrations*

Materlik and Zegenhagen (1984), Hertel *et al.* (1985) and Bedzyk and Materlik (1985) have shown that the standing wavefield structure factor should include a Debye–Waller factor as does the usual scattering structure factor:

$$F_{SW} = \frac{1}{N} \sum_{j=1}^{N} \exp(-M_j) \exp(2\pi \mathbf{h} \cdot \mathbf{r_j}),$$

where

$$M_j = \tfrac{1}{2} \langle 2\pi \mathbf{h} \cdot \mathbf{u_j} \rangle^2$$

and $\mathbf{u_j}$ is the displacement of atom j due to the thermal vibrations. For instance, the projection on [001] of the mean amplitude of the thermal vibrations of arsenic adsorbed on silicon, $\sqrt{\langle u_{004}^2 \rangle}$, is 0.14 Å (Franklin *et al.* 1994). It is interesting to compare this value with the height of the arsenic above the last plane of silicon atoms, $\Delta d = 1.40$ Å, as measured with the standing wave experiment; this shows that the thermal vibrations are quite important.

A careful analysis of the thermal vibrations requires the measurement of several Fourier coefficients of the atomic density distribution and, in particular, of higher-order coefficients. The rocking curve widths of the high-order reflections are very narrow, as shown in Section 7.1.2, which makes the standing wave measurements very difficult. To obviate this difficulty Huang *et al.* (1989), in a study of an epitaxial epilayer $NiSi_2$ on Si (111), have used reflections with Bragg angles close to 90° for which the Darwin width, $2\delta_{os}$, is very wide. Using synchrotron radiation, it is possible to tune the wavelength and therefore to choose the Bragg angle at will. Huang *et al.* used the 555 reflection from silicon at a wavelength for which the Bragg angle was $\theta_B = 89°$. The Darwin width is then $2\delta_{os} = 20$ arcsec. At this Bragg angle, the usual dynamical theory where the Ewald sphere is approximated by its tangential plane is still valid. For angles $89° < \theta_B < 90°$, a more rigorous theory must be used, such as that developed by Brümmer *et al.* (1979) or by Caticha and Caticha-Ellis (1982)—see Section 5.9.

For a review of the studies of thermal vibrations with the standing wave method, see Patel and Fontes (1996).

16.3 Bulk crystals

16.3.1 *Extinction effect*

The secondary emission at the standing wave antinodes comes from foreign atoms as well as from the atoms of the crystal itself. By selecting the energy of this secondary emission one may study either type. In the case of the atoms of the crystal itself, the secondary emission comes not only from the surface of the crystal but also from the bulk. The yield (16.4) must therefore be integrated over the depth t inside the crystal. It follows (Lagomarsino 1996*a*) that

$$Y_T(\Delta\theta) = \int_0^{t_o} \exp\Big[-\Big(\frac{\mu_e}{\gamma_o} + \frac{\mu_F}{\sin\beta}\Big)t\Big] \times \Big[1 + |\xi|^2 + 2|\xi|\frac{1}{N}\sum_{j=1}^{N}\cos(2\pi z_j + \psi)\Big]\,\mathrm{d}t$$
$$= \frac{1 - \exp[-(\mu_e/\gamma_o + \mu_F/\sin\beta)t_o]}{[\mu_e/\gamma_o + \mu_F/\sin\beta]} \times \Big[1 + |\xi|^2 + 2|\xi|\frac{1}{N}\sum_{j=1}^{N}\cos(2\pi z_j + \psi)\Big], \qquad (16.7)$$

where μ_e is the effective absorption coefficient of the diffracted beam (given by (5.46)) which, like ξ and ψ, depends on $\Delta\theta$, γ_o has its usual meaning, μ_F is the absorption coefficient of the secondary radiation (fluorescence radiation, photoelectrons, etc.), β is the angle between the crystal surface and the direction in which the secondary radiation is measured and t_o is the crystal thickness.

The result of the integration is that the signal from the surface layer is masked by the signal from the bulk as shown by Lagomarsino *et al.* (1984) in a study of bulk garnet crystals. In order to palliate this drawback and to be able to look at the near-surface layers only, it is necessary to have a situation where $\mu_F/\sin\beta$ is very large. The measured radiation then only comes from layers close to the surface.

One way to do this is to use a very small takeoff angle for the secondary emission. This is what was done for the first time by Patel and Golovchenko (1983) who used a takeoff angle of 7 mrad. Another way is to measure the photoelectrons which have a much higher absorption coefficient than fluorescence radiation (Takahashi and Kikuta 1979). Bedzyk *et al.* (1984) and Maslov *et al.* (1985) showed that it is possible to analyse the depth from which the photoelectrons are emitted. Afanas'ev *et al.* (1989) have suggested a *double-channel* technique in which photoelectrons and fluorescence radiation are simultaneously recorded, which enables one to limit the escape depth of the fluorescence radiation to that of the photoelectrons. Extinction effects can thus be limited while taking advantage of the high resolution of solid-state detectors to determine the nature of the atoms emitting the secondary radiation. This technique was applied by Mukhamedzhanov *et al.* (1991) to analyse the positions occupied by gallium and gadolinium in GGG garnets.

16.3.2 *Determination of the polarity of heteropolar crystals*

In a non-centrosymmetric crystal a face which is not parallel to an axis of even order (namely of order 2, $\bar{2} = m$, 4, $\bar{4}$, 6 or $\bar{6} = 3/m$) is polar; in other words the atoms occupying the top and bottom surfaces of the crystal are different. A typical example is given by the (111) and $(\bar{1}\bar{1}\bar{1})$ surfaces of GaAs, which are occupied by Ga and As

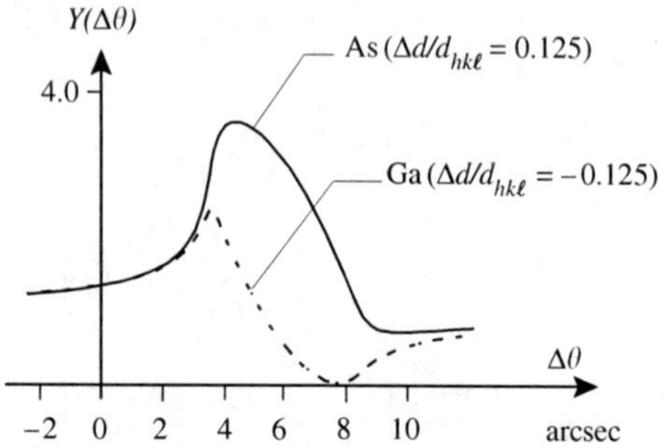

Fig. 16.6. Variations of the fluorescence yield of gallium (dotted curve) and arsenic atoms (solid curve). GaAs, 111 reflection, Mo$K\alpha$.

atoms, respectively (Fig. 7.9); a similar situation holds for GaP, InP, etc., It has been seen in Sections 2.7.2 and 7.1.2 that the intensities of the hkl and $\bar{h}\bar{k}\bar{l}$ reflections are different (Fig. 7.7). This expresses the non-validity of the so-called Friedel law for non-centrosymmetric crystals.

The fluorescence signal excited by standing waves gives the possibility to identify the nature of the atoms which occupy the surface of the crystal and provides a direct and independent way of determining the polarity. Let us consider the case of GaAs represented in Fig. 7.9. The positional parameter P for the atoms occupying the top plane of the top layer is 0.125 and the variations of the corresponding fluorescence signal are given by the solid curve, presenting a maximum, in Fig. 16.6. In the experiment by Patel and Golovchenko (1983), the analysis of this signal with an energy detector showed that these atoms were arsenic. The positional parameter P for the atoms occupying the bottom plane of the top layer is -0.125 and the variations of the signal from these atoms correspond to the dashed curve, presenting a dip, in Fig. 16.6. Its analysis showed that the atoms were gallium atoms. There was thus an unambiguous identification of the nature of the surface under study. A similar identification had been performed by Trucano (1976) and by Takahashi and Kikuta (1979) on GaP. They got round the problem of extinction, Trucano by taking advantage of the high absorption of the phosphorus $K\alpha$ wavelength and therefore of its short absorption depth, and Takahashi and Kikuta by using photoelectrons. As recent examples, Kazimirov *et al.* (1998) determined the polarity of GaN thin films on sapphire and Bedzyk *et al.* (2000) that of ferroelectric thin films of $PbTiO_3$ on $SrTiO_3$.

16.4 Solution to the surface registration problem

As mentioned in Section 16.2.1, the information provided by a standing wave measurement concerns the distance of the emitting atom from the origin of the unit cell in the direction normal to the reflecting planes. In order to locate its lateral position, at least one more measurement must be made with a set of reflecting planes lying in a different direction. By measuring the distance of the atom from two different sets of

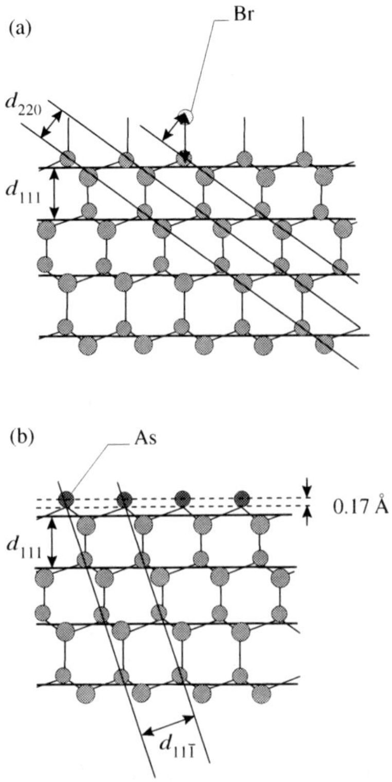

Fig. 16.7. Determination by triangulation of the position of an atom adsorbed on a (111) silicon surface (projection on a (1$\bar{1}$0) plane). (a) Bromine (Patel and Golovchenko 1983). (b) Arsenic (Patel *et al.* 1987).

lattice planes, its position is found in two dimensions, this is the *triangulation* method (Golovchenko *et al.* 1982).

For instance, Golovchenko *et al.* (1982) determined the position of bromine atoms adsorbed on a silicon (111) surface with two successive measurements: one in the direction perpendicular to the surface with the 111 reflection and another inclined with respect to the surface with the 220 reflection (Fig. 16.7(a)). They could show in that way that the bromine atom lies exactly at a silicon site. In another experiment, Patel *et al.* (1987), using the 11$\bar{1}$ reflection, could show that arsenic atoms adsorbed on a silicon (111) surface lie at *on top* substitutional positions located 0.17 Å above the normal positions (Fig. 16.7(b)).

More than two sets of reflecting planes can also be used; for instance, Berman *et al.* (1988) used the 111, 220 and 11$\bar{1}$ reflections to determine the position of the gold atoms adsorbed on a Si (111) surface and Eteläniemi *et al.* (1995) the 111, 220 and 422 reflections to study the adsorption of bromine on a Si (211) surface.

A very interesting possibility occurs if one uses a set of lattice planes perpendicular or nearly perpendicular to the surface and a grazing incidence (this geometry is studied in Chapter 8). In that case the position of the emitting atom parallel to the crystal surface can be determined (Cowan 1985). The first standing-waves effects at grazing incidence were observed in silicon by Afanas'ev *et al.* (1984, 1986*a*,*b*,*c*, 1989*a*)—see also the review by Afanas'ev *et al.* (1992)—and in germanium by Cowan *et al.* (1986) and Hashizume and Sakata (1989*b*). The first actual determination of the position of an adsorbate atom was by Jach and Bedzyk (1990) for iodine atoms adsorbed on a germanium (111) surface. The corresponding theory was developed by the same authors (Jach and Bedzyk 1993). The same type of measurements were performed by Sakata *et al.* (1993) and by Sakata and Hashizume (1995) for arsenic atoms adsorbed on a silicon (111) surface and by Sakata and Hashizume (1997) for the determination of the in-plane structure of a 100 Å $Ca_{0.39}Sr_{0.61}F_2$ epilayer film on a GaAs(111) substrate.

In a three-beam diffraction experiment, three sets of reflecting planes are simultaneously excited (see Chapter 9 for details). The intensity of the wavefield is

$$|D|^2 = |\mathbf{D_o} + \mathbf{D_h} \exp[2\pi i \mathbf{h} \cdot \mathbf{r}] + \mathbf{D_g} \exp[2\pi i \mathbf{g} \cdot \mathbf{r}]|^2.$$

Greiser and Materlik (1987) have shown that the normalized fluorescence yield (16.5) can be written as

$$\begin{aligned}\frac{Y(\Delta\theta, z, \Psi)}{Nc} &= 1 + |\xi_h|^2 + |\xi_g|^2 + 2|\xi_h| C_h F_h \cos(2\pi P_h + \psi_h)|\xi_g|^2 \\ &\quad + 2|\xi_g| C_g F_g \cos(2\pi P_g + \psi_g) \\ &\quad + 2|\xi_g \xi_h| C_{g\text{–}h}\, F_{g\text{–}h} \cos(2\pi P_{g\text{–}h} + \psi_g - \psi_h),\end{aligned}$$

where Ψ is the rotation angle around $\mathbf{h}$, ψ_h and ψ_g are the phases of $\xi_h = D_h/D_o$ and $\xi_g = D_g/D_o$, respectively, C_h, C_g and $C_{g\text{–}h}$ are the cosines of the angles between $\mathbf{D_h}$ and $\mathbf{D_o}$, $\mathbf{D_g}$ and $\mathbf{D_o}$, $\mathbf{D_h}$ and $\mathbf{D_g}$, respectively, and F_h, P_h, F_g, P_g and $F_{g\text{–}h}$, $P_{g\text{–}h}$ are the coherent fraction and positional parameters associated with the three reflections, **h**, **g** and **g–h**, respectively.

Greiser and Materlik used the 000, 333, 115 three-beam case in germanium and showed that two independent Fourier coefficients of the atomic density function can be determined. The possibilities of multiple reflection standing wave experiments were further discussed by Koval'chuk *et al.* (1987), Kohn (1988) and by Kazimirov *et al.* (1992, 1993).

Another, entirely different, way of solving the registration problem was suggested by Materlik *et al.* (1984), who coupled a measurement in the reflection geometry with a measurement in the transmission geometry where the sample was actually the analyser of an *LLL* interferometer. They applied it to the determination of bromine adsorbed on a (111) silicon surface.

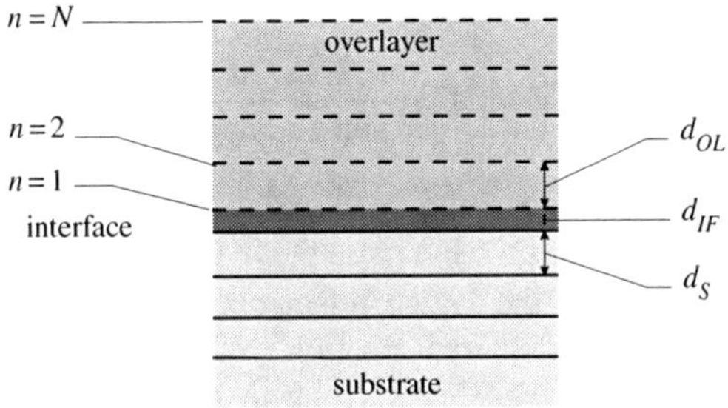

Fig. 16.8. Overlayer of lattice spacing d_{OL} above a substrate of lattice spacing d_s. d_{IF}: distance between the top plane of the substrate and the bottom plane of the overlayer.

16.5 Thin films and buried interfaces

16.5.1 *Simple model*

Epilayers, buried thin layers and interfaces have been for many years, and are still, a topic of major interest for technical applications: heterojunctions of III/V or II/VI compounds, metal (alkali, nickel, cobalt, iron, etc.) or fluoride overlayers epitaxially grown on silicon or gallium arsenide, etc. Standing wave techniques have proved to be a useful tool to investigate the structure and distortions of these layers and of their interface and are unique in the case of buried layers. After pioneering work by Akimoto *et al.* (1985, 1986) on the $NiSi_2$/Si system, Vlieg *et al.* (1986) proposed a simple method to analyse the structure of an overlayer. They assumed that the overlayer has a lattice spacing, d_{OL}, different from that, d_S, of the substrate and that the interface is sharp, the distance between the top plane of the substrate and the bottom plane of the overlayer being d_{IF} (Fig. 16.8). The standing wave pattern is that generated by the substrate, and the total fluorescence yield by the overlayer is

$$Y(\Delta\theta, z) = c \int_{\text{overlayer}} I_{SW}(\Delta\theta, z)\rho(z)\,\mathrm{d}z, \tag{16.8}$$

where

$$\rho(z) = \sum_{n=1}^{N} \delta\left(z - \frac{d_{IF}}{d_S} - n\frac{d_{OL}}{d_S}\right) \tag{16.9}$$

assuming there are N layers and that all the atoms occupy lattice sites.

Using (16.1), (16.8) and (16.9), it follows, for the total yield by the overlayer, that

$$\frac{Y(\Delta\theta)}{Nc} = 1 + |\xi|^2 + 2|\xi|F\cos(2\pi P + \psi), \tag{16.10}$$

where F and P are now given by

$$F = f_c \frac{\sin(\pi N \delta d/d_S)}{N \sin(\pi \delta d/d_S)},$$

$$P = \frac{d_{IF}}{d_S} + \frac{N-1}{2}\frac{\delta d}{d_S}.$$

$\delta d/d_S = (d_{IF} - d_S)/d_S$ is the lattice mismatch between the overlayer and the substrate, and f_c is called the coherent fraction.

This analysis was applied by Vlieg's group to Ni_2Si/Si (Vlieg *et al.* 1986) and to Co_2Si/Si (Fischer *et al.* 1987) and then by many authors to a variety of systems, such as $Ca_xSr_{1-x}F_2$ epilayers on GaAs(111) (Saitoh *et al.* 1988), Fe epilayer on Si(111) (Boulliard *et al.* 1992), interface of Cs or Rb adlayers on Si(111) (Lagomarsino *et al.* 1992), GaAs/InAs/GaAs(100) heterointerface (Giannini *et al.* 1993), buried ultrathin layer of InAs in GaAs(001) (Woicik *et al.* 1995; Gupta *et al.* 1999), CdTe/MnTe/CdTe(001) heterointerfaces (Boulliard *et al.* 1997), GaSe/Si(111) heteroepitaxy (Koëbel *et al.* 1997) among others.

Actually, as noticed by Zegenhagen *et al.* (1989) and Huang *et al.* (1989), in a study of epilayers of $NiSi_2$ on silicon, the atoms never occupy all lattice sites and a fraction, U, of them occupies randomly distributed sites. They are also submitted to the thermal vibrations. The coherent fraction, f_c, should therefore include a term $(1-U)\exp -M$.

If the number, N, of planes of the overlayer is known (as it can be when it is vapour deposited, for instance), there are three parameters to be determined, d_{IF}, d_{OL} and f_c, but a standing wave experiment only enables one to determine two parameters, F and P. To overcome this difficulty, Vlieg *et al.* (1986) have proposed to perform two experiments, with two different overlayer thicknesses. Saitoh *et al.* (1988) proposed two other ways: one, which also requires to know the number of planes, is to make two measurements on the same specimen, with two different reflections, for instance 111 and 333; the other one, which does not require this information, is to make two measurements, with 111 and 333, on a first specimen and one measurement with 111 on a second specimen of different thickness; it is then possible to determine d_{IF}, d_{OL}, f_c and the thicknesses of the two specimens.

In many cases more than one site is occupied and the parameters F and P of the total yield are expressed in terms of the positional parameters, P_i, of sites i and of their probabilities of occupation, f_i (Lagomarsino *et al.* 1992):

$$F \exp(2\pi \mathrm{i} P) = f_c \sum_i f_i \exp(2\pi \mathrm{i} P_i),$$

hence

$$F = f_c \left\{ \left[\sum_i f_i \sin(2\pi P_i) \right]^2 + \left[\sum_i f_i \cos(2\pi P_i) \right]^2 \right\}^{1/2},$$

$$\tan(2\pi P) = \frac{\sum_i f_i \sin(2\pi P_i)}{\sum_i f_i \cos(2\pi P_i)}.$$

16.5.2 *Calculation of the standing pattern in an overlayer with the dynamical theory*

In the simple model described in Section 16.5.1, the assumption is made that the standing wave pattern generated in the substrate extends unchanged in the overlayer. The situation is in general more complicated and the two parts, the bulk and the overlayer, cannot be considered independently, as is shown by the calculations of the intensity of the wavefield with the dynamical theory.

1. *Strained overlayer*: Authier *et al.* (1989) have considered the case of an overlayer with a continuous variation of the lattice parameter given by

$$\frac{\mathrm{d}u(z)}{\mathrm{d}z} = -\frac{\Delta d_S}{d_S}\frac{1}{1+\exp[(z-z_I)/C]}$$

following Bensoussan *et al.* (1987*a*). The interface is not sharp; it is characterized by its midpoint, of ordinate z_I, the origin being taken at the surface, and its steepness by the parameter C. The interface has a thickness of the order of $4.4 \times C$ and the variation of the lattice parameter in the overlayer above the interface is nearly saturated. Authier *et al.* calculated the intensity of the standing wavefield by solving Takagi–Taupin's equations numerically (the principle of the numerical integration is given in Section 14.3.3). Figure 16.9 shows its variations with the deviation parameter, η, and with depth inside the crystal. The result of the study is that the nodes and antinodes of the standing wave pattern can only be assumed to be hooked to the perfect bulk lattice planes for the first 10 or so atomic layers above the interface. Above these few lattice planes the nodes are no longer hooked to the bulk, except at the exact Bragg

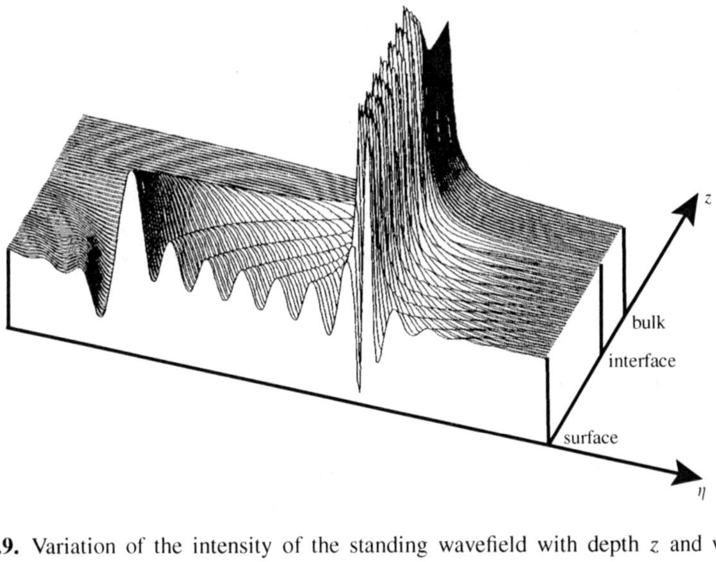

Fig. 16.9. Variation of the intensity of the standing wavefield with depth z and with the deviation parameter, η, in an overlayer, calculated assuming a continuous variation of the lattice parameter in an interface 0.05 mm thick; epilayer thickness 1.31 mm; lattice mismatch $\delta d/d_S = 7.8 \times 10^{-4}$; GaAs, 400 reflection, Cu$K\alpha$, extinction distance $\Lambda_B = 5.01$ μm.

angle for the substrate, and they start to drift, their spacing being neither that of the bulk lattice planes nor that of the epilayer lattice planes. This shows that the simple model cannot be applied for strained epilayers, but if an appropriate elastic model exists, the value of the fluorescence signal can be calculated by the dynamical theory for deformed crystals and compared to the experimental values.

2. *Thin buried layer*: Zheng *et al.* (1998) have analysed the case of an ultrathin buried layer where the thin crystal layer above the buried layer is perfect and identical to the bulk crystal; this upper thin layer is simply shifted with respect to the bulk and dynamical theory for perfect crystals can be applied. In the geometry which is used here and which is the reflection geometry, only one type of wavefield is normally excited, that one which propagates towards the inside of the crystal (of tiepoint P_1' or P_2' in Fig. 4.19), but Zheng *et al.* have shown that in this particular case the backwards wavefields propagating towards the outside of the crystal (of tiepoint P_1'' or P_2'' in Fig. 4.19) are excited in the upper thin part of the crystal. This induces anomalous transmission of the X-rays in the crystal and oscillations in the reflectivity profile. These effects must be taken into account when calculating the total fluorescence yield. Zheng *et al.* (2004) have applied dynamical theory to evaluate the yield in a thin $Zn_{1-x}Co_xO$ epilayer and in the $ZnO(00.\bar{1})$–O substrate.

16.6 Standing waves in deformed crystals

The analysis of the secondary emission due to the standing wave pattern according to the scheme described in Section 16.2 is only possible if the crystal is highly perfect. This is indeed the case in many applications where the adsorption of impurities at the surface of crystals such as silicon, germanium or gallium arsenide is studied. It has been seen in Section 16.5.2 that it is not true in the case of strained overlayers. In the case of rather imperfect crystals the solution proposed by Woodruff *et al.* (1987, 1988) for the study of chlorine chemisorption on copper (111) crystals is to use Bragg angles close to 90° for which the Darwin width is large. The sensitivity to defects is then much smaller. In addition, they scanned the rocking curve in energy rather than in angle. Backreflection was also used by Hashizume *et al.* (1992) to study epilayers of CaF_2 on GaAs and by Sugiyama *et al.* (1995) to investigate a reconstructed Sb/GaAs surface. As mentioned in Section 16.2.2, for angles $89° < \theta_B < 90°$, a more rigorous dynamical theory must be used, (Brümmer *et al.* 1979; Caticha and Caticha-Ellis 1982; Stepanov 1992; Grehk *et al.* 2001 – see Section 5.9).

When the whole crystal is submitted to a simple elastic deformation, the standing wave pattern can be calculated using Takagi's equations. This was done by Vartanyants *et al.* (1993*a*,*b*) for a crystal with a linear gradient of the parameter normal to the crystal surface, for bent crystals by Vartanyants *et al.* (1994) and Chukhovskii *et al.* (1996*b*) and, more generally, for a crystal with a uniform strain gradient by Chukhovskii *et al.* (1996*a*) who gave an exact analytical solution. They showed that one has to take into account the interferences with the wavefields whose paths have been bent back towards the incident surface (see Section 13.6.5 and Fig. 13.12); this results in fringes in the rocking curves and in oscillations in the intensity of the standing wavefield which will be reflected in the fluorescence yield.

16.7 Standing waves due to specular reflection

The formation of X-ray standing waves at a mirror has been discussed by Bedzyk *et al.* (1989). Using the notations of Section 2.10 where specular reflection of X-rays was presented, let $E_o^{(a)}$ and E_s be the amplitude of the incident and reflected waves, respectively. Their ratio is, from (2.61), and for the σ-component:

$$R = \frac{E_o^{(a)}}{E_s} = \frac{\Delta\psi - \sqrt{\chi_o + \Delta\psi^2}}{\Delta\psi + \sqrt{\chi_o + \Delta\psi^2}},$$

where $\Delta\psi$ is the glancing angle; it can also be written, neglecting absorption in the mirror, as

$$R = \frac{x - \mathrm{i}\sqrt{1 - x^2}}{x + \mathrm{i}\sqrt{1 - x^2}} \quad (0 < x < 1)$$

$$= \frac{x - \sqrt{x^2 - 1}}{x + \sqrt{x^2 - 1}} \quad (x < 1),$$

with $x = \Delta\psi/\omega_c$; $\omega_c = \sqrt{-\chi_o}$ is the critical angle. For $0 < x < 1$, the reflectivity can be written as

$$R = \exp(-\mathrm{i}\phi_s),$$

with $\cos(\phi_s/2) = x$. Figure 16.10(a) represents the variations of the reflectivity and of its phase with glancing angle.

The normalized intensity of the standing wavefield is given by

$$I_{SW} = \left| \exp\left(- 2\pi \mathrm{i}\mathbf{K_o^{(a)}} \cdot \mathbf{r}\right) + R \exp\left(- 2\pi \mathrm{i}\mathbf{K_s} \cdot \mathbf{r}\right)\right|^2,$$

where $\mathbf{K_o^{(a)}}$ and $\mathbf{K_s}$ are the wavevectors of the incident and specularly reflected waves, respectively. Their difference is normal to the crystal surface and

$$\left(\mathbf{K_s} - \mathbf{K_o^{(a)}}\right) \cdot \mathbf{r} = \frac{2 \sin \Delta\psi}{\lambda} z.$$

Setting

$$\Lambda = \frac{\lambda}{2 \sin \Delta\psi},$$

the intensity of the standing wavefield can be written as

$$I_{SW} = \left[2 + 2 \cos\left(\phi_s + 2\pi \frac{z}{\Lambda}\right)\right] \quad (0 < x < 1)$$

$$= \left[1 + R^2 + 2R \cos\left(2\pi \frac{z}{\Lambda}\right)\right] \quad (x > 1),$$

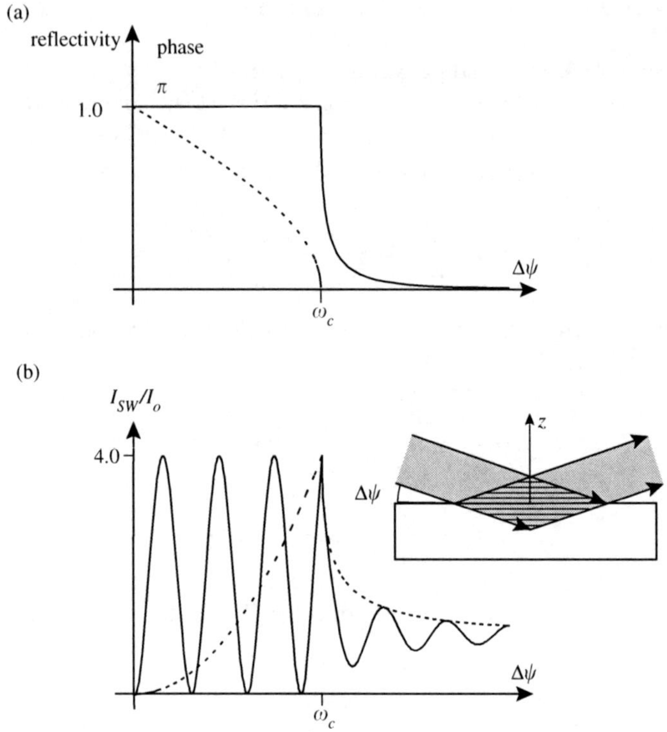

Fig. 16.10. Standing waves at a reflecting mirror. (a) Reflectivity and phase of the reflected wave for zero absorption. (b) Intensity of the standing wavefield at a height $z = 3\Lambda_c$ (solid curve) and at the surface (dashed curve); Λ_c is the period at the critical angle.

which shows that the period of the standing waves depends on the angle of incidence. Letting $\Lambda_c = \lambda/2\sin\omega_c$ be the period at the critical angle and putting $n = z/\Lambda_c$, the intensity can be written, following Bedzyk *et al.* (1989), as

$$\frac{I_{SW}}{I_o} = [2 + 2\cos(\phi_s - 2\pi nx)] \quad (0 < x < 1)$$

$$= [1 + R^2 + 2R\cos(2\pi nx)] \quad (x > 1),$$

with $x = \Lambda_c/\Lambda$.

Figure 16.10(b) represents the variations of the intensity of the standing wavefield for $z = 3\Lambda_c$ (solid line) and $z = 0$ (dashed line). The curve for $z = 3\Lambda_c$ shows the angle dependence of the standing wavefield; the maximum presented for $\Delta\psi = \omega_c$ at the surface ($z = 0$) is due to the generation of the evanescent wave (Marra *et al.* 1979; Becker *et al.* 1983). For a recent review, see Zheludeva *et al.* (1997).

17

X-ray diffraction topography

17.1 Introduction

Diffraction topography is an imaging technique visualizing the variations of intensity diffracted by various points of the surface or the bulk of a crystal. These variations may be due to the fact that one region satisfies Bragg's condition for diffraction while another does not—this is called *orientation* contrast; they may also be due to the fact that the local crystal perfection and therefore the local reflectivity varies from point to point—this is called *extinction* or *diffraction* contrast. Since a topograph is an image, its most important property is the *spatial resolution* that depends on various parameters and which will be discussed further down; it must be as good as possible, of the order of a few microns or less. To take full advantage of the spatial resolution of the method, the resolution of the detector must be as good. Two types of detectors are used: fine-grain films or plates and direct-display systems used for real-time observations. The best resolution is obtained with thick emulsions nuclear plates (25 μm or 50 μm thick); their processing technique is described in Lang (1978) and Bowen and Tanner (1998). The developments of direct-display systems up to 1977 are described in Hartmann (1977). Modern developments are reviewed by Arndt (1986, 1990, 1999*b*) for position-sensitive detectors, by Chikawa (1999) for TV cameras and by Amemiya and Chikawa (1999) for storage phosphors.

As it is based on Bragg diffraction, a topograph is also sensitive to misorientations and it must have a very good *angular resolution*; spatial and angular resolutions are not always compatible, in particular for images of dislocations. One distinguishes different types of topographic settings. On the one hand, there is:

(1) *reflection topography* if the sample is set for the reflection geometry or

(2) *transmission topography* if the sample is set for the transmission geometry

and, on the other hand, whether the setting is by reflection or by transmission:

(1) *single-crystal topography* if there is no monochromator and the diffracted intensity is the integrated intensity, because the angular or the spectral divergence of the incident beam is wider than the width of the rocking curve (the latter case is called *white beam topography*);

(2) *double- or multiple-crystal topography* if the angular and spectral widths of the incident beam issued from an appropriate beam conditioner are significantly smaller than the width of the rocking curve.

After the pioneering work by Berg (1931), Ramachandran (1944), Barrett (1945) and Wooster and Wooster (1945), diffraction topography really started when perfect crystals became available in which single dislocation lines could be observed (Lang 1957, 1958; Barth and Hosemann 1958; Borrmann *et al.* 1958; Bonse 1958; Bonse and Kappler 1958; Newkirk 1958*a*,*b*). It is a versatile, non-destructive characterization technique which has been applied to very many materials and many types of problems. More than 2000 papers have been devoted to it, including many reviews and several books. A few of them are selected here: Authier (1977*c*, 1978), Miltat (1980), Authier *et al.* (1996), Baruchel *et al.* (2002), Bowen and Tanner (1992, 1998), Dudley (1997), Hildebrandt (1975), Lang (1978, 1999), Miltat and Dudley (1990), Sauvage and Pétroff (1980), Steiner *et al.* (1990), Tanner (1996), Tanner and Bowen (1980, 1992), Tuomi (2002), Weissmann *et al.* (1984) and Baruchel and Schlenker (1984) and Schlenker and Baruchel (1986, 1996, 1999) for neutron topography. It should also be mentioned that many papers devoted to diffraction topography can be found in the Proceedings of the XTOP meetings: *J. Physics D (Applied Physics)*, 1993, Vol. 26 (ed. J. Baruchel); 1995, Vol. 28 (ed. R. Köhler and H. Klapper); 1999, Vol. 32 (ed. B. K. Tanner); 2001, Vol. 34 (ed. J. Gronkowski); 2003, Vol. 36 (ed. J. Baruchel) and *Il Nuovo Cimento*, 1997, Vol. 19D (ed. C. Bocchi, C. Ferrari, S. Lagomarsino and L. Tapfer). The reader may also consult the ESRF ID19 beamline home page at http://www.esrf.fr/.

17.2 Single-crystal reflection topography (Berg–Barrett technique)

17.2.1 *Principle*

This method was first used by Berg (1931), who studied sodium chloride crystals. He showed the presence of scratches on the surface, observed the subgrain structure of the crystal and separated the effects due to lattice parameter variations from those due to rotations of the lattice planes. The method was applied by Barrett (1945) to metallography and a similar setting using a simultaneous oscillation of crystal and film was applied by Wooster and Wooster (1945) to the study of imperfections in diamond. A slightly different technique using white radiation was developed by Schulz (1954) to measure tilts between grains in polygonized structures.

The Berg-Barrett technique was improved by Newkirk by using a microfocus and a small sample-to-specimen distance in order to increase the resolution. He could thus observe subgrain boundaries in iron–silicon (Newkirk 1958*a*) and reveal the outcrops of single dislocations at the surface of silicon crystals (Newkirk 1958*b*, 1959) and dislocation outcrops and subgrain boundaries in lithium fluoride crystals (Newkirk 1959). Soon after, Yoshimatsu and Kohra (1959, 1960) also observed dislocation outcrops and subgrain boundaries in lithium fluoride crystals. Reviews of the method are given in Austerman and Newkirk (1967), Newkirk (1970), Armstrong and Wu (1973), Armstrong (1980), Bowen and Tanner (1998) and Lang (1999).

The principle of the method is very simple and is described in Fig. 17.1(a). In general no monochromator is used. An incident beam coming from a fine focus or

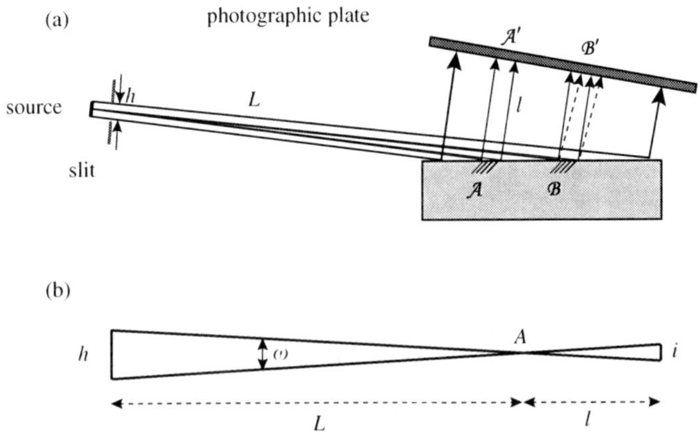

Fig. 17.1. Principle of the Berg–Barrett technique. (a) Setup: $\mathcal{A}$ is an area which has a different reflecting power from that of the surrounding regions; $\mathcal{B}$ is an area where the lattice planes have a different orientation or a different lattice spacing. (b) Geometrical resolution; i is the image on the photographic plate of a point A of the sample.

from a synchrotron source and collimated by a slit impinges on a crystal set for Bragg diffraction. The diffracted beam is received on a photographic plate. The resolution of the method, namely the size of the image of a given defect in the sample, depends on several factors:

1. *The geometrical resolution*: The size, i, of the image of a point A of the crystal on the detector is (Fig. 17.1(b))

$$i = \frac{hl}{L} = \omega l, \tag{17.1a}$$

   where h is the size of the source in the direction perpendicular to that of the diffraction plane or that of the limiting slit, L is the source-to-sample distance, l is the sample-to-plate distance and ω is the divergence of the beam received by A. If, however, the natural divergence of the beam, Ω, is smaller than ω, as can be the case with synchrotron radiation, (17.1a) is replaced by

$$i = \Omega l. \tag{17.1b}$$

2. *The photographic resolution*: It depends on the projection of the trace of the reflected beam on the detector and on the photographic process or on the pixel size of a direct viewing detector. For a 25 μm thick photographic emulsion, the trace of the beam reflected by a point of the crystal is 1 μm wide if the beam is 2° off the perpendicular to the plate. The plate must therefore be placed perpendicular to the reflected beam.
3. *The diffraction process*: which will be discussed later.

For the best geometrical resolution one should therefore have:

1. A small source size: In the laboratory it is recommended to use a microfocus tube; third-generation-synchrotron source sizes are very small (see Section 15.1.2.3).
2. A small sample-to-plate distance and a small penetration depth inside the crystal. To meet these conditions, one should use:
   - a grazing incidence
   - a long wavelength
   - a high Bragg angle, so that the photographic plate can be put normal to the reflected beam while being as parallel as possible to the crystal surface, in order to reduce distortions; the best results are obtained if the Bragg angle is close to $\pi/4$. If a thin emulsion is used, the film can be placed parallel to the crystal surface to avoid distortions or parallel to the incident beam to increase resolution.

With a synchrotron source having a natural divergence $\Omega < \omega$, the sample-to-detector distance, l, can be large, which is very convenient for the experimental setup.

17.2.2 *Image formation*

17.2.2.1 *Main origins of the contrast*

Each point of the crystal receives a beam whose angular and spectral widths are larger than the width of the rocking curve and the reflected intensity is integrated over the area of the rocking curve. The expressions of the integrated intensity for an ideally imperfect and for a perfect crystal have been given in Sections 3.5 and 4.9, respectively. Two extreme cases can be distinguished for the origin of the contrast:

1. *Diffraction, or extinction, contrast.* Let us assume that the crystal is nearly perfect and that $\mathcal{A}$ (Fig. 17.1) is an area where the crystal perfection is less than that of the surrounding regions. It acts as a small imperfect crystal and its image $\mathcal{A}'$ is an area of higher diffracted intensity than the surrounding regions, giving rise to *diffraction* contrast. As an example, Fig. 17.2 compares a Berg–Barrett topograph and an optical micrograph of the surface of a lithium fluoride crystal. The etch pits on the optical micrograph are at the outcrops of dislocations. One can see a few isolated dislocations, a roughly horizontal very low-angle subgrain boundary in which the dislocations can be resolved, and a roughly vertical low-angle grain boundary in which the dislocations cannot be resolved. The angle between subgrains is smaller than the divergence of the incident beam and all the subgrains are visible on the X-ray topographs. The contrast is due to the strong reflectivity of the distorted areas around the dislocations; it is pure diffraction contrast. The image of the vertical subgrain boundary is darker because it contains a higher dislocation density and the deformations around it are stronger; its left edge is sharp, corresponding to the intersection of the boundary with the crystal surface, and its right edge is fuzzy as it comes from the inner part

(a) (b)

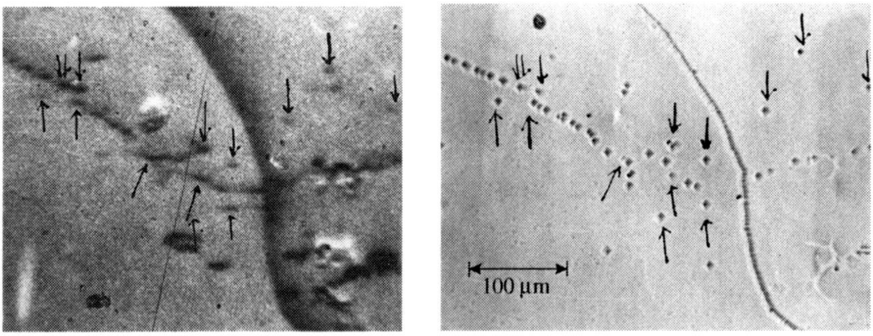

Fig. 17.2. Dislocation outcrops at the surface of a LiF crystal. (a) Berg–Barrett topograph; Cr$K\alpha$, 220 reflection. (b) Optical micrograph; arrows indicate etch pits at the outcrop of the dislocations. Courtesy K. Kohra.

of the crystal. The width of the image is determined by the penetration depth and the angle between the subgrain boundary and the surface. Surface damage is also observed.

2. *Orientation contrast.* Let $\mathcal{B}$ (Fig. 17.1) be a region where the orientation or spacing of the lattice planes is different from that of the surrounding regions. If the angular and spectral widths of the incident beam are smaller than the effective misorientation or the difference in Bragg angle, this area is out of reflection and its image $\mathcal{B}'$ on the photographic plate is white (Fig. 17.1); this is *orientation* contrast.

If the incident beam is monochromatic, the crystal must be rotated in order to bring this area $\mathcal{B}$ in reflection and this provides a means to measure the misorientation. If one wishes to have an image of all the areas at the same time, or if the crystal is bent or warped, crystal and plate can be oscillated simultaneously. This has been proposed for instance by Auleytner and Zielinska-Rohozinska (1963). Their method was improved by Kohra and Takano (1968), who used a monochromatic divergent beam produced with a Johansson monochromator (described in Section 15.4.4.1). They applied it to the study of lithium fluoride crystals with very misoriented subgrains and silicon wafers with an oxide layer. In order to measure misorientations they took several topographs after successive rotations of the crystal. The same setting was used by Polcarová and Brádler (1987, 1988) to measure strain fields in deformed crystals of Fe-6 at.% Si.

If the incident beam is polychromatic as in the method developed by Schulz (1954) or at a synchrotron, the misoriented areas or subgrains are diffracted by slightly different wavelengths and therefore at different Bragg angles. Depending on the relative values of the Bragg angle for two neighbouring subgrains, their images can slightly overlap or be separated by a gap. The image formation by neighbouring subgrains and the measurement of their misorientations was discussed by various authors, for instance Aristov *et al.* (1974), Aristov and Shulakov (1975), Wu and

Armstrong (1975), Armstrong *et al.* (1980) and Kulda and Polcarová (1981). Different techniques are possible, such as taking successive topographs with different sample-to-plate distances or different reflections or placing an array of vertical wires in the incident beam.

17.2.2.2 *Other possible origins of the contrast*

Two examples will be given.

One is purely geometric. If the surface is rough, any irregularity will cast a shadow, giving rise to a black and white contrast. It will be the more pronounced, the smaller the incidence angle of the beam on the surface and the smaller the penetration depth of the X-rays in the crystal.

The other example is related to the breakdown of Friedel's law in absorbing crystals and is observed with domains of different polarity. For instance, barium titanate ($BaTiO_3$) is ferroelectric below 120 °C. It presents 90° domains where the directions of the electric polarization are at 90° in neighbouring domains and 180° domains where the directions of the electric polarization are at 180° in neighbouring domains. In the former case, the lattice parameter is slightly different in each domain, but in the latter case they are identical. Since the polarity is inverse in each domain, the reflection is hkl in one domain and $\bar{h}\bar{k}\bar{l}$ in the neighbouring domain. If the crystal is highly absorbing, as in barium titanate, Friedel's law does not apply and there is a strong contrast between the two domains. This effect was observed by Niizeki and Hasegawa (1964). It was also observed by Huang *et al.* (1996) on KTA with white beam synchrotron topography.

17.2.3 *Penetration depth*

The defects observed by reflection topography lie within a certain depth inside the crystal. This penetration depth depends on the degree of perfection of the crystal. If the crystal is perfect, the penetration depth, z_e, is given by the extinction distance within the total reflection domain. It was shown in Section 4.9.2 that it is the distance from the crystal surface for which the attenuation factor $\exp(-\mu_e z/\gamma_o) = \exp(-2\pi z/\Lambda_B)$ is equal to $1/e$:

$$z_e = \frac{\Lambda_B}{2\pi}, \tag{17.2}$$

where $\Lambda_B = \lambda\sqrt{\gamma_o|\gamma_h|}/\sqrt{\chi_h\chi_{\bar{h}}}$ with $\gamma_o\gamma_h = \cos(\psi_n + \theta_B)\cos(\psi_n - \theta_B)$. ψ_n is the angle between the normal to the crystal surface and the lattice planes (see Section 4.7.3).

The penetration depth decreases as the incidence angle decreases; it is represented in Fig. 17.3 for silicon, 220 and Cu$K\alpha$. When $\psi_n \to -\pi/2 \mp \theta_B$, one has a glancing incidence or emergence, respectively, and evanescent waves are generated (Cowan 1985).

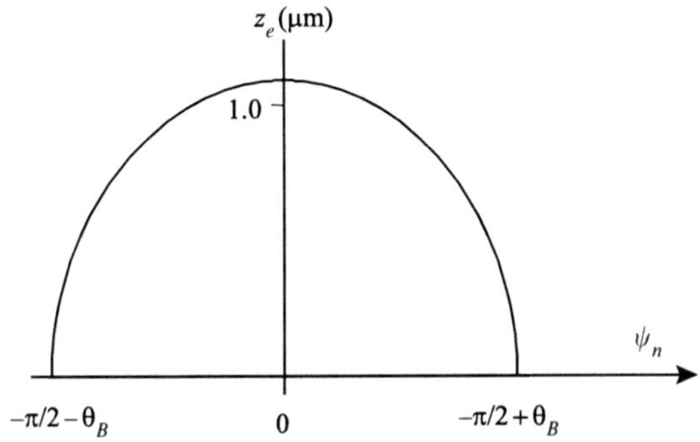

Fig. 17.3. Variation of the penetration depth with the angle ψ_n between the normal to the crystal surface and the lattice planes. Silicon, 220 reflection, Cu$K\alpha$.

For that part of the beam which is out of the diffraction range of the crystal, or if the crystal is imperfect, the penetration depth is

$$z_e = \frac{\gamma_o}{\mu_o},$$

where μ_o is the normal absorption coefficient. The penetration depth due to absorption is always much larger than that due to extinction. In the case of silicon and Cu$K\alpha$, it is 28 μm. The relation between the visibility of dislocations and these two depths has been discussed, for instance, by Halliwell *et al.* (1973) and by Shul'pina and Argunova (1995).

17.2.4 *Stereographic views*

The images of defects, dislocations or subgrain boundaries come from a certain depth inside the crystals. If one takes two images with different reflections, the defects are viewed from two different angles and the two images can thus constitute a 'stereographic pair'. This is common practice for transmission topographs (Lang 1959*a*,*b*); in that case the two reflections chosen are hkl and $\bar{h}\bar{k}\bar{l}$ and the stereo 'convergence' angle is twice the Bragg angle. It is also possible to take two topographs with the same reflection but with two different orientations of the crystal around the reciprocal lattice vector **h** (Haruta 1965). In the case of reflection topographs one can take two topographs with the hkl and $\bar{h}\bar{k}l$ reflections or two reflections with two orientations of the crystal around **h** (Tu and Vreeland 1974; Vreeland 1976). In the latter case, the convergence angle is $180^\circ - 2\theta_B - \phi$, where θ_B is the Bragg angle and ϕ is the angle between the lattice planes of the two reflections. For instance, Vreeland (1976) took stereo pictures of dislocations in zinc using the $10\bar{1}3$ and $\bar{1}013$ reflections. An important difficulty in reflection topography arises from the fact that the photographic plate or film is usually not parallel to the crystal surface and the images are distorted.

This distortion must be removed before observing the stereo pair. This was done by Vreeland (1976) by computer image processing of the topographs.

17.2.5 *Applications*

Although reflection topography provides images of the outcrops of dislocations, it is best adapted to measure lattice parameter variations and misorientations. A few recent examples are given here of the most frequent applications:

1. *Real-time observation of crystal growth using synchrotron radiation*: germanium epilayer on GaAs, silicon, Fe-6 at.% Si (Queisser *et al.* 1981).
2. *Characterization of absorbing materials*: $BaTiO_3$ (Petrov *et al.* 1989); supraconductors (Vakhrushev *et al.* 1991*a*,*b*), tungsten (Katoh *et al.* 1991), $Hg_{1-x}Cd_xTe$ (Williams and Vere 1987; Sharma *et al.* 1993).
3. *Study of the surface structures of semiconductors*: (Argunova *et al.* 1987*a*,*b*; Imamov *et al.* 1989; Kshevetskii *et al.* 1990; InSb (Fodchuk and Kshevetskii 1992); Sharma *et al.* 1993); GaAs (Minari and Billia 1993).
4. *Plastic deformation*: HgI_2 (Gits and Authier 1982); MgO (Armstrong 1984), niobium and zinc (Bilello 1984); NaCl (Fries *et al.* 1984); molybdenum and niobium (Hmelo *et al.* 1984); Fe-6 at.% Si (Polcarová and Brádler 1987, 1988); zinc (Blish and Vreeland 1969) and ZrO_2 (Fries *et al.* 1989; Guiberteau *et al.* 1994).
5. *Ferromagnetic domains*: Cobalt zinc ferrite (Merz 1960); NiO (Blech and Meieran 1966; Yamada *et al.* 1966; Kranjc 1968, 1969; Kurosawa *et al.* 1974; Nakahigashi *et al.* 1975); CoO (Saito *et al.* 1966; Kurosawa *et al.* 1974; Nakahigashi *et al.* 1975); Fe 3% Si (Stephenson *et al.* 1980).
6. *Ferroelectric domains*: $BaTiO_3$ (Niizeki and Hasegawa, 1964; Kranjc, 1969).

17.3 Single-crystal transmission topography

17.3.1 *Early history*

The early history of topography is recorded in Lang (1993, 1995). The expression *diffraction topograph* was introduced by Ramachandran (1944, 1946*a*,*b*), who used a simple transmission technique to study defects in diamond crystals. The method was improved by Lang (1957), who used a collimated beam and characteristic radiation (Fig. 17.4). He could thus observe the image of single dislocations in a thin silicon crystal with Ag$K\alpha$ radiation (Lang 1958). A correlation with an infrared micrograph of the same silicon crystal after copper decoration enabled the unambiguous identification of the observed lines as dislocations. Independently, Borrmann *et al.* (1958) also observed the images of single dislocation lines in silicon, in the reflected beam of a wide-angle diagram taken with Cu$K\alpha$ (the method is described in Section 1.10.1).

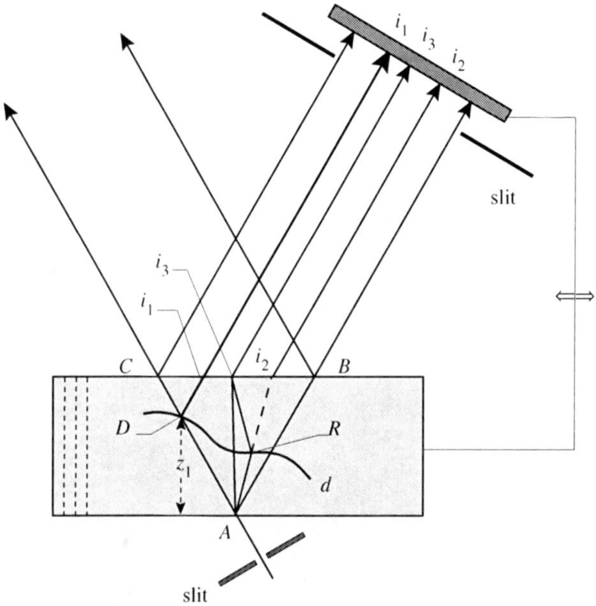

Fig. 17.4. Principle of single-crystal transmission topography. d: projection of a dislocation line on the plane of incidence; D: intersection of the dislocation with the plane of incidence; i_1: direct image; i_2: dynamical image; i_3: intermediary image. Section topographs: the crystal is immobile; projection topographs: the crystal and the photographic plate are traversed simultaneously.

While in Lang's case the crystal was not very absorbing, it was very absorbing in Borrmann's case and the origin of the image contrast was completely different, as will be explained in the following subsections. Various variants of these techniques were published soon afterwards (Barth and Hosemann 1958; Gerold and Meier 1959; Ishii and Kohra 1959).

In the topographs obtained by Lang and Borrmann that have been just mentioned, a small volume only of the crystal was observed, a *section*, as it was described by Lang. This latter author then devised an ingenious system in order to visualize big portions of the crystal (Lang 1959*a*): a simultaneous translation of the crystal and the photographic plate (see Section 17.3.3). The images thus obtained were simply projections in the reflected directions of the three-dimensional distribution of defects in the crystal, and for that reason this type of topograph is called *projection topograph*; it is also called *traverse topograph* in reference to the way in which it is obtained. The contrast results from an integration over the angular width of the rocking curve.

An entirely new method appeared with the advent of synchrotron radiation. Because of the polychromatic nature of the source a crystal placed in the beam will generate Laue spots (Fig. 17.5(a)), and because the synchrotron radiation beams are highly

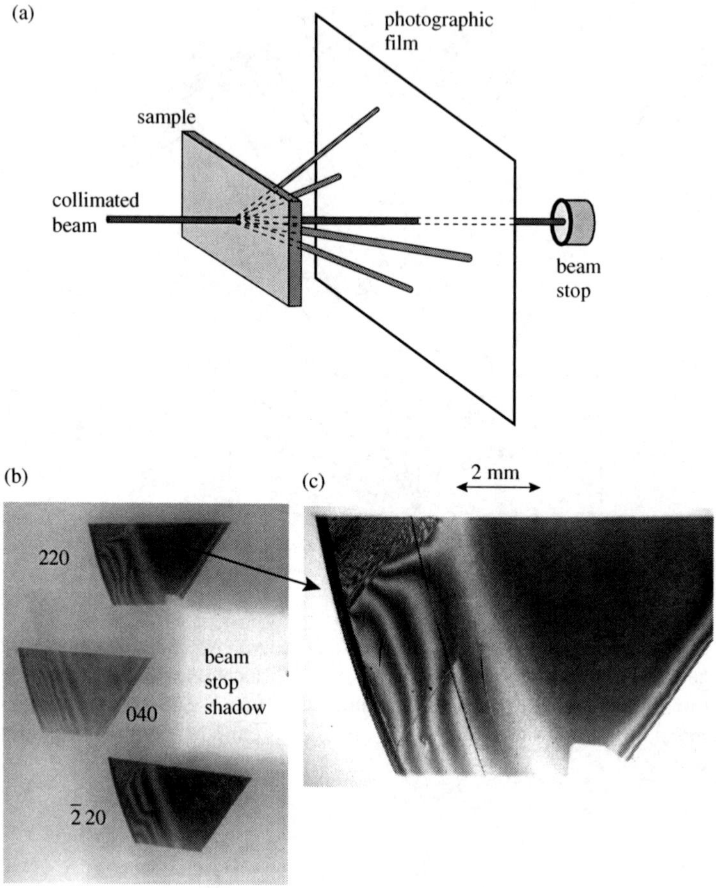

Fig. 17.5. Synchrotron white beam topography. (a) Principle. (b) Laue spots from a white beam topograph of a low-dose SIMOX silicon wafer (superposition of a thin silicon layer on a silicon substrate with a thin silicon oxide layer in between them, produced by implantation of oxygen). (c) Enlargement of one of the spots. Dislocations, scratches, moiré fringes, and Pendellösung fringes are visible. Courtesy W. Ludwig and J. Härtwig.

collimated the fine structure of these Laue spots can be resolved (Fig. 17.5(b)). It is the equivalent of a projection topograph where the integration is over the spectral width of the rocking curve; this type of topograph is therefore called *white beam topograph.* The first topographs obtained in this way were reported by Tuomi *et al.* (1973, 1974) and the first review of the possibilities of synchrotron radiation topography was given by Hart (1975*a*). As an example, Fig. 17.6 shows a synchrotron white beam topograph of misfit dislocations between epilayer and substrate in silicon.

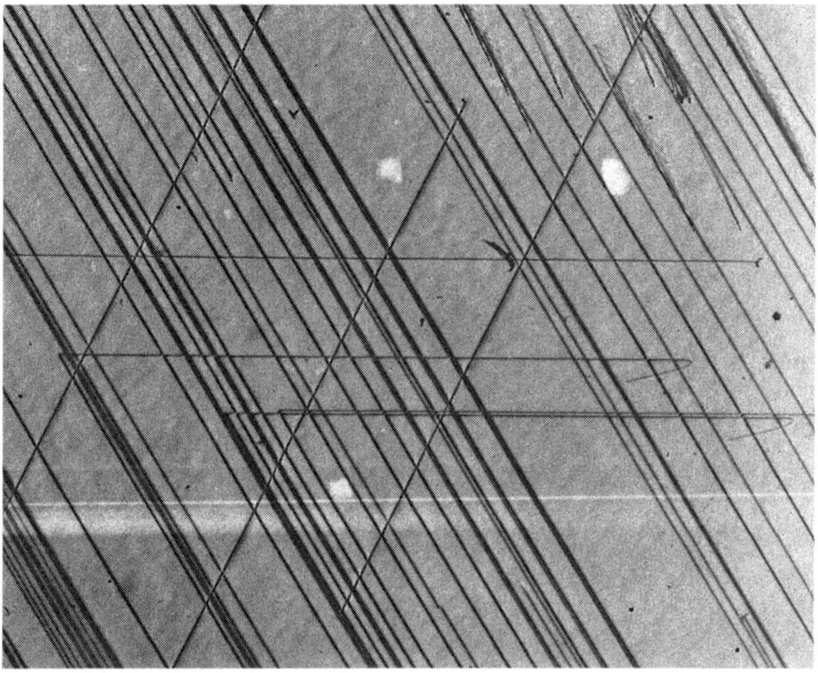

Fig. 17.6. X-ray white beam synchrotron topograph of misfit dislocations between epilayer and substrate. Silicon, 220, $\lambda \approx 1$ Å; crystal thickness 0.35 mm; $\mu_o t \approx 1.4$. Courtesy M. Sauvage.

17.3.2 *Principle of section topographs*

Figure 17.4 shows the principle of Lang section topography. The main factors governing the spatial resolution are, again, the geometrical resolution, the photographic resolution and the diffraction process. The discussion of Section 17.2.1 relative to the geometrical and the photographic resolutions applies here as well. The incident beam from a microfocus X-ray tube is collimated by a slit about 10 μm wide. The divergence of the incident beam is small enough for the characteristic $K\alpha_1$ and $K\alpha_2$ lines not to be simultaneously reflected and the crystal acts as its own monochromator. Section topographs can also be taken with synchrotron radiation, but with a special arrangement (Mai *et al.* 1980; Sauvage *et al.* 1982; Lang 1983*b*). The impact of the beam on the crystal can be considered, as a first approximation, as a point source, A, which will generate wavefields propagating within the Borrmann triangle, ABC, as has been described in Chapter 12. Actually, one should speak of a Borrmann prism since the slit is perpendicular to the plane of incidence and the dispersion surface is approximated by a cylindrical surface. The influence of the slit width on the topograph is, however, non-negligible and should be taken into account in the computer simulations of images (Aristov *et al.* 1982; Epelboin and Authier 1983). This is done by adding five or eight simulated section topographs. As shown in Chapter 10, these wavefields interfere, giving rise to spherical-wave hyperbolic Pendellösung fringes which appear as parallel lines on the topograph if the crystal is plane parallel (Fig. 17.7(a)) and have the shape of hyperbolae if the crystal is wedge shaped (Fig. 10.6). The intensity distribution of the reflected beam along the base *BC* of the Borrmann triangle has been given in Section 10.7 for the Pendellösung fringes and in Section 12.7 for each type of wavefield, separately. It was seen that there is a concentration of intensity along the edges of the Borrmann triangle, due to the high density of wavefields paths, for values $\mu_o t$ up to about 5 (Fig. 12.12). This effect, called the margin effect, was predicted qualitatively by Borrmann (1959*a*) and by Laue (1960). It was first observed by Kato and Lang (1959) on section topographs and is now a well-known feature of these images.

Section topographs have the big advantage of containing a large amount of information about the distribution of strain fields inside the crystal, but they are more complicated to interpret than the reflection topographs. One should distinguish the effects due to:

1. *Long-range strains*: their influence on the intensities and on the trajectories of the wavefields has been discussed in Chapter 13; it results in particular in modifications of the Pendellösung fringes (see Section 13.8.4); an example is given in Fig. 13.22(a).

2. *Microdefects*: if they are too small to be resolved they will appear as small dots. If their density is not too high, the Pendellösung fringes will not be disturbed, but as their density increases, the Pendellösung fringes become blurred as shown in Fig. 17.7 representing a Czochralski-grown silicon crystal containing oxygen that precipitates when the crystal is annealed: before annealing, the crystal appears as perfect (Fig. 17.7(a)), after 1 h annealing at 1000 °C, silicon oxide starts to precipitate, but the Pendellösung fringes are still visible (Fig. 17.7(b)), after 6 h annealing at the same temperature the Pendellösung fringes are no longer visible (Fig. 17.7(c)), but growth striations made visible by decoration with the tiny particles of silicon oxide which have precipitated can be observed (Patel 1973).

3. *Localized defects such as dislocations and precipitates*: their image is composite and the various components have been identified by Authier (1967). Let *D* be the intersection of the defect with the incident beam, also called the direct beam (Fig. 17.4). The strongly deformed regions around the defect diffract that part of the incident beam which is far from the reflection domain and give rise to a reflected beam which reaches the exit surface at i_1 (Figs 17.4 and 17.8). The corresponding image is called the *direct image*. It is obvious from the figure that the distance of the direct image, i_1, from the edge of the topograph is directly related to the depth at which the dislocation intersects the direct beam.

As an illustration, Fig. 17.9 shows the variation of the position of the direct image for different depths of the dislocation on experimental section topographs (Fig. 17.9(a)) and on their computer simulations (Fig. 17.9(b)) in a 440 μm thick silicon crystal

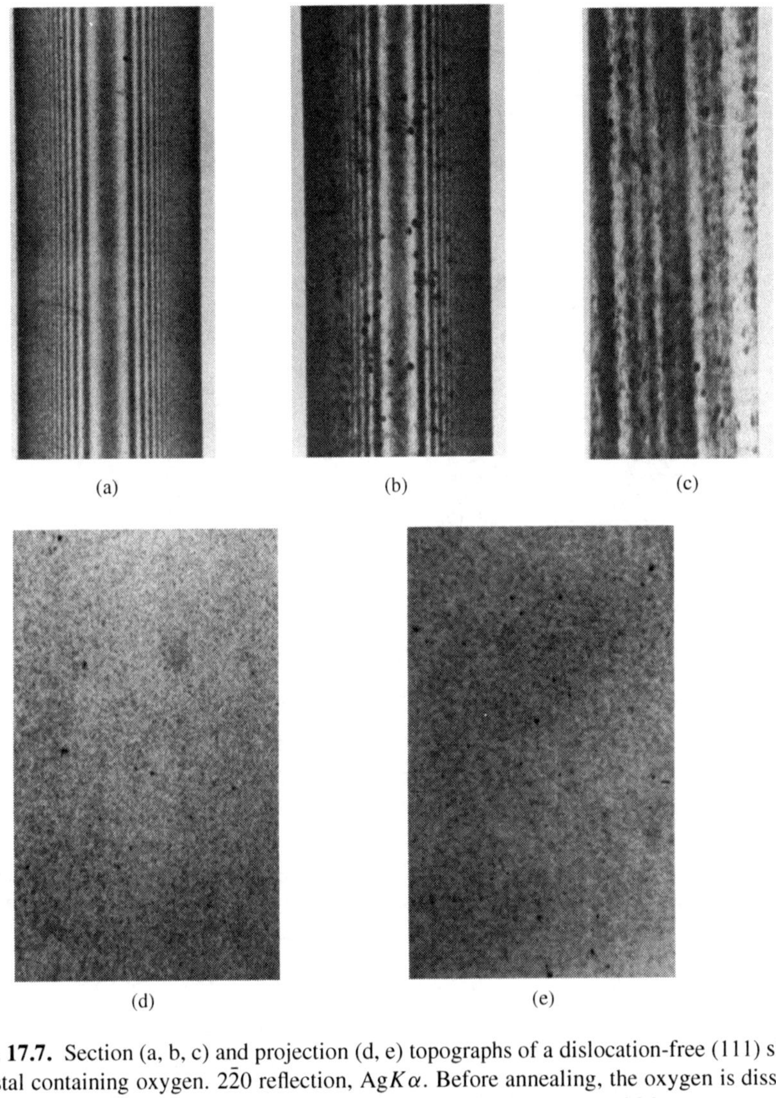

Fig. 17.7. Section (a, b, c) and projection (d, e) topographs of a dislocation-free (111) silicon crystal containing oxygen. $2\bar{2}0$ reflection, Ag$K\alpha$. Before annealing, the oxygen is dissolved and the crystal appears perfect (a and d); after 1 h annealing at 1000 °C, oxygen starts to precipitate in the form of small silicon oxide inclusions (b); after 6 h annealing at 1000 °C, the density of precipitates is so high that the Pendellösung fringes are blurred (c) and growth striations become visible, but no change is visible in the projection topograph (e). Courtesy J. R. Patel.

(after Authier *et al.* 1985); the simulations were performed using Epelboin's DEFV variable-step algorithm (Epelboin 1983).

Independently of the formation of the direct image, wavefields propagating within the Borrmann triangle are deviated or stopped when they arrive at the defect. Let R

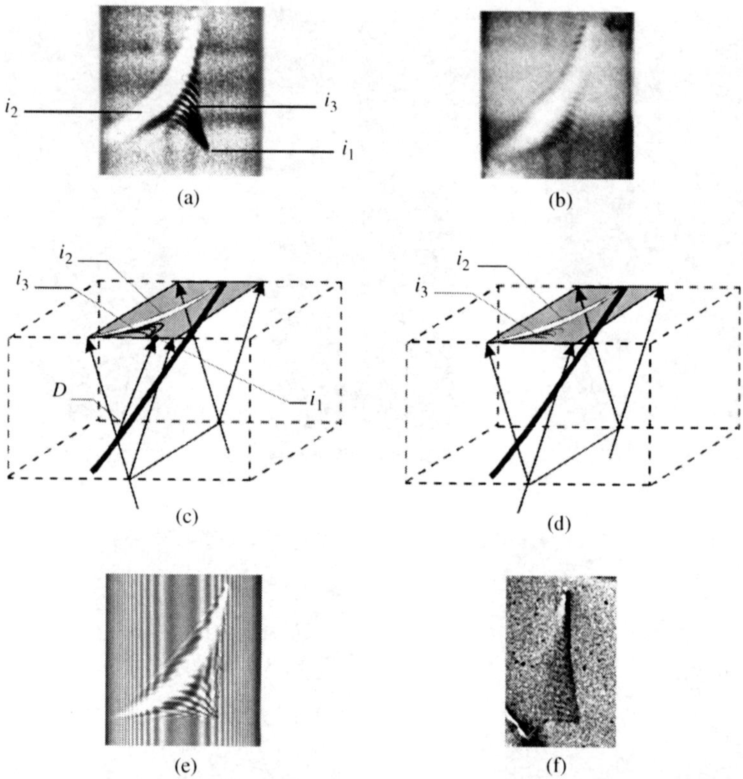

Fig. 17.8. Image of a dislocation in a section topograph. Silicon, Mo$K\alpha$, crystal thickness 0.8 mm, $\mu_o t \approx 1.12$. (a) $2\bar{2}0$ reflection; i_1: direct image; i_2: dynamical image; i_3: intermediary image. (b) Same dislocation as in (a), $\bar{2}20$ reflection, the dislocation does not intersect the direct beam, there is no direct image. (c) Image formation of (a). (d) Image formation of (b). (e) Computer simulation of (a) (reproduced with permission from Epelboin and Authier 1983). (f) Projection topograph of the dislocation in (a), same reflection.

be a point on the defect, dislocation line for instance, whose projection on the plane of incidence is marked d in Fig. 17.4. There is a lack of intensity in the direction AR after the defect, which will thus cast a shadow at i_2 on the exit surface (Fig. 17.8); the term 'shadow' was introduced by Borrmann *et al.* (1958) and the corresponding image is called *dynamical image* (Authier 1967). When the wavefields propagating along AR fall on the most strongly deformed regions around the defect, interbranch scattering occurs and new wavefields are generated which reach the exit surface at i_3 in Figs 17.4 and 17.8. They are coherent with the wavefields which have propagated along Ai_3 in the perfect crystal and interfere with them, giving rise to fringes. This third component of the image is called the *intermediary image* (Authier 1967). In addition, the long-range strain field from the defect induces a deformation of the spherical-wave Pendellösung fringes, an effect first observed by Kato (1963*a*).

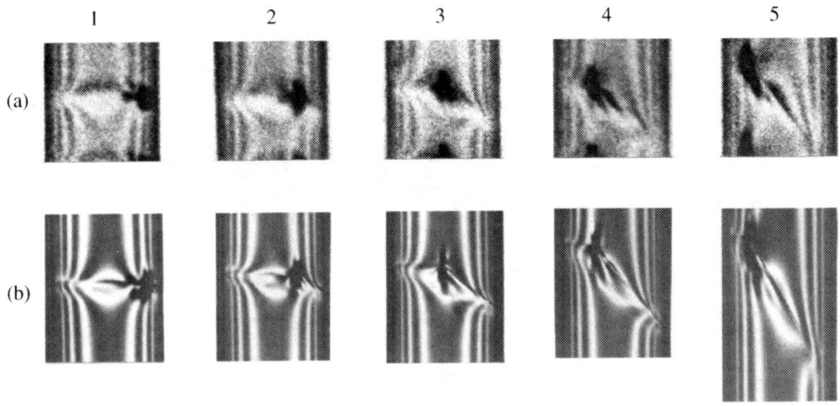

Fig. 17.9. Experimental (a) and computer (b) simulated section topographs of a silicon crystal. $\bar{3}\bar{3}3$ reflection, Mo$K\alpha$, total crystal thickness, 440 μm ($\mu_o t \approx 0.62$). The depths of the intersection of the dislocation with the direct beam are, respectively, 44, 108, 223, 299 and 352 μm for images 1, 2, 3, 4, and 5. (Reproduced with permission from Authier *et al.* 1985.)

The mechanisms of the image formation are discussed in Section 17.3.4 for dislocations.

The images of precipitates follow roughly the same principles. As an example, a section topograph showing precipitates is presented in Fig. 17.10. The computer simulations of the images of precipitate are discussed, for instance, in Cui *et al.* (1989) and Green *et al.* (1990) for hydrogen precipitates in silicon and in Holland and Tanner (1995) for oxygen precipitates in silicon; in both cases a spherical strain distribution is assumed.

The relative importance of the various components of a dislocation image depends on the product, $\mu_o t$, of the absorption coefficient, μ_o, by the crystal thickness, t. If it is smaller than 2 or 3, the direct image of defects is predominant. If it is larger than 6 or 8, the direct image and branch 2 wavefields are absorbed out. Only branch 1 wavefields whose tiepoints are close to the apex of the dispersion surface can propagate. When they arrive in the vicinity of a dislocation, their path is deviated and they are absorbed out; the dynamical image is predominant and is due to a breakdown of the Borrmann effect. This was the case of the images observed by Borrmann *et al.* (1958) and topographs taken in these conditions are sometimes called Borrmann topographs (see, for instance, Klapper and Smolsky 1998). If $1 < \mu_o t < 6$, the three types of images are observed simultaneously.

4. *Planar defects*: their direct image is the projection of their intersection with the direct beam, the remainder of the image is due to the interferences of the wavefields generated at the defect; the resulting image presents typically the shape of an hourglass (Fig. 17.11) and is discussed in Section 17.3.5.

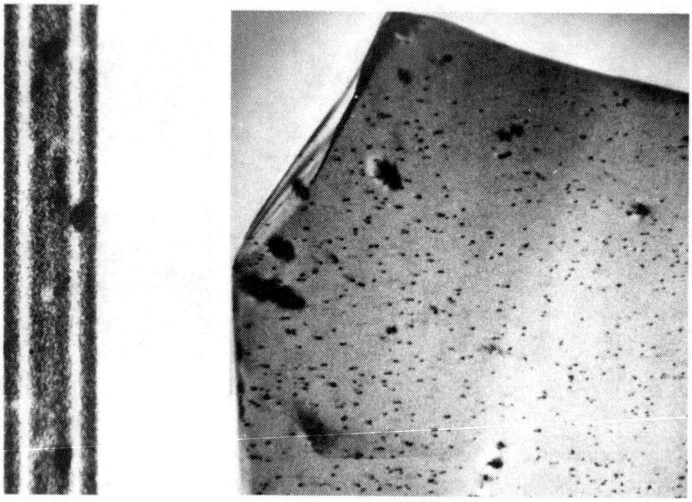

(a) (b)

Fig. 17.10. Section (a) and projection (b) topographs of precipitates in triglycine sulphate. Mo$K\alpha$; 100 reflection. Courtesy J.-F. Pétroff.

17.3.3 *Projection topographs*

17.3.3.1 *Traverse topographs*

In order to view the whole crystal, Lang (1959*a*) suggested to traverse the crystal and the photographic plate simultaneously (Fig. 17.4). In a laboratory setup, the focus is a microfocus and the distance to the sample should be at least 50 cm. The incident beam is limited by a slit, as for section patterns, but it does not need to be as narrow; a slit 100 μm wide is sufficient. The reflected beam is also isolated by a slit, as shown in the figure and the sample to photographic plate distance should not exceed 10 mm. For best resolution a thick emulsion should be used and the photographic plate must be set normal to the direction of the reflected beam. In practice, the crystal and the photographic plate move back and forth by means of a gliding table. The mechanism must be accurate enough so that the crystal orientation remains constant within 1 or 2 seconds of arc. The entrance slit and the rotation axis which serve to set the crystal to the Bragg angle for $K\alpha_1$ radiation must be parallel and perpendicular to the direction of the translation. For a laboratory setup, the rotation axis is vertical and the translation horizontal. In order to have the maximum field of view in the vertical direction, the lattice planes must be vertical. If α is their departure from verticality, one must have

$$\alpha < \frac{\omega_h \cos\theta_B}{\omega_v},$$

where ω_h is the total horizontal divergence and ω_v the vertical divergence. In practice, α should be at most half a degree.

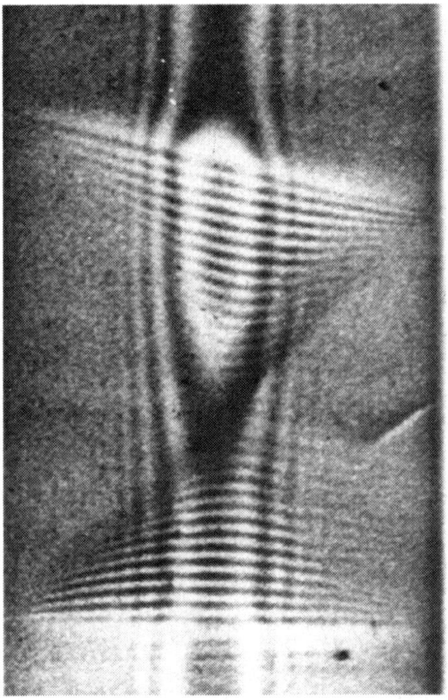

Fig. 17.11. Section topograph of a planar defect. Dolomite ($CaMg(CO_3)_2$), 100 reflection, Mo$K\alpha$. Courtesy A. Zarka.

For more details on the experimental setup, see Lang (1978, 1999) and Bowen and Tanner (1998).

Other setups have been proposed, such as using a linear focus and Soller slits, but the resolution is never as good as in Lang's traverse topography technique. A curved monochromator and an oscillation of crystal and photographic plate, as suggested by Kohra and Takano (1968), is useful when the sample is bent or warped.

Recently, Bowen *et al.* (2003) have suggested a very ingenious system in which the photographic plate is replaced by a high speed digital imaging detector and the mechanical translation of the photographic plate is replaced by a virtual scan of the detector: at each step of the translation of the sample, the diffracted image is stored in a memory and the integration of all the images thus obtained is performed by the computer.

17.3.3.2 *Synchrotron white beam topographs*

Thanks to the properties of synchrotron radiation described in Section 15.1.2, in particular the long source-to-sample distances (145 m and 1000 m at the ESRF and SPring-8 topography beamlines, respectively) and the small source size, the fine structure of Laue spots can be recorded with a very good resolution; each spot is obtained with a different part of the white spectrum of the incident beam. Two methods are possible, one is simply to record many Laue spots at the same time (Fig. 17.5); this is very convenient because there is no need for exact positioning and because one can compare at one glance the contrast variations from one reflection to another and determine the Burgers vectors of dislocations. But to set up a particular reflection for an arbitrary wavelength, one needs to be able to orient both the crystal and the detector in at least two directions. This is done by remote control and there are special programs to calculate these orientations (see, for instance, Miltat and Dudley 1980). As noticed in Section 17.2.1 in relation to the Berg–Barrett technique, because of the very small divergence of the synchrotron beam, much larger sample-to-detector distances than in the laboratory can be afforded and this is very useful for the applications, in particular when special stages are required for real-time experiments or experiments at high or low temperature, at high pressure or under an applied stress, under an applied electric or magnetic field, etc.

The advantages of using synchrotron radiation are:

1. *The special properties of the white beam*
   (a) No need to traverse the crystal (the field of view is 2–3 cm^2).
   (b) Short wavelengths which undergo very small absorption so that materials containing heavy atoms can be investigated.
   (c) Possibility to image simultaneously distorted areas or the whole volume of a bent or warped crystal (as in a laboratory oscillation technique).
   (d) Easiness to determine Burgers vectors, using the simultaneous Laue spots.
2. *The high flux*
   (a) Very short exposure times, a few hundredths of a second to a few seconds even for thick samples; for instance at $\lambda = 0.21$ Å, $\mu_o t \approx 0.7$ for a 1 cm thick silicon crystal. As an example, Medrano *et al.* (1999) studied domains in a 0.77 mm thick magnetite (Fe_2O_3) crystal for which $\mu_o t \approx 1.5$ at $\lambda = 0.34$ Å. The very short exposure times have also been taken advantage of by Ludwig *et al.* (2001) to develop a new technique, *topo-tomography*, in which a large number of topographs are taken after successive rotations of a diamond sample by small angular steps, enabling a three-dimensional tomographic reconstruction. A similar technique has been applied to silicon by Kawado *et al.* (2004).
   (b) Suitability to perform *in situ* real-time dynamic experiments.
3. *The pulse structure of the beam*
   (a) Possibility to perform stroboscopic experiments.

The contrast of defects is on the whole similar to that observed in traverse patterns taken with characteristic radiation in the laboratory, as can be seen in Fig. 17.6: for values of $\mu_o t$ up to 2 or 3, the direct image due to the diffraction of the incident beam by the highly distorted areas around the dislocation core is predominant. There are, however, a certain number of differences due to the continuous nature of the spectrum,

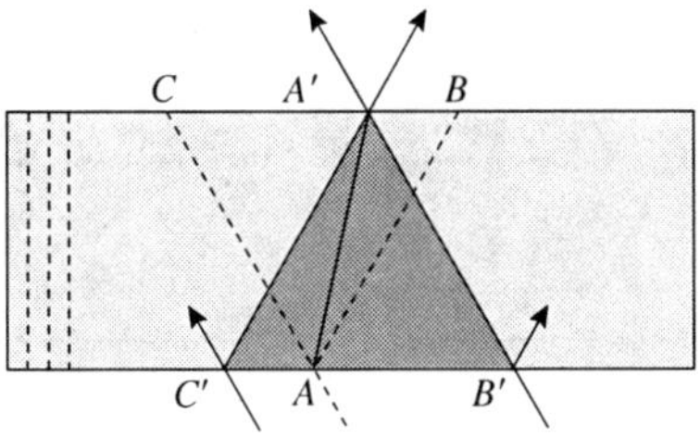

Fig. 17.12. In a projection topograph, the intensity received by a point of the exit surface is the result of an integration over the base of an inverted Borrmann triangle.

the presence of harmonics and the long sample-to-detector distances which make contrast interpretation more complex. The simulation of images with white radiation is a formidable task, but it can be done using parallel computers or multiprocessors (Epelboin 1996; Epelboin *et al.* 1998).

For reviews of synchrotron topography see, for instance, Miltat (1980), Sauvage and Pétroff (1980), Steiner *et al.* (1990), Bowen and Tanner (1992, 1998), Baruchel (1996), Baruchel *et al.* (2002), Tuomi (2002); for the contrast of dislocation images, Zontone *et al.* (1996).

17.3.3.3 *Contrast features of projection topographs*

1. *Diffracted intensity along the exit surface*: when taking a projection topograph, a point A' of the exit surface receives wavefields which have followed all the possible paths, AA', from $B'A'$ to $C'A'$ within the inverted Borrmann triangle, $B'C'A'$, either simultaneously in a synchrotron white beam topograph, or successively in a traverse topograph (Fig. 17.12).

The diffracted intensity at A' is therefore an integrated intensity. The reciprocity theorem, extended by Kato (1968*c*) from the case of optics to the case of X-rays, shows that this intensity is equal to the intensity integrated over the basis of the inverted Borrmann triangle, $B'C'A'$. As seen in Section 10.7, it is also equal to the integrated intensity under the rocking curve in the plane-wave case (Kato 1968*c*; Authier and Simon 1968), integration performed over the angular range of reflection for traverse topographs and over the spectral range of reflection for synchrotron white beam topographs. If the crystal thickness varies from point to point, the diffracted intensity will vary periodically according to (6.17) as shown in Fig. 6.8. This results in equal-thickness fringes on a projection topograph. An example is given in Fig. 9.7.

2. *Long-range strains*: the influence of long-range strains on the propagation of wavefields was discussed in Chapter 13. If $\mu_o t$ is small and branch 2 wavefields are not absorbed out, interferences occur between branch 1 and 2 wavefields whose path and phase have been modified, which generates additional Pendellösung fringes on section patterns. On projection topographs this results in equal-strain fringes. An example is given in Fig. 13.22(b). At the same time, the intensities of the wavefields

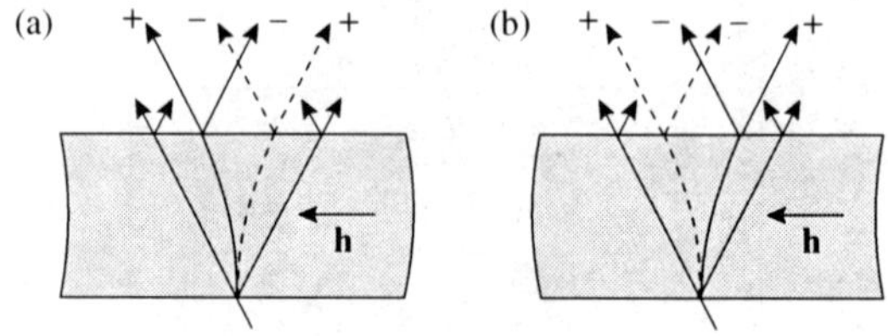

Fig. 17.13. Variations of the reflected and refracted intensities of branch 1 (solid lines) and branch 2 (dashed lines) wavefields in a crystal deformed by a uniform strain gradient. The + and − signs indicate an increase or decrease of intensity, respectively. (a) The directions of the curvature and of the diffraction vector coincide. (b) The directions of the curvature and of the diffraction vector are opposite. If absorption is important (branch 2 wavefields absorbed out), the intensity of the reflected beam is higher in case (b) than in case (a); this can be related to the fact that the trajectories of branch 1 wavefields are then bent towards the reflected direction.

are modified as was first shown by Penning and Polder (1961) and as was discussed in Section 13.7. The results are summarized in Fig. 17.13. There are intensity transfers between branch 1 and branch 2 wavefields and between the reflected and refracted beams:

(i) The directions of the curvature and of the diffraction vector coincide, the intensity is transferred from the reflected to the refracted beam for branch 1 and it is the contrary for branch 2.

(ii) The directions of the curvature and of the diffraction vector are opposite, the intensity is transferred from the refracted to the reflected beam for branch 1 and it is the contrary for branch 2.

These rules are only useful when $\mu_o t$ is larger than 3 or 4, and branch 2 wavefields are absorbed out. Regions of opposite misorientations will then appear with a black or white contrast on an hkl topograph. These contrasts will be reversed on $\bar{h}\bar{k}\bar{l}$ topographs. This remark is extremely useful for the determination of the sign of misorientations. As an example, Fig. 17.14 shows $2\bar{2}0$ and $\bar{2}20$ topographs of a rectangular silicon bar, $2.8 \times 3.3\ \mathrm{mm}^2$ in section. The axis of the bar was [111] and the side faces were the pairs $(1\bar{1}0)$, $(\bar{1}10)$ and $(11\bar{2})$, $(\bar{1}\bar{1}2)$. The bar had been deformed by torsion about its long axis at approximately 900 °C, in a lathe.[1] The four {111} slip systems were activated and several Frank–Read dislocation sources were generated in planes parallel to these directions, in particular in planes perpendicular to the axis of the bar; there is an important strain concentration in these planes which results in a strong black and white contrast above and below them, as can be seen in the figures. This contrast is reversed in the $2\bar{2}0$ and $\bar{2}20$ topographs, as predicted. A Frank–Read source lying in a $(\bar{1}11)$ glide plane is faintly to be seen on the right-hand side of the topographs and is much better seen in Figs 9.7 and 17.26 (see Section 17.3.4.2).

[1] See footnote 2 on p. 544.

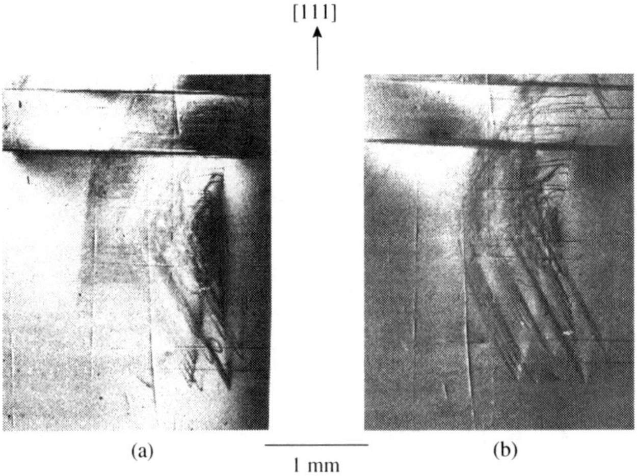

Fig. 17.14. Projection topographs of a silicon crystal deformed by torsion. (a) $2\bar{2}0$ reflection. (b) $\bar{2}20$ reflection; Ag$K\alpha$; crystal thickness 3.3 mm ($\mu_o t \approx 2.1$). Note the contrast reversal between the two topographs. Part of the same crystal is also represented in the topographs of Figs 9.7 and 17.26.

This method of determining the sense of the distortions is used to determine the sign of the Burgers vector of dislocations, for instance by studying the black–white contrast at the outcrop of the dislocation at the surface (Dunia *et al.* 1980).

1. *Dislocations*

(i) *Direct image.* Every point of the dislocation is crossed in turn by the direct beam in traverse topographs, or simultaneously in white beam topographs, and one obtains on the topograph a direct image along the projection of the dislocation in the reflected direction (Figs 17.4 and 17.8). Their contrast is normally black (more intensity than the background), but occasionally it happens that a dislocation lies in a highly strained region, a curved crystal for instance, and is actually out of contrast; its image is then white (Meieran and Blech 1972). If the crystal is very thin, the expressions given by the geometric and dynamical theories for the reflected intensity are not very different; the contrast of the dislocations is then weak (Penning and Goemans 1968). Direct images are predominant for values of $\mu_o t$ up to 2 or 3 as can be seen in Fig. 17.6.

(ii) *Dynamical image.* For each point of a dislocation, the dynamical image on a section topograph is a curved segment (Fig. 17.8). On a projection topograph it becomes a blurred white line after integration. Qualitatively, one may say that the main reason for its origin is a breakdown of the Borrmann effect. As the Borrmann effect is maximum for the wavefields propagating along the lattice planes, it follows that the central part of the dynamical image is a white line along the projection of the dislocation in the direction parallel to the lattice planes.

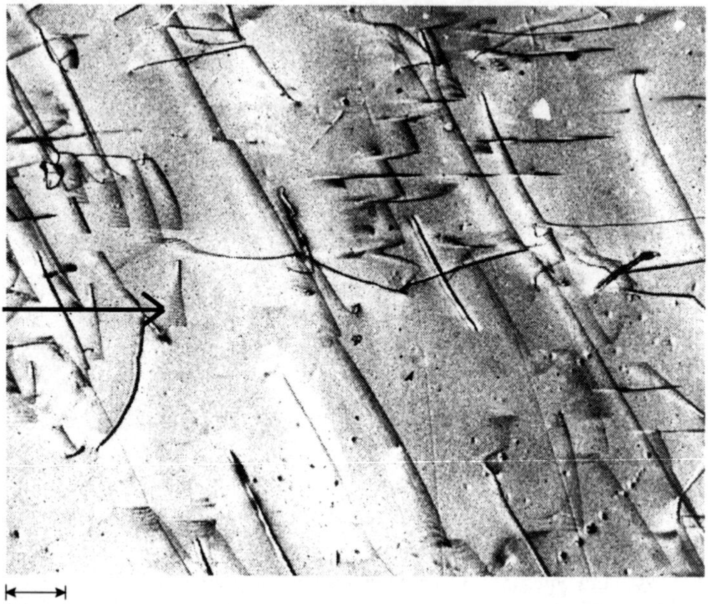

Fig. 17.15. Projection topograph of a silicon crystal. Crystal surface (111), $2\bar{2}0$ reflection, Mo$K\alpha$; crystal thickness 800 μm ($\mu_o t \approx 1.1$). The arrow indicates the dislocation reproduced in Fig. 17.8.

If $\mu_o t$ is between 1 and 5, roughly, both images will be seen simultaneously, the dynamical image being a white line along the direct image, always on the same side. This is particularly well seen in Fig. 17.15 for dislocations that run more or less parallel to the crystal surface. The distance between the two images depends on the depth of the dislocation and, by following its variations along the dislocation line, one gets an idea about the depth variations of the dislocation. If $\mu_o t$ is larger than 5 or 6, the direct image is absorbed out and only a rather well-defined dynamical image along the projection of the dislocation in the direction parallel to the lattice planes can be seen. This is the case, for instance, of Fig. 17.16.

(iii) *Intermediary image*. The intermediary image gives rise to a set of fringes which are particularly well visible when the dislocation is inclined with respect to the surface (Figs 17.8 and 17.15) and crosses both surfaces of the crystal. At the intersection of the dislocation with the exit surface the three images (direct, dynamical and intermediary) merge while the width of the intermediary image is maximum at the intersection of the dislocation with the entrance surface.

2. *Microdefects*. Very small defects which appear as tiny dots on section topographs do not present resolvable contrast on the projection topograph and the overall contrast of the projection topograph may not be changed in a visible way. This can be observed

Fig. 17.16. X-ray white beam synchrotron topograph of a heteroepitaxy of $Ga_{1-x}Al_xAs_{1-y}P_y$ on a GaAs substrate corresponding to the 220 reflection and $\lambda \approx 1.3$ Å. Epilayer thickness 1.3 μm, substrate thickness 342 μm ($\mu_o t \approx 6.5$). Courtesy J.-F. Pétroff and M. Sauvage.

by comparing Figs 17.7(d) and (e). The former is a projection topograph of a perfect silicon crystal corresponding to the section topograph represented in Fig. 17.7(a); the latter is a projection topograph of the same crystal after a 6 h anneal. The corresponding section topograph, Fig. 17.7(c), shows no Pendellösung fringes but growth striations. The crystal perfection is now degraded but no visible change can be seen between Figs 17.7(d) and (e). It is important to remember that section topographs are much more sensitive to imperfections than projection topographs and that a crystal which looks perfect from its projection topograph may in reality be imperfect. The good visibility of Pendellösung fringes in a section topograph is always a good test of crystal perfection.

3. *Precipitates*. Small precipitates appear on projection topographs as unresolved dots as can be seen in Fig. 17.10(b), but larger precipitates usually have a black–white–black contrast with the white line parallel to the trace of the reflecting planes.

4. *Planar defects*. See Section 17.3.5.

17.3.3.4 *Special setups*

1. *Stereographic views*. As mentioned in Section 17.2.4, one obtains stereographic views by taking successively two topographs, one with an *hkl* reflection and one

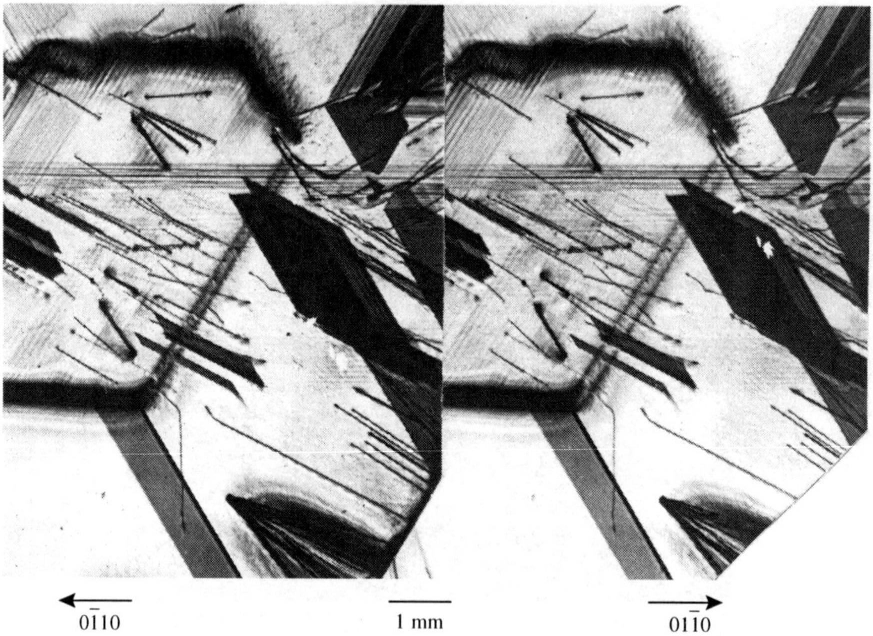

Fig. 17.17. Stereographic pair of topographs representing dislocations, fault planes and growth sector boundaries in natural quartz. Mo$K\alpha$. Courtesy A. Zarka.

with an $\bar{h}\bar{k}\bar{l}$ reflection (Lang 1959*a*,*b*); they are called a stereo pair. The visibility conditions of defects are the same for the two reflections, but the dynamical image lies on the other side of the direct image and the contrast of long-range strains is opposite. It works best with a low value of $\mu_o t$ and symmetric reflections. As an example, a stereo pair of topographs is shown in Fig. 17.17; the two topographs were taken with the $0\bar{1}10$ and $01\bar{1}0$ reflections, respectively. One can see dislocations, fault planes and growth sector boundaries in a 0001 quartz plate.

As was mentioned in Section 17.2.4, another method to obtain a stereo pair consists in taking two topographs for two different orientations of the crystal around the diffraction vector (Haruta 1965). A 5° rotation is usually enough. It is the best solution in the case of a very asymmetric reflection, such as a $\bar{1}11$ reflection from a (111) oriented silicon wafer because the projection angles and image distortions would be too different for the $\bar{1}11$ and $1\bar{1}\bar{1}$ reflections.

2. *Limited projection topographs (Fig. 17.18).* It is often useful to have a projection topograph of the defects contained in a layer thinner than the whole crystal, for instance if there is surface damage that cannot be removed. This can be done simply by modifying the position of the two lips of the slit limiting the reflected beam (Lang 1963; Caslavsky 1970). If the two lips intercept part of the reflected beam (Fig. 17.18), only the defects contained in the volume that lies between the depths z_1 and z_2 and

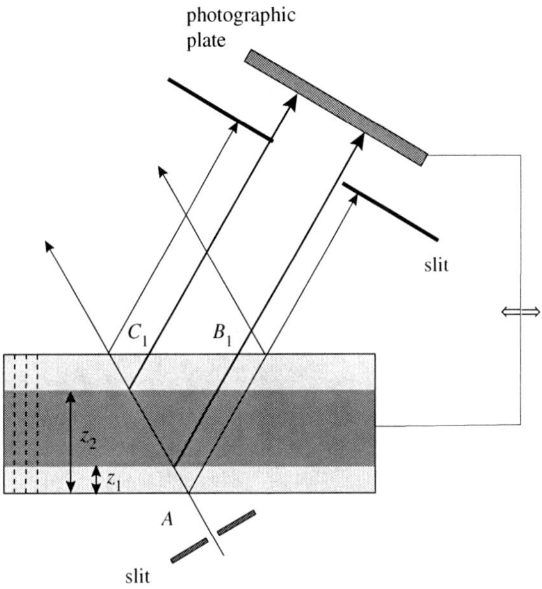

Fig. 17.18. Limited projection topography.

is represented in dark grey in the figure will give rise to direct images; these lie along a portion B_1C_1 of the basis of the Borrmann triangle. This technique only works with traverse topographs and is most satisfactory for small values of $\mu_o t$ when direct images are predominant.

3. *Refracted beam topographs (Fig. 17.19)*. It is also possible to take topographs in the refracted beam (Lang 1963). For this one must isolate the refracted beam by means of a slit, taking care to stop the direct beam, namely that part of the beam that does not satisfy Bragg's condition (Fig. 17.19). This adjustment is rather delicate and is easier for thicker crystals. The reflection of the direct beam by the distorted regions around a defect that gives rise to the direct image on a usual topograph gives no contribution to the refracted beam topograph. This type of topograph is most convenient when $\mu_o t$ is not too small. The intensity distribution in the refracted beam is given in Section 12.7. For values of $\mu_o t$ larger than 6, it presents a maximum in the centre, as does the reflected beam, because of the Borrmann effect. The dynamical image is predominant and has a white contrast as in a usual topograph. The contrast of long-range strains is opposite to that on the reflected beam topograph. Refracted beam topographs are useful in comparison with reflected beam topographs for the determination of the sign of misorientations or of the phase change associated with a stacking fault (Jiang and Lang 1983).

4. *Reticulography*. A new, simple, system has been proposed recently by Lang and Makepeace (1996, 1999*a*,*b*) to measure local distortions. A fine-scale X-ray

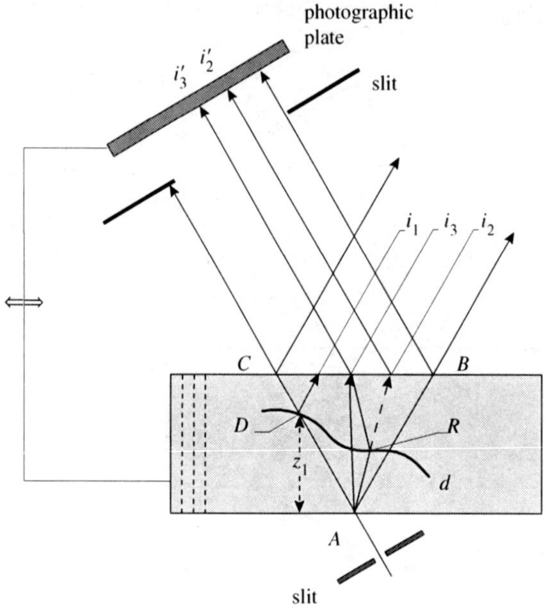

Fig. 17.19. Refracted beam topography.

absorbing mesh is placed between the sample and the photographic plate. The mesh splits the diffracted beam into an array of microbeams; the period of the mesh is of the order of a few tens of microns. The orientations of these microbeams vary with the local distortions of the crystal. They can be measured by varying the distance between the mesh and the photographic plate. A map of the crystal distortions can thus be obtained (see, for instance, Arridge *et al.* 2002).

17.3.4 *Dislocation images*

17.3.4.1 *Direct images*

The direct image is due to the reflection of the direct beam by the more distorted areas close to the dislocation core. A simple kinematical model was proposed by Lang and Polcarová (1965) and developed independently by Chikawa (1965) and Authier (1967). It was improved by Miltat and Bowen (1975) and was used by various authors for Burgers vector determinations by comparison between experimental and computer-simulated topographs (see, for instance, Epelboin and Ribet 1974; G'Sell and Epelboin 1979). It was also used more recently by Zontone *et al.* (1996) to interpret features in third-generation synchrotron radiation topographs.

The principle of this crude model is to assume that the regions which contribute to the direct image are those which are contained in the volumes around the dislocation core where the effective misorientation (13.7b) is larger than a certain value:

$$\delta\theta = -\frac{1}{k \sin 2\theta_B} \frac{\partial(\mathbf{h} \cdot \mathbf{u})}{\partial x_h} \geq K \times 2\delta_{os}, \tag{17.3}$$

where $\mathbf{u}$ is the displacement vector, $2\delta_{os}$ is the width of the rocking curve (FWHM) and K is an arbitrary coefficient. This relation defines an equal-strain surface around the dislocation. This surface presents in general two or more lobes which are well visible in the experimental and simulated topographs of Fig. 17.9. They are marked i_1 in Fig. 17.20 where the equal-strain contour, represented by a white line, has been drawn by the computer. It is the simulation of a 333 topograph for identical conditions as in Fig. 17.9 and a dislocation depth of 240 μm. The contour is around a region where the reflected intensity is larger than an arbitrary value and it gives an idea of the shape of the direct image.

The reflection of the direct beam by the two-lobe volume gives rise to the double contrast of the direct image on projection topographs (Authier 1967) as shown in Fig. 17.21. This double contrast has been frequently observed (see, for instance, Pétroff and Authier 1964; Lang 1965; Willaime and Authier 1966; Miltat and Bowen 1975; Klapper 1976, 1991; Zontone *et al.* 1996). Four lobes have been observed by Epelboin and Ribet (1974) in the case of packets of dislocations on traverse topographs of lithium formiate monohydrate and interpreted quantitatively using this simple theory.

The width of the direct image, namely the distance between the two maxima in Fig. 17.21, can be estimated by calculating the effective misorientation. The components of the displacement vector of a dislocation are, assuming elastic isotropy (see

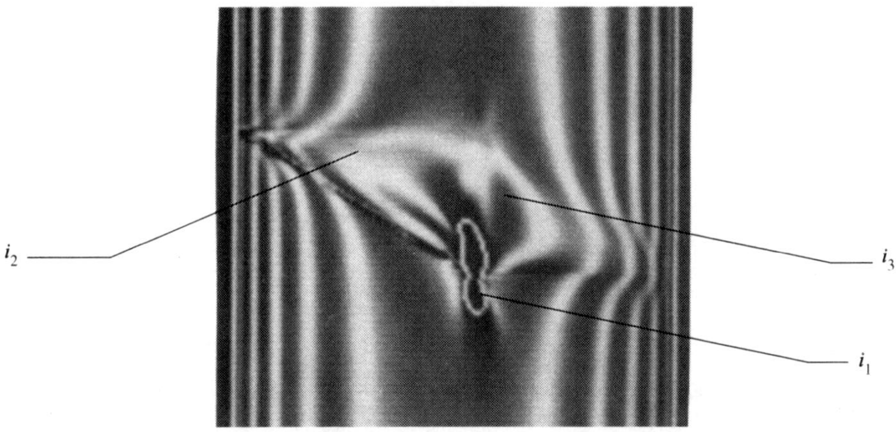

Fig. 17.20. Various features of a dislocation image on a computer-simulated section topograph. Silicon, 333 reflection; Mo$K\alpha$; depth of the dislocation: 240 μm. i_1: direct image; i_2: dynamical image; i_3: intermediary image.

any textbook on dislocation theory, for instance Hirth and Lothe 1967),

$$u_1 = \frac{b_e}{2\pi}\left[\arctan\frac{x_2}{x_1} + \frac{x_1 x_2}{2(1-\nu)(x_1^2+x_2^2)}\right],$$

$$u_2 = -\frac{b_e}{8\pi(1-\nu)}\left[2(1-2\nu)\ln\frac{\sqrt{x_1^2+x_2^2}}{r_o} + \frac{x_1^2-x_2^2}{x_1^2+x_2^2}\right], \qquad (17.4)$$

$$u_3 = \frac{b_s}{2\pi}\arctan\frac{x_2}{x_1},$$

where ν is the Poisson ratio, b_e and b_s are the edge and screw components of the Burgers vector, respectively, r_o is the radius of the core of the dislocation, x_1 is a coordinate in the direction parallel to the edge component of the dislocation, x_3 is a coordinate in the direction parallel to the dislocation line and x_2 is a coordinate in the direction perpendicular to the other two. When the dislocation is close to the crystal surface, surface relaxation must be taken into account, as was done for instance by Dunia *et al.* (1980).

For the calculation of the effective misorientation, it is more convenient to choose a coordinate system related to the diffraction experiment: x axis perpendicular to the diffraction plane, y axis parallel to the reciprocal lattice vector, **h**, z axis perpendicular to the crystal surface, oriented towards the inside of the crystal (Fig. 17.21), limiting ourselves here to symmetrical reflections. The effective misorientation is given by

$$\delta\theta = -\left[\frac{\partial}{\partial z} - \tan\theta_B\frac{\partial}{\partial y}\right](u_h), \qquad (17.5)$$

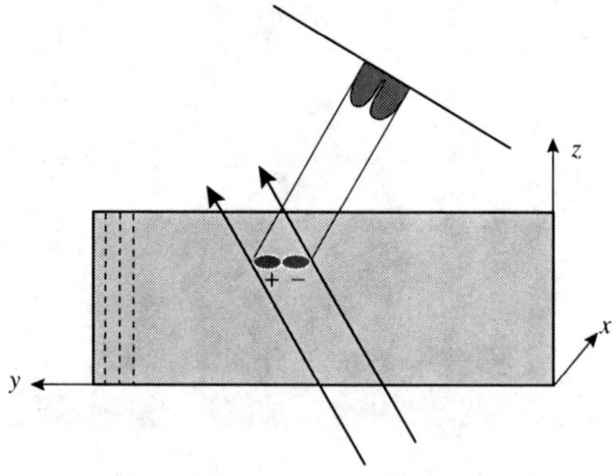

Fig. 17.21. Origin of the double contrast of direct dislocation images. The + and − signs indicate the sign of the effective misorientation.

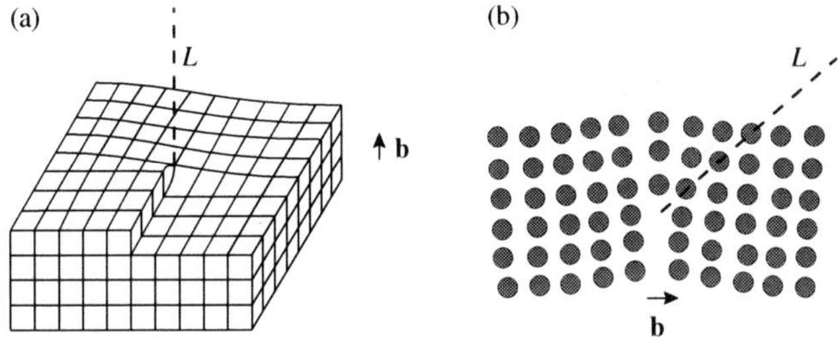

Fig. 17.22. (a) Screw dislocation. (b) Edge dislocation. **b**: Burgers vector; L: dislocation line.

where u_h is the component of **u** in the direction of **h**. Let us consider the cases of a pure screw (Fig. 17.22(a)) and a pure edge (Fig. 17.22(b)) dislocation in a few simple orientations.

1. *Screw dislocation parallel to the reciprocal lattice vector (Fig. 17.23(a))*: $\mathbf{h} \cdot \mathbf{b} = hb$; using (17.4) and (17.5), it follows for a symmetrical reflection that

$$\delta\theta = -\frac{b}{2\pi} \frac{x}{x^2 + z^2}, \tag{17.6}$$

where b is the Burgers vector. The width $2x_s$ of the image is calculated in the plane $z = 0$ using (17.3) and (17.6); it is

$$2x_s = \frac{1}{K} \frac{b}{2\pi} \frac{1}{2\delta_{os}}. \tag{17.7a}$$

It is inversely proportional to the width of the rocking curve. Its width may also be expressed in terms of the Pendellösung distance, Λ_L. Using (4.26), the width $2\delta_{os}$ of the rocking curve may be written, for a symmetrical reflection, as

$$2\delta_{os} = \frac{\lambda}{\Lambda_L \sin\theta_B} = \frac{2}{h\Lambda_L},$$

where $h = 2\sin\theta_B/\lambda$ is the modulus of the reciprocal lattice vector. It follows, for the width of the image of the screw dislocation, that

$$2x_s = \frac{1}{K} \frac{hb\Lambda_L}{\pi}. \tag{17.7b}$$

It is proportional to $\mathbf{h} \cdot \mathbf{b}$ and inversely proportional to the width of the rocking curve. The image is wider, the larger the value of $\mathbf{h} \cdot \mathbf{b}$ and the smaller the rocking curve width.

2. *Screw dislocation parallel to the lattice planes (Fig. 17.23(b))*: $\mathbf{h} \cdot \mathbf{b} = 0$ and $\delta\theta = 0$, there is no contrast as is normal since the lattice planes parallel to a screw dislocation are not deformed (Fig. 17.22(a)).

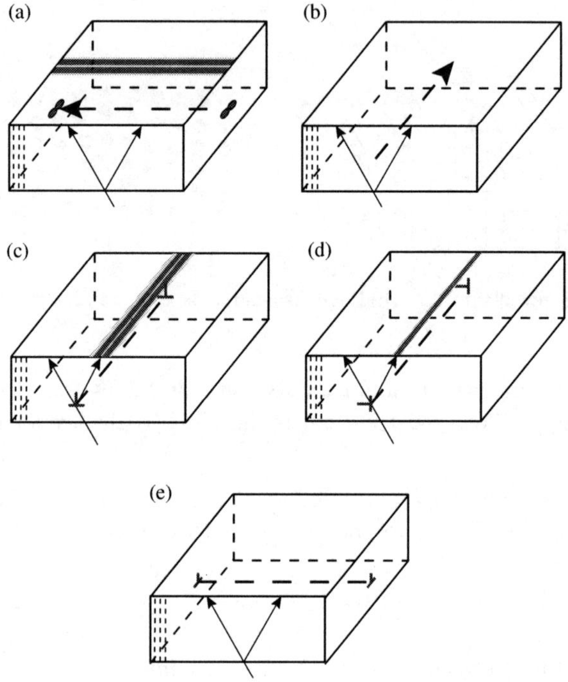

Fig. 17.23. Variation of the contrast of a dislocation line according to its orientation. (a) screw dislocation perpendicular to the lattice planes, contrast maximum. (b) Screw dislocation parallel to the lattice planes, no contrast. (c) Edge dislocation parallel to the lattice planes, **b** parallel to **h**, contrast maximum. (d) Edge dislocation parallel to the lattice planes, **b** perpendicular to **h**, medium contrast. (e) Edge dislocation perpendicular to the lattice planes, no contrast.

3. *Edge dislocation parallel to the lattice planes* ($\mathbf{L} \cdot \mathbf{h} = 0$, *where* **L** *is the direction of the dislocation line), Burgers vector parallel to the reciprocal lattice vector* ($\mathbf{h} \cdot \mathbf{b} = hb$) *(Fig. 17.23(c))*:

$$\delta\theta = -\frac{b}{2\pi}\frac{y + z\tan\theta_B}{y^2 + z^2}\left[1 + \frac{y^2 - z^2}{2(1-\nu)(y^2+z^2)}\right], \tag{17.8}$$

where b is the Burgers vector. The width $2y_e$ of the image is obtained for $z = 0$ and, using (17.3) and (17.8), it follows that

$$2y_e = \frac{1}{K}\frac{b}{2\pi}\left[1 + \frac{1}{2(1-\nu)}\right]\frac{1}{2\delta_{os}}. \tag{17.9a}$$

Poisson's ratio is never very different from 0.3 so that

$$2y_e \approx 1.7\frac{1}{K}\frac{b}{2\pi}\frac{1}{2\delta_{os}} = \frac{1.7}{K}\frac{hb\Lambda_L}{\pi}, \tag{17.9b}$$

which shows that the images of edge dislocations are wider than the images of screw dislocations; as for screw dislocations, the image is wider, the larger the value of $\mathbf{h} \cdot \mathbf{b}$ and the smaller the rocking curve width.

4. *Edge dislocation parallel to the lattice planes* ($\mathbf{L} \cdot \mathbf{h} = 0$)*, Burgers vector perpendicular to the reciprocal lattice vector* ($\mathbf{h} \cdot \mathbf{b} = 0$) *(Fig. 17.23(d))*: It follows, for the width of the image, that

$$\begin{aligned} 2y_e &= \frac{1}{K}\frac{b}{2\pi}\frac{1-2\nu}{2(1-\nu)}\frac{1}{2\delta_{os}}\tan\theta_B \\ &\approx 0.3\frac{1}{K}\frac{b}{2\pi}\frac{1}{2\delta_{os}}\tan\theta_B = \frac{0.3}{K}\frac{hb\Lambda_L}{\pi}\tan\theta_B. \end{aligned} \tag{17.9c}$$

The image is much thinner than when the Burgers vector is parallel to $\mathbf{h}$, which is normal because the reflecting planes are in this case parallel to the dislocation line and to the Burgers vector and are only slightly deformed (Fig. 17.22(b)).

5. *Edge dislocation perpendicular to the lattice planes (Fig. 17.23(e))*: $\mathbf{h} \cdot \mathbf{b} = 0$, $\delta\theta = 0$, there is no contrast as is normal since the lattice planes perpendicular to an edge dislocation are not deformed (Fig. 17.22(b)).

Equations (17.7) and (17.9) show that the image width is inversely proportional to the width of the rocking curve. This is very easy to understand: with a weak reflection the width of the rocking curve is smaller and one is sensitive to smaller distortions and therefore to regions which are further away from the dislocation core. This is illustrated in Fig. 17.24 which compares dislocation image widths in topographs taken with different orders of reflection. Figures 17.24(a) and (b) are topographs of a dislocation in mica taken with 400 and 200 reflections, respectively (after Willaime and Authier 1966), and Figs 17.24(c) and (d) are topographs of a dislocation loop in silicon taken with 333 and 111 reflections, respectively. In each case, the contrast is double for the higher-order reflection and single for the low-order reflection.

The image widths (17.7) and (17.9) are only valid if isotropic elasticity can be assumed. Otherwise, elastic anisotropy must be taken into account as was shown, for instance, by Klapper (1976) for the images of screw dislocations in potassium hydrogen oxylate.

Double contrast direct images have also been observed by Dudley *et al.* (1995) for giant superscrew dislocations in silicon carbide with large Burgers vectors (75 Å), but they found that the contrast is better explained with an orientation contrast model based on the variation of the direction of the reflected beam as the distance from the dislocation core increases (Dudley *et al.* 1999, 2003).

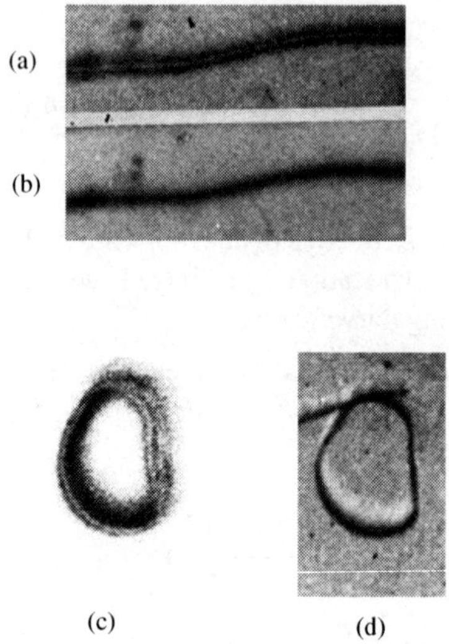

Fig. 17.24. Influence of the order of the reflection on the contrast of direct images. (a) and (b) Mica, Mo$K\alpha$, crystal thickness 100 μm: 400 reflection for (a), 200 reflection for (b) (after Willaime and Authier 1966). (c) and (d) Silicon, Mo$K\alpha$, crystal thickness 500 μm: 333 reflection for (c), 111 reflection for (d).

17.3.4.2 *Burgers vector determination from direct images*

The preceding subsection has given the explanation of the well-known rules for the determination of the Burgers vector of a dislocation: the contrast is

- *maximum*: if $\mathbf{h} \cdot \mathbf{b} = hb$ ($\mathbf{h}$ parallel to $\mathbf{b}$)
- *zero*: if $\mathbf{h} \cdot \mathbf{b} = 0$ for screw dislocations and if both $\mathbf{h} \cdot \mathbf{b} = 0$ and $\mathbf{L} \cdot \mathbf{b} = 0$ for edge dislocations
- *small*: if $\mathbf{h} \cdot \mathbf{b} = 0$ and $\mathbf{L} \cdot \mathbf{b} \neq 0$ for edge dislocations.

These rules will be illustrated with two examples:

1. *Emitter-edge dislocations in phosphorus-diffused silicon (after Sauvage and Simon 1969)*. This silicon crystal is a (111) wafer (Fig. 17.25). It has been diffused with phosphorus in certain areas and emitter-edge dislocations have been generated at the border between diffused and non-diffused areas, as can be seen in Figs 17.25(a)–(d). In silicon, the directions of Burgers vectors are ⟨110⟩. Figures 17.25(a)–(d) are topographs taken with $2\bar{2}0$, $\bar{1}11$, $11\bar{1}$ and $1\bar{1}1$ reflections, respectively. Three families of dislocation lines can be identified from the contrast they present in the four topographs: 1 (thin solid line) visible in Figs 17.25(a), (b) and (d), 2 (dashed line) visible in Figs 17.25(a), (c) and (d), and 3 (thick solid line) visible in Figs 17.25(a)–(c). From this information their Burgers vectors can be identified unambiguously as $\frac{1}{2}[1\bar{1}0]$, $\frac{1}{2}[01\bar{1}]$ and $\frac{1}{2}[\bar{1}01]$, respectively. Their sum is zero, as it should be, since the three types of dislocations are joined by nodes. Dislocation reactions were first described by Jenkinson and Lang

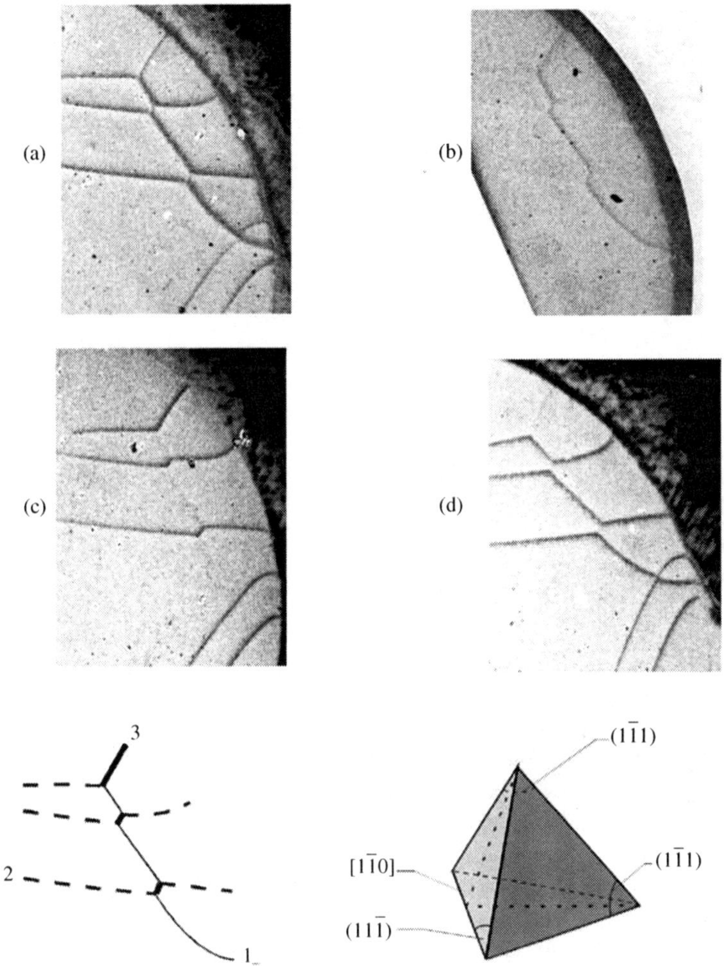

Fig. 17.25. Determination of the direction of the Burgers vector. Figure (a)–(d) are topographs of a silicon (111) wafer taken with reflections $2\bar{2}0$, $\bar{1}11$, $11\bar{1}$ and $1\bar{1}1$ respectively; Mo$K\alpha$. Dislocations of type 1 (thin solid line), 2 (dashed line) and 3 (thick solid line) have $\frac{1}{2}[1\bar{1}0]$, $\frac{1}{2}[01\bar{1}]$ and $\frac{1}{2}[\bar{1}01]$ as Burgers vectors, respectively. Magnification ×17.5. After Sauvage and Simon (1969).

(1962) and were also observed recently in silicon web dendrites (Morelhão *et al.* 2000).

2. *Frank–Read source in silicon (after Authier and Lang 1964).* Figure 17.26 shows topographs of a Frank–Read source that was generated by torsion at high temperature of the silicon crystal represented in Fig. 17.14 and described in the accompanying

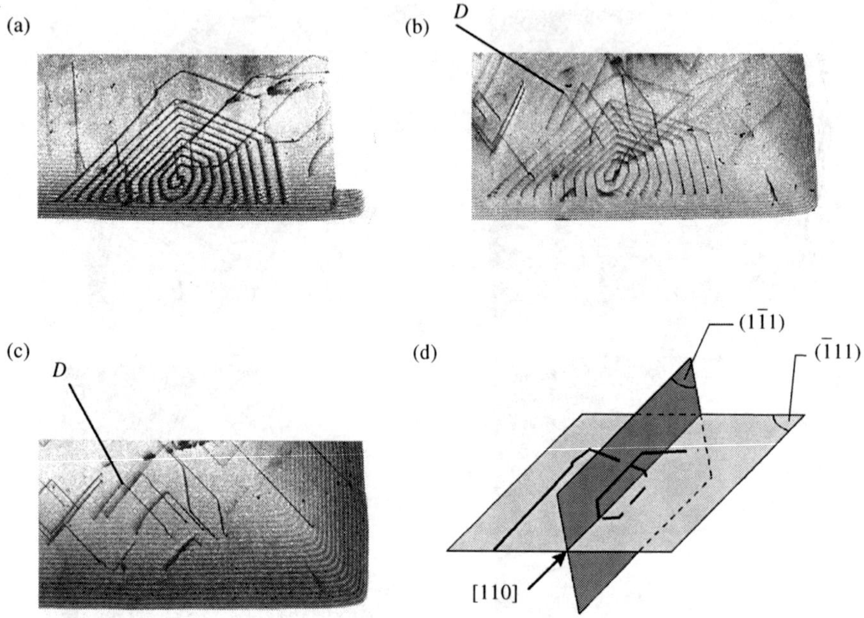

Fig. 17.26. Determination of the direction of the Burgers vector of a Frank–Read source in a $(\bar{1}11)$ glide plane in silicon. (a), (b), (c) Topographs taken with reflections, 111, $0\bar{2}2$, $1\bar{1}1$ respectively, Mo$K\alpha$. The Burgers vector of the Frank–Read source is $\frac{1}{2}[110]$; that of the dislocations marked D is $\frac{1}{2}[01\bar{1}]$. (d) Diagram explaining the cross-slip of the last loop of the source. After Authier and Lang (1964).

text.[2] To facilitate the study, the part containing the source was cut away from the silicon bar. The cut was in a plane parallel to the plane of the source and about half a millimetre from it. The piece cut away had roughly the shape of a flat pyramid and is represented in Fig. 9.7. Figure 17.26(a) is an enlargement of that topograph taken with the 111 reflection, and Figs 17.26(b) and (c) are topographs of the same area taken with the $0\bar{2}2$ and $1\bar{1}1$ reflections, respectively. The plane of the source is $(\bar{1}11)$ and eight separated half-loops can be seen. The last loop underwent cross-slip and glided in the $(1\bar{1}1)$ plane (another very nice example of dislocation multiplication by cross-slip is given in Vallino *et al.* 2000). The Burgers vector of the dislocations of the source is therefore necessarily parallel to the intersection of the $(\bar{1}11)$ and $(1\bar{1}1)$ planes and is $\frac{1}{2}[110]$. This is confirmed by comparing their contrast in the topographs of Figs 17.26(a)–(c).

[2] The history of this crystal is interesting; it was prepared by W. C. Dash who was the first to observe Frank–Read sources (Dash 1956, 1957). For this he used copper decoration and infrared optical observation. He gave one such crystal to A. R. Lang before decoration. It was studied by A. R. Lang and A. Authier and theirs were the first X-ray topographs of Frank–Read sources.

The rules recalled at the beginning of this subsection were derived assuming elastic isotropy of the crystal. They fail when there is a strong elastic anisotropy as was observed, for instance, by Klapper and Küppers (1973) in ammonium hydrogen oxalate hemihydrate ($NH_4C_2O_4.\frac{1}{2}H_2O$) of space group $Pnma$ and by Epelboin and Patel (1982) in quartz of space group $P32_1$. In such cases Burgers vectors determination requires the comparison of experimental section topographs and computer simulations of the full image performed for several possible Burgers vectors (Epelboin and Patel 1982; Authier *et al.* 1985).

17.3.4.3 *Dynamical images*

The dynamical image, which presents a characterestic black-white-black contrast in thick crystals (Borrmann 1959*b*; Authier 1961*b*, 1967; Ishii 1962) is a shadow cast along the direction of propagation of the wavefields impinging on the dislocation. It is due, on the one hand, to the curvature of these wavefields away from their normal path by the strain field surrounding the dislocation (Kambe 1963) and, on the other hand, to interbranch scattering (Authier *et al.* 1970) in the regions closer to the core where the strain gradient $\partial^2(\mathbf{h}\cdot\mathbf{u})/\partial x_o\,\partial x_h$ takes values that are larger than a critical value (Authier and Balibar 1970) as shown in Section 14.4. The relative importance of these two effects has been discussed by Epelboin (1979*a*) and the mechanism of interbranch scattering has been explained theoretically by Balibar *et al.* (1983). The dynamical image appears as a white line which is much wider in the centre of the section pattern than near the edges (Figs 17.8, 17.9, 17.20 and 17.27) and its visibility increases with increasing values of $\mu_o t$.

It can be seen in the experimental and simulated section topographs of a 4 mm thick silicon crystal (Fig. 17.27) that the dynamic image presents a fine structure and that there are additional fringes on each side of the main black–white–black contrast. They can be interpreted roughly as Fraunhofer diffraction fringes, the dislocation being assimilated to a diffracting object. The curved shape of the dynamical image can be explained geometrically. Let us consider two axes, Bx and By, on the exit surface of the crystal, parallel to the axis of the section topograph and to the base,

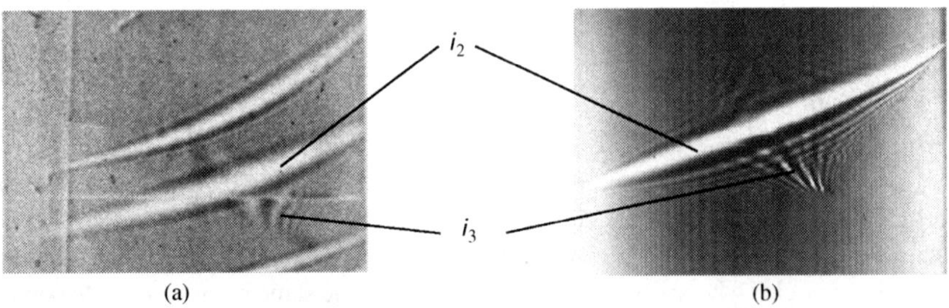

Fig. 17.27. Experimental (a) and simulated (b) section topographs of a 4 mm thick silicon crystal. 220 reflection, Ag$K\alpha$; $\mu_o t \approx 2.5$; i_2: dynamical image; i_3: intermediary image.

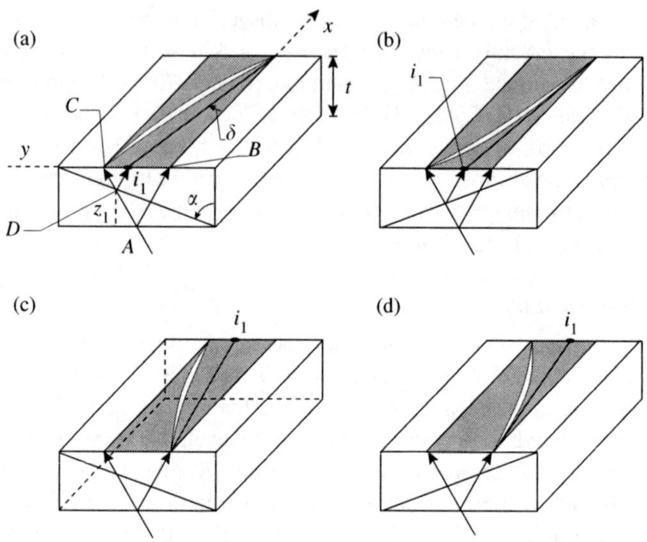

Fig. 17.28. Variation of the shape of the dynamical image of a dislocation with its orientation in the crystal. (a) $\alpha > 0$, $\delta < 0$; i_1: direct image; z_1: depth of the dislocation; t: crystal thickness. (b) $\alpha < 0, \delta < 0$; (c) $\alpha > 0, \delta > 0$; (d) $\alpha < 0, \delta > 0$.

BC, of the Borrmann triangle, respectively (Fig. 17.28(a)). The equation of the trace of the dynamic image on the exit surface is (Authier *et al.* 1985)

$$x = \frac{yz_1(\tan\alpha - \tan\theta_B)\tan^{-1}(\delta)}{t(\tan\alpha + \tan\theta_B) - y},$$

where α is the angle between the projection of the dislocation on the plane of incidence and the normal to the crystal surface, δ is the angle between the projection of the dislocation along the reflected direction on the exit surface and the axis of the section topograph, z_1 is the depth of the intersection of the dislocation with the direct beam and t is the crystal thickness. Figures 17.28(a)–(d) show how the curvature of the line changes according to the signs of α and δ.

17.3.4.4 *Intermediary images*

The intermediary image was first described by Authier (1967) on experimental topographs and is well visible on the computer simulations (Epelboin 1975, 1979*a*; Authier 1977*c*; Kowalski and Gronkowski 1982; Authier *et al.* 1985). It is due to interference effects between new wavefields resulting from interbranch scattering and normal wavefields whose tiepoints have remained on the same branch of the dispersion surface. It merges into the direct image and there is no discontinuity between the two images. The paths of the wavefields generated at the dislocation by interbranch scattering have been studied by Epelboin (1975, 1979*a*) using computer simulations. The interference effects which are characteristic of the intermediary image give rise to flat fringes (Authier 1967, 1977*c*) which extend over the whole width of the section pattern when the dislocation lies close to the entrance surface of the crystal and which remain visible even for large values of $\mu_o t$. They are in many ways similar to the fringes which appear in the image of a stacking fault and which are also due to interferences between wavefields which have and have not undergone interbranch scattering (Authier 1968). This image is well visible in Figs 17.8, 17.9, 17.20 and 17.27.

Another very interesting feature that can be observed in some cases is a line of high intensity along the projection of the dislocation line on the photographic plate, that is along what would be the direct image on a traverse pattern. Such an accumulation of intensity along the reflected direction from the dislocation was first to be observed on simulations of energy propagation in the plane of incidence (Epelboin 1975) and it was mentioned on a comparison of simulated and experimental topographs (Gronkowski and Kowalski 1981). It is very clearly seen on practically all simulated topographs of Fig. 17.9 and quite well on many of the experimental ones. Its contrast is the strongest when the dislocation lies closest to the direct beam side of the Borrmann prism.

17.3.4.5 *Deformation of the Pendellösung fringes due to the long-range strain field of the dislocation*

The normal spherical-wave Kato Pendellösung fringes are deformed due to the long-range strain field of dislocations, as can be seen in the experimental and simulated topographs of Fig. 17.9 and in the simulated topograph of Fig. 17.20. This effect is due to a decrease of the Pendellösung distance, itself a consequence of the bending of wavefields in the slightly deformed areas far from the dislocation core. It can obviously only be seen for values of $\mu_o t$ for which Pendellösung fringes are visible. It was first observed by Authier and Lang (unpublished, cited in Kato 1963*a*) and explained by Kato (1963*a*) and was further discussed by Suvorov *et al.* (1974) in terms of the phase shift to be associated with the relative translation of lattice planes above and below a dislocation. The paths of the wavefields can be calculated by means of the theory developed in Chapter 13.

Let us consider the simple case of a screw dislocation parallel to the plane of incidence, lying at depth z_1 (Fig. 17.29). The effective misorientation in the vicinity of the dislocation is given by (17.6) which can be rewritten, taking the origin at depth z_1, as

$$\delta\theta = -\frac{b}{2\pi}\frac{x}{x^2 + (z - z_1)^2}.$$

The slope $\mathrm{d}y/\mathrm{d}z$ of the tangent to the path of the wavefields in the plane of incidence is given by (5.21b) for non-deformed crystals:

$$\frac{\mathrm{d}y}{\mathrm{d}z} = \frac{\mp\eta}{\sqrt{\eta^2 + 1}} \tan\theta_B,$$

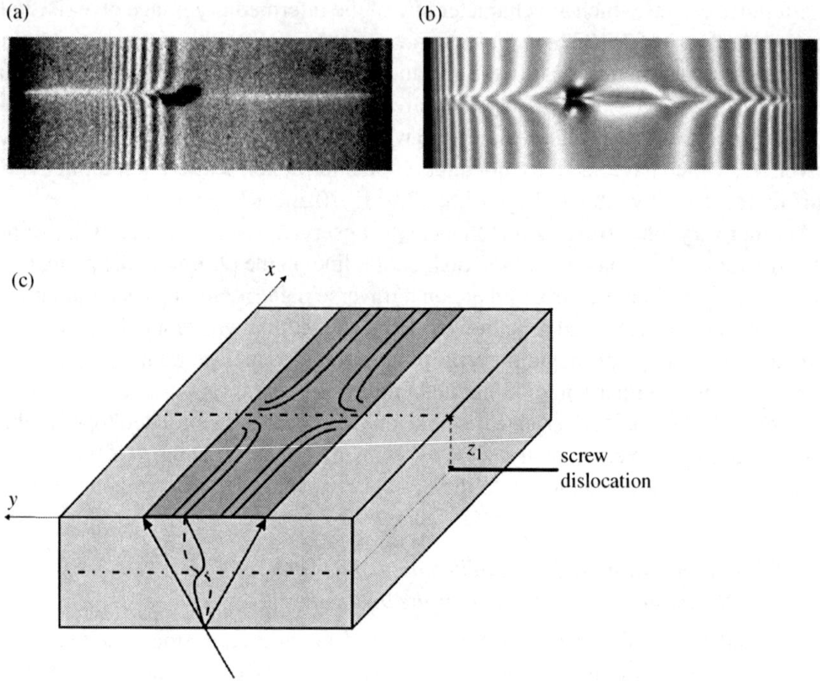

Fig. 17.29. Deformation of the Pendellösung fringes due to the long-range strains of a screw dislocation. The black dot is the direct image. (a) Experimental topograph of a wedge-shaped silicon crystal, which is why the fringes are only seen in the left-hand part of the crystal, 220 reflection, Mo$K\alpha$. (b) Computer simulation performed for a plane-parallel crystal. (c) Wavefield paths in the strain field of the dislocation (solid line: branch 1; dashed line: branch 2).

where the $-$ sign corresponds to branch 1 and the $+$ sign to branch 2 wavefields, and η is the usual deviation parameter defined in (5.18) which reduces to, for a symmetric reflection,

$$\eta = \frac{\Delta\theta \sin 2\theta_B}{|C|\sqrt{|\chi_h \chi_{\bar{h}}|}}.$$

When the wavefield penetrates in the strain field of the dislocation, η varies by a quantity:

$$\delta\eta = \frac{\delta\theta \sin 2\theta_B}{|C|\sqrt{|\chi_h \chi_{\bar{h}}|}}.$$

It follows, using (17.10), that

$$\frac{\mathrm{d}y}{\mathrm{d}z} = \mp\left(\frac{\eta}{\sqrt{\eta^2+1}} + \frac{\delta\eta}{(\eta^2+1)^{3/2}}\right)\tan\theta_B.$$

It follows, after integration, that

$$y = \mp\left[z\eta + \frac{A}{\eta^2+1}\left(\arctan\frac{z-z_1}{x} + \arctan\frac{z}{x}\right)\frac{\tan\theta_B}{\sqrt{\eta^2+1}}\right],$$

with

$$A = \frac{b\sin^2\theta_B}{\pi|C|\sqrt{|\chi_h\chi_{\bar{h}}|}}.$$

The paths of the two wavefields reaching a point of the exit surface are represented schematically in Fig. 17.29(c). The fringes which are observed on the exit surface and which can be seen in the experimental and simulated topographs of Figs 17.29(a) and (b) are hyperbolae whose asymptotes are the axis of the section topograph and the projection of the dislocation.

The modification of the Pendellsung fringes around a dislocation outcrop have been observed in the plane wave case by Ishida *et al.* (1976) and simulated using Takagi's equations by Ishida (1980).

17.3.5 *Images of planar defects*

17.3.5.1 *Classification of planar defects*

A planar defect is a surface of discontinuity separating two regions of perfect crystal. Different cases may be distinguished:

1. *Stacking fault*: The contrast of a stacking fault on a section topograph was first considered by Kato *et al.* (1967) in the non-absorbing case, but the first complete calculation showing all the features of the image and taking into account absorption is by Authier (1968). Many studies by computer simulations have been performed; a review is given in Epelboin (1985, 1987); a more recent study comparing simulations and synchrotron radiation topographs is given in Kowalski *et al.* (1989). The determination of the sign of the fault vector is discussed in Lang (1972, 1983*a*), Authier and Patel (1975) and Yang *et al.* (1990). All these studies refer to Laue–Laue case geometries. The Laue–Bragg case geometry is discussed by Indenbom and Chukhovskii (1974), Indenbom and Slobodetskii (1974) and Epelboin (1979*b*) in the case of a stacking fault and by Schlenker *et al.* (1968) in the case of walls between ferromagnetic domains in Fe 3% at. Si. The Bragg–Laue geometry was studied by Wierzchowski and Moore (1995). The nature of the stacking faults in growth striations in silicon was determined by double crystal topography in the Bragg case by Gabrielyan *et al.* (1986).
2. *Antiphase domain boundaries*: The contrast due to antiphase domain boundaries in ferroelastic, ferroelectric, gadolinium molybdate, $GdDy(MoO_4)_3$, was observed experimentally and compared to computer simulations by Capelle and Malgrange (1982, 1984*a*,*b*) and Capelle *et al.* (1982).
3. *Bicrystal*: The two regions are separated by a gap. This case was considered by Authier *et al.* (1968*b*), Bonse and Hart (1968), Tanemura and Lang (1973) and Bezirganyan *et al.* (1981).
4. *Moiré*: If the two parts of a bicrystal or two separate crystals are slightly rotated or have a different lattice parameter, moiré patterns can be observed. The first

observations were by Authier *et al.* (1964) on a silicon crystal bombarded with high-energy protons, Chikawa (1965*b*) from superposed CdS crystals, Lang and Miuscov (1965) with a quartz crystal containing a crack, Bonse and Hart (1966*b*) with an interferometer (see Subsection 15.5.2.1), Brádler and Lang (1968) and Lang (1968) from two separate silicon and quartz crystals, respectively and, as a recent example, by Fedorov *et al.* (2003) in the case of a non diffracting buried layer near the surface of a silicon crystal (other examples due to implanted layers are given in Section 17.3.6). The theory was first developed for electron diffraction by Hashimoto *et al.* (1961) and Gevers (1962) and extended to the X-ray case by Chikawa (1967). Simon and Authier (1968) calculated the contrast by deriving Takagi's equations and the theory was further developed by Tanemura and Lang (1973), Polcarová (1978*a*,*b*, 1980), Yoshimura (1996), Haroutyunyan and Sedrakyan (1997) and Ohler and Härtwig (1999).

5. *Growth twins*: Several cases have to be considered, depending on whether the absolute value of the structure factor is different in the two regions or simply the phase of the structure factor. Among the first observations of twins by topography were those of Lang (1967) and Phakey (1969) in quartz and of Chikawa and Austerman (1968) in BeO. The first discussions relative to their images were by Chikawa and Austerman (1968), Kato (1974) and Katagawa *et al.* (1975). A general review of the observations of twins by diffraction topography is given in Klapper (1991). Simulations of the contrast of twin boundaries in section topographs have been performed by Yang *et al.* (1986, 1991*a*) and González-Mañas *et al.* (1993).
6. *Misorientation boundaries*: The orientation and/or the lattice spacing are different in the two regions. This is frequently the case for growth sector boundaries and many planar defects such as twin boundaries when they are decorated by impurities, for instance. This case was discussed by Authier and Sauvage (1966), Kato (1974), Katagawa *et al.* (1975), Polcarová (1978*a*,*b*, 1980) and, more recently, by Yang *et al.* (1991*b*).
7. *Mechanical twins*: Mechanical twin lamellae were observed by Authier and Sauvage (1966) in calcite; their contrast is similar to that of a crystal with a gap to which is superimposed the effect misorientations when the density of dislocations in the boundary is very high.
8. *Growth bands and growth sector boundaries*: Growth bands and growth sector boundaries in solution-grown crystals are usually decorated by impurities and constitute discontinuity surfaces for the crystal. They give rise in section topographs to a contrast similar to that of misorientation boundaries (Yoshimatsu 1965; Authier 1977*c*; Sauvage *et al.* 1982; Léon *et al.* 1989).

17.3.5.2 *Study of planar defects in the plane-wave case*

1. *Wavefield multiplication at a planar defect—new wavefields*: There are essentially two ways of calculating the contrast of a planar defect. The first one follows the

principle of Kato's spherical-wave theory (see Chapter 10): one first calculates the contrast in the plane-wave case; the contrast on section topographs is then calculated by performing a Fourier transform. The second one solves Takagi's equations (see Chapters 11 and 14) in a crystal containing a planar defect (Authier and Simon 1968). This latter method is that which will be best suited for comparison with experiment because it takes into account the direct image and because it can include deformations as well as a phase shift or a modification of the structure factor. However, it does not provide a physical interpretation of the phenomena taking place at the fault plane. For this reason the former method will be briefly described here. For the details of the calculation, see Authier (1968).

Figure 17.30 explains the wavefield multiplication at the surface of the fault. Even if the fault plane is very thin, it is as if the two parts of the crystal on each side of the fault were separated by a gap. The two wavefields generated by the incident wave at the entrance surface split at the boundary into their two component waves, o_1, h_1 and o_2, h_2, respectively (Fig. 17.30(a)). Each one is an incident wave on part *II*

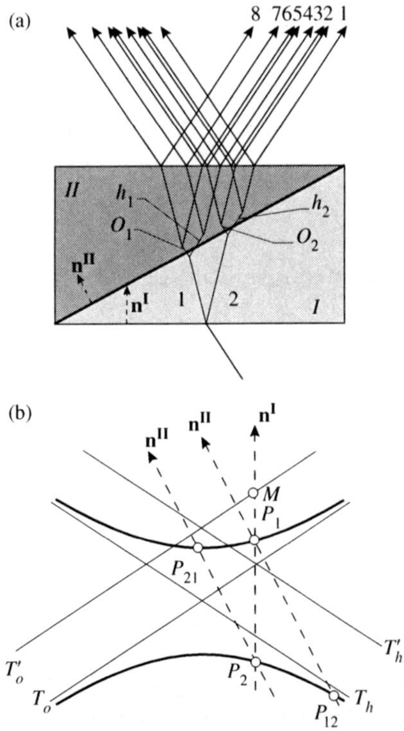

Fig. 17.30. Wavefield multiplication at the fault plane and paths of the wavefields in the plane-wave case; $\mathbf{n}^I$: normal to the crystal surface; $\mathbf{n}^{II}$: normal to the fault surface. (a) Real space; (b) Reciprocal space.

of the crystal and generates two wavefields whose tiepoints lie on the normal $\mathbf{n}^{II}$ to the boundary (Fig. 17.30(b)). Eight wavefields are thus generated: four have the same tiepoints as the wavefields propagating in part I (P_1 and P_2) and are called *normal* wavefields; the four others have as tiepoints P_{12} and P_{21} and are called *new* wavefields. When they reach the exit surface of the crystal, they in turn split into their o and h components. The reflected and refracted waves are therefore the sum of eight waves each, which interfere. Let $\mathcal{A}'$ and $\mathcal{A}''$ be the total complex amplitudes of the normal and new wavefields, respectively. The total intensity of the reflected beam at the exit surface is

$$I = I_1 + I_2 + I_3 = |\mathcal{A}'|^2 + |\mathcal{A}''|^2 + 2\mathcal{R}e(\mathcal{A}'\mathcal{A}''), \tag{17.11}$$

where $\mathcal{R}e(\,)$ means real part of.

2. *Stacking faults*: The influence of the stacking fault is to introduce a phase shift between the two parts of the crystal. The polarizability of part I is expanded in Fourier series (2.36):

$$\chi^I(\mathbf{r}) = \sum_{\mathbf{h}} \chi_h^I \exp(2\pi i \mathbf{h} \cdot \mathbf{r}). \tag{17.12}$$

Part II is translated with respect to part I through a translation $\mathbf{u}$, so that the position vector of a point of part II is

$$\mathbf{r}' = \mathbf{r} + \mathbf{u}.$$

By substitution in (17.12), it follows, for the second part, that

$$\chi^{II}(\mathbf{r}') = \sum_{\mathbf{h}} \chi_h^I \exp[2\pi i \mathbf{h} \cdot (\mathbf{r}' - \mathbf{u})],$$

which can also be written as

$$\chi^{II}(\mathbf{r}) = \sum_{\mathbf{h}} \chi_h^{II} \exp(2\pi i \mathbf{h} \cdot \mathbf{r})$$

with

$$\chi_h^{II} = \chi_h^I \exp(-2\pi i \mathbf{h} \cdot \mathbf{u}) = \chi_h^I \exp(i\delta), \tag{17.13}$$

where δ is the phase shift introduced by the stacking fault.

The expression of the reflected intensity will not be developed here. There are few experimental studies of the contrast of stacking faults in the plane-wave case (see for instance Sauvage and Malgrange (1970), who observed stacking faults in silicon in the transmission geometry, and Wierzchowski and Moore (1995), who observed stacking faults in diamond in the reflection geometry).

3. *Twins*: A similar treatment can be followed with two possibilities: either the absolute value of the structure factor is the same in the two parts and (17.13) applies, or it is not and χ_h^{II} and χ_h^{I} differ both in magnitude and phase. In the latter case, the dispersion surface has a different diameter in the two parts of the crystal.

17.3.5.3 *Contrast of planar defects in the spherical-wave case*

1. *Geometrical study, focalization of the new wavefields*: Figure 17.31(a) shows the paths of the normal and new wavefields in the Borrmann triangle and Fig. 17.31(b) shows their tiepoints in reciprocal space. It is assumed here that there is no difference in the magnitude of the structure factor in the two parts. Let us consider a path Aq in part I; it is followed by two wavefields, of tiepoints Q_1 and Q_2. At the fault plane they generate *normal* wavefields with the same tiepoints which propagate along the same direction in part II and *new* wavefields of tiepoints P_2 and P_1, respectively, which propagate along path qp in part II. In part I, there are also wavefields with tiepoints P_1 and P_2; they propagate along Ap'. When they reach the fault plane they generate in part II *normal* wavefields with the same tiepoints and *new* wavefields with tiepoints Q_2 and Q_1. These wavefields propagate along $p'q'$ in part II. Let A' be the intersection of qp and $p'q'$. Depending on the depth of the fault in the crystal, it lies

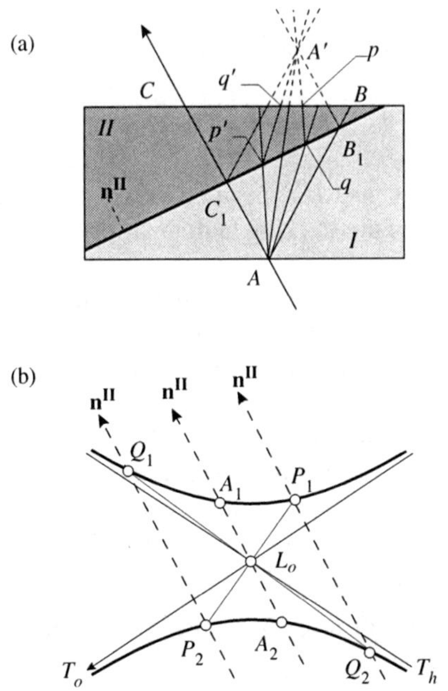

Fig. 17.31. Paths and tiepoints of the wavefields generated at the fault plane in the spherical-wave case; $\mathbf{n}^{II}$: normal to the fault surface. (a) Real space. (b) Reciprocal space.

either inside or outside the crystal. Let A_1 and A_2 be the extremities of the diameter of the dispersion surface which is parallel to the normal $\mathbf{n}^{//}$ to the fault plane (and to Q_2P_1 and P_2Q_1). It is a property of hyperbolae that the wavefields of tiepoints A_1 and A_2 propagate along the same direction, AA', in both parts of the crystal. In this manner the paths of all the *new* wavefields will focalize at A'. The new wavefields generated at the intersection C_1 of the direct beam with the fault plane produce the direct image of the fault plane.

In the case of a twin boundary (Yang *et al.* 1986; Yang *et al.* 1991*a*) or of a misorientation boundary (Authier and Sauvage 1966; Kato 1974; Yang *et al.* 1991*b*), the focalization effect breaks down.

2. *Intensity distribution—Pendellösung fringes*: The calculation of the diffracted amplitudes is performed by a Fourier transform of $\mathcal{A}'$ and $\mathcal{A}''$. The integration is done using the stationary phase method. The details and the full expressions are given in Authier (1968). As in the plane-wave case, the intensity distribution on the exit surface is given by a sum of three terms:

$$I = I_1 + I_2 + I_3,$$

where the three terms are of the form

$$I_1 = I_{\text{perfect}}\left[1 - A\sin^2\left(\frac{\delta}{2}\right)\right],$$

$$I_2 = B\sin^2\left(\frac{\delta}{2}\right),$$

$$I_3 = C\sin^2\left(\frac{\delta}{2}\right) + D\sin(\delta),$$

δ is the phase shift, A, B, C and D depend on the conditions of the reflection, on the depth and the orientation of the stacking fault in the crystal and on the position on the exit surface.

The first term, I_1, corresponds to the interference of the *normal* wavefields and gives rise to the usual hyperbolic Pendellösung fringes (Fig. 17.32). It reduces to the expression, I_{perfect}, of the intensity for the perfect crystal when $\delta = 0$. The second term, I_2, corresponds to the interference of the *new* wavefields and gives rise to hyperbolic fringes asymptotic to $A'B_1$ and $A'C_1$. The third term, I_3, gives rise to interferences between *new* and *normal* wavefields. The corresponding fringes are 'flat' fringes which pass through the intersections of the I_1 and I_2 fringes (Figs 17.32 and 17.33). The contrast of the first fringe depends on the value of the phase shift, δ, and provides a means to determine it (Lang 1972, 1983*a*; Authier and Patel 1975; Yang *et al.* 1990).

Examples of the images of stacking faults, twin boundaries and misorientation boundaries are given below.

1. *Stacking faults*

Figure 17.34(a) shows an experimental section topograph of a stacking fault in silicon. The corresponding crystal was a Czochralski-grown (111) silicon crystal containing

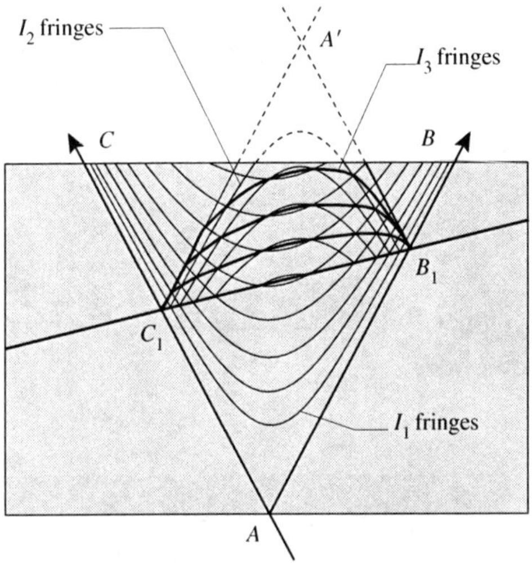

Fig. 17.32. Stacking fault: fringes due to I_1, asymptotic to AB and AC; fringes due to I_2, asymptotic to $A'B_1$ and $A'C_1$; fringes due to I_3, flat fringes.

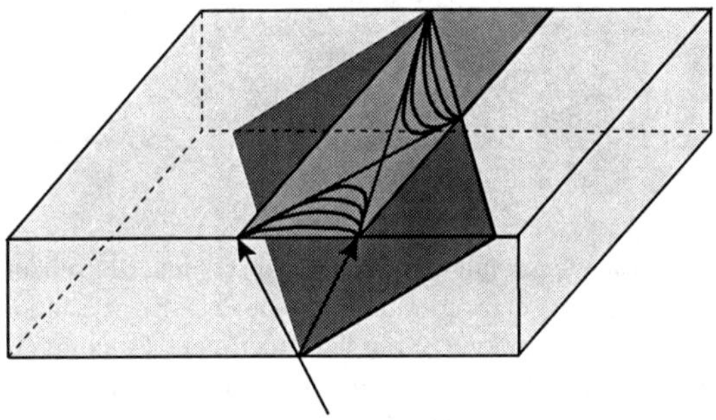

Fig. 17.33. Stacking fault contrast: fringes due to I_3 or 'flat' fringes; dark grey: plane of the fault; medium grey: trace of the reflected beam on the exit surface.

oxygen, annealed at 1200 °C. During the annealing, silicon oxide precipitated and stacking fault loops were formed (Patel and Authier 1975; Authier and Patel 1975). These stacking faults lie in the $\{111\}$ planes (Fig. 17.35). They are bordered by partial dislocations, and by taking topographs with different reflections it was possible to show that these dislocations are of the *sessile* or Frank type, with a Burgers vector $\frac{1}{3}\langle 111\rangle$, perpendicular to the plane of the fault.

The figure also gives an example of the determination of the sign of the phase shift associated with the stacking fault. Depending on the sign of the fault vector, the fault is either extrinsic (if the fault is a disk of interstitials) or intrinsic (if the fault is a disk of vacancies). Figures 17.34(b) and (c) are computer simulations (Authier and Patel 1975; Wonsiewicz and Patel 1976) of the same fault as in Fig. 17.34(a),

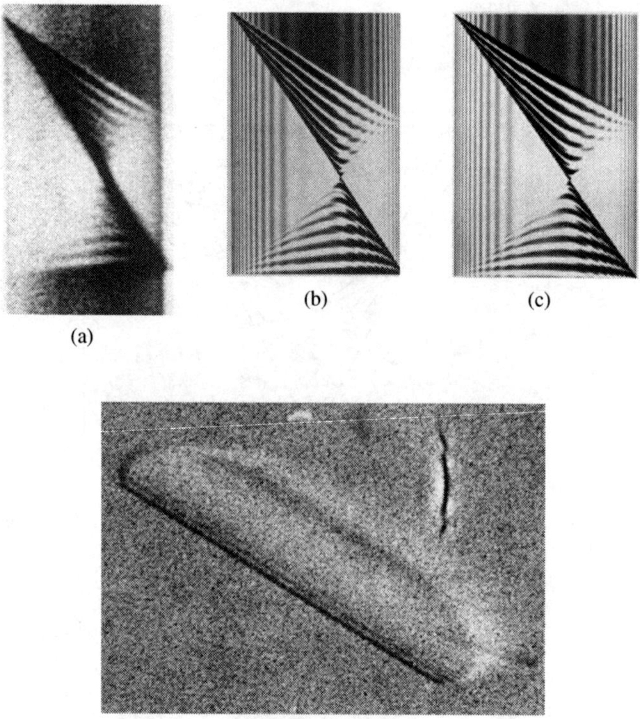

(a) (b) (c)

(d)

Fig. 17.34. Contrast of a stacking fault in a (111) silicon crystal, $1\bar{1}\bar{1}$ reflection, Cu$K\alpha$. (a), (b) Section topographs, (a) experimental; (b) simulation for an extrinsic stacking fault, the contrast of the first fringe agrees with that observed in the experimental topograph; (c) simulation for an intrinsic stacking fault; (d) projection topograph. (a) and (d) after Authier and Patel (1975), (b) and (c) after Wonsiewicz and Patel (1976).

but with opposite signs of the fault vector, and therefore opposite signs of the phase $\delta = -2\pi \mathbf{h} \cdot \mathbf{u}$ where $\mathbf{u}$ is the fault vector. It can be seen that there is a reversal of the contrast of the first fringe between the two simulations and that the agreement with the experimental topograph is very good if one assumes that the fault is extrinsic (simulation of Fig. 17.34(b)). Figure 17.34(d) shows the projection topograph of the same stacking fault as in Fig. 17.34(a). One can see that the integration in the projection topograph results in equal-depth fringes.

2. *Twin boundaries*

Figure 17.36 shows computer simulations of section patterns of a hypothetical crystal containing a twin. The structure factors of the two parts are assumed to be different:

$$F_h^I = 1.5 F_h^{II} \exp(-2\pi \mathrm{i}\delta).$$

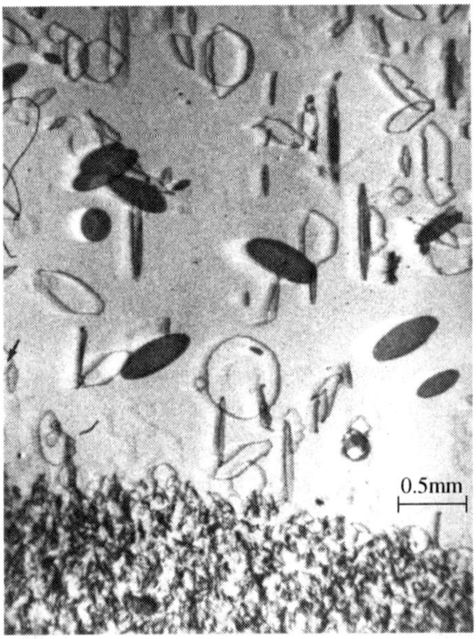

Fig. 17.35. Traverse topograph of a (111) silicon crystal; $1\bar{1}\bar{1}$ reflection, Mo$K\alpha$, crystal thickness 1 mm. The images of stacking faults are grey circles or ellipses, depending on the orientation of the plane of the fault. Courtesy J. R. Patel.

The simulations (Kato 1974; Katagawa *et al.* 1975) are performed for three different values of the phase shift: $\delta = 0$ (Fig. 17.36(a)), $\delta = \pi/2$ (Fig. 17.36(b)), $\delta = \pi$ (Fig. 17.36(c)). The effect of the variation of the Pendellösung distance on the I_1 fringes is most visible for $\delta = 0$ while the I_3 fringes are most visible for $\delta = \pi$. Authier (1977*c*) has given a geometrical construction of the shape of the I_1 and I_3 fringes.

3. *Misorientation boundaries*

Kato (1974) and Katagawa *et al.* (1975) have simulated section patterns of an hypothetical crystal containing a misorientation boundary, assuming that there is a phase shift $\delta = \pi$ between the two parts (Fig. 17.37). The defocalization effect, which was first observed by Authier and Sauvage (1966), increases with increasing misorientations.

As an example, Fig. 17.38 is a section topograph of a boundary between two neighbouring growth sectors in potash alum, typical of a misorientation boundary. A geometrical interpretation of the defocalization effect has been given by Authier (1977*c*). It enables one to evaluate the misorientation between the two growth sector boundaries to about 0.45 arcsec.

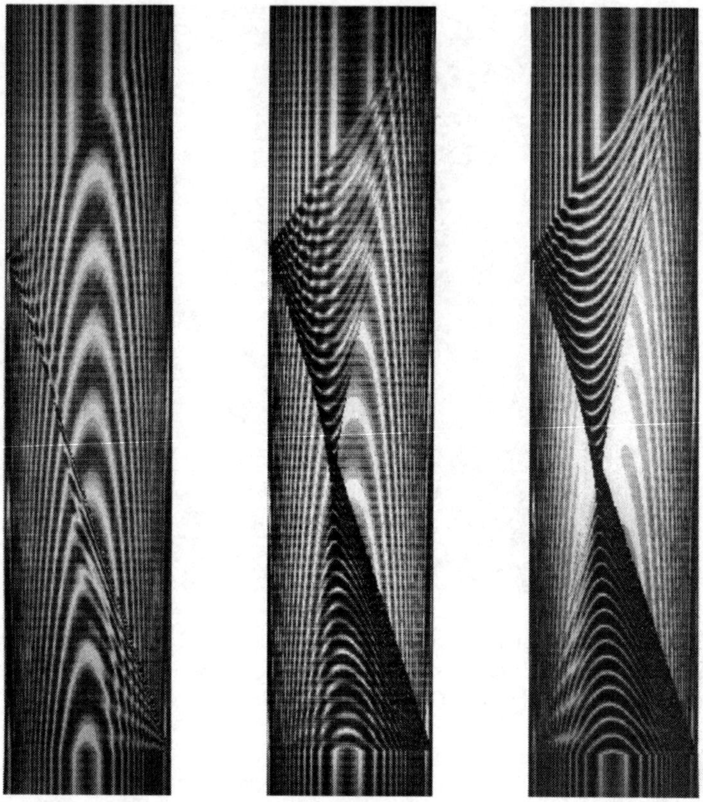

(a) (b) (c)

Fig. 17.36. Computer simulations of the section patterns of a hypothetical crystal containing a twin with $F_h^I = 1.5F_h^{II}$, where F_h^I and F_h^{II} are the structure factors of parts *I* and *II*, respectively. (a) $\delta = 0$; (b) $\delta = \pi/2$; (c) $\delta = \pi$. Courtesy N. Kato.

17.3.5.4 *Contrast of planar defects in projection topographs*

The expression of the reflected intensity on a projection topograph is obtained by integration. One obtains (Authier and Simon 1968)

$$I^i \propto \int_0^{2\pi t/\Lambda_L} J_0(\zeta)\,\mathrm{d}\zeta + 4Fz^j \sin^2\left(\frac{\delta}{2}\right)\left[\int_0^{2\pi t/\Lambda_L} J_1(\zeta)\,\mathrm{d}\zeta - 1\right],$$

where z^j is the depth of the fault from the surface ($j = I$, entrance surface, or *II*, exit surface) and $F = (2\pi k|C|(\sqrt{\chi_h\chi_{\bar{h}}}\sqrt{\gamma_o\gamma_h})/\sin^2\theta_B$.

The intensity on a projection topograph depends on the value of the phase shift, but not on its sign, and section topographs are necessary to determine it. The first projection topographs of stacking faults were observed in silicon by Kohra and Yoshimatsu

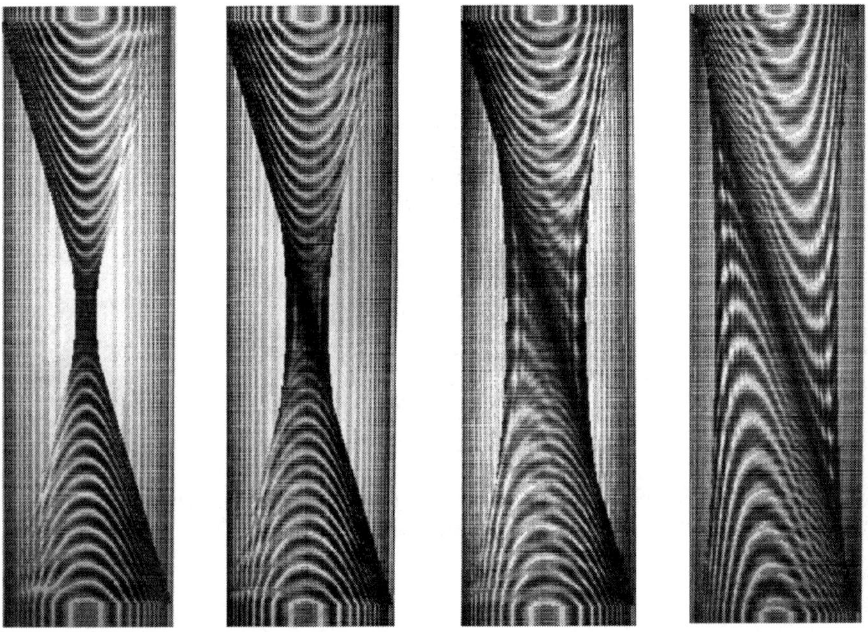

(a) (b) (c) (d)

Fig. 17.37. Computer simulations of the section patterns of a hypothetical crystal containing misorientation boundaries with a phase shift $\delta = \pi$ between the two parts. (a) $x = 1/8$; (b) $x = 1/4$; (c) $x = \frac{1}{2}$; (d) $x = 1$($x = \Delta\theta/2\delta_{os}$, where $\Delta\theta$ is the misorientation between the two parts and $2\delta_{os}$ is the width of the rocking curve). Courtesy N. Kato.

(1962), Schwuttke and Sils (1963), Yoshimatsu (1964), Meieran and Blech (1967) and Authier (1968) and in ice by Oguro and Higashi (1973).

17.3.6 *Applications*

The number of papers devoted to the applications of transmission single-crystal topography is so large that it is only possible to list here some of the problems for which diffraction topography has proved to be very useful. The first purpose is the overall characterization of the defects in crystals:

- Dislocations
- Planar defects and domain walls
- Precipitates and inclusions
- Radiation damage
- Long-range strains.

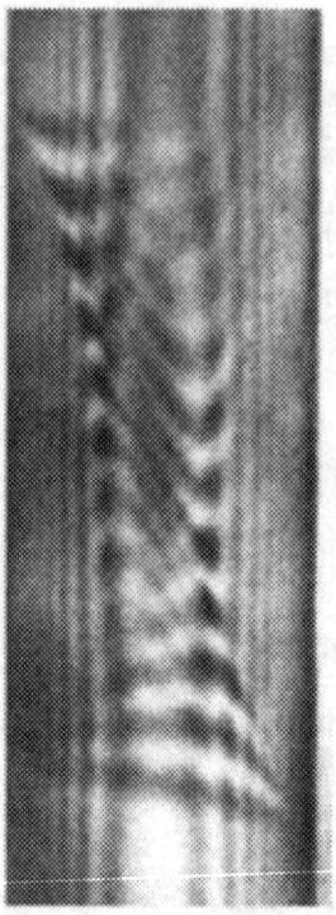

Fig. 17.38. Section topograph of a growth sector boundary in potash alum; 220 reflection, Mo$K\alpha$; crystal thickness 1.5 mm. The contrast is typical of a misorientation boundary. Courtesy F. Lefaucheux and M. C. Robert.

As examples of the type of problems studied by means of X-ray topography, one may mention:

(1) *Generation of defects during growth*: For minerals see the review by Authier and Zarka (1992); for synthetic crystals see the reviews by Authier (1972, 1977*a*,*b*), Bhat (1985), Klapper (1990, 1991, 1996); for the characterization of protein crystals see, for instance, Dobrianov *et al.* (1998, 1999), Lorber *et al.* (1999), Boggon *et al.* (2000), Hu *et al.* (2001, 2004), Vetter *et al.* (2002), Tachibana *et al.* (2003) and Capelle *et al.* (2004); for the characterization of space-grown crystals, see for instance Authier *et al.* (1987), Robert *et al.* (1988), Steiner *et al.* (1991), Calzadilla *et al.* (1991) and Raghothamachar *et al.* (1998); for recent, real time, *in situ* studies, see below. It should be noted that, owing to the high flux of synchrotron sources it is possible to characterize highly absorbing crystals with white beam topography such as, for instance, mercury iodide (HgI_2) (Rémy *et al.* 1994; Rossberg *et al.* 1996).

(2) *Phase transitions*: For reviews see Bordas (1975); Aleshko-Ozhevskii (1982); Klapper *et al.* (1983) and Baruchel and Schlenker (1996); for recent, real time, *in situ* studies, see below.

(3) *Magnetic domain walls*: Neutron topography is of course the ideal tool for the observation of magnetic domains. The first observation of antiferromagnetic domains was reported by Ando and Hosoya (1972*b*) in chromium, of ferromagnetic domains by Schlenker and Shull (1973) and Schlenker *et al.* (1977, 1978) in Fe–Si and of chirality domains by Baruchel *et al.* (1981) in the helimagnetic phase of terbium; for reviews see Schlenker and Baruchel (1978, 1986, 1999), Guigay *et al.* (1984), Baruchel and Schlenker (1984, 1996). Domain walls are also visible with X-rays because of the distortions around them and many papers were devoted to that topic; pioneer work was done on Fe3% Si by Polcarová and

Lang (1962), Roessler *et al.* (1965), Frank *et al.* (1966, 1993) and Schlenker *et al.* (1968), on terbium iron garnet by Basterfield and Prescott (1967) and on yttrium iron garnet (YIG) by Patel *et al.* (1968); reviews are given in Polcarová (1969), Schlenker and Kleman (1972), Tanner (1981), Miltat (1984*a*), Baruchel (1993), Baruchel and Schlenker (1996); for recent studies with synchrotron radiation, see Baruchel *et al.* (1988); Clark *et al.* (1983); Medrano *et al.* (1998, 1999); Polcarová *et al.* (1999*a*); Baruchel *et al.* (2001); Schetinkin *et al.* (2003).

(4) *Ferroelectric domain walls*: Early work was done on triglycine sulphate (TGS) by Authier and Pétroff (1964), Pétroff (1969), Izrael *et al.* (1972), Takahashi and Takagi (1978*a*,*b*) and Parpia (1982*a*,*b*), barium titanate ($BaTiO_3$) by Caslavsky and Polcarová (1964), Akaba *et al.* (1979) and Takagi *et al.* (1979) and on sodium nitrite ($NaNO_2$) by Suzuki and Takagi (1966); for recent studies with synchrotron radiation, see, for instance, Ribet *et al.* (1990) on HTMA ($[N(CH_3)_4]_2ZnCl_4$); Robert *et al.* (1991) on GASH (guanisine aluminium sulfate hexahydrate); Aoyama *et al.* (1992) on thiourea; Huang *et al.* (1996) and Pernot-Rejmánková *et al.* (2000) on $KTiOAsO_4$ (KTA) and $KTa_{1-x}Nb_xO_3$ (KTN) Park *et al.* (1997) on $BaTiO_3$; Hu *et al.* (1996), Rejmánková *et al.* (1998) and Baruchel *et al.* (2000) on $LiNbO_3$.

(5) *Ferroelastic domains*: Barium sodium niobate (Jiang *et al.* 1988; Jiang *et al.* 1993); HTMA ($[N(CH_3)_4]_2ZnCl_4$) (Ribet *et al.* 1990); Neodymium pentaphosphate (NPP) (Huang *et al.* 1996).

(6) *Plastic deformation*: Early observations of dislocation sources and dislocation movement are reported in germanium by Gerold and Meier (1962), Meier (1962) and Datsenko and Skorokhod (1967), in silicon by Authier and Lang (1964), Foct and Champier (1965), Meieran (1965) and Suzuki and Kojima (1966), in copper by Elistratov and Madzhitov (1965) and Clarebrough *et al.* (1967), in lithium fluoride by Forwood and Lawn (1966) and Pétroff and Authier (1966), in diamond by Wild *et al.* (1967). Twinning dislocations and mechanical twinning were studied in calcite by Sauvage and Authier (1965), Authier and Sauvage (1966), Sauvage (1968) and Jourdan and Sauvage (1970). Dislocation velocities were measured in silicon by Patel and Freeland (1967, 1970) and Patel *et al.* (1976) and in ice by Shearwood and Whitworth (1991, 1992*a*, *b*). Major developments in the study of plastic deformation can be found, among others, for silicon in Miltat and Bowen (1970, 1979) and Jung and Lefeld-Sosnowska (1984*a*,*b*), for copper in Young and Sherrill (1971, 1972) and Pichaud and Minari (1976, 1984), for ice in Fukuda *et al.* (1987) and Ahmad *et al.* (1992), for benzil in Scheffen-Lauenroth *et al.* (1981) and Klapper (1996). For recent, *in situ*, real time experiments with synchrotron radiation, see below.

(7) *Implantation*: See, for instance, Authier *et al.* (1963); Simon and Authier, 1968; Bonse *et al.* 1969; Tomimitsu, 1983; Ohler *et al.* 1997, 1999; Wieteska *et al.* 1997, 2000; Lang and Makepeace 1999*a*).

In particular since the advent of synchrotron radiation, it has been possible to do *in situ*, real-time experiments for:

(1) *Crystal growth*: Silicon (Chikawa 1974, 1978, 1980); aluminium (Gastaldi *et al.* 1988, 1992; Gastaldi *et al.*, 1988; Grange *et al.* 1987, 1990); aluminium–copper alloys (Grange *et al.* 1995); InGaAs epilayers on GaAs (Tanner *et al.* 2001); directional solidification of NiAl alloys (Nguyen *et al.* 2003); annealing of helium crystals at 0.3 K (Nakajima *et al.* 1995).

(2) *Plastic deformation*: In silicon (Michot *et al.* 1984; Vallino *et al.* 2000); in ice (Ahmad *et al.* 1986, 1987; Hondoh 1986; Ahmad and Whitworth 1988; Hondoh *et al.* 1989; Hondoh *et al.* 1990; Shearwood and Whitworth 1989, 1992*a*,*b*; Liu *et al.* 1993*a*,*b*, 1995); in Fe-4at% Si (Polcarová *et al.* 1998, 1999*b*, *c*, 2003; Gemperlová *et al.* 1999); dislocation generation in InGaAsP/AlGaAs/GaAs quantum well lasers (Mazuelas *et al.* 2001).

(3) *Phase transitions*: α–β phase transition in quartz (Zarka 1983); orthorhombic to tetragonal transition in non-linear-optical $KNbO_3$ (Zhao *et al.* 1991; Wang *et al.* 1998); orthorhombic to rhombohedral and rhombohedral to cubic transitions in $LaGaO_3$ and $LaAlO_3$, respectively, which serve as substrates for high T_c superconductor epilayer growth (Dudley and Yao 1993); hexagonal to orthorhombic transition in potassium lithium sulphate (Scherf *et al.* 1997); cubic to tetragonal transition in strontium titanate (Ozaki *et al.* 2003) paraelectric to ferroelectric phase transition in ammonium sulphate (Bhat *et al.* 1995); ferroelectric phase transitions in DKDP and KDP (Aleshko-Ozhevskii *et al.* 1985); hcp to bcc phase transition of titanium (Jourdan *et al.* 1991); titanium orientation memory effect (Jourdan *et al.* 1992); martensitic transformation of the CuZnAl shape memory alloy (Jourdan *et al.* 1995*a*,*b*).

(4) *movement of magnetic domains under the influence of an applied field* (see, for instance, Dudley and Miltat 1989);

(5) *influence of an applied electric field on the defects* (Baruchel *et al.* 1991; Rejmánková *et al.* 1995, 1996);

(6) *stroboscopic study of the propagation of acoustic waves* (see, for piezoelectric materials, Whatmore *et al.* 1982; Gluer *et al.* 1983; Cerva and Graeff 1984, 1985; Zarka *et al.* 1987; Capelle *et al.* 1992, 1998, 2001; Détaint *et al.* 1998; Zolotoyabko *et al.* 1998, 1999; Shilo and Zolotoyabko 2003; for magnetic materials, Miltat and Kléman 1979; Matsouli *et al.* 1998, 1999; Lorut *et al.* 2003).

The types of materials to which diffraction topography has been applied are very varied: semiconductors, metals, non-metallic elements, oxides and ionic compounds, organic materials, minerals and even, recently, quasicrystals (Wang *et al.* 1994; Baruchel *et al.* 1994, 2000; Gastaldi *et al.* 1995, 1999; Mancini *et al.* 1998*a*,*b*).

17.4 Double- or multiple-crystal topography

17.4.1 *Principle of double-crystal topography*

Double-crystal topography was introduced as early as 1958 by Bonse (Bonse and Kappler 1958; Bonse 1958) and was one of the first topographic methods for the observation of single dislocations. It was used in particular for the study of the contrast of dislocation outcrops or of dislocations lying close to the surface by Bonse (1962, 1964*b*,*c*), Renninger (1962, 1963, 1964, 1965*a*,*b*, 1967), Renninger and Theis (1969), Wierzchowski (1985). In order to reveal very fine microdefects such as swirls, Renninger (1976, 1981) adapted the "kinematical image technique" invented by Chikawa *et al.* (1970) to enhance the direct images. Chikawa *et al.* screened off the dynamically diffracted rays of the Borrmann fan and Renninger preferred offsetting the sample with respect to the monochromator. Double-crystal topography was also used to measure the concentration of impurities in single crystals, for instance of various dopants in gallium arsenide by Baker *et al.* (1976), of carbon in silicon by Hart (1968*b*), of various dopants in silicon by Buschert *et al.* (1980) or to analyse the variations of the lattice parameter in relation with the growth conditions of the crystal: for instance, in quartz by Grosswig *et al.* (1983), in yttrium alumnium garnet (YAG) by Gao *et al.* (1987), in CdTe by Mühlberg *et al.* (1990), in diamond by Wierzchowski *et al.* (1991), in gallium arsenide by Kitano *et al.* (1991). A review of the possibilities of double crystal topography for accurate lattice parameter mapping is given in Hart (1981). Improvements of the technique have been reported recently by Kudo *et al.* (2000) and applied to the case of silicon.

The main applications of double-crystal topography are related to its sensitivity to small misorientations. For small distortions to be easily detected, the reflected intensity should be a rapidly varying function of the local orientation and spacing of the lattice planes, the divergence of the beam should be at most of the order of the width of the rocking curve and its spectral width should introduce no broadening. This is achieved by the double-crystal setting represented in Fig. 17.39(a). In the ideal situation, two crystals of the same nature are set in the non-dispersive $(+, -)$ arrangement which has been described in Section 15.2.3. The first crystal, or reference crystal, should be as perfect as possible. In the laboratory a line focus parallel to the plane of incidence is used: because of the small value of the divergence the beam can be wide. The lattice planes of the reference crystal and the sample must be well aligned to the vertical direction, otherwise the field of view is restricted to a horizontal line. This can be, however, useful in some cases to measure local distortions (see Section 17.4.3).

The rocking curve of the second crystal is represented schematically in Fig. 17.39(b) and has been discussed in Section 15.2.3. If the crystal is set at a position for which the intensity is at half maximum (positions 1 or 2), any local distortion will induce a rapid increase or decrease of intensity, depending on the sense of the local rotations of the lattice planes, as indicated by the arrows. As has been seen in Section 15.2.3, the curve is close to an isosceles triangle in the low-absorption case. Around positions 1 and 2 the intensity varies linearly with angle if the distortions are small. The local value of the distortion can thus be measured directly from the intensity. Distortions corresponding to a small fraction of the rocking curve width can be evaluated. The sense of the intensity variations are opposite for positions 1 and 2 and one observes contrast reversal between the two as shown in Fig. 17.40 for dislocation outcrops in silicon (Renninger 1965*a*).

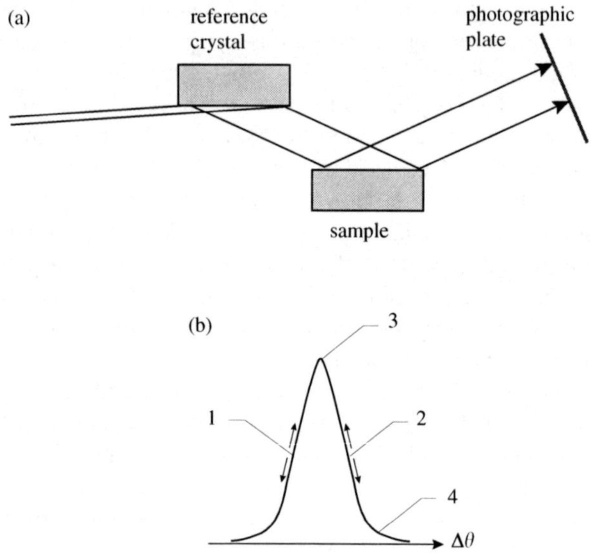

Fig. 17.39. Principle of double-crystal topography for a high sensitivity to small strains. (a) Experimental setup. (b) Rocking curve.

If the crystal is set at position 3, the intensity variation will be a decrease, whatever the sense of local lattice rotations. It is also very interesting to set the crystal at position 4. In that case the average reflectivity of the crystal is low, but for distortions large enough, and of the right sense, the local diffracted intensity may reach the maximum value. This occurs for regions which are closer to the core of dislocations, and narrow images of dislocations can be obtained. This is equivalent to the *weak beam* technique for electron diffraction (see for example Fig. 17.41).

With a two-crystal setting the non-dispersive character of the $(+, -)$ arrangement requires that the reference crystal and the sample be of the same nature. This is a severe restriction and it is not always possible to find a sufficiently perfect crystal to use as a reference. What is actually necessary is that the lattice spacing of the first crystal should be close to that of the sample. Deslattes and Paretzkin (1968) have collected for that purpose a set of silicon reflections which can be used for various samples. This requirement is, however, not so stringent for synchrotron radiation, as will be mentioned in Section 17.4.2. Chukhovskii *et al.* (1992) have suggested that it is also possible to go round the restriction by bending the two crystals to appropriate radii.

Overviews of the experimental arrangements and diffraction conditions for double-crystal topography are given, for instance, in Sauvage (1978, 1980), Kuo and Bilello (1984) and Matsouli *et al.* (2000).

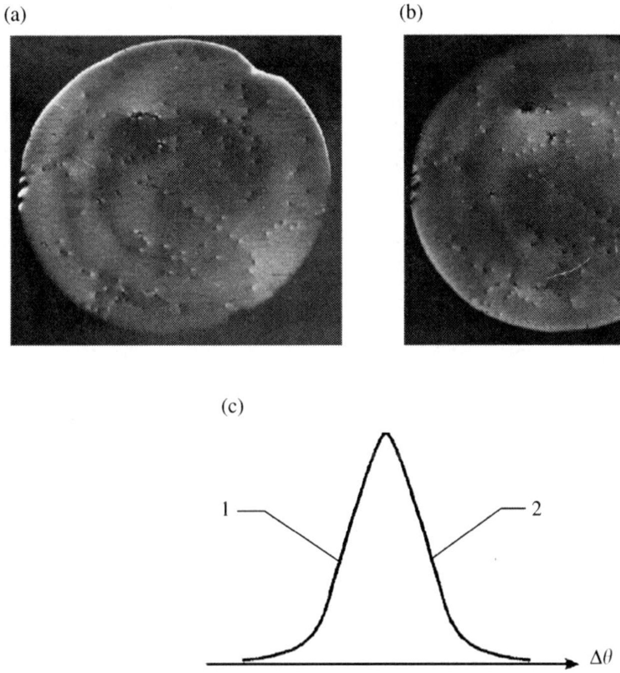

Fig. 17.40. Double-crystal topographs of dislocations in a (111) silicon crystal, $1\bar{1}\bar{1}$ reflection, Cu$K\alpha$. (a), (b) Topographs taken at positions 1 and 2 of the rocking curve, respectively. (c) Rocking curve. Courtesy M. Renninger.

17.4.2 *Plane-wave topography*

From the preceding subsection it is clear that the sensitivity of the method is higher, the closer the sample rocking curve is to the intrinsic rocking curve. The rocking curve of the first crystal must be as narrow as possible. For this reason, Renninger (1961, 1965, 1967) recommends the use of an asymmetric reflection for the first crystal, as shown in Fig. 17.39a (see also Kohra 1972; Kuriyama and Boettinger 1976; Boettinger *et al.* 1976, 1979, 1982). This is the arrangement which is widely used now. An even better situation can be achieved with an incident pseudo-plane wave. This was already done by Chikawa and Austerman (1968) using the same setting as Authier (1960*a*, 1961*a*) for observing double refraction of X-rays: a slit at the centre of the beam diffracted by a crystal in the transmission geometry (see Fig. 12.14). Nowadays, plane-wave monochromators (Pétroff *et al.* 1980; Sauvage and Pétroff 1980) or multiple-crystal arrangements producing pseudo-plane waves are used with synchrotron radiation; for instance, for the study of microdefects or swirls in silicon (Chikaura *et al.* 1987, 1990; Chikaura and Imai 1990; Kimura *et al.* 1992, 1994*b*), for the characterization of a six-inch silicon wafer (Kitano *et al.* 1987), for the study of growth striations in

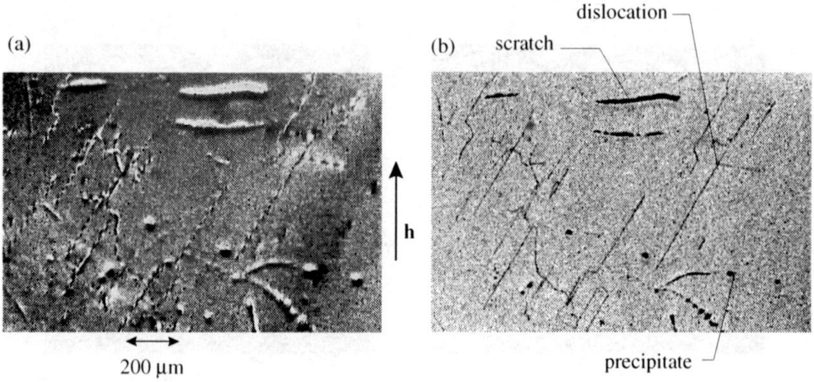

Fig. 17.41. Synchrotron double-crystal topographs of a germanium crystal (transmission geometry); 111 reflection, $\lambda \approx 0.35$ Å; the monochromator is a silicon (111) crystal. Dislocations, scratches and precipitates can be seen. Courtesy J. Baruchel. (a) Topograph taken with the crystal set at the top of the rocking curve. The dynamical and intermediary images predominate. (b) Topograph of the same area taken with the crystal set at the high angle flank of the rocking curve. The images of dislocations are narrower and direct images predominate.

quartz (Bernhardt *et al.* 1992) or in silicon (Kawado *et al.*,1989, 1991*a*,*b*) or of high resolution images of dislocations in silicon (Kimura *et al.* 1998; Iida *et al.* 2002). One then speaks of plane-wave topography.

Asymmetric reflections can be made use of to magnify the image and increase the resolution. This was done for one-dimensional magnifying by Boettinger *et al.* (1979) and by Spal (2001), Schäfer and Köhler (2003) and Korytár *et al.* (2003) for two-dimensional magnification.

Sometimes the sample is bent and distorted in such a way that the whole crystal cannot be pictured at the same time. As will be mentioned in Section 17.4.3 orientation contours may then be observed. In order to see the whole crystal, Jenichen *et al.* (1982, 1985, 1988) and Köhler (1994) have bent the first crystal so as to match the curvature of the sample.

As in the case of single-crystal topographs, the images of defects in double-crystal topographs have been simulated by solving numerically Takagi's equations with a computer, of dislocations (Gronkowski 1980; Ishida 1980; Riglet *et al.* 1980; Indenbom *et al.* 1984; Spirkl *et al.* 1994*a*,*b*) and of microdefects (Indenbom and Kaganer 1985).

The contrast of defects in GaAs near the absorbtion edges of Ga and As (resonant scattering) has been studied by Negishi *et al.* (2001, 2004).

17.4.3 *Synchrotron double-crystal topography*

Double-crystal topography and plane-wave topography benefit very much from the use of synchrotron radiation. The first double-crystal topographic studies with synchrotron radiation were made by M. Hart in W. Parrish's laboratory (see Fig. 14

of Sauvage 1980) and by Pétroff *et al.* (1980) and Riglet *et al.* (1980). The small source size, the long source-to-sample distance, the high flux make third-generation synchrotron radiation a remarkable improvement over laboratory sources for double-crystal topography. The problems related with dispersion are much less important because of the very small divergence and it is not compulsory to use the same material for the reference crystal and the sample. It is not necessary to put the detector very close to the sample; this has as a consequence that the defects of the monochromator are as visible as those of the sample itself; it is therefore preferable to use a perfect silicon sample as reference crystal. The alignment constraint of the two crystals is not as important as with an X-ray tube.

Because of the high flux of the synchrotron radiation, exposure times are short and it is possible to do real-time experiments. For instance, Barnett *et al.* (1995) have followed *in situ* molecular beam epitaxy of InGaAs on GaAs, George *et al.* (1987) have measured the velocity of dislocations in sulphur-doped InP, and Michot and George (1989) have studied the plastic deformation at the tips of cracks in silicon under an applied stress.

17.4.4 *Mapping of distortions and of lattice parameter variations*

Double-crystal topography and plane-wave topography are ideal tools to map minute variations of the lattice parameter, as was shown by Hart (1968*b*, 1981) in the case of silicon. In solution-grown crystals the various growth sectors usually have slightly different lattice parameters and are slightly misoriented. These variations in lattice parameter and orientation can be mapped and measured using plane-wave topography. The method is described by Yoshimura *et al.* (1979) and Robert and Lefaucheux (1983). One takes a series of plane-wave topographs for various settings of the sample corresponding to different positions on the rocking curve in Fig. 17.39(b). This first series is denoted azimuth 0° and the operation is repeated for azimuths 90°, 180° and 270° obtained by rotating the sample by the same angles around the reciprocal lattice vector. The local differences in lattice parameter and orientation of a given area with respect to a reference region are deduced from the differences between the angles for which the area and the reference region present maximum reflected intensity. This method was applied, for instance, by Yoshimura *et al.* (1979), Zarka and Liu Lin (1983) and Zarka *et al.* (1993) to synthetic quartz samples, by Robert *et al.* (1981) to strontium and barium nitrates ($Sr(NO_3)_2$ and $Ba(NO_3)_2$), by Belouet *et al.* (1983) to KDP (KH_2PO_4) and by Léon *et al.* (1989) to ADP ($NH_4H_2PO_4$).

17.4.5 *Equal-strain or equal-lattice parameter contours*

When the distortions and variations in lattice parameter are large, double-crystal topography can be adapted so that equal-strain or equal-lattice parameter contours can

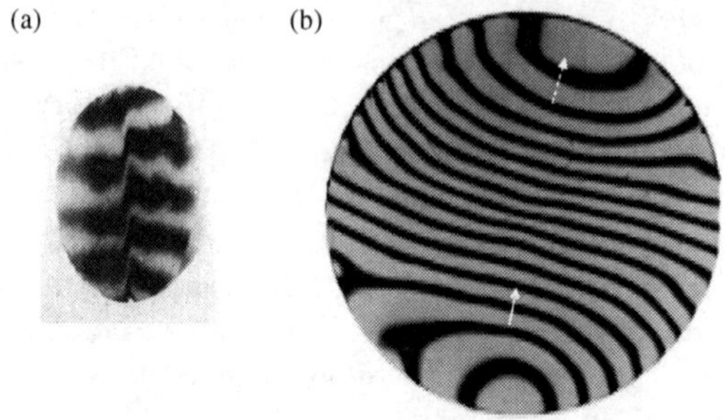

Fig. 17.42. (a) silicon (111) crystal with a scratch on one of its surfaces; 333 reflection, CuKα. Courtesy M. Renninger; (b) Reconstucted step-scanned X-ray synchrotron plane-wave reflection topograph, (+n, −n, +m) setting, of a 300-mm diameter CZ-Si wafer, reflection 115, 21 keV radiation. Step interval: 20 arcseconds. Courtesy S. Kawado.

be pictured. The idea was first proposed by Renninger (1962, 1963, 1965). If the lattice planes of the first and second crystal are not perfectly parallel to the rotation axis, only a horizontal stripe of the second crystal is visible at one time. The vertical position of this stripe varies if the crystal is rotated. By taking topographs on the same photographic film or plate after successive rotations of the sample, a series of horizontal stripes will be observed, which Renninger called *zebra* pattern. This technique has been recently applied to the characterization of 200 mm and 300 mm diameter silicon wafers (Kawado *et al.* 2002; Kawado 2003) at the SPring-8 topography station (Chikaura *et al.* 2001; Iida *et al.* 2002). Fig. 17.42 shows two examples of zebra patterns: In Fig. 17.42(a) the stripes reveal the distortions of the crystal due a local scratch and in Fig. 17.42(b) they show the overall warping of a very large silicon wafer (300 mm diameter).

In a similar way, if the crystal is curved, its whole surface will not be in reflection at the same time and an equal-strain contour will be observed; by rotating the crystal a series of contours will be obtained and it is possible to map the distortions of the crystal surface in this way. This technique has been applied and improved by Jacobs and Hart (1977), Hart (1981), Brown *et al.* (1984), Ishikawa *et al.* (1985), Minato *et al.* (1986) and Bilello *et al.* (1989).

Another technique to map equi-lattice spacing contours consists in using a fixed analyser after the sample. When rotating the crystal, only diffraction from constant lattice spacing areas will be recorded (Ishikawa *et al.* 1986, 1987; Kitano *et al.* 1991).

17.4.6 *Double-crystal setting for high spatial resolution topography*

The double crystal can also be used in a different mode in order to achieve high spatial resolution instead of a high sensitivity to small distortions. For this reason the sense of the asymmetric cuts of the first and second crystals (Fig. 17.43) is opposite to that shown in Fig. 17.39(a) (Beard and Armstrong 1989): grazing emergence from the reference crystal and grazing incidence on the sample. The beam impinging on the

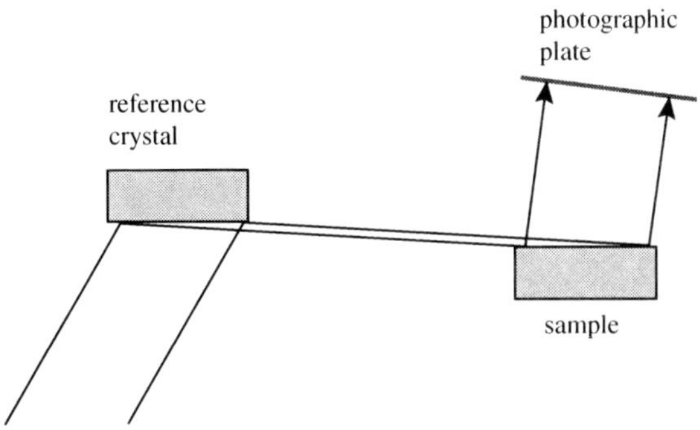

Fig. 17.43. Double-crystal setting for high spatial resolution topography.

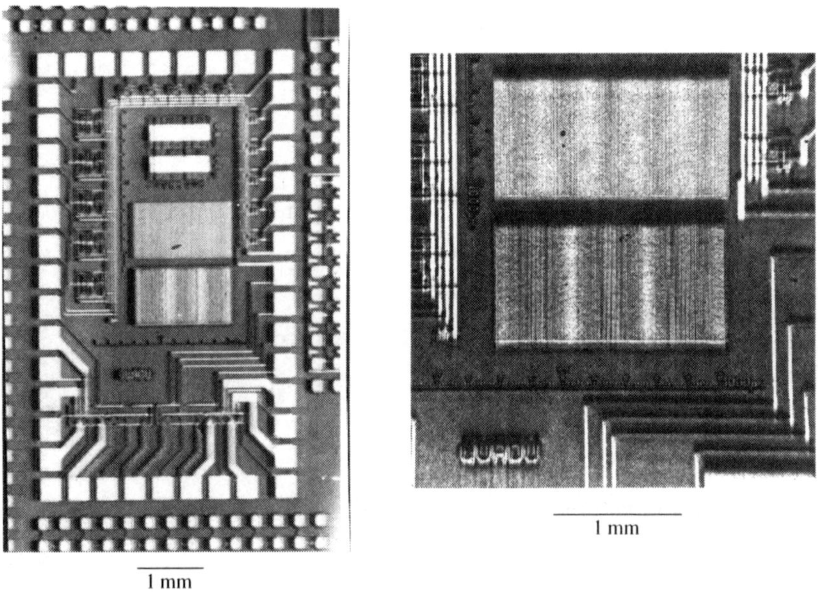

Fig. 17.44. Double-crystal high-resolution topograph of a microelectronic device on a (111) silicon crystal, $4\bar{2}\bar{2}$ reflection, Cu$K\alpha$. The thin vertical lines on the right magnified picture are 1 mm wide metallic strips. Courtesy R. W. Armstrong.

second crystal has then a wide angular spread. Regions with a high distortion and which are closest to the core of dislocations contribute most to the image. A very good resolution can thus be achieved (Beard *et al.* 1996; Armstrong *et al.* 1997) and an example is given in Fig. 17.44 which is the topograph of a microelectronic device.

Appendix 1

Useful formulae

A1.1 Fourier expansion of the dielectric susceptibility (or polarizability)

$$\chi = \sum_{\mathbf{h}} \chi_h \exp(2\pi \mathrm{i}\mathbf{h}\cdot\mathbf{r}) = -\frac{R\lambda^2}{\pi V}\sum_{\mathbf{h}} F_h \exp(2\pi \mathrm{i}\mathbf{h}\cdot\mathbf{r}), \qquad (2.36, 2.37)$$

where R is the classical radius of the electron, λ the wavelength, V the volume of the unit cell and F_h the structure factor.

A1.2 Index of refraction (see Section 2.7.3)

$$n = \sqrt{1+\chi_o} \approx 1 + \tfrac{1}{2}\chi_{ro} + \mathrm{i}\tfrac{1}{2}\chi_{io},$$

with $\chi_o = -R\lambda^2 F_o/(\pi V)$ and χ_{ro} and χ_{io} are the real and imaginary parts of χ_o, respectively (F_o/V is the electronic density).

A1.3 Critical angle for specular reflection (see Section 2.7.3)

$$\omega_c = \sqrt{-\chi_{ro}} = \sqrt{\frac{RF_{ro}}{\pi V}}\,\lambda.$$

A1.4 Linear absorption coefficient (normal)

$$\mu_o = -2\pi k \chi_{io} = \frac{2R\lambda}{V}F_{io}. \qquad (2.42b)$$

A1.5 Asymmetry ratio (see Section 4.7.3)

$$\gamma = \frac{\gamma_h}{\gamma_o} = \frac{\cos(\psi_n - \theta_B)}{\cos(\psi_n + \theta_B)},$$

where

$$\psi_o = (\mathbf{n}, \mathbf{s_o}) = \psi_n + \theta_B,$$
$$\psi_h = (\mathbf{n}, \mathbf{s_h}) = \psi_n - \theta_B,$$
$$\psi_n = (\mathbf{n}, \mathbf{s}).$$

n, $\mathbf{s_o}$, $\mathbf{s_h}$ and **s** are unit vectors normal to the crystal surface and parallel to the incident and reflected directions and the trace of the lattice planes in the diffraction plane, respectively. γ_o is positive in the transmission and reflection geometries; γ and γ_h are positive in the transmission geometry and negative in the reflection geometry.

A1.6 Deviation from Bragg's angle of the middle of the reflection domain (refraction effect)

$$\Delta\theta_{os} = -\frac{\chi_o(1-\gamma)}{2\sin 2\theta_B} = \frac{R\lambda^2 F_o}{2\pi V}\frac{1-\gamma}{\sin 2\theta_B}. \tag{4.26}$$

A1.7 Deviation from Bragg's angle of the middle of the reflected beam (refraction effect)

$$\Delta\theta_{hs} = -\frac{\chi_o}{2\sin 2\theta_B}\left(1-\frac{1}{\gamma}\right) = -\frac{R\lambda^2 F_o}{2\pi V\sin 2\theta_B}\left(1-\frac{1}{\gamma}\right). \tag{4.49b}$$

A1.8 Deviation parameter

$$\eta = \frac{(\Delta\theta - \Delta\theta_{os})\sin 2\theta_B}{|C|\sqrt{|\gamma|}\sqrt{\chi_h\chi_{\bar{h}}}} = \frac{\Delta\theta - \Delta\theta_{os}}{\delta_{os}} = \frac{(\Delta\theta - \Delta\theta_{os})\sin 2\theta_B\,\Lambda_o}{\lambda|\gamma_h|}, \tag{4.25}$$

where $2\delta_{os}$ is the Darwin width and Λ_o the Pendellösung or extinction distance.

A1.9 Width of the rocking curve (FWHM in the transmission geometry, Darwin width in the reflection geometry)

$$2\delta_{os} = 2\frac{|C|\sqrt{|\gamma|}\sqrt{\chi_h\chi_{\bar{h}}}}{\sin 2\theta_B} = \frac{2R\lambda^2}{\pi V}\frac{|C|\sqrt{|\gamma|}\sqrt{F_h F_{\bar{h}}}}{\sin 2\theta_B}. \tag{4.26}$$

A1.10 Full width at half maximum of the reflected beam

$$2\delta_{hs} = 2\frac{|C|\sqrt{\chi_h\chi_{\bar{h}}}}{\sin 2\theta_B}\frac{1}{\sqrt{|\gamma|}} = \frac{2R\lambda^2}{\pi V\sin 2\theta_B}\frac{1}{\sqrt{|\gamma|}}|C|\sqrt{F_h F_{\bar{h}}}. \tag{4.49a}$$

A1.11 Pendellösung distance (transmission geometry)—extinction distance in the reflection geometry

$$\Lambda_o = \frac{\lambda\sqrt{\gamma_o|\gamma_h|}}{|C|\sqrt{\chi_h\chi_{\bar{h}}}} = \frac{\pi V\sqrt{\gamma_o|\gamma_h|}}{R|C|\lambda\sqrt{F_h F_{\bar{h}}}} = \frac{\lambda|\gamma_h|}{\delta_{os}\sin 2\theta_B}. \tag{4.26}$$

Λ_o is called Λ_L in the transmission geometry and Λ_B in the reflection geometry.

A1.12 Amplitude ratio of the reflected and refracted waves

$$\xi_j = \frac{D_{hj}}{D_{oj}} = \frac{S(C)S(\gamma_h)}{\sqrt{|\gamma|}}\frac{\sqrt{\chi_h\chi_{\bar{h}}}}{\chi_{\bar{h}}}\left[\eta \pm \sqrt{\eta^2 + S(\gamma_h)}\right], \tag{5.34}$$

where the + sign corresponds to branch 1 wavefields ($j = 1$) and the − sign to branch 2 wavefields ($j = 2$) and $S(\ )$ means sign of.

A1.13 Equation of the dispersion surface—excitation errors

$$X_{oj}X_{hj} = \frac{k^2C^2\chi_h\chi_{\bar{h}}}{4} = \frac{\gamma_o|\gamma_h|}{4\Lambda_o^2}, \tag{4.15}$$

where X_{oj} and X_{hj} are the excitation errors:

$$\begin{aligned} X_{oj} &= \frac{kC\chi_{\bar{h}}\xi}{2} = \frac{\gamma_o S(\gamma_h)}{2\Lambda_o}\left[\eta \pm \sqrt{\eta^2 + S(\gamma_h)}\right], \\ X_{hj} &= \frac{kC\chi_h}{2\xi} = \frac{|\gamma_h|}{2\Lambda_o}\left[-\eta \pm \sqrt{\eta^2 + S(\gamma_h)}\right] \end{aligned} \tag{5.20}$$

are the distances of the tiepoint from the spheres of centres O and H and of radius nk.

A1.14 Diameter of the dispersion surface, or Bragg gap

$$G_B = \frac{k|C|\sqrt{\chi_h\chi_{\bar{h}}}}{\cos\theta_B} = \frac{R\lambda|C|\sqrt{F_h F_{\bar{h}}}}{\pi V\cos\theta_B}. \tag{4.16}$$

A1.15 Slope of Poynting vector S

$$\tan\alpha = \frac{1 - |\xi|^2}{1 + |\xi|^2}\tan\theta_B, \tag{4.21}$$

where $\alpha = (\mathbf{s}, \mathbf{S})$ and $\mathbf{s}$ is defined in A1.5.

A1.16 Anomalous absorption coefficient in the symmetric transmission geometry (Section 4.8.4)

$$\mu_j = \mu_o\left[1 \mp \frac{|C\chi_{ih}/\chi_{io}|\cos\varphi}{\sqrt{1+\eta_r^2}}\right],$$

where $\varphi = \varphi_{rh} - \varphi_{ih}$ is the phase difference between χ_{rh} and χ_{ih} and η_r is the real part of η. The $-$ sign corresponds to branch 1 wavefields ($j = 1$) and the $+$ sign to branch 2 wavefields ($j = 2$).

A1.17 Penetration depth (reflection geometry)

(i) *In the absence of absorption*:

$$z_e = \frac{\Lambda_B}{2\pi\sqrt{1-\eta^2}}. \tag{5.30a}$$

(ii) *In the presence of absorption*:

$$z_e = \frac{1}{\mu_o/\gamma_o + 2\pi\mathcal{I}m\big[\big(\eta + \sqrt{\eta^2-1}\big)/\Lambda_B\big]}. \tag{5.30b}$$

(iii) *Far from the reflection domain (due to absorption)*:

$$z_o = \frac{\gamma_o}{\mu_o}.$$

A1.18 Integrated intensity (transmission geometry)

(i) *Non-absorbing crystals*:

$$I_{hi} = \frac{\pi|C\chi_h|\sqrt{\gamma}}{2\sin 2\theta_B}\int_0^{2\pi t\Lambda_L^{-1}} J_0(z)\,\mathrm{d}z, \tag{4.40}$$

where $J_0(z)$ is the zeroth-order Bessel function.

(ii) *Absorbing crystals*:

$$I_{hi} = A\left[\int_0^{2\pi t\Lambda_L^{-1}} J_0(z)\,\mathrm{d}z - 1 + I_0(\zeta)\right], \tag{6.17}$$

where

$$A = \frac{\pi |C\chi_h| \sqrt{\gamma}}{2 \sin 2\theta_B} \left| \frac{\chi_h}{\chi_{\bar{h}}} \right| \exp \left[-\frac{\mu_o t}{2} \left(\frac{1}{\gamma_o} + \frac{1}{\gamma_h} \right) \right],$$

$$\zeta = \frac{\mu_o t}{\sqrt{\gamma_o \gamma_h}} \sqrt{|C|^2 \left| \frac{\chi_{ih}}{\chi_{io}} \right|^2 \cos^2 \varphi + \frac{\gamma_h - \gamma_o}{4 \gamma_o \gamma_h}}$$

and $I_0(\zeta)$ is the modified Bessel function of zeroth-order.

A1.19 Integrated intensity (reflection geometry)

Non–absorbing crystals:

$$I_{hi} = \pi \frac{|C| \sqrt{|\gamma|} \sqrt{\chi_h \chi_{\bar{h}}}}{\sin 2\theta_B} \tanh \left[\frac{\pi t}{\Lambda_B} \right], \tag{4.47}$$

where t is the crystal thickness.

A1.20 Intensity distribution on the exit surface (incident spherical wave, transmission geometry)

$$\begin{aligned} I_h(Y) &= \pi^2 |\mathcal{A}_h|^2 \exp(-\mu_s t) \left| J_0 \left[\pi \frac{t}{\Lambda_L} \sqrt{1 - Y^2} \right] \right|^2, \\ I_o(Y) &= \pi^2 |\mathcal{A}_o|^2 \exp(-\mu_s t) \frac{1 + Y}{1 - Y} \left| J_1 \left[\pi \frac{t}{\Lambda_L} \sqrt{1 - Y^2} \right] \right|^2, \end{aligned} \tag{10.25a}$$

with

$$\mu_s = \frac{1}{2} \mu_o \left[\frac{1}{\gamma_o} + \frac{1}{\gamma_h} + Y \left(\frac{1}{\gamma_o} - \frac{1}{\gamma_h} \right) \right]. \tag{10.25b}$$

where Y is the reduced coordinate of a point p of the exit surface (Fig. 10.2):

$$Y = \frac{\overline{Ep}}{\overline{EC}} \tag{10.16}$$

$$= \mp \frac{\eta}{\sqrt{1 + \eta^2}} \tag{10A.4}$$

$$\eta = \mp \frac{Y}{\sqrt{1 - Y^2}} \tag{10A.5}$$

Appendix 2

The early days of dynamical theory

By Paul P. Ewald

Transcript[1] of an evening lecture given at the
Summer School on X-ray Topography and Dynamical Theory
(Limoges, France, August 1975)

I was asked to say something about the origins of dynamical theory, and what I would like to do is to stress that dynamical theory is part of General Optics. It started out from Optics and the formulae for the Dynamical Theory were really ready in 1910 or 1911 before Laue's discovery. So, let us find out what this means. I was a student in Munich and I decided to do Theoretical Physics for Professor Sommerfeld. When I came to ask him for a thesis, he had a long list of papers and at that time he worked mainly on boundary problems, on questions of self-inductance, on the propagation of wireless waves which was new at the time and, quite at the end of the list, there was to explain double refraction in crystals by an anisotropic arrangement of isotropic dipoles. Isotropic dipoles means that there is no preferential direction in the dipole for its vibrations and an anisotropic arrangement, the simplest you can think of, is that they are arranged in an orthorhombic simple lattice like match boxes packed in a package. Well, I told Sommerfeld that I chose that subject in spite of his warning that he did not know how to tackle that problem and that he could not be of as much help, while the boundary problems were all more or less routine and he could tell me exactly what to do. So, I insisted on the problem which he did not know the solution of. How did Sommerfeld come to take this interest?

In Munich there were several famous professors at the time. Röntgen led a laboratory of his own and actually did not take much part in the discovery of X-ray interference; that is to say, when Laue, Friedrich and Knipping had done the experiments, he came over and looked at the diagrams they had, shook his head and said: well this is very beautiful but certainly it is not diffraction! He had tried himself, many years ago, to obtain diffraction, without success of course, for various reasons. There was also a mineralogist, Groth. Sommerfeld was a friend both of Schoenflies and Groth. Groth was his colleague in Munich and the two, Schoenflies and Groth, were very interested in crystal structure theory but they could not come to terms, they could not understand one another. Schönfliess did it from a mathematical point

[1] Slightly edited from the oral version; the help of D. W. J. Cruickshank is gratefully acknowledged.

of view—in fact he had been a student and a Privatdozent of mathematics and had worked out all the space groups and brought the manuscript to Felix Klein. Felix Klein looked at the manuscript and told him: what you are doing is crystallography. Felix Klein was a famous mathematician in Göttingen at the time, with very wide interests and an enormous survey of all branches of Mathematics, Physics and Technology. Schoenflies really looked at the space groups as something as operators which operate on something and that duplicated and multiplied, whereas Groth had a much more physical approach. He thought of the Bravais space lattices and he said a crystal consists of a Bravais lattice or rather of the groups which his friend and colleague Sohncke had established—the 65 or 66 groups which are not as general as the Schoenflies or Fedorov ones—and that each kind of atom forms one of what they called space groups. You see, in German, it is all Gitter, Kristall Gitter and there is not the difference between lattice and structure which has been introduced by the English. These atomic space groups are intercalated, they penetrate one another.

Laue made his discovery in 1912, early 1912, at about Easter time. It was published. I was in Göttingen at the time and I heard of the experiment only when Sommerfeld went to Göttingen and gave a lecture to the Physical Society there at Whitsuntide, and then I sat down and discussed the formulae which I had developed in my thesis and that gave the sphere—the usual construction of which I am not really very proud of because it was obvious whereas the dynamical theory was not quite obvious! Well, I have to come back on my thesis, you see. If you have dipoles which are in this anisotropic array and if they swing this way, they will interact in a different way when they swing along the long distance of separation than if they swing normal to it, that is to say in the direction of the small separation. And therefore, you can expect that by the coupling of the dipoles a difference in refractive index will be produced, that is to say double refraction. To establish the theory, it was first of all necessary to find out how to sum the dipoles which issue from the various points of the lattice.

Each dipole emits a spherical wavelet and you have to sum them. This took about a year and when one day I had found a solution, I wanted to tell it to Sommerfeld. He said 'wait a moment before you start, I have something to tell you' and he showed me that he had found exactly the same mode, how to transform the sum of spherical waves into a sum of plane waves. The interesting point in this is that neither Sommerfeld nor I ever thought of Fourier transforms. In fact what was usual at the time was to speak of Fourier developments. Sommerfeld's own thesis was on the development of arbitrary functions according to certain types of other functions to orthogonal sets of functions. But the idea of Fourier transforms came much later.

I remember when I first read the words Fourier transform. I was so horrified because I thought it was such bad English. I remember very well that this was quite new to me but I cannot remember the name of the author of the paper. I have asked several mathematicians and people interested in the history of mathematics to find out who invented the point of view of a transformation. Of course Hilbert had it, but that was in very abstract and highbrow mathematics and it hadn't penetrated to us. This summation is now done in a few lines; if you know Fourier transform theory, there is nothing to it. At that time, I had to do it first by inventing complex integration; it was quite a job and it took me quite long. But it didn't solve the problem because if you want to find out the double refraction, you have to find out the field that is active on one particular dipole, that is to say, from the transformation of the sum of all the spherical waves you have to take out one spherical wave and see what remains, that is the so-called field of excitation. While for X-rays this does not play any role, for visible optics and for any static problem like for instance the anisotropy of the dielectric constant, it is most important to consider the field of excitation. The whole theory resulted in a formula expressing the double refraction in terms of the parameters of the crystal, the axial ratio etc. and the polarizability of the dipole of course.

When I was that far, I went to Groth and asked him: well, I had done this work, could you tell me of a crystal which is built up according to this simple orthorhombic lattice which I took as a model? And so Groth was silent for a while and, suddenly, his face lit up and he said: yes, there is one crystal where I can guarantee that it is built according to this simple lattice, that is anhydrite, a form of gypsum, without water. Anhydrite! I said, why? Well, it has three cleavage planes which are orthogonal to one another and which are fairly good. One is excellent and the others are still fairly good. So I calculated the double refraction for the anhydrite model and I can only say, it came out quite wrong, compared to the experiment! The only thing I could conclude is that the order of magnitude was correct, so it was correct to assume that by the influence of the coupling through radiation of the dipoles there would result double refraction of an acceptable order, but the sign was wrong. It was positive when it should be negative and it was 10 times bigger or twice smaller, so there was no correlation and I concluded that this was evidently because the structure was not the correct one!

Well, the structure of anhydrite was determined many years after and it has a face-centred lattice or I have forgotten what, but it is entirely different, has nothing to do with Groth's prediction. That was the state of structure theory at the time before Laue. Nobody could tell what was the object of the repetition and how it came about. Well now, in doing my work, I came across a difficulty. I could find the radiation coupling of the dipoles and obtained, let us say, a determinant which contained the refractive index and that could be calculated from the determinant having to be zero. But in all this calculation, I never used any incident radiation and in the usual theory of dispersion, as it was called at the time, which had been started by Drude and other people, there always played a role the incident beam, the incident radiation. I could really not do anything with it. The whole system was a linear system so you could superimpose solutions and if you wanted the incident ray to continue in the crystal, this upset everything so I left out the incident ray and I did not like to do that, you see, being an inexperienced young graduate student. I thought, all these big venerable people have spoken of an incident radiation, so how could I leave it out?

The man to whom I went was Laue because he had a real feeling for fundamental things, but in this case he was not very attentive to what I told him. I began telling him when we left the University, and he had invited me for supper at his house, and so we went through the Änglische Garten and I began telling him of my work of which he did not seem to know anything. He always asked what happens if you have very small waves, what are the distances of the atoms? I said the distances of the atoms I can't tell, it depends on the structure you assume, whatever you put in, but what I know is that the atomic distances must be about 10 000 times smaller than the wavelength of the light I am considering and that is quite sufficient because it tells me that my approximations are correct. So Laue repeated his question and I got no answer from him on the problem on which I wanted to consult him, but I left out the incident wave and it was always a sore point in my mind how this could be justified. Actually, I think I have told Laue: well here is the transformation of these spherical wavelets into plane waves; this is a strict transformation, you can discuss it yourself if you just make the wavevector k long instead of very small as in the optical case. But he of course never did it. It would have been a geometrical theory, at all events a kinematical theory of the diffraction effects.

So, after a while, the war broke out—First World War by the way—and the X-ray installation which we had in Sommerfeld's Institute was taken to an improvised hospital,[2] and I went with the equipment and learned to do medical work. Then after a while I thought the war would end victoriously and I would come late, not seeing anything of it. So I was eager and besides I might have been drafted, so I was quite eager to see something and I applied for a position as Feld-Röntgen Mechaniker, field X-ray mechanic. I was ordered to Russia;[3] at first to assist the building of the whole equipment in Berlin and then it was resent to Russia. They wouldn't divulge where the Headquarters in Russia were but I was told, well, take a ticket to Königsberg, that's good enough. So I took a ticket to Königsberg and, being of very non-military nature, I never reported, of course! I took a room in the best hotel, by making very long steps in front of several generals and other high brass. I got the last room and there I was. I went every day to the station to see whether my wagon, my X-ray wagon, had appeared, which came there by train, but I always got the answer 'no'.

So, I had plenty of time and enjoyed Königsberg and the good marzipan and worked out the theory of reflection and refraction on a half crystal, that is to say a crystal which has a boundary, because no boundary had appeared in my theory. And, you see, this is the point which I think the optical people have never acknowledged, that the problem which usually was called theory of dispersion really consists of two entirely different parts. It is first of all the propagation of a set of waves or a single wave in the interior of the crystal to determine under which conditions such a wavefield can propagate unchanged through the crystal, and then a set of problems is how such an allowed field is generated in the crystal by the incident wave. So these were clearly separated I think for the first time in my thesis and it led in my thesis already to the statement that at the surface there must be generated waves which annihilate the incident wave, which is quite true. In the work I did in Königsberg, I proved this to be correct for the half crystal. At nearly the same time, Oseen in Stockholm also deduced the same

[2] In 1914 (mentioned in *Ewald Special Lecture*, Kyoto, 1961, communicated by D. W. J. Cruickshank).
[3] In 1915 (mentioned in *Ewald Special Lecture*, Kyoto, 1961, communicated by D. W. J. Cruickshank).

result from Maxwell's equations but he could not really describe the field he wanted annihilated—it was more an assumption that this field would be annihilated. So this extinction, so-called Ewald–Oseen extinction theorem, has been very much discussed recently because you have now much more complicated theories of dispersion, with polarons and other things moving along and transformation of radiation into other forms of energy, and so it has become quite a subject of discussion again.

Finally, my money ran out and I phoned headquarters not knowing where they were. But then came the connection and I asked for some more money to be sent, and then I got a telephone back and I nearly fell flat on my back because I had never heard such cursing and such strong language. I should take the next train, it would be at 7 o'clock in the evening, I could take a train to Tilsitt and report the next morning at 7 o'clock at the Sanität Depot to which I belonged by formation. I was there punctually, again taking long steps in order to secure a room and, the next morning, I got up and went there and saw the sergeant who had telephoned me. It so happened that the chef of the formation was a physicist and we became very good friends. He took me out riding and we had a very nice time. It took another three weeks, I believe, in Tilsitt, until my wagon finally appeared. Well, then I was taken through Russian landscape which was beautiful—the first snow falling—and in a beautiful train, in a hospital train that had been established by the town of Bremen and was full of good wines and good cigars and a cook who was so fat that he could only go outside the train when it stopped in order to report what meals he would prepare for the next day. And in the little place where I finally ended, I always had a room for myself. My X-ray equipment was there and I had a table on which to lay the patients but you could also write on that table. There was not very much to be done because the fighting had stopped, and the only casualties were the casualties of the Russian winter and the old people who were transferred from our western front to the eastern front in order to recover.

So I had a quiet time, and I could go on with the theory of dispersion and that is where it originated. It was not quite easy and I would like to tell you some of the problems. At first I tried, quite in vain of course, to find out, with an incident ray, how many interferences would be produced. This is a problem of the theory of numbers and I think it is insoluble and so it took some time until I finally said: well, if I can't find that answer, let us assume that there are a certain number of waves generated and from there on an important role was played by the reciprocity theorem. That is to say, if you have a ray falling in this direction and generating a secondary ray in another direction, then, if you let the incident ray take the direction of the secondary one, the primary ray will also be generated again. So this showed that you cannot consider a single wave but you have to consider a whole bundle of waves or, as we now say, a wave field propagating in the crystal and then, of course, the problem was what is the velocity of one of the rays, because if you fix that, then the others are automatically fixed. The whole thing is based on Lorentz' theory of electrons which was then the outstanding theory. It was not very old, Lorentz' book appeared I think in about 1908 or 1910.

This is just to illustrate the reciprocity theorem: if you have the origin there and you get interference things there, if you change the direction of the primary ray to one of the 'h' rays, then of course the whole picture will repeat because the vectors all end in lattice points so that shows that you have to consider a bundle of waves instead of a single wave which is sufficient in Optics.

It took quite some hard work to get the ideas correct, I mean to treat the problem of what you assume to find the idea of the dispersion surface if you vary the direction of incidence without really getting very far away from the case where there are only a certain number of secondary rays split up. But all that can be solved and also I solved the problem of what happens at a boundary because again this problem of propagation of X-rays is divided up into one internal problem and of how its internal field is related to the incident outside field. So when all this work was ready, I wrote it up and sent it to Sommerfeld as a thesis for becoming a lecturer in Munich. Sommerfeld looked at it and said it seemed very nice work, but of course entirely unpractical, it would never find any application! But in spite of that, he agreed that it be handled for my Habilitation so, in 1917, I became Privatdozent and I had to prepare my theses. Now they were very disappointed that I did not come in uniform! but I preferred to be in mastic[4] for the lecture I had to give—a trial lecture—and I also had to prepare the theses, which in the old days were to be defended against somebody who opposed them, the opponent. In my father's day this was in Latin, but in my days it was in German. I would like to show you this excerpt: *If the absorption of X-rays is similar to that of optical waves, and is caused by loss of energy in the vibrations of the dipoles, then, under certain conditions, X-rays should find no absorption in an absorbing crystal.* Well, that was the origin. I had forgotten all about it until I found—40 years later—a copy of this thesis. By that time Borrmann had made his experiments, but it was obvious from my work that if there was a full resonance between the wavefield and the dipoles, then the dipoles were not required to vibrate at all and the wavefield could pass through unmolested as it were.

This dynamical theory was rigged up only for a simple lattice—no structure factor—but from the beginning I stressed to have n-beams. This is the difference with Darwin's theory of which I must have known because I remember that I had given a report of Darwin's theory in a colloquium but I had forgotten everything about it. For the simple reason that we considered the phenomenon as a phenomenon of diffraction (and we knew that, if you have a plane filled regularly with atoms, besides the specularly reflected ray there are a number of other rays), I did not believe in Darwin's theory at all. Darwin had the argument that these other rays are not in phase and so need not be regarded but, of course, this has to be proved and it is not true in various cases where you have multiple reflections, so I always treated the case of n rays but I could not do it including the structure factor. This was done later and I gave a lecture at the Institut Henri Poincaré on that in 1932[5] where the full structure factor is there.

So I developed the theory during my lifetime from 1910 when I started work on the optical problem and I always took it up again... I am still interested in it!

[4] Civilian clothes.

[5] Published in 1938: Ewald, P. P. (1938). Optique cristalline (lumière et rayons X), interaction des atomes par rayonnement. *Annales de l'Institut Henri Poincaré*, **VIII**, 79–110 (in French).

Question (M. Kuriyama): I would like to know from you, personally, what was the interaction between you and Professor Bloch?

Answer (P. P. Ewald): Professor Bloch at the time was probably in the cradle.[6] Actually, I think it was the Solid State people who invented the name Bloch wave. If anybody of the Solid State people had known about my work, they would have known that this was not new.

[6] The actual reference to Bloch's work is: Bloch, F. (1928). *Z. Physik*, **52**, 555 (Note by the present author).

References

Abdul Gani, S. M., Clark, G. F. and Tanner, B. K. (1981). *Inst. Phys. Conf. Ser. No. 60*, 259–64.

Abrosimov, N. V., Rossolenko, S. N., Alex, V., Gerhardt, A. and Schröder, W. (1996). *J. Crystal Growth*, **166**, 657–62.

Abruña, H. D., Bommarito, G. M. and Acevedo, D. (1990). *Science*, **250**, 69–74.

Adams, B. W. (2004*a*). *Acta Cryst. A*, **60**, 120–33.

Adams, B. W. (2004*b*). *Chem. Phys.*, **299**, 193–201.

Adams, B. W. (2004*c*). *Proc. SPIE*, **5194**, 46–55.

Afanas'ev, A. M. and Esayan, A. V. (1991). *Phys. Stat. Sol. (a)*, **126**, 303–11.

Afanas'ev, A. M. and Kagan, Yu. (1968). *Acta Crystallogr. A*, **24**, 163–70.

Afanas'ev, A. M. and Kohn, V. G. (1971). *Acta Crystallogr. A*, **27**, 421–30.

Afanas'ev, A. M. and Kohn, V. G. (1975). *Phys. Stat. Sol. (a)*, **28**, 61–70.

Afanas'ev, A. M. and Kohn, V. G. (1976). *Acta Crystallogr. A*, **32**, 308–10.

Afanas'ev, A. M. and Kohn, V. G. (1977). *Acta Crystallogr. A*, **33**, 178–84.

Afanas'ev, A. M. and Melikyan, O. G. (1990). *Phys. Stat. Sol. (a)*, **122**, 459–68.

Afanas'ev, A. M. and Melkonyan, M. K. (1983). *Acta Crystallogr. A*, **39**, 207–10.

Afanas'ev, A. M. and Pertsnev, I. P. (1969). *Acta Crystallogr. A*, **25**, 520–3.

Afanas'ev, A. M., Baturin, V. E., Imamov, R. M., Koval'chuk, E. K., Kovev, E. K., Kohn, V. G. and Semiletov, S. A. (1977). In *Proc. 7th Int. Vac. Cong.*, Vienna, pp. 2209–12.

Afanas'ev, A. M., Afanas'ev, S. M., Aleksandrov, P. A., Imamov, R. M. and Pashaev, E. M. (1984). *Phys. Stat. Sol. (a)*, **86**, K1–K5.

Afanas'ev, A. M., Imamov, R. M., Maslov, A. V., and Pashaev, E. M. (1984). *Phys. Stat. Sol. (a)*, **84**, 73–8.

Afanas'ev, A. M., Aleksandrov, P. A., Fanchenko, S. S., Chaplanov, V. A. and Yakimov, S. S. (1986*a*). *Acta Crystallogr. A*, **42**, 116–22.

Afanas'ev, A. M., Imamov, R. M., Mukhamedzhanov, E. Kh. and Chuzo, A. N. (1986*b*). *Acta Crystallogr. A*, **42**, 24–9.

Afanas'ev, A. M., Imamov, R. M., Mukhamedzhanov, E. Kh. and Peregudov, V. N. (1986*c*). *Phys. Stat. Sol. (a)*, **98**, 367–75.

Afanas'ev, A. M., Imamov, R. M., Maslov, A. V. and Mukhamedzhanov, E. Kh. (1989*a*). *Phys. Stat. Sol. (a)*, **113**, K153–K156.

Afanas'ev, A. M., Imamov, R. M., Mukhamedzhanov, E. Kh. and Bzheumikhov, A. A. (1989*b*). *Phys. Stat. Sol. (a)*, **116**, 197–201.

Afanas'ev, A. M., Imamov, R. M. and Mukhamedzhanov, E. Kh. (1992). *Crystal. Rev.*, **3**, 157–230.

Ahmad, S. and Whitworth, R. W. (1988). *Phil. Mag. A*, **57**, 749–66.

Ahmad, S., Ohtomo, M. and Whitworth, R. W. (1986). *Nature*, **319**, 659–60.

Ahmad, S., Ohtomo, M. and Whitworth, R. W. (1987). *J. Physique*, **48**, C1-175–81.

Ahmad, S., Shearwood, C. and Whitworth, R. W. (1992). In *Physics and chemistry of ice* (ed. N. Maeno and T. Hondoh), pp. 492–6. Hokkaido University Press, Sapporo.

Akaba, N., Suzuki, S. and Takagi, M. (1979). *J. Phys. Soc. Jap.*, **46**, 1583–93.

Akimoto, K., Ishikawa, T., Takahashi, T. and Kikuta, S. (1985). *Japan J. Appl. Phys.*, **24**, 1425–31.

Akimoto, K., Ishikawa, T., Takahashi, T. and Kikuta, S. (1986). *Nucl. Instrum. Methods A*, **246**, 755–9.

Albertini, G., Boeuf, A., Cesini, G., Mazkedian, S., Melone, S. and Rustichelli, F. (1976). *Acta Crystallogr. A*, **32**, 863–8.

Albertini, G., Boeuf, A., Klar, B., Lagomarsino, S., Mazkedian, S., Medone, S., Puliti, P. and Rustichelli, F. (1977). *Phys. Stat. Sol. (a)*, **44**, 127–36.

Aldred, P. J. E. and Hart, M. (1973*a*). *Proc. Roy. Soc. A*, **332**, 223–38.

Aldred, P. J. E. and Hart, M. (1973*b*). *Proc. Roy. Soc. A*, **332**, 239–54.

Aleksandrov, P. A., Afanas'ev, A. M. and Stepanov, S. A. (1984*a*). *Phys. Stat. Sol. (a)*, **86**, 143–54.

Aleksandrov, P. A., Afanas'ev, A. M. and Stepanov, S. A. (1984*b*). *Sov. Phys. Crystallogr.* **29**, 119–22.

Aleshko–Ozhevskii, O. P. (1982). *Kristallografiya*, **27**, 1125–30.

Aleshko–Ozhevskii, O. P., Bowen, K. and Davies, S. T. (1985). *Ferroelectrics*, **62**, 53–8.

Al Haddad, M. and Becker, P. J. (1988). *Acta Crystallogr. A*, **44**, 262–70.

Allison, S. K. and Williams, J. H. (1930). *Phys. Rev.*, **35**, 149–54.

Als-Nielsen, J. and McMorrow, D. (2001). *Elements of Modern X-ray Physics.* J. Wiley and Sons, Ltd, New York.

Alstrup, A. (1978). *Acta Crystallogr. A*, **34**, 362–7.

Altin, D., Härtwig, J., Köhler, R., Ludwig, W., Ohler, M. and Klein, H. (2002). *J. Synchrotron Rad.*, **9**, 282–6.

Amemiya, Y. and Chikawa, J. (1999). In *International tables of crystallography*, Vol. C (ed. A. J. C. Wilson and E. Prince), pp. 629–32. Kluwer Academic Publishers, Dordrecht.

An, H., Li, M., Yang, S., Mai, Z. and Liu, S. (1995). *J. Cryst. Growth*, **148**, 31–4.

Andersen, S. K., Golovchenko, J. A. and Mair, G. (1976). *Phys. Rev. Lett.*, **37**, 1141–5.

Ando, M. and Hosoya, S. (1971). *Techn. Rep. ISSP A*, **483**, 1–16.

Ando, M. and Hosoya, S. (1972*a*). In *Proc. 6th Int. Conf. on X-ray Optics and Microanalysis* (ed. G. Shinoda, K. Kohra and T. Ichinokawa), pp. 63–8. University of Tokyo Press, Tokyo.

Ando, M. and Hosoya, S. (1972*b*). *Phys. Rev. Letts.*, **29**, 281–5.

Ando, M., Bailey, D. and Hart, M. (1978). *Acta Crystallogr. A*, **34**, 484–9.

Ando, Y. and Kato, N. (1966). *Acta Crystallogr.*, **21**, 284–5.

Ando, Y. and Kato, N. (1970). *J. Appl. Crystallogr.*, **3**, 74–89.

Ando, Y., Patel, J. R. and Kato, N. (1973). *J. Appl. Phys.*, **44**, 4405–12.

Andreev, A. V., Gorshkov, M. E. and Ilinskii, Yu. A. (1985). *Dokl. Akad. Nauk SSSR*, **282**, 69–72.

Andreeva, M. A., Rocete, K. and Khapachev, Y. P. (1985). *Phys. Stat. Sol. A*, **88**, 455–62.

Andronova, N. V., Kohn, V. G. and Chechin, A. I. (1995). *Nucl. Instr. & Methods in Phys. Res. A*, **359**, 131–4.

Andronova, N. V., Kohn, V. G., Chechin, A. I., Esheng, T., Miao, J. and Cui, M. (1995). *Nucl. Instr. & Methods in Phys. Res. A*, **359**, 135–7.

Annaka, S. (1982). *J. Phys. Soc. Japan*, **51**, 1927–31.

Annaka, S., Kikuta, S. and Kohra, K. (1965). *J. Phys. Soc. Jap.*, **20**, 2093.

Annaka, S., Kikuta, S. and Kohra, K. (1966). *J. Phys. Soc. Jap.*, **21**, 1559–64.

Annaka, S., Kohra, K. and Ando, M. (1981). *Jap. J. Appl. Phys.*, **20**, L463–6.

Aoyama, J., Suzuki, S. and Takagi, M. (1992). *J. Phys. Soc. Jap.*, **61**, 3613–24.

Appel, A. and Bonse, U. (1991). *Phys. Rev. Letters*, **67**, 1673–6.

Argunova, T. S., Baranov, A. N., Djourtanov, B. E., Mosina, G. N., Shul'pina, I. L. and Yakovlev, Yu. P. (1987*a*). *Sov. Phys. Techn. Phys.*, **32**, 19–22.

Argunova, T. S., Kusov, A. A., Petrashen, P. V. and Shu'lpina, I. L. (1987*b*). *Sov. Phys. Techn. Phys.*, **32**, 646–9.

Aristov, V. V. and Shulakov, E. V. (1975). *J. Appl. Crystallogr.*, **8**, 445–51.

Aristov, V. V., Shmytko, I. M. and Shulakov, E. V. (1974). *J. Appl. Cryst.*, **7**, 409–13 and 413–6.

Aristov, V. V., Shmytko, I. M. and Shuvalov, E. V. (1977*a*). *Acta Crystallogr. A*, **33**, 412–18.

Aristov, V. V., Shmytko, I. M. and Shuvalov, E. V. (1977*b*). *Acta Crystallogr. A*, **33**, 418–23.

Aristov, V. V., Kohn, V. G. and Polovinkina, V. I. (1980). *Phys. Stat. Sol. (a)*, **62**, 431–40.

Aristov, V. V., Polovinkina, V. I., Afanas'ev, A. M. and Kohn, V. G. (1980). *Acta Crystallogr. A*, **36**, 1002–13.

Aristov, V. V., Kohn, V. G., Polovinkina, V. I. and Snigirev, A. A. (1982). *Phys. Stat. Sol. (a)*, **72**, 483–91.

Aristov, V. V., Snigirev, A. A., Basov, Yu. A. and Nikulin, A. Yu. (1986). In *Short wavelength coherent radiation: generation and application* (ed. D.T. Attwood and J. Bokor), Conference Proceedings, Vol. 147, pp. 253–9. American Institute of Physics.

Aristov, V. V., Basov, Yu. A., Redkin, S. V., Snigirev, A. A. and Yunkin, V. A. (1987). *Nucl. Instrum. Methods A*, **261**, 72–4.

Aristov, V. V., Basov, Yu. A., Kulipanov, G. N., Pindyurin, V. F., Snigirev, A. A., and Sokolov, A. S. (1989). *Nucl. Instrum. Methods A*, **274**, 390–3.

Aristov, V. V., Gureyev, T. E., Nikulin, Yu. and Snigirev, A. A. (1991*a*). *Phys. Stat. Sol. (a)*, **127**, 33–42.

Aristov, V. V., Kuznetsov, S. M., Kouyumchian, A. V. and Snigirev, A. A., (1991*b*). *Phys. Stat. Sol. (a)*, **125**, 57–66.

Aristov, V. V., Basov, Yu. A., Gureyev, T. E., Snigirev, A. A., Ishikawa, T., Izumi, K. and Kikuta, S. (1992). *Jpn. J. Appl. Phys.*, **31**, 2616–20.

Aristov, V. V., Gureyev, T. E., Nikulin, Yu., Petrashen, P. V. and Snigirev, A. A. (1992). *Semiconductor Sci. & Technology*,**7**, 1109–12.

Aristov, V. V., Gureyev, T. E., Nikulin, Yu., Petrashen, P. V. and Snigirev, A. A. (1993). *J. Phys. D: Appl. Phys.*, **26**, A29–A31.

Aristov, V. V., Grigoriev, M., Kuznetsov, S., Shabelnikov, L., Yunkin, V., Weitkamp, T., Rau, C., Snigireva, I., Snigirev, A., Hoffmann, M. and Voges, E. (2000). *Appl. Phys. Letters*, **77**, 4058–60.

Aristov, V. V., Grigoriev, M., Kuznetsov, S., Shabelnikov, L., Yunkin, V., Rau, C., Snigirev, A., Snigireva, I., Weitkamp, T., Hoffmann, M. and Voges, E. (2001). *Proc. SPIE – The International Society for Optical Engineering.*, **4145**, 285–93.

Armstrong, R. W. (1980). In *Characterization of crystal growth defects by X-ray methods*, NATO ASI Series, Series B, Physics, Vol. 63, pp. 349–67. Plenum Press, New York.

Armstrong, R. W. (1984). In *Applications of X-ray topographic methods to material science* (ed. S. Weissmann, F. Balibar and J. F. Pétroff), pp. 295–300. Plenum Press, New York.

Armstrong, R. W. and Wu, C. C. (1973). In *Tools and techniques for microstructural analysis* (ed. J. L. McCawll and W. M. Mueller), pp. 169–219. Plenum Press, New York.

Armstrong, R. W., Boettinger, W. J. and Kuriyama, M. (1980). *J. Appl. Crystallogr.*, **13**, 417–24.

Armstrong, R. W., Beard, W. T., Green, K. A. and Zhang, X. J. (1997). *Il Nuovo Cimento*, **19D**, 147–52.

Arndt, U. W. (1986). *J. Appl. Crystallogr.*, **19**, 145–63.

Arndt, U. W. (1990). *Synchrotron Radiat. News*, **3**, 17–22.

Arndt, U. W. (1999*a*). In *International tables of crystallography*, Vol. C (ed. A. J. C. Wilson and E. Prince), pp. 191–200. Kluwer Academic Publishers, Dordrecht.

Arndt, U. W. (1999*b*). In *International tables of crystallography*, Vol. C (ed. A. J. C. Wilson and E. Prince), pp. 617–27. Kluwer Academic Publishers, Dordrecht.

Arridge, R. G. C ., Lang, A. R. and Makepeace, A. P. W. (2002). *Proc. Roy. Soc. London A*, **458**, 2623–43.

Ashkin, M. and Kuriyama, M. (1966). *J. Phys. Soc. Jap.*, **21**, 1549–58.

Aslanov, L. A., Fetisov, G. V. and Howard J. A. K. (1998). *Crystallographic instrumentation*. Oxford University Press, Oxford.

Auleytner, J. and Zielinska-Rohozinska, E. (1963). *Phys. Stat. Sol.*, **3**, 1846–50.

Austerman, S. B. and Miller, K. T. (1965). *Phys. Stat. Sol.*, **11**, 241.

Austerman, S. B. and Newkirk, J. B. (1967). *Adv. X-ray Anal.*, **10**, 134–52.

Authier, A. (1960*a*). *C. R. Acad. Sci. Paris*, **251**, 2003–5.

Authier, A. (1960*b*). *C. R. Acad. Sci. Paris*, **251**, 2502–4.

Authier, A. (1961*a*). *Bull. Soc. Fr. Minér. Crist.*, **84**, 51–89.

Authier, A. (1961*b*). *Bull. Soc. Fr. Minér. Crist.*, **94**, 115–27.

Authier, A. (1962). *J. Phys. Rad.*, **23**, 961–9.

Authier, A. (1966). *J. Physique (France)*, **27**, 57–60.

Authier, A. (1967). *Adv. X-ray Anal.*, **10**, 9–31.

Authier, A. (1968). *Phys. Stat. Sol.*, **27**, 77–93.

Authier, A. (1970). In *Advances in structure research by diffraction methods*, Vol. 3 (ed. Brill and Mason), pp. 1–15. Vieweg, Braunschweig.

Authier, A. (1972). *J. Cryst. Growth*, **13/14**, 34–8.

Authier, A. (1977*a*). In *Crystal growth and materials* (ed. E. Kaldis and H. J. Scheel), pp. 516–48. North-Holland, Amsterdam.

Authier, A. (1977*b*). *J. Cryst. Growth*, **42**, 612–20.

Authier, A. (1977*c*). In *X-ray optics*. Topics in Applied Physics (ed. H.-J. Queisser), Vol. 22, pp. 145–88. Springer-Verlag, Berlin.

Authier, A. (1978). In *Diffraction and imaging techniques in material science* (ed. S. Amelinckx, R. Gevers, G. Remaut and J. Van Landuy), pp. 715–56. North-Holland, Amsterdam.

Authier, A. (1986). *Acta Crystallogr. A*, **42**, 414–26.

Authier, A. (1989). *J. Physique*, **50**, C7-215–C7-224.

Authier, A. (1996). In *X-ray and neutron dynamical diffraction: theory and applications* (ed. A. Authier, S. Lagomarsino and B. K. Tanner), NATO Series, Series B, Physics, Vol. 357, pp. 1–31. Plenum Press, New York and London.

Authier, A. (1998). *Cryst. Res. Technol.*, **33**, 517–33.

Authier, A. (2001). In *International tables of crystallography*, Vol. B (ed. U. Shmueli), pp. 534–51, Kluwer Academic Publishers, Dordrecht.

Authier, A. (2003). *J. Phys. D: Appl. Physics*, **36**, A1–A4.

Authier, A. (2008), *Acta Crystallogr A*, **64**, 337–340.

Authier, A. and Balibar, F. (1970). *Acta Crystallogr. A*, **26**, 647–54.

Authier, A. and Capelle, B. (1992). In *P. P. Ewald and his dynamical theory of X-ray diffraction* (ed. D. W. J. Cruickshank, H. J. Juretschke and N. Kato), pp. 64–70. Oxford University Press, Oxford.

Authier, A. and Lang, A.R. (1964). *J. Appl. Phys.*, **35**, 1956–9.

Authier, A. and Malgrange, C. (1966). *C. R. Acad. Sci. Paris*, **262**, 429–32.

Authier, A. and Malgrange, C. (1998). *Acta Crystallogr. A*, **54**, 806–19.

Authier, A. and Patel, J. R. (1975). *Phys. Stat. Sol. (a)*, **27**, 213–22.

Authier, A. and Pétroff, J. F. (1964). *C. R. Acad. Sci. Paris*, **258**, 4238.

Authier, A. and Sauvage, M. (1966). *J. Phys. Rad. (France)*, **27**, 137–42.

Authier, A. and Simon, D. (1968). *Acta Crystallogr. A*, **24**, 517–26.

Authier, A. and Zarka, A. (1992). In *Advanced mineralogy* (ed. A. S. Marfunin), pp. 290–302. Springer-Verlag, Heidelberg.

Authier, A., Lallemand, P. and Pfister, J. C. (1963). *J. Physique, Paris*, **24**, 467–70.

Authier, A., Malgrange, C. and Tournarie, M. (1968*a*). *Acta Crystallogr. A*, **24**, 126–36.

Authier, A., Milne, D. and Sauvage, M. (1968*b*). *Phys. Stat. Sol.*, **26**, 469–84.

Authier, A., Balibar, F. and Epelboin, Y. (1970). *Phys. Stat. Sol.*, **41**, 225–38.

Authier, A., Lefeld-Sosnowska, M., Epelboin, Y. and Soyer, A. (1985). *J. Appl. Crystallogr.*, **18**, 93–105.

Authier, A., Benz, K. W. and Wallrafen, F. (1987). In *Fluid Sciences and Materials Science in Space* (ed. H. U. Walter), pp. 405–49. Springer, Berlin.

Authier, A., Gronkowski, J. and Malgrange, C. (1989). *Acta Crystallogr. A*, **45**, 432–41.

Authier, A., Lagomarsino, S. and Tanner, B.K. (ed.) (1996). *X-ray and neutron dynamical diffraction: theory and applications*. NATO Series, Series B, Physics, Vol. 357. Plenum Press, New York and London.

Avanci, L. H. and Morelhão, S. L. (2000). *Acta Crystallogr. A*, **56**, 507–8.

Avanci, L. H., Cardoso, L. P., Girdwood, S. E., Pugh, D., Sherwood, J. N. and Roberts, K. J. (1998). *Phys. Rev. Lett.*, **81**, 5426–9.

Avanci, L. H., Cardoso, L. P., Sasaki, J. M., Girdwood, S. E., Roberts, K. J., Pugh, D. and Sherwood, J. N. (2000). *Phys. Rev. B*, **61**, 6507–14.

Awaji, M., Suzuki, Y., Takeuchi, A., Takano, H., Kamijo, N., Tamura, S. and Yasumoto, M. (2002). *J. Synchrotron Rad.*, **9**, 125–7.

Ayvazyan, V. *et al.* (2002). *Phys. Rev. Lett.*, **88**, 104802.

Bäcklin, E. (1935). *Z. Physik*, **93**, 450.

Baez, A. V. (1952). *J. Opt. Soc. Am.*, **42**, 756.

Baker, J. F. C. and Hart, M. (1975). *Acta Crystallogr. A*, **31**, 364–7.

Baker, J. F. C., Hart, M., Halliwell, M. A. G. and Heckingbotton, R. (1976). *Solid Stat. Elect.*, **19**, 331–4.

Bąk-Misiuk, J., Gronkowski, J., Härtwig, J. and Wierzchowski, W. (1987). *Phys. Stat. Sol. (a)*, **99**, 345–51.

Baldwin, T. O., Sherrill, F. A. and Young, F. W. (1968). *J. Appl. Phys.*, **39**, 1541–50.

Balibar, F. (1968). *Acta Crystallogr. A*, **24**, 666–76.

Balibar, F. (1969). *Acta Crystallogr. A*, **25**, 650–8.

Balibar, F. and Authier, A. (1967). *Phys. Stat. Sol.*, **21**, 413–22.

Balibar, F. and Malgrange, C. (1975). *Acta Crystallogr. A*, **31**, 425–34.

Balibar, F., Epelboin, Y. and Malgrange, C. (1975). *Acta Crystallogr. A*, **31**, 836–40.

Balibar, F., Chukhovskii, F. N. and Malgrange, C. (1983). *Acta Crystallogr. A*, **39**, 387–99.

Bancel, P. A. (1989). *Phys. Rev. Lett.*, **63**, 2741–4.

Baranova, N. B. and Zel'dovich, B. Ya. (1980). *Zhurnal Exper. Teor. Fiziki*, **79**, 1779–86.

Barbee, T. W. Jr (1985). *Superlattices Microstruct.*, **1**, 311.

Barbee, T. W. Jr and Warburton, W. K. (1984). *Mater. Lett.*, **3**, 17–23.

Barbee Jr, T. W., Warburton, W. K. and Underwood, J. H. (1984). *J. Opt. Soc. Am. B*, **1**, 691–8.

Barkla, C. G. (1905). *Phil. Trans. Roy. Soc. London A*, **204**, 467–79.

Barkla, C. G. and Sadler, C. A. (1909). *Phil. Mag.*, **17**, 739.

Barnett, S. J., Keir, A. M., Cullis, A. G., Johnson, A. D., Jefferson, J., Smith, G. W., Martin, T., Whitehouse, C. R., Lacey, G., Clark, G. F., Tanner, B. K., Spirkl, W.,

Lunn, B., Hogg, J. C. H., Ashu, P., Hagston, W. E. and Castelli, C. M. (1995). *J. Phys. D: Appl. Phys.*, **28**, A17–A22.

Baron, A. Q. R., Tanaka, Y., Ishikawa, D., Miwa, D., Yabashi, M. and Ishikawa, T. (2001). *J. Synchrotron Rad.*, **8**, 1127–30.

Barrett, C. S. (1945). *Trans. Metal. Soc. AIME*, **161**, 15–64.

Bartels, W. J. (1983). *J. Vac. Sci. Technol. B*, **1**, 338–45.

Bartels, W. J., Hornstra, J. and Lobeek, D. J. W. (1986). *Acta Crystallogr. A*, **42**, 539–45.

Barth, H. and Hosemann, R. (1958). *Z. Naturforsch. (a)*, **13**, 792–4.

Bartscher, M. and Bonse, U. (1998). *Cryst. Res. Technol.*, **33**, 535–41.

Baruchel, J. (1993). *Physica B*, **192**, 79–93.

Baruchel, J. (1996). In *X-ray and neutron dynamical diffraction: theory and applications* (ed. A. Authier, S. Lagomarsino and B. K. Tanner), NATO ASI Series, Series B, Physics, Vol. 357, pp. 199–210. Plenum Press, New York and London.

Baruchel, J. and Schlenker, M. (1984). In *Applications of X-ray topographic methods to materials science* (ed. S. Weissmann, F. Balibar and J.-F. Pétroff), pp. 59–74. Plenum Press, New York.

Baruchel, J. and Schlenker (1996). in *X-ray and neutron dynamical diffraction: theory and applications*. ed. A. Authier, S. Lagomarsino and B. K. Tanner, NATO ASI Series, Series B: Physics, Plenum Press, New York and London, **357**, 177–86.

Baruchel, J., Palmer, S. B., and Schlenker, M. (1981). *J. Physique*, **42**, 1279–83.

Baruchel, J., Guigay, J.-P., Mazuré–Espejo, C., Schlenker, M. and Schweizer, J. (1982). *J. Physique*, **43**, C7-101–6.

Baruchel, J., Patterson, C. and Palmer, S. B. (1988). *Phil. Mag. B*, **57**, 505–19.

Baruchel, J., Bouillot, J. and Coquet, E. (1991). *Phil. Mag. B*, **63**, 1051–62.

Baruchel, J., Epelboin, Y., Gastaldi, J., Härtwig, J., Kulda, J., Rejmánková, P., Schlenker, M. and Zontone, F. (1994). *Phys. Stat. Sol. (a)*, **141**, 59–69.

Baruchel, J., Buffiere, J. Y., Cloetens, P., Guigay, J. P., Mancini, L., Peix, G., Peyrin, F., Salomé, M., Schlenker, M. and Spanne, P. (1997). *SPIE*, **3154**, 51–62.

Baruchel, J., Cloetens, P., Härtwig, J., Ludwig, W., Mancini, L., Pernot, P. and Schlenker, M. (2000). *J. Synchrotron Rad.*, **7**, 196–201.

Baruchel, J., de Camargo, P. C., Klein, H., Mazzaro, I., Nogues, J. and de Oliveira, A. (2001). *J. Phys. D: Appl. Phys.*, **34**, A114–A116.

Baruchel, J., Härtwig, J. and Pernot-Rejmánková, P. (2002). *J. Synchrotron Rad.*, **9**, 107–14.

Basile, G., Bergamin, A., Cavagnero, G., Mana, G. and Zosi, G. (1989). *IEEE Trans. Instrum. Meas.*, **38**, 210–6.

Basile, G., Bergamin, A., Cavagnero, G., Mana, G., Vittone, E. and Zosi, G. (1995*a*). *IEEE Trans. Instrum. Meas.*, **44**, 526–9.

Basile, G., Becker, P., Bergamin, A., Bettin, H., Cavagnero, G., de Bievre, P., Kütgens, U., Mana, G., Mosca, M., Pajot, B., Panciera, R., Pasin, W., Pettoruso, S., Peuto, A., Sacconi, A., Stümpel, J., Valkiers, S., Vittone, E. and Zosi, G. (1995*b*). *IEEE Trans. Instrum. Meas.*, **44**, 538–41.

Basov, Yu. A., Pravdivtseva, T. L., Snigirev, A. A., Belakhovsky, M., Dhez, R. and Freund, A. (1991). *Nucl. Instrum. Methods A*, **308**, 363–6.
Basterfield, J. and Prescott, M. J. (1967). *J. Appl. Phys.*, **38**, 3190.
Bastie, D. and Dolino, G. (1985). *Phys. Rev. B*, **31**, 2857–61.
Batterman, B. W. (1962*a*). *Phys. Rev.*, **126**, 1461–9.
Batterman, B. W. (1962*b*). *Phys. Rev.*, **127**, 686–90.
Batterman, B. W. (1964*a*). *Phys. Rev. A*, **133**, 759–64.
Batterman, B. W. (1964*b*). *Phys. Rev. A*, **134**, 1354–6.
Batterman, B. W. (1969). *Phys. Rev. Lett.*, **22**, 703–5.
Batterman, B. W. (1992). *Phys. Rev. B*, **22**, 12677–81.
Batterman, B. W. and Berman, L. E. (1983). *Nucl. Instrum. Methods Phys. Res.*, **208**, 327–31.
Batterman, B. W. and Bilderback, D. H. (1991). In *Handbook on Synchrotron Radiation*, Vol. 3 (ed. G. Brown and D. E. Moncton), pp. 105–53. Elsevier Science B.V., Amsterdam.
Batterman, B. W. and Cole, H. (1964). *Rev. Mod. Phys.*, **36**, 681–717.
Batterman, B. W. and Hildebrandt, G. (1967). *Phys. Stat. Sol.*, **23**, K147–K149.
Batterman, B. W. and Hildebrandt, G. (1968). *Acta Crystallogr. A*, **24**, 150–7.
Batterman, B.W. and Patel, J. R. (1968). *J. Appl. Phys.*, **39**, 1882–7.
Baumbach, T., Brühl, H.-G., Rhan, H. and Pietsch, U. (1988). *J. Appl. Cryst.*, **21**, 386–92.
Baumbach, T., Tixier, S., Pietsch, U. and Holý, V. (1995). *Phys. Rev. B*, **51**, 16848–59.
Bauspiess, W., Bonse, U., Rauch, H. and Treimer, W. (1974). *Z. Physik*, **271**, 177–82.
Bauspiess, W., Bonse, U. and Graeff, W. (1976). *J. Appl. Crystallogr.*, **9**, 68–80.
Bauspiess, W., Bonse, U. and Rauch, H. (1977). In *Proc. Conference on Neutron Scattering*, pp. 1094–102. US Dept of Commerce, Springfield, VA, USA.
Beard, W. T. Jr and Armstrong, R. W. (1989). *Adv. X-ray Analysis*, **32**, 659.
Beard, W. T. Jr, Green, K. A., Zhang, X-J. and Armstrong, R. W. (1996). *Appl. Phys. Lett.*, **69**, 488–90.
Beaumont, J. H. and Hart, M. (1974). *J. Phys. E*, **7**, 823–9.
Becker, P. and Bonse, U. (1974). *J. Appl. Crystallogr.*, **7**, 593–8.
Becker, P. J. and Dunstetter, F. (1984). *Acta Crystallogr. A*, **40**, 241–51.
Becker, P., Dorenwendt, K., Ebeling, G., Lauer, R., Lucas, W., Probst, R., Rademacher, H.-J., Reim, G., Seyfired, P. and Siegert, H. (1981). *Phys. Rev. Lett.*, **46**, 1540–3.
Becker, P. J. and Al Haddad, M. (1989). *Acta Crystallogr. A*, **45**, 333–7.
Becker, P. J. and Al Haddad, M. (1990). *Acta Crystallogr. A*, **46**, 123–9.
Becker, P. J. and Al Haddad, M. (1992). *Acta Crystallogr. A*, **48**, 121–34.
Becker, P. J. and Coppens, P. (1974). *Acta Crystallogr. A*, **30**, 129–47 and 148–53.
Becker, P. J. and Coppens, P. (1975). *Acta Crystallogr. A*, **31**, 417–25.
Becker, R. S., Golovchenko, J. A. and Patel, J. R. (1983). *Phys. Rev. Lett.*, **50**, 153–6.

Becker, P., Bettin, H., Danzebrink, H. U., Glaser, M., Kuetgens, U., Nicolaus, A., Schiel, D., De Bivre, P., Valkiers, S. and Taylor, P. (2003). *Metrologia*, **40**, 271–87.

Beckman, F., Bonse, U., Busch, F. and Günnewig, O. (1997). *J. Comput. Assist. Tomography*, **21**, 539–53.

Beckman, F., Bonse, U. and Biermann, T. (1999). *Proc. SPIE*, **3772**, 179–87.

Bedynska, T. (1973*a*). *Phys. Stat. Sol. (a)*, **18**, 147–54.

Bedynska, T. (1973*b*). *Phys. Stat. Sol. (a)*, **19**, 365–72.

Bedynska, T. (1974). *Phys. Stat. Sol. (a)*, **25**, 405–11.

Bedynska, T., Bubáková, R. and Šourek, Z. (1976). *Phys. Stat. Sol. (a)*, **36**, 509–16.

Bedzyk, M. J. (1988). *Nucl. Instrum. Methods A*, **266**, 679–83.

Bedzyk, M. J. and Materlik, G. (1985). *Phys. Rev. B*, **32**, 6456–63.

Bedzyk, M. J., Materlik, G. and Koval'chuk, M. V. (1984). *Phys. Rev. B*, **30**, 4881–4.

Bedzyk, M. J., Bilderback, D. H., White, J., Abruña, H. D. and Bommarito, M. J. (1986). *J. Phys. Chem.*, **90**, 4926–8.

Bedzyk, M. J., Bilderback, D. H., Bommarito, G. M., Caffrey, M. and Schildkraut, J. S. (1988). *Science*, **241**, 1788–91.

Bedzyk, M. J., Bommarito, G. M. and Schildkraut, J. S. (1989). *Phys. Rev. Lett.*, **62**, 1376–9.

Bedzyk, M. J., Kazimirov, A., Marasco, D. L., Lee, T.-L., Foster, C. M., Bai, G.-R., Lyman, P. F. and Keane, D. T. (2000). *Phys. Rev. B*, **61**, R7873–6.

Begum, R., Hart, M., Lea, K. R. and Siddons, D. P. (1986). *Acta Crystallogr. A*, **42**, 456–64.

Belouet, C., Dunia, E., Pétroff, J. F. and Sauvage, M. (1983). *J. Cryst. Growth*, **61**, 383–92.

Belyakov, V. A. and Dmitrienko, V. E. (1989). *Sov. Phys. Usp.*, **32**, 697–719.

Bensoussan, S., Malgrange, C. and Sauvage-Simkin, M. (1987*a*). *J. Appl. Crystallogr.*, **20**, 222–9.

Bensoussan, S., Malgrange, C., Sauvage-Simkin, M., N'Guessan, K. and Gibart, P. (1987*b*). *J. Appl. Crystallogr.*, **20**, 230–4.

Berenson, R. and Birman, J. L. (1986). *Phys. Rev. B*, **34**, 8926–9.

Berg, O. (1926). *Wiss. Veröff. a.d. Siemens-Konz.*, **5**, 89.

Berg, W. F. (1931). *Naturwissenshaften*, **19**, 391–6.

de Bergevin, F., and Brunel, M. (1972). *Phys. Lett. A*, **39**, 141–2.

de Bergevin, F. and Brunel, M. (1981). *Acta Crystallogr. A*, **37**, 314–24.

de Bergevin, F. and Brunel, M. (1991). In *Handbook of synchrotron radiation*, Vol. 3, pp. 535–64. North-Holland, Amsterdam.

Berman, L. E. and Hart, M. (1991*a*). *Nucl. Instr. & Methods in Phys. Res. A*, **300**, 415–21.

Berman, L. E. and Hart, M. (1991*b*). *Nucl. Instr. & Methods in Phys. Res. A*, **302**, 558–62.

Berman, L. E. and Hart, M. (1993). *Nucl. Instrum. Methods Phys. Res. A*, **334**, 617–20.

Berman, L. E., Durbin, S. M. and Batterman, B. W. (1985). *Nucl. Instrum. Methods A*, **241**, 295–301.

Berman, L. E., Batterman, B. W. and Blakely, J. M. (1988). *Phys. Rev. B*, **38**, 5397–405.

Berman, L. E., Hart, M. and Sharma, S. (1992). *Nucl. Instr. & Methods in Phys. Res. A*, **321**, 617–28.

Berman, L. E., Shen, Q., Finkelstein, K. D., Park, D., Zhijian-Yin and Guoqiang-Pan (2002). *Rev. Sci. Instrum.*, **73**, 1502–04.

Bernhardt, H. J., Härtwig, J. and Lerche, V. (1992). *Progress in Crystal Growth and Characterization*, **24**, 1–51.

Berreman, D. W. and Macrander, A. T. (1988). *Phys. Rev. B*, **37**, 6030–40.

Bethe, H. A. and Hildebrandt, G. (1988). *Biographical Memoirs of Fellows of the Royal Society*, **34**, 135–76.

Bezirganyan, P. Kh., Kocharyan, A. K. and Truni, K. G. (1981). *Phys. Stat. Sol. (a)*, **64**, 431–4.

Bhat, H. L. (1985). in *Progress in Crystal Growth and Characterization*. ed. B. R. Pamplin, Pergamon, Oxford, **11**, 57–87.

Bhat, H. L., Clark, G.F., Klapper, H. and Roberts, K. J. (1995). *J. Phys. D: Appl. Phys.*, **28**, A23–6.

Bijvoet, J. M. and MacGillavry, C. H. (1939). *Chem. Weekbl.*, **36**, 330–1.

Bijvoet, J. M., Burgers, W. G. and Häggs, G. (1969). *Early papers on diffraction of X-rays by crystals*. Published for the IUCr by A. Oosthoek's Uitgeversmaatschappij N.V., Utrecht.

Bilderback, D. H. (1986). *Nucl. Instrum. Methods Phys. Res. A*, **302**, 558–62.

Bilderback, D. H., Freund, A., Knapp, G. S. and Mills, D. M. (2000). *J. Synchrotron Rad.*, **7**, 53–60.

Bilello, J. C. (1984). In *Applications of X-ray topographic methods to material science* (ed. S. Weissmann, F. Balibar and J.F. Pétroff), pp. 333–42. Plenum Press, New York.

Bilello, J. C., Schmitz, H. A. and Dew-Hughes, D. (1989). *J. Appl. Phys.*, **65**, 2282–8.

Bionta, R. M., Ables, E., Clamp, O., Edwards, O. D., Gabriele, P. C., Miller, K., Ott, L. L., Skulima, K. M., Yilley, R. and Viada, T. (1989). *Proc. SPIE*, **1160**, 12–18.

Bionta, R. M., Ables, E., Clamp, O., Edwards, O. D., Gabriele, P. C., Miller, K., Ott, L. L., Skulima, K. M., Yilley, R. and Viada, T. (1990). *Opt. Eng.*, **29**, 576–80.

Bionta, R. M., Skulina, K. M. and Weinberg, J. (1994). *Appl. Phys. Lett.*, **64**, 945–7.

Blech, I. A. and Meieran, E. S. (1966). *Phil. Mag.*, **14**, 275–88.

Blech, I. A. and Meieran, E. S. (1969). *Phys. Rev.*, **179**, 731–5.

Blewett, J. P. (1998). *J. Synchrotron Rad.*, **5**, 135–9.

Blish, R. C. II and Vreeland, T. Jr (1969). *J. Appl. Phys.*, **40**, 884–90.

Bloch, F. (1928). *Z. Physik*, **52**, 555.

Blume, M. (1985). *J. Appl. Phys.*, **57**, 3615–18.

Blume, M. (1994). In *Resonant anomalous X-ray scattering* (ed. G. Materlik, C. J. Sparks and K. Fischer), pp. 495–515. Elsevier Science B.V., Amsterdam.
Blume, M. and Gibbs, D. (1988). *Phys. Rev. B*, **37**, 1779–89.
Boettinger, W. J., Burdette, H. E., Kuriyama, M. and Green, R. E. (1976). *Rev. Sci. Instrum.*, **47**, 906–11.
Boettinger, W. J., Burdette, H. E. and Kuriyama, M. (1979). *Rev. Sci. Instrum.*, **50**, 26–30.
Boettinger, W. J., Dobbyn, R. C., Burdette, H. E. and Kuriyama, M. (1982). *Nucl. Instrum. Methods A*, **195**, 355–61.
Boeuf, A., Lagomarsino, S., Mazkedian, S., Melone, S., Puliti, P. and Rustichelli, F. (1978). *J. Appl. Crystallogr.*, **11**, 442–9.
Boggon, T. J., Helliwell, J. R., Judge, R. A., Olczak, A., Siddons, D. P., Snell, E. H. and Stojanoff, V. (2000). *Acta Crystallogr. D*, **56**, 868–80.
Bond, W. L. (1960). *Acta Crystallogr.*, **13**, 814–17.
Bongaerts, J. H. H., David, C., Drakopoulos, M., Zwanenburg, M. J., Wegdam, G. H., Lackner, T., Keymeulen, H. and van der Veen, J. F. (2002). *J. Synchrotron Rad.*, **9**, 383–93.
Bonse, U. (1958). *Z. Phys.*, **153**, 278–96.
Bonse, U. (1962). In *Direct observation of imperfection in crystals* (ed. J. B. Newkirk and J. H. Wernick), pp. 431–60. Interscience, New York.
Bonse, U. (1964*a*). *Z. Phys.*, **177**, 385–423.
Bonse, U. (1964*b*). *Z. Phys.*, **177**, 529–42.
Bonse, U. (1964*c*). *Z. Phys.*, **177**, 543–61.
Bonse, U. (1969). In *Proc. of the Vth Int. Cong. on X-ray Optics and Microanalysis* (ed. G. Möllenstedt and K. H. Gaukler), pp. 1–10. Springer, Berlin.
Bonse, U. (1988). *Physica B*, **151**, 7–21.
Bonse, U. and Beckmann, F. (2001). *J. Synchrotron Rad.*, **8**, 1–5.
Bonse, U. and Busch, F. (1996). *Prog. Biophys. Molec. Biol.*, **65**, 133–69.
Bonse, U. and Graeff, W. (1973). *Z. Naturforsch. (a)*, **28**, 558–64.
Bonse, U. and Graeff, W. (1977). In *X-ray optics. Applications to solids*. Topics in Applied Physics (ed. H. J. Queisser), pp. 93–143. Springer, Berlin.
Bonse, U. and Hart, M. (1965*a*). *Appl. Phys. Lett.*, **6**, 155–6.
Bonse, U. and Hart, M. (1965*b*). *Appl. Phys. Lett.*, **7**, 99–100.
Bonse, U. and Hart, M. (1965*c*). *Appl. Phys. Lett.*, **7**, 238–40.
Bonse, U. and Hart, M. (1965*d*). *Z. Phys.*, **188**, 154–64.
Bonse, U. and Hart, M. (1966*a*). *Z. Phys.*, **189**, 151–62.
Bonse, U. and Hart, M. (1966*b*). *Z. Phys.*, **190**, 455–67.
Bonse, U. and Hart, M. (1966*c*). *Z. Phys.*, **194**, 1–17.
Bonse, U. and Hart, M. (1968). *Acta Crystallogr. A*, **24**, 240–5.
Bonse, U. and Hart, M. (1969). *Phys. Stat. Sol.*, **33**, 351–9.
Bonse, U. and Hellköter, H. (1969). *Z. Phys.*, **223**, 345–52.
Bonse, U. and Kappler, E. (1958). *Z. Naturforsch.*, **13a**, 348–9.
Bonse, U. and Materlik, G. (1972). *Z. Phys.*, **253**, 232–9.

Bonse, U. and Materlik, G. (1974). In *Anomalous scattering* (ed. S. Ramaseshan and S. C. Abrahams), pp. 107–9. Munjsgaard, Copenhagen.
Bonse, U. and Materlik, G. (1976). *Z. Phys. B*, **24**, 189–91.
Bonse, U. and te Kaat, E. (1968). *Z. Phys.*, **214**, 16–21.
Bonse, U. and te Kaat, E. (1971). *Z. Phys.*, **243**, 14–45.
Bonse, U. and Rauch, H. ed. (1979). *Neutron Interferometry*. Clarendon, Oxford.
Bonse, U. and Teworte, R. (1980). *J. Appl. Crystallogr.*, **13**, 410–16.
Bonse, U., Hart, M. and Schwuttke, G. H. (1969). *Phys. Stat. Sol.*, **33**, 361–74.
Bonse, U., Graeff, W. and Materlik, G. (1976*a*). *Rev. Physique Appl.*, **11**, 83.
Bonse, U., Materlik, G. and Schröder, W. (1976*b*). *J. Appl. Crystallogr.*, **9**, 223–30.
Bonse, U., Graeff, W., Teworte, R. and Rauch, H. (1977). *Phys. Stat. Sol. (a)*, **43**, 487–92.
Bonse, U., Spieker, P., Hein, J. T. and Materlik, G. (1980). *Nuclear Instrum. & Methods*, **172**, 223–6.
Bonse, U., Olthoff-Münter, K. A. and Rumpf, A. (1983). *J. Appl. Crystallogr.*, **9**, 524–31.
Bonse, U., Riekel, C. and Snigirev, A. A. (1992). *Rev. Sci. Instrum.*, **63**, 622–4.
Bordas, J., Glazer, A. M., Howard, C. J. and Bourdillon, A. (1977). *Phil. Mag.*, **35**, 311–23.
Borie, B. (1957). *Acta Crystallogr.*, **10**, 89–96.
Borie, B. (1959). *Acta Crystallogr.*, **12**, 280–2.
Borie, B. (1967). *Acta Crystallogr.*, **23**, 210–16.
Born, M. and Wolf, E. (1993). *Principles of optics*. 6th Edition, with corrections, Pergamon Press, Oxford; reissued by Cambridge Press, Cambridge, 1999.
Borrmann, G. (1936). *Ann. Physik, Lpz*, **27**, 669–93.
Borrmann, G. (1941). *Physik Z.*, **42**, 157–62.
Borrmann, G. (1950). *Z. Phys.*, **127**, 297–323.
Borrmann, G. (1954). *Z. Krist.*, **106**, 109–21.
Borrmann, G. (1955). *Naturwiss.*, **42**, 67–8.
Borrmann, G. (1959*a*). *Beitr. Phys. Chem. 20 Jahrhunderts*, pp. 262–82. Vieweg & Sohn, Braunschweig.
Borrmann, G. (1959*b*). *Phys. Bl.*, **15**, 508–9.
Borrmann, G. (1964). *Z. Krist.*, **120**, 143–81.
Borrmann, G. (1971). *Wissenschaft. Zeit. Tech. Univ. Dresden*, **20**, 415.
Borrmann, G. and Hartwig, W. (1965). *Z. Krist.*, **121**, 401–9.
Borrmann, G. and Lehmann, K. (1962). In *Direct observation of imperfections in crystals*, pp. 409–14. Interscience Publ., New York.
Borrmann, G. and Lehmann, K. (1963). In *Crystallography and crystal perfection*, pp. 101–8 (ed. G. N. Ramachandran). Academic Press, London.
Borrmann, G. and Wagner, H. (1955). *Naturwissenshaften*, **42**, 68.
Borrmann, G., Hildebrandt, G. and Wagner, H. (1955). *Z. Phys.*, **142**, 406–14.
Borrmann, G., Hartwig, W. and Irmler, H. (1958). *Z. Naturforsch.*, **13a**, 423–5.

Bouchard, R., Hupfeld, D., Lippmann, T., Neuefeind, J., Neumann, H.-B., Poulsen, H.F., Rütt, U., Schmidt, T., Schneider, J. R., Süssenbach, J. and von Zimmermann, M. (1998). *J. Synchrotron Rad.*, **5**, 90–101.

Boulliard, J. C., Capelle, B., Ferret, D., Lifchitz, A., Malgrange, C., Pétroff, J. F., Taccoen, A. and Zheng, Y. L. (1992). *J. Phys. I France*, **2**, 1215–32.

Boulliard, J. C., Capelle, B., Gualandris, S., Lifchitz, A., Cibert, J. and Tatarenko, S. (1997). *J. Appl. Phys.*, **81**, 1173–9.

Bowen, D. K. (1996). In *X-ray and neutron dynamical diffraction: theory and applications* (ed. A. Authier, S. Lagomarsino and B. K. Tanner), NATO ASI Series, Series B, Physics, Vol. 357, pp. 381–410. Plenum Press, New York and London.

Bowen, D. K. and Miltat, J. (1976). *J. Phys. E: Sci. Instr.*, **9**, 868–70.

Bowen, D. K. and Tanner, B. K. (1992). *Mater. Sci. Rep.*, **8**, 369–407.

Bowen, D. K. and Tanner, B. K. (1998). *High resolution X-ray diffraction and topography*. Taylor & Francis Ltd, London.

Bowen, D. K., Chetwynd, D. G. and Davies, S. T. (1985). In *29th Symposium on Optical and Electro-optical Engineering, SPIE Proc.*, Vol. 563, pp. 412–19.

Bowen, D. K., Chetwynd, D. G. and Schwarzenberger, D. R. (1990*a*). *Meas. Sci. Technol.*, **1**, 107–19.

Bowen, D. K., Chetwynd, D. G., Schwarzenberger, D. R. and Smith, S. T. (1990*b*). *Precision Engineering*, **12**, 165–71.

Bowen, D. K., Wormington, M. and Feichtinger, P. (2003). *J. Phys. D: Appl. Physics*, **36**, A17–A23.

Bradaczek, H., Hildebrandt, G. and Uebach, W. (1993). *Acta Crystallogr. A*, **49**, C374.

Brádler, J. and Lang, A. R. (1968). *Acta Crystallogr. A*, **24**, 246–7.

Bragg, W.H. (1914). *Phil. Mag.*, **27**, 881–99; reprinted in *Acta Crystallogr. A* (1969), **25**, 3–11.

Bragg, W. H. and Bragg, W. L. (1913*a*). *Proc. Roy. Soc. (London) A*, **88**, 428–38.

Bragg, W. H. and Bragg, W. L. (1913*b*). *Proc. Roy. Soc. (London) A*, **89**, 248–77.

Bragg, W. L. (1913). *Proc. Cambridge Phil. Soc.*, **17**, 43–57.

Bragg, W. L., James, R. W. and Bosanquet, C. H. (1921). *Phil. Mag.*, **41**, 309–37.

Bragg, W. L., Darwin, C.G. and James, R. W. (1926). *Phil. Mag.*, **1**, 897–922.

Bravin, A. (2003). *J. Phys. D: Appl. Physics*, **36**, A24–A29.

Bremer, J. (1984). *Acta Crystallogr. A*, **42**, 283–91.

Breuer, D., Klimanek, J. and Pantleon, W. (2000). *J. Appl. Crystallogr.*, **33**, 1284–94.

Brogren, G. (1954). *Acta Crystallogr.*, **7**, 211–13.

Brogren, G. and Adell, O. (1954*a*). *Arkiv Fys.*, **8**, 97–112.

Brogren, G. and Adell, O. (1954*b*). *Arkiv Fys.*, **9**, 1.

Brown, G. T., Skolnick, M. S., Jones, G. R., Tanner, B. K. and Barnett, S. J. (1984). In *Semi-insulating III–V materials*, pp. 76–82 (ed. D. C. Look and J. S. Blakemore). Shiva, Nantwich.

Brown, B., Clark, G. F., Dineen, C., Isherwood, B. J., Pantos, E., Roberts, K. J. and Scheffen–Lauenroth, T. (1989). *J. Appl. Crystallogr.*, **22**, 201–4.

Brühl, H.-G., Poecker, A., Nickel, H., Lösch, R. and Schlapp, W. (1993). *J. Appl. Crystallogr.*, **26**, 645–9.

Brümmer, O. and Stephanik, H. (1976). *Dynamische interferenztheorie*. Akademische Verlagsgesellschaft, Leipzig.

Brümmer, O., Höche, H. R. and Nieber, J. (1976*a*). *Phys. Stat. Sol. (a)*, **33**, 587–93.

Brümmer, O., Höche, H. R. and Nieber, J. (1976*b*). *Phys. Stat. Sol. (a)*, **37**, 529–32.

Brümmer, O., Höche, H. R. and Nieber, J. (1978). *Phys. Stat. Sol. (a)*, **46**, K131–K133.

Brümmer, O., Höche, H. R. and Nieber, J. (1979). *Phys. Stat. Sol. (a)*, **53**, 565–70.

Brümmer, O., Eisenschmidt, Ch. and Höche, H. R. (1982*a*). *Z. Naturforsch. (a)*, **37**, 524–7.

Brümmer, O., Höche, H. R. and Nieber, J. (1982*b*). *Z. Naturforsch. (a)*, **37**, 519–23.

Brümmer, O., Eisenschmidt, Ch. and Höche, H. R. (1984). *Acta Crystallogr. A*, **40**, 394–8.

Brunel, M. and de Bergevin, F. (1974). In *Proc. Int. Conf. Magn. ICM-73*, Vol. II, pp. 127–30. Nauka, Moscow.

Bubáková, R. (1962). *Czech. J. Phys. B*, **12**, 776–83.

Bubáková, R. and Šourek, Z. (1976). *Czech. J. Phys. B*, **26**, 863–4.

Bubáková, R., Drahokoupil, J. and Fingerland, A. (1961). *Czech. J. Phys. B*, **11**, 205–22.

Bubáková, R., Drahokoupil, J. and Fingerland, A. (1962*a*). *Czech. J. Phys. B*, **12**, 538–41.

Bubáková, R., Drahokoupil, J. and Fingerland, A. (1962*b*). *Czech. J. Phys. B*, **12**, 764–75.

Bucksch, R., Otto, J. and Renninger, M. (1967). *Acta Crystallogr.*, **23**, 507–11.

Buffière, J.-Y., Maire, E., Cloetens, P., Lormand, G. and Fougères, R. (1999). *Acta Met.*, **47**, 1613–25.

Bullough, R. K. and Hynne, F. (1992). In *P. P. Ewald and his dynamical theory of X-ray diffraction* (ed. D. W. J. Cruickshank, H. J. Juretschke and N. Kato), pp. 98–110. Oxford University Press, Oxford.

Burzlaff, H., Hümmer, K., Rothammel, W., Schäfer, G., Weckert, E., Bestmann, H. J. and Roth, D. (1989). *Acta Cryst. B*, **45**, 303–6.

Buschert, R. C. (1965). *Bull. Am. Phys. Soc.*, **10**, 125–33.

Buschert R. C., Meyer A. J., Kauffman D. S. and Gotwals J. K. (1983). *J. Appl. Crystallogr.*, **16**, 599–605.

Buschert, R. C., Pace, S., Inzaghi, D. and Merlini, A. E. (1980). *J. Appl. Crystallogr.*, **13**, 207–11.

Bushuev, V. A. (1989*a*). *Sov. Phys. Crystallograph.*, **34**, 163–7.

Bushuev, V. A. (1989*b*). *Sov. Phys. Solid State*, **31**, 1877–82.

Bushuev, V. A., Ingal, V. N. and Beliaevskaya, E. A. (1996). *Crystallogr. Repts*, **41**, 766–74.

Bushuev, V. A., Beliaevskaya, E. A. and Ingal, V. N. (1997). *Il Nuovo Cimento*, **19D**, 513–20.

Bushuev, V. A., Ingal, V. N. and Beliaevskaya, E. A. (1998). *Crystallogr. Repts*, **43**, 538–47.

Caciuffo, R., Melone, S., Rustichelli, F. and Boeuf, A. (1987). *Physics Reports*, **152**, 1–71.

Calzadilla, A. O., Mosina, G. N., Ratnikov, V. V., Sorokin, L. M., Trebugova, A. S., Fuentes, J., Shul'pina, I. L. and Shcheglov, M. P. (1991). *Microgravity Science & Technology*, **IV**, 172–8.

Campbell, H. N. (1951*a*). *Acta Crystallogr.*, **4**, 180–1.

Campbell, H. N. (1951*b*). *J. Appl. Phys.*, **22**, 1139–42.

Campos, C. and Chang, S.-L. (1986). *Acta Crystallogr. A*, **42**, 348–53.

Capelle, B. and Malgrange, C. (1982). *J. Appl. Phys.*, **53**, 6762–6.

Capelle, B. and Malgrange, C. (1984*a*). in *Applications of X-ray topographic methods to material science*. ed. S. Weissmann, F. Balibar and J.-F. Pétroff, Plenum Press, New York, 511–22.

Capelle, B. and Malgrange, C. (1984*b*). *J. Physique*, **45**, 1827–34.

Capelle, B., Epelboin, Y. and Malgrange, C. (1982). *J. Appl. Phys.*, **53**, 6767–71.

Capelle, B., Zarka, A., Zheng, Y., Détaint, J. and Schwartzel, J. (1989). *Proc. 43rd Annual Frequency Control Symposium*, pp. 470–6.

Capelle, B., Détaint, J., Schwartzel, J., Zheng, Y. and Zarka, A. (1992). *Proc. 1992 IEEE Frequency Control Symposium*, pp. 553–60.

Capelle, B., Détaint, J., Zarka, A. and Epelboin, Y. (1998). *Proc. 1998 IEEE Frequency Control Symposium*, pp. 816–22.

Capelle, B., Détaint, J. and Epelboin, Y. (2001). *J. Appl. Crystallogr.*, **34**, 625–9.

Capelle, B., Epelboin, Y., Härtwig J., Moraleda, A. B., Otálora, F. and Stojanoff, V. (2004). *J. Appl. Crystallogr.*, **37**, 67–71.

Capitan, M. J. and Alvarez, J. (2000). *J. Appl. Crystallogr.*, **33**, 888–92.

Cappello, G., Dechelette, A., Schmithusen, F., Decossas, S., Chevrier, J., Comin, F., Formoso, V., De Boissieu, M., Jach, T., Colella, R., Lograsso, T. A., Jenks, C. and Delaney, D. (2000). *Mat. Sci. & Engineering A*, **294/296**, 863–6.

Carvalho, C. A. M. and Epelboin, Y. (1993*a*). *Acta Crystallogr. A*, **49**, 460–7.

Carvalho, C. A. M. and Epelboin, Y. (1993*b*). *Acta Crystallogr. A*, **49**, 467–73.

Caslavsky, J. L. (1970). *Rev. Sci. Instrum.*, **41**, 517–18.

Caslavsky, J. L. and Polcarová, M. (1964). *Czech. J. Phys. B*, **14**, 454–8.

Caticha, A. (1993). *Phys. Rev. B*, **47**, 76–83.

Caticha, A. (1994). *Phys. Rev. B*, **49**, 33–8.

Caticha, A. and Caticha-Ellis, S. (1982). *Phys. Rev. B*, **25**, 971–83.

Caticha, A. and Caticha-Ellis, S. (1990*a*). *Phys. Stat. Sol. (a)*, **119**, 47–54.

Caticha, A. and Caticha-Ellis, S. (1990*b*). *Phys. Stat. Sol. (a)*, **119**, 643–54.

Caticha, A. and Caticha-Ellis, S. (1996). *Phys. Stat. Sol. (a)*, **153**, 29–46.

Caticha, A., Caticha, N. and Caticha-Ellis, S. (1989). *Appl. Phys. Lett.*, **54**, 887–9.

Caticha, A., Aliberti, K. and Caticha-Ellis, S. (1990). *Rev. Sci. Instrum.*, **67**, 3380–1.

Cauchois, Y., Hulubei, H. and Weigle, J. (1937). *Helv. Physica Acta*, **10**, 218–24.

Cedola, A., Di Fonzo, S., Jark, W., Lagomarsino, S. and Soullié, G. (1999). *J. Phys. D: Appl. Phys.*, **32**, A179–A183.

Cembali, F., Servidori, M., Gabili, E. and Lotti, R. (1985). *Phys. Stat. Sol. (a)*, **87**, 225–33.

Cembali, F., Fabri, R., Servidori, M., Zani, A., Basile, G., Cavegnero, G., Bergamin, A. and Zosi, G. (1992). *J. Appl. Crystallogr.*, **25**, 424–31.

Cerva, H. and Graeff, W. (1984). *Phys. Stat. Sol. (a)*, **82**, 35–45.

Cerva, H. and Graeff, W. (1985). *Phys. Stat. Sol. (a)*, **87**, 507–16.

Chan, C. T., Ho, K. M. and Soukoulis, C. M. (1992). *Modern Physics Letters B*. **6**, 139–44.

Chandrasekaran, K. S. (1959). *Acta Crystallogr.*, **12**, 916–22.

Chang, S.-L. (1979). *Acta Crystallogr. A*, **35**, 543–7.

Chang, S.-L. (1982). *Phys. Rev. Lett.*, **48**, 163–6.

Chang, S.-L. (1986). *Phys. Rev. B*, **33**, 5848–50.

Chang, S.-L. (1987). *Cryst. Rev.*, **1**, 87–189.

Chang, S.-L. (1992). *Int. J. Mod. Phys. B*, **6**, 2987–3020.

Chang, S.-L. (1998). *Acta Crystallogr. A*, **54**, 886–94.

Chang, S.-L. (2004). *X-Ray Multiple-Wave Diffraction: Theory and Application*. Springer Series in Solid-State Sciences, **143**, Springer-Verlag, Berlin.

Chang, S.-L. and Caticha-Ellis, S. (1978). *Acta Crystallogr. A*, **34**, 825–7.

Chang, S.-L. and Tang, M.-T. (1988). *Acta Crystallogr. A*, **44**, 1065–72.

Chang, S.-L., Huang, M.-T., Tang, M.-T. and Lee, C.-H. (1989). *Acta Crystallogr. A*, **45**, 870–8.

Chang, S.-L., King, H. E. Jr, Huang, M.-T. and Gao, Y. (1991). *Phys. Rev. Lett.*, **67**, 3113–16.

Chang, S.-L., Huang, Y.-S., Chao, C. H., Tang, M. T. and Stetsko, Y. (1998). *Phys. Rev. Lett.*, **80**, 301–4.

Chang, S.-L., Stetsko, Yu. P., Huang, Y.-S., Chao, C.-H., Liang, F.-J. and Chen, C.-K. (1999). *Phys. Lett. A*, **264**, 328–33.

Chang, S.-L., Stetsko, Yu. P. and Lee, Y.-R. (2002). *Z. Krist.*, **217**, 662–7.

Chang, W. Z. and Förster, E. (1997). *J. Opt. Soc. Am. A*, **14**, 1647–53.

Chang, W. Z., Kley, E. B., Fuchs, H. J., Schnabel, B., Förster, E. and Chukhovskii, F. N. (1995). *Proc. SPIE – The International Society for Optical Engineering.*, **2516**, 69–76.

Chantler, C. T. (1994). In *Resonant anomalous scattering, theory and applications* (ed. G. Materlik, C. J. Sparks and K. Fischer), pp. 61–78. Elsevier Science B.V., Amsterdam.

Chantler, C. T. (1995). *J. Phys. Chem. Ref. Data*, **24**, 71–643.

Chapman, D., Thomlinson, W., Johnston, R. E., Washburn, D., Pisano, E., Gmur, N., Zhong, Z., Menk, R., Arfelli, F. and Sayers, D. (1997). *Physics in Medicine and Biolology*, **42**, 2015–25.

Chetwynd, D. G., Cockerton, S. C., Smith, S. T. and Fung, W. W. (1991). *Nanotechnol.*, **2**, 1–10.

Chevallier, P., Dhez, P., Legrand, F., Idir, M., Soullié, G., Mirone, A., Erko, A., Snigirev, A., Snigireva, I., Souvorov, A., Freund, A., Engström, P., Nielsen, J. A. and Grubel, A. (1995). *Nucl. Instrum. & Methods Phys. Res. A*, **354**, 584–7.

Chikaura, Y. and Imai, M. (1990). *Jap. J. Appl. Phys.*, **29**, 221–5.

Chikaura, Y., Imai, M. and Ishikawa, T. (1987). *Jap. J. Appl. Phys.*, **26**, L889–L892.

Chikaura, Y., Imai, M., Suzuki, Y. and Yatsurugi, Y. (1990). *J. Cryst. Growth*, **103**, 141–9.

Chikaura, Y., Iida, S., Kawado, S., Mizuno, K., Kimura, S., Matsui, J., Umeno, M., Ozaki, K., Shimura, T., Suzuki, Y., Izumi, K., Kawasaki, K., Kajiwara, K. and Ishikawa, T. (2001). *J. Phys. D: Appl. Phys.*, **34**, A158–A162.

Chikawa, J.-I. (1965*a*). *J. Appl. Phys.*, **36**, 3496–3502.

Chikawa, J.-I. (1965*b*). *Appl. Phys. Lett.*, **7**, 193–5.

Chikawa, J.-I. (1967). *J. Phys. Chem. Solids*, **28 (Suppl. 1)**, 817–23.

Chikawa, J.-I. (1974). *J. Cryst. Growth*, **24/25**, 61–8.

Chikawa, J.-I. (1978). *J. Jap. Assoc. Cryst. Growth*, **5**, 141–63.

Chikawa, J.-I. (1980). In *Characterization of crystal growth defects by X-ray methods* (ed. B. K. Tanner and D. K. Bowen), NATO ASI Series, Series B, Physics, Vol. 63, pp. 368–400. Plenum Press, New York.

Chikawa, J.-I. (1999). In *International tables of crystallography*, Vol. C (ed. A. J. C. Wilson and E. Prince), pp. 627–9. Kluwer Academic Publishers, Dordrecht.

Chikawa, J.-I. and Austerman, S. B. (1968). *J. Appl. Crystallogr.*, **1**, 165–71.

Chikawa, J.-I., Asaeda, Y. and Fujimoto, I. (1970). *J. Appl. Phys.*, **41**, 1922–5.

Christiansen, G., Gerward, L. and Lindegaard-Andersen, A. (1971). *J. Appl. Crystallogr.*, **4**, 370–6.

Cho, W., Huang, X. R. and Dudley, M. (2004). *Acta Crystallogr. A*, **60**, 195–7.

Chukhovskii, F. N. (1974). *Kristallografiya*, **19**, 482–8.

Chukhovskii, F. N. (1981). *Metallofizika*, **3**, 3–30.

Chukhovskii, F. N. and Förster, E. (1995). *Acta Crystallogr. A*, **51**, 668–72.

Chukhovskii, F. N. and Khapachev, Yu. P. (1985). *Phys. Stat. Sol. (a)*, **88**, 69–76.

Chukhovskii, F. N. and Malgrange, C. (1989). *Acta Crystallogr. A*, **45**, 732–8.

Chukhovskii, F. N. and Petrashen, P. V. (1977). *Acta Crystallogr. A*, **33**, 311–19.

Chukhovskii, F. N. and Petrashen, P. V. (1988). *Acta Crystallogr. A*, **44**, 8–14.

Chukhovskii, F. N., Gabrielyan, K. T. and Petrashen, P. V. (1978). *Acta Crystallogr. A*, **34**, 610–21.

Chukhovskii, F.N., Krisch, M. and Freund, A. (1992). *Rev. Sci. Instrum.*, **63**, 920–3.

Chukhovskii, F. N., Chang, W. Z. and Förster, E. (1995*a*). *J. Appl. Phys.*, **77**, 1843–8.

Chukhovskii, F. N., Chang, W. Z. and Förster, E. (1995*b*). *J. Appl. Phys.*, **77**, 1849–54.

Chukhovskii, F. N., Malgrange, C. and Gronkowski, J. (1996). *Acta Crystallogr. A*, **52**, 47–55.

Chukhovskii, F. N., Hölzer, G., Wehrhan, O. and Förster, E. (1996*b*). *J. Appl. Crystallogr.*, **29**, 438–45.

Chung, J.-S. and Durbin, S. M. (1995). *Phys. Rev. B*, **51**, 14976–9.

Chung, J.-S. and Durbin, S. M. (1999). *Acta Crystallogr. A*, **55**, 14–19.

Clarebrough, C. M., Michell, D. and Smith, A.P. (1967). *Phys. Stat. Sol.*, **21**, 369–73.

Clark, G. F., Goddard, P. A., Nicholson, J. R. S., Tanner, B. K. and Wanklyn, B. M. (1983). *Phil. Mag. B*, **47**, 307–13.

Cloetens, P., Barrett, R., Baruchel, J., Guigay, J.-P. and Schlenker, M. (1996). *J. Phys. D: Appl. Phys.*, **29**, 133–46.

Cloetens, P., Guigay, J. P., De Martino, C. and Baruchel, J. (1997*a*). *Opt. Lett.*, **22-14**, 1059–61.

Cloetens, P., Guigay, J.-P., De Martino, C., Salomé, M., Schlenker, M. and Van Dyck, D. (1997*b*). *SPIE Proc.*, **3154**, 72–82.

Cloetens, P., Pateyron-Salomé, M., Buffière, J. Y., Peix, G., Baruchel, J., Peyrin, F. and Schlenker, M. (1997*c*). *J. Appl. Phys.*, **81**, 5878–86.

Cloetens, P., Ludwig, W., Baruchel, J., Guigay, J.-P., Pernot-Rejmánková, P., Salomé-Pateyron, M., Schlenker, M., Buffière, J.-Y., Maire, E. and Peix, G. (1999*a*). *J. Phys. D: Appl. Phys.*, **32**, A145–A151.

Cloetens, P., Ludwig, W., Baruchel, J., Van Dyck, D., Van Landuyt, J., Guigay, J.-P. and Schlenker, M. (1999*b*). *Appl. Phys. Lett.*, **75**, 2912–14.

Cochran, W. (1956). *Acta Crystallogr.*, **9**, 259–62.

Coïsson, R. (1995). *Appl. Optics*, **34**, 904–8.

Cole, H., Chambers, F. W. and Wood, C. G. (1961). *J. Appl. Phys.*, **32**, 1942–5.

Cole, H., Chambers, F. W. and Dunn, H. M. (1962). *Acta Crystallogr.*, **15**, 138–44.

Colella, R. (1974). *Acta Crystallogr. A*, **30**, 413–23.

Colella, R. (1980). *Acta Crystallogr. A*, **36**, 827.

Colella, R. (1988). *Acta Crystallogr. A*, **44**, 1055–9.

Colella, R. (1991). *Phys. Rev. B*, **43**, 13827–32.

Colella, R. (1995). *Comments Cond. Mat. Phys.*, **17**, 175–98 and 199–215.

Colella, R. (1996). In *X-ray and neutron dynamical diffraction: theory and applications* (ed. A. Authier, S. Lagomarsino and B. K. Tanner), NATO ASI Series, Series B, Physics, Vol. 357, pp. 369–860. Plenum Press, New York and London.

Colella, R. and Luccio, A. (1984). *Optics Commun.*, **50**, 41–4.

Colella, R. and Merlini, A. (1966). *Phys. Stat. Sol.*, **14**, 81–90.

Colella, R., Bessiere, M., Lee, H., Motsch, T., Lefebvre, S. and Denoyer, F. (1994). *Z. für Krist.*, **209**, 390–4.

Colella, R., Zhang, Y., Sutter, J. P., Ehrlich, S. N. and Kycia, S. W. (2001). *Phys. Rev. B*, **63**, 14202-1-5.

Compton, A. H. (1917). *Phys. Rev.*, **9**, 29–57.

Compton, A. H. and Allison, S. K. (1935). *X-rays in theory and experiment*. van Nostrand, New York; reprinted by Van Nostrand, Princeton, 1963.

Cottrell, S., Spirkl, W. and Tanner, B. K. (1993). *J. Phys. D: Appl. Phys.*, **26**, A126–A130.

Courant, R. and Hilbert, D. (1962). *Methods of mathematical physics*. Interscience Publ., New York.

Cowan, P. L. (1985). *Phys. Rev. B*, **32**, 5437–9.

Cowan, P. L., Golovchenko, J. A. and Robbins, M. F. (1980). *Phys. Rev. Lett.*, **44**, 1680–3.

Cowan, P. L., Brennan, S., Jach, T., Bedzyk, M. J. and Materlik, G. (1986). *Phys. Rev. Lett.*, **57**, 2399–2402.

Cowley, J. M. (1964). *Acta Crystallogr.*, **17**, 33–40.

Cowley, J. M. (ed.) (1992). *Electron diffraction techniques*. Oxford University Press, Oxford.

Cowley, J. M. and Rees, A. L. G. (1947). *Nature (London)*, **158**, 550.

Creagh, D. C. (1970). *Austr. J. Phys.*, **23**, 99–103.

Creagh, D. C. (1985). *Austr. J. Phys.*, **38**, 371–404.

Creagh, D. C. (1999). In *International tables of crystallography*, Vol. C (ed. A. J. C. Wilson and E. Prince), pp. 245–58. Kluwer Academic Publishers, Dordrecht.

Creagh, D. C. and Hart, M. (1970). *Phys. Stat. Sol.*, **37**, 753–8.

Cromer, D. T. (1983). *J. Appl. Crystallogr.*, **16**, 437–8.

Cromer, D. T. and Liberman, D. A. (1970). *J. Chem. Phys.*, **53**, 1891–8.

Cromer, D. T. and Liberman, D. A. (1981). *Acta Crystallogr. A*, **37**, 267–8.

Cruickshank, D. W. J. (1992). In *P. P. Ewald and his dynamical theory of X-ray diffraction* (ed. D. W. J. Cruickshank, H. J. Juretschke and N. Kato), pp. 53–60. Oxford University Press, Oxford.

Cruickshank, D. W. J., Juretschke, H. J. and Kato, N. (ed.) (1992). *P. P. Ewald and his dynamical theory of X-ray diffraction.* Oxford University Press, Oxford.

Cui, S. F. and Mai, Z. (1990). *J. Appl. Crystallogr.*, **23**, 147–50.

Cui, S. F., Green, G. S. and Tanner, B. K. (1989). *Mater. Res. Soc. Symp. Proc.*, **139**, 71–6.

Cusatis, C. and Hart, M. (1974). In *Anomalous scattering* (ed. S. Ramaseshan and S.C. Abrahams), pp. 57–68. Munksgaard, Copenhagen.

Cusatis, C. and Hart, M. (1976). *Proc. Roy. Soc. London A*, **354**, 291–301.

Dalisa, A. L., Zajac, A. and Ng, C. H. (1968). *Phys. Rev.*, **168**, 859–66.

Darbinyan, S. P., Chukhovskii, F. N. and Voronkov, S. N. (1991). *Phys. Stat. Sol. (a)*, **125**, 441–9.

Darwin, C. G. (1914*a*). *Phil. Mag.*, **27**, 315–33.

Darwin, C. G. (1914*b*). *Phil. Mag.*, **27**, 675–90.

Darwin, C. G. (1922). *Phil. Mag.*, **43**, 800–29.

Dash, W. C. (1956). *J. Appl. Phys.*, **27**, 1193.

Dash, W. C. (1957). In *Dislocations and mechanical properties of crystals* (ed. L. Fischer), p. 57. J. Wiley and Sons, New York.

Datsenko, L. I. and Skorokhod, M. J. A. (1967). *Kristallografiya*, **12**, 480–2.

Datsenko, L. I., Khrupa, V. I. and Kislovskii, E. N. (1981). *Phys. Stat. Sol. (a)*, **68**, 399–404.

Davis, B. and von Nardroff, R. (1924). *Proc. Nat. Acad. Amer.*, **10**, 384.

Davis, B. and Slack, C. M. (1926). *Phys. Rev.*, **26**, 18–22.

Davis, B. and Stempel, W. M. (1921). *Phys. Rev.*, **17**, 608–23.

Davis, B. and Stempel, W. M. (1922). *Phys. Rev.*, **19**, 504–11.

Davis, B. and Terrill, H. M. (1922). *Proc. Nat. Acad. Amer.*, **8**, 357.

Davis, T. J. (1991). *Austr. J. Phys.*, **44**, 693–704.

Davis, T. J. (1994). *Acta Crystallogr. A*, **50**, 224–31.
Davis, T. J. (1996). *J. X-ray Sci. Technol.*, **6**, 317–42.
Davis, T. J. and Stevenson, A. W. (1996). *J. Opt. Soc. Am. A*, **13**, 1193–8.
Davis, T. J., Gao, D., Gureyev, T. E., Stevenson, A. W. and Wilkins, S. W. (1995*a*). *Nature*, **373**, 595–8.
Davis, T. J., Gureyev, T. E., Gao, D., Stevenson, A. W. and Wilkins, S. W. (1995*b*). *Phys. Rev. Lett.*, **74**, 3173–6.
Debye, P. (1913). *Verh. Dtsch. Phys. Gesell.*, **15**, 678–89.
Debye, P. (1914). *Ann. d. Physik*, **43**, 49.
Debye, P. (1930). *Phys. Z.*, **31**, 419.
De Caro, L. and Tapfer, L. (1997). *Phys. Rev. B*, **55**, 105–12.
De Caro, L., Giannini, C. and Tapfer, L. (1997). *Phys. Rev. B*, **56**, 9744–52.
Dederichs, P. H. (1967). *Phys. Stat. Sol.*, **23**, 377–86.
Dederichs, P. H. (1970). *Phys. Rev. B*, **1**, 1306–17.
Dederichs, P. H. (1971). *Phys. Rev. B*, **4**, 1041–50.
Dederichs, P. H. (1973). *J. Phys. F: Metal Physics*, **3**, 471–96.
Deslattes, R. D. (1968). *Appl. Phys. Lett.*, **12**, 133.
Deslattes, R. D. (1969). *Appl. Phys. Lett.*, **15**, 386–8.
Deslattes, R. D. and Henins, A. (1969). *Phys. Rev. Lett.*, **15**, 386–8.
Deslattes, R. D. and Henins, A. (1973). *Phys. Rev. Lett.*, **31**, 972–5.
Deslattes, R. D. and Paretzkin, B. (1968). *J. Appl. Crystallogr.*, **1**, 176–8.
Deslattes, R. D., Henins, A., Bowman, H. A., Schoonover, R. M., Carroll, C. L., Barnes, I. L., Machlan, L. A., Moore, L. J. and Shields, W. R. (1974). *Phys. Rev. Lett.*, **33**, 463–6.
Deslattes, R. D., Tanaka, M., Greene, G. L., Henins, A. and Kessler, E. G. (1987). *IEEE Trans. Instrum. Meas.*, **IM-36**, 166–9.
Deslattes, R. D., Kessler, E. G. Jr, Indelicato, P. and Lindroth, E. (1999*a*). In *International tables of crystallography*, Vol. C (ed. A. J. C. Wilson and E. Prince), pp. 200–13. Kluwer Academic Publishers, Dordrecht.
Deslattes, R. D., Kessler, E. G. Jr, Owens, S., Black, D. and Henins, A. (1999*b*). *J. Phys. D: Appl. Phys.*, **32**, A3–A7.
DESY (2001). TESLA Technical Design Report 001-011. ECFA 2001-209 DESY, Hamburg, Germany.
Détaint, J., Zarka, A., Capelle, B. and Epelboin, Y. (1998). *Proc. 1998 Int. Frequency Control Symp.*, pp. 882–90.
Deutsch, M. (1980*a*). *J. Appl. Crystallogr.*, **13**, 252–5.
Deutsch, M. (1980*b*). *J. Appl. Crystallogr.*, **13**, 256–8.
Deutsch, M. (1987). *Appl. Phys. Lett.*, **51**, 1410–12.
Deutsch, M. and Hart, M. (1982). *Phys. Rev. B*, **26**, 5558–67.
Deutsch, M. and Hart, M. (1984). *Phys Rev. B*, **30**, 643–6.
Deutsch, M. and Hart, M. (1985*a*). *Acta Crystallogr. A*, **41**, 48–55.
Deutsch, M. and Hart, M. (1985*b*). *Phys Rev. B*, **31**, 3846–58.
Deutsch, M. and Hart, M. (1988). *Phys Rev. B*, **37**, 2701–3.
Deutsch, M., Hart, M. and Cummings, S. (1990). *Phys Rev. B*, **42**, 1248–53.

Deutsch, M., Hart, M. and Sommer-Larsen, P. (1989). *Phys. Rev. B*, **40**, 11666–9.

Dewey, M. S., Kessler Jr, E. G., Greene, G. L., Deslattes, R. D., Sacchetti, F., Petrillo, C. and Freund, A. (1994). *Phys. Rev. B*, **50**, 2800–8.

Di Fabrizio, E., Gentili, M., Grella, L., Baciocchi, M., Krasnoperova, A., Cerrina, F., Yun, W., Lai, B. and Gluskin, E. (1994). *J. Vacuum Sci. & Technol. B*, **12**, 3979–85.

Di Fabrizio, E., Romanato, F. and Gentili, M. (1999). In *X-ray Microscopy: Proceedings of the Sixth International Conference*, pp. 635–40. American Institute of Physics, New York.

Di Fonzo, S., Jark W., Soullié, G., Cedola, A., Lagomarsino, S., Cloetens, P. and Riekel, C. (1998). *J. Synchrotron Rad.*, **5**, 376–8.

Di Fonzo, S., Jark, W., Lagomarsino, S., Giannini, C., De Caro, L., Cedola, A. and Müller, M. (2000). *Nature (London)*, **403**, 638–40.

Dmitrienko, V. E. (1982). *Sov. Phys. Crystallogr.*, **27**, 131–5.

Dmitrienko, V. E. (1983). *Acta Crystallogr. A*, **39**, 29–37.

Dmitrienko, V. E. (1984). *Acta Crystallogr. A*, **40**, 89–95.

Dmitrienko, V. E. and Belyakov, V. A. (1980*a*). *Acta Crystallogr. A*, **36**, 1044–50.

Dmitrienko, V. E. and Belyakov, V. A. (1980*b*). *Sov. Tech. Phys. Lett.*, **6**, 621–2.

Dmitrienko, V. E. and Ovchinnikova, E. N. (2000). *Acta Crystallogr. A*, **56**, 340–7.

Dmitrienko, V. E. and Ovchinnikova, E. N. (2001). *Acta Crystallogr. A*, **57**, 642–8.

Dmitrienko, V. E., Ovchinnikova, E. N. and Ishida, K. (1999). *JETP Lett.*, **69**, 938–42.

Dobrianov, I., Finkelstein, K. D., Lemay, S. G. and Thorne, R. E. (1998). *Acta Cryst. D*, **54**, 922–37.

Dobrianov, I., Caylor, C., Lemay, S. G., Finkelstein, K. D. and Thorne, R. E. (1999). *J. Cryst. Growth*, **196**, 511–23.

Doniach, S. (2000). *J. Synchrotron Rad.*, **7**, 116–20.

Dorner, B., Burkel, E. and Peisl, J. (1986). *Nucl. Instrum. & Methods in Phys. Res. A*, **246**, 450–1.

Dub, P. and Litzman, O. (1999). *Acta Crystallogr. A*, **55**, 613–20.

Dub, P. and Litzman, O. (2001). *Acta Crystallogr. A*, **57**, 212–16.

Dudley, M. (1997). In *Encyclopedia of applied physics*, Vol. 21, pp. 533–47. Wiley-VCH Verlag GmbH.

Dobrianov, I., Caylor, C., Lemay, S. G., Finkelstein, K. D. and Thorne, R. E. (1999). *J. Cryst. Growth*, **196**, 511–23.

Dobrianov, I., Finkelstein, K. D., Lemay, S. G. and Thorne, R. E. (1998). *Acta Cryst. D*, **54**, 922–37.

Dudley, M. and Miltat, J. (1989). *Nucl. Instrum. Methods B*, **40/41**, 393–7.

Dudley, M. and Yao, G.-D. (1993). *J. Phys. D: Appl. Phys.*, **26**, A120–5.

Dudley, M., Wang, S., Huang, W., Carter, C. H. Jr, Tsvetkov, V. F. and Fazi, C. (1995). *J. Phys. D: Appl. Phys.*, **28**, A63–A68.

Dudley, M., Huang, X. R. and Huang, W. (1999). *J. Phys. D: Appl. Phys.*, **32**, A139–A144.

Dudley, M., Huang, X.-R. and Vetter, W. M. (2003). *J. Phys. D*, **36**, A30–A36.

Duke, P. J. (2000). *Synchrotron Radiation (Production and Properties)*. Oxford Science Publications.

DuMond, J. W. M. (1937). *Phys. Rev.*, **52**, 872–83.

Dunia, E., Malgrange, C. and Pétroff, J. F. (1980). *Phil. Mag. A*, **41**, 291–306.

Durbin, S. M. (1995). *Acta Crystallogr. A*, **51**, 258–68.

Durbin, S. M. and Follis, G. C. (1995). *Phys. Rev. B*, **51**, 10127–33.

Durbin, S. M. and Gog, T. (1989). *Acta Crystallogr. A*, **45**, 132–41.

Durbin, S. M., Berman, L. E., Batterman, B. W. and Blakely, J. M. (1985). *Scann. Electr. Micr.*, **3**, 1099–1102.

Eisenberger, P. and Marra, W. C. (1981). *Phys. Rev. Lett.*, **16**, 1081–4.

Eisenhower, R., Colella, R. and Grushko, B. (1998). *Phys. Rev. B*, **57**, 8218–22.

Egert, G. and Dachs, H. (1970). *J. Appl. Crystallogr.*, **3**, 214–20.

Elder, F. R., Gurewitsch, A. M., Langmuir, R. V. and Pollock, H. C. (1947). *Phys. Rev.*, **71**, 829–30.

Elder, F. R., Langmuir, R. V. and Pollock, H. C. (1948). *Phys. Rev.*, **74**, 52–6.

Elistratov, A. M. and Madzhitov (1965). *Fiz. Metallov Metallov.*, **19**, 349.

Eng, P. J., Rivers, M. L., Yang, B. X. and Schildkamp, W. (1995). *Proc. SPIE – The International Society for Optical Engineering.*, **2516**, 41–51.

Eng, P. J., Newville, R., Rivers, M. L. and Sutton, S. R. (1998). *Proc. SPIE – The International Society for Optical Engineering.*, **3449**, 145–56.

Entin, I. R. and Assur, K. P. (1981). *Acta Cryst. A.* **37**, 769–74.

Epelboin, Y. (1974). *J. Appl. Crystallogr.*, **7**, 372–7.

Epelboin, Y. (1975). *Acta Crystallogr. A*, **31**, 591–600.

Epelboin, Y. (1979*a*). *Acta Crystallogr. A*, **35**, 38–44.

Epelboin, Y. (1979*b*). *J. Appl. Phys.*, **50**, 1312–17.

Epelboin, Y. (1983). *Acta Crystallogr. A*, **39**, 761–7.

Epelboin, Y. (1985). *Mater. Sci. Eng.*, **73**, 1–43.

Epelboin, Y. (1987). *Progr. Crystal Growth Characterization*, **14**, 465–506.

Epelboin, Y. (1996). *J. Appl. Crystallogr.*, **29**, 331–40.

Epelboin, Y. and Authier, A. (1983). *Acta Crystallogr. A*, **39**, 767–72.

Epelboin, Y. and Patel, J.R. (1982). *J. Appl. Phys.*, **53**, 271–5.

Epelboin, Y. and Ribet, M. (1974). *Phys. Stat. Sol. (a)*, **25**, 507–13.

Epelboin, Y. and Riglet, P. (1979). *Phys. Stat. Sol. (a)*, **54**, 547–56.

Epelboin, Y. and Soyer, A. (1984). In *Applications of X-ray topographic methods to materials science* (ed. S. Weissmann, F. Balibar and J.-F. Pétroff), pp. 161–70. Plenum Press, New York.

Epelboin, Y. and Soyer, A. (1985). *Acta Crystallogr. A*, **41**, 67–72.

Epelboin, Y., Détaint, J. and Capelle, B. (1998). *J. Appl. Crystallogr.*, **31**, 574–82.

Erko, A., Schäfers, F., Gudat, W., Abrosimov, N. V., Rossolenko, S. N., Alex, V., Groth, S. and Schröder, W. (1996*a*). *Nucl. Instrum. Methods Phys. Res. A*, **374**, 408–12.

Erko, A., Schäfers, F., Gudat, W., Sawney, K. J., Abrosimov, N. V., Rossolenko, S. N., Alex, V., Groth, S. and Schröder, S. W. (1996*b*). *Proc. SPIE—The International Society for Optical Engineering.*, **2856**, 110–9.

Erko, A., Veldkamp, M., Gudat, W., Abrosimov, N. V., Rossolenko, S. N., Shekhtman, V., Khasanov, S., Alex, V., Groth, S., Schröder, W., Vidal, B. and Yashkin, A. (1998). *J. Synchrotron Rad.*, **5**, 239–45.

Erola, E., Eteläniemi, V., Suortti, P., Pattison, P. and Thomlinson, W. (1990). *J. Appl. Crystallogr.*, **23**, 35–42.

Espeso, J. I., Cloetens, P., Baruchel, J., Härtwig, J., Mairs, T., Biasci, J.-C., Marot, G., Salomé-Pateyron, M. and Schlenker, M. (1998). *J. Synchrotron Rad.*, **5**, 1243–9.

Eteläniemi, V., Michel, E. G. and Materlik, G. (1995). *J. Vac. Sci. & Technol. A*, **13**, 1583–8.

Ewald, P. P. (1913). *Physik. Z.*, **14**, 465–72.

Ewald, P. P. (1916*a*). *Ann. Physik*, **49**, 1–38.

Ewald, P. P. (1916*b*). *Ann. Physik*, **49**, 117–43.

Ewald, P. P. (1917). *Ann. Physik*, **54**, 519–97.

Ewald, P. P. (1920). *Z. Phys.*, **2**, 332–42.

Ewald, P. P. (1921). *Z. Kristall.*, **56**, 129–56.

Ewald, P. P. (1924). *Z. Phys.*, **30**, 1–13.

Ewald, P. P. (1925). *Phys. Zeit.*, **26**, 29–32.

Ewald, P. P. (1937). *Z. Kristall.*, **A97**, 1–27.

Ewald, P. P. (1938). *Annales de l'Institut Henri Poincaré*, **VIII**, 79–110.

Ewald, P. P. (1958). *Acta Crystallogr.*, **11**, 888–91.

Ewald, P. P. (1961). *J. Phys. Soc. Jap.*, **17** (Suppl. B-II), 48–52.

Ewald, P. P. (ed.) (1962). *Fifty years of X-ray diffraction.* Published for the IUCr by A. Oosthoek's Uitgeversmaatschappij N.V., Utrecht.

Ewald, P. P. (1965). *Rev. Mod. Phys.*, **37**, 46–56.

Ewald, P. P. (1968). *Acta Crystallogr. A*, **24**, 1–3.

Ewald, P. P. (1969). *Acta Crystallogr. A*, **25**, 103–8.

Ewald, P. P. (1979). *Acta Crystallogr. A*, **35**, 1–9.

Ewald, P. P. (1986). *Acta Crystallogr. A*, **42**, 411–13.

Ewald, P. P. and Heno, Y. (1968). *Acta Crystallogr. A*, **24**, 5–15.

Faleev, N. N., Flaks, L. I., Konnikov, S. G., Solomin, I. K. and Batashova, S. V. (1990). *Phys. Stat. Sol. (a)*, **120**, 327–37.

Faleev, N. N., Pavlov, K., Tabuchi, M. and Takeda, Y. (1999*a*). *Jpn J. Appl. Phys.*, **38**, 818–21.

Faleev, N. N., Pavlov, K., Tabuchi, M. and Takeda, Y. (1999*b*). *Proceedings of the Second International Conference on Synchrotron Radiation in Materials Science (SRMS-2)*, Kobe, Japan, *Jpn J. Appl. Phys.*, **38** (Suppl. 38-1), 277.

Farwig, P. and Schürmann, H. W. (1967). *Z. Physik*, **204**, 489–500.

Fedorov, V. A., Ganshin, V. A. and Korkishko, Yu. N. (1993). *Phys. Stat. Sol. (a)*, **135**, 493–505.

Fedorov, A. A., Trukhanov, E. M., Vasilenko, A. P., Kolesnikov, A. V. and Revenko, M. A. (2003). *J. Phys. D: Appl. Physics*, **36**, A44–A48.

Feng, Y., Gao, D., Yuan, Y. and Eda, K. (1988). *J. Lumin.*, **40/41**, 757–8.

Feng, Y. P., Sinha, S. K., Fullerton, E. E., Grübel, G., Abernathy, D., Siddons, D. P. and Hastings, J. B. (1995). *Appl. Phys. Lett.*, **67**, 3647–9.

Ferrari, C. and Korytár, D. (2001). *J. Appl. Crystallogr.*, **34**, 608–12.
Fewster, P. F. (1989). *J. Appl. Crystallogr.*, **22**, 64–9.
Fewster, P. F. (1991). *Appl. Surf. Sci.*, **50**, 9–18.
Fewster, P. F. (1992). *J. Appl. Crystallogr.*, **25**, 714–23.
Fewster, P. F. (1999). *J. Mater. Science: Materials in Electronics*, **10**, 175–83.
Fewster, P. F. (2000). *X-ray scattering from semiconductors*. Imperial College Press. London.
Fewster, P. F. and Andrew, N.L. (1995*a*). *J. Appl. Crystallogr.*, **28**, 451–8.
Fewster, P. F. and Andrew, N.L. (1995*b*). *J. Phys. D: Appl. Phys.*, **28**, A97–A103.
Fezzaa, K. and Lee, W.-K. (2001). *J. Appl. Crystallogr.*, **34**, 166–71.
Fingerland, A. (1971). *Acta Crystallogr. A*, **27**, 280–4.
Finkelstein, K. D., Shen, Q. and Shastri, S. (1992). *Phys. Rev. Lett.*, **69**, 1612–15.
Fischer, A. E. M. J., Vlieg, E., van der Veen, J. F., Clausnitzer, M. and Materlik, G. (1987). *Phys. Rev. B*, **36**, 4769–73.
Floquet, G. (1883). *Annales Ecole Normale Sup.*, **12**, 47.
Foct, J. and Champier, G. (1966). *C. R. Acad. Sci., Paris*, **263**, 908–11.
Fodchuk, I. M. and Kshevetskii, O. S. (1992). *Sov. Phys. Metallophys.*, **14**, 57–62.
Fodchuk, I. M. and Raransky, N. D. (2003). *J. Phys. D: Appl. Physics*, **36**, A55–A59.
Fontaine, A., Warburton, W. K. and Ludwig Jr, K. F. (1985). *Phys. Rev. B*, **31**, 3599–605.
Fornari, R., Franzosi, P., Salviati, G., Ferrari, C. and Ghezzi, C. (1985). *J. Cryst. Growth*, **72**, 717–25.
Förster, E., Goetz, K. and Zaumseil, P. (1980). *Kristall und Technik*, **1**, 937–45.
Forwood, A. and Lawn, B. R. (1966). *Phil. Mag.*, **13**, 395.
Françon, M. (1950). *Le contraste de phase en optique et en microscopie*. Revue d'Optique, Paris.
Frank, F.C., Lang, A. R., Kaczer, J. and Polcarová, M. (1966). *IEEE Trans. Magn., Mag. -2*, **3**, 511.
Frank, F. C., Kaczer, J., Lang, A. R. and Polcarová, M. (1993). *Proc. Roy. Soc., London, A*, **442**, 161–75.
Frank, F. C., Harris, J. W., Kaneko, K. and Lang, A. R. (1994). *J. Crystal Growth*, **143**, 46–57.
Franklin, G. E., Fontes, E., Qian, Y., Bedzyk, M., Golovchenko, J. A. and Patel, J. R. (1994). *Phys. Rev. B*, **50**, 7483–7.
Freund, A. (1973). Thesis. Technische Universität München.
Freund, A. (1989). In *Synchrotron radiation in structural biology* (ed. R. M. Sweet and A. D. Woodhead), pp. 255–92. Plenum Press, New York.
Freund, A. (1990). In *SPIE Conference on 'Advanced X-ray-UV Radiation Sources and Applications', SPIE Proc.* Vol. , 1345, pp. 234–44.
Freund, A. (1993). In *Neutron and synchrotron radiation for condensed matter studies*, Vol. 1 (ed. J. Baruchel, J.-L. Hodeau, M. S. Lehmann, J.-R. Regnard and C. Schlenker), pp.79–93. Springer-Verlag, Berlin and Les Editions de Physique, Les Ullis.

Freund, A. (1995). *Opt. Eng.*, **34**, 432–40.

Friedel, G. (1913). *C. R. Acad. Sci. Paris*, **157**, 1533.

Friedrich, W., Knipping, P. and von Laue, M. (1912). *Sitzungsberichte der Kgl. Bayer. Akad. der Wiss.*, 303–322, reprinted in *Ann. Phys.* (1913) **41**, 971.

Fries, E., Dolin, C., Spendel, M. and Castaing, J. (1984). In *Applications of X-ray topographic methods to material science* (ed. S. Weissmann, F. Balibar and J. F. Pétroff), pp. 359–65. Plenum Press, New York.

Fries, E., Guiberteau, F., Dominguez-Rodriguez, A., Cheong, D. S. and Heuer, A. (1989). *Phil. Mag. A*, **60**, 107–21.

Fues, E. (1939). *Ann. Physik Lpz.*, **36**, 209–25.

Fujimoto, Z., Fujii, K., Tanaka, M. and Nakayama, K. (1997). *IEEE Trans. Instrum. Meas.*, **50**, 123–4.

Fukamachi, T. and Kawamura, T. (1993). *Acta Crystallogr. A*, **49**, 384–8.

Fukamachi, T., Negishi, R., Yoshizawa, M., Ehara, K., Kawamura, T., Nakajima, T. and Zhao, Z. (1993). *Acta Crystallogr. A*, **49**, 573–5.

Fukamachi, T., Negishi, R. and Kawamura, T. (1995). *Acta Crystallogr. A*, **51**, 253–8.

Fukamachi, T., Negishi, R., Zhou, S., Yoshizawa, M., Sakamaki, T., Kawamura, T. and Nakajima, T. (1996). *Acta Crystallogr. A*, **52**, 669–74.

Fukamachi, T., Negishi, R., Zhou, S.-M., Yoshizawa, M. and Kawamura, T. (2002). *Acta Crystallogr. A*, **58**, 552–8.

Fukuda, A., Hondoh, T., and Higashi, A. (1987). *J. Physique*, **48**, Cl-183-7.

Fukuhara, A. and Takano, Y. (1977*a*). *Acta Crystallogr. A*, **33**, 137–42.

Fukuhara, A. and Takano, Y. (1977*b*). *J. Appl. Crystallogr.*, **10**, 287–90.

Fukumachi, T., Hosoya, S. and Terasaki, O. (1973). *J. Appl. Crystallogr.*, **6**, 117–22.

Gabrielyan, K. T. and Aslanian, H. A. (1988). *Phys. Stat. Sol. (a)*, **108**, K85–K88.

Gabrielyan, K. T., Kuběna, J. and Holý, V. (1986). *Phys. Stat. Sol. (a)*, **95**, 579–88.

Gao, D., Li, M., Cui, Y., Feng, Y. and Zhang, Z. (1987). *Journal of the Chinese Rare Earth Society*, **5**, 33–7.

Gao, D., Davis, T. J. and Wilkins, S. W. (1995). *Austr. J. Phys.*, **48**, 103–11.

Gao, D., Gureyev, T. E., Pogany, A., Stevenson, A. W. and Wilkins, S. W. (1998). In *Medical applications of synchrotron radiation* (ed. M. Ando and C. Uyama), pp. 63–70. Springer-Verlag, Tokyo.

Gastaldi, J. and Jourdan, C. (1976). *J. Cryst. Growth*, **35**, 17–27.

Gastaldi, J., Jourdan, C. and Grange, G. (1988*a*). *Phil. Mag. A*, **57**, 971–81.

Gastaldi, J., Jourdan, C., Grange, G. and Bauer, C. L. (1988*b*). *Phys. Stat. Sol. (a)*, **109**, 403–11.

Gastaldi, J., Jourdan, C. and Grange, G. (1992). In *Materials science forum*, Vol. 94–96 (ed. G. Abruzzesse and P. Brozzo), pp. 17–25.

Gastaldi, J., Reinier, E., Jourdan, C., Grange, G., Quivy, A. and Boudard, M. (1995). *Phil. Mag. Lett.*, **72**, 311–21.

Gastaldi, J., Mancini, L., Reinier, E., Cloetens, P., Ludwig, W., Janot, C., Baruchel, J., Härtwig, J. and Schlenker, M. (1999). *J. Phys. D: Appl. Phys.* **32**, A152–A159.

Gastaldi, J., Agliozzo, S., Letoublon, A., Wang, J., Mancini, L., Klein, H., Härtwig, J., Baruchel, J., Fisher, I. R., Sato, T., Tsai, A. P. and De Boissieu, M. (2003). *Phil. Mag.*, **83**, 1–29.

Gemperlová, J., Jacques, A., Polcarová, M., Brádler, J., Novak, V. and George, A. (1999). *Materials Science Forum*, **294–296**, 373–6.

George, A., Jacques, A. and Coquille, R. (1987). *Inst. Phys. Conf. Series*, **87**, 615.

Gerold, V. and Meier, F. (1959). *Z. Phys.*, **155**, 387–94.

Gerold, V. and Meier, F. (1962). In *Direct observations of imperfections in crystals* (ed. J. B. Newkirk and J. H. Wernick), p. 509. Interscience Publ., New York.

Gevers, R. (1962). *Phil. Mag.*, **7**, 1681–720.

Ghezzi, C., Merlini, A. and Pace, S. (1971). *Phys. Rev. B*, **4**, 1833–42.

Giannini, C., Tapfer, L., Lagomarsino, S., Boulliard, J. C., Taccoen, A., Capelle, B., Ilg, M., Brandt, O. and Ploog, K. H. (1993). *Phys. Rev. B, Rapid Comm.*, **48**, 11496–9.

Giannini, C., De Caro, L., Lagomarsino, S., Cedola, A., Di Fonzo, S. and Jark, W. (2001). *J. Phys. D: Appl. Phys.*, **34**, A40–A43.

Giardina, M. D. and Merlini, A. (1973). *Z. Naturforsch.*, **28a**, 1360–5.

Gibbs, D., Harshman, D. R., Isaacs, E. D., Mcwhan, D. B., Mills, D. and Vettier, C. (1988). *Phys. Rev. Lett.*, **61**, 1241–4.

Giles, C. and Cusatis, C. (1991). *Appl. Phys. Lett.*, **59**, 641–3.

Giles, C., Malgrange, C., Goulon, J., de Bergevin, F., Vettier, C., Dartyge, E., Fontaine, A., Giorgetti, C. and Pizzini, S. (1994*a*). *J. Appl. Crystallogr.*, **27**, 232–40.

Giles, C., Malgrange, C., Goulon, J., de Bergevin, F., Vettier, C., Fontaine, A., Dartyge, E. and Pizzini, S. (1994*b*). *Nucl. Instrum. Methods A*, **349**, 622–5.

Giles, C., Malgrange, C., Goulon, J., de Bergevin, F., Vettier, C., Dartyge, E., Fontaine, A., Pizzini, S., Baudelet, F. and Freund, A. (1995*a*). *Physica B*, **208 & 209**, 784–6.

Giles, C., Malgrange, C., Goulon, J., de Bergevin, F., Vettier, C., Fontaine, A., Dartyge, E., Pizzini, S., Baudelet, F. and Freund, A. (1995*b*). *Rev. Sci. Instrum.*, **66**, 1549–53.

Giles, C., Vettier, C., Bergevin, de F., Malgrange, C., Grübel, G. and Grossi, F. (1995*c*). *Rev. Sci. Instrum.*, **66**, 1518–21.

Gits, S. and Authier, A. (1982). *J. Cryst. Growth*, **58**, 473–85.

Gluer, C. C., Graeff, W. and Moller, H. (1983). *Nucl. Instrum. Methods A*, **208**, 701–4.

Golovchenko, J. A., Batterman, B. W. and Brown, W. L. (1974). *Phys. Rev. B*, **10**, 4239–43.

Golovchenko, J. A., Patel, J. R., Kaplan, D. R., Cowan, P. L. and Bedzyk, M. J. (1982). *Phys. Rev. Lett.*, **49**, 560–3.

Golovchenko, J. A., Kincaid, B. M., Levesque, R. A., Meixner, A. E. and Kaplan, D. R. (1986). *Phys. Rev. Lett.*, **57**, 202–5.

Golovin, A. L., Imamov, R. M. and Stepanov, S. A. (1984). *Acta Crystallogr. A*, **40**, 225–8.

Golovin, A. L., Imamov, R. M., Kondrashikina, E. A. and Novikov, V. (1985). *Sov. Phys. Technical Physics*, **30**, 834–5.

Golovin, A. L., Imamov, R. M. and Melikyan, O. G. (1989). *J. Appl. Cryst.*, **22**, 406–9.

González-Mañas, M., Caballero, M. A., Capelle, B. and Epelboin, Y. (1993). *J. Appl. Crystallogr.*, **26**, 122–7.

Goulon, J., Malgrange, C., Giles, C., Neumann, C., Rogalev, A., Moguiline, E., De Bergevin, F. and Vettier, C. (1996). *J. Synchrotron Rad.*, **3**, 272–81.

Graeff, W. (2002*a*). *J. Synchrotron Rad.*, **9**, 82–5.

Graeff, W. (2002*b*). *J. Synchrotron Rad.*, **9**, 293–97.

Graeff, W. (2004). *J. Synchrotron Rad.*, **11**, 261–5.

Graeff, W. and Bonse, U. (1977). *Z. Phys. B*, **27**, 19–32.

Graeff, W. and Engelke, K. (1991). in *Handbook of Synchrotron Radiation*. ed. S. Ebashi, M. Koch and E. Rubinstein, Elsevier Scence, Amsterdam, 361–405.

Graeff, W. and Materlik, G. (1982). *Nucl. Instrum. Methods*, **195**, 97–103.

Graf, H. A. and Schneider, J. R. (1986). *Phys. Rev. B*, **34**, 8629–38.

Grange, G., Gastaldi, J. and Jourdan, C. (1987). *J. Appl. Phys.*, **62**, 1202–7.

Grange, G., Gastaldi, J. and Jourdan, C. (1990). In *Proc. 2nd European Conference of Progress in X-ray Synchrotron Research*, pp. 457–9. SIF, Bologna.

Grange, G., Gastaldi, J., Jourdan, C. and Billia, B. (1995). *J. Cryst. Growth*, **151**, 192–9.

Green, G. S. and Tanner, B. K. (1987). *Inst. Phys. Conf. Series*, **87**, 627–32.

Green, G. S., Cui, S. F. and Tanner, B. K. (1990). *Phil. Mag. A*, **61**, 23–33.

Green, G. S., Loxley, N. and Tanner, B. K. (1991). *J. Appl. Crystallogr.*, **24**, 304–11.

Grehk, T. M., Drube, W., Kipp, L. and Materlik, G. (2001). *J. Synchrotron Rad.*, **8**, 1015–20.

Greiser, N. and Materlik, G. (1987). *Z. Phys. B*, **66**, 83–9.

Gronkowski, J. (1980). *Phys. Stat. Sol. (a)*, **57**, 105–12.

Gronkowski, J. (1991). *Physics Reports*, **206**, 1–41.

Gronkowski, J. and Kowalski, G. (1981). *Acta Crystallogr. A*, **37**, C251.

Gronkowski, J. and Malgrange, C. (1984*a*). *Acta Crystallogr. A*, **40**, 507–14.

Gronkowski, J. and Malgrange, C. (1984*b*). *Acta Crystallogr. A*, **40**, 515–22.

Grosswig, S., Härtwig, J., Alter, U., and Christoph, A. (1983). *Crystal Res. & Technol.*, **18**, 501–11.

Guiberteau, F., Cumbrera, F. L., Dominguez-Rodriguez, A., Fries, E. and Castaing, J. (1994). *J. Appl. Crystallogr.*, **27**, 406–10.

Guigay, J.-P. (1989). *Acta Crystallogr. A*, **45**, 241–4.

Guigay, J.-P. (1999). *Acta Crystallogr. A*, **55**, 561–3.

Guigay, J.-P. and Chukhovskii, F. N. (1992). *Acta Crystallogr. A*, **48**, 819–26.

Guigay, J.-P. and Chukhovskii, F. N. (1995). *Acta Crystallogr. A*, **51**, 288–95.

Guigay, J.-P. and Vartanyants, I. A. (1999). *J. Phys. D: Appl. Phys.*, **32**, A84–A87.

Guigay, J.-P., Schlenker, M. and Baruchel, J. (1984). in *Applications of X-ray topographic methods to material science*. ed. S. Weissmann, F. Balibar and J.-F. Pétroff, Plenum Press, New York, 171–83.

Guinier, A. and Tennevin, J. (1949). *Acta Crystallogr.*, **2**, 133–8.

Gupta, J. A., Woicik, J., Watkind, S. P., Harrison, D. A., Crozier, E. D. and Karlin, B. A. (1999). *J. Synchrotron Rad.*, **6**, 500–2.

Gureyev, T. E. and Wilkins, S. W. (1997). *Il Nuovo Cimento*, **19D**, 545–52.

Gureyev, T. E. and Wilkins, S. W. (1998). *Opt. Commun.*, **147**, 229–32.

Gureyev, T. E., Roberts, A. and Nugent, K. A. (1995). *J. Opt. Soc. Am. A*, **12**, 1942–6.

Gureyev, T. E., Raven, C., Snigirev, A., Snigireva, I. and Wilkins, S. W. (1999). *J. Phys. D: Appl. Phys.*, **32**, 563–7.

Gureyev, T. E., Mayo, S., Wilkins, S. W., Paganin, D. M. and Stevenson, A. W. (2001). *Phys. Rev. Lett.*, **86**, 5827–30.

Gureyev, T. E., Stevenson, A. W., Paganin, D. M., Weitkamp, T., Snigirev, A., Snigireva, I. and Wilkins, S. W. (2002). *J. Synchrotron Rad.*, **9**, 148–53.

Guryan, C. A., Goldman, A. I., Stephens, P. W., Hiraga, K., Tsai, A. P., Inoue, A. and Masumoto, T. (1989). *Phys. Rev. Lett.*, **63**, 2409–12.

G'Sell, C. and Epelboin, Y. (1979). *J. Appl. Crystallogr.*, **12**, 110–15.

Habenschuss, A., Ice, E., Sparks, C. J. Jr and Neiser, R. A. (1988). *Nucl. Instrum. Methods Phys. Res. A*, **266**, 215–19.

Haga, H. and Wind, C. H. (1903). *Annalen der Physik*, **10**, 305–12.

Halliwell, M. A. G., Childs, J. B. and O'Hara, S. (1973). In *GaAs and related compounds*, p. 98. IOP Publishing, Bristol.

Halliwell, M. A. G., Lyons, M. H. and Hill, M. J. (1984). *J. Cryst. Growth*, **68**, 523–31.

Hamilton, W. C. (1957). *Acta Crystallogr.*, **10**, 629–34.

Hannon, J. P. (1994). In *Resonant anomalous X-ray scattering* (ed. G. Materlik, C. J. Sparks and K. Fischer), p. 565. Elsevier Science B.V., Amsterdam.

Hannon, J. P., Trammell, G. T., Blume, M. and Gibbs, D. (1988). *Phys. Rev. Lett.*, **61**, 1245–8.

Haroutyunyan, V. S. and Sedrakyan, A. G. (1997). *Acta Cryst. A*, **53**, 410–4.

Hart, M. (1965). *Appl. Phys. Lett.*, **7**, 96–8.

Hart, M. (1966). *Z. Phys.*, **189**, 269–91.

Hart, M. (1968*a*). *Brit. J. Appl. Phys. (J. Phys. D)*, **1**, 1405–8.

Hart, M. (1968*b*). *Sci. Progr. (Oxford)*, **56**, 429–47.

Hart, M. (1969). *Proc. Roy. Soc. A*, **309**, 281–96.

Hart, M. (1972). *Phil. Mag.*, **26**, 821–31.

Hart, M. (1975*a*). *J. Appl. Crystallogr.*, **8**, 436–44.

Hart, M. (1975*b*). *Proc. Roy. Soc. A*, **346**, 1–22.

Hart, M. (1978). *Phil. Mag. B*, **38**, 41–56.

Hart, M. (1979). *J. Phys. E (Sci. Instr.)*, **12**, 911–12.

Hart, M. (1980). *Nucl. Instrum. Methods*, **172**, 209–14.

Hart, M. (1981). *J. Cryst. Growth*, **55**, 409–27.

Hart, M. (1990). *Nucl. Instr. & Methods in Phys. Res. A*, **297**, 306–11.

Hart, M. (1996). In *X-ray and neutron dynamical diffraction: theory and applications* (ed. A. Authier, S. Lagomarsino and B. K. Tanner), NATO ASI Series, Series B, Physics, Vol. 357, pp. 73–90. Plenum Press, New York and London.

Hart, M. and Berman, L. E. (1998). *Acta Crystallogr. A*, **54**, 850–8.

Hart, M. and Bonse, U. (1970). *Phys. Today*, **23**, 26–31.

Hart, M. and Lang, A. R. (1961). *Phys. Rev. Lett.*, **4**, 120–1.

Hart, M. and Lang, A. R. (1965). *Acta Crystallogr.*, **19**, 73–7.

Hart, M. and Milne, A. D. (1968). *Phys. Stat. Sol.*, **26**, 185–9.

Hart, M. and Milne, A. D. (1969*a*). *Acta Crystallogr. A*, **25**, 134–8.

Hart, M. and Milne, A. D. (1969*b*). *J. Phys. E: Sci. Instr.*, **2**, 646–8.

Hart, M. and Milne, A. D. (1970). *Acta Crystallogr. A*, **26**, 223–9.

Hart, M. and Milne, A. D. (1971). *Acta Crystallogr. A*, **27**, 430–5.

Hart, M. and Morgan, I. (1971). *Proceedings of the 4th International Conference on Atomic Masses and Fundamental Constants*, pp. 509–15. Plenum Press, New York.

Hart, M. and Rodrigues, A. R. D. (1978). *J. Appl. Crystallogr.*, **11**, 248–53.

Hart, M. and Rodrigues, A. R. D. (1979). *Phil. Mag.*, **40**, 149–57.

Hart, M. and Rodrigues, A. R. D. (1981). *Phil. Mag. B*, **43**, 321–32.

Hart, M. and Siddons, D. P. (1981). *Proc. Roy. Soc. London A*, **376**, 465–82.

Hart, M., Sauvage, M. and Siddons, D. P. (1980). *Acta Crystallogr. A*, **36**, 947–51.

Hart, M., Rodrigues, A. R. D. and Siddons, D. P. (1984). *Acta Crystallogr. A*, **40**, 502–7.

Hart, M., Siddons, D. P., Amemiya, Y. and Stojanoff, V. (1991). *Rev. Sci. Instrum.*, **62**, 2540–4.

Hartman, Y., Kohn, V., Kuznetsov, S., Snigirev, A. and Snigireva, I. (1997). *Il Nuovo Cimento*, **19D**, 571–6.

Hartmann, W. (1977). In *X-ray optics*. Topics in Applied Physics (ed. H.-J. Queisser), Vol. 22, pp.191–219. Springer-Verlag, Berlin.

Härtwig, J. (1976). *Phys. Stat. Sol. (a)*, **37**, 417–25.

Härtwig, J. (1977). *Phys. Stat. Sol. (a)*, **42**, 495–500.

Härtwig, J. (1978*a*). *Exper. Technik der Physik, Berlin*, **26**, 131–4.

Härtwig, J. (1978*b*). *Exper. Technik der Physik, Berlin*, **26**, 447–56.

Härtwig, J. (1978*c*). *Exper. Technik der Physik, Berlin*, **26**, 535–46.

Härtwig, J. (1978*d*). *Kristall und Technik*, **13**, 1117–26.

Härtwig, J. (1981). *Acta Crystallogr. A*, **37**, 802–4.

Härtwig, J. (2001). *J. Phys. D: Appl. Phys.*, **34**, A70–A77.

Härtwig, J., Jäckel, K. H. and Lerche, V. (1987). *Crystal Res. Technol.*, **22**, 951–9.

Härtwig, J., Grosswig, S., Becker, P. and Windisch, P. (1991). *Phys. Stat. Sol. (a)*, **125**, 79–89.

Härtwig, J., Hölzer, G., Wolf, J. and Förster E. (1993). *J. Appl. Crystallogr.*, **26**, 539–48.

Härtwig, J., Hölzer, G., Förster, E., Goetz, K., Wokulska, K. and Wolf, J. (1994). *Phys. Stat. Sol. (a)*, **143**, 23–34.

Härtwig, J., Agliozzo, S., Baruchel, J., Colella, R., de Boissieu, M., Gastaldi, J., Klein, H., Mancini, L. and Wang, J. (2001). *J. Phys. D: Appl. Phys.*, **34**, A103–A108.

Haruta, K. (1965). *J. Appl. Phys.,* **36**, 1789–90.

Hasegawa, Y., Ueji, Y., Okitsu, K., Ablett, J. M., Siddons, D. P. and Amemiya, Y. (1998). *J. Synchrotron Rad.*, **5**, 738–40.

Hasegawa, Y., Ueji, Y., Okitsu, K., Ablett, J. M., Siddons, D. P. and Amemiya, Y. (1999). *Acta Crystallogr. A*, **55**, 955–62.

Hashimoto, H., Mannami, M. and Naiki, T. (1961). *Philoph. Trans. Roy. Soc. london*, **253**, 490–516.

Hashizume, H. (1983). *J. Appl. Crystallogr.*, **16**, 420–7.

Hashizume, H. and Nakahata, T. (1988). *Jap. J. Appl. Phys.*, **27**, L1568–L1571.

Hashizume, H. and Sakata, O. (1989*a*). *Colloques de Physique*, **50**, C7-225–C7-229.

Hashizume, H. and Sakata, O. (1989*b*). *Rev. Sci. Instrum.*, **60**, 2373–5.

Hashizume, H., Nakayama, K., Matsushita, T. and Kohra, K. (1970). *J. Phys. Soc. Jap.*, **29**, 806.

Hashizume, H., Sugitama, M., Niwa, T., Sakata, O. and Cowan, P. L. (1992). *Rev. Sci. Instrum.*, **63**, 1142–5.

Hattori, H. and Kato, N. (1966). *J. Phys. Soc. Jap.*, **21**, 1772–7.

Hattori, H., Kuriyama, H. and Kato, N. (1965*a*). *J. Phys. Soc. Jap.*, **20**, 1047–50.

Hattori, H., Kuriyama, H., Katagawa, T. and Kato, N. (1965*b*). *J. Phys. Soc. Jap.*, **20**, 988–96.

Häusermann, D. and Hart, M. (1990). *J. Appl. Crystallogr.*, **23**, 63–9.

Hayashi, M. A., Morelheão, S. L., Avanci, L. H., Cardoso, L. P., Sasaki, J. M., Kretly, L. C. and Chang, S. L. (1997). *Appl. Phys. Lett.*, **71**, 2614–16.

Helliwell, J. R. (1998). *Acta Crystallogr. A*, **54**, 738–49.

Heno, Y. and Ewald, P. P. (1968). *Acta Crystallogr. A*, **24**, 17–24.

Hertel, N., Materlik, G. and Zegenhagen, J. (1985). *Z. Phys. B*, **58**, 199–204.

Hertz, H. (1889). *Ann. der Physik*, **36**, 1.

Heyroth, F., Höche, H.-R. and Eisenschmidt, C. (1999). *J. Phys.D: Appl. Phys.*, **32**, A133–8.

Heyroth, F., Zellner, J., Höche, H. R., Eisenschmidt, C., Weckert, E. and Drakopoulos, M. (2001). *J. Phys. D: Appl. Physics*, **34**, A151–A157.

Hildebrandt, G. (1967). *Phys. Stat. Sol.*, **24**, 245–61.

Hildebrandt, G. (1975). *Fortschr. Miner.*, **53**, 19–78.

Hildebrandt, G. (1978). *Kristall und Technik*, **13**, 1095–1104.

Hildebrandt, G. (1979*a*). *Acta Crystallogr. A*, **35**, 696–7.

Hildebrandt, G. (1979*b*). *Phys. Bl.*, **35**, 55–64.

Hildebrandt, G. (1979*c*). *Phys. Bl.*, **35**, 103–18.

Hildebrandt, G. (1992). In *P. P. Ewald and his dynamical theory of X-ray diffraction* (ed. D. W. J. Cruickshank, H. J. Juretschke and N. Kato), pp. 27–34. Oxford University Press, Oxford.

Hildebrandt, G. (1993). *Cryst. Res. Technol.*, **28**, 747–66.

Hildebrandt, G. (1995). *J. Phys. D, Appl. Phys.*, **28**, A8–A16.

Hildebrandt, G. (1997). *Z. Krist.*, **212**, 618–27.
Hildebrandt, G. (2002). *Crystal Res. and Technol.*, **37**, 777–82.
Hildebrandt, G. and Bradaczek, H. (2000). *Rigaku J.*, **17**, 13–22.
Hildebrandt, G., Stephenson, J. D. and Wagenfeld, H. (1973). *Z. Naturforsch.*, **28a**, 588–600.
Hildebrandt, G., Stephenson, J. D. and Wagenfeld, H. (1975). *Z. Naturforsch.*, **30a**, 697–707.
Hirano, K. and Momose, A. (1996). *Phys. Rev. Lett.*, **76**, 3735–7.
Hirano, K., Izumi, K., Ishikawa, T., Annaka, S. and Kikuta, S. (1991). *Jpn. J. Appl. Phys. Lett.*, **30**, L407–L410.
Hirano, K., Ishikawa, T., Koreeda, S., Fuchigami, K., Kanzaki, K. and Kikuta, S. (1992*a*). *Jpn. J. Appl. Phys. Lett.*, **31**, L1209–L1211.
Hirano, K., Kanzaki, K., Mikami, M., Miura, M., Tamasaku, K., Ishikawa, T. and Kikuta, S. (1992*b*). *J. Appl. Crystallogr.*, **25**, 531–5.
Hirano, K., Ishikawa, T. and Kikuta, S. (1993). *Nucl. Instrum. & Methods in Phys. Res. A*, **336**, 343–53.
Hirano, K., Ishikawa, T. and Kikuta, S. (1995). *Rev. Sci. Instrum.*, **66**, 1604–9.
Hirsch, P. B. (1952). *Acta Crystallogr.*, **5**, 176–81.
Hirsch, P. B. and Ramachandran, G. N. (1950). *Acta Crystallogr.*, **3**, 187–94.
Hirsch, P. B., Howie, A. and Whelan, M. J. (1960). *Trans. Roy. Soc. A*, **252**, 499–529.
Hirth, J. P. and Lothe, J. (1967). *Theory of dislocations*. McGraw-Hill Series on Materials Science and Engineering. McGraw-Hill, New York. 2nd edn. (1992), Krieger Pub. Co., Malabar.
Hjalmar, E. (1920). *Z. Physik*, **1**, 439.
Hmelo, A. B., Bilello, J. C., Davies, S. T. and Bowen, D. K. (1984). In *Applications of X-ray topographic methods to material science* (ed. S. Weissmann, F. Balibar and J. F. Pétroff), pp. 343–57. Plenum Press, New York.
Höche, H. R. and Nieber, J. (1988). *Phys. Stat. Sol. (a)*, **108**, 489–93.
Höche, H. R., Brümmer, O. and Nieber, J. (1986). *Acta Crystallogr. A*, **42**, 585–6.
Höche, H. R., Nieber, J., Clausnitzer, M. and Materlik, G. (1988). *Phys. Stat. Sol. (a)*, **105**, 53–60.
Høier, R. and Aanestad, A. (1981). *Acta Crystallogr. A*, **37**, 787–94.
Høier, R. and Marthinsen, K. (1983). *Acta Crystallogr. A*, **39**, 854–60.
Holland, A. J. and Tanner, B. K. (1995). *J. Phys. D: Appl. Phys.*, **28**, A27–A32.
Holt, D. B., Fornari, R., Franzosi, P., Kumar, J. and Salviati, G. (1989). In *Microscopy of semiconducting materials, Inst. Phys. Conf. Series*, Vol. 100, pp. 421–6.
Holý, V. (1980). *Phys. Stat. Sol. (b)*, **101**, 575–83.
Holý, V. (1982*a*). *Phys. Stat. Sol. (b)*, **111**, 341–51.
Holý, V. (1982*b*). *Phys. Stat. Sol. (b)*, **112**, 161–9.
Holý, V. (1984). *Acta Crystallogr. A*, **40**, 675–9.
Holý, V. (1996). In *X-ray and neutron dynamical diffraction: theory and applications* (ed. A. Authier, S. Lagomarsino and B. K. Tanner), NATO ASI Series, Series B, Physics, Vol. 357, pp. 33–42. Plenum Press, New York and London.
Holý, V. and Fewster, P. F. (2003). *J. Phys. D: Appl. Phys.*, **36**, A5-A8.

Holý, V. and Gabrielyan, K. T. (1987). *Phys. Stat. Sol. (b)*, **140**, 39–50.

Holý, V. and Kuběna, J. (1987). *Phys. Stat. Sol. (b)*, **141**, 35–45.

Holý, V. and Kuběna, J. (1989*a*). *Phys. Stat. Sol. (b)*, **151**, 23–8.

Holý, V. and Kuběna, J. (1989*b*). *Phys. Stat. Sol. (b)*, **155**, 339–47.

Holý, V. and Mikulik, P. (1996). in *X-ray and neutron dynamical diffraction: theory and applications*. ed. A. Authier, S. Lagomarsino and B. K. Tanner, NATO ASI Series, Series B: Physics, Plenum Press, New York and London, **357**, 259–68.

Holý, V., Cummings, S. and Hart, M. (1988). *J. Appl. Crystallogr.*, **21**, 516–20.

Holý, V., Kuběna, J. and Bochnicek, Z. (1991). *J. Appl. Phys.*, **70**, 3537–41.

Holý, V., Darhuber, A. A., Stangl, J., Zerlauth, S., Schäffer, F., Bauer, G., Darowski, N., Lübbert, D., Pietsch, U. and Vávra, I. (1998). *Phys. Rev. B*, **58**, 7934–43.

Holý, V., Pietsch, U. and Baumbach, G. T. (1999). *High-resolution X-ray scattering from thin films and multilayers*. Springer Tracts in Modern Physics, Berlin Heidelberg and New York.

Hölzer, G., Werhan, O. and Förster, E. (1998). *Cryst. Res. Technol.*, **33**, 555–67.

Homma, S., Ando, Y. and Kato, N. (1966). *J. Phys. Soc. Jap.*, **21**, 1160–5.

Hondoh, T. (1986). *Proc. JIMIS-4, Suppl. Transactions of the Japan. Inst. of Metals*, pp. 723–9.

Hondoh, T., Goto, A., Hoshi, R., Ono, T., Anzai, H., Kawase, R., Pimienta, P. and Mae, S. (1989). *Rev. Sci. Instrum.*, **60**, 2494–7.

Hondoh, T., Iwamatsu, H. and Mae, S. (1990). *Phil. Mag. A*, **62**, 89–102.

Hönl, H. (1933). *Z. Phys.*, **84**, 1–16.

Hoszowska, J., Freund, A., Boller, E., Sellschop, J. P. F., Level, G., Härtwig J., Burns, R. C., Rebak, M. and Baruchel, J. (2001). *J. Phys. D: Appl. Phys.*, **34**, A47–A51.

Howie, A. and Whelan, M. J. (1960). *Proc. Eur. Reg. Conf. on Electron Microscopy, Delft*, **1**, 194.

Howie, A. and Whelan, M. J. (1961). *Proc. Roy. Soc. A*, **263**, 217–37.

Howie, A. and Whelan, M. J. (1962). *Proc. Roy. Soc. A*, **267**, 206–30.

Hu, H.-C. (1997*a*). *Acta Crystallogr. A*, **53**, 484–92.

Hu, H.-C. (1997*b*). *Acta Crystallogr. A*, **53**, 493–504.

Hu, S. W., Thomas, P. A. and Webjörn, J. (1996). *J. Appl. Crystallogr.*, **29**, 279–84.

Hu, Z. H., Thomas, P. A., Snigirev, A., Snigireva, I., Souvorov, A., Smith, P. G. R., Ross, G. W. and Teat, S. (1998). *Nature*, **392**, 690–3.

Hu, Z.W., Thomas, B. R. and Chernov, A. A. (2001). *Acta Crystallogr. D*, **57**, 840–6.

Hu, M.-Y., Sinn, H., Alatas, A., Stuhrhahn, W., Alp, E. E., Wille, H.-C., Shvyd'ko, Yu. V., Sutter, J. P., Bandaru, J., Haller, E. E., Ozhogin, V. I., Rodriguez, S., Colella, R., Kartheuser, E. and Villeret, M. A. (2003). *Phys. Rev. B*, **67**, 113306–1-4.

Hu, Z.-W., Chu, Y.-S., Lai, B., Thomas, B. R. and Chernov, A. A. (2004). *Acta Crystallogr. D*, **60**, 621–9.

Huang, K. (1947). *Proc. Roy. Soc. (London) A*, **190**, 122.

Huang, X. R. and Dudley, M. (2003). *Acta Crystallogr. A*, **59**, 163–7.

Huang, K. G., Gibson, W. M. and Zegenhagen, J. (1989). *Phys. Rev. B*, **40**, 4216–9.

Huang, M.-T., Wang, C.-M. and Chang, S.-L. (1994). *Acta Crystallogr. A*, **50**, 342–50.

Huang, X. R., Jiang, S.-S., Liu, W.-J., Wu, X.-S. Feng, D., Wang, Z.-G., Han, Y. and Wang, J.-Y. (1996). *J. Appl. Crystallogr.*, **29**, 371–7.

Hümmer, K. and Billy, H. W. (1982). *Acta Crystallogr. A*, **38**, 841–8.

Hümmer, K. and Billy, H. W. (1986). *Acta Crystallogr. A*, **42**, 127–33.

Hümmer, K. and Weckert, E. (1995). *Acta Crystallogr. A*, **51**, 431–8.

Hung, H.-H. and Chang, S.-L. (1989). *Acta Crystallogr. A*, **45**, 823–33.

Ice, G. E. and Sparks, C. J. Jr (1994). *J. Opt. Soc. Amer.*, **11**, 1265–71.

Iida, A. and Kohra, K. (1979). *Phys. Stat. Sol. (a)*, **51**, 533–42.

Iida, A., Matsushita, T. and Ishikawa, T. (1985). *Jap. J. Appl. Phys.*, **24**, L675–L678.

Iida, S., Chikaura, Y., Kawado, S., and Kimura, S. (2002). *J. Synchrotron Rad.*, **9**, 169–73.

Imamov, R. M., Lomov, A. A. and Novikov, D. V. (1989). *Phys. Stat. Sol. (a)*, **115**, K133–K134.

Indenbom, V. L. (1979). *JETP Lett.*, **29**, 5–8.

Indenbom, V. L. and Chukhovskii, F. N. (1974). *Kristallografiya*, **19**, 35–41.

Indenbom, V. L. and Kaganer, V. M. (1985). *Phys. Stat. Sol. (a)*, **87**, 253–65.

Indenbom, V. L. and Slobodetskii, I. Sh. (1974). *Kristallografiya*, **19**, 42–53.

Indenbom, V. L. and Slobodetskii, I. Sh. (1976). *Phys. Stat. Sol. (b)*, **73**, K9–K12.

Indenbom, V. L., Kaganer, V. M., Möhling, W. and Suvorov, E. V. (1984). *Phys. Stat. Sol. (a)*, **83**, 195–205.

Ingal, V. N. and Beliaevskaya, E. A. (1995). *J. Phys. D: Appl. Phys.*, **28**, 2314–17.

Ingal, V. N. and Beliaevskaya, E. A. (1996). *Physica Medica*, **XII**, 74–81.

Ingal, V. N. and Beliaevskaya, E. A. (1997*a*). *Il Nuovo Cimento*, **19D**, 553–60.

Ingal, V. N. and Beliaevskaya, E. A. (1997*b*). *Sov. Phys. Technical Physics*, **42**, 59–67.

Ishida, H. (1980). *J. Appl. Crystallogr.*, **13**, 58–64.

Ishida, H., Miyamoto, N. and Kohra, K. (1976). *J. Appl. Crystallogr.*, **9**, 240–1.

Ishii, Z. (1962). *J. Phys. Soc. Jap.*, **17**, 838–46.

Ishii, Z. and Kohra, K. (1959). *J. Phys. Soc. Jap.*, **14**, 1250–1.

Ishikawa, T. (1988). *Acta Crystallogr. A*, **44**, 496–9.

Ishikawa, T., Kitano, T. and Matsui, J. (1985). *Jap. J. Appl. Phys.*, **24**, L968–L971.

Ishikawa, T., Matsui, J. and Kitano, T. (1986). *Nucl. Instrum. Methods A*, **246**, 613–6.

Ishikawa, T., Kitano, T. and Matsui, J. (1987). *J. Appl. Crystallogr.*, **20**, 344–8.

Ishikawa, T., Hirano, K. and Kikuta, S. (1991). *J. Appl. Crystallogr.*, **24**, 982–6.

Izrael, A., Pétroff, J.-F. and Authier, A. (1972). *J. Physique—C2*, **33**, 206–8.

Izumi, K., Mitsui, T., Seto, M., Yoda, Y., Ishikawa, T., Zhang, X. W., Ando, M. and Kikuta, S. (1995). *Jap. J. Appl. Phys.*, **34**, 5862–8.

Jach, T. and Bedzyk, M. J. (1990). *Phys. Rev. B*, **42**, 5399–402.

Jach, T. and Bedzyk, M. J. (1993). *Acta Crystallogr. A*, **49**, 346–50.

Jach, T., Cowan, P. L., Shen, Q. and Bedzyk, M. J. (1989). *Phys. Rev. B*, **39**, 5739–47.

Jach, T., Thurgate, S. M., Zhang, Y., Colella, R., Goldman, A. I., Kycia, S., de Boissieu, M. and Boudard, M. (1998). *Proceedings of the 6th International Conference on Quasicrystals* (ed. S. Takeuchi and T. Fujiwara), pp. 43–6. World Scientific, Singapore.

Jach, T., Zhang, Y., Colella, R., de Boissieu, M., Boudard, M., Goldman, A. I., Lograsso, T. A., Delaney, D. W. and Kycia, S. (1999). *Phys. Rev. Lett.*, **82**, 2904–7.

Jackson, J. D. (1999). *Classical electrodynamics*, 3rd edn. John Wiley and Sons, New York.

Jacobs, L. and Hart, M. (1977). *Nucl. Instrum. Methods A*, **173**, 319–25.

James, R. W. (1950). *The optical principles of the diffraction of X-rays*. G. Bell & Sons, London.

James, R. W. (1963). In *Solid state physics*, Vol. 15 (ed. F. Seitz and D. Turnbull), p. 53. Academic Press, New York.

Jark, W. and Di Fonzo, S. (2004). *J. Synchrotron Rad.*, **11**, 386–92.

Jark, W., Di Fonzo, S., Lagomarsino, S., Cedola, A., Di Fabrizio, E., Bram, A. and Riekel, C. (1996). *J. Appl. Phys.*, **80**, 4831–36.

Jark, W., Cedola, A., Di Fonzo, S., Fiordelisi, M., Lagomarsino, S., Kovalenko, N. V. and Chernov, V. A. (2001). *Appl. Phys. Lett.*, **78**, 1192–4.

Jenichen, B., Köhler, R. and Möhling, W. (1982). *Z. Naturforsch. (a)*, **37**, 405–10.

Jenichen, B., Köhler, R. and Möhling, W. (1985). *Phys. Stat. Sol. (a)*, **89**, 79–87.

Jenichen, B., Köhler, R. and Möhling, W. (1988). *J. Phys. E: Sci. Instr.*, **21**, 1062–6.

Jenkinson, A. E. and Lang, A. R. (1962). In *Direct observation of imperfections in crystals* (ed. J. B. Newkirk and J. H. Wernick), pp. 471–95. Interscience, New York.

Jergel, M., Anopchenko, A., Holý, V., Majková, E., Luby, S. and Senderák, R. (2000). *J. Appl. Crystallogr.*, **33**, 753–7.

Jiang, S.-S. and Lang, A. R. (1983). *Proc. Roy. Soc. A*, **388**, 249–71.

Jiang, S.-S., Hu, Z.-W., Yang, P., Feng, D., Zhao, J.-Y. and Jiang, J.-H. (1993). *Ferroelectrics*, **140**, 71–80.

Jiang, S.-S., Surowiec, M. and Tanner, B. K. (1988). *J. Appl. Crystallogr.*, **21**, 145–50.

Joensen, K. D., Voutov, P., Szentgyorgyi, A., Roll, J., Gorenstein, P., Høghøj, P. and Christensen, F. E. (1995). *Applied Optics*, **34**, 7935–44.

Johann, H. H. (1931). *Z. Phys.*, **69**, 185–206.

Johansson, T. (1933). *Z. Phys.*, **82**, 507–28.

Joko, T. and Fukuhara, A. (1967). *J. Phys. Soc.*, **22**, 597.

Joksch, S., Marot, G., Freund, A. and Krisch, M. (1991). *Nucl. Instr. & Methods in Phys. Res. A*, **306**, 386–90.

Jourdan, C., Gastaldi, J. and Grange, G. (1991). *Phase Transitions B*, **31**, 137–43.

Jourdan, C., Gastaldi, J. and Grange, G. (1992). *J. Physique III*, **2**, 343–53.

Jourdan, C., Gastaldi, J., Grange, G., Belkahla, S. and Guenin, G. (1995*b*). *Acta Met. et Mat.*, **43**, 4227–34.

Jourdan, C., Gastaldi, J., Roques, V., Grange, G., Belkahla, S. and Guenin, G. (1995*a*). *Acta Met. et Mat.*, **43**, 4213–25.

Jourdan, C. and Sauvage, M. (1970). *Phys. Stat. Sol. (a)*, **3**, 343–52.

Jung, J. and Lefeld-Sosnowska, M. (1984*a*). *Phil. Mag. A*, **184**, 233–55.

Jung, J. and Lefeld-Sosnowska, M. (1984*b*). *Phil. Mag. A*, **184**, 257–74.

Juretschke, H. J. (1982*a*). *Phys. Rev. Lett.*, **48**, 1487–9.

Juretschke, H. J. (1982*b*). *Phys. Lett. A*, **92**, 183–5.

Juretschke, H. J. (1984). *Acta Crystallogr. A*, **40**, 379–89.
Juretschke, H. J. (1998). *Cryst. Res. Technol.*, **33**, 569–81.
Kaganer, V. M. and Möhling, W. (1991). *Phys. Stat. Sol. (a)*, **123**, 379–92.
Kaganer, V. M., Köhler, R., Schmidbauer, M. and Opitz, R. (1997*a*). *Phys. Rev. B*, **55**, 1793–1810.
Kaganer, V. M., Opitz, R., Schmidbauer, M., Köhler, R. and Jenichen, B. (1997*b*). *Il Nuovo Cimento*, **19D**, 285–92.
Kagishima, Y., Yokoyama, Y., Ibuki, T., Niimi, T., Tsusaka, Y., Takai, K. and Matsui, J. (2002). *J. Synchrotron Rad.*, **9**, 132–5.
Kalman, Z. H. (1979). *Acta Crystallogr. A*, **35**, 634–41.
Kambe, K. (1963). *Z. Naturforsch. (a)*, **18**, 1010–11.
Kambe, K. (1965). *Z. Naturforsch. (a)*, **20**, 1730–2.
Kambe, K. (1968). *Z. Naturforsch. (a)*, **23**, 25–43.
Kambe, K. and Moliere, K. (1970). in *Advances in Structure Research by Diffraction Methods*. ed. Brill and Mason, Vieweg, Braunschweig, **3**, 53–100.
Kamijo, N., Tamura, S., Suzuki, Y. and Kihara, H. (1995). *Rev. Sci. Instrum.*, **66**, 2132–4.
Kamijo, N., Suzuki, Y., Awaji, M., Takeuchi, A., Takano, H., Ninomiya, T., Tamura, S. and Yasumoto, M. (2002). *J. Synchrotron Rad.*, **9**, 182–6.
Kanamaru, T., Kuribayashi, M. and Ishida, K. (1995). *J. Phys. Soc. Jap.*, **64**, 3529–36.
Kane, P. P., Kissel, L., Pratt, R. H. and Roy, S. C. (1986). *Phys. Rep.*, **140**, 75–159.
Katagawa, T. and Kato, N. (1974). *Acta Crystallogr. A*, **30**, 830–6.
Katagawa, T., Ishikawa, H. and Kato, N. (1975). *Acta Crystallogr. A*, **31**, S246.
Kato, N. (1949). *Proc. Jap. Acad.*, **25**, 41–4.
Kato, N. (1952*a*). *J. Phys. Soc. Jap.*, **7**, 397–406.
Kato, N. (1952*b*). *J. Phys. Soc. Jap.*, **7**, 406–14.
Kato, N. (1953). *J. Phys. Soc. Jap.*, **8**, 350–9.
Kato, N. (1955). *J. Phys. Soc. Jap.*, **10**, 46–55.
Kato, N. (1958). *Acta Crystallogr.*, **11**, 885–7.
Kato, N. (1960a). *Acta Crystallogr.*, **13**, 349–56.
Kato, N. (1960*b*). *Z. Naturforsch.*, **15a**, 369–70.
Kato, N. (1961*a*). *Acta Crystallogr.*, **14**, 526–32.
Kato, N. (1961*b*). *Acta Crystallogr.*, **14**, 627–36.
Kato, N. (1963*a*). *In Crystallography and crystal perfection* (ed. G. N. Ramachandran), pp. 153–73. Academic Press, London.
Kato, N. (1963*b*). *Acta Crystallogr.*, **16**, 276–81.
Kato, N. (1963*c*). *Acta Crystallogr.*, **16**, 282–90.
Kato, N. (1963*d*). *J. Phys. Soc. Jap.*, **18**, 1785–91.
Kato, N. (1964*a*). *J. Phys. Soc. Jap.*, **19**, 67–77.
Kato, N. (1964*b*). *J. Phys. Soc. Jap.*, **19**, 971–85.
Kato, N. (1968*a*). *J. Appl. Phys.*, **39**, 2225–30.
Kato, N. (1968*b*). *J. Appl. Phys.*, **39**, 2231–7.
Kato, N. (1968*c*). *Acta Crystallogr. A*, **24**, 157–60.

Kato, N. (1969). *Acta Crystallogr. A*, **25**, 119–28.

Kato, N. (1974). In *X-Ray diffraction* (ed. L. V. Azároff, R. Kaplow, N. Kato, R. J. Weiss, A. J. C. Wilson and R. A. Young), pp. 176–438. McGraw-Hill, New York.

Kato, N. (1976). *Acta Crystallogr. A*, **32**, 453–7.

Kato, N. (1980*a*). *Acta Crystallogr. A*, **36**, 763–9.

Kato, N. (1980*b*). *Acta Crystallogr. A*, **36**, 770–8.

Kato, N. (1991). *Acta Crystallogr. A*, **47**, 1–11.

Kato, N. (1992). *Acta Crystallogr. A*, **48**, 834–41.

Kato, N. (1994). *Acta Crystallogr. A*, **50**, 17–22.

Kato, N. (1996). In *X-ray and neutron dynamical diffraction: theory and applications* (ed. A. Authier, S. Lagomarsino and B. K. Tanner), NATO ASI Series, Series B, Physics, Vol. 357, pp. 111–35. Plenum Press, New York and London.

Kato, N. (1998). *Cryst. Res. Technol.*, **33**, 583–92.

Kato, N. and Ando, Y. (1966). *J. Phys. Soc. Jap.*, **21**, 964–8.

Kato, N. and Lang, A. R. (1959). *Acta Crystallogr.*, **12**, 787–94.

Kato, N. and Patel, J. R. (1973). *J. Appl. Phys.*, **44**, 965–70.

Kato, N., Usami, K. and Katagawa, T. (1967). *Adv. X-ray Anal.*, **10**, 46–66.

Katoh, M., Iida, S., Sugita, Y. and Okamoto, K.-I. (1991). *J. Cryst. Growth*, **112**, 368–72.

Kaulich, B., Oestreich, S., Salomé, M., Barrett, R., Susini, J., Wilhein, T., Di Fabrizio, E., Gentili, M. and Charalambous, P. (1999). *Appl. Phys. Lett.*, **75**, 4061–3.

Kawado, S. (2003). *Materials Science in Semiconductor Processing*, **5**, 435–44.

Kawado, S., Kojima, S., Kato, Y., Hayashi, H. and Ishikawa, T. (1989). *Proc. Intern. Conf. on Sci. & Technol. of Defect Control in Semiconductors*, 17–22.

Kawado, S., Kojima, S., Maekawa, I. and Ishikawa, T. (1991*a*). in *Defects in Silicon II*. ed. W.M. Bullis, U. Gösele and F. Shimura, Electrochemical Society, Pennington, 65.

Kawado, S., Kojima, S., Maekawa, I. and Ishikawa, T. (1991*b*). *Diffusion and Defect Data—Solid State Data, Part B (Solid State Phenomena)*, **19–20**, 429–38.

Kawado, S., Iida, S., Yamagughi, S., Kimura, S., Hirose, Y., Kajiwara, K., Chikaura, Y. and Umeno, M. (2002). *J. Synchrotron Rad.*, **9**, 166–68.

Kawado, S., Taishi, T., Iida, S., Suzuki, Y., Chikaura, Y. and Kajiwara, K. (2004). *J. Synchrotron Rad.*, **11**, 304–8.

Kazimirov, A. Yu., Koval'chuk, M. V. and Chukhovskii, F. N. (1987). *Kristallografiya*, **32**, 776–8.

Kazimirov, A. Yu., Koval'chuk, M. V., Kharitonov, I. Yu., Samoilova, L. V., Ishikawa, T. and Kikuta, S. (1992). *Rev. Sci. Instrum.*, **63**, 1019–22.

Kazimirov, A. Yu., Koval'chuk, M. V., Kohn, V. G., Kharitonov, I. Yu., Samoilova, L. V., Ishikawa, T., Kikuta, S. and Hirano, K. (1993). *Phys. Stat. Sol. (a)*, **135**, 507–12.

Kazimirov, A., Scherb, G., Zegenhagen, J., Lee, T.-L., Bedzyk, M. J., Kelly, M. K., Angerer, H. and Ambacher, O. (1998). *J. Appl. Phys.*, **84**, 1703–5.

Keitel, S., Retsch, C. C., Niemöller, T., Schneider, J., Abrosimov, N. V., Rossolenko, S. N. and Riemann, H. (1998). *Nucl. Instrum. Methods Phys. Res. A*, **414**, 427–30.

Keitel, S., Malgrange, C., Niemöller, T. and Schneider, J. (1999). *Acta Crystallogr. A*, **55**, 855–63.

Kellström, G. (1932). *Nova Acta R. Soc. Scient. Uppsala*, **8 (5)**.

Kies, J., Köhler, R., Möhling, W. and Peibst, H. (1973). *Z. Naturforsch. (a)*, **28**, 610–4.

Kiflawi, I. and Lang, A. R. (1974). *Phil. Mag.*, **30**, 219–23.

Kiflawi, I. and Lang, A. R. (1976). *Phil. Mag.*, **33**, 697–701.

Kiflawi, I. and Lang, A. R. (1977). *Nature*, **267**, 36–7.

Kikuta, S. (1971). *J. Phys. Soc. Jap.*, **30**, 222–7.

Kikuta, S. and Kohra, K. (1968*a*). *J. Phys. Soc. Jap.*, **25**, 923–4.

Kikuta, S. and Kohra, K. (1968*b*). *J. Phys. Soc. Jap.*, **25**, 924.

Kikuta, S. and Kohra, K. (1970). *J. Phys. Soc. Jap.*, **29**, 1322–8.

Kikuta, S., Matsushita, T. and Kohra, K. (1970). *Phys. Lett. A*, **33**, 151–2.

Kikuta, S., Kohra, K., Minakawa, N. and Doi, K. (1971). *J. Phys. Soc. Jap.*, **31**, 954–5.

Kikuta, S., Takahashi, T. and Tuzi, Y. (1975). *Phys. Lett. A*, **50**, 453–4.

Kikuta, S., Imai, Y., Iizuka, T., Yoda, Y., Zhang, X.-W. and Hirano, K. (1998). *J. Synchrotron Rad.*, **5**, 670–2.

Kilcoyne, A. L. D., Tyliszczak, T., Steele, W. F., Fakra, S., Hitchcock, P., Franck, K., Anderson, E., Harteneck, B., Rightor, E. G., Mitchell, G. E., Hitchcock, A. P., Yang, L., Warwick, T. and Ade, H. (2003). *J. Synchrotron Rad.*, **10**, 125–36.

Kim, K.-J. (1989). In *Proceedings of the US Particle Accelerator School*, Vol. 184, pp. 567–632. American Institute of Physics, New York.

Kim, C.-Y., Bedzyk, M. J., Nelson, E. J., Woicic, J. C. and Berman, L. E. (2002). *Phys. Rev. B*, **66**, 85115–1–4.

Kimura, S., Ishikawa, T., Mizuki, J. and Matsui, J. (1992). *J. Cryst. Growth*, **116**, 22–6.

Kimura, S., Harada, J. and Ishikawa, T. (1994*a*). *Acta Crystallogr. A*, **50**, 337–42.

Kimura, S., Ishikawa, T. and Matsui, J. (1994*b*). *Phil. Mag. A*, **69**, 1179–87.

Kimura, S., Matsumara, T., Kinoshita, K., Hirano, K. and Kihara, H. (1998). *J. Synchrotron Rad.*, **5**, 1079–81.

Kirfel, A. and Petcov, A. (1991). *Z. Kristallogr.*, **195**, 1–15.

Kirkpatrick, P. and Baez, A. V. (1948). *J. Opt. Soc. Am.*, **38**, 766–74.

Kishino, S. (1971). *J. Phys. Soc. Jap.*, **31**, 1168–73.

Kishino, S. and Kohra, K. (1971). *Jap. J. Appl. Phys.*, **10**, 551–7.

Kishino, S., Noda, A. and Kohra, K. (1972). *J. Phys. Soc. Jap.*, **33**, 158–66.

Kissel, L. and Pratt, R. H. (1990). *Acta Crystallogr. A*, **46**, 170–5.

Kissel, L., Pratt, R. H. and Roy, S. C. (1980). *Phys. Rev. A*, **22**, 1970–2004.

Kissel, L., Zhou, B., Roy, S. C., Sen Gupta, S. K. and Pratt, R. H. (1995). *Acta Crystallogr. A*, **51**, 271–88.

Kitano, T., Ishikawa, T., Matsui, J., Akimoto, K., Mizuki, J. and Kawase, Y. (1987). *Jap. J. Appl. Phys.*, **26**, L108–L110.

Kitano, T., Ishikawa, T. and Matsui, J. (1989). *Phys. Stat. Sol. (a)*, **115**, 383–7.

Kitano, T., Ishikawa, T. and Matsui, J. (1991). *Phil. Mag. A*, **63**, 95–109.

Klappe, J. G. and Fewster, P. F. (1994). *J. Appl. Crystallogr.*, **27**, 103–10.

Klapper, H. (1976). *J. Appl. Crystallogr.*, **9**, 310–7.

Klapper, H. (1987). *Progr. in Crystal Growth Characterization*, **14**, 367–401.

Klapper, H. (1990). In *Defect control in semiconductors* (ed. K. Suomino), pp. 1641–52. North-Holland, Amsterdam.

Klapper, H. (1991). In *Crystals: growth, properties and applications*, Vol. 13 (ed. N. Karl), pp. 109–62. Springer-Verlag, Berlin.

Klapper, H. (1996). In *X-ray and neutron dynamical diffraction: theory and applications* (ed. A. Authier, S. Lagomarsino and B. K. Tanner), NATO ASI Series, Series B, Physics, Vol. 357, pp. 167–76. Plenum Press, New York and London.

Klapper, H. and Küppers, H. (1973). *Acta Crystallogr. A*, **29**, 495–503.

Klapper, H. and Smolsky, I. L. (1998). *Cryst. Res. & Technol.*, **33**, 605–11.

Klapper, H., Roberts, K. J., Götz, D. and Herres, N. (1983). *J. Cryst. Growth*, **65**, 621–36.

Klar, B. and Rustichelli, F. (1973). *Il Nuovo Cimento B.*, **13B**, 249–71.

Klein, A. G. and Werner, S. A. (1983). *Rep. Progr. Phys.*, **46**, 259–335.

Klein, H., Agliozzo, S., Mancini, L., Gastaldi, J., Härtwig, J. and Baruchel, J. (2001). *J. Phys. D: Appl. Phys.*, **34**, A98–A102.

Knowles, J. W. (1956). *Acta Crystallogr.*, **9**, 61–9.

Koch, P. P. (1912). *Ann. Physik*, **38**, 507.

Kochelab, V. V., Molodkin, V. B., Olikhovskii, S. I. and Osinovskii, M. E. (1988). *Phys. Stat. Sol. (a)*, **108**, 67–79.

Koëbel, A., Zheng, Y., Pétroff, J. F., Boulliard, J. C., Capelle, B. and Eddrief, M. (1997). *Phys. Rev. B*, **56**, 12296–302.

Kohler, M. (1933). *Ann. Phys. (Leipzig)*, **18**, 265.

Kohler, M. (1935). *Sitzungsber. Preuss. Akad. Math. Physikal. Kl.*, 334.

Köhler, R. (1994). *Appl. Phys. A*, **58**, 149.

Kohn, V. G. (1979). *Phys. Stat. Sol. A*, **54**, 375–84.

Kohn, V. G. (1980). *Phys. Stat. Sol. A*, **60**, K93–K94.

Kohn, V. G. (1988). *Phys. Stat. Sol. (a)*, **106**, 31–9.

Kohn, V. G. (1994). *JETP*, **78**, 357–65.

Kohn, V. G. and Chumakov, A. I. (2002). *J. Physics: Condensed matter*, **14**, 11875–87.

Kohn, V. G. and Shvyd'ko, Yu. V. (1995). *J. Phys.: Condensed Matter*, **7**, 7589–600.

Kohn, V. G., Kohn, I. V. and Manykin, E. A. (1999). *JETP*, **89**, 500–7.

Kohn, V. G., Snigireva, I. and Snigirev, A. (2000). *Phys. Rev. Lett.*, **85**, 2745–8.

Kohn, V. G., Snigireva, I. and Snigirev, A. (2001). *Optics Com.*, **198**, 293–309.

Kohn, V. G., Shvyd'ko, Yu. V. and Gerdau, E. (2000). *Phys. Stat. Sol. (b)*, **221**, 597–615.

Kohra, K. (1962). *J. Phys. Soc. Jap.*, **17**, 589–90.

Kohra, K. (1972). In *Proceedings of the Sixth International Conference on X-ray Optics and Microanalysis* (ed. G. Shinoda, K. Kohra and T. Ichinokawa), pp. 35–45. University of Tokyo Press, Tokyo.

Kohra, K. and Kikuta, S. (1968). *Acta Crystallogr. A*, **24**, 200–5.

Kohra, K. and Matsushita, T. (1972). *Z. Naturforsch. (a)*, **27**, 484–7.

Kohra, K. and Takano, Y. (1968). *Jap. J. Appl. Phys.*, **7**, 982–8.

Kohra, K. and Yoshimatsu, M. (1962). *J. Phys. Soc. Jap.*, **17**, 1041–5.

Kohra, K., Kikuta, S. and Nakano, S. (1965). *J. Phys. Soc. Jap.*, **20**, 1959–60.

Kohra, K., Kikuta, S., Annaka, S. and Nakano, S. (1966). *J. Phys. Soc. Jap.*, **21**, 1565–72.

Kohra, K., Ando, M., Matsushita, T. and Hashizume, H. (1978). *Nucl. Instr. Methods*, **152**, 161–6.

Korytár, D., Mikulik, P., Ferrari, C., Hrdý, J., Baumbach, T., Freund, A. and Kuběna, A. (2003). *J. Phys. D: Appl. Physics*, **36**, A65–A68.

Kostroun, V. O. and Materlik, G. (1980). *Nucl. Instrum. Methods*, **172**, 215–22.

Koval'chuk, M. V. and Kohn, V. G. (1986). *Sov. Phys. Usp.*, **29**, 426–46.

Koval'chuk, M. V., Vartanyantz, I. A. and Kohn, V. G. (1987). *Acta Crystallogr. A*, **43**, 180–7.

Kowalski, G. and Gronkowski, J. (1982). *Phys. Stat. Sol. (a)*, **71**, 611–17.

Kowalski, G. and Lang, A. R. (1986). *J. Appl. Crystallogr.*, **19**, 224–8.

Kowalski, G., Lang, A. R., Makepeace, A. P. W. and Moore, M. (1989). *J. Appl. Crystallogr.*, **22**, 410–30.

Koz'mik, V. D., Kshevetskii, S. A., Kshevetskaya, M. L., Mikhailyuk, I. P. and Ostapovich, M. V. (1976). *Sov. Phys. Crystallogr.*, **21**, 514–18.

Kranjc, K. (1968). *J. Appl. Crystallogr.*, **1**, 227–33.

Kranjc, K. (1969). *J. Appl. Crystallogr.*, **2**, 262–73.

Krivoglaz, M. A. (1961). *Fiz. Metalllov. Metalloved.*, **12**, 465–75.

Krivoglaz, M. A. (1969). *Theory of X-ray and thermal neutron scattering by real crystals*. Plenum Press, New York.

Krivoglaz, M. A. (1996). *X-ray and neutron diffraction in nonideal crystals*. Springer-Verlag, Berlin.

Kruglov, M. V., Sozontov, E. A., Shchemelev, V. N. and Zakharov, B. G. (1977). *Sov. Phys. Crystallogr.*, **22**, 397–400.

Kruglov, M. V., Shchemelev, V. N. and Kareva, G. G. (1978). *Phys. Stat. Sol. (a)*, **46**, 343–50.

Kshevetskii, S. A., Stetsko, Yu. P. and Shelud'ko, S. A. (1985). *Sov. Phys. Crystallogr.*, **30**, 270–2.

Kshevetskii, S. A., Stetsko, Yu. P., Fodchuk, I. M. and Polyanko, V. S. (1990). *Ukrainian J. Phys.*, **35**, 444–8.

Kudo, Y., Liu, K.-Y., Kawado, S. and Hirano, K. (2000). *J. Appl. Crystallogr.*, **33**, 226–33.

Kulda, J. (1987). *Acta Crystallogr. A*, **43**, 167–73.

Kulda, J. (1988). *Acta Crystallogr. A*, **44**, 283–5.

Kulda, J. (1991). *Acta Crystallogr. A*, **47**, 775–9.
Kulda, J. and Polcarová, M. (1981). *J. Appl. Crystallogr.*, **14**, 265–9.
Kuo, C. L. and Bilello, J. C. (1984). *J. Appl. Crystallogr.*, **17**, 442–50.
Kurbakov, A. and Sokolov, A. (1999). *Acta Crystallogr. A*, **55**, 259–66.
Kuribayashi, M., Takumi, K., Kanamaru, T., Katoh, H. and Ishida, K. (1997). *J. Phys. Soc. Jap.*, **66**, 859–63.
Kuriyama, M. (1967). *J. Phys. Soc. Jap.*, **23**, 1369–79.
Kuriyama, M. (1970). *Acta Crystallogr. A*, **26**, 56–9.
Kuriyama, M. (1971). *Acta Crystallogr. A*, **27**, 273–80.
Kuriyama, M. (1972). *Acta Crystallogr. A*, **28**, 588–93.
Kuriyama, M. (1973). *Z. Naturforsch. (a)*, **28**, 622–6
Kuriyama, M. (1975). *Acta Crystallogr. A*, **31**, 774–9.
Kuriyama, M. and Boettinger, W. J. (1976). *Acta Crystallogr. A*, **32**, 511–12.
Kuriyama, M. and Early, J.G. (1974). *Acta Crystallogr. A*, **30**, 525–35.
Kurosawa, K., Saito, S. and Takemoto, S. (1974). *Jap. J. Appl. Phys.*, **13**, 804–11.
Kushnir, V. I. and Suvorov, E. V. (1986). *JETP*, **44**, 205–7.
Kushnir, I., Quintana, J. P. and Georgopoulos, P. (1993). *Nucl. Instrum. Methods Phys. Res. A*, **328**, 588–91.
Kuznetsov, S. M., Snigereva, I. I., Snigirev, A. A., Engtröm, P. and Riekel, C. (1994). *Appl. Phys. Lett.*, **65**, 827–9.
Kvardakov, V. V. and Somenkov, V. A. (1992). *Acta Crystallogr. A*, **48**, 423–30.
Kycia, S. W., Goldman, A. I., Lograsso, T. A., Delaney, D. W., Black, D., Sutton, M., Dufresne, E., Brüning, R. and Rodricks, B. (1993). *Phys. Rev. B*, **48**, 3544–7.
Kyutt, R. N. and Efimov, O. N. (1969). *Sov. Phys. Sol. State*, **11**, 228–32.
Lagomarsino, S. (1996*a*). In *X-ray and neutron dynamical diffraction: theory and applications* (ed. A. Authier, S. Lagomarsino and B. K. Tanner), NATO ASI Series, Series B, Physics, Vol. 357, pp. 225–34. Plenum Press, New York and London.
Lagomarsino, S. (1996*b*). In *X-ray and neutron dynamical diffraction: theory and applications* (ed. A. Authier, S. Lagomarsino and B. K. Tanner), NATO ASI Series, Series B, Physics, Vol. 357, pp. 249–58. Plenum Press, New York and London.
Lagomarsino, S., Scarinci, F. and Tucciarone, A. (1984). *Phys. Rev. B*, **29**, 4859–63.
Lagomarsino, S., Scarinci, F., Castrucci, P., Giannini, C., Fontes, E. and Patel, J. R. (1992). *Phys. Rev. B*, **46**, 13631–4.
Lagomarsino, S., Jark, W., Di Fonzo, S., Cedola, A., Mueller, B., Engström, P. and Riekel, C. (1996). *J. Appl. Phys.*, **79**, 4471–3.
Lagomarsino, S., Cedola, A., Cloetens, P., Di Fonzo, S., Jark, W., Soullié, G. and Riekel, C. (1997). *Appl. Phys. Lett.*, **71**, 2557–9.
Lagomarsino, S., Di Fonzo, S., Jark, W., Giannini, C., De Caro, L. and Cedola, A. (2000). *Microelectronic-Engineering*, **53**, 645–8.
Lai, B., Yun, B., Legnini, D., Xiao, Y., Chrzas, J., Viccaro, P. J., White, V., Bajikar, S., Denton, D., Cerrina, F., Di Fabrizio, E., Gentili, M., Grella, L. and Baciocchi, M. (1992). *Appl. Phys. Lett.*, **61**, 1877–9.
Lai, B., Yun, B., Xiao, Y., Yang, L., Legnini, D., Cai, Z., Krasnoperova, A., Cerrina, F. and Gentili, M. (1995). *Rev. Sci. Instrum.*, **66**, 2287–9.

Lang, A. R. (1957). *Acta Met.*, **5**, 358–64.
Lang, A. R. (1958). *J. Appl. Phys.*, **29**, 597–8.
Lang, A. R. (1959*a*). *Acta Crystallogr.*, **12**, 249–50.
Lang, A. R. (1959*b*). *J. Appl. Phys.*, **30**, 1748–55.
Lang, A. R. (1963). *Brit. J. Appl. Phys.*, **14**, 904–7.
Lang, A. R. (1965). *Z. Naturforsch. (a)*, **20**, 636–8.
Lang, A. R. (1967). *Adv. X-ray Anal.*, **10**, 91–107.
Lang, A.R. (1968). *Nature, London*, **220**, 652–7.
Lang, A. R. (1972). *Z. Naturforsch. (a)*, **27**, 461–8.
Lang, A. R. (1978). In *Diffraction and imaging techniques in material science* (ed. S. Amelinckx, R. Gevers, Remaut G. and J. Van Landuy), pp. 715–56. North-Holland, Amsterdam.
Lang, A. R. (1983*a*). *Phys. Stat. Sol. (a)*, **76**, 595–9.
Lang, A. R. (1983*b*). *Rev. Sci. Instrum.*, **54**, 897–9.
Lang, A. R. (1993). *J. Phys. D: Appl. Phys.*, **26**, A1–A8.
Lang, A. R. (1995). *Z. Krist.*, **210**, 394–7.
Lang, A. R. (1999). In *International tables of crystallography*, Vol. C (ed. A. J. C. Wilson and E. Prince), pp. 113–23. Kluwer Academic Publishers, Dordrecht.
Lang, A. R. and Mai, Z.-H. (1979). *Proc. Roy. Soc. A*, **368**, 313–29.
Lang, A. R. and Makepeace, A. P. W. (1996). *J. Synchrotron Rad.*, **3**, 313–15.
Lang, A. R. and Makepeace, A. P. W. (1999*a*). *J. Appl. Crystallogr.*, **32**, 1119–26.
Lang, A. R. and Makepeace, A. P. W. (1999*b*). *J. Phys. D: Appl. Phys.*, **32**, A97–A103.
Lang, A. R. and Makepeace, A. P. W. (1999*c*). *J. Synchrotron Rad.*, **6**, 59–61.
Lang, A. R. and Miuscov, V. F. (1965). *Appl. Phys. Lett.*, **7**, 213–16.
Lang, A. R. and Moore, M. (1991). *Proc. 2nd Int. Conf. on New Diamond Science & Technology (MRS)*, pp. 683–94.
Lang, A. R. and Polcárová, M. (1965). *Proc. Roy. Soc. A*, **285**, 297–311.
Lang, A. R., Kowalski, G., Makepeace, A. P. W. and Moore, M. (1986*a*). *Phil. Mag. B*, **53**, L53–L58.
Lang, A. R., Kowalski, G., Makepeace, A. P. W. and Moore, M. (1986*b*). *Acta Crystallogr. A*, **42**, 501–10.
Lang, A. R., Kowalski, G., Makepeace, A. P. W., Moore, M. and Clackson, S. G. (1987). *J. Phys. D: Appl. Phys.*, **20**, 541–4.
Lang, A. R., Kowalski, G. and Makepeace, A. P. W. (1990). *Acta Crystallogr. A*, **46**, 215–27.
Lang, A. R., Moore, M., Makepeace, A. P. W., Wierzchowski, W. and Welbourn, C. M. (1991). *Phil. Trans. Roy. Soc. A*, **337**, 497–520.
Lang, A. R., Yelisseyev, A. P., Pokhilenko, N. P., Steeds, J. W. and Wotherspoon, A. (2004). *J. Crystal Growth*, **263**, 575–89.
Larsen, H. B. and Thorkildsen, G. (1998*a*). *Acta Crystallogr. A*, **54**, 129–36.
Larsen, H. B. and Thorkildsen, G. (1998*b*). *Acta Crystallogr. A*, **54**, 137–45.
Larsen, H. B. and Thorkildsen, G. (1998*c*). *Acta Crystallogr. A*, **54**, 511–12.
Larson, B. C. and Schmatz, W. (1974). *Phys. Rev. B*, **10**, 2307–14.

Larsson, A., Siegbahn, M. and Waller, I. (1924). *Naturwissenshaften*, **52**, 1212.
von Laue, M. (1912). *Sitzungsberichte der Kgl. Bayer. Akad. der Wiss.*, pp. 363–73; reprinted in *Ann. Phys.* (1913), **41**, 989.
von Laue, M. (1914). *Jahrb. der Radioaktivität und Elektronik*, **11**, 308–45.
von Laue, M. (1931). *Ergeb. Exakt. Naturwiss.*, **10**, 133–58.
von Laue, M. (1935). *Ann. Physik*, **23**, 705–746.
von Laue, M. (1937). *Ann. Physik*, **28**, 528–32.
von Laue, M. (1941). *Röntgenstrahl-Interferenzen*. Akademische Verlagsgesellshaft, Beckert & ErlerKolm.-Ges., Leipzig.
von Laue, M. (1949). *Acta Crystallogr.*, **2**, 106–13.
von Laue, M. (1952). *Acta Crystallogr.*, **5**, 619–25.
von Laue, M. (1959). *Sitzungsberichte der Deutschen Akademie der Wissenschaften zu Berlin. Klasse für Mathematik, Physik und Technik.*, **1**, 1–25.
von Laue, M. (1960). *Röntgenstrahl-Interferenzen*. Akademische Verlagsgesellschaft, Frankfurt am Main.
von Laue, M. (1962). *Gesammelte Schriften und Vorträge*. Vieweg & Sohn, Braunschweig.
Laundy, D., Brown, S., Cooper, M. J., Bowyer, D. Thompson, P., Paul, D. F. and Stirling, W. G. (1998). *J. Synchrotron Rad.*, **5**, 1235–9.
Laval, J. (1945). *Bull. Soc. Fr. Minér.*, **68**, 211–79.
Laval, J. (1958). *Rev. Mod. Phys.*, **30**, 222–7.
Lee, H., Colella, R. and Chapman, L. D. (1993). *Acta Crystallogr. A*, **49**, 600–5.
Lee, H., Colella, R. and Shen, Q. (1996). *Phys. Rev. B*, **54**, 214–22.
Lefeld-Sosnowska, M. and Malgrange, C. (1968). *Phys. Stat. Sol.*, **30**, K23–K25.
Lefeld-Sosnowska, M. and Malgrange, C. (1969). *Phys. Stat. Sol.*, **34**, 635–47.
Lehmann, K. and Borrmann, G. (1967). *Z. Krist.*, **125**, 234–48.
Leitenberger, W., Kuznetsov, S. M. and Snigirev, A. (2001). *Optics Comm.*, **191**, 91–6.
Lengeler, B., Tümmler, J., Snigirev, A., Snigireva, I. and Raven, C. (1998). *J. Appl. Phys.*, **84**, 5855–61.
Lengeler, B., Schroer, C., Richwin, M., Tümmler, J., Drakopoulos, M., Snigirev, A. and Snigireva, I. (1999*a*). *Appl. Phys. Lett.*, **74**, 3924–6.
Lengeler, B., Schroer, C., Tümmler, J., Benner, B., Richwin, M., Snigirev, A., Snigireva, I. and Drakopoulos, M. (1999*b*). *J. Synchrotron Rad.*, **6**, 1153–67.
Lengeler, B., Schroer, C. G., Benner, B., Gerhardus, A., Günzler, T. F., Kuhlmann, M., Meyer, J. and Zimprich, C. (2002). *J. Synchrotron Rad.*, **9**, 119–24.
Léon, S., Lefaucheux, F., Robert, M. C., Malgrange, C. and Lorin, J. C. (1989). *J. Cryst. Growth*, **97**, 631–44.
Li, M., Mai, Z.-H., Li, J., Li, C. and Cui, S. (1995). *Acta Crystallogr. A*, **51**, 350–4.
Lienert, U., Schulze, C., Honkimäki, V., Tschentscher, Th., Garbe, S., Hignette, O., Horsewell, A., Lingham, M., Poulsen, H. F., Thomsen, N. B. and Ziegler, E. (1998). *J. Synchrotron Rad.*, **5**, 226–31.
Lin, B., Schlossman, M. L., Meron, M., Williams, S. M. and Viccaro, P. J. (1996). *Rev. Sci. Instrum.*, **67**, 3363
Ling, D. and Wagenfeld, H. (1965). *Phys. Lett.*, **15**, 8–10.
Lipscomb, W. N. (1949). *Acta Crystallogr.*, **2**, 193–4.

Liss, K.-D., Royer, A., Tschentscher, T., Suortti, P. and Williams, A. P. (1998). *J. Synchrotron Rad.*, **5**, 82–9.

Litzman, O. and Janacek, Z. (1974). *Phys. Stat. Sol. (a)*, **25**, 663–6.

Liu, F., Baker, I. and Dudley, M. (1993*a*). *Phil. Mag. A*, **67**, 1261–76.

Liu, F., Baker, I. and Dudley, M. (1993*b*). *Phil. Mag. A*, **71**, 15–42.

Liu, F., Baker, I. and Dudley, M. (1995). *Phil. Mag. A*, **71**, 1–14.

Lomov, A. A., Zaumseil, P. and Winter, U. (1985). *Acta Crystallogr. A*, **41**, 223–7.

Lorber, B., Sauter, C., Ng, J. D., Giegé, R., Vidal, O., Robert, M. C. and Capelle, B. (1999). *J. Cryst. Growth*, **204**, 357–68.

Lorut, F., Pernot, E., Pernot-Rejmánková, P. and Baruchel, J. (2003). *J. Phys. D: Appl. Physics*, **36**, A69–A73.

Lovesey, S. W. (1987). *J. Phys. C*, **20**, 5625–39.

Lovesey, S. W. and Collins, S. P. (1996). *X-ray scattering and absorption by magnetic materials*, Oxford Series on Synchrotron Radiation. Clarendon Press, Oxford.

Lovesey, S. W. and Collins, S. P. (2001). *J. Synchrotron Rad.*, **8**, 1065–77.

Ludwig, W., Cloetens, P., Härtwig, J., Baruchel, J., Hamelin, B. and Bastie, P. (2001). *J. Appl. Crystallogr.*, **34**, 602–7.

Ludewig, J. (1969). *Acta Crystallogr. A*, **25**, 116–18.

Lukáš, P., Kulda, J., Mikula, P., Sedláková, L., Alexandrov, Yu. A. and Vrána, M. (1991). *Acta Crystallogr. A*, **47**, 166–9.

Lund, M. W. (1997). *J. X-ray Sci. Technol.*, **7**, 265–70.

Mai, Z.-H., Mardix, S. and Lang, A. R. (1980). *J. Appl. Crystallogr.*, **13**, 180–7.

Mai, Z.-H. and Zhao, H. (1989). *Acta Crystallogr. A*, **45**, 602–9.

Malgrange, C. (1969). *Acta Crystallogr. A*, **25**, 356–63.

Malgrange, C. (1992). *Acta Physica Polonica A*, **82**, 13–32.

Malgrange, C. and Authier, A. (1965). *C. R. Acad. Sci. Paris*, **261**, 3774–7.

Malgrange, C. and Ferret, D. (1992). *Nucl. Instrum. Methods Phys. Res. A*, **314**, 285–96.

Malgrange, C. and Graeff, W. (2003). *J. Synchrotron Rad.*, **10**, 248–54.

Malgrange, C. and Gronkowski, J. (1984). *Phys. Stat. Sol. (a)*, **85**, 389–97.

Malgrange, C., Velu, E. and Authier, A. (1968). *J. Appl. Crystallogr.*, **1**, 181–4.

Malgrange, C., Pétroff, J.-F., Sauvage, M., Zarka, A. and Englander, M. (1976). *Phil. Mag.*, **33**, 743–51.

Malgrange, C., Varga, L., Giles, C., Rogalev, A. and Goulon, J. (1999). *Proc. SPIE - The International Society for Optical Engineering.*, **3773**, 326–9.

Mana, G., Palmisano, C. and Zosi, G. (2004*a*). *J. Appl. Crystallogr.*, **37**, 773–7.

Mana, G., Palmisano, C. and Zosi, G. (2004*b*). *Metrologia*, **41**, 238–45.

Mancini, D. C. and Bilderback, D. H. (1983). *Nucl. Instrum. Methods*, **208**, 263–72.

Mancini, L., Janot, C., Loreto, L., Farinato, R., Gastaldi, J. and Baruchel, J. (1998*a*). *Phil. Mag. Lett.*, **78**, 159–67.

Mancini, L., Reinier, E., Cloetens, P., Gastaldi, J., Härtwig, J., Schlenker, M. and Baruchel, J. (1998*b*). *Phil. Mag. A*, **78**, 1175–94.

Manninen, S., Hämäläinen, K., Schneider, J. R., Rollason, A. J. and Drube, W. (1990). *Nucl. Instrum. & Methods in Phys. Res. A*, **290**, 242–7.

Margaritondo, G. (1995). *J. Synchrotron Rad.*, **2**, 148–54.
Margaritondo, G. (2002). *Elements of Synchrotron Light - for Biology, Chemistry and Medical Research*. Oxford University Press.
Marot, G. (1995). *Opt. Eng.*, **34**, 426–31.
Marot, G., Freund, A. and Krisch, M. (1991). *Nucl. Instr. & Methods in Phys. Res. A*, **306**, 386–90.
Marot, G., Rossat, M., Freund, A., Jocksch, S., Kawata, H., Zhang, L., Ziegler, E., Berman, L. E., Chapman, D., Hastings, J. B. and Iaroci, M. (1992). *Rev. Sci. Instrum.*, **63**, 477–80.
Marra, W. C., Eisenberger, P. and Cho, A. Y. (1979). *J. Appl. Phys.*, **50**, 6927–33.
Maskil, N. and Deutsch, M. (1988). *Phys. Rev. A*, **37**, 2947–52.
Maslov, A. V., Mukhamedzhanov, E. Kh., Imamov, R. M., Man, L. I. and Le Cong Qui (1985). *Phys. Stat. Sol. (a)*, **90**, 439–44.
Materlik, G. (1985). *Z. Phys. B*, **61**, 405–9.
Materlik, G. and Kostroun, V. O. (1980). *Rev. Sci. Instrum.*, **51**, 86–94.
Materlik, G. and Zegenhagen, J. (1984). *Phys. Lett. A*, **104**, 47–50.
Materlik, G., Frahm, A. and Bedzyk, M. J. (1984). *Phys. Rev. Lett.*, **52**, 441–6.
Mathiesen, R. H., Arnberg, L., Mo, F., Weitkamp, T. and Snigirev, A. (1999). *Phys. Rev. Lett.*, **83**, 5062–5.
Matsouli, I., Kvardakov, V., Espeso, J., Chabert, L. and Baruchel, J. (1998). *J. Phys. D: Appl. Phys.*, **31**, 1478–88.
Matsouli, I., Pernot, E., Baruchel, J., Kvardakov, V., Chabert, L. and Palmer, S. B. (1999). *J. Phys. D: Appl. Phys.*, **32**, A104–A108.
Matsouli, I., Kvardakov, V. and Baruchel, J. (2000). *J. Appl. Crystallogr.*, **33**, 1051–8.
Matsushita, T. (1974). *J. Appl. Crystallogr.*, **7**, 254–9.
Matsushita, T. and Hashizume, H. (1983). In *Handbook on synchrotron radiation*, Vol. 1 (ed. E. Koch), pp. 261–314. North-Holland, Amsterdam.
Matsushita, T. and Kaminaga, U. (1980*a*). *J. Appl. Crystallogr.*, **13**, 465–71.
Matsushita, T. and Kaminaga, U. (1980*b*). *J. Appl. Crystallogr.*, **13**, 472–8.
Matsushita, T., Kikuta, S. and Kohra, K. (1971). *J. Phys. Soc. Jap.*, **30**, 1136–44.
Matsushita, T., Ishikawa, T. and Oyanagi, H. (1986). *Nucl. Instrum. Methods Phys. Res. A*, **246**, 377–9.
Matveev, B. A., Petrov, V. I., Stus', N. M., Talalakin, G. N. and Shabalin, A. V. (1988). *Sov. Phys. Semicond.*, **22**, 788–90.
Mayer, G. (1928). *Z. Krist.*, **66**, 585–636.
Mazuelas, A., Dotor, M. I., Golmayo, D., Zeimer, U., Baumbach, T., Luebbert, D., Grenzer, J. and Baruchel, J. (2001). *J. Phys. D: Appl. Phys.*, **34**, A117–A121.
Medrano, C., Heyroth, F., Schlenker, M., Baruchel, J. and Espeso, J. (1998). *J. Appl. Crystallogr.*, **31**, 726–32.
Medrano, C., Schlenker, M., Baruchel, J., Espeso, J. and Miyamoto, Y. (1999). *Phys. Rev. B*, **59**, 1185–95.
Meier, F. (1962). *Z. Phys.*, **168**, 29–41.
Meieran, E. S. (1965). *J. Appl. Phys.*, **36**, 1497.
Meieran, E. S. and Blech, I. A. (1967). *J. Appl. Phys.*, **38**, 3495–508.

Meieran, E. S. and Blech, I. A. (1972). *J. Appl. Phys.*, **43**, 265–9.

Merz, K. M. (1960). *J. Appl. Phys.*, **31**, 147.

Michot, G. and George, A. (1989). *Inst. Phys. Conf. Series*, **104**, 385.

Michot, G., George, A. and Champier, G. (1984). In *Applications of X-ray topographic methods to materials science* (ed. S. Weissmann, F. Balibar and J.-F. Pétroff), pp. 325–32. Plenum Press, New York.

Mikhailyuk, I. P., Kshevetskii, S. A., Ostapovich, M. V. and Koz'mik, V. D. (1978). *Sov. Phys. Crystallography*, **23**, 226.

Mikula, P., Kulda, J., Vrana, M. and Chalupa, B. (1984). *J. Appl. Crystallogr.*, **17**, 189–95.

Mikula, P., Lukáš, P. and Eichhorn, F. (1988). *J. Appl. Crystallogr.*, **21**, 981–3.

Mikulik, P., Holý, V., Kuběna, J. and Ploog, K. (1995). *Acta Crystallogr. A*, **51**, 825–30.

Mills, D. M. (1987). *Phys. Rev. B*, **36**, 6178–81.

Mills, D. M. (1988). *Nucl. Instrum. Methods A*, **266**, 531–7.

Mills, D. M. (ed) (2002). *Third-generation hard X-ray synchrotron radiation sources*, John Wiley & Sons, New York.

Miltat, J. (1980). In *Characterization of crystal growth defects by X-ray methods* (ed. B. K. Tanner and D. K. Bowen), NATO ASI Series, Series B, Physics, Vol. 63, pp. 401–20. Plenum Press, New York.

Miltat, J. (1984*a*). In *Applications of X-ray topographic methods to material science* (ed. S. Weissmann, F. Balibar and J. F. Pétroff), pp. 436–56. Plenum Press, New York.

Miltat, J. (1984*b*). *IEEE Trans. Magn.*, **20**, 1114–16.

Miltat, J. and Bowen, D. K. (1970). *Phys. Stat. Sol. (a)*, **3**, 431–45.

Miltat, J. and Bowen, D. K. (1975). *J. Appl. Crystallogr.*, **8**, 657–69.

Miltat, J. and Bowen, D. K. (1979). *J. Physique*, **60**, 389–401.

Miltat, J. and Dudley, M. (1980). *J. Appl. Crystallogr.*, **13**, 555–62.

Miltat, J. and Dudley, M. (1990). In *Applications of synchrotron radiation* (ed. C. R. A. Catlow and G. N. Greaves), pp. 65–99. Blackie & Sons, Glasgow.

Miltat, J. and Kléman, M. (1979). *J. Appl. Phys.*, **50**, 7695–7.

Minari, F. and Billia, B. (1993). *J. Phys. D: Appl. Phys.*, **26**, A73–A75.

Minato, I., Hashizume, H., Watanabe, H. and Matsui, J. (1986). *Jap. J. Appl. Phys.*, **25**, 1485–9.

Miyake, S. and Kambe, K. (1954). *Acta Crystallogr.*, **7**, 220.

Miyake, S. and Ohtsuki, Y. H. (1974). *Acta Crystallogr. A*, **30**, 103–4.

Miyake, S. and Uyeda, R. (1955). *Acta Crystallogr.*, **8**, 335–42.

Mo, F., Mathiesen, R. H., Alzari, P. M., Lescar, J. and Rasmussen, B. (2002). *Acta Cryst. D*, **58**, 1780–6.

Mocella, V., Guigay, J.-P., Epelboin, Y., Härtwig, J. and Baruchel, J. (1999). *J. Phys. D: Appl. Phys.*, **32**, A88–A91.

Mocella, V., Epelboin, Y. and Guigay, J. P. (2000). *Acta Crystallogr. A*, **56**, 308–16.

Mocella, V., Ferrero, C., Freund, A. K., Hoszowska, J., Zhang, L. and Epelboin, Y. (2001). *Nucl. Instrum. & Methods in Physics Research A*, **467–468**, 414–7.

Mocella, V., Lee, W.-K., Tajiri, G., Mills, D., Ferrero, C. and Epelboin, Y. (2003). *J. Appl. Crystallogr.*, **36**, 129–36.

Mohr, P. J. and Taylor, B. N. (2000). *Rev. Modern Phys.*, **72**, 351–495.

Moliere, G. (1939*a*). *Ann. Physik*, **35**, 272–96.

Moliere, G. (1939*b*). *Ann. Physik*, **35**, 297–313.

Moliere, G. (1939*c*). *Ann. Physik*, **36**, 265–74.

Molodkin, V. B., Olikhovskii, S. I., Osinovskii, M. E., Gureev, A. N., Datsenko, L. I., Nizkova, A. I. and Zhuravlev, B. F. (1985). *Phys. Stat. Sol. (a)*, **87**, 597–608.

Momose, A. (1995). *Nucl. Instrum. Methods A*, **352**, 622–8.

Momose, A. (2002). *J. Synchrotron Rad.*, **9**, 136–42.

Momose, A. and Fukuda, J. (1995). *Med. Phys.*, **22**, 375–80.

Momose, A., Takeda, T. and Itai, Y. (1995). *Rev. Sci. Instrum.*, **66**, 1434–6.

Momose, A., Takeda, T., Itai, Y. and Hirano, K. (1996*a*). *Nature (London) Medecine*, **2**, 473–5.

Momose, A., Takeda, T., Itai, Y. and Hirano, K. (1996*b*). *Proc. SPIE*, **2708**, 674–84.

Momose, A., Yoneyama, A. and Hirano, K. (1997). *J. Synchrotron Rad.*, **4**, 311–12.

Momose, A., Takeda, T., Itai, Y., Yoneyama, A. and Hirano, K. (1998). *J. Synchrotron Rad.*, **5**, 309–14.

Moodie, A. F., Cowley, J. C. and Goodman, P. (1997). In *International tables of crystallography*, Vol. B (ed. U. Shmueli), pp. 481–6. Kluwer Academic Publishers, Dordrecht.

Morawe, C., Peffen, J. C., Ziegler, E. and Freund, A. K. (2001). *Proc. SPIE – The International Society for Optical Engineering.*, **4145**, 61–71.

Morelhão, S. L. and Abramof, E. (1999). *J. Appl. Crystallogr.*, **32**, 871–7.

Morelhão, S. L. and Avanci, L. H. (2001). *Acta Crystallogr. A*, **57**, 192–6.

Morelhão, S. L. and Cardoso, L. P. (1996). *J. Appl. Crystallogr.*, **29**, 446–56.

Morelhão, S. L. and Kycia, S. (2002). *Phys. Rev. Letters*, **89**, 15501-1-4.

Morelhão, S. L., Avanci, L. H., Hayashi, M. A. and Cardoso, L. P. (1999). *Mater. Res. Soc. Proc.*, **561**, 87–92.

Morelhão, S. L., Härtwig, J. and Meier, D. L. (2000). *J. Cryst. Growth*, **213**, 288–98.

Mori, K., Sekine, N., Sato, H., Shimano, D., Shiwaku, H., Hyodo, K., Sugiyama, H., Ando, M., Ohashi, K., Koyama, M. and Nakajima, Y. (2002). *J. Synchrotron Rad.*, **9**, 143–47.

Morse, P. M. (1930). *Phys. Rev.*, **35**, 1310–24.

Moseley, H. G. J. and Darwin, C. G. (1913). *Phil. Mag.*, **26**, 210.

Mühlberg, M., Rudolph, P., Genzel, C., Wermke, B. and Becker, U. (1990). *J. Cryst. Growth*, **101**, 275–80.

Mukhamedzhanov, E. Kh. and Bocchi, C. (2000). *J. Appl. Crystallogr.*, **33**, 1430–3.

Mukhamedzhanov, E. Kh., Maslov, A. V., Imamov, R. M., Bzhaumikhov, A. A., Fedorov, E. A. and Kobzareva, S. A. (1991). *J. Appl. Crystallogr.*, **24**, 6–13.

Müller, M., Burghammer, M., Flot, D., Riekel, C., Morawe, C., Murphy, B. and Cedola, A. (2000). *J. Appl. Crystallogr.*, **33**, 1231–40.

Nagel, D. J., Gilfrich, J. V. and Barbee, T. W. Jr (1982). *Nucl. Instrum. Methods A*, **124**, 83–91.

Nakahigashi, K., Fukuoka, N. and Shimomura, Y. (1975). *J. Phys. Soc. Jap.*, **38**, 1634–40.

Nakajima, T., Ohta, J., Yonenaga, I., Koizumi, H., Iwasa, I., Suzuki, H., Suzuki, T. and Suzuki, H. (1995). *J. Appl. Crystallogr.*, **28**, 375–84.

Nakatani, S. and Takahashi, T. (1994). *Surf. Sci.*, **311**, 433–9.

Nakayama, K., Hashizume, H. and Kohra, K. (1970). *J. Phys. Soc. Jap.*, **30**, 893–7.

Nakayama, K., Hashizume, H., Miyoshi, A., Kikuta, S. and Kohra, K. (1973). *Z. Naturforsch. (a)*, **28**, 632–8.

Negishi, R., Fukumachi, T. and Kawamura, T. (1999). *Acta Crystallogr. A*, **55**, 267–73.

Negishi, R., Fukamachi, T., Yoshizawa, M., Zhou, S., Xu, Z., Kawamura, T., Matsumoto, I., Sakamaki, T. and Nakajima, T. (1998). *J. Appl. Crystallogr.*, **31**, 351–5.

Negishi, R., Yoshizawa, M., Zhou, S., Matsumoto, I., Fukamachi, T. and Kawamura, T. (2001). *Jpn. J. Appl. Phys.*, **40**, L884–7.

Negishi, R., Yoshizawa, M., Zhou, S., Matsumoto, I., Fukamachi, T. and Kawamura, T. (2004). *J. Synchrotron Rad.*, **11**, 266–71.

Nesterets, Ya. I., Punegov, V.I., Pirshin, I. V., Touryanski, A. G., Vinogradov, A. V., Förster, E. and Podorov, S. G. (2000). *Phys. Stat. Sol. (a)*, **179**, 311–7.

Neuhäusler, U., Schneider, G., Ludwig, W. and Hambach, D. (2003). *J. Phys. D: Appl. Physics*, **36**, A79–A82.

Neumann, H. B., Rütt, U., Bouchard, R., Schneider, J. R. and Nagasawa, H. (1994). *J. Appl. Crystallogr.*, **27**, 1030–8.

Neumann, H. B., Schneider, J. R., Suessenbach, J., Stock, S. R. and Rek, Z. U. (1996). *Nucl. Instrum. & Methods in Phys. Res. A*, **372**, 551–5.

Newkirk, J. B. (1958*a*). *J. Appl. Phys.*, **29**, 996.

Newkirk, J. B. (1958*b*). *Phys. Rev.*, **110**, 1465–6.

Newkirk, J. B. (1959). *Trans. Metal. Soc. AIME*, **215**, 483–97.

Newkirk, J. B. (1970). In *Applications of modern metallographic techniques*, Vol. 480, p. 152. ASTM STP, New York.

Nguyen, Thi H., Jamgotchian, H., Dabo, Y., Billia, B., Gastaldi, J., Härtwig, J., Baruchel, J., Schenk, T. and Klein, H. (2003). *J. Phys. D: Appl. Physics*, **36**, A83–A86.

Niizeki, N. and Hasegawa, M. (1964). *J. Phys. Soc. Jap.*, **19**, 550–4.

Nikulin, A. Yu. (1997). *Appl. Phys. Letters*, **70**, 1545–7.

Nikulin, A. Y., Davis, J. R., Jones, N. T., Usher, B. F., Souvorov, A. Yu. and Freund, A. K. (2000). *Phys. Stat. Sol. A*, **179**, 103–8.

Nikulin, A. Y., Davis, J. R., Usher, B. F., Freund, A. K. and Ishikawa, T. (2001). *Proc. SPIE – The International Society for Optical Engineering.*, **4145**, 129–39.

Nikulin, A. Y., Souvorov, A., Goodden, K., Tamasaku, K. and Ishikawa, T. (2003). *J. Phys. D: Appl. Physics*, **36**, A87–A92.

Nöhammer, B., Hoszowska, J., Freund, A. K. and David, C. (2003). *J. Synchrotron Rad.*, **10**, 168–71.

Nourtier, C., Taupin, D., Kléman, M., Labrune, M. and Miltat, J. (1977). *J. Appl. Crystallogr.*, **10**, 328–37.
Nourtier, C., Kleman, M., Taupin, D., Miltat, J., Labrune, M. and Epelboin, Y. (1979). *J. Appl. Phys.*, **50**, 2143–5.
Nugent, K. A., Gureyev, T. E., Cookson, D. F., Paganin, D. and Barnea, Z. (1996). *Phys. Rev. Lett.*, **77**, 2961–4.
Nusshardt, M. and Bonse, U. (2003). *J. Appl. Crystallogr.*, **36**, 269–79.
Oestreich, S., Kaulich, B. and Susini, J. (1999). *Rev. Sci. Instrum.*, **70**, 1921–5.
Oguro, M. and Higashi, A. (1973). *Intern. Symposium on Physics and Chemistry of Ice, Ottawa, Canada*, 338–43.
Ohama, N., Sakashita, H. and Okazaki, A. (1979). *J. Appl. Crystallogr.*, **12**, 455–9.
Ohler, M. and Härtwig, J. (1999). *Acta Crystallogr. A*, **55**, 413–22.
Ohler, M., Härtwig, J. and Prieur, E. (1997). *Acta Crystallogr. A*, **53**, 199–201.
Ohler, M., Köhler, S. and Härtwig, J. (1999). *Acta Crystallogr. A*, **55**, 423–32.
Ohtsuki, Y. H. (1964). *J. Phys. Soc. Jap.*, **19**, 2285–92.
Ohtsuki, Y. H. (1966). *J. Phys. Soc. Jap.*, **21**, 2300–6.
Okazaki, A. and Ohama, N. (1979). *J. Appl. Crystallogr.*, **12**, 450–4.
Okazaki, A. and Soejima, Y. (2001). *Acta Crystallogr. A*, **57**, 708–12.
Okitsu, K. (2003). *Acta Crystallogr. A*, **59**, 235–44.
Okitsu, K., Iida, S., Sugita, Y., Takeno, H., Yagou, Y. and Kawata, H. (1992). *Jap. J. Appl. Phys.*, **31**, 3779–85.
Okitsu, K., Ueji, Y., Sato, K. and Amemiya, Y. (2001). *J. Synchrotron Rad.*, **8**, 33–7.
Okitsu, K., Ueji, Y., Sato, K. and Amemiya, Y. (2002). *Acta Crystallogr. A*, **58**, 146–54.
Okkerse, B. (1962). *Philips Res. Repts*, **17**, 464–78.
Okkerse, B. (1963). *Philips Res. Repts*, **18**, 413–31.
Okkerse, B. and Penning, P. (1963). *Philips Res. Repts*, **18**, 82–94.
Olekhnovich, N. M., Karpei, A. L., Olekhnovich, A. I. and Puzenkova, L. D. (1983). *Acta Crystallogr. A*, **39**, 116–22.
Ovchinnikova, E. N. and Dmitrienko, V. E. (2000). *Acta Crystallogr. A.*, **56**, 2–10.
Ozaki, T., Fujimoto, I., Mizuno, K., Iida, S., Kajiwara, K., Taira, T., Yoshimura, J., Shimura, T. and Chikaura, Y. (2003). *Nucl. Instrum. & Methods in Physics Research B*, **199**, 81–4.
Palmer, A. and Jauch, W. (1995). *Acta Crystallogr. A*, **51**, 662–7.
Panzner, T., Leitenberger, W., Grenzer, J., Bodenthin, Y., Geue, Th., Pietsch, U. and Möhwald, H. (2003). *J. Phys. D: Appl. Physics*. **36**, A93–A97.
Parente, C. B. R., Mazzochi, V. L. and Pimentel, F. J. F. (1994). *J. Appl. Crystallogr.*, **27**, 463–74.
Park, B. M., Chung, S. J., Kim, H. S., Si, W. and Dudley, M. (1997). *Phil. Mag. A*, **75**, 611–20.
Parpia, D. Y. (1982*a*). *Phil. Mag. A*, **45**, 593–605.
Parpia, D. Y. (1982*b*). *Phil. Mag. A*, **46**, 691–7.

Parrish, W. and Langford, J. I. (1999). In *International tables of crystallography*, Vol. C (ed. A. J. C. Wilson and E. Prince), pp. 42–72. Kluwer Academic Publishers, Dordrecht.

Parthasarathy, R. (1960). *Acta Crystallogr.*, **13**, 802–6.

Parthasarathy, D., Ramachandran, G. N. and Mallikarjunan, M. (1963). In *Crystallography and crystal perfection* (ed. G. N. Ramachandran), pp. 133–43. Academic Press, London.

Patel, J. R. (1973). *J. Appl. Phys.*, **44**, 3903–6.

Patel, J. R. (1975). *J. Appl. Crystallogr.*, **8**, 186–91.

Patel, J. R. (1996). In *X-ray and neutron dynamical diffraction: theory and applications* (ed. A. Authier, S. Lagomarsino and B. K. Tanner), NATO ASI Series, Series B, Physics, Vol. 357, pp. 211–24. Plenum Press, New York and London.

Patel, J. R. and Authier, A. (1975). *J. Appl. Phys.*, **46**, 118–25.

Patel, J. R. and Fontes, E. (1996). In *X-ray and neutron dynamical diffraction: theory and applications* (ed. A. Authier, S. Lagomarsino and B. K. Tanner), NATO ASI Series, Series B, Physics, Vol. 357, pp. 235–48. Plenum Press, New York and London.

Patel, J. R. and Freeland, P. E. (1967). *Phys. Rev. Lett.*, **18**, 833–5.

Patel, J. R. and Freeland, P. E. (1970). *J. Appl. Phys.*, **41**, 2814–7.

Patel, J. R. and Golovchenko, J. A. (1983). *Phys. Rev. Lett.*, **50**, 1858–61.

Patel, J. R. and Kato, N. (1968). *Appl. Phys. Lett.*, **13**, 40–2.

Patel, J. R. and Kato, N. (1973). *J. Appl. Phys.*, **44**, 971–7.

Patel, J. R., Jackson, K. A. and Dillon, J. F. (1968). *J. Appl. Phys.*, **39**, 3767–70.

Patel, J. R., Testardi, L. R. and Freeland, P. E. (1976). *Phys. Rev. B*, **13**, 3548–57.

Patel, J. R., Freeland, P. E., Golovchenko, J. A., Kortan, A. R., Chadi, D. J. and Qian, G.-X. (1986). *Phys. Rev. Lett.*, **57**, 3077–80.

Patel, J. R., Golovchenko, J. A., Freeland, P. E. and Gossmann, H.-J. (1987). *Phys. Rev. B*, **36**, 7715–17.

Pavlov, K. M. and Punegov, V. I. (1998). *Acta Crystallogr. A*, **54**, 214–18.

Pavlov, K. M. and Punegov, V. I. (2000). *Acta Crystallogr. A*, **56**, 227–34.

Pavlov, K. M., Punegov, V. I. and Faleev, N. N. (1995). *JETP*, **80**, 1090–7.

Pavlov, K. M., Faleev, N., Tabuchi, M. and Takeda, Y. (1999). *Proceedings of the Second International Conference on Synchrotron Radiation in Materials Science (SRMS-2)*, Kobe, Japan, *Jpn. J. Appl. Phys.*, **38** (Suppl. 38-1), 269–72.

Pavlov, K. M., Kewish, C. M., Davis, J. R. and Morgan, M. J. (2000). *J. Phys. D: Appl. Phys.* **33**, 1596–605.

Penning, P. (1966). Theory of X-ray diffraction in unstrained and lightly strained perfect crystals. Ph. D. Thesis, University of Delft. Published in *Philips Res. Repts. Suppl.*, **5**, 1–109.

Penning, P. and Goemans, A. H. (1968). *Phil. Mag.*, **18**, 297–300.

Penning, P. and Polder, D. (1961). *Philips Res. Repts*, **16**, 419–40.

Pernot-Rejmánková, P., Thomas, P. A., Cloetens, P., Lorut, F., Baruchel, J., Hu, Z. W., Urenski, P. and Rosenman, G. (2000). *J. Appl. Crystallogr.*, **33**, 1149–53.

Perkins, R. T. and Knight, L. V. (1984). *Acta Cryst. A*, **40**, 617–19.

Perutz, M. F. (1990). *Acta Crystallogr. A*, **46**, 633–43.
Petrascheck, D. and Folk, R. (1976). *Phys. Stat. Sol. (a)*, **36**, 147–59.
Petrashen, P. V. (1973). *Fiz. Tverd. Tela*, **15**, 3131–2.
Petrashen, P. V. (1986). *Metallofizika*, **8**, 35–43.
Petrashen, P. V. and Chukhovskii, F. N. (1976). *Sov. Phys. JETP*, **42**, 243–8.
Petrashen, P. V., Chukhovskii, F. N. and Shul'pina, I. L. (1980). *Acta Crystallogr. A*, **36**, 287–95.
Pétroff, J.-F. (1969). *Phys. Stat. Sol.*, **31**, 285–95.
Pétroff, J.-F. and Authier, A. (1966). *Phys. Stat. Sol.*, **13**, 373–86.
Pétroff, J.-F. and Sauvage, M. (1978). *J. Cryst. Growth*, **43**, 628–36.
Pétroff, J.-F., Sauvage, M., Riglet, P. and Hashizume, H. (1980). *Phil. Mag. A*, **42**, 319–38.
Petrov, A. A., Afanas'ev, Yu. B., Shu'lpina, I. L. and Kartenko, N. F. (1989). *Sov. Phys. Crystallogr.*, **34**, 271–2.
Pfeiffer, F., Salditt, T., Høghøj, P., Andersen, I. and Schell, N. (2000). *Phys. Rev. B*, **62**, 16939–16943.
Pfeiffer, F., Salditt, T. and David, C. (2002). *J. Appl. Crystallogr.*, **35**, 430–3.
Phakey, P. P. (1969). *Phys. Stat. Sol.*, **34**, 105–19.
Pichaud, B. and Minari, F. (1976). *Phil. Mag.*, **34**, 1121–8.
Pichaud, B. and Minari, F. (1984). in *Applications of X-ray topographic methods to material science*. ed. S. Weissmann, F. Balibar and J.-F. Pétroff, Plenum Press, New York, 385–92.
Pinsker, Z. G. (1978). *Dynamical scattering of X-rays in crystals*, Springer Series in Solid-State Physics, Vol. 3. Springer-Verlag, Berlin.
Platzmann, P. M. and Tzoar, N. (1970). *Phys. Rev. B*, **2**, 3556–59.
Podorov, S. G. and Förster, E. (2000). *Phys. Stat. Sol. (b)*, **220**, 829–36.
Pogany, A., Gao, D. and Wilkins, S. W. (1997). *Rev. Sci. Instrum.*, **68**, 2774–82.
Polcarová, M. (1969). *IEEE Trans. Magnetics*, **5**, 536–44.
Polcarová, M. (1978*a*). *Phys. Stat. Sol. (a)*, **46**, 567–75.
Polcarová, M. (1978*b*). *Phys. Stat. Sol. (a)*, **47**, 179–86.
Polcarová, M. (1980). *Phys. Stat. Sol. (a)*, **59**, 779–85.
Polcárová, M. and Brádler, J. (1987). *J. Appl. Crystallogr.*, **20**, 374–8.
Polcárová, M. and Brádler, J. (1988). *J. Appl. Crystallogr.*, **21**, 169–75.
Polcarová, M. and Lang, A. R. (1962). *Appl. Phys. Lett.*, **1**, 13–15.
Polcarová, M., Gemperlová, J., Brádler, J., Jacques, A., George, A. and Priester, L. (1998). *Phil. Mag. A*, **78**, 105–30.
Polcarová, M., Brádler, J., Tomas, I., Jacques, A. and George, A. (1999*a*). *Phil. Trans. Roy. Soc. (London) A*, **357**, 2689–700.
Polcarová, M., Brádler, J., Gemperlová, J., Jacques, A. and George, A. (1999*b*). *J. Phys. D: Applied Physics*, **32**, A109–13.
Polcarová, M., Jacques, A., Brádler, J., Vallino, F., Gemperlová, J. and George, A. (1999*c*). *Materials Science Forum*, **294–296**, 377–80.
Polcarová, M., Brádler, J., Jacques, A., Lejcek, P., George, A. and Ferry, O. (2003). *J. Phys. D: Appl. Physics*, **36**, A98-A101.

Prins, J. A. (1930). *Z. Phys.*, **63**, 477–93.

Punegov, V. I. (1990). *Sov. Phys. Crystallogr.*, **35**, 336–40.

Punegov, V. I. (1991). *Sov. Phys. Solid State*, **33**, 136–40.

Punegov, V. I. (1993). *Phys. Stat. Sol. (a)*, **136**, 9–19.

Punegov, V. I., Pavlov, K. M., Podorov, S. G. and Faleev, N. N. (1996). *Phys. Solid State*, **38**, 148–52.

Qian, Y., Sturchio, N. C., Chiarello, R. P., Lyman, P. F., Lee, T.-L. and Bedzyk, M. J. (1994). *Science*, **265**, 1555–7.

Queisser, H. J., Hartmann, W. and Hagen, W. (1981). *J. Cryst. Growth*, **52**, 897–906.

Quintana, J. P. and Hart, M. (1995). *J. Synchrotron Rad.*, **2**, 119–23.

Quintana, J. P., Hart, M., Bilderback, D. H., Henderson, C., Richter, D., Setterson, T., White, J., Hausermann, D., Krumrey, M. and Schulte-Schrepping, H. (1995). *J. Synchrotron Rad.*, **2**, 1–5.

Raghothamachar, B., Chung, H., Dudley, M. and Larson, D. J. (1998). *J. Electronic Materials*, **27**, 556–63.

Ramachandran, G. N. (1944). *Proc. Ind. Acad. Sci. A*, **19**, 280–92.

Ramachandran, G. N. (1946*a*). *Proc. Ind. Acad. Sci. A*, **24**, 58–65.

Ramachandran, G. N. (1946*b*). *Proc. Ind. Acad. Sci. A*, **24**, 81.

Raoux, D. (1993). In *Neutron and synchrotron radiation for condensed matter studies*, Vol. 1 (ed. J. Baruchel, J.-L. Hodeau, M. S. Lehmann, J.-R. Regnard and C. Schlenker), pp. 37–78. Springer-Verlag Berlin and Les Editions de Physique, Les Ullis.

Rauch, H. and Petrascheck, D. (1978). In *Neutron diffraction* (ed. H. Dachs), pp. 305–51. Springer-Verlag, Berlin.

Rauch, H. and Suda, M. (1974). *Phys. Stat. Sol. (a)*, **25**, 495–505.

Rauch, H. and Werner, S. (2000). *Neutron interferometry. Lessons in experimental quantum mechanics*. Clarendon Press, Oxford.

Rauch, H., Treimer, W. and Bonse, U. (1974). *Phys. Lett.*, **47A**, 369–71.

Raven, C., Snigirev, A., Snigireva, I., Spanne, P., Souvorov, A. and Kohn, V. (1996). *Appl. Phys. Lett.*, **69**, 1826–8.

Raven, C., Snigirev, A., Koch, A., Snigireva, I. and Kohn, V. (1997). *SPIE*, **3149**, 140–8.

Reinier, E., Mancini, L., Gastaldi, J., Baluc, N., Härtwig, J. and Baruchel, J. (1998). *Physica B*, **253**, 61–7.

Rejmánková, P., Baruchel, J., Kulda, J., Calemczuk, R. and Salce, B. (1995). *J. Phys. D: Appl. Phys.*, **28**, A69–A73.

Rejmánková, P., Baruchel, J. and Moretti, P. (1996). *Physica B*, **226**, 293–303.

Rejmánková, P., Baruchel, J., Moretti, P., Arbore, M., Fejer, M. and Foulon, G. (1998). *J. Appl. Crystallogr.*, **31**, 106–8.

Rejmánková-Pernot, P., Cloetens, P., Baruchel, J. and Guigay, J.-P. (1998). *Phys. Rev. Lett.*, **81**, 3435–8.

Rémy, F., Gastaldi, J., Le Lay G., Rossberg, A., Piechotka, M., Wetzel, G. and Kaldis, E. (1994). *J. Physique (Colloque)*, **4**, C9/203–7.

Renninger, M. (1934). *Z. Krist.*, **89**, 344–74.

Renninger, M. (1937*a*). *Z. Phys.*, **106**, 141–76.
Renninger, M. (1937*b*). *Z. Krist.*, **97**, 107–21.
Renninger, M. (1938). *Z. Krist.*, **99**, 181–7.
Renninger, M. (1955). *Acta Crystallogr.*, **8**, 597–606.
Renninger, M. (1961). *Z. Naturforsch.*, **16a**, 1110–11.
Renninger, M. (1962). *Phys. Lett.*, **1**, 106–9.
Renninger, M. (1963). In *Crystallography and crystal perfection* (ed. G. N. Ramachandran), pp. 145–52. Academic Press, London.
Renninger, M. (1964). *Z. Naturforsch. (a)*, **19**, 783–7.
Renninger, M. (1965*a*). *Z. Angewandte Phys.*, **19**, 20–33.
Renninger, M. (1965*b*). *Z. Angewandte Phys.*, **19**, 34–5.
Renninger, M. (1967). *Adv. X-ray Anal.*, **10**, 32–41.
Renninger, M. (1968). *Acta Crystallogr. A*, **24**, 143–9.
Renninger, M. (1976). *J. Appl. Crystallogr.*, **9**, 178–80.
Renninger, M. (1981). *J. Appl. Crystallogr.*, **14**, 59.
Renninger, M. and Theis, W. (1969). *J. Appl. Crystallogr.*, **2**, 48–52.
Ribet, M., Leon, S., Lefaucheux, F. and Robert, M. C. (1990). *J. Appl. Crystallogr.*, **23**, 277–81.
Rieutord, F. (1990). *Acta Crystallogr. A*, **46**, 526–34.
Riglet, P., Sauvage, M., Pétroff, J. F. and Epelboin, Y. (1980). *Phil. Mag. A*, **42**, 339–58.
Rigon, L., Besch, H-J., Arfelli, F., Menk, R-H., Heitner, G. and Plothow-Besch, H. (2003). *J. Phys. D: Appl. Physics*. **36**, A107–A112.
Robert, M. C. and Lefaucheux, F. (1983). *J. Cryst. Growth*, **65**, 637–50.
Robert, M. C., Lefaucheux, F., Sauvage, M. and Ribet, M. (1981). *J. Cryst. Growth*, **52**, 976–82.
Robert, M. C., Lefaucheux, F., Jannot, B., Godefroy, G. and Garnier, E. (1988). *J. Cryst. Growth*, **88**, 499–510.
Robert, M. C., Lefaucheux, F., Le Bihan, R. and Rafii, N. (1991). *Crystal Res. Technol.*, **26**, 59–67.
Roessler, B., Kramer, J. J. and Kuriyama, M. (1965). *Phys. Stat. Sol.*, **11**, 117–20.
Rogers, C. S., Mills, D. M., Lee, W. K., Knapp, G. S., Holmberg, J., Freund, A., Wulff, M., Rossat, M., Hanfland, M. and Yamaoka, H. (1995). *Rev. Sci. Instrum.*, **66**, 3494–9.
Rogosa, G. L. and Schwarz, G. (1952). *Phys. Rev.*, **87**, 995–8.
Röntgen, W. (1896). *Beiblätter zu den Annalen der Physik und Chemie*, 401–6.
Rossberg, A., Piechotka, M., Wetzel, G., Gastaldi, J., Magerl, A. and Kaldis, E. (1996). *J. Crystal Growth*, **166**, 354–60.
Rossmanith, E. (1992). *Acta Crystallogr. A*, **48**, 596–610.
Rugel, S., Wallner, G., Metzger, H. and Peisl, J. (1993). *J. Appl. Crystallogr.*, **26**, 34–40.
Rustichelli, F. (1975). *Phil. Mag.*, **31**, 1–12.
Rütt, U., Neumann, H. B., Poulsen, H. F. and Schneider, J. R. (1995). *J. Appl. Crystallogr.*, **28**, 729–37.

Sabine, T. M. (1988). *Acta Crystallogr. A*, **44**, 368–73.
Sabine, T. M. (1999). In *International tables of crystallography*, Vol. C (ed. A. J. C. Wilson and E. Prince), pp. 603–7. Kluwer Academic Publishers, Dordrecht.
Saito, S., Nakahigashi, K. and Shimomura, Y. (1966). *J. Phys. Soc. Jap.*, **21**, 850–60.
Saitoh, Y., Hashizume, H. and Tsutsui, K. (1988). *Jap. J. Appl. Phys.*, **27**, 1386–96.
Saitoh, K., Inagawa, K., Kohra, K., Hayashi, C., Iida, A. and Kato, N. (1989). *Rev. Sci. Instrum.*, **60**, 1519–23.
Saka, T. and Kato, N. (1986). *Acta Crystallogr. A*, **42**, 469–78.
Saka, T. and Kato, N. (1987). *Acta Crystallogr. A*, **43**, 252–4.
Saka, T., Katagawa, T. and Kato, N. (1972*a*). *Acta Crystallogr. A*, **28**, 102–13.
Saka, T., Katagawa, T. and Kato, N. (1972*b*). *Acta Crystallogr. A*, **28**, 113–20.
Saka, T., Katagawa, T. and Kato, N. (1973). *Acta Crystallogr. A*, **29**, 192–200.
Sakai, N. and Ono, K. (1976). *Phys. Rev. Lett.*, **37**, 351–3.
Sakata, O. and Hashizume, H. (1987). *Rep. Res. Lab. Eng. Mat. Tokyo Inst. Techn.*, **12**, 45–57.
Sakata, O. and Hashizume, H. (1988). *Jap. J. Appl. Phys.*, **27**, L1976–L1979.
Sakata, O. and Hashizume, H. (1991). *Rep. Res. Lab. Eng. Mat. Tokyo Inst. Techn.*, **16**, 27–38.
Sakata, O. and Hashizume, H. (1995). *Acta Crystallogr. A*, **51**, 375–84.
Sakata, O. and Hashizume, H. (1997). *Acta Crystallogr. A*, **53**, 781–8.
Sakata, O., Hashizume, H. and Kurashina, H. (1993). *Phys. Rev. B*, **48**, 11408–11.
Salles da Costa, C. A. B., Cardoso, L. P., Mazzochi, V. L. and Parente, C. B. R. (1990). In *Defect control of semiconductors*, Vol. II (ed. K. Sumino), pp. 1535–9. Elsevier, Amsterdam.
Sano, H., Ohtaka, K. and Ohtsuki, Y. (1969). *J. Phys. Soc. Jap.*, **27**, 1254–61.
Šaroun, J., Lukáš, P., Mikula, P. and Alefeld, B. (1994). *J. Appl. Crystallogr.*, **27**, 80–8.
Sasaki, J. M., Cardoso, L. P., Campos, C., Roberts, K. J., Clark, G. F., Pantos, E. and Sacilotti, M. A. (1996). *J. Appl. Crystallogr.*, **29**, 325–30.
Sato, K., Okitsu, K., Ueji, Y., Matsushita, T. and Amemiya, Y. (2000). *J. Synchrotron Rad.*, **7**, 368–73.
Sauvage, M. (1968). *Phys. Stat. Sol.*, **29**, 725–36.
Sauvage, M. (1978). *Nucl. Instrum. Methods*, **152**, 313–17.
Sauvage, M. (1980). In *Characterization of crystal growth defects by X-ray methods* (ed. B. K. Tanner and D. K. Bowen,) NATO ASI Series, Series B, Physics, Vol. 63, pp. 433–55. Plenum Press, New York.
Sauvage, M. and Authier, A. (1965). *Bull. Soc. Fr. Minér. Crist.*, **88**, 379–88.
Sauvage, M. and Malgrange, C. (1970). *Phys. Stat. Sol.*, **37**, 759–71.
Sauvage, M. and Pétroff, J.-F. (1980). In *Synchrotron radiation research* (ed. H. Winick and S. Doniach), pp. 607–38. Plenum Press, New York.
Sauvage, M. and Simon, D. (1969). *Phys. Stat. Sol.*, **35**, 173–76.
Sauvage, M., Lefaucheux, F., Robert, M. C. and Ribet, M. (1982). *Phys. Stat. Sol. (a)*, **71**, K5–K7.

Sauvage, M., Malgrange, C. and Pétroff, J.-F. (1983). *J. Appl. Crystallogr.*, **16**, 14–20.

Schäfer, P. and Köhler, R. (2003). *J. Phys. D: Appl. Physics*, **36**, A113–A117.

Scheffen–Lauenroth, Th., Klapper, H. and Becker, R. A. (1981). *J. Cryst. Growth*, **55**, 557–70.

Schelokov, I. A., Roshchupkin, D. V., Irzhaka, D. V. and Tucoulou, R. (2004). *J. Appl. Crystallogr.*, **37**, 52–61.

Scherf, C., Hahn, T., Heger, G., Becker, R. A., Wunderlich, W. and Klapper, H. (1997). *Ferroelectrics*, **191**, 171–7.

Schetinkin, S. A., Wheeler, E., Kvardakov, V. V., Schlenker, M. and Baruchel, J. (2003). *J. Phys. D: Appl. Physics*, **36**, A118–A121.

Schlangenotto, H. (1967). *Z. Phys.*, **203**, 17–36.

Schlenker, M. and Baruchel, J. (1986). *Physica B*, **137**, 309–19.

Schlenker, M. and Baruchel, J. (1996). in *X-ray and neutron dynamical diffraction: theory and applications*. ed. A. Authier, S. Lagomarsino and B. K. Tanner, NATO ASI Series, Series B: Physics, Plenum Press, New York and London, **357**, 177–86

Schlenker, M. and Baruchel, J. (1999). In *International tables of crystallography*, Vol. C (ed. A. J. C. Wilson and E. Prince), pp. 124–5. Kluwer Academic Publishers, Dordrecht.

Schlenker, M. and Guigay, J.-P. (1996). In *X-ray and neutron dynamical diffraction: theory and applications* (ed. A. Authier, S. Lagomarsino and B. K. Tanner), NATO ASI Series, Series B, Physics, Vol. 357, pp. 63–72. Plenum Press, New York and London.

Schlenker, M. and Guigay, J.-P. (2001). In *International tables of crystallography*, Vol. B (ed. U. Shmueli) pp. 557–646. Kluwer Academic Publishers, Dordrecht.

Schlenker, M. and Kleman, M. (1972). *J. Physique*, **32**, C1-256–8.

Schlenker, M. and Shull, C. G. (1973). *J. Appl. Phys.*, **44**, 4181–4.

Schlenker, M., Brissonneau, P. and Perrier, J. P. (1968). *Bull. Soc. Fr. Minér. Cristal.*, **91**, 653–65.

Schlenker, M., Linares–Galvez, J. and Baruchel, J. (1977). *Physica B*, **86-88**, 1339–40.

Schlenker, M., Linares–Galvez, J. and Baruchel, J. (1978). *Phil. Mag. B*, **37**, 1–11.

Schmahl, G., Rudolph, D., Niemann, B., Guttmann, J., Schneider, G., David, C., Diehl, M. and Wilhein, T. (1994). *Optik (Stuttgart)*, **93**, 95–102.

Schmahl, G., Rudolph, D., Schneider, G., Guttmann, J. and Niemann, B. (1994). *Optik (Stuttgart)*, **97**, 181–2.

Schneider, J. R., Gonçalves, O. D., Rollason, A. J., Bonse, U., Lauer, J. and Zulehner, J. (1988). *Nucl. Instrum. Methods Phys. Res. B*, **29**, 661–74.

Schneider J. R., Nagasawa, H., Berman, L. E., Hastings, J. B., Siddons, D. P. and Zulehner, J. (1989). *Nucl. Instrum. & Methods Phys. Res. A*, **276**, 636–42.

Schneider, J. R., Bouchard, R., Graf, H. A. and Nagasawa, H. (1992). *Acta Crystallogr. A*, **48**, 804–19.

Schulte–Schrepping, H., Heuer, J. and Huckelmann, B. (1998). *J. Synchrotron Rad.*, **5**, 682–84.

Schulte–Schrepping, H., Materlik, G., Heuer, J. and Teichmann, T. (1995). *Rev. sci. Instrum.*, **66**, 2217–9.

Schulz, L. G. (1954). *J. Metals Trans. AIME*, **200**, 1082–3.

Schulze, C., Lienert, U., Hanfland, M., Lorenzen, M. and Zontone, F. (1998). *J. Synchrotron Rad.*, **5**, 77–81.

Schuster, M. (1991). *Spectrochim. Acta B*, **46**, 1341–9.

Schuster, M. and Göbel, H. (1995). *J. Appl. Phys.*, **28**, A270–A275.

Schuster, M., Müller, L., Mauser, K. E. and Straub, R. (1988). *Thin Solid Films*, **157**, 325–36.

Schütz, G., Frahm, R., Mautner, P., Wienke, R., Wagner, W. and Kniele, P. (1989). *Phys. Rev. Lett.*, **62**, 2620–3.

Schwarz, G. and Rogosa, G. L. (1954). *Phys. Rev.*, **95**, 950–3.

Schwarzschild, M. (1928). *Phys. Rev.*, **32**, 162–71.

Schwinger, J. (1949). *Phys. Rev.*, **75**, 1912–25.

Schwuttke, G. H. and Sils, V. (1963). *J. Appl. Phys.*, **34**, 3127.

Sears, V. F. (1979). *Can. J. Phys.*, **56**, 1261–88.

Sears, V. F. (1989). *Neutron optics*. Oxford University Press, New York.

Servidori, M., Šourek, Z. and Solmi, S. (1987). *J. Appl. Phys.*, **62**, 1723–8.

Servidori, M., Cembali, F., Fabri, R. and Zani, A. (1992). *J. Appl. Crystallogr.*, **25**, 46–51.

Servidori, M., Cembali, F. and Milita, S. (1996). In *X-ray and neutron dynamical diffraction: theory and applications* (ed. A. Authier, S. Lagomarsino and B. K. Tanner), NATO ASI Series, Series B, Physics, Vol. 357, pp. 301–22. Plenum Press, New York and London.

Sharma, R. K., Singh, V. K., Nayyar, N. K., Gupta, S. R. and Sharma, B. B. (1993). *J. Cryst. Growth*, **131**, 565–73.

Shastri, S. D., Zambianchi, P. and Mills, D. M. (2001*a*). *J. Synchrotron Rad.*, **8**, 1131–5.

Shastri, S.D., Zambianchi, P. and Mills, D. M. (2001*b*). *Proc. SPIE – The International Society for Optical Engineering.*, **4143**, 69–77.

Shearwood, C. and Whitworth, R. W. (1989). *J. Glaciol.*, **35**, 281–3.

Shearwood, C. and Whitworth, R. W. (1991). *Phil. Mag. A*, **64**, 289–302.

Shearwood, C. and Whitworth, R. W. (1992*a*). *Phil. Mag. A*, **65**, 85–9.

Shearwood, C. and Whitworth, R. W. (1992*b*). in *Physics and Chemistry of Ice*. ed. N. Maeno and T. Hondo, Hokkaido University Press, Sapporo, 488–91.

Shechtman, D., Blech, I. A., Gratias, D. and Cahn, J. W. (1984). *Phys. Rev. Lett.*, **53**, 1951–3.

Sheludko, S. (2004). *Acta Crystallogr. A*, **60**, 281–2.

Shen, Q. (1986). *Acta Crystallogr. A*, **42**, 525–33.

Shen, Q. (1993). *Acta Crystallogr. A*, **49**, 605–13.

Shen, Q. (1998). *Phys. Rev. Lett.*, **80**, 3268–71.

Shen, Q. (1999*a*). *Phys. Rev. Lett.*, **83**, 4784–7.

Shen, Q. (1999*b*). *Phys. Rev. B*, **59**, 11109–12.

Shen, Q. (2000). *Phys. Rev. B*, **61**, 8593–7.

Shen, Q. and Colella, R. (1988). *Acta Crystallogr. A*, **44**, 17–21.

Shen, Q. and Finkelstein, K. D. (1990). *Phys. Rev. Lett.*, **65**, 3337–40.

Shen, Q. and Finkelstein, K. D. (1992). *Phys. Rev. B*, **45**, 5075–8.

Shen, Q. and Huang, X.-R. (2001). *Phys. Rev. B*, **63**, 174102–1–8.

Shen, Q. and Wang, J. (2003). *Acta Crystallogr. D*, **59**, 809–14.

Shen, Q., Shastri, S. and Finkelstein, K. D. (1995). *Rev. Sci. Instrum.*, **66**, 1610–13.

Shen, Q., Pringle, D., Szebenyi, M. and Wang, J. (2002). *Rev. Sci. Instrum.*, **73**, 1646–8.

Shen, Q., Kycia, S. and Dobrianov, I. (2000*a*). *Acta Crystallogr. A*, **56**, 264–7.

Shen, Q., Kycia, S. and Dobrianov, I. (2000*b*). *Acta Crystallogr. A*, **56**, 268–79.

Shevchenko, M. B. (2003). *Acta Cryst. A.* **59**, 481–6.

Shevchenko, M. B. and Pobydaylo, O. V. (2003). *Acta Crystallogr. A*, **59**, 45–47.

Shilo, D. and Zolotoyabko, E. (2003). *J. Phys. D: Appl. Physics*, **36**, A122–A127.

Shull, C. G. (1968). *Phys. Rev. Lett.*, **21**, 1585–9.

Shull, C. G. and Oberteuffer, J. A. (1972). *Phys. Rev. Lett.*, **28**, 871–4.

Shull, C. G. and Shaw, W. N. (1973). *Z. Naturforsch.*, **28a**, 657–61.

Shul'pina, I. L. and Argunova, T. S. (1995). *J. Phys. D: Appl. Phys.*, **28**, A47–A49.

Shul'pina, I. L., Petrashen, P. N., Chukhovskii, F. N. and Gabrielyan, K. T. (1984). *Tezisi Dokl. IV Vsesoyuz. Sovetch. Defecti Strukturi v Poluprovodniakh, Novosibirsk 23–25 Oktyabr 1984*, **2**, 114.

Shvyd'ko, Yu. V., Gerdau, E., Jäschke, J., Leupold, O., Lucht, M. and Rüter, H. D. (1998). *Phys. Rev. B*, **57**, 4968–71.

Shvyd'ko, Yu. V., Lerche, M., Jaschke, J., Lucht, M., Gerdau, E., Gerken, M., Ruter, H. D., Wille, H.-C., Becker, P., Alp, E. E., Stuhrhan, W., Sutter, J. and Toellner, T. S. (2000). *Phys. Rev. Lett.*, **85**, 495–8.

Siddons, D. P. and Hart, M. (1983). in *Proc. Intern. Conf. on EXAFS and Near Edge Structures*. Springer–Verlag, Berlin, 373–5.

Siddons, D. P., Hart, M., Amemiya, Y. and Hastings, J. B. (1990). *Phys. Rev. Letters*, **64**, 1967–70.

Siegbahn, M. (1931). *Spektroskopie der Röntgenstrahlen*. Berlin.

Simon, D. and Authier, A. (1968). *Acta Crystallogr. A*, **24**, 527–34.

Sippel, D., Kleinstück, K. and Schulze, G. E. R. (1962). *Physica Stat. Sol.*, **2**, K104–K105.

Sippel, D., Kleinstück, K. and Schulze, G. E. R. (1964). *Phys. Lett.*, **8**, 241–2.

Sippel, D., Kleinstück, K. and Schulze, G. E. R. (1965). *Phys. Lett.*, **14**, 174–5.

Skalicky, P. and Malgrange, C. (1972). *Acta Crystallogr. A*, **28**, 501–7.

SLAC (1998). Report SLAC-R-521, UC-414 (1998). Stanford Linear Accelerator Center.

Smith, D. Y. (1987). *Phys. Rev. A*, **35**, 3381–7.

Smith, S. H. and Chetwynd, D. G. (1992). *Foundations of ultraprecision mechanism and design*. Gordon and Breach, London.

Smither, R. K. (1982). *Rev. Sci. Instrum.*, **53**, 131–41.

Smither, R. K. and Fernandez, P. B. (1992). *Rev. Sci. Instrum.*, **63**, 1755–62.

Sneddon, I. N. (1951). *Fourier transforms*. McGraw-Hill, New York.

Snigirev, A. (1995). *Rev. Sci. Instrum.*, **66**, 2053–8.

Snigirev, A. (1996). *SPIE Proc.*, **2856**, 26–33.

Snigirev, A. and Kohn, V. (1995). In *X-ray microbeam technology and applications* (ed. W. Yun), *Proc. SPIE*, **2516**, 27–37.

Snigirev, A. and Raven, C. (1997). *Bull. de la S. F. P.*, **110**, 10–14.

Snigirev, A. and Suvorov, A. Yu. (1993). *Acta Cryst. A*, **49**, 818–24.

Snigirev, A., Snigireva, I., Engström, P., Lequien, S., Souvorov, A., Hartman, Ya., Chevallier, P., Idir, M., Legrand, F., Soullie, G. and Engrand, S. (1995*a*). *Rev. Sci. Instrum.*, **66**, 1461–3.

Snigirev, A., Snigireva, I., Kohn, V., Kuznetsov, S. and Schelokov, I. (1995*b*). *Rev. Sci. Instrum.*, **66**, 5486–92.

Snigirev, A., Kohn, V., Snigireva, I. and Lengeler, B. (1996). *Nature*, **384**, 49–51.

Snigirev, A., Snigireva, I., Bösecke, P., Lequien, S. and Schelokov, I. (1997). *Opt. Commun.*, **135**, 378–84.

Snigirev, A., Kohn, V., Snigireva, I., Souvorov, A. and Lengeler, B. (1998). *Appl. Opt.*, **37**, 653–62.

Snigireva, I., Souvorov, A. and Snigirev, A. (1998). In *X-ray microscopy and spectroscopy*, pp. IV37–IV44. Springer-Verlag, Heidelberg.

Somenkov, V. A., Tkalich, A. K. and Shil'stein, S. Sh. (1991). *Sov. Phys. Tech. Phys.*, **36**, 1309–11.

Sommerfeld, A. (1912). *Ann. Physik*, **38**, 473.

Sommerfeld, A. (1949). *Partial differential equations in physics*. Academic Press, New York.

Sondhauss, P. and Wark, J. S. (2003). *Acta Crystallogr. A*, **59**, 7–13.

Souvorov, A., Snigirev, A., Snigireva, I. and Aristova, E. (1996). *Rev. Sci. Instrum.*, **67**, 1733–6.

Souvorov, A., Snigireva, I., Snigirev, A. A., Aristova, E. and Hartman, Ya. (1997). *Appl. Phys. Lett.*, **70**, 829–31.

Sozontov, E., Cao, L.-X., Kazimirov, A., Kohn, V., Konuma, M., Cardona, M. and Zegenhagen, J. (2001). *Phys. Rev. Lett.*, **86**, 5329–32.

Spal, R. D. (2001). *Phys. Rev. Lett.*, **86**, 3044–7.

Spanne, P., Raven, C., Snigireva, I. and Snigirev, A. (1999). *Phys. Med. Biol.*, **44**, 741–9.

Sparks, C. J. Jr, Borie, B. S. and Hastings, J. B. (1980). *Nucl. Instrum. Methods Phys. Res.*, **172**, 237–42.

Sparks, C. J. Jr, Ice, G. E., Wong, J. and Batterman, B. W. (1982). *Nucl. Instrum. Methods Phys. Res.*, **194**, 73–8.

Spiller, E. and Segmüller, A. (1974). *Appl. Phys. Lett.*, **24**, 60–1.

Spirkl, W., Tanner, B. K., Whitehouse, C., Barnett, S. J., Cullis, A. G., Johnson, A. D., Keir, A., Usher, B., Clark, G. E., Hagston, W., Hogg, C. R. and Lunn, B. (1994*a*). *Phil. Mag. A*, **69**, 221–36.

Spirkl, W., Tanner, B. K., Whitehouse, C., Barnett, S. J., Cullis, A. G., Johnson, A. D., Keir, A., Usher, B., Clark, G. E., Hagston, W., Hogg, C. R. and Lunn, B. (1994*b*). *Phil. Mag. A*, **70**, 531–48.

Stahn, J., Pucher, A., Pietsch, U., Zellner, J. and Weckert, E. (1999). *Acta Crystallogr. A*, **55**, 1034–7.

Stangl, J., Zerlauth, S., Holý, V., Faschinger, W. and Bauer, G. (1997). *Il Nuovo Cimento*, **19D**, 355–60.

Stanglmeier, F., Lengeler, B., Weber, W., Göbel, H. and Schuster, M. (1992). *Acta Crystallogr. A*, **48**, 626–39.

Staudenmann, J.-L. and Chapman, L. D. (1989). *J. Appl. Crystallogr.*, **22**, 162–72.

Staudenmann, J. L., Werner, S. A., Colella, R. and Overhauser, A. W. (1980). *Phys. Rev.*, **21**, 1419–38.

Staudenmann, J.-L., Chapman, L. D., Murphy, W. J., Horning, R. D. and Liedl, G. L. (1985). *J. Appl. Crystallogr.*, **18**, 519–27.

Steiner, B., Kuriyama, M., Dobbyn, R. C., Laor, U., Larson, D. and Brown, M. (1989). *J. Appl. Phys.*, **66**, 559–68.

Steiner, B., Kuriyama, M. and Dobbyn, R. C. (1990). *Progr. Crystal Growth Characterization*, **20**, 189–216.

Steiner, B., Dobbyn, R. C., Black, D., Burdette, H., Kuriyama, M., Spal, R. D., van der Berg, L., Fripp, A., Simchik, R., Lal, R. B., Batra, A., Matthiesen, D. and Ditchek, B. (1991). *J. Research NIST*, **96**, 305–31.

Stepanov, S. A. (1992). *Phys. Stat. Sol. (a)*, **132**, 9–16.

Stepanov, S. A. (1994). *Crystallography Reports*, **39**, 182–7.

Stepanov, S. A. and Köhler, R. (1994*a*). *J. Appl. Phys.*, **76**, 7809–15.

Stepanov, S. A. and Köhler, R. (1994*b*). *J. Phys. D: Appl. Phys.*, **27**, 1922–8.

Stepanov, S.A. and Kondrashkina, E. A. (1991). *Phys. Chem. and Mech. of Surfaces*, **6**, 1152–61.

Stepanov, S. A. and Ulyanenkov, A. P. (1994). *Acta Crystallogr.*, **A50**, 579–85.

Stepanov, S. A., Kondrashikina, E. A. and Novikov, D. V. (1991). *Nucl. Instrum. Methods A*, **301**, 350–7.

Stepanov, S. A., Pietsch, U. and Baumbach, G. T. (1995). *Z. Phys. B—Condens. Matter*, **96**, 341–7.

Stepanov, S.A., Kondrashkina, E. A., Köhler, R., Novikov, D. V., Materlik, G. and Durbin, S. (1998). *Phys. Rev. B*, **57**, 4829–41.

Stephenson, J. D. (1992). *Phys. Stat. Sol. (a)*, **130**, 75–82.

Stephenson, J. D. (1997). *Phys. Stat. Sol. (a)*, **159**, 317–22.

Stephenson, J. D., Tuomi, T. and Kelhä, V. (1980). *Phys. Stat. Sol. (a)*, **57**, 191–202.

Stern, R. M., Perry, J. J. and Bourdreux, D. S. (1969). *Rev. Mod. Phys.*, **41**, 275–95.

Stetsko, Yu. P. and Chang, S.-L. (1997). *Acta Crystallogr.*, **53**, 28–34.

Stetsko, Yu. P. and Chang, S.-L. (1999*a*). *Acta Crystallogr. A*, **55**, 457–65.

Stetsko, Yu. P. and Chang, S.-L. (1999*b*). *Acta Crystallogr. A*, **55**, 683–94.

Stetsko, Yu. P., Juretschke, H. J., Huang, Y.-S., Chao, C.-H., Chen, C.-K. and Chang, S.-L. (2000). *Acta Crystallogr. A*, **56**, 394–400.

Stetsko, Yu. P., Juretschke, H. J., Huang, Y.-S., Lee, Y.-R., Lin, T.-C. and Chang, S.-L. (2001). *Acta Crystallogr. A*, **57**, 359–367.

Stetsko, Yu. P., Lee, Y.-R., Tang, M.-T. and Chang, S.-L. (2004). *Acta Crystallogr. A*, **60**, 64–74.

Stevenson, A. W., Gao, D., Gureyev, T. E., Pogany, A. and Wilkins, S. W. (1997). *Materials Australia*, May/June, pp. 14–17.

Steyerl, A. and Steinhauser, K. A. (1979). *Zeit. Phys. B*, **34**, 221–7.

Sturkey, L. (1948). *Phys. Rev.*, **73**, 183.

Sugii, K., Matsushita, T. and Kohra, K. (1971). *J. Phys. Soc. Jap.*, **31**, 1295–6.

Sugiyama, M., Maeyama, S., Maeda, F., Oshima, M. (1995). *Phys. Rev. B*, **52**, 2678–81.

Suortti, P. (1992). *Rev. Sci. Instrum.*, **63**, 942–5.

Suortti, P. and Freund, A. (1989). *Rev. Sci. Instrum.*, **60**, 2579–85.

Suortti, P. and Schulze, C. (1995). *J. Synchrotron Rad.*, **2**, 6–12.

Suortti, P., Chapman, D., Schneider, J. R. and Tschentscher, T. (1992). *J. Appl. Crystallogr.*, **25**, 432–8.

Suortti, P., Fiedler, S., Bravin, S., Brochard, T., Mattenet, M., Renier, M., Spanne, P., Thomlinson, W., Charvet, A. M., Elleaume, H., Schulze-Briese C. and Thompson A. C. (2000). *J. Synchrotron Rad.*, **7**, 340–7.

Suortti, P., Hastings, J. B., and Cox, D. E. (1985). *Acta Crystallogr.*, **41**, 417–20.

Suortti, P., Thomlinson, W. Chapman, D., Gmur, N., Greene, R. and Lazarz, N. (1990). *Nucl. Instrum. & Methods in Phys. Res. A*, **297**, 68–274.

Sutter, J. P., Alp, E. E., Hu, M .Y., Sinn, H., Stuhrbahn, W., Toellner, T. S., Bortel, G. and Colella, R. (2001). *Phys. Rev. B*, **63**, 94111-1-12.

Sutter, J. P., Kuetgens, U., Materlik, G. and Rostomyan, A. (2003). *J. Appl. Crystallogr.*, **36**, 1432–9.

Sutter, J. P., Ishikawa, T., Kuetgens, U., Materlik, G., Nishino, Y., Rostomyan, A., Tamasaku, K. and Yabashi, M. (2004). *J. Synchrotron Rad.*, **11**, 378–85.

Suvorov, E. V., Polovinkina, V. I., Nikitenko, V. I. and Indenbom, V. L. (1974). *Phys. Stat. Sol. (a)*, **26**, 385–95.

Suzuki, S. and Takagi, M. (1966). *J. Phys. Soc. Jap.*, **21**, 554.

Suzuki, T. and Kojima, H. (1966). *Acta Met.*, **14**, 913.

Suzuki, Y., Kamijo, N., Tamura, S., Handa, K., Takeuchi, A., Yamamoto, S., Sugiyama, H., Ohsumi, K. and Ando, M. (1997). *J. Synchrotron Rad.*, **4**, 60–3.

Suzuki, Y., Awaji, M., Kohmura, Y., Takeuchi, A., Takano, H., Kamijo, N., Tamura, S., Yasumoto, M. and Handa, K. (2001). *Nucl. Instrum. & Methods A*, **467/468**, 951–3.

Tachibana, M., Koizumi, H., Izumi, K., Kajiwara, K. and Kojima, K. (2003). *J. Synchrotron Rad.*, **10**, 416–20.

Takagi, S. (1962). *Acta Crystallogr.*, **15**, 1311–12.

Takagi, S. (1969). *J. Phys. Soc. Jap.*, **26**, 1239–53.

Takagi, M., Akaba, N. and Suzuki, S. (1979). *J. Phys. Soc. Jap.*, **46**, 1811–18.

Takahashi, K. and Takagi, M. (1978*a*). *J. Phys. Soc. Jap.*, **44**, 1266–74.

Takahashi, K. and Takagi, M. (1978*b*). *J. Phys. Soc. Jap.*, **44**, 1664–9.

Takahashi, T. and Hashimoto, M. (1995). *Physics Lett. A*, **200**, 73–5.
Takahashi, T. and Kikuta, S. (1977). *J. Phys. Soc. Jap.*, **42**, 1433–4.
Takahashi, T. and Kikuta, S. (1979). *J. Phys. Soc. Jap.*, **47**, 620–4.
Takahashi, T. and Nakatani, S. (1995). *Surf. Sci.*, **326**, 347–60.
Takama, T. and Harima, H. (1994). *Acta Crystallogr. A*, **50**, 239–46.
Takama, T., Iwasaki, M. and Sato, S. (1980). *Acta Crystallogr. A*, **36**, 1025–30.
Takeda, T. and Kato, N. (1978). *Acta Crystallogr. A*, **34**, 43–7.
Takeda, T., Momose, A., Ueno, E. and Itai, Y. (1998). *J. Synchrotron Rad.*, **5**, 1133–65.
Takeda, T., Momose, A., Yu, Q., Wu, J., Hirano, K. and Itai, Y. (2000). *J. Synchrotron Rad.*, **7**, 280–2.
Takeuchi, T., Ohta, N., Sugita, Y. and Fukuhara, A. (1983). *J. Appl. Phys.*, **54**, 715–21.
Tamasaku, K. and Ishikawa, T. (2001). *Acta Crystallogr. A*, **57**, 197–200.
Tamasaku, K., Yabashi, M. and Ishikawa, T. (2002). *Phys. Rev. Lett.*, **88**, 044801-1-4.
Tamasaku, K., Ishikawa, T. and Yabashi, M. (2003). *Appl. Phys. Letters*, **83**, 2994–6.
Tamura, S., Ohtani, K. and Kamijo, N. (1994). *Appl. Surf. Sci.*, **79/80**, 514–18.
Tanemura, S. and Kato, N. (1972). *Acta Cryst. A*, **28**, 69–80.
Tanemura, S. and Lang, A. R. (1973). *Z. Naturforsch. (a)*, **28**, 668–76.
Tang, M.-T. and Chang, S.-L. (1988). *Acta Crystallogr. A*, **44**, 1073–8.
Tanner, B. K. (1976). *X-ray diffraction topography*. Pergamon Press, Oxford.
Tanner, B. K. (1981). *Science Progress, Oxford*, **67**, 411–40.
Tanner, B. K. (1996). In *X-ray and neutron dynamical diffraction: theory and applications* (ed. A. Authier, S. Lagomarsino and B. K. Tanner), NATO ASI Series, Series B, Physics, Vol. 357, pp. 147–66. Plenum Press, New York and London.
Tanner, B. K. and Bowen, D. K. (ed). (1980). *Characterization of crystal growth defects by X-ray methods*, NATO ASI Series, Series B, Physics, Vol. 63. Plenum Press, New York.
Tanner, B. K. and Bowen, D. K. (1992). *Mater. Sci. Rept.*, **8**, 369–407.
Tanner, B. K., Safa, M., Midgley, D. and Bordas, J. (1976). *J. Magn. Mater.*, **7**, 337–41.
Tanner, B. K., Parbrook, P. J., Whitehouse, C. R., Keir, A. M., Johnson, A. D., Jones, J., Wallis, D., Smith, I. M., Lunn, B. and Hogg, J. H. C. (2001). *J. Phys. D: Appl. Phys.*, **34**, A109–A113.
Taupin, D. (1964). *Bull. Soc. Fr. Minér. Crist.*, **87**, 469–511.
Taupin, D. (1967). *Acta Crystallogr.*, **23**, 25–35.
Tchen, T. (2002). *J. Appl. Crystallogr.*, **35**, 713–19.
Templeton, D. H. (1994). In *Resonant anomalous X-ray scattering* (ed. G. Materlik, C. J. Sparks and K. Fischer), pp. 1–7. Elsevier Science B.V., Amsterdam.
Templeton, D. H. and Templeton, L. K. (1980). *Acta Crystallogr. A*, **36**, 237–41.
Templeton, D. H. and Templeton, L. K. (1985). *Acta Crystallogr. A*, **41**, 133–42.
Templeton, D. H. and Templeton, L. K. (1986). *Acta Crystallogr. A*, **42**, 478–81.
Templeton, D. H. and Templeton, L. K. (1988). *J. Appl. Crystallogr.*, **21**, 151–3.
Templeton, D. H. and Templeton, L. K. (1994). *Phys. Rev. B*, **49**, 14850–3.
Templeton, D. H., Templeton, L. K., Phillips, J. C. and Hodgson, K. O. (1980). *Acta Crystallogr. A*, **36**, 436–42.

Templeton, D. H., Templeton, L. K., Phizackerley, R. P. and Hodgson, K. O. (1982). *Acta Crystallogr. A*, **38**, 74–8.
Teworte, R. and Bonse, U. (1984). *Phys. Rev. B*, **29**, 2102–8.
Teworte, R. and Bonse, U. (1987). *Z. Phys. B*, **65**, 275–89.
Thomas, J. E., Baldwin, T. O. and Dederichs, P. H. (1971). *Phys. Rev. B*, **3**, 1167–73.
Thomson, J. J. (1898). *Proc. Camb. Phil. Soc.*, **9**, 393–7.
Thorkildsen, G. (1987). *Acta Crystallogr. A*, **43**, 361–9.
Thorkildsen, G. and Larsen, H. B. (1998*a*). *Acta Crystallogr. A*, **54**, 120–8.
Thorkildsen, G. and Larsen, H. B. (1998*b*). *Acta Crystallogr. A*, **54**, 172–85.
Thorkildsen, G. and Larsen, H. B. (1998*c*). *Acta Crystallogr. A*, **54**, 186–90.
Thorkildsen, G. and Larsen, H. B. (1998*d*). *Acta Crystallogr. A*, **54**, 416–29.
Thorkildsen, G. and Larsen, H. B. (1999*a*). *Acta Crystallogr. A*, **55**, 1–13.
Thorkildsen, G. and Larsen, H. B. (1999*b*). *Acta Crystallogr. A*, **55**, 840–54.
Thorkildsen, G. and Larsen, H. B. (1999*c*). *Acta Crystallogr. A*, **55**, 1000–13.
Thorkildsen, G. and Larsen, H. B. (2002). *Acta Crystallogr. A*, **58**, 252–8.
Thorkildsen, G., Larsen, H. B., Semmingsen, D. and Bjaanes, Ø. (2001). *Acta Crystallogr. A*, **57**, 201–11.
Thorkildsen, G., Larsen H. B. and Weckert, E. (2001). *Acta Crystallogr. A*, **57**, 389–94.
Tomie, T. (1997) US Patent Number 5,594,773.
Tomimitsu, H. (1983). *Jap. J. Appl. Phys.*, **22**, L674–76.
Tomiyoshi, S., Yamada, M. and Watanabe, H. (1980). *Acta Crystallogr. A*, **36**, 600–4.
Tomov, I. V., Chen, P. and Rentzepis, P. M. (1998). *J. Appl. Phys.*, **83**, 5546–8.
Treimer, W. (1978). *Kristall und Technik*, **13**, 1105–16.
Treimer, W. (1998). *Cryst. Res. Technol.*, **33**, 643–52.
Treimer, W. and Hildebrandt, G. (1989). *J. Appl. Crystallogr.*, **22**, 84–6.
Trucano, P. (1976). *Phys. Rev. B*, **13**, 2524–31.
Tseng, T.-P. and Chang, S.-L. (1990). *Acta Crystallogr. A*, **46**, 567–76.
Tu, L. K. and Vreeland, T. Jr (1974). *Acta Crystallogr. A*, **30**, 852–3.
Tuomi, T. (2002). *J. Synchrotron Rad.*, **9**, 174–8.
Tuomi, T., Naukkarinen, K., Laurila, E. and Rabe, P. (1973). *Acta Polytechn. Scand.*, **Ph 100**, 1–4.
Tuomi, T., Naukkarinen, K. and Rabe, P. (1974). *Phys. Stat. Sol. (a)*, **25**, 93–106.
Uebach, W. (1973). *Z. Naturforsch. (a)*, **28 a**, 1214–20.
Uebach, W. and Hildebrandt, G. (1969). *Z. Krist.*, **129**, 1–8.
Ulyanenkov, A. P., Stepanov, S. A., Pietsch, U. and Köhler, R. (1995). *J. Phys. D: Appl. Phys.*, **28**, 2522–8.
Umeno, M. and Hildebrandt, G. (1975). *Phys. Stat. Sol. (a)*, **31**, 583–94.
Underwood, J. H. (1995). *Proc. SPIE – The International Society for Optical Engineering.*, **2523**, 51–7.
Underwood, J. H. and Barbee, T. W. Jr (1981). *AIP Conf. Proc.*, **75**, 170–8.
Underwood, J. H., Barbee, T. W. Jr and Frieber, C. (1986). *Appl. Optics*, **25**, 1730–2.
Underwood, J. H., Thomson, A. C., Kortright, J. B., Chapman, K. C. and Lunt, D. (1996). *Rev. Sci. Instrum.*, **67**, 3359.

Underwood, J. H., Thompson, A. C., Wu, Y. and Giauque, R. D. (1998). *Nucl. Instrum. Methods A*, **266**, 296–302.

Uragami, T. (1969). *J. Phys. Soc. Jap.*, **27**, 147–54.

Uragami, T. (1970). *J. Phys. Soc. Jap.*, **28**, 1508–27.

Uragami, T. (1971). *J. Phys. Soc. Jap.*, **31**, 1141–61.

Uschmann, I., Förster, E., Gäbel, K., Hölzer, G. and Ensslen, M. (1993). *J. Appl. Crystallogr.*, **26**, 405–12.

Uschmann, I., Malgrange, C. and Förster, E. (1997). *J. Appl. Crystallogr.*, **30**, 1150–1.

Vakhrushev, S. B., Kolla, E. V., Okuneva, N. M., Shul'pina, I. L. and Shcheglov, M. P. (1991*a*). *Supercond. Phys. Chem. Techn.*, **4**, 1110–13.

Vakhrushev, S. B., Kolla, E. V., Okuneva, N. M., Shul'pina, I. L., Shcheglov, M. P., Osiko, V. V., Gamayunov, K. V., Ivanov, A. L. and Tatarntsev, V. M. (1991*b*). *Supercond. Phys. Chem. Techn.*, **4**, 1017–20.

Vallino, F., Jacques, A. and George, A. (2000). *Phys. Stat. Sol. (b)*, **222**, 51–62.

Vartanyants, I. A. and Koval'chuk, M. V. (2001). *Rep. Prog. Phys.*, **64**, 1009–84.

Vartanyants, I. A. and Zegenhagen, J. (1999). *Solid State Comm.*, **113**, 299–320.

Vartanyants, I. A., Koval'chuk, M. V. and Beresovskii, V. M. (1993*a*). *J. Phys. D: Appl. Phys.*, **26**, A197–A201.

Vartanyants, I. A., Koval'chuk, M. V. and Beresovskii, V. M. (1993*b*). *Phys. Stat. Sol. (a)*, **135**, 513–17.

Vartanyants, I. A., Koval'chuk, M. V. and Ustimenko, I. N. (1994). *J. de Physique IV*, **C9**, 191–4.

Vetter, W. M., Gallagher, D. T. and Dudley, M. (2002). *Acta Crystallogr. D*, **58**, 579–84.

Villa, M., Baron, M., Hainbuchner, M., Jericha, E., Leiner, V., Schwahn, D., Seidl, E., Stahn, J. and Rauch, H. (2003). *J. Appl. Cryst.*, **36**, 769–73.

Vineyard, G. H. (1982). *Phys. Rev. B*, **26**, 4146–59.

Vlieg, E., Fischer, A. E. M. J., van der Veen, J. F., Dev, B. N. and Materlik, G. (1986). *Surf. Sci.*, **178**, 36–46.

Vlieg, E., Fontes, E. and Patel, J. R. (1991). *Phys. Rev. B*, **43**, 7185–93.

Voronkov, S. N., Maksimov, S. K. and Chukhovskii, F. N. (1984). *Fiz. Tverd. Tela*, **26**, 2019.

Vreeland, T. Jr (1976). *J. Appl. Crystallogr.*, **9**, 34–8.

Wabnitz, H. *et al.* (2002). *Nature*, **420**, 482.

Wada, M. and Kato, N. (1977). *Acta Crystallogr. A*, **33**, 161–8.

Wagenfeld, H. (1966). *Phys. Rev.*, **144**, 216–24.

Wagenfeld, H. (1968). *Acta Crystallogr. A*, **24**, 170–4.

Wagenfeld, H. (1975). In *Anomalous scattering* (ed. S. C. Abrahams and S. Ramaseshan), pp. 12–23. Munksgaard, Copenhagen.

Wagner, E. (1920). *Physik. Z.*, **21**, 632.

Wagner, E. and Kulenkampf, H. (1922). *Ann. der Physik*, **68**, 369.

Wagner, E. H. (1959*a*). *Acta Crystallogr.*, **12**, 345–6.

Wagner, E. H. (1959*b*). *Z. Phys.*, **154**, 352–60.

Wagner, H. (1956). *Z. Phys.*, **146**, 127–68.

Waller, I. (1923). *Z. Phys.*, **17**, 398–408.

Walter, B. and Pohl, R. (1908). *Ann. d. Physik*, **25**, 715–24.

Walter, B. and Pohl, R. (1909). *Ann. d. Physik*, **29**, 331–54.

Wang, R., Yang, X., Zou, W., Gui, J. and Jiang, J. (1994). *J. Phys.: Condens. Matter.*, **6**, 9009–16.

Wang, G.-L., Jiang, J.-H., Tian, Y.-L., Han, Y. and Wang Z.-G. (1998). *J. Synchrotron Rad.*, **5**, 1082–84.

Warburton, W. K. and Ludwig Jr, K. F. (1986). *Phys. Rev. B*, **33**, 8424–32.

Wark, J. S. and He, H. (1994). *Laser Part. Beams*, **12**, 507–13.

Wark, J. S. and Lee, R.W. (1999). *J. Appl. Crystallogr.*, **32**, 692–703.

Wark, J. S., Allen, A. M., Ansbro, P. C., Bucksbaum, P. H., Chang, Z., Decamp, M., Falcone, R. W., Heimann, P. A., Johnson, S. L., Kang, I., Kapteyn, H. C., Larsson, J., Lee, R. W., Lindenberg, A. M., Merlin, R., Missalla, T., Naylor, G., Padmore, A., Reis, D. A., Scheidt, K., Sjoegren, A., Sondhauss, P. C. and Wulff, M. (2001). *Proc. SPIE – The International Society for Optical Engineering.*, **4143**, 26–37.

Warren, B. E. (1969). *X-ray diffraction.* Addison-Wesley, Reading; second edition: 1990, Dover, New York, Addison-Wesley, Reading.

Weckert, E. and Hümmer, K. (1990). *Acta Crystallogr. A*, **46**, 387–93.

Weckert, E. and Hümmer, K. (1996). In *X-ray and neutron dynamical diffraction: theory and applications* (ed. A. Authier, S. Lagomarsino and B. K. Tanner), NATO ASI Series, Series B, Physics, Vol. 357, pp. 345–67. Plenum Press, New York and London.

Weckert, E. and Hümmer, K. (1997). *Acta Crystallogr. A*, **53**, 108–43.

Weckert, E. and Hümmer, K. (1998). *Crystal Res. Technol.*, **33**, 653–74.

Weckert, E., Schwegle, W. and Hümmer, K. (1993). *Proc. Roy. Soc. London A*, **442**, 33–46.

Weckert, E., Müller, R., Zellner, J., Zegers, I. and Loris, R. (2002). *Z. Krist.*, **217**, 651–61.

Weeks, C. M., Xu, H., Hauptman, H. A. and Shen, Q. (2000). *Acta Crystallogr. A*, **56**, 280–3.

Weiss, R. J. (1965). *Acta Crystallogr.*, **18**, 814.

Weissmann, S., Balibar, F. and Pétroff, J.-F. (ed.) (1984). *Applications of X-ray topographic methods to materials science.* Plenum Press, New York.

Werner, S. A. (1974). *J. Appl. Phys.*, **45**, 3246–54.

Werner, S. A. (1980). *Phys. Today*, **33**, 24–30.

Werner, S. A., Berliner, R. R. and Arif, M. (1986). *Physica*, **173B**, 245–55.

Whatmore, R. W., Goddard, P. A., Tanner, B. K. and Clark, G. F. (1982). *Nature*, **299**, 44–6.

Whelan, M. J. (1986). *Acta Crystallogr. A*, **42**, 493–501.

Wierzchowski, W. (1985). *Phys. Stat. Sol. (a)*, **88**, 77–87.

Wierzchowski, W. and Moore, M. (1995). *Acta Crystallogr. A*, **51**, 831–40.

Wierzchowski, W., Moore, M., Makepeace, A. P. W. and Yacoot, A. (1991). *J. Cryst. Growth*, **114**, 209–27.

Wierzchowski, W., Mazur, K. and Wieteska, K. (1995). *J. Phys. D: Appl. Phys.*, **28**, A33–A37.

Wierzchowski, W., Wieteska, K. and Graeff, W. (1997). *Il Nuovo Cimento*, 227–32.

Wieteska, K., Wierzchowski, W. and Graeff, W. (1997). *Il Nuovo Cimento*, **19D**, 233–39.

Wieteska, K., Wierzchowski, W. Graeff, W., Turos, A. and Grötzschel, R. (2000). *J. Synchrotron Rad.*, **7**, 318–25.

Wild, R. K., Evans, T. and Lang, A. R. (1967). *Phil. Mag.*, **15**, 267–79.

Wilkens, M. (1970). *Phys. Stat. Sol. (a)*, **2**, 359–70.

Wilkens, M. (1976). *Kristall und Technik*, **11**, 1159–69.

Wilkins, S. W. (1978*a*). *Proc. Roy. Soc. A*, **364**, 569–89.

Wilkins, S. W. (1978*b*). *Acta Crystallogr. A*, **34**, 343–4.

Wilkins, S. W. (1978*c*). *Phys. Lett. A*, **66**, 134–6.

Wilkins, S. W. (1981). *Phil. Trans. Roy. Soc. (London)*, **299**, 275–317.

Wilkins, S. W., Gureyev, T. E., Gao, D., Pogany, A. and Stevenson, A. W. (1996). *Nature*, **384**, 335–8.

Willaime, C. and Authier, A. (1966). *Bull. Soc. Fr. Minér. Crist.*, **89**, 269–70.

Williams, D. J. and Vere, A. W. (1987). *J. Cryst. Growth*, **83**, 341–52.

Windish, D. and Becker, P. (1988). *Phil. Mag. A*, **58**, 435–43.

Windish, D. and Becker, P. (1990). *Phys. Stat. Sol. (a)*, **118**, 379–88.

Windish, D. and Becker, P. (1992). *J. Appl. Crystallogr.*, **25**, 377–83.

Winick, H. (1980). In *Synchrotron radiation research* (ed. H. Winick and S. Doniach), pp. 11–25. Plenum Press, New York.

Winick, H. (1988). In *Applications of synchrotron radiation* (ed. H. Winick, D. Xian, M.-H. Ye and T. Huang), pp. 1–75. Gordon and Breach Science Publishers, New York.

Winick, H. (1998). *J. Synchrotron Rad.*, **5**, 168–75.

Winick, H., Brown, G., Halbach, K. and Harris, J. (1981). *Phys. Today*, **34**(5), 50–63.

Wittry, D. B. (1990). *J. Appl. Physics*, **67**, 1633–8.

Wittry, D. B. and Chang, W. Z. (1993). *J. Appl. Physics*, **73**, 601–7.

Wittry, D. B. and Sun, S. (1990). *J. Appl. Physics*, **68**, 387–91.

Wittry, D. B., Chang, W. Z. and Li R.,Y. (1993). *J. Appl. Physics*, **74**, 3534–40.

Woicik, J. C., Pellegrino, J. G., Southworth, S. H., Shaw, P. S., Karlin, B. A. and Bouldin, C. E. (1995). *Phys. Rev. B*, **52**, R2281–R2284.

Woicik, J. C., Heskett, D., Warner, J., Berman, L. E., Karlin, B. A., Vartanyants, I. A., Hasan, M. Z., Kendelewicz, T., Shen, Z. X. and Pianetta, P. (2001). *Phys. Rev. B*, **64**, 125115–1–15.

Woicik, J. C., Heskett, D., Warner, J., Berman, L. E., Karlin, B. A., Vartanyants, I. A., Hasan, M. Z., Kendelewicz, T., Shen, Z. X. and Pianetta, P. (2001). *Phys. Rev. B*, **64**, 125115–1–15.

Wonsiewicz, B. C. and Patel, J. R. (1976). *J. Appl. Phys.*, **47**, 1837–45.

Woodruff, D. P., Seymour, D. L., Mc Conville, C. F., Riley, C. E., Crapper, M. D., Prince, N. P. and Jones, R. G. (1987). *Phys. Rev. Lett.*, **58**, 1460–2.

Woodruff, D. P., Seymour, D. L., Mc Conville, C. F., Riley, C. E., Crapper, M. D., Prince, N. P. and Jones, R. G. (1988). *Surf. Sci.*, **195**, 237–55.

Wooster, N. and Wooster, W. A. (1945). *Nature (London)*, **155**, 786.

Wu, C. Cm. and Armstrong, R. W. (1975). *J. Appl. Crystallogr.*, **8**, 29–36.

Yabashi, M., Tamasaku, K. and Ishikawa, T. (2001). *Phys. Rev. Lett.*, **87**, 140801–1-4.

Yamada, T., Saito, S. and Shimomura, Y. (1966). *J. Phys. Soc. Jap.*, **21**, 672–80.

Yamaoka, H., Ohtomo, K. and Ishikawa, T. (1998). *J. Synchrotron Rad.*, **5**, 687–9.

Yamazaki, H. and Ishikawa, T. (1996). *Proc. SPIE – The International Society for Optical Engineering.*, **2856**, 279–88.

Yamazaki, H. and Ishikawa, T. (2002). *J. Appl. Crystallogr.*, **35**, 314–8.

Yamazaki, H. and Ishikawa, T. (2003). *J. Appl. Crystallogr.*, **36**, 213–9.

Yamazaki, H. and Ishikawa, T. (2004). *J. Appl. Crystallogr.*, **37**, 48–51.

Yang, P., Jiang, S. S., Zhou, H.-N. and Feng, D. (1986). *Phys. Stat. Sol. (a)*, **97**, 411–20.

Yang, P., Jiang, S. S. and Feng, D. (1990). *Phys. Stat. Sol. (a)*, **122**, 475–9.

Yang, P., Jiang, S. S. and Feng, D. (1991*a*). *Phys. Stat. Sol. (a)*, **123**, 51–60.

Yang, P., Jiang, S. S. and Feng, D. (1991*b*). *Phys. Stat. Sol. (a)*, **124**, 57–66.

Yang, B. X., Rivers, M., Schildkamp, W. and Eng, P. J. (1993). *Rev. Sci. Instrum.*, **66**, 2278–80.

Yashiro, W. and Takahashi, T. (2000). *Acta Crystallogr. A*, **56**, 163–7.

Yokosuka, H., Watanabe, N., Ohigashi, T., Yoshida, Y., Maeda, S., Aoli, S., Suzuki, Y., Takeuchi, A. and Takano, H. (2002). *J. Synchrotron Rad.*, **9**, 179–81.

Yoneyama, A., Momose, A., Koyama, I., Seya, E., Takeda, T., Itai, Y., Hirano, K. and Hyodo, K. (2002). *J. Synchrotron Rad.*, **9**, 277–81.

Yoshimatsu, M. (1964). *Jap. J. Appl. Phys.*, **3**, 94–103.

Yoshimatsu, M. (1965). *Jap. J. Appl. Phys.*, **4**, 619–20.

Yoshimatsu, M. and Kohra, K. (1959). *J. Phys. Soc. Jap.*, **14**, 1249–50.

Yoshimatsu, M. and Kohra, K. (1960). *J. Phys. Soc. Jap.*, **15**, 1760–70.

Yoshimura, J. (1996). *Acta Crystallogr. A*, **52**, 312–25.

Yoshimura, J., Miyazaki, T., Wada, T., Kohra, K., Hosaka, M., Ogawa, T. and Taki, S. (1979). *J. Cryst. Growth*, **46**, 691–700.

Young, F. W. and Sherrill, F. A. (1971). *J. Appl. Phys.*, **42**, 230–7.

Young, F. W. and Sherrill, F. A. (1972). *J. Appl. Phys.*, **43**, 2949–50.

Yun, W. B., Lai, B., Legnini, D., Xiao, Y. H., Chrzas, J., Skulima, K. M., Bionta, M., White, V. and Cerrina, F. (1992). *SPIE Proc.*, **1740**, 117.

Yun, W., Lai, B., Cai, Z., Maser, J., Legnini, D., Gluskin, E., Chen, Z., Krasnoperova, A. A., Vladimirsky, Y., Cerrina, F., Di Fabrizio, E. and Gentili, M. (1999). *Rev. Sci. Instrum.*, **70**, 2238–41.

Zachariasen, W. H. (1945). *Theory of X-ray diffraction in crystals.* John Wiley and Sons, Inc., New York. Reprinted by Dover, New York, 1967.

Zachariasen, W. H. (1967). *Acta Crystallogr.*, **23**, 558–64.

Zachariasen, W. H. (1968). *Acta Crystallogr. A*, **24**, 421–4.

Zarka, A. (1983). *J. Appl. Crystallogr.*, **16**, 354–6.

Zarka, A. and Liu Lin (1983). *J. Cryst. Growth*, **61**, 397–405.

Zarka, A., Liu Lin and Sauvage, M. (1983). *J. Cryst. Growth*, **62**, 409–24.

Zarka, A., Capelle, B., Détaint, J. and Schartzel, J. (1987). *Proc. of the 41st Annual Frequency Control Symp.*, Philadelphia, USA, pp. 236–40.

Zarka, A., Capelle, B., Détaint, J. and Schwartzel, J. (1988*a*). *J. Appl. Crystallogr.*, **21**, 967–71.

Zarka, A., Capelle, B., Zheng, Y., Détaint, J. and Schwartzel, J. (1988*b*). *Proc. of the 42nd Annual Frequency Control Symp.*, Denver, USA, pp. 85–92.

Zaumseil, P. (1997). *Il Nuovo Cimento*, **19D**, 361–7.

Zaumseil, P. and Winter, U. (1982). *Phys. Stat. Sol. (a)*, **70**, 497–505.

Zaumseil, P. and Winter, U. (1990). *Phys. Stat. Sol. (a)*, **120**, 67–75.

Zaumseil, P., Winter, U., Cembali, F., Servidori, M. and Šourek, Z. (1987). *Phys. Stat. Sol. (a)*, **100**, 95–104.

Zaus, R. (1993). *J. Appl. Crystallogr.*, **26**, 801–11.

Zegenhagen, J. (1993). *Surf. Sci. Reports*, **18**, 199–271.

Zegenhagen, J. (2004). *Surface Science*, **554**, 77–9.

Zegenhagen, J., Huang, K.-G., Gibson, W. M., Hunt, B. D. and Schowalter, L. J. (1989). *Phys. Rev. B*, **39**, 10254–60.

Zegenhagen, J., Lyman, P. F., Böhringer, M. and Bedzyk, M. J. (1997). *Phys. Stat. Sol. (b)*, **204**, 587–616.

Zernike, F. (1935). *Z. Tech. Phys.*, **16**, 454.

Zernike, F. (1938). *Physica*, **5**, 785.

Zhang, L., Hoszowska, J., Migliore, J. S., Mocella, V., Ferrero, C. and Freund, A. K. (2001). *Nucl. Instrum. & Methods in Physics Research A*, **467–468**, 409–13.

Zhang, X., Sugiyama, H., Ando, M., Imai, Y. and Yoda, Y. (2003). *J. Appl. Crystallogr.*, **36**, 188–92.

Zhang, Y., Colella, R., Shen, Q. and Kycia, S. W. (1998). *Acta Crystallogr. A*, **54**, 411–15.

Zhang, Z., Fenter, P., Cheng, L., Sturchio, N. C., Bedzyk, M. J., Machewski, M. L. and Wesolowski, D. J. (2004). *Surf. Sci. Letters*, **554**, L95–100.

Zhao, J.-Y., Yang, P., Jiang, S.-S., Jiang, X.-M., Jiang, J.-H., Xian, D.-C., Gen, Z.-H. and Zhou, Q. (1991). *Appl. Phys. Lett.*, **59**, 1952–3.

Zheludeva, S. I., Koval'chuk, M. V., Novikova, N. N., Sosphenov, A. N., Salaschenko, N. N., Shamov, E. A., Prokhorov, K. A., Burattini, E. and Cappuccio, G. (1997). *J. Appl. Crystallogr.*, **30**, 833–8.

Zheng, Y., Zarka, A., Capelle, B., Détaint, J. and Schwartzel, J. (1989). *Acta Crystallogr. A*, **45**, 275–85.

Zheng, Y., Boulliard, J. C., Capelle, B. and Malgrange, C. (1998). *Acta Crystallogr. A*, **54**, 74–82.

Zheng, Y., Boulliard, J. C., Soyer, A. and Pétroff, J. F. (2004). *Acta Crystallogr. A*, **60**, 339–50.

Zhong, Z., Kao, C. C., Siddons, D. P. and Hastings, J. B. (2002). *Acta Crystallogr. A*, **58**, 487–93.

Zhong, Z., Kao, C. C., Siddons, D. P., Zhong, H. and Hastings, J. B. (2003). *Acta Crystallogr. A*, **59**, 1–6.

Ziegler, E. (1995). *Opt. Eng.*, **34**, 445–52.

Zolotoyabko, E. and Panov, V. (1992). *Acta Crystallogr. A*, **48**, 225–31.

Zolotoyabko, E., Shilo, D., Sauer, W., Pernot, E. and Baruchel, J. (1998). *Appl. Phys. Lett.*, **66**, 2278–80.

Zolotoyabko, E., Shilo, D., Sauer, W., Pernot, E. and Baruchel, J. (1999). *Rev. Sci. Instrum.*, **70**, 3341–5.

Zontone, F. and Comin, F. (1992). *Rev. Sci. Instrum.*, **63**, 501–4.

Zontone, F., Mancini, L., Barrett, R., Baruchel, J., Härtwig, J. and Epelboin, Y. (1996). *J. Synchrotron Rad.*, **3**, 173–84.

Zwanenburg, M. J., Peters, J. F., Bongaerts, J. H. H., de Vries, S. A., Abernathy, D. L. and van der Veen, J. F. (1999). *PRL*, **82**, 1696–9.

Zwanenburg, M. J., Peters, J. F., Bongaerts, J. H. H., Riese, D. O. and van der Veen, J. F. (2000). *Phys. Rev. Letters*, **85**, 5154–7.

Zwillinger, D. (1997). *Handbook of differential equations*. 3rd ed. Academic Press, Boston, MA, USA.

List of symbols

$\mathcal{A}$	Amplification ratio
$\mathbf{A}$	Vector potential
$\mathbf{B}$	Magnetic induction
$\mathbf{B_o}$, $\mathbf{B_h}$, $\mathbf{B_g}$	Fourier components of the magnetic induction
c	Velocity of mass in vacuum
$C = 1$ or $\cos 2\theta_B$	Polarization factor
d_{hkl}	Reflecting-plane spacing ($= d/n$, where d is lattice spacing, and n is order of the reflection)
$\mathbf{D}$	Electric displacement
$\mathbf{D_o}$, $\mathbf{D_h}$, $\mathbf{D_g}$	Fourier components of the electric displacement
$\mathbf{D_{h'[h]}}$	Projection of $\mathbf{D_{h'}}$ on the plane normal to $\mathbf{K_h}$
$\mathbf{D_o^{(a)}}$	Incident wave
$\mathbf{D_h^{(a)}}$	Reflected wave (Bragg geometry)
$\mathbf{D_o^{(d)}}$	Refracted wave (Bragg and Laue geometries)
$\mathbf{D_h^{(d)}}$	Reflected wave (Laue geometry)
e	Charge of the electron
$\mathbf{E}$	Electric field
$\mathbf{E_o}$, $\mathbf{E_h}$, $\mathbf{E_g}$	Fourier components of the electric field
f	Form factor
f', f''	Real and imaginary part of the anomalous dispersion corrections
$F_h = F_{rh} + \mathrm{i}F_{ih}$	Structure factor associated with reciprocal lattice node hkl, taking into account the Debye–Waller factor and the anomalous dispersion corrections
F_{rh}	Structure factor calculated from the real part of the form factor taking into account the anomalous dispersion corrections
F_{ih}	Structure factor calculated from the imaginary part of the form factor
$\mathbf{h}$, $\mathbf{g}$	Reciprocal lattice vector
$\mathbf{H}$	Magnetic field
$\mathbf{H_o}$, $\mathbf{H_h}$, $\mathbf{H_g}$	Fourier components of the magnetic field
G_B	Bragg gap (diameter of the dispersion surface)
$I_h^{(a)}$, $I_h^{(d)}$	Reflectivity (Bragg and Laue geometries, respectively)
$I_{hi}^{(a)}$, $I_{hi}^{(d)}$	Integrated intensity (Bragg and Laue geometries, respectively)
$\mathcal{I}m(x)$	Imaginary part of x
$\mathbf{j}$	Current density
$k = 1/\lambda$	Wave number in vacuum

$\mathbf{K_o}, \mathbf{K_h}, \mathbf{K_g}$	Wavevectors of a wavefield
$\mathbf{K_o^i}, \mathbf{K_h^i}, \mathbf{K_g^i}$	Imaginary part of the wavevectors
$\mathbf{K_o^{(a)}}$	Wavevector of the incident wave
$\mathbf{K_o^{(s)}}$	Wavevector of the specularly reflected wave
$\mathbf{K_h^{(a)}}$	Wavevector of the reflected wave, Bragg geometry
$\mathbf{K_o^{(d)}}, \mathbf{K_h^{(d)}}$	Wavevectors of the refracted and reflected waves, Laue geometry
m	Rest mass of the electron
$\mathbf{M}$	Magnetization
$\exp(-M)$	Debye–Waller factor
n	Index of refraction
$\mathbf{n}$	Normal to the crystal surface, drawn towards the inside of the crystal
N	Number of scattering units per unit volume
$p = \tan\alpha/\tan\theta_B$	Slope of the wavefield trajectory
$\mathbf{P}$	Electric polarization
$\mathbf{r}$	Position vector in real space
R	Classical radius of the electron
$\mathcal{R}$	Radius of curvature
$\mathcal{R}e(x)$	Real part of x
$\mathbf{s}$	Unit vector parallel to the lattice planes
$\mathbf{s_o}, \mathbf{s_h}$	Unit vectors in the incident and reflected directions
$\mathbf{S}$	Poynting vector
$S(x)$	Sign of x
t	Crystal thickness
t	Time
v	Velocity of light in the medium
V	Volume of the unit cell
x_o, x_h	Coordinates in the incident and reflected directions (direct space)
X_o, X_h	Excitation errors; they reduce to the distances of the tiepoint from the spheres passing through the Lorentz point when the incidence and reflection angles are not close to 0 or $\pi/2$
Y	Reduced coordinate along the base of the Borrmann triangle
z_e	Penetration depth
$\alpha = (\mathbf{s}, \mathbf{S})$	Angle between the Poynting vector and the lattice planes
α	Coefficient of linear expansion
β	Phase of $\sqrt{\chi_h\chi_{\bar{h}}}$
β'	Phase of $\sqrt{\chi_h/\chi_{\bar{h}}}$
β	Strain gradient

β_h	In Takagi's theory, parameter proportional to the departure from Bragg's incidence of the arbitrarily chosen incident wavevector—usually taken equal to zero
β'_h	In Takagi's theory, parameter equal to the sum of β_h and a term proportional to the local deformation
χ	Dielectric susceptibility, or polarizability
χ_r, χ_i	Real and imaginary part of the dielectric susceptibility
χ_h	Fourier component of the dielectric susceptibility,
χ_{rh}, χ_{ih}	Fourier components of the real and imaginary parts of the dielectric susceptibility
$2\delta_{os}$	Darwin width or FWHM, neglecting curvature of the Ewald sphere
$2\delta_{oc}$	Darwin width or FWHM, taking curvature of the Ewald sphere into account
$2\delta_{hs}$	Angular width of the reflected beam, neglecting curvature of the Ewald sphere
$2\delta_{hc}$	Angular width of the reflected beam, taking curvature of the Ewald sphere into account
ϵ	Dielectric constant, or permittivity
ϵ_0	Permittivity of vacuum
ϕ	Scalar potential
$\gamma = \gamma_h/\gamma_o$	Asymmetry ratio
$\gamma_h = \cos(\psi_h)$	Cosinus of the angle between the normal to the crystal surface and the reflected direction
$\gamma_o = \cos(\psi_o)$	Cosinus of the angle between the normal to the crystal surface and the incident direction
η	Deviation parameter
η_r, η_i	Real and imaginary parts of the deviation parameter
$\varphi_h, \varphi(\mathbf{h})$	Phase of the structure factor, F_h
$\varphi_{rh}, \varphi_{ih}$	Phases of F_{rh} and F_{ih}, respectively
$\varphi = \varphi_{rh} - \varphi_{ih}$	Difference between the phases of F_{rh} and F_{ih}
Φ_3	Triplet phase
$\kappa = \lvert\chi_{ih}/\chi_{rh}\rvert$	Magnitude of the ratio of the Fourier components of the imaginary and real parts of the polarizability
λ	Wavelength
$\Lambda = \Lambda_B/\sqrt{\eta^2 - 1}$	Period of the Pendellösung oscillations (Bragg geometry)
$\Lambda = \Lambda_L/\sqrt{\eta^2 + 1}$	Period of the Pendellösung oscillations (Laue geometry)
Λ_o	Extinction distance (Bragg geometry), Pendellösung distance (Laue geometry)
Λ_B	Extinction distance (Bragg geometry)
Λ_L	Pendellösung distance (Laue geometry)—interpreted as a diameter of the dispersion surface
μ	Magnetic permeability of the medium
μ_0	Magnetic permeability of vacuum

μ_o	Linear absorption coefficient (normal)
μ_e	Linear absorption coefficient (anomalous, or effective)
$\mu_{\mathbf{S}}$	Linear absorption coefficient along the direction of propagation
ν	Frequency of light
π-polarization	Polarization parallel to the diffraction plane
$\Pi_{\mathbf{e}}$, $\Pi_{\mathbf{m}}$	Hertz vectors
θ	Angle of incidence
θ_B	Bragg angle
$\Delta\theta$	Departure from Bragg's angle of the incident beam
$\Delta\theta_{os}$	Deviation from Bragg's angle of the middle of the reflection domain, neglecting curvature of the Ewald sphere
$\Delta\theta_{oc}$	Deviation from Bragg's angle of the middle of the reflection domain, taking curvature of the Ewald sphere into account
$\Delta\theta_{hs}$	Deviation from Bragg's angle of the middle of the reflection domain, neglecting curvature of the Ewald sphere
$\Delta\theta_{hc}$	Deviation from Bragg's angle of the middle of the reflection domain, taking curvature of the Ewald sphere into account
$\rho(\mathbf{r})$	Electronic density
σ	Specific conductivity
$\sigma = \chi_h/\chi_{\bar{h}}$	Ratio of the *hkl* and the $\bar{h}\bar{k}\bar{l}$ components of the polarizability
σ-polarization	Polarization perpendicular to the diffraction plane
$\mathrm{d}\tau$	Volume element in direct space
ω_c	Critical angle for specular reflection
$\xi = D_h/D_o$	Amplitude ratio
ψ	Phase of the amplitude ratio, ξ
$\psi_h = (\mathbf{n}, \mathbf{s_h})$	Angle between the normal to the crystal surface and the lattice planes
$\psi_o = (\mathbf{n}, \mathbf{s_o})$	Angle between the normal to the crystal surface and the incident direction
$\psi_n = (\mathbf{n}, \mathbf{s})$	Angle between the normal to the crystal surface and the lattice planes
Ψ	Rotation angle around the reciprocal lattice vector

Index of names

Italic numbers denote reference to illustrations; n. denotes reference to a footnote

Index of Subjects

Italic numbers denote reference to illustrations; n. denotes reference to a footnote; **Bold** type denotes most important page numbers

Lightning Source UK Ltd.
Milton Keynes UK
UKOW040741140612

194390UK00002B/99/P

9 780198 528920